Undergraduate Texts in Mathematics

Undergraduate Texts in Mathematics

Undergraduate Texts in Mathematics are generally aimed at third- and fourth-year undergraduate mathematics students at North American universities. These texts strive to provide students and teachers with new perspectives and novel approaches. The books include motivation that guides the reader to an appreciation of interrelations among different aspects of the subject. They feature examples that illustrate key concepts as well as exercises that strengthen understanding.

For further volumes:
http://www.springer.com/series/666

Peter J. Olver

Introduction to Partial Differential Equations

Peter J. Olver
School of Mathematics
University of Minnesota
Minneapolis, MN
USA

ISSN 0172-6056 ISSN 2197-5604 (electronic)
ISBN 978-3-319-34744-8 ISBN 978-3-319-02099-0 (eBook)
DOI 10.1007/978-3-319-02099-0
Springer Cham Heidelberg New York Dordrecht London

Library of Congress Control Number: 2013954394

Mathematics Subject Classification: 35-01, 42-01, 65-01

Springer is part of Springer Science+Business Media (www.springer.com)

To the memory of my father, Frank W.J. Olver (1924-2013) and mother, Grace E. Olver (née Smith, 1927-1980), whose love, patience, and guidance formed the heart of it all.

Preface

The momentous revolution in science precipitated by Isaac Newton's calculus soon revealed the central role of partial differential equations throughout mathematics and its manifold applications. Notable examples of fundamental physical phenomena modeled by partial differential equations, most of which are named after their discoverers or early proponents, include quantum mechanics (Schrödinger, Dirac), relativity (Einstein), electromagnetism (Maxwell), optics (eikonal, Maxwell–Bloch, nonlinear Schrödinger), fluid mechanics (Euler, Navier–Stokes, Korteweg–de Vries, Kadomstev–Petviashvili), superconductivity (Ginzburg–Landau), plasmas (Vlasov), magneto-hydrodynamics (Navier–Stokes + Maxwell), elasticity (Lamé, von Karman), thermodynamics (heat), chemical reactions (Kolmogorov–Petrovsky–Piskounov), finance (Black–Scholes), neuroscience (FitzHugh–Nagumo), and many, many more. The challenge is that, while their derivation as physical models — classical, quantum, and relativistic — is, for the most part, well established, [**57, 69**], most of the resulting partial differential equations are notoriously difficult to solve, and only a small handful can be deemed to be completely understood. In many cases, the only means of calculating and understanding their solutions is through the design of sophisticated numerical approximation schemes, an important and active subject in its own right. However, one cannot make serious progress on their numerical aspects without a deep understanding of the underlying analytical properties, and thus the analytical and numerical approaches to the subject are inextricably intertwined.

This textbook is designed for a one-year course covering the fundamentals of partial differential equations, geared towards advanced undergraduates and beginning graduate students in mathematics, science, and engineering. No previous experience with the subject is assumed, while the mathematical prerequisites for embarking on this course of study will be listed below. For many years, I have been teaching such a course to students from mathematics, physics, engineering, statistics, chemistry, and, more recently, biology, finance, economics, and elsewhere. Over time, I realized that there is a genuine need for a well-written, systematic, modern introduction to the basic theory, solution techniques, qualitative properties, and numerical approximation schemes for the principal varieties of partial differential equations that one encounters in both mathematics and applications. It is my hope that this book will fill this need, and thus help to educate and inspire the next generation of students, researchers, and practitioners.

While the classical topics of separation of variables, Fourier analysis, Green's functions, and special functions continue to form the core of an introductory course, the inclusion of nonlinear equations, shock wave dynamics, dispersion, symmetry and similarity methods, the Maximum Principle, Huygens' Principle, quantum mechanics and the Schrödinger equation, and mathematical finance makes this book more in tune with recent developments and trends. Numerical approximation schemes should also play an essential role in an introductory course, and this text covers the two most basic approaches: finite differences and finite elements.

On the other hand, modeling and the derivation of equations from physical phenomena and principles, while not entirely absent, has been downplayed, not because it is unimportant, but because time constraints limit what one can reasonably cover in an academic year's course. My own belief is that the primary purpose of a course in partial differential equations is to learn the principal solution techniques and to understand the underlying mathematical analysis. Thus, time devoted to modeling effectively lessens what can be adequately covered in the remainder of the course. For this reason, modeling is better left to a separate course that covers a wider range of mathematics, albeit at a more cursory level. (Modeling texts worth consulting include [**57**, **69**].) Nevertheless, this book continually makes contact with the physical applications that spawn the partial differential equations under consideration, and appeals to physical intuition and familiar phenomena to motivate, predict, and understand their mathematical properties, solutions, and applications. Nor do I attempt to cover stochastic differential equations — see [**83**] for this increasingly important area — although I do work through one important by-product: the Black–Scholes equation, which underlies the modern financial industry. I have tried throughout to balance rigor and intuition, thus giving the instructor flexibility with their relative emphasis and time to devote to solution techniques versus theoretical developments.

The course material has now been developed, tested, and revised over the past six years here at the University of Minnesota, and has also been used by several other universities in both the United States and abroad. It consists of twelve chapters along with two appendices that review basic complex numbers and some essential linear algebra. See below for further details on chapter contents and dependencies, and suggestions for possible semester and year-long courses that can be taught from the book.

Prerequisites

The initial prerequisite is a reasonable level of mathematical sophistication, which includes the ability to assimilate abstract constructions and apply them in concrete situations. Some physical insight and familiarity with basic mechanics, continuum physics, elementary thermodynamics, and, occasionally, quantum mechanics is also very helpful, but not essential.

Since partial differential equations involve the partial derivatives of functions, the most fundamental prerequisite is calculus — both univariate and multivariate. Fluency in the basics of differentiation, integration, and vector analysis is absolutely essential. Thus, the student should be at ease with limits, including one-sided limits, continuity, differentiation, integration, and the Fundamental Theorem. Key techniques include the chain rule, product rule, and quotient rule for differentiation, integration by parts, and change of variables in integrals. In addition, I assume some basic understanding of the convergence of sequences and series, including the standard tests — ratio, root, integral — along with Taylor's theorem and elementary properties of power series. (On the other hand, Fourier series will be developed from scratch.)

When dealing with several space dimensions, some familiarity with the key constructions and results from two- and three-dimensional vector calculus is helpful: rectangular (Cartesian), polar, cylindrical, and spherical coordinates; dot and cross products; partial derivatives; the multivariate chain rule; gradient, divergence, and curl; parametrized curves and surfaces; double and triple integrals; line and surface integrals, culminating in Green's Theorem and the Divergence Theorem — as well as very basic point set topology: notions of open, closed, bounded, and compact subsets of Euclidean space; the boundary of a domain and its normal direction; etc. However, all the required concepts and results will be quickly reviewed in the text at the appropriate juncture: Section 6.3 covers the two-dimensional material, while Section 12.1 deals with the three-dimensional counterpart.

Many solution techniques for partial differential equations, e.g., separation of variables and symmetry methods, rely on reducing them to one or more ordinary differential equations. In order to make progress, the student should therefore already know how to find the general solution to first-order linear equations, both homogeneous and inhomogeneous, along with separable nonlinear first-order equations, linear constant-coefficient equations, particularly those of second order, and first-order linear systems with constant-coefficient matrices, in particular the role of eigenvalues and the construction of a basis of solutions. The student should also be familiar with initial value problems, including statements of the basic existence and uniqueness theorems, but not necessarily their proofs. Basic references include [**18**, **20**, **23**], while more advanced topics can be found in [**52**, **54**, **59**]. On the other hand, while boundary value problems for ordinary differential equations play a central role in the analysis of partial differential equations, the book does not assume any prior experience, and will develop solution techniques from the beginning.

Students should also be familiar with the basics of complex numbers, including real and imaginary parts; modulus and phase (or argument); and complex exponentials and Euler's formula. These are reviewed in Appendix A. In the numerical chapters, some familiarity with basic computer arithmetic, i.e., floating-point and round-off errors, is assumed. Also, on occasion, basic numerical root finding algorithms, e.g., Newton's Method; numerical linear algebra, e.g., Gaussian Elimination and basic iterative methods; and numerical solution schemes for ordinary differential equations, e.g., Runge–Kutta Methods, are mentioned. Students who have forgotten the details can consult a basic numerical analysis textbook, e.g., [**24**, **60**], or reference volume, e.g., [**94**].

Finally, knowledge of the basic results and conceptual framework provided by modern linear algebra will be essential throughout the text. Students should already be on familiar terms with the fundamental concepts of vector space, both finite- and infinite-dimensional, linear independence, span, and basis, inner products, orthogonality, norms, and Cauchy–Schwarz and triangle inequalities, eigenvalues and eigenvectors, determinants, and linear systems. These are all covered in Appendix B; a more comprehensive and recommended reference is my previous textbook, [**89**], coauthored with my wife, Cheri Shakiban, which provides a firm grounding in the key ideas, results, and methods of modern applied linear algebra. Indeed, Chapter 9 here can be viewed as the next stage in the general linear algebraic framework that has proven to be so indispensable for the modern analysis and numerics of not just linear partial differential equations but, indeed, all of contemporary pure and applied mathematics.

While applications and solution techniques are paramount, the text does not shy away from precise statements of theorems and their proofs, especially when these help shed light on the applications and development of the subject. On the other hand, the more advanced results that require analytical sophistication beyond what can be reasonably assumed at this level are deferred to a subsequent, graduate-level course. In particular, the book does *not* assume that the student has taken a course in real analysis, and hence, while the basic ideas underlying Hilbert space are explained in the context of Fourier analysis, knowledge of measure theory and Lebesgue integration is neither assumed nor used. Consequently, the precise definitions of Hilbert space and generalized functions (distributions) are necessarily left somewhat vague, with the level of detail being similar to that found in a basic physics course on quantum mechanics. Indeed, one of the goals of the course is to inspire mathematics students (and others) to take a rigorous real analysis course, because it is so indispensable to the more advanced theory and applications of partial differential equations that build on the material presented here.

Outline of Chapters

The first chapter is brief and serves to set the stage, introducing some basic notation and describing what is meant by a partial differential equation and a (classical) solution thereof. It then describes the basic structure and properties of linear problems in a general sense, appealing to the underlying framework of linear algebra that is summarized in Appendix B. In particular, the fundamental superposition principles for both homogeneous and inhomogeneous linear equations and systems are employed throughout.

The first three sections of Chapter 2 are devoted to first-order partial differential equations in two variables — time and a single space coordinate — starting with simple linear cases. Constant-coefficient equations are easily solved, leading to the important concepts of characteristic and traveling wave. The method of characteristics is then extended, initially to linear first-order equations with variable coefficients, and then to the nonlinear case, where most solutions break down into discontinuous shock waves, whose subsequent dynamics relies on the underlying physics. The material on shocks may be at a slightly higher level of difficulty than the instructor wishes to deal with this early in the course, and hence may be downplayed or even omitted, perhaps returned to at a later stage, e.g., when studying Burgers' equation in Section 8.4, or when the concept of weak solution is introduced in Chapter 10. The final section of Chapter 2 is essential, and shows how the second-order wave equation can be reduced to a pair of first-order partial differential equations, thereby producing the celebrated solution formula of d'Alembert.

Chapter 3 covers the essentials of Fourier series, which is *the* most important tool in our analytical arsenal. After motivating the subject by adapting the eigenvalue method for solving linear systems of ordinary differential equations to the heat equation, the remainder of the chapter develops basic Fourier series analysis, in both real and complex forms. The final section investigates the various modes of convergence of Fourier series: pointwise, uniform, in norm. Along the way, Hilbert space and completeness are introduced, at an appropriate level of rigor. Although more theoretical than most of the material, this section is nevertheless strongly recommended, even for applications-oriented students, and can serve as a launching pad for higher-level analysis.

Chapter 4 immediately delves into the application of Fourier techniques to construct solutions to the three paradigmatic second-order partial differential equations in two independent variables — the heat, wave, and Laplace/Poisson equations — via the method of separation of variables. For dynamical problems, the separation of variables approach reinforces the importance of eigenfunctions. In the case of the Laplace equation, separation is performed in both rectangular and polar coordinates, thereby establishing the averaging property of solutions and, consequently, the Maximum Principle as important by-products. The chapter concludes with a short discussion of the classification of second-order partial differential equations, in two independent variables, into parabolic, hyperbolic, and elliptic categories, emphasizing their disparate natures and the role of characteristics.

Chapter 5 is the first devoted to numerical approximation techniques for partial differential equations. Here the emphasis is on finite difference methods. All of the preceding cases are discussed: heat equation, transport equations, wave equation, and Laplace/Poisson equation. The student learns that, in contrast to the field of ordinary differential equations, numerical methods must be specially adapted to the particularities of the partial differential equation under investigation, and may well not converge unless certain stability constraints are satisfied.

Chapter 6 introduces a second important solution method, founded on the notion of a Green's function. Our development relies on the use of distributions (generalized functions), concentrating on the extremely useful "delta function", which is characterized both as an unconventional limit of ordinary functions and, more rigorously but more abstractly, by duality in function space. While, as with Hilbert space, we do not assume familiarity with the analysis tools required to develop the fully rigorous theory of such generalized functions, the aim is for the student to assimilate the basic ideas and comfortably work with them in the context of practical examples. With this in hand, the Green's function approach is then first developed in the context of boundary value problems for ordinary differential equations, followed by consideration of elliptic boundary value problems for the Poisson equation in the plane.

Chapter 7 returns to Fourier analysis, now over the entire real line, resulting in the Fourier transform. Applications to boundary value problems are followed by a further development of Hilbert space and its role in modern quantum mechanics. Our discussion culminates with the Heisenberg Uncertainty Principle, which is viewed as a mathematical property of the Fourier transform. Space and time considerations persuaded me not to press on to develop the Laplace transform, which is a special case of the Fourier transform, although it can be profitably employed to study initial value problems for both ordinary and partial differential equations.

Chapter 8 integrates and further develops several different themes that arise in the analysis of dynamical evolution equations, both linear and nonlinear. The first section introduces the fundamental solution for the heat equation, and describes applications in mathematical finance through the celebrated Black–Scholes equation. The second section is a brief discussion of symmetry methods for partial differential equations, a favorite topic of the author and the subject of his graduate-level monograph [**87**]. Section 8.3 introduces the Maximum Principle for the heat equation, an important tool, inspired by physics, in the advanced analysis of parabolic problems. The last two sections study two basic higher-order nonlinear equations. Burgers' equation combines dissipative and nonlinear effects, and can be regarded as a simplified model of viscous fluid mechanics. Interestingly, Burgers' equation can be explicitly solved by transforming it into the linear heat equation. The convergence of its solutions to the shock-wave solutions of the limiting nonlinear transport equation underlies the modern analytic method of viscosity solutions. The final section treats basic third-order linear and nonlinear evolution equations arising, for example, in the modeling of surface waves. The linear equation serves to introduce the phenomenon of dispersion, in which different Fourier modes move at different velocities, producing common physical effects observed in, for instance, water waves. We also highlight the recently discovered and fascinating Talbot effect of dispersive quantization and fractalization on periodic domains. The nonlinear Korteweg–de Vries equation has many remarkable properties, including localized soliton solutions, first discovered in the 1960s, that result from its status as a completely integrable system.

Before proceeding further, Chapter 9 takes time to formulate a general abstract framework that underlies much of the more advanced analysis of linear partial differential equations. The material is at a slightly higher level of abstraction (although amply illustrated by concrete examples), so the more computationally oriented reader may wish to skip ahead to the last two chapters, referring back to the relevant concepts and general results in particular contexts as needed. Nevertheless, I strongly recommend covering at least some of this chapter, both because the framework is important to understanding the commonalities among various concrete instantiations, and because it demonstrates the pervasive power of mathematical analysis, even for those whose ultimate goal is applications. The development commences with the adjoint of a linear operator between inner product spaces — a powerful and far-ranging generalization of the matrix transpose — which naturally leads to consideration of self-adjoint and positive definite operators, all illustrated by finite-dimensional linear algebraic systems and boundary value problems governed by ordinary and partial differential equations. A particularly important construction, forming the foundation of the finite element numerical method, is the characterization of solutions to positive definite boundary value problems via minimization principles. Next, general results concerning eigenvalues and eigenfunctions of self-adjoint and positive definite operators are established, which serve to explain the key features of reality, orthogonality, and completeness that underlie Fourier and more general eigenfunction series expansions. A general characterization of complete eigenfunction systems based on properties of the Green's function nicely ties together two of the principal themes of the text.

Chapter 10 returns to the numerical analysis of partial differential equations, introducing the powerful finite element method. After outlining the general construction based on the preceding abstract minimization principle, we present its practical implementation, first for one-dimensional boundary value problems governed by ordinary differential equations and then for elliptic boundary value problems governed by the Laplace and Poisson equations in the plane. The final section develops an alternative approach, based on the idea of a weak solution to a partial differential equation, a concept of independent interest. Indeed, the nonclassical shock-wave solutions encountered in Section 2.3 are properly characterized as weak solutions.

The final two Chapters, 11 and 12, survey the analysis of partial differential equations in, respectively, two and three space dimensions, concentrating, as before, on the Laplace, heat, and wave equations. Much of the analysis relies on separation of variables, which, in curvilinear coordinates, leads to new classes of special functions that arise as solutions to certain linear second-order non-constant-coefficient ordinary differential equations. Since we are not assuming familiarity with this subject, the method of power series solutions to ordinary differential equations is developed in some detail. We also present the methods of Green's functions and fundamental solutions, including their qualitative properties and various applications. The material has been arranged according to spatial dimension rather than equation type; thus Chapter 11 deals with the planar heat and wave equations (the planar Laplace and Poisson equations having been treated earlier, in Chapters 4 and 6), while Chapter 12 covers all their three-dimensional counterparts. This arrangement allows a more orderly treatment of the required classes of special functions; thus, Bessel functions play the leading role in Chapter 11, while spherical harmonics, Legendre/Ferrers functions, and Laguerre polynomials star in Chapter 12. The last chapter also presents the Kirchhoff formula that solves the wave equation in three-dimensional space, an important consequence being the validity of Huygens' Principle concerning the localization of disturbances in space, which, surprisingly, does not hold in a two-dimensional universe. The book culminates with an analysis of the Schrödinger equation for the hydrogen atom, whose bound states are the atomic energy levels underlying the periodic table, atomic spectroscopy, and molecular chemistry.

Course Outlines and Chapter Dependencies

With sufficient planning and a suitably prepared and engaged class, most of the material in the text can be covered in a year. The typical single-semester course will finish with Chapter 6. Some pedagogical suggestions:

Chapter 1: Go through quickly, the main take-away being linearity and superposition.

Chapter 2: Most is worth covering and needed later, although Section 2.3, on shock waves, is optional, or can be deferred until later in the course.

Chapter 3: Students that have already taken a basic course in Fourier analysis can move directly ahead to the next chapter. The last section, on convergence, is important, but could be shortened or omitted in a more applied course.

Chapter 4: The heart of the first semester's course. Some of the material at the end of Section 4.1 — Robin boundary conditions and the root cellar problem — is optional, as is the very last subsection, on characteristics.

Chapter 5: A course that includes numerics (as I strongly recommend) should start with Section 5.1 and then cover at least a couple of the following sections, the selection depending upon the interests of the students and instructor.

Chapter 6: The material on distributions and the delta function is important for a student's general mathematical education, both pure and applied, and, in particular, for their role in the design of Green's functions. The proof of Green's representation formula (6.107) might be heavy going for some, and can be omitted by just covering the preceding less-rigorous justification of the logarithmic formula for the free-space Green's function.

Chapter 7: Sections 7.1 and 7.2 are essential, and convolution in Section 7.3 is also important. Section 7.4, on Hilbert space and quantum mechanics, can easily be omitted.

Chapter 8: All five sections are more or less independent of each other and, except for the fundamental solution and maximum principle for the heat equation, not used subsequently. Thus, the instructor can pick and choose according to interest and time alotted.

Chapter 9: This chapter is at a more abstract level than the bulk of the text, and can be skipped entirely (referring back when required), although if one intends to cover the finite element method, the material in the first three sections leading to minimization principles is required. Chapters 11 and 12 can, if desired, be launched into straight after Chapter 8, or even Chapter 7 plus the material on the heat equation in Chapter 8.

Chapter 10: Again, for a course that includes numerics, finite elements is extremely important and well worth covering. The final Section 10.4, on weak solutions, is optional, particularly the revisiting of shock waves, although if this was skipped in the early part of the course, now might be a good time to revisit Section 2.3.

Chapters 11 and 12: These constitute another essential component of the classical partial differential equations course. The detour into series solutions of ordinary differential equations is worth following, unless this is done elsewhere in the curriculum. I recommend trying to cover as much as possible, although one may well run out of time before reaching the end, in which case, consider omitting the end of Section 11.6, on Chladni figures and nodal curves, Section 12.6, on Kirchhoff's formula and Huygens' Principle, and Section 12.7, on the hydrogen atom. Of course, if Chapter 6, on Green's functions, and Section 8.1, on fundamental solutions, were omitted, those aspects will also presumably be omitted here; even if they were covered, there is not a compelling reason to revisit these topics in higher dimensions, and one may prefer to jump ahead to the more novel material appearing in the final sections.

Exercises and Software

Exercises appear at the end of almost every subsection, and come in a variety of genres. Most sets start with some straightforward computational problems to develop and reinforce the principal new techniques and ideas. Ability to solve these basic problems is a minimal requirement for successfully assimilating the material. More advanced exercises appear later on. Some are routine, but others involve challenging computations, computer-based projects, additional practical and theoretical developments, etc. Some will challenge even the most advanced reader. A number of straightforward technical proofs, as well as interesting and useful extensions of the material, particularly in the later chapters, have been relegated to the exercises to help maintain continuity of the narrative.

Don't be afraid to assign only a few parts of a multi-part exercise. I have found the True/False exercises to be particularly useful for testing of a student's level of understanding. A full answer is not merely a T or F, but must include a detailed explanation of the reason, e.g., a proof or a counterexample, or a reference to a result in the text. Many computer projects are included, particularly in the numerical chapters, where they are essential for learning the practical techniques. However, computer-based exercises are not tied to any specific choice of language or software; in my own course, MATLAB is the preferred programming platform. Some exercises could be streamlined or enhanced by the use of computer algebra systems, such as MATHEMATICA and MAPLE, but, in general, I have avoided assuming access to any symbolic software.

As a rough guide, some of the exercises are marked with special signs:

$\diamondsuit$ indicates an exercise that is referred to in the body of the text, or is important for further development or applications of the subject. These include theoretical details, omitted proofs, or new directions of importance.

$\heartsuit$ indicates a project — usually a longer exercise with multiple interdependent parts.

$\spadesuit$ indicates an exercise that requires (or at least strongly recommends) use of a computer. The student could be asked either to write their own computer code in, say, MATLAB, MAPLE, or MATHEMATICA, or to make use of pre-existing packages.

$\clubsuit = \spadesuit + \heartsuit$ indicates a more extensive computer project.

Movies

In the course of writing this book, I have made a number of movies to illustrate the dynamical behavior of solutions and their numerical approximations. I have found that they are an extremely effective pedagogical tool and strongly recommend showing them in the classroom with appropriate commentary and discussion. They are an ideal medium for fostering a student's deep understanding and insight into the phenomena exhibited by the at times indigestible analytical formulas — much better than the individual snapshots that appear in the figures in the printed book.

While it is clearly impossible to include the movies directly in the printed text, the electronic e-book version will contain direct links. In addition, I have posted all the movies on my own web site, along with the MATHEMATICA code used to generate them:

http://www.math.umn.edu/~olver/mov.html

When a movie is available, the sign ⊎ appears in the figure caption.

Conventions and Notation

A complete list of symbols employed can be found in the Symbol Index that appears at the end of the book.

Equations are numbered consecutively within chapters, so that, for example, (3.12) refers to the 12th equation in Chapter 3, irrespecive of which section it appears in.

Theorems, lemmas, propositions, definitions, and examples are also numbered consecutively within each chapter, using a single scheme. Thus, in Chapter 1, Definition 1.2 follows Example 1.1, and precedes Proposition 1.3 and Theorem 1.4. I find this numbering system to be the most helpful for speedy navigation through the book.

References (books, papers, etc.) are listed alphabetically at the end of the text, and are referred to by number. Thus, [**89**] is the 89th listed reference, namely my *Applied Linear Algebra* text.

Q.E.D. signifies the end of a proof, an acronym for "quod erat demonstrandum", which is Latin for "which was to be demonstrated".

The variables that appear throughout will be subject to consistent notational conventions. Thus t always denotes time, while x, y, z represent (Cartesian) space coordinates. Polar coordinates r, θ, cylindrical coordinates r, θ, z, and spherical coordinates r, θ, φ, will also be used when needed, and our conventions appear at the appropriate places in the exposition; be especially careful with the last case, since the angular variables θ, φ are subject to two contradictory conventions in the literature. The above are almost always independent variables in the partial differential equations under study; the dependent variables or unknowns will mostly be denoted by u, v, w, while f, g, h and F, G, H represent known functions, appearing as forcing terms or in boundary data. See Chapter 4 for our convention, used in differential geometry, used to denote functions in different coordinate systems, i.e., $u(x, y)$ versus $u(r, \theta)$.

In accordance with standard contemporary mathematical notation, the "blackboard bold" letter $\mathbb{R}$ denotes the real number line, $\mathbb{C}$ denotes the field of complex numbers, $\mathbb{Z}$ denotes the set of integers, both positive and negative, while $\mathbb{N}$ denotes the natural numbers, i.e., the nonnegative integers, including 0. Similarly, $\mathbb{R}^n$ and $\mathbb{C}^n$ denote the corresponding n-dimensional real and complex vector spaces consisting of n–tuples of elements of $\mathbb{R}$ and $\mathbb{C}$, respectively. The zero vector in each is denoted by $\mathbf{0}$.

Boldface lowercase letters, e.g., $\mathbf{v}, \mathbf{x}, \mathbf{a}$, usually denote vectors (almost always column vectors), whose entries are indicated by subscripts: v_1, x_i, etc. Matrices are denoted by ordinary capital letters, e.g., A, C, K, M — but not all such letters refer to matrices; for instance, V often refers to a vector space, while F is typically a forcing function. The entries of a matrix, say A, are indicated by the corresponding subscripted lowercase letters: a_{ij}, with i the row index and j the column index.

Angles are *always* measured in radians, although occasionally degrees will be mentioned in descriptive sentences. All trigonometric functions are evaluated on radian angles. Following the conventions advocated in [**85, 86**], we use $\operatorname{ph} z$ to denote the *phase* of a complex number $z \in \mathbb{C}$, which is more commonly called the *argument* and denoted by $\arg z$. Among the many reasons to prefer "phase" are to avoid potential confusion with the argument x of a function $f(x)$, as well as to be in accordance with the "Method of Stationary Phase" mentioned in Chapter 8.

We use $\{ f \,|\, C \}$ to denote a set, where f gives the formula for the members of the set and C is a (possibly empty) list of conditions. For example, $\{ x \mid 0 \le x \le 1 \}$ means the closed unit interval from 0 to 1, also written $[0,1]$, while $\{ a x^2 + b x + c \mid a,b,c \in \mathbb{R} \}$ is the set of real quadratic polynomials, and $\{0\}$ is the set consisting only of the number 0. We use $x \in S$ to indicate that x is an element of the set S, while $y \notin S$ says that y is not an element. Set theoretic union and intersection are denoted by $S \cup T$ and $S \cap T$, respectively. The subset sign $S \subset U$ includes the possibility that the sets S and U might be equal, although for emphasis we sometimes write $S \subseteq U$. On the other hand, $S \subsetneq U$ specifically implies that the two sets are not equal. We use $U \setminus S = \{ x \,|\, x \in U, x \notin S \}$ to denote the set-theoretic difference, meaning all elements of U that do not belong to S. We use the abbreviations max and min to denote the maximum and minimum elements of a set of real numbers, or of a real-valued function.

The symbol $\equiv$ is used to emphasize when two functions are identically equal, so $f(x) \equiv 1$ means that f is the constant function, equal to 1 at all values of x. It is also occasionally used in modular arithmetic, whereby $i \equiv j \bmod n$ means $i - j$ is divisible by n. The symbol $:=$ will define a quantity, e.g., $f(x) := x^2 - 1$. An arrow is used in two senses: first, to indicate convergence of a sequence, e.g., $x_n \to x^\star$ as $n \to \infty$, or, alternatively, to indicate a function, so $f\colon X \to Y$ means that the function f maps the domain set X to the image or target set Y, with formula $y = f(x)$. Composition of functions is denoted by $f \circ g$, while f^{-1} indicates the inverse function. Similarly, A^{-1} denotes the inverse of a matrix A.

By an *elementary function* we mean a combination of rational, algebraic, trigonometric, exponential, logarithmic, and hyperbolic functions. Familiarity with their basic properties is assumed. We always use $\log x$ for the natural (base e) logarithm — avoiding the ugly modern notation $\ln x$. On the other hand, the required properties of the various special functions — the error and complementary error functions, the gamma function, Airy functions, Bessel and spherical Bessel functions, Legendre and Ferrers functions, Laguerre functions, spherical harmonics, etc. — will be developed as needed.

Summation notation is used throughout, so $\displaystyle\sum_{i=1}^{n} a_i$ denotes the finite sum $a_1 + a_2 + \cdots + a_n$ or, if the upper limit is $n = \infty$, an infinite series. Of course, the lower limit need not be 1; if it is $-\infty$ and the upper limit is $+\infty$, the result is a doubly infinite series, e.g., the complex Fourier series in Chapter 3. We use $\displaystyle\lim_{n \to \infty} a_n$ to denote the usual limit of a sequence a_n. Similarly, $\displaystyle\lim_{x \to a} f(x)$ denotes the limit of the function $f(x)$ at a point a, while $f(x^-) = \displaystyle\lim_{x \to a^-} f(x)$ and $f(x^+) = \displaystyle\lim_{x \to a^+} f(x)$ are the one-sided (left- and right-hand, respectively) limits, which agree if and only if $\displaystyle\lim_{x \to a} f(x)$ exists.

We will employ a variety of standard notations for derivatives. In the case of ordinary derivatives, the most basic is the Leibniz notation $\frac{du}{dx}$ for the derivative of u with respect to x. As for partial derivatives, both the full Lebiniz notation $\frac{\partial u}{\partial t}, \frac{\partial u}{\partial x}, \frac{\partial^2 u}{\partial x^2}, \frac{\partial^3 u}{\partial t\, \partial x^2}$, and the more compact subscript notation $u_t, u_x, u_{xx}, u_{txx}$, etc. will be interchangeably employed throughout; see also Chapter 1. Unless specifically mentioned, all functions are assumed to be sufficiently smooth that any indicated derivatives exist and the relevant mixed partial derivatives are equal. Ordinary derivatives can also be indicated by the Newtonian notation u' instead of $\frac{du}{dx}$ and u'' for $\frac{d^2u}{dx^2}$, while $u^{(n)}$ denotes the nth order derivative $\frac{d^n u}{dx^n}$. If the variable is time, t, instead of space, x, then we may employ dots, $\dot{u}, \ddot{u}$, instead of primes.

Definite integrals are denoted by $\int_a^b f(x)\, dx$, while $\int f(x)\, dx$ is the corresponding indefinite integral or anti-derivative. We assume familiarity only with the Riemann theory of integration, although students who have learned Lebesgue integration may wish to take advantage of that on occasion, e.g., during the discussion of Hilbert space.

Historical Matters

Mathematics is both a historical and a social activity, and many notable algorithms, theorems, and formulas are named after famous (and, on occasion, not-so-famous) mathematicians, scientists, and engineers — usually, but not necessarily, the discoverer(s). The text includes a succinct description of many of the named contributors. Readers who are interested in more extensive historical details, complete biographies, and, when available, portraits or photos, are urged to consult the informative University of St. Andrews Mactutor web site:

`http://www-history.mcs.st-andrews.ac.uk/history/index.html`

Early prominent contributors to the subject include the Bernoulli family, Euler, d'Alembert, Lagrange, Laplace, and, particularly, Fourier, whose remarkable methods in part sparked the nineteenth century's rigorization of mathematical analysis and then mathematics in general, as pursued by Cauchy, Riemann, Cantor, Weierstrass, and Hilbert. In the twentieth century, the subject of partial differential equations reached maturity, producing an ever-increasing number of research papers, both theoretical and applied. Nevertheless, it remains one of the most challenging and active areas of mathematical research, and, in some sense, we have only scratched the surface of this deep and fascinating subject.

Textbooks devoted to partial differential equations began to appear long ago. Of particular note, Courant and Hilbert's monumental two-volume treatise, [**34, 35**], played a central role in the development of applied mathematics in general, and partial differential equations in particular. Indeed, it is not an exaggeration to state that all modern treatments, including this one, as well as large swaths of research, have been directly influenced by this magnificent text. Modern undergraduate textbooks worth consulting include [**50, 91, 92, 114, 120**], which are more or less at the same mathematical level but have a variety of points of view and selection of topics. The graduate-level texts [**38, 44, 61, 70, 99**] are recommended starting points for the more advanced reader and beginning researcher. More specialized monographs and papers will be referred to at the appropriate junctures.

This book began life in 1999 as a part of a planned comprehensive introduction to applied math, inspired in large part by Gilbert Strang's wonderful text, [**112**]. After some time and much effort, it was realized that the original vision was much too ambitious a goal, so my wife, Cheri Shakiban, and I recast the first part as our applied linear algebra textbook, [**89**]. I later decided that a large fraction of the remainder could be reworked into an introduction to partial differential equations, which, after some time and classroom testing, resulted in the book you are now reading.

Some Final Remarks

To the student: You are about to delve into the vast and important field of partial differential equations. I hope you enjoy the experience and profit from it in your future studies and career, wherever they may take you. Please send me your comments. Did you find the explanations helpful or confusing? Were enough examples included? Were the exercises of sufficient variety and appropriate level to enable you to learn the material? Do you have suggestions for improvements to be incorporated into a new edition?

To the instructor: Thank you for adopting this text! I hope you enjoy teaching from it as much as I enjoyed writing it. Whatever your experience, I want to hear from you. Let me know which parts you liked and which you didn't. Which sections worked and which were less successful. Which parts your students enjoyed, which parts they struggled with, and which parts they disliked. How can it be improved?

To all readers: Like every author, I sincerely hope that I have eliminated all errors in the text. But, more realistically, I know that no matter how many times one proofreads, mistakes still manage to squeeze through (or, worse, be generated during the editing process). Please email me your questions, typos, mathematical errors, comments, suggestions, and so on. The book's dedicated web site

`http://www.math.umn.edu/~olver/pde.html`

will actively maintain a comprehensive list of known corrections, commentary, feedback, and resources, as well as links to the movies and MATHEMATICA code mentioned above.

Acknowledgments

I have immensely profited from the many comments, corrections, suggestions, and remarks by students and mathematicians over the years. I would like to particularly thank my current and former colleagues at the University of Minnesota — Markus Keel, Svitlana Mayboroda, Willard Miller, Jr., Fadil Santosa, Guillermo Sapiro, Hans Weinberger, and the late James Serrin — for their invaluable advice and help. Over the past few years, Ariel Barton, Ellen Bao, Stefanella Boatto, Ming Chen, Bernard Deconinck, Greg Pierce, Thomas Scofield, and Steven Taylor all taught from these notes, and alerted me to a number of errors, made valuable suggestions, and shared their experiences in the classroom. I would like to thank Kendall Atkinson, Constantine Dafermos, Mark Dunster, and Gil Strang, for references and answering questions. Others who sent me commentary and corrections are Steven Brown, Bruno Carballo, Gong Chen, Neil Datta, René Gonin, Zeng Jianxin, Ben Jordan, Charles Lu, Anders Markvardsen, Cristina Santa Marta, Carmen Putrino, Troy Rockwood, Hullas Sehgal, Lubos Spacek, Rob Thompson, Douglas Wright, and Shangrong Yang. The following students caught typos during various classes: Dan Brinkman, Haoran Chen, Justin Hausauer, Matt Holzer, Jeff Gassmann, Keith Jackson, Binh Lieu, Dan Ouellette, Jessica Senou, Mark Stier, Hullas Seghan, David Toyli, Tom Trogdon, and Fei Zheng. While I didn't always agree with or follow their suggestions, I particularly want to thank the many reviewers of the book for their insightful comments on earlier drafts and valuable suggestions.

I would like to thank Achi Dosanjh for encouraging me to publish this book with Springer and for her enthusiastic encouragement and help during the production process. I am grateful to David Kramer for his thorough job copyediting the manuscript. While I did not always follow his suggested changes (and, somethimes, chose to deliberately go against certain grammatical and stylistic conventions in the interests of clarity), they were all seriously considered and the result is a much-improved exposition.

And last, but far from least, my mathematical family — my wife, Cheri Shakiban, my father, Frank W.J. Olver, and my son, Sheehan Olver — had a profound impact with their many comments, help, and advice over the years. Sadly, my father passed away at age 88 on April 23, 2013, and so never got to see the final printed version. I am dedicating this book to him and to my mother, Grace, who died in 1980, for their amazing influence on my life.

Peter J. Olver
September 2013

Corrected Printing

This is the corrected printing of the book that fixes all currently known typos and errors. Some minor improvements of the exposition, typesetting, and figures have also been incorporated. Again, I thank the staff at Springer for all their help and understanding during the production process. I would also like to thank to Adrian Fellhauer, Samuel Fleischer, Ulrich Gerlach, Christopher Grant, Joost Hulshof, Qunli Ji, Ted Kroon, Christoph Leuenberger, Artem Novozhilov, Ercüment Ortaccgil, Paul Georg Papatzacos, and Mikhail Shvartsman for their helpful comments on and corrections to the first printing.

Peter J. Olver
University of Minnesota
olver@umn.edu
http://www.math.umn.edu/~olver
May 2016

Table of Contents

Preface vii

Chapter 1. What Are Partial Differential Equations? 1

Classical Solutions 4

Initial Conditions and Boundary Conditions 6

Linear and Nonlinear Equations 8

Chapter 2. Linear and Nonlinear Waves 15

2.1. Stationary Waves 16

2.2. Transport and Traveling Waves 19

Uniform Transport 19

Transport with Decay 22

Nonuniform Transport 24

2.3. Nonlinear Transport and Shocks 31

Shock Dynamics 37

More General Wave Speeds 46

2.4. The Wave Equation: d'Alembert's Formula 49

d'Alembert's Solution 50

External Forcing and Resonance 56

Chapter 3. Fourier Series 63

3.1. Eigensolutions of Linear Evolution Equations 64

The Heated Ring 69

3.2. Fourier Series 72

Periodic Extensions 77

Piecewise Continuous Functions 79

The Convergence Theorem 82

Even and Odd Functions 85

Complex Fourier Series 88

3.3. Differentiation and Integration 92

Integration of Fourier Series 92

Differentiation of Fourier Series 94

3.4. Change of Scale 95

3.5. Convergence of Fourier Series 98

Pointwise and Uniform Convergence 99

Smoothness and Decay 104

Hilbert Space 106

Convergence in Norm 109

Completeness 112

Pointwise Convergence 115

Chapter 4. Separation of Variables 121
4.1. The Diffusion and Heat Equations 122
The Heat Equation 124
Smoothing and Long Time Behavior 126
The Heated Ring Redux 130
Inhomogeneous Boundary Conditions 133
Robin Boundary Conditions 134
The Root Cellar Problem 136
4.2. The Wave Equation 140
Separation of Variables and Fourier Series Solutions 140
The d'Alembert Formula for Bounded Intervals 146
4.3. The Planar Laplace and Poisson Equations 152
Separation of Variables 155
Polar Coordinates 160
Averaging, the Maximum Principle, and Analyticity 167
4.4. Classification of Linear Partial Differential Equations 172
Characteristics and the Cauchy Problem 175

Chapter 5. Finite Differences 181
5.1. Finite Difference Approximations 182
5.2. Numerical Algorithms for the Heat Equation 186
Stability Analysis 188
Implicit and Crank–Nicolson Methods 190
5.3. Numerical Algorithms for First Order Partial Differential Equations . 195
The CFL Condition 196
Upwind and Lax–Wendroff Schemes 198
5.4. Numerical Algorithms for the Wave Equation 201
5.5. Finite Difference Algorithms for the Laplace and Poisson Equations . 207
Solution Strategies 211

Chapter 6. Generalized Functions and Green's Functions 215
6.1. Generalized Functions 216
The Delta Function 217
Calculus of Generalized Functions 221
The Fourier Series of the Delta Function 229
6.2. Green's Functions for One–Dimensional Boundary Value Problems . . 234
6.3. Green's Functions for the Planar Poisson Equation 242
Calculus in the Plane 242
The Two–Dimensional Delta Function 246
The Green's Function 248
The Method of Images 256

Chapter 7. Fourier Transforms 263
7.1. The Fourier Transform 263
Concise Table of Fourier Transforms 272
7.2. Derivatives and Integrals 275
Differentiation 275
Integration 276
7.3. Green's Functions and Convolution 278
Solution of Boundary Value Problems 278
Convolution 281
7.4. The Fourier Transform on Hilbert Space 284
Quantum Mechanics and the Uncertainty Principle 286

Chapter 8. Linear and Nonlinear Evolution Equations 291
8.1. The Fundamental Solution to the Heat Equation 292
The Forced Heat Equation and Duhamel's Principle 296
The Black–Scholes Equation and Mathematical Finance 299
8.2. Symmetry and Similarity 305
Similarity Solutions 308
8.3. The Maximum Principle 312
8.4. Nonlinear Diffusion 315
Burgers' Equation 315
The Hopf–Cole Transformation 317
8.5. Dispersion and Solitons 323
Linear Dispersion 324
The Dispersion Relation 330
The Korteweg–de Vries Equation 333

Chapter 9. A General Framework for Linear Partial Differential Equations 339
9.1. Adjoints 340
Differential Operators 342
Higher–Dimensional Operators 345
The Fredholm Alternative 350
9.2. Self–Adjoint and Positive Definite Linear Functions 353
Self–Adjointness 354
Positive Definiteness 355
Two–Dimensional Boundary Value Problems 359
9.3. Minimization Principles 362
Sturm–Liouville Boundary Value Problems 363
The Dirichlet Principle 368
9.4. Eigenvalues and Eigenfunctions 371
Self–Adjoint Operators 371
The Rayleigh Quotient 375
Eigenfunction Series 378
Green's Functions and Completeness 379

9.5. A General Framework for Dynamics 385
Evolution Equations 386
Vibration Equations 388
Forcing and Resonance 389
The Schrödinger Equation 394

Chapter 10. Finite Elements and Weak Solutions 399
10.1. Minimization and Finite Elements 400
10.2. Finite Elements for Ordinary Differential Equations 403
10.3. Finite Elements in Two Dimensions 410
Triangulation 411
The Finite Element Equations 416
Assembling the Elements 418
The Coefficient Vector and the Boundary Conditions 422
Inhomogeneous Boundary Conditions 424
10.4. Weak Solutions 427
Weak Formulations of Linear Systems 428
Finite Elements Based on Weak Solutions 430
Shock Waves as Weak Solutions 431

Chapter 11. Dynamics of Planar Media 435
11.1. Diffusion in Planar Media 435
Derivation of the Diffusion and Heat Equations 436
Separation of Variables 439
Qualitative Properties 440
Inhomogeneous Boundary Conditions and Forcing 442
The Maximum Principle 443
11.2. Explicit Solutions of the Heat Equation 445
Heating of a Rectangle 446
Heating of a Disk — Preliminaries 450
11.3. Series Solutions of Ordinary Differential Equations 453
The Gamma Function 453
Regular Points 456
The Airy Equation 459
Regular Singular Points 463
Bessel's Equation 467
11.4. The Heat Equation in a Disk, Continued 474
11.5. The Fundamental Solution to the Planar Heat Equation 481
11.6. The Planar Wave Equation 486
Separation of Variables 487
Vibration of a Rectangular Drum 488
Vibration of a Circular Drum 490
Scaling and Symmetry 494
Chladni Figures and Nodal Curves 497

Chapter 12. Partial Differential Equations in Space 503
12.1. The Three–Dimensional Laplace and Poisson Equations 504
Self–Adjoint Formulation and Minimum Principle 505
12.2. Separation of Variables for the Laplace Equation 506
Laplace's Equation in a Ball 507
The Legendre Equation and Ferrers Functions 510
Spherical Harmonics 517
Harmonic Polynomials 519
Averaging, the Maximum Principle, and Analyticity 521
12.3. Green's Functions for the Poisson Equation 527
The Free–Space Green's Function 528
Bounded Domains and the Method of Images 531
12.4. The Heat Equation for Three–Dimensional Media 535
Heating of a Ball 537
Spherical Bessel Functions 538
The Fundamental Solution to the Heat Equation in Space . . . 543
12.5. The Wave Equation for Three–Dimensional Media 545
Vibration of Balls and Spheres 547
12.6. Spherical Waves and Huygens' Principle 551
Spherical Waves 551
Kirchhoff's Formula and Huygens' Principle 558
Descent to Two Dimensions 561
12.7. The Hydrogen Atom 564
Bound States 565
Atomic Eigenstates and Quantum Numbers 567

Appendix A. Complex Numbers 571

Appendix B. Linear Algebra 575
B.1. Vector Spaces and Subspaces 575
B.2. Bases and Dimension 576
B.3. Inner Products and Norms 578
B.4. Orthogonality 581
B.5. Eigenvalues and Eigenvectors 582
B.6. Linear Iteration 583
B.7. Linear Functions and Systems 585

References 589

Symbol Index 595

Author Index 603

Subject Index 607

Chapter 1
What Are Partial Differential Equations?

Let us begin by delineating our field of study. A *differential equation* is an equation that relates the derivatives of a (scalar) function depending on one or more variables. For example,

$$\frac{d^4u}{dx^4} + \frac{d^2u}{dx^2} + u^2 = \cos x \tag{1.1}$$

is a differential equation for the function $u(x)$ depending on a single variable x, while

$$\frac{\partial u}{\partial t} = \frac{\partial^2 u}{\partial x^2} + \frac{\partial^2 u}{\partial y^2} - u \tag{1.2}$$

is a differential equation involving a function $u(t,x,y)$ of three variables.

A differential equation is called *ordinary* if the function u depends on only a single variable, and *partial* if it depends on more than one variable. Usually (but not quite always) the dependence of u can be inferred from the derivatives that appear in the differential equation. The *order* of a differential equation is that of the highest-order derivative that appears in the equation. Thus, (1.1) is a fourth-order ordinary differential equation, while (1.2) is a second-order partial differential equation.

Remark: A differential equation has order 0 if it contains no derivatives of the function u. These are more properly treated as *algebraic equations*,† which, while of great interest in their own right, are not the subject of this text. To be a bona fide *differential equation*, it must contain at least one derivative of u, and hence have order ≥ 1.

There are two common notations for partial derivatives, and we shall employ them interchangeably. The first, used in (1.1) and (1.2), is the familiar Leibniz notation that employs a d to denote ordinary derivatives of functions of a single variable, and the ∂ symbol (usually also pronounced "dee") for partial derivatives of functions of more than one variable. An alternative, more compact notation employs subscripts to indicate partial derivatives. For example, u_t represents $\partial u/\partial t$, while u_{xx} is used for $\partial^2 u/\partial x^2$, and $\partial^3 u/\partial x^2 \partial y$ for u_{xxy}. Thus, in subscript notation, the partial differential equation (1.2) is written

$$u_t = u_{xx} + u_{yy} - u. \tag{1.3}$$

† Here, the term "algebraic equation" is used only to distinguish such equations from true "differential equations". It does not mean that the defining functions are necessarily algebraic, e.g., polynomials. For example, the transcendental equation $\tan u = u$, which appears later in (4.50), is still regarded as an algebraic equation in this book.

We will similarly abbreviate partial differential operators, sometimes writing $\partial/\partial x$ as ∂_x, while $\partial^2/\partial x^2$ can be written as either ∂_x^2 or ∂_{xx}, and $\partial^3/\partial x^2 \partial y$ becomes $\partial_{xxy} = \partial_x^2 \partial_y$.

It is worth pointing out that the preponderance of differential equations arising in applications, in science, in engineering, and within mathematics itself are of either first or second order, with the latter being by far the most prevalent. Third-order equations arise when modeling waves in dispersive media, e.g., water waves or plasma waves. Fourth-order equations show up in elasticity, particularly plate and beam mechanics, and in image processing. Equations of order ≥ 5 are very rare.

A basic prerequisite for studying this text is the ability to solve simple ordinary differential equations: first-order equations; linear constant-coefficient equations, both homogeneous and inhomogeneous; and linear systems. In addition, we shall assume some familiarity with the basic theorems concerning the existence and uniqueness of solutions to initial value problems. There are many good introductory texts, including [**18, 20, 23**]. More advanced treatises include [**31, 52, 54, 59**]. Partial differential equations are considerably more demanding, and can challenge the analytical skills of even the most accomplished mathematician. Many of the most effective solution strategies rely on reducing the partial differential equation to one or more ordinary differential equations. Thus, in the course of our study of partial differential equations, we will need to develop, ab initio, some of the more advanced aspects of the theory of ordinary differential equations, including boundary value problems, eigenvalue problems, series solutions, singular points, and special functions.

Following the introductory remarks in the present chapter, the exposition begins in earnest with simple first-order equations, concentrating on those that arise as models of wave phenomena. Most of the remainder of the text will be devoted to understanding and solving the three essential linear second-order partial differential equations in one, two, and three space dimensions:† the *heat equation*, modeling thermodynamics in a continuous medium, as well as diffusion of animal populations and chemical pollutants; the *wave equation*, modeling vibrations of bars, strings, plates, and solid bodies, as well as acoustic, fluid, and electromagnetic vibrations; and the *Laplace equation* and its inhomogeneous counterpart, the *Poisson equation*, governing the mechanical and thermal equilibria of bodies, as well as fluid-mechanical and electromagnetic potentials.

Each increase in dimension requires an increase in mathematical sophistication, as well as the development of additional analytic tools — although the key ideas will have all appeared once we reach our physical, three-dimensional universe. The three starring examples — heat, wave, and Laplace/Poisson — are not only essential to a wide range of applications, but also serve as instructive paradigms for the three principal classes of linear partial differential equations — parabolic, hyperbolic, and elliptic. Some interesting nonlinear partial differential equations, including first-order transport equations modeling shock waves, the second-order Burgers' equation governing simple nonlinear diffusion processes, and the third-order Korteweg–de Vries equation governing dispersive waves, will also be discussed. But, in such an introductory text, the further reaches of the vast realm of nonlinear partial differential equations must remain unexplored, awaiting the reader's more advanced mathematical excursions.

More generally, a *system of differential equations* is a collection of one or more equations relating the derivatives of one or more functions. It is essential that all the functions

† For us, *dimension* always refers to the number of space dimensions. Time, although theoretically also a dimension, plays a very different physical role, and therefore (at least in nonrelativistic systems) is to be treated on a separate footing.

occurring in the system *depend on the same set of variables*. The symbols representing these functions are known as the *dependent variables*, while the variables that they depend on are called the *independent variables*. Systems of differential equations are called *ordinary* or *partial* according to whether there are one or more independent variables. The *order* of the system is the highest-order derivative occurring in any of its equations.

For example, the three-dimensional *Navier–Stokes equations*

$$\begin{aligned}
\frac{\partial u}{\partial t} + u\,\frac{\partial u}{\partial x} + v\,\frac{\partial u}{\partial y} + w\,\frac{\partial u}{\partial z} &= -\,\frac{\partial p}{\partial x} + \nu\left(\frac{\partial^2 u}{\partial x^2} + \frac{\partial^2 u}{\partial y^2} + \frac{\partial^2 u}{\partial z^2}\right),\\
\frac{\partial v}{\partial t} + u\,\frac{\partial v}{\partial x} + v\,\frac{\partial v}{\partial y} + w\,\frac{\partial v}{\partial z} &= -\,\frac{\partial p}{\partial y} + \nu\left(\frac{\partial^2 v}{\partial x^2} + \frac{\partial^2 v}{\partial y^2} + \frac{\partial^2 v}{\partial z^2}\right),\\
\frac{\partial w}{\partial t} + u\,\frac{\partial w}{\partial x} + v\,\frac{\partial w}{\partial y} + w\,\frac{\partial w}{\partial z} &= -\,\frac{\partial p}{\partial z} + \nu\left(\frac{\partial^2 w}{\partial x^2} + \frac{\partial^2 w}{\partial y^2} + \frac{\partial^2 w}{\partial z^2}\right),\\
\frac{\partial u}{\partial x} + \frac{\partial v}{\partial y} + \frac{\partial w}{\partial z} &= 0,
\end{aligned} \tag{1.4}$$

is a second-order system of differential equations that involves four functions, $u(t,x,y,z)$, $v(t,x,y,z)$, $w(t,x,y,z)$, $p(t,x,y,z)$, each depending on four variables, while $\nu \geq 0$ is a fixed constant. (The function p necessarily depends on t, even though no t derivative of it appears in the system.) The independent variables are t, representing time, and x, y, z, representing space coordinates. The dependent variables are u, v, w, p, with $\mathbf{v} = (u, v, w)$ representing the velocity vector field of an incompressible fluid flow, e.g., water, and p the accompanying pressure. The parameter ν measures the viscosity of the fluid. The Navier–Stokes equations are fundamental in fluid mechanics, [**12**], and are notoriously difficult to solve, either analytically or numerically. Indeed, establishing the existence or nonexistence of solutions for all future times remains a major unsolved problem in mathematics, whose resolution will earn you a $1,000,000 prize; see `http://www.claymath.org` for details. The Navier–Stokes equations first appeared in the early 1800s in works of the French applied mathematician/engineer Claude-Louis Navier and, later, the British applied mathematician George Stokes, whom you already know from his eponymous multivariable calculus theorem.[†] The inviscid case, $\nu = 0$, is known as the *Euler equations* in honor of their discoverer, the incomparably influential eighteenth-century Swiss mathematician Leonhard Euler.

We shall be employing a few basic notational conventions regarding the variables that appear in our differential equations. We always use t to denote time, while x, y, z will represent (Cartesian) space coordinates. Polar coordinates r, θ, cylindrical coordinates r, θ, z, and spherical coordinates[‡] r, θ, φ, will also be used when needed. An *equilibrium equation* models an unchanging physical system, and so involves only the space variable(s). The time variable appears when modeling *dynamical*, meaning time-varying, processes. Both time and space coordinates are (usually) independent variables. The dependent variables will mostly be denoted by u, v, w, although occasionally — particularly in representing

[†] Interestingly, Stokes' Theorem was taken from an 1850 letter that Lord Kelvin wrote to Stokes, who turned it into an undergraduate exam question for the Smith Prize at Cambridge University in England. However, unbeknownst to either, the result had, in fact, been discovered earlier by George Green, the father of Green's Theorem and also the Green's function, which will be the subject of Chapter 6.

[‡] See Section 12.2 for our notational convention.

particular physical quantities — other letters may be employed, e.g., the pressure p in (1.4). On the other hand, the letters f, g, h typically represent specified functions of the independent variables, e.g., forcing or boundary or initial conditions.

In this introductory text, we must confine our attention to the most basic analytic and numerical solution techniques for a select few of the most important partial differential equations. More advanced topics, including all systems of partial differential equations, must be deferred to graduate and research-level texts, e.g., [**35, 38, 44, 61, 99**]. In fact, many important issues remain incompletely resolved and/or poorly understood, making partial differential equations one of the most active and exciting fields of contemporary mathematical research. One of my goals is that, by reading this book, you will be both inspired and equipped to venture much further into this fascinating and essential area of mathematics and/or its remarkable range of applications throughout science, engineering, economics, biology, and beyond.

Exercises

1.1. Classify each of the following differential equations as ordinary or partial, and equilibrium or dynamic; then write down its order. (a) $\dfrac{du}{dx} + x\,u = 1$, (b) $\dfrac{\partial u}{\partial t} + u\,\dfrac{\partial u}{\partial x} = x$, (c) $u_{tt} = 9u_{xx}$, (d) $\dfrac{\partial u}{\partial t} = \dfrac{\partial^2 u}{\partial x^2} + \dfrac{\partial u}{\partial x}$, (e) $-\dfrac{\partial^2 u}{\partial x^2} - \dfrac{\partial^2 u}{\partial y^2} = x^2 + y^2$, (f) $\dfrac{d^2 u}{dt^2} + 3u = \sin t$, (g) $u_{xx} + u_{yy} + u_{zz} + (x^2 + y^2 + z^2)u = 0$, (h) $u_{xx} = x + u^2$, (i) $\dfrac{\partial u}{\partial t} + \dfrac{\partial^3 u}{\partial x^3} + u\,\dfrac{\partial u}{\partial x} = 0$, (j) $\dfrac{\partial^2 u}{\partial x^2} + \dfrac{\partial^2 u}{\partial y\,\partial z} = u$, (k) $u_{tt} = u_{xxxx} + 2u_{xxyy} + u_{yyyy}$.

1.2. In two space dimensions, the *Laplacian* is defined as the second-order partial differential operator $\Delta = \partial_x^2 + \partial_y^2$. Write out the following partial differential equations in (*i*) Leibniz notation; (*ii*) subscript notation: (a) the Laplace equation $\Delta u = 0$; (b) the Poisson equation $-\Delta u = f$; (c) the two-dimensional heat equation $\partial_t u = \Delta u$; (d) the von Karman plate equation $\Delta^2 u = 0$.

1.3. Answer Exercise 1.2 for the three-dimensional Laplacian $\Delta = \partial_x^2 + \partial_y^2 + \partial_z^2$.

1.4. Identify the independent variables, the dependent variables, and the order of the following systems of partial differential equations: (a) $\dfrac{\partial u}{\partial x} = \dfrac{\partial v}{\partial y}$, $\dfrac{\partial u}{\partial y} = -\dfrac{\partial v}{\partial x}$; (b) $u_{xx} + v_{yy} = \cos(x + y)$, $u_x v_y - u_y v_x = 1$; (c) $\dfrac{\partial u}{\partial t} = \dfrac{\partial v}{\partial x}$, $\dfrac{\partial^2 v}{\partial t^2} = \dfrac{\partial^2 u}{\partial x^2}$; (d) $u_t + u\,u_x + v\,u_y = p_x$, $v_t + u\,v_x + v\,v_y = p_y$, $u_x + v_y = 0$; (e) $u_t = v_{xxx} + v(1 - v)$, $v_t = u_{xxy} + v\,w$, $w_t = u_x + v_y$.

Classical Solutions

Let us now focus our attention on a single differential equation involving a single, scalar-valued function u that depends on one or more independent variables. The function u

is usually real-valued, although complex-valued functions can, and do, play a role in the analysis. Everything that we say in this section will, when suitably adapted, apply to systems of differential equations.

By a *solution* we mean a sufficiently smooth function u of the independent variables that satisfies the differential equation at every point of its domain of definition. We do not necessarily require that the solution be defined for all possible values of the independent variables. Indeed, usually the differential equation is imposed on some domain D contained in the space of independent variables, and we seek a solution defined only on D. In general, the *domain* D will be an open subset, usually connected and, particularly in equilibrium equations, often bounded, with a reasonably nice boundary, denoted by ∂D.

We will call a function *smooth* if it can be differentiated sufficiently often, at least so that all of the derivatives appearing in the equation are well defined on the domain of interest D. More specifically, if the differential equation has order n, then we require that the solution u be of *class* C^n, which means that it and all its derivatives of order $\leq n$ are continuous functions in D, and such that the differential equation that relates the derivatives of u holds throughout D. However, on occasion, e.g., when dealing with shock waves, we will consider more general types of solutions. The most important such class consists of the so-called "weak solutions" to be introduced in Section 10.4. To emphasize the distinction, the smooth solutions described above are often referred to as *classical solutions*. In this book, the term "solution" without extra qualification will usually mean "classical solution".

Example 1.1. A classical solution to the heat equation

$$\frac{\partial u}{\partial t} = \frac{\partial^2 u}{\partial x^2} \tag{1.5}$$

is a function $u(t,x)$, defined on a domain $D \subset \mathbb{R}^2$, such that all of the functions

$$u(t,x), \quad \frac{\partial u}{\partial t}(t,x), \quad \frac{\partial u}{\partial x}(t,x), \quad \frac{\partial^2 u}{\partial t^2}(t,x), \quad \frac{\partial^2 u}{\partial t\,\partial x}(t,x) = \frac{\partial^2 u}{\partial x\,\partial t}(t,x), \quad \frac{\partial^2 u}{\partial x^2}(t,x),$$

are well defined and continuous[†] at every point $(t,x) \in D$, so that $u \in \mathrm{C}^2(D)$, and, moreover, (1.5) holds at every $(t,x) \in D$. Observe that, even though only u_t and u_{xx} explicitly appear in the heat equation, we require continuity of *all* the partial derivatives of order ≤ 2 in order that u qualify as a classical solution. For example,

$$u(t,x) = t + \tfrac{1}{2}x^2 \tag{1.6}$$

is a solution to the heat equation that is defined on the full domain $D = \mathbb{R}^2$ because it is[‡] C^2, and, moreover,

$$\frac{\partial u}{\partial t} = 1 = \frac{\partial^2 u}{\partial x^2}.$$

Another, more complicated but extremely important, solution is

$$u(t,x) = \frac{e^{-x^2/(4t)}}{2\sqrt{\pi\, t}}. \tag{1.7}$$

[†] The equality of the mixed partial derivatives follows from a general theorem in multivariable calculus, [**8**, **97**, **108**]. Classical solutions automatically enjoy equality of all their relevant mixed partial derivatives.

[‡] In fact, the function (1.6) is C^∞, meaning infinitely differentiable, on all of $\mathbb{R}^2$.

One easily verifies that $u \in \mathrm{C}^2$ and, moreover, solves the heat equation on the domain $D = \{ (t,x) \mid t > 0 \} \subset \mathbb{R}^2$. The reader is invited to verify this by computing $\partial u / \partial t$ and $\partial^2 u / \partial x^2$, and then checking that they are equal. Finally, with $\mathrm{i} = \sqrt{-1}$ denoting the imaginary unit, we note that

$$u(t,x) = e^{-t + \mathrm{i}\,x} = e^{-t} \cos x + \mathrm{i}\, e^{-t} \sin x, \tag{1.8}$$

the second expression following from Euler's formula (A.11), defines a complex-valued solution to the heat equation. This can be verified directly, since the rules for differentiating complex exponentials are identical to those for their real counterparts:

$$\frac{\partial u}{\partial t} = -e^{-t + \mathrm{i}\,x}, \qquad \frac{\partial u}{\partial x} = \mathrm{i}\, e^{-t + \mathrm{i}\,x}, \qquad \text{and so} \qquad \frac{\partial^2 u}{\partial x^2} = -e^{-t + \mathrm{i}\,x} = \frac{\partial u}{\partial t}.$$

It is worth pointing out that both the real part, $e^{-t} \cos x$, and the imaginary part, $e^{-t} \sin x$, of the complex solution (1.8) are individual real solutions, which is indicative of a fairly general property.

Incidentally, most partial differential equations arising in physical applications are real, and, although complex solutions often facilitate their analysis, at the end of the day we require real, physically meaningful solutions. A notable exception is quantum mechanics, which is an inherently complex-valued physical theory. For example, the one-dimensional *Schrödinger equation*

$$\mathrm{i}\,\hbar \,\frac{\partial u}{\partial t} = -\frac{\hbar^2}{2m} \frac{\partial^2 u}{\partial x^2} + V(x)\, u, \tag{1.9}$$

with $\hbar$ denoting *Planck's constant*, which is real, governs the dynamical evolution of the complex-valued wave function $u(t,x)$ describing the probabilistic distribution of a quantum particle of mass m, e.g., an electron, moving in the force field prescribed by the (real) potential function $V(x)$. While the solution u is complex-valued, the independent variables t, x, representing time and space, remain real.

Initial Conditions and Boundary Conditions

How many solutions does a partial differential equation have? In general, lots. Even ordinary differential equations have infinitely many solutions. Indeed, the general solution to a single n^{th} order ordinary differential equation depends on n arbitrary constants. The solutions to partial differential equations are yet more numerous, in that they depend on *arbitrary functions*. Very roughly, we can expect the solution to an nth order partial differential equation involving m independent variables to depend on n arbitrary functions of $m - 1$ variables. But this must be taken with a large grain of salt — only in a few special instances will we actually be able to express the solution in terms of arbitrary functions.

The solutions to dynamical ordinary differential equations are singled out by the imposition of initial conditions, resulting in an *initial value problem*. On the other hand, equations modeling equilibrium phenomena require boundary conditions to specify their solutions uniquely, resulting in a *boundary value problem*. We assume that the reader is already familiar with the basics of initial value problems for ordinary differential equations. But we will take time to develop the perhaps less familiar case of boundary value problems for ordinary differential equations in Chapter 6.

A similar specification of auxiliary conditions applies to partial differential equations. Equations modeling equilibrium phenomena are supplemented by boundary conditions imposed on the boundary of the domain of interest. In favorable circumstances, the boundary conditions serve to single out a unique solution. For example, the equilibrium temperature of a body is uniquely specified by its boundary behavior. If the domain is unbounded, one must also restrict the nature of the solution at large distances, e.g., by asking that it remain bounded. The combination of a partial differential equation along with suitable boundary conditions is referred to as a *boundary value problem*.

There are three principal types of boundary value problems that arise in most applications. Specifying the value of the solution along the boundary of the domain is called a *Dirichlet boundary condition*, to honor the nineteenth-century analyst Johann Peter Gustav Lejeune Dirichlet. Specifying the normal derivative of the solution along the boundary results in a *Neumann boundary condition*, named after his contemporary Carl Gottfried Neumann. Prescribing the function along part of the boundary and the normal derivative along the remainder results in a *mixed boundary value problem*. For example, in thermal equilibrium, the Dirichlet boundary value problem specifies the temperature of a body along its boundary, and our task is to find the interior temperature distribution by solving an appropriate partial differential equation. Similarly, the Neumann boundary value problem prescribes the heat flux through the boundary. In particular, an insulated boundary has no heat flux, and hence the normal derivative of the temperature is zero on the boundary. The mixed boundary value problem prescribes the temperature along part of the boundary and the heat flux along the remainder. Again, our task is to determine the interior temperature of the body.

For partial differential equations modeling dynamical processes, in which time is one of the independent variables, the solution is to be specified by one or more initial conditions. The number of initial conditions required depends on the highest-order time derivative that appears in the equation. For example, in thermodynamics, which involves only the first-order time derivative of the temperature, the initial condition requires specifying the temperature of the body at the initial time. Newtonian mechanics describes the acceleration or second-order time derivative of the motion, and so requires two initial conditions: the initial position and initial velocity of the system. On bounded domains, one must also impose suitable boundary conditions in order to uniquely characterize the solution and hence the subsequent dynamical behavior of the physical system. The combination of the partial differential equation, the initial conditions, and the boundary conditions leads to an *initial-boundary value problem*. We will encounter, and solve, many important examples of such problems during the course of this text.

Remark: An additional consideration is that, besides any smoothness required by the partial differential equation within the domain, the solution and any of its derivatives specified in any initial or boundary condition should also be continuous at the initial or boundary point where the condition is imposed. For example, if the initial condition specifies the function value $u(0,x)$ for $a < x < b$, while the boundary conditions specify the derivatives $\dfrac{\partial u}{\partial x}(t,a)$ and $\dfrac{\partial u}{\partial x}(t,b)$ for $t > 0$, then, in addition to any smoothness required inside the domain $\{a < x < b,\ t > 0\}$, we also require that u be continuous at all initial points $(0,x)$, and that its derivative $\dfrac{\partial u}{\partial x}$ be continuous at all boundary points (t,a) and (t,b), in order that $u(t,x)$ qualify as a *classical solution* to the initial-boundary value problem.

Exercises

1.5. Show that the following functions $u(x,y)$ define classical solutions to the two-dimensional Laplace equation $\dfrac{\partial^2 u}{\partial x^2} + \dfrac{\partial^2 u}{\partial y^2} = 0$. Be careful to specify an appropriate domain.
(a) $e^x \cos y$, (b) $1+x^2-y^2$, (c) x^3-3xy^2, (d) $\log(x^2+y^2)$, (e) $\tan^{-1}(y/x)$, (f) $\dfrac{x}{x^2+y^2}$.

1.6. Find all solutions $u = f(r)$ of the two-dimensional Laplace equation $u_{xx} + u_{yy} = 0$ that depend only on the radial coordinate $r = \sqrt{x^2+y^2}$.

1.7. Find all (real) solutions to the two-dimensional Laplace equation $u_{xx}+u_{yy} = 0$ of the form $u = \log p(x,y)$, where $p(x,y)$ is a quadratic polynomial.

1.8. (a) Find all quadratic polynomial solutions of the three-dimensional Laplace equation $\dfrac{\partial^2 u}{\partial x^2} + \dfrac{\partial^2 u}{\partial y^2} + \dfrac{\partial^2 u}{\partial z^2} = 0$. (b) Find all the homogeneous cubic polynomial solutions.

1.9. Find all polynomial solutions $p(t,x)$ of the heat equation $u_t = u_{xx}$ with $\deg p \leq 3$.

1.10. Show that each of the following functions $u(t,x)$ is a solution to the wave equation $u_{tt} = 4u_{xx}$: (a) $4t^2 - x^2$; (b) $\cos(x+2t)$; (c) $\sin 2t \cos x$; (d) $e^{-(x-2t)^2}$.

1.11. Find all polynomial solutions $p(t,x)$ of the wave equation $u_{tt} = u_{xx}$ with
(a) $\deg p \leq 2$, (b) $\deg p = 3$.

1.12. Suppose $u(t,x)$ and $v(t,x)$ are C^2 functions defined on $\mathbb{R}^2$ that satisfy the first-order system of partial differential equations $u_t = v_x$, $v_t = u_x$.
(a) Show that both u and v are classical solutions to the wave equation $u_{tt} = u_{xx}$. Which result from multivariable calculus do you need to justify the conclusion?
(b) Conversely, given a classical solution $u(t,x)$ to the wave equation, can you construct a function $v(t,x)$ such that $u(t,x), v(t,x)$ form a solution to the first-order system?

1.13. Find all solutions $u = f(r)$ of the three-dimensional Laplace equation
$u_{xx} + u_{yy} + u_{zz} = 0$ that depend only on the radial coordinate $r = \sqrt{x^2+y^2+z^2}$.

1.14. Let $u(x,y)$ be defined on a domain $D \subset \mathbb{R}^2$. Suppose you know that all its second-order partial derivatives, $u_{xx}, u_{xy}, u_{yx}, u_{yy}$, are defined and continuous on all of D. Can you conclude that $u \in C^2(D)$?

1.15. Write down a partial differential equation that has
(a) no real solutions; (b) exactly one real solution; (c) exactly two real solutions.

1.16. Let $u(x,y) = xy\,\dfrac{x^2-y^2}{x^2+y^2}$ for $(x,y) \neq (0,0)$, while $u(0,0) = 0$. Prove that

$$\frac{\partial^2 u}{\partial x\,\partial y}(0,0) = 1 \neq -1 = \frac{\partial^2 u}{\partial y\,\partial x}(0,0).$$

Explain why this example does not contradict the theorem on the equality of mixed partials.

Linear and Nonlinear Equations

As with algebraic equations and ordinary differential equations, there is a crucial distinction between linear and nonlinear partial differential equations, and one must have a firm grasp of the linear theory before venturing into the nonlinear wilderness. While linear algebraic equations are (modulo numerical difficulties) eminently solvable by a variety of techniques, linear ordinary differential equations, of order ≥ 2, already present a challenge, as most cannot be solved in terms of elementary functions. Indeed, as we will learn in Chapter 11, solving many of those equations that arise in applications requires introducing new types of "special functions" that are typically not encountered in a basic calculus course. Linear partial differential equations are of a yet higher level of difficulty, and only a small handful of specific equations can be completely solved. Moreover, explicit solutions tend to be expressible only in the form of infinite series, requiring subtle analytic tools to understand their convergence and properties. For the vast majority of partial differential equations, the only feasible means of producing general solutions is through numerical approximation. In this book, we will study the two most basic numerical schemes: finite differences and finite elements. Keep in mind that, in order to develop and understand numerics for partial differential equations, one must already have a good understanding of their analytical properties.

The distinguishing feature of linearity is that it enables one to straightforwardly combine solutions to form new solutions, through a general Superposition Principle. Linear superposition is universally applicable to all linear equations and systems, including linear algebraic systems, linear ordinary differential equations, linear partial differential equations, linear initial and boundary value problems, as well as linear integral equations, linear control systems, and so on. Let us introduce the basic idea in the context of a single differential equation.

A differential equation is called *homogeneous linear* if both sides are sums of terms, each of which involves the dependent variable u *or* one of its derivatives to the first power; on the other hand, there is no restriction on how the terms involve the independent variables. Thus,

$$\frac{d^2 u}{dx^2} + \frac{u}{1+x^2} = 0$$

is a homogeneous linear second-order ordinary differential equation. Examples of homogeneous linear partial differential equations include the heat equation (1.5), the partial differential equation (1.2), and the equation

$$\frac{\partial u}{\partial t} = e^x \frac{\partial^2 u}{\partial x^2} + \cos(x-t)\, u.$$

On the other hand, Burgers' equation

$$\frac{\partial u}{\partial t} + u\,\frac{\partial u}{\partial x} = \frac{\partial^2 u}{\partial x^2} \tag{1.10}$$

is not linear, since the second term involves the product of u and its derivative u_x. A similar terminology is applied to systems of partial differential equations. For example, the Navier–Stokes system (1.4) is not linear because of the terms $u u_x, v u_y$, etc. — although its final constituent equation is linear.

A more precise definition of a homogeneous linear differential equation begins with the concept of a *linear differential operator* L. Such operators are assembled by summing the basic partial derivative operators, with either constant coefficients or, more generally, coefficients depending on the independent variables. The operator acts on sufficiently smooth functions depending on the relevant independent variables. According to Definition B.32, *linearity* imposes two key requirements:

$$L[u+v] = L[u] + L[v], \qquad L[cu] = c\,L[u], \tag{1.11}$$

for any two (sufficiently smooth) functions u, v, and any constant c.

Definition 1.2. A *homogeneous linear differential equation* has the form

$$L[u] = 0, \tag{1.12}$$

where L is a linear differential operator.

As a simple example, consider the second-order differential operator

$$L = \frac{\partial^2}{\partial x^2}, \qquad \text{whereby} \qquad L[u] = \frac{\partial^2 u}{\partial x^2}$$

for any C^2 function $u(x, y)$. The linearity requirements (1.11) follow immediately from basic properties of differentiation:

$$\begin{aligned} L[u+v] &= \frac{\partial^2}{\partial x^2}(u+v) = \frac{\partial^2 u}{\partial x^2} + \frac{\partial^2 v}{\partial x^2} = L[u] + L[v], \\ L[cu] &= \frac{\partial^2}{\partial x^2}(cu) = c\,\frac{\partial^2 u}{\partial x^2} = c\,L[u], \end{aligned}$$

which are valid for any C^2 functions u, v and any constant c. The corresponding homogeneous linear differential equation $L[u] = 0$ is

$$\frac{\partial^2 u}{\partial x^2} = 0.$$

The heat equation (1.5) is based on the linear partial differential operator

$$L = \partial_t - \partial_x^2, \qquad \text{with} \qquad L[u] = \partial_t u - \partial_x^2 u = u_t - u_{xx} = 0. \tag{1.13}$$

Linearity follows as above:

$$\begin{aligned} L[u+v] &= \partial_t(u+v) - \partial_x^2(u+v) = (\partial_t u - \partial_x^2 u) + (\partial_t v - \partial_x^2 v) = L[u] + L[v], \\ L[cu] &= \partial_t(cu) - \partial_x^2(cu) = c\,(\partial_t u - \partial_x^2 u) = c\,L[u]. \end{aligned}$$

Similarly, the linear differential operator

$$L = \partial_t^2 - \partial_x\,\kappa(x)\,\partial_x = \partial_t^2 - \kappa(x)\,\partial_x^2 - \kappa'(x)\,\partial_x,$$

where $\kappa(x)$ is a prescribed C^1 function of x alone, defines the homogeneous linear partial differential equation

$$L[u] = \partial_t^2 u - \partial_x(\kappa(x)\,\partial_x u) = u_{tt} - \partial_x(\kappa(x)\,u_x) = u_{tt} - \kappa(x)\,u_{xx} - \kappa'(x)\,u_x = 0,$$

which is used to model vibrations in a nonuniform one-dimensional medium.

The defining attributes of linear operators (1.11) imply the key properties shared by all homogeneous linear (differential) equations.

Proposition 1.3. *The sum of two solutions to a homogeneous linear differential equation is again a solution, as is the product of a solution with any constant.*

Proof: Let u_1, u_2 be solutions, meaning that $L[u_1] = 0$ and $L[u_2] = 0$. Then, thanks to linearity,

$$L[u_1 + u_2] = L[u_1] + L[u_2] = 0,$$

and hence their sum $u_1 + u_2$ is a solution. Similarly, if c is any constant and u any solution, then

$$L[cu] = c\,L[u] = c\,0 = 0,$$

and so the constant multiple cu is also a solution. *Q.E.D.*

As a result, starting with a handful of solutions to a homogeneous linear differential equation, by repeating these operations of adding solutions and multiplying by constants, we are able to build up large families of solutions. In the case of the heat equation (1.5), we are already in possession of two solutions, namely (1.6) and (1.7). Multiplying each by a constant produces two infinite families of solutions:

$$u(t,x) = c_1 (t + \tfrac{1}{2} x^2) \qquad \text{and} \qquad u(t,x) = \frac{c_2\, e^{-x^2/(4t)}}{2\sqrt{\pi\, t}},$$

where c_1, c_2 are arbitrary constants. Moreover, one can add the latter solutions together, producing a two-parameter family of solutions

$$u(t,x) = c_1 (t + \tfrac{1}{2} x^2) + \frac{c_2\, e^{-x^2/(4t)}}{2\sqrt{\pi\, t}},$$

valid for any choice of the constants c_1, c_2.

The preceding construction is a special case of the general *Superposition Principle* for homogeneous linear equations:

Theorem 1.4. *If $u_1, \ldots, u_k$ are solutions to a common homogeneous linear equation $L[u] = 0$, then the* linear combination, *or* superposition, $u = c_1 u_1 + \cdots + c_k u_k$ *is a solution for any choice of constants $c_1, \ldots, c_k$.*

Proof: Repeatedly applying the linearity requirements (1.11), we find

$$\begin{aligned} L[u] = L[c_1 u_1 + \cdots + c_k u_k] &= L[c_1 u_1 + \cdots + c_{k-1} u_{k-1}] + L[c_k u_k] \\ = \cdots &= L[c_1 u_1] + \cdots + L[c_k u_k] = c_1 L[u_1] + \cdots + c_k L[u_k]. \end{aligned} \tag{1.14}$$

In particular, if the functions are solutions, so $L[u_1] = 0, \ \ldots\ , L[u_k] = 0$, then the right-hand side of (1.14) vanishes, proving that u also solves the equation $L[u] = 0$. *Q.E.D.*

In the linear algebraic language of Appendix B, Theorem 1.4 tells us that the solutions to a homogeneous linear partial differential equation form a vector space. The same holds true for linear algebraic equations, [**89**], and linear ordinary differential equations, [**18, 20, 23, 52**]. In the latter two situations, once one finds a sufficient number of independent solutions, the general solution is obtained as a linear combination thereof. In the language of linear algebra, the solution space is finite-dimensional. In contrast, most linear systems of partial differential equations admit an infinite number of independent solutions, meaning that the solution space is infinite-dimensional, and, as a consequence, one cannot hope to build the general solution by taking *finite* linear combinations. Instead, one requires the far more delicate operation of forming infinite series involving the basic solutions. Such considerations will soon lead us into the heart of Fourier analysis, and require spending an entire chapter developing the required analytic tools.

Definition 1.5. An *inhomogeneous linear differential equation* has the form

$$L[v] = f, \tag{1.15}$$

where L is a linear differential operator, v is the unknown function, and f is a prescribed nonzero function of the independent variables alone.

For example, the inhomogeneous form of the heat equation (1.13) is

$$L[v] = \partial_t v - \partial_x^2 v = v_t - v_{xx} = f(t,x), \tag{1.16}$$

where $f(t,x)$ is a specified function. This equation models the thermodynamics of a one-dimensional medium subject to an external heat source.

You already learned the basic technique for solving inhomogeneous linear equations in your study of elementary ordinary differential equations. Step one is to determine the general solution to the homogeneous equation. Step two is to find a particular solution to the inhomogeneous version. The general solution to the inhomogeneous equation is then obtained by adding the two together. Here is the general version of this procedure:

Theorem 1.6. *Let $v_\star$ be a particular solution to the inhomogeneous linear equation $L[v_\star] = f$. Then the general solution to $L[v] = f$ is given by $v = v_\star + u$, where u is the general solution to the corresponding homogeneous equation $L[u] = 0$.*

Proof: Let us first show that $v = v_\star + u$ is also a solution whenever $L[u] = 0$. By linearity,

$$L[v] = L[v_\star + u] = L[v_\star] + L[u] = f + 0 = f.$$

To show that every solution to the inhomogeneous equation can be expressed in this manner, suppose v satisfies $L[v] = f$. Set $u = v - v_\star$. Then, by linearity,

$$L[u] = L[v - v_\star] = L[v] - L[v_\star] = 0,$$

and hence u is a solution to the homogeneous differential equation. Thus, $v = v_\star + u$ has the required form. *Q.E.D.*

In physical applications, one can interpret the particular solution $v_\star$ as a response of the system to the external forcing function. The solution u to the homogeneous equation represents the system's internal, unforced behavior. The general solution to the inhomogeneous linear equation is thus a combination, $v = v_\star + u$, of the external and internal responses.

Finally, the *Superposition Principle* for *inhomogeneous* linear equations allows one to combine the responses of the system to different external forcing functions. The proof of this result is left to the reader as Exercise 1.26.

Theorem 1.7. *Let $v_1, \ldots, v_k$ be solutions to the inhomogeneous linear systems $L[v_1] = f_1, \ \ldots \ , L[v_k] = f_k$, involving the same linear operator L. Then, given any constants $c_1, \ldots, c_k$, the linear combination $v = c_1 v_1 + \cdots + c_k v_k$ solves the inhomogeneous system $L[v] = f$ for the combined forcing function $f = c_1 f_1 + \cdots + c_k f_k$.*

The two general Superposition Principles furnish us with powerful tools for solving linear partial differential equations, which we shall repeatedly exploit throughout this text. In contrast, nonlinear partial differential equations are much tougher, and, typically, knowledge of several solutions is of scant help in constructing others. Indeed, finding even one solution to a nonlinear partial differential equation can be quite a challenge. While this text will primarily concentrate on analyzing the solutions and their properties to some of the most basic and most important linear partial differential equations, we will have occasion to briefly venture into the nonlinear realm, introducing some striking recent developments in this fascinating arena of contemporary research.

Exercises

1.17. Classify the following differential equations as either
(i) homogeneous linear; (ii) inhomogeneous linear; or (iii) nonlinear:
(a) $u_t = x^2 u_{xx} + 2x u_x$, (b) $-u_{xx} - u_{yy} = \sin u$; (c) $u_{xx} + 2y u_{yy} = 3$;
(d) $u_t + u u_x = 3u$; (e) $e^y u_x = e^x u_y$; (f) $u_t = 5u_{xxx} + x^2 u + x$.

1.18. Write down all possible solutions to the Laplace equation you can construct from the various solutions provided in Exercise 1.5 using linear superposition.

1.19. (a) Show that the following functions are solutions to the wave equation $u_{tt} = 4u_{xx}$:
(i) $\cos(x - 2t)$, (ii) e^{x+2t}; (iii) $x^2 + 2xt + 4t^2$.
(b) Write down at least four other solutions to the wave equation.

1.20. The displacement $u(t,x)$ of a forced violin string is modeled by the partial differential equation $u_{tt} = 4u_{xx} + F(t,x)$. When the string is subjected to the external forcing $F(t,x) = \cos x$, the solution is $u(t,x) = \cos(x - 2t) + \frac{1}{4}\cos x$, while when $F(t,x) = \sin x$, the solution is $u(t,x) = \sin(x - 2t) + \frac{1}{4}\sin x$. Find a solution when the forcing function $F(t,x)$ is
(a) $\cos x - 5\sin x$, (b) $\sin(x - 3)$.

1.21. (a) Show that the partial derivatives $\partial_x[f] = \dfrac{\partial f}{\partial x}$ and $\partial_y[f] = \dfrac{\partial f}{\partial y}$ both define linear operators on the space of continuously differentiable functions $f(x,y)$. (b) For which values of a, b, c, d is the differential operator $L[f] = a\,\dfrac{\partial f}{\partial x} + b\,\dfrac{\partial f}{\partial y} + c f + d$ linear?

1.22. (a) Prove that the Laplacian $\Delta = \partial_x^2 + \partial_y^2$ defines a linear differential operator.
(b) Write out the Laplace equation $\Delta[u] = 0$ and the Poisson equation $-\Delta[u] = f$.

1.23. Prove that, on $\mathbb{R}^3$, the gradient, curl, and divergence all define linear operators.

1.24. Let L and M be linear partial differential operators. Prove that the following are also linear partial differential operators: (a) $L - M$, (b) $3L$, (c) fL, where f is an arbitrary function of the independent variables; (d) $L \circ M$.

1.25. Suppose L and M are linear differential operators and let $N = L + M$.
(a) Prove that N is a linear operator. (b) *True or false*: If u solves $L[u] = f$ and v solves $M[v] = g$, then $w = u + v$ solves $N[w] = f + g$.

♢ 1.26. Prove Theorem 1.7.

1.27. Solve the following inhomogeneous linear ordinary differential equations:
(a) $u' - 4u = x - 3$, (b) $5u'' - 4u' + 4u = e^x \cos x$, (c) $u'' - 3u' = e^{3x}$.

1.28. Use superposition to solve the following inhomogeneous ordinary differential equations:
(a) $u' + 2u = 1 + \cos x$, (b) $u'' - 9u = x + \sin x$, (c) $9u'' - 18u' + 10u = 1 + e^x \cos x$,
(d) $u'' + u' - 2u = \sinh x$, where $\sinh x = \frac{1}{2}(e^x - e^{-x})$, (e) $u''' + 9u' = 1 + e^{3x}$.

Chapter 2
Linear and Nonlinear Waves

Our initial foray into the vast mathematical continent that comprises partial differential equations will begin with some basic first-order equations. In applications, first-order partial differential equations are most commonly used to describe dynamical processes, and so time, t, is one of the independent variables. Our discussion will focus on dynamical models in a single space dimension, bearing in mind that most of the methods we introduce can be extended to higher-dimensional situations. First-order partial differential equations and systems model a wide variety of wave phenomena, including transport of pollutants in fluids, flood waves, acoustics, gas dynamics, glacier motion, chromatography, traffic flow, and various biological and ecological systems.

A basic solution technique relies on an inspired change of variables, which comes from rewriting the equation in a moving coordinate frame. This naturally leads to the fundamental concept of characteristic curve, along which signals and physical disturbances propagate. The resulting method of characteristics is able to solve a first-order *linear* partial differential equation by reducing it to one or more first-order *nonlinear* ordinary differential equations.

Proceeding to the nonlinear regime, the most important new phenomenon is the possible breakdown of solutions in finite time, resulting in the formation of discontinuous shock waves. A familiar example is the supersonic boom produced by an airplane that breaks the sound barrier. Signals continue to propagate along characteristic curves, but now the curves may cross each other, precipitating the onset of a shock discontinuity. The ensuing shock dynamics is *not* uniquely specified by the partial differential equation, but relies on additional physical properties, to be specified by an appropriate conservation law along with a causality condition. A full-fledged analysis of shock dynamics becomes quite challenging, and only the basics will be developed here.

Having attained a basic understanding of first-order wave dynamics, we then focus our attention on the first of three paradigmatic second-order partial differential equations, known as the wave equation, which is used to model waves and vibrations in an elastic bar, a violin string, or a column of air in a wind instrument. Its multi-dimensional versions serve to model vibrations of membranes, solid bodies, water waves, electromagnetic waves, including light, radio waves, microwaves, acoustic waves, and many other physical phenomena. The one-dimensional wave equation is one of a small handful of physically relevant partial differential equations that has an explicit solution formula, originally discovered by the eighteenth-century French mathematician (and encyclopedist) Jean d'Alembert. His solution is the result of being able to "factorize" the second-order wave equation into a pair of first-order partial differential equations, of a type solved in the first part of this

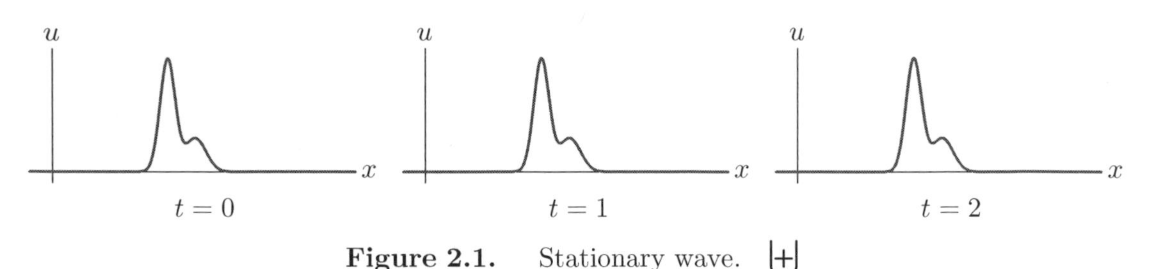

Figure 2.1. Stationary wave.

chapter. We investigate the consequences of d'Alembert's solution formula for the initial value problem on the entire real line; solutions on bounded intervals will be deferred until Chapter 4. Unfortunately, d'Alembert's method is of rather limited scope, and does not extend beyond the one-dimensional case, nor to equations modeling vibrations of nonuniform media. The analysis of the wave equation in more than one space dimension can be found in Chapters 11 and 12.

2.1 Stationary Waves

When entering a new mathematical subject — in our case, partial differential equations — one should first analyze and fully understand the very simplest examples. Indeed, mathematics is, at its core, a bootstrapping enterprise, in which one builds on one's knowledge of and experience with elementary topics — in the present case, ordinary differential equations — to make progress, first with the simpler types of partial differential equations, and then, by developing and applying each newly gained insight and technique, to more and more complicated situations.

The simplest partial differential equation, for a function $u(t,x)$ of two variables, is

$$\frac{\partial u}{\partial t} = 0. \tag{2.1}$$

It is a first-order, homogeneous, linear equation. If (2.1) were an ordinary differential equation[†] for a function $u(t)$ of t alone, the solution would be obvious: $u(t) = c$ must be constant. A proof of this basic fact proceeds by integrating both sides with respect to t and then appealing to the Fundamental Theorem of Calculus. To solve (2.1) as a partial differential equation for $u(t,x)$, let us similarly integrate both sides of the equation from, say, 0 to t, producing

$$0 = \int_0^t \frac{\partial u}{\partial t}(s,x)\,ds = u(t,x) - u(0,x).$$

Therefore, the solution takes the form

$$u(t,x) = f(x), \qquad \text{where} \qquad f(x) = u(0,x), \tag{2.2}$$

and hence is a function of the space variable x alone. The only requirement is that $f(x)$ be continuously differentiable, so $f \in \mathrm{C}^1$, in order that $u(t,x)$ be a bona fide classical

† Of course, in this situation, we would write the equation as $du/dt = 0$.

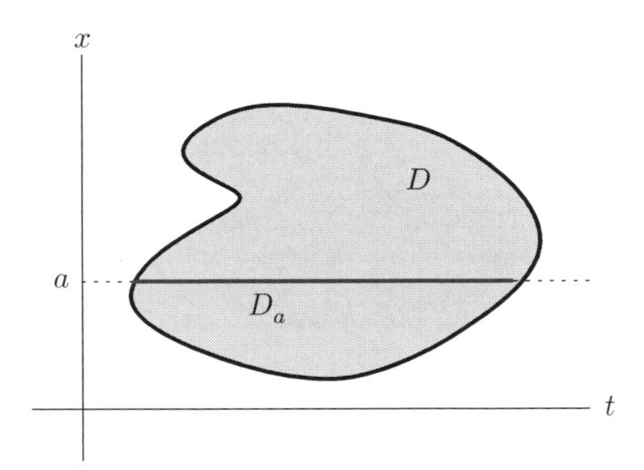

Figure 2.2. Domain for stationary-wave solution.

solution of the first-order partial differential equation (2.1). The solution (2.2) represents a *stationary wave*, meaning that it does not change in time. The initial profile stays frozen in place, and the system remains in equilibrium. Figure 2.1 plots a representative solution as a function of x at three successive times.

The preceding analysis seems very straightforward and perhaps even a little boring. But, to be completely rigorous, we need to take a bit more care. In our derivation, we implicitly assumed that the solution $u(t,x)$ was defined everywhere on $\mathbb{R}^2$. And, in fact, the solution formula (2.2) is *not* completely valid as stated if the solution $u(t,x)$ is defined only on a subdomain $D \subset \mathbb{R}^2$.

Indeed, a solution $u(t)$ to the corresponding ordinary differential equation $du/dt = 0$ is constant, *provided it is defined on a connected subinterval* $I \subset \mathbb{R}$. A solution that is defined on a disconnected subset $D \subset \mathbb{R}$ need only be constant on each connected subinterval $I \subset D$. For instance, the nonconstant function

$$u(t) = \begin{cases} 1, & t > 0, \\ -1, & t < 0, \end{cases} \qquad \text{satisfies} \qquad \frac{du}{dt} = 0$$

everywhere on its domain of definition, that is, $D = \{ t \neq 0 \}$, but is constant only on the connected positive and negative half-lines.

Similar counterexamples can be constructed in the case of the partial differential equation (2.1). If the domain of definition is disconnected, then we do not expect $u(t,x)$ to depend only on x if we move from one connected component of D to another. Even that is not the full story. For example, the function

$$u(t,x) = \begin{cases} 0, & x > 0, \\ x^2, & x \leq 0, \quad t > 0, \\ -x^2, & x \leq 0, \quad t < 0, \end{cases} \tag{2.3}$$

is continuously differentiable[†] on its domain of definition, namely $D = \mathbb{R}^2 \setminus \{ (0,x) \mid x \leq 0 \}$, satisfies $\partial u/\partial t = 0$ everywhere in D, but, nevertheless, is not a function of x alone, because, for example, $u(1,x) = x^2 \neq u(-1,x) = -x^2$.

[†] You are asked to rigorously prove differentiability in Exercise 2.1.10.

A completely correct formulation can be stated as follows: If $u(t,x)$ is a classical solution to (2.1), defined on a domain $D \subset \mathbb{R}^2$ whose intersection with any horizontal[‡] line, namely $D_a = D \cap \{ (t,a) \,|\, t \in \mathbb{R} \}$, for each fixed $a \in \mathbb{R}$, is either empty or a connected interval, then $u(t,x) = f(x)$ is a function of x alone. An example of such a domain is sketched in Figure 2.2. In Exercise 2.1.9, you are asked to justify these statements.

We are thus slightly chastened in our dismissal of (2.1) as a complete triviality. The lesson is that, in future, one must *always* be careful when interpreting such "general" solution formulas — since they often rely on unstated assumptions on their underlying domain of definition.

Exercises

2.1.1. Solve the partial differential equation $\dfrac{\partial u}{\partial t} = x$ for $u(t,x)$.

2.1.2. Solve the partial differential equation $\dfrac{\partial^2 u}{\partial t^2} = 0$ for $u(t,x)$.

2.1.3. Find the general solution $u(t,x)$ to the following partial differential equations: (a) $u_x = 0$, (b) $u_t = 1$, (c) $u_t = x - t$, (d) $u_t + 3u = 0$, (e) $u_x + t\,u = 0$, (f) $u_{tt} + 4u = 1$.

2.1.4. Suppose $u(t,x)$ is defined for all $(t,x) \in \mathbb{R}^2$ and solves $\partial u/\partial t + 2u = 0$. Prove that $\lim_{t \to \infty} u(t,x) = 0$ for all x.

2.1.5. Write down the general solution to the partial differential equation $\partial u/\partial t = 0$ for a function of three variables $u(t,x,y)$. What assumptions should be made on the domain of definition for your solution formula to be valid?

2.1.6. Solve the partial differential equation $\dfrac{\partial^2 u}{\partial x\, \partial y} = 0$ for $u(x,y)$.

2.1.7. Answer Exercise 2.1.6 when $u(x,y,z)$ depends on the three independent variables x, y, z.

♡ 2.1.8. Let $u(t,x)$ solve the initial value problem $\dfrac{\partial u}{\partial t} + u^2 = 0$, $u(0,x) = f(x)$, where $f(x)$ is a bounded C^1 function of $x \in \mathbb{R}$. (a) Show that if $f(x) \geq 0$ for all x, then $u(t,x)$ is defined for all $t > 0$, and $\lim_{t \to \infty} u(t,x) = 0$. (b) On the other hand, if $f(x) < 0$, then the solution $u(t,x)$ is not defined for all $t > 0$, but in fact, $\lim_{t \to \tau^-} u(t,x) = -\infty$ for some $0 < \tau < \infty$. Given x, what is the corresponding value of τ? (c) Given $f(x)$ as in part (b), what is the longest time interval $0 < t < t_\star$ on which $u(t,x)$ is defined for all $x \in \mathbb{R}$?

♢ 2.1.9. Justify the claim in the text that if $u(t,x)$ is a solution of $\partial u/\partial t = 0$ that is defined on a domain $D \subset \mathbb{R}^2$ with the property that $D_a = D \cap \{ (t,a) \mid t \in \mathbb{R} \}$ is either empty or a connected interval, then $u(t,x) = v(x)$ depends only on $x \in D$.

♢ 2.1.10. Prove that the function in (2.3) is continuously differentiable at all points (t,x) in its domain of definition.

[‡] *Important*: We will adopt the (slightly unusual) convention of displaying the (t,x)–plane with time t along the horizontal axis and space x along the vertical axis — which also conforms with our convention of writing t before x in expressions like $u(t,x)$. Later developments will amply vindicate our adoption of this convention.

2.2 Transport and Traveling Waves

In many respects, the stationary-wave equation (2.1) does not quite qualify as a partial differential equation. Indeed, the spatial variable x enters only parametrically in the solution to what is, in essence (ignoring technical difficulties with domains), a very simple ordinary differential equation.

Let us then turn to a more "genuine" example. Consider the linear, homogeneous first-order partial differential equation

$$\frac{\partial u}{\partial t} + c\,\frac{\partial u}{\partial x} = 0, \tag{2.4}$$

for a function $u(t,x)$, in which c is a fixed, nonzero constant, known as the *wave speed* for reasons that will soon become apparent. We will refer to (2.4) as the *transport equation*, because it models the transport of a substance, e.g., a pollutant, in a uniform fluid flow that is moving with velocity c. In this model, the solution $u(t,x)$ represents the concentration of the pollutant at time t and spatial position x. Other common names for (2.4) are the *first-order* or *unidirectional wave equation*. But for brevity, as well as to avoid any confusion with the second-order, bidirectional wave equation discussed extensively later on, we will stick with the designation "transport equation" here. Solving the transport equation is slightly more challenging, but, as we will see, not difficult.

Since the transport equation involves time, its solutions are distinguished by their initial values. As a first-order equation, we need only specify the value of the solution at an initial time t_0, leading to the initial value problem

$$u(t_0,x) = f(x) \qquad \text{for all} \qquad x \in \mathbb{R}. \tag{2.5}$$

As we will show, as long as $f \in \mathrm{C}^1$, i.e., is continuously differentiable, the initial conditions serve to specify a unique classical solution. Also, by replacing the time variable t by $t - t_0$, we can, without loss of generality, set $t_0 = 0$.

Uniform Transport

Let us begin by assuming that the wave speed c is constant. In general, when one is confronted with a new equation, one solution strategy is to try to convert it into an equation that you already know how to solve. In this case, we will introduce a simple change of variables that effectively rewrites the equation in a moving coordinate system, inspired by the interpretation of c as the overall transport speed.

If x represents the position of an object in a fixed coordinate frame, then

$$\xi = x - ct \tag{2.6}$$

represents the object's position relative to an observer who is uniformly moving with velocity c. Think of a passenger in a moving train to whom stationary objects appear to be moving *backwards* at the train's speed c. To formulate a physical process in the reference frame of the passenger, we replace the stationary space-time coordinates (t,x) by the moving coordinates (t,ξ).

Remark: These are the same changes of reference frame that underlie Einstein's special theory of relativity. However, unlike Einstein, we are working in a purely classical,

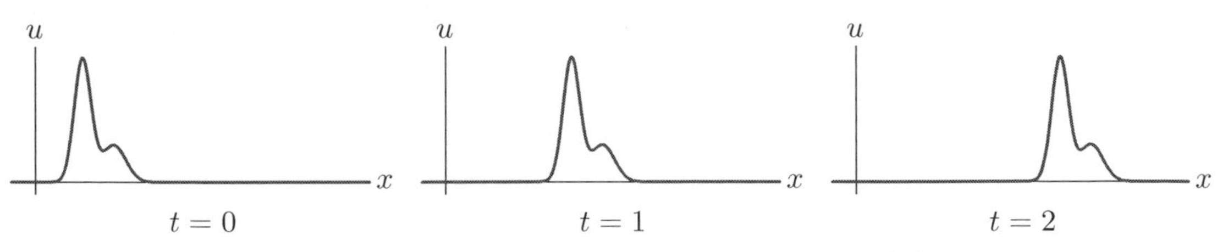

Figure 2.3. Traveling wave with $c > 0$.

nonrelativistic universe here. Such changes to moving coordinates are, in fact, of a much older vintage, and named *Galilean boosts* in honor of Galileo Galilei, who was the first to champion such "relativistic" moving coordinate systems.

Let us see what happens when we re-express the transport equation in terms of the moving coordinate frame. We rewrite

$$u(t,x) = v(t, x - ct) = v(t, \xi) \tag{2.7}$$

in terms of the *characteristic variable* $\xi = x - ct$, along with the time t. To write out the differential equation satisfied by $v(t, \xi)$, we apply the chain rule from multivariable calculus, [**8**, **108**], to express the derivatives of u in terms of those of v:

$$\frac{\partial u}{\partial t} = \frac{\partial v}{\partial t} - c\,\frac{\partial v}{\partial \xi}\,, \qquad \frac{\partial u}{\partial x} = \frac{\partial v}{\partial \xi}\,.$$

Therefore,

$$\frac{\partial u}{\partial t} + c\,\frac{\partial u}{\partial x} = \frac{\partial v}{\partial t} - c\,\frac{\partial v}{\partial \xi} + c\,\frac{\partial v}{\partial \xi} = \frac{\partial v}{\partial t}\,. \tag{2.8}$$

We deduce that $u(t,x)$ solves the transport equation (2.4) if and only if $v(t,\xi)$ solves the stationary-wave equation

$$\frac{\partial v}{\partial t} = 0. \tag{2.9}$$

Thus, the effect of using a moving coordinate system is to convert a wave moving with velocity c into a stationary wave. Think again of the passenger in the train — a second train moving at the same speed appears as if it were stationary.

According to our earlier discussion, the solution $v = v(\xi)$ to the stationary-wave equation (2.9) is a function of the characteristic variable alone. (For simplicity, we assume that $v(t,\xi)$ has an appropriate domain of definition, e.g., it is defined everywhere on $\mathbb{R}^2$.) Recalling (2.7), we conclude that the solution

$$u = v(\xi) = v(x - ct)$$

to the transport equation must be a function of the characteristic variable only. We have therefore proved the following result:

Proposition 2.1. *If $u(t,x)$ is a solution to the partial differential equation*

$$u_t + c\,u_x = 0, \tag{2.10}$$

which is defined on all of $\mathbb{R}^2$, then

$$u(t,x) = v(x - ct), \tag{2.11}$$

where $v(\xi)$ is a C^1 function of the characteristic variable $\xi = x - ct$.

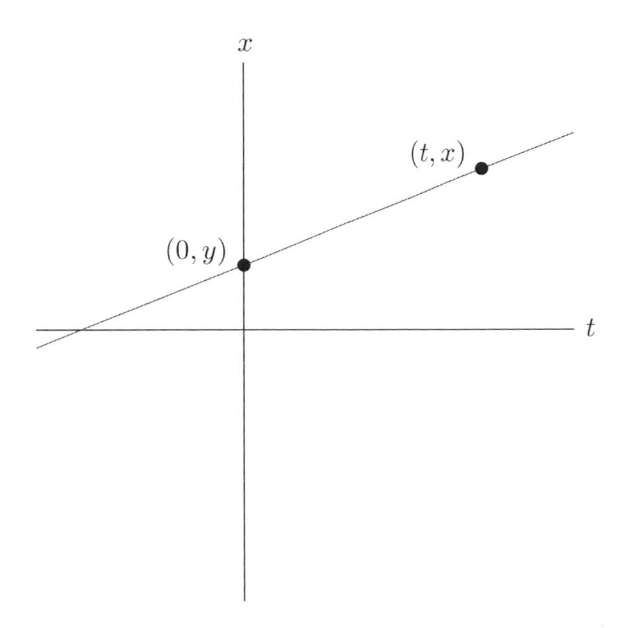

Figure 2.4. Characteristic line.

In other words, *any* (reasonable) function of the characteristic variable, e.g., ξ^2+1, or $\cos\xi$, or e^ξ, will produce a corresponding solution, $(x-ct)^2+1$, or $\cos(x-ct)$, or e^{x-ct}, to the transport equation with constant wave speed c. And, in accordance with the counting principle of Chapter 1, the general solution to this first-order partial differential equation in two independent variables depends on one arbitrary function of a single variable.

To a stationary observer, the solution (2.11) appears as a *traveling wave* of unchanging form moving at constant velocity c. When $c > 0$, the wave translates to the right, as illustrated in Figure 2.3. When $c < 0$, the wave translates to the left, while $c = 0$ corresponds to a stationary wave form that remains fixed at its original location, as in Figure 2.1.

At $t = 0$, the wave has the initial profile

$$u(0,x) = v(x), \tag{2.12}$$

and so (2.11) provides the (unique) solution to the initial value problem (2.4, 12). For example, the solution to the particular initial value problem

$$u_t + 2u_x = 0, \qquad u(0,x) = \frac{1}{1+x^2}, \qquad \text{is} \qquad u(t,x) = \frac{1}{1+(x-2t)^2}.$$

Since it depends only on the characteristic variable $\xi = x - ct$, every solution to the transport equation is constant on the *characteristic lines* of slope† c, namely

$$x = ct + k, \tag{2.13}$$

where k is an arbitrary constant. At any given time t, the value of the solution at position x depends only on its original value on the characteristic line passing through (t,x).

† This makes use of our convention that the t–axis is horizontal and the x–axis is vertical. Reversing the axes will replace the slope by its reciprocal.

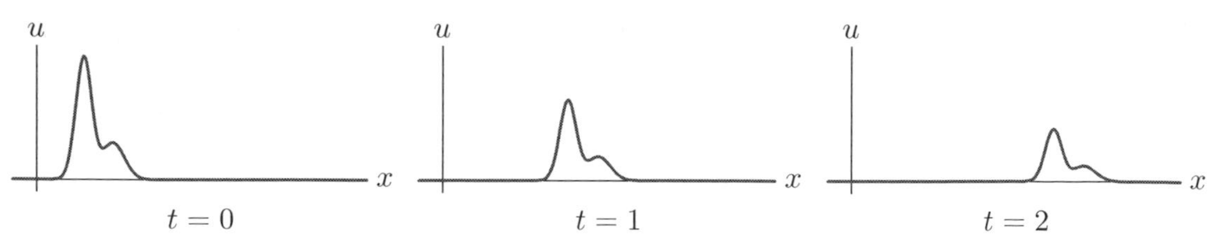

Figure 2.5. Decaying traveling wave.

This is indicative of a general fact concerning such wave models: *Signals propagate along characteristics*. Indeed, a disturbance at an initial point $(0, y)$ only affects the value of the solution at points (t, x) that lie on the characteristic line $x = c\,t + y$ emanating therefrom, as illustrated in Figure 2.4.

Transport with Decay

Let $a > 0$ be a positive constant, and c an arbitrary constant. The homogeneous linear first-order partial differential equation

$$\frac{\partial u}{\partial t} + c\,\frac{\partial u}{\partial x} + a\,u = 0 \tag{2.14}$$

models the transport of, say, a radioactively decaying solute in a uniform fluid flow with wave speed c. The coefficient a governs the rate of decay. We can solve this variant of the transport equation by the self-same change of variables to a uniformly moving coordinate system.

Rewriting $u(t, x)$ in terms of the characteristic variable, as in (2.7), and then recalling our chain rule calculation (2.8), we find that $v(t, \xi) = u(t, \xi + c\,t)$ satisfies the partial differential equation

$$\frac{\partial v}{\partial t} + a\,v = 0.$$

The result is, effectively, a homogeneous linear first-order ordinary differential equation, in which the characteristic variable ξ enters only parametrically. The standard solution technique learned in elementary ordinary differential equations, [**20**, **23**], tells us to multiply the equation by the exponential *integrating factor* $e^{a\,t}$, leading to

$$e^{a\,t}\left(\frac{\partial v}{\partial t} + a\,v\right) = \frac{\partial}{\partial t}\left(e^{a\,t} v\right) = 0.$$

We conclude that $w = e^{a\,t} v$ solves the stationary-wave equation (2.1). Thus,

$$w = e^{a\,t} v = f(\xi), \qquad \text{and hence} \qquad v(t, \xi) = f(\xi)\, e^{-a\,t},$$

where $f(\xi)$ is an arbitrary function of the characteristic variable. Reverting to physical coordinates, we produce the solution formula

$$u(t, x) = f(x - c\,t)\, e^{-a\,t}, \tag{2.15}$$

which solves the initial value problem $u(0, x) = f(x)$. It represents a wave that is moving along with fixed velocity c while simultaneously decaying at an exponential rate as prescribed by the coefficient $a > 0$. A typical solution, for $c > 0$, is plotted at three successive times in Figure 2.5. While the solution (2.15) is no longer constant on the characteristics, signals continue to propagate along them, since a solution's initial value at a point $(0, y)$ will only affect its subsequent (decaying) values on the associated characteristic line $x = ct + y$.

Exercises

2.2.1. Find the solution to the initial value problem $u_t + u_x = 0, \; u(1, x) = x/(1 + x^2)$.

2.2.2. Solve the following initial value problems and graph the solutions at times $t = 1, 2,$ and 3:
(a) $u_t - 3u_x = 0, \; u(0, x) = e^{-x^2}$; (b) $u_t + 2u_x = 0, \; u(-1, x) = x/(1 + x^2)$;
(c) $u_t + u_x + \frac{1}{2}u = 0, \; u(0, x) = \tan^{-1} x$; (d) $u_t - 4u_x + u = 0, \; u(0, x) = 1/(1 + x^2)$.

2.2.3. Graph some of the characteristic lines for the following equations, and write down a formula for the general solution:
(a) $u_t - 3u_x = 0$, (b) $u_t + 5u_x = 0$, (c) $u_t + u_x + 3u = 0$, (d) $u_t - 4u_x + u = 0$.

2.2.4. Solve the initial value problem $u_t + 2u_x = 1, \; u(0, x) = e^{-x^2}$.
Hint: Use characteristic coordinates.

2.2.5. Answer Exercise 2.2.4 for the initial value problem $u_t + 2u_x = \sin x, \; u(0, x) = \sin x$.

♢ 2.2.6. Let c be constant. Suppose that $u(t, x)$ solves the initial value problem $u_t + cu_x = 0$, $u(0, x) = f(x)$. Prove that $v(t, x) = u(t - t_0, x)$ solves the initial value problem $v_t + cv_x = 0$, $v(t_0, x) = f(x)$.

2.2.7. Is Exercise 2.2.6 valid when the transport equation is replaced by the damped transport equation (2.14)?

2.2.8. Let $c \neq 0$. Prove that if the initial data satisfies $u(0, x) = v(x) \to 0$ as $x \to \pm\infty$, then, for each fixed x, the solution to the transport equation (2.4) satisfies $u(t, x) \to 0$ as $t \to \infty$.

2.2.9. (a) Prove that if the initial data is bounded, $|\,f(x)\,| \leq M$ for all $x \in \mathbb{R}$, then the solution to the damped transport equation (2.14) with $a > 0$ satisfies $u(t, x) \to 0$ as $t \to \infty$.
(b) Find a solution to (2.14) that is defined for all (t, x) but docs not satisfy $u(t, x) \to 0$ as $t \to \infty$.

2.2.10. Let $F(t, x)$ be a C^1 function of $(t, x) \in \mathbb{R}^2$. (a) Write down a formula for the general solution $u(t, x)$ to the inhomogeneous partial differential equation $u_t = F(t, x)$.
(b) Solve the inhomogeneous transport equation $u_t + c\,u_x = F(t, x)$.

♡ 2.2.11. (a) Write down a formula for the general solution to the nonlinear partial differential equation $u_t + u_x + u^2 = 0$. (b) Show that if the initial data is positive and bounded, $0 \leq u(0, x) = f(x) \leq M$, then the solution exists for all $t > 0$, and $u(t, x) \to 0$ as $t \to \infty$.
(c) On the other hand, if the initial data is negative somewhere, so $f(x) < 0$ at some $x \in \mathbb{R}$, then the solution *blows up* in finite time: $\lim_{t \to \tau^-} u(t, y) = -\infty$ for some $\tau > 0$ and some $y \in \mathbb{R}$. (d) Find a formula for the earliest blow-up time $\tau_\star > 0$.

2.2.12. A sensor situated at position $x = 1$ monitors the concentration of a pollutant $u(t, 1)$ as a function of t for $t \geq 0$. Assuming that the pollutant is transported with wave speed $c = 3$, at what locations x can you determine the initial concentration $u(0, x)$?

2.2.13. Write down a solution to the transport equation $u_t + 2u_x = 0$ that is defined on a connected domain $D \subset \mathbb{R}^2$ and that is *not* a function of the characteristic variable alone.

2.2.14. Let $c > 0$. Consider the uniform transport equation $u_t + c u_x = 0$ restricted to the quarter-plane $Q = \{x > 0,\ t > 0\}$ and subject to initial conditions $u(0,x) = f(x)$ for $x \geq 0$, along with boundary conditions $u(t,0) = g(t)$ for $t \geq 0$. (a) For which initial and boundary conditions does a classical solution to this initial-boundary value problem exist? Write down a formula for the solution. (b) On which regions are the effects of the initial conditions felt? What about the boundary conditions? Is there any interaction between the two?

2.2.15. Answer Exercise 2.2.14 when $c < 0$.

Nonuniform Transport

Slightly more complicated, but still linear, is the *nonuniform transport equation*

$$\frac{\partial u}{\partial t} + c(x)\,\frac{\partial u}{\partial x} = 0, \tag{2.16}$$

where the wave speed $c(x)$ is now allowed to depend on the spatial position. Characteristics continue to guide the behavior of solutions, but when the wave speed is not constant, we can no longer expect them to be straight lines. To adapt the method of characteristics, let us look at how the solution varies along a prescribed curve in the (t,x)–plane. Assume that the curve is identified with the graph of a function $x = x(t)$, and let

$$h(t) = u\big(t, x(t)\big)$$

be the value of the solution on it. We compute the rate of change in the solution along the curve by differentiating h with respect to t. Invoking the multivariable chain rule, we obtain

$$\frac{dh}{dt} = \frac{d}{dt}\,u\big(t,x(t)\big) = \frac{\partial u}{\partial t}\big(t,x(t)\big) + \frac{\partial u}{\partial x}\big(t,x(t)\big)\,\frac{dx}{dt}\,. \tag{2.17}$$

In particular, if $x(t)$ satisfies

$$\frac{dx}{dt} = c\big(x(t)\big), \qquad \text{then} \qquad \frac{dh}{dt} = \frac{\partial u}{\partial t}\big(t,x(t)\big) + c\big(x(t)\big)\,\frac{\partial u}{\partial x}\big(t,x(t)\big) = 0,$$

since we are assuming that $u(t,x)$ solves the transport equation (2.16) for all values of (t,x), including those points $\big(t,x(t)\big)$ on the curve. Since its derivative is zero, $h(t)$ must be a constant, which motivates the following definition.

Definition 2.2. The graph of a solution $x(t)$ to the autonomous ordinary differential equation

$$\frac{dx}{dt} = c(x) \tag{2.18}$$

is called a *characteristic curve* for the transport equation with wave speed $c(x)$.

In other words, at each point (t,x), the slope of the characteristic curve equals the wave speed $c(x)$ there. In particular, if c is constant, the characteristic curves are straight lines of slope c, in accordance with our earlier construction.

Proposition 2.3. *Solutions to the linear transport equation* (2.16) *are constant along characteristic curves.*

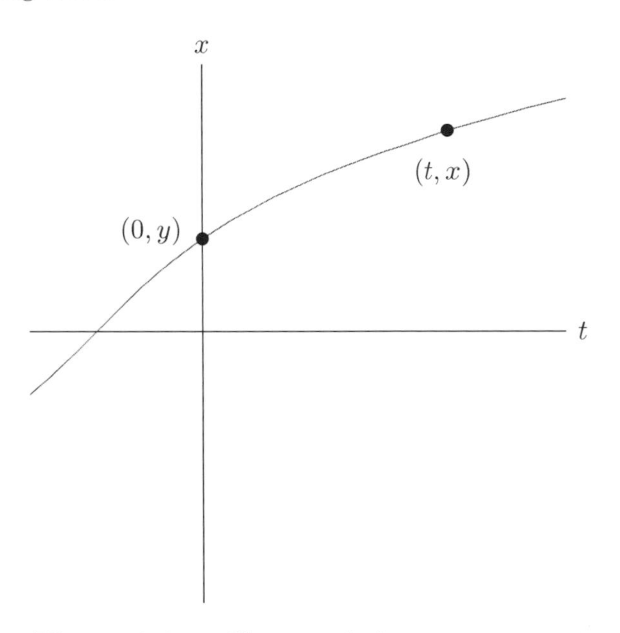

Figure 2.6. Characteristic curve.

The characteristic curve equation (2.18) is an autonomous first-order ordinary differential equation. As such, it can be immediately solved by separating variables, [**20, 23**]. Assuming $c(x) \neq 0$, we divide both sides of the equation by $c(x)$, and then integrate the resulting equation:

$$\frac{dx}{c(x)} = dt, \qquad \text{whereby} \qquad \beta(x) := \int \frac{dx}{c(x)} = t + k, \tag{2.19}$$

with k denoting the integration constant. For each fixed value of k, (2.19) serves to implicitly define a characteristic curve, namely,

$$x(t) = \beta^{-1}(t + k),$$

with β^{-1} denoting the inverse function. On the other hand, if $c(x_\star) = 0$, then $x_\star$ is a *fixed point* for the ordinary differential equation (2.18), and the horizontal line $x \equiv x_\star$ is a stationary characteristic curve.

Since the solution $u(t, x)$ is constant along the characteristic curves, it must therefore be a function of the *characteristic variable*

$$\xi = \beta(x) - t \tag{2.20}$$

alone, and hence of the form

$$u(t, x) = v\big(\beta(x) - t\big), \tag{2.21}$$

where $v(\xi)$ is an arbitrary C^1 function. Indeed, it is easy to check directly that, provided $\beta(x)$ is defined by (2.19), $u(t, x)$ solves the partial differential equation (2.16) for *any* choice of C^1 function $v(\xi)$. (But keep in mind that the algebraic solution formula (2.21) may fail to be valid at points where the wave speed vanishes: $c(x_\star) = 0$.)

Warning: The definition of characteristic variable used here is slightly different from that in the constant wave speed case, which, by (2.20), would be $\xi = x/c - t = (x - ct)/c$. Clearly, rescaling the characteristic variable by $1/c$ is an inessential modification of our original definition.

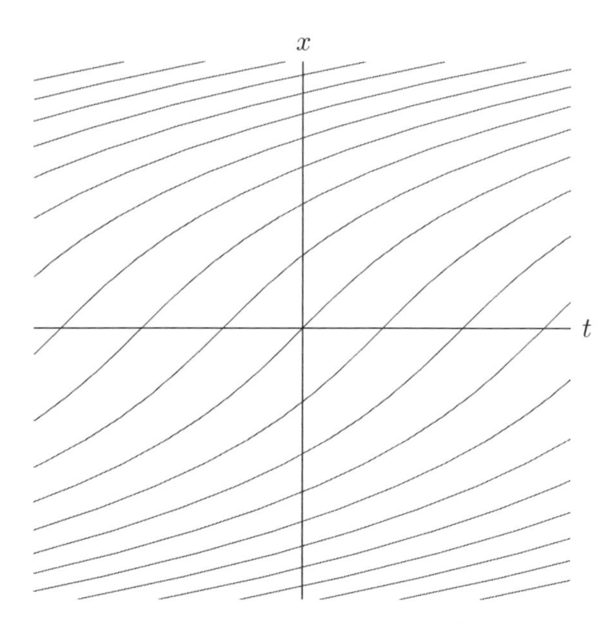

Figure 2.7. Characteristic curves for $u_t + (x^2+1)^{-1} u_x = 0$.

To find the solution that satisfies the prescribed initial conditions

$$u(0,x) = f(x), \tag{2.22}$$

we merely substitute the general solution formula (2.21). This leads to the implicit equation $v(\beta(x)) = f(x)$ for the function $v(\xi) = f \circ \beta^{-1}(\xi)$. The resulting solution formula

$$u(t,x) = f \circ \beta^{-1}\big(\beta(x) - t\big) \tag{2.23}$$

is not particularly enlightening, but it does have a simple graphical interpretation: To find the value of the solution $u(t,x)$, we look at the characteristic curve passing through the point (t,x). If this curve intersects the x–axis at the point $(0,y)$, as in Figure 2.6, then $u(t,x) = u(0,y) = f(y)$, since the solution must be constant along the curve. On the other hand, if the characteristic curve through (t,x) doesn't intersect the x–axis, the solution value $u(t,x)$ is *not* prescribed by the initial data.

Example 2.4. Let us solve the nonuniform transport equation

$$\frac{\partial u}{\partial t} + \frac{1}{x^2+1}\,\frac{\partial u}{\partial x} = 0 \tag{2.24}$$

by the method of characteristics. According to (2.18), the characteristic curves are the graphs of solutions to the first-order ordinary differential equation

$$\frac{dx}{dt} = \frac{1}{x^2+1}\,.$$

Separating variables and integrating, we obtain

$$\beta(x) = \int (x^2+1)\,dx = \tfrac{1}{3}x^3 + x = t + k, \tag{2.25}$$

where k is the integration constant. Representative curves are plotted in Figure 2.7. (In this case, inverting the function β, i.e., solving (2.25) for x as a function of t, is not particularly enlightening.)

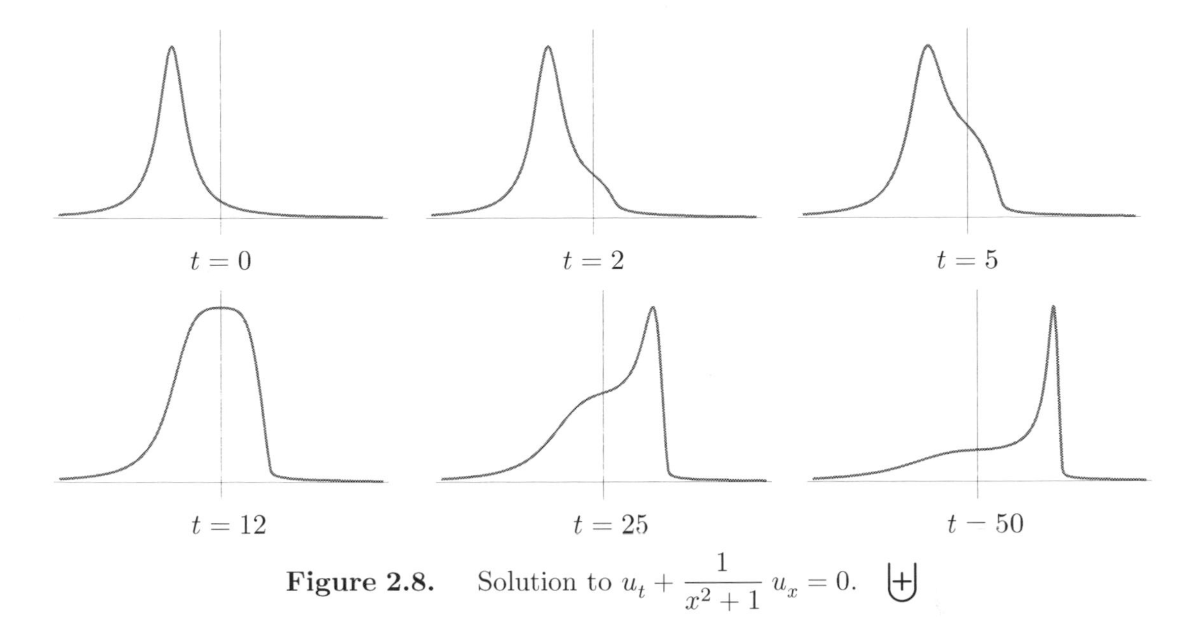

Figure 2.8. Solution to $u_t + \dfrac{1}{x^2+1}\, u_x = 0.$

According to (2.20), the characteristic variable is $\xi = \frac{1}{3}x^3 + x - t$, and hence the general solution to the equation takes the form

$$u = v\left(\tfrac{1}{3}x^3 + x - t\right), \tag{2.26}$$

where $v(\xi)$ is an arbitrary C^1 function. A typical solution, corresponding to initial data

$$u(0,x) = \frac{1}{1+(x+3)^2}, \tag{2.27}$$

is plotted[†] at the indicated times in Figure 2.8. Although the solution remains constant along each individual curve, a stationary observer will witness a dynamically changing profile as the wave moves through the nonuniform medium. In this example, since $c(x) > 0$ everywhere, the wave always moves from left to right; its speed as it passes through a point x determined by the magnitude of $c(x) = (x^2+1)^{-1}$, with the consequence that each part accelerates as it approaches the origin from the left, and then slows back down once it passes by and $c(x)$ decreases in magnitude. To a stationary observer, the wave spreads out as it speeds through the origin, and then becomes progressively narrower and slower as it gradually moves off to $+\infty$.

Example 2.5. Consider the nonuniform transport equation

$$u_t + (x^2-1)\, u_x = 0. \tag{2.28}$$

[†] The required function $v(\xi)$ in (2.26) is implicitly given by the equation $v\left(\frac{1}{3}x^3 + x\right) = u(0,x)$, and so the explicit formula for $u(t,x)$ is not very instructive or useful. Indeed, to make the plots, we instead sampled the initial data (2.27) at a collection of uniformly spaced points $y_1 < y_2 < \cdots < y_n$. Since the solution is constant along the characteristic curve (2.25) passing through each sample point $(0, y_i)$, we can find nonuniformly spaced sample values for $u(t, x_i)$ at any later time. The smooth solution curve $u(t,x)$ is then approximated using spline interpolation, [**89**; §11.4], on these sample values.

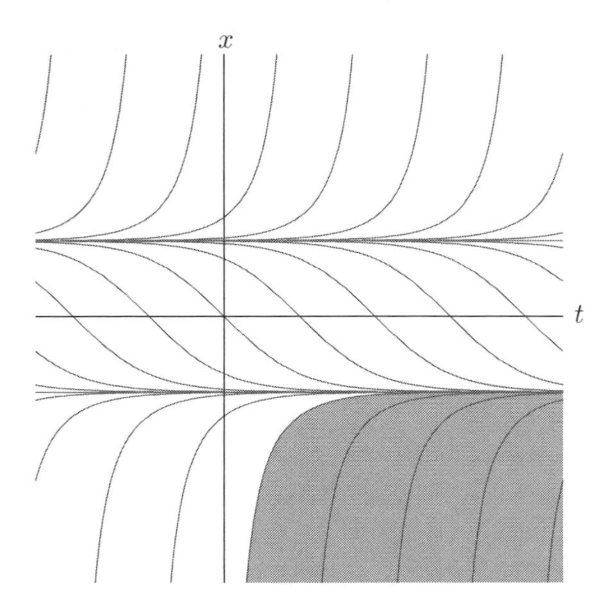

Figure 2.9. Characteristic curves for $u_t + (x^2 - 1)u_x = 0$.

In this case, the characteristic curves are the solutions to

$$\frac{dx}{dt} = x^2 - 1,$$

and so

$$\beta(x) = \int \frac{dx}{x^2 - 1} = \frac{1}{2} \log \left| \frac{x - 1}{x + 1} \right| = t + k. \tag{2.29}$$

One must also include the horizontal lines $x = x_\pm = \pm 1$ corresponding to the roots of $c(x) = x^2 - 1$. The curves are graphed in Figure 2.9. Note that those curves starting below $x_+ = 1$ converge to $x_- = -1$ as $t \to \infty$, while those starting above $x_+ = 1$ veer off to ∞ in finite time. Owing to the sign of $c(x) = x^2 - 1$, points on the graph of $u(0, x)$ lying over $|\, x \,| < 1$ will move to the left, while those over $|\, x \,| > 1$ will move to the right.

In Figure 2.10, we graph several snapshots of the solution whose initial value is a bell-shaped Gaussian profile

$$u(0, x) = e^{-x^2}.$$

The initial conditions uniquely prescribe the value of the solution along the characteristic curves that intersect the x–axis. On the other hand, if

$$x \le \frac{1 + e^{2t}}{1 - e^{2t}} \qquad \text{for} \qquad t > 0,$$

the characteristic curve through (t, x) does not intersect the x–axis, and hence the value of the solution at such points, lying in the shaded region in Figure 2.9, is *not* prescribed by the initial data. Let us arbitrarily assign the solution to be $u(t, x) = 0$ at such points. At other values of (t, x) with $t \ge 0$, the solution (2.23) is

$$u(t, x) = \exp\left[-\left(\frac{x + 1 + (x - 1)\, e^{-2t}}{x + 1 - (x - 1)\, e^{-2t}} \right)^2 \right]. \tag{2.30}$$

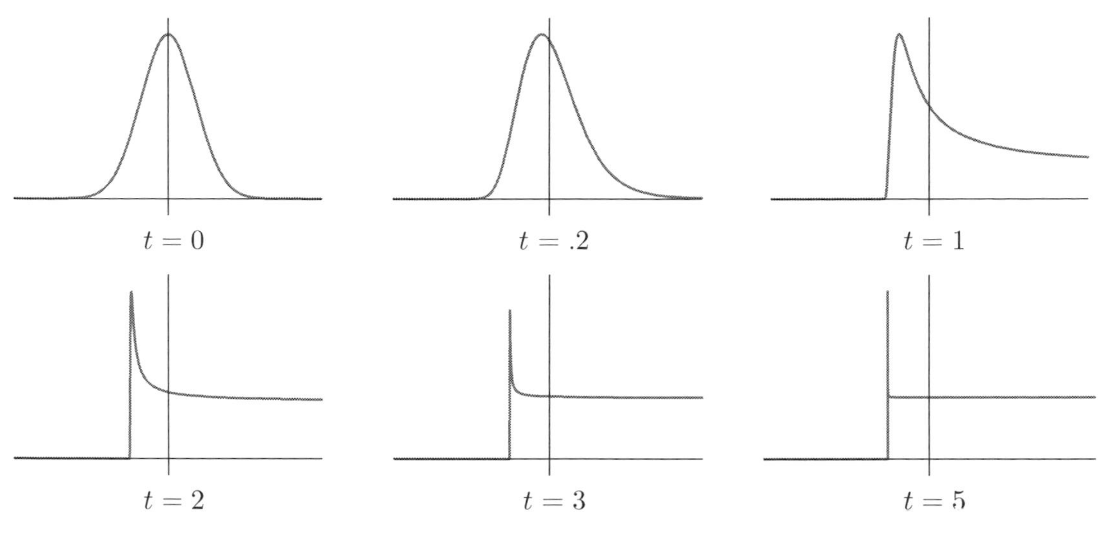

Figure 2.10. Solution to $u_t + (x^2 - 1)u_x = 0$.

(The derivation of this solution formula is left as Exercise 2.2.23.) As t increases, the solution's peak becomes more and more concentrated near $x_- = -1$, while the section of the wave above $x > x_+ = 1$ rapidly spreads out to ∞. In the long term, the solution converges (albeit nonuniformly) to a step function of height $1/e$:

$$u(t,x) \longrightarrow s(x) = \begin{cases} 1/e \approx .367879, & x \geq -1, \\ 0, & x < -1, \end{cases} \qquad \text{as} \qquad t \longrightarrow \infty.$$

Let us finish by making a few general observations concerning the characteristic curves of transport equations whose wave speed $c(x)$ depends only on the position x. Using the basic existence and uniqueness theory for such autonomous ordinary differential equations, [**20**, **23**, **52**], and assuming that $c(x)$ is continuously differentiable:†

- There is a unique characteristic curve passing through each point $(t,x) \in \mathbb{R}^2$.
- Characteristic curves cannot cross each other.
- If $t = \beta(x)$ is a characteristic curve, then so are all its horizontal translates: $t = \beta(x) + k$ for any k.
- Each non-horizontal characteristic curve is the graph of a strictly monotone function. Thus, each point on a wave always moves in the same direction, and can never reverse its direction of propagation.
- As t increases, the characteristic curve either tends to a fixed point, $x(t) \to x_\star$ as $t \to \infty$, with $c(x_\star) = 0$, or goes off to $\pm\infty$ in either finite or infinite time.

Proofs of these statements are assigned to the reader in Exercise 2.2.25.

† For those who know about such things, [**18**, **52**], this assumption can be weakened to just Lipschitz continuity.

Exercises

2.2.16. (a) Find the general solution to the first-order equation $u_t + \frac{3}{2} u_x = 0$.
(b) Find a solution satisfying the initial condition $u(1,x) = \sin x$. Is your solution unique?

2.2.17. (a) Solve the initial value problem $u_t - x\,u_x = 0$, $u(0,x) = (x^2+1)^{-1}$.
(b) Graph the solution at times $t = 0, 1, 2, 3$. (c) What is $\lim_{t\to\infty} u(t,x)$?

2.2.18. Suppose the initial data $u(0,x) = f(x)$ of the nonuniform transport equation (2.28) is continuous and satisfies $f(x) \to 0$ as $|\,x\,| \to \infty$. What is the limiting solution profile $u(t,x)$ as (a) $t \to \infty$? (b) $t \to -\infty$?

♡ 2.2.19. (a) Find and graph the characteristic curves for the equation $u_t + (\sin x)\,u_x = 0$.
(b) Write down the solution with initial data $u(0,x) = \cos \frac{1}{2}\pi x$. (c) Graph your solution at times $t = 0, 1, 2, 3, 5$, and 10. (d) What is the limiting solution profile as $t \to \infty$?

2.2.20. Consider the linear transport equation $u_t + (1+x^2)\,u_x = 0$. (a) Find and sketch the characteristic curves. (b) Write down a formula for the general solution. (c) Find the solution to the initial value problem $u(0,x) = f(x)$ and discuss its behavior as t increases.

2.2.21. Prove that, for $t \gg 0$, the speed of the wave in Example 2.4 is asymptotically proportional to $t^{-2/3}$.

2.2.22. Verify directly that formula (2.21) defines a solution to the differential equation (2.16).

♢ 2.2.23. Explain how to derive the solution formula (2.30). Justify that it defines a solution to equation (2.28).

2.2.24. Let $c(x)$ be a bounded C^1 function, so $|\,c(x)\,| \le c_\star < \infty$ for all x. Let $f(x)$ be any C^1 function. Prove that the solution $u(t,x)$ to the initial value problem $u_t + c(x)\,u_x = 0$, $u(0,x) = f(x)$, is uniquely defined for all $(t,x) \in \mathbb{R}^2$.

♡ 2.2.25. Suppose that $c(x) \in C^1$ is continuously differentiable for all $x \in \mathbb{R}$. (a) Prove that the characteristic curves of the transport equation (2.16) cannot cross each other. (b) A point where $c(x_\star) = 0$ is known as a *fixed point* for the characteristic equation $dx/dt = c(x)$. Explain why the characteristic curve passing through a fixed point $(t, x_\star)$ is a horizontal straight line. (c) Prove that if $x = g(t)$ is a characteristic curve, then so are all the horizontally translated curves $x = g(t+\delta)$ for any δ. (d) *True or false*: Every characteristic curve has the form $x = g(t+\delta)$, for some fixed function $g(t)$. (e) Prove that each non-horizontal characteristic curve is the graph $x = g(t)$ of a strictly monotone function. (f) Explain why a wave cannot reverse its direction. (g) Show that a non-horizontal characteristic curve starts, in the distant past, $t \to -\infty$, at either a fixed point or at $-\infty$ and ends, as $t \to +\infty$, at either the next-larger fixed point or at $+\infty$.

♡ 2.2.26. Consider the transport equation $\dfrac{\partial u}{\partial t} + c(t,x)\dfrac{\partial u}{\partial x} = 0$ with time-varying wave speed. Define the corresponding characteristic ordinary differential equation to be $\dfrac{dx}{dt} = c(t,x)$, the graphs of whose solutions $x(t)$ are the *characteristic curves*. (a) Prove that any solution $u(t,x)$ to the partial differential equation is constant on each characteristic curve. (b) Suppose that the general solution to the characteristic equation is written in the form $\xi(t,x) = k$, where k is an arbitrary constant. Prove that $\xi(t,x)$ defines a *characteristic variable*, meaning that $u(t,x) = f(\xi(t,x))$ is a solution to the time-varying transport equation for any continuously differentiable scalar function $f \in C^1$.

2.2.27. (a) Apply the method in Exercise 2.2.26 to find the characteristic curves for the equation $u_t + t^2\,u_x = 0$. (b) Find the solution to the initial value problem $u(0,x) = e^{-x^2}$, and discuss its dynamic behavior.

2.2.28. Solve Exercise 2.2.27 for the equation $u_t + (x - t)\, u_x = 0$.

♡ 2.2.29. Consider the first-order partial differential equation $u_t + (1 - 2t)\, u_x = 0$. Use Exercise 2.2.26 to: (a) Find and sketch the characteristic curves. (b) Write down the general solution. (c) Solve the initial value problem with $u(0,x) = \dfrac{1}{1+x^2}$. (d) Describe the behavior of your solution $u(t,x)$ from part (c) as $t \to \infty$. What about $t \to -\infty$?

2.2.30. Discuss which of the conclusions of Exercise 2.2.25 are valid for the characteristic curves of the transport equation with time-varying wave speed, as analyzed in Exercise 2.2.26.

◇ 2.2.31. Consider the two-dimensional transport equation $\dfrac{\partial u}{\partial t} + c(x,y)\,\dfrac{\partial u}{\partial x} + d(x,y)\,\dfrac{\partial u}{\partial y} = 0$, whose solution $u(t,x,y)$ depends on time t and space variables x, y. (a) Define a characteristic curve, and prove that the solution is constant along it. (b) Apply the method of characteristics to solve the initial value problem $u_t + y\, u_x - x\, u_y$, $u(0,x,y) = e^{-(x-1)^2-(y-1)^2}$. (c) Describe the behavior of your solution.

2.3 Nonlinear Transport and Shocks

The first-order nonlinear partial differential equation

$$u_t + u\, u_x = 0 \tag{2.31}$$

has the form of a transport equation (2.4), but the wave speed $c = u$ now depends, not on the position x, but rather on the size of the disturbance u. Larger waves will move faster, and overtake smaller, slower-moving waves. Waves of elevation, where $u > 0$, move to the right, while waves of depression, where $u < 0$, move to the left. This equation is considerably more challenging than the linear transport models analyzed above, and was first systematically studied in the early nineteenth century by the influential French mathematician Siméon–Denis Poisson and the great German mathematician Bernhard Riemann.† It and its multi-dimensional and multi-component generalizations play a crucial role in the modeling of gas dynamics, acoustics, shock waves in pipes, flood waves in rivers, chromatography, chemical reactions, traffic flow, and so on. Although we will be able to write down a solution formula, the complete analysis is far from trivial, and will require us to confront the possibility of discontinuous shock waves. Motivated readers are referred to Whitham's book, [**122**], for further details.

Fortunately, the method of characteristics that was developed for linear transport equations also works in the present context and leads to a complete mathematical solution. Mimicking our previous construction, (2.18), but now with wave speed $c = u$, let us define a *characteristic curve* of the nonlinear wave equation (2.31) to be the graph of a solution $x(t)$ to the ordinary differential equation

$$\frac{dx}{dt} = u(t,x). \tag{2.32}$$

† In addition to his fundamental contributions to partial differential equations, complex analysis, and number theory, Riemann also was the inventor of Riemannian geometry, which turned out to be absolutely essential for Einstein's theory of general relativity some 70 years later!

As such, the characteristics depend upon the solution u, which, in turn, is to be specified by its characteristics. We appear to be trapped in a circular argument.

The resolution of the conundrum is to argue that, as in the linear case, the solution $u(t,x)$ remains constant along its characteristics, and this fact will allow us to simultaneously specify both. To prove this claim, suppose that $x = x(t)$ parametrizes a characteristic curve associated with the given solution $u(t,x)$. Our task is to show that $h(t) = u\big(t, x(t)\big)$, which is obtained by evaluating the solution along the curve, is constant, which, as usual, is proved by checking that its derivative is identically zero. Repeating our chain rule computation (2.17), and using (2.32), we deduce that

$$\frac{dh}{dt} = \frac{d}{dt}\,u\big(t,x(t)\big) = \frac{\partial u}{\partial t}\big(t,x(t)\big) + \frac{dx}{dt}\,\frac{\partial u}{\partial x}\big(t,x(t)\big) = \frac{\partial u}{\partial t}\big(t,x(t)\big) + u\big(t,x(t)\big)\,\frac{\partial u}{\partial x}\big(t,x(t)\big) = 0,$$

since u is assumed to solve the nonlinear transport equation (2.31) at all values of (t,x), including those on the characteristic curve. We conclude that $h(t)$ is constant, and hence u is indeed constant on the characteristic curve.

Now comes the clincher. We know that the right-hand side of the characteristic ordinary differential equation (2.32) is a constant whenever $x = x(t)$ defines a characteristic curve. This means that the derivative dx/dt is a constant — namely the fixed value of u on the curve. Therefore, the characteristic curve must be a *straight line*,

$$x = u\,t + k, \tag{2.33}$$

whose slope equals the value assumed by the solution u on it.

And, as before, since the solution is constant along each characteristic line, it must be a function of the *characteristic variable*

$$\xi = x - t\,u \tag{2.34}$$

alone, and so

$$u = f(x - t\,u), \tag{2.35}$$

where $f(\xi)$ is an arbitrary C^1 function. Formula (2.35) should be viewed as an algebraic equation that implicitly defines the solution $u(t,x)$ as a function of t and x. Verification that the resulting function is indeed a solution to (2.31) is the subject of Exercise 2.3.14.

Example 2.6. Suppose that

$$f(\xi) = \alpha\,\xi + \beta,$$

with α, β constant. Then (2.35) becomes

$$u = \alpha(x - t\,u) + \beta, \qquad \text{and hence} \qquad u(t,x) = \frac{\alpha\,x + \beta}{1 + \alpha\,t} \tag{2.36}$$

is the corresponding solution to the nonlinear transport equation. At each fixed t, the graph of the solution is a straight line. If $\alpha > 0$, the solution flattens out: $u(t,x) \to 0$ as $t \to \infty$. On the other hand, if $\alpha < 0$, the straight line rapidly steepens to vertical as t approaches the critical time $t_\star = -1/\alpha$, at which point the solution ceases to exist. Figure 2.11 graphs two representative solutions. The top row shows the solution with $\alpha = 1,\ \beta = .5$, plotted at times $t = 0, 1, 5$, and 20; the bottom row takes $\alpha = -.2,\ \beta = .1$, and plots the solution at times $t = 0, 3, 4$, and 4.9. In the second case, the solution *blows up* by becoming vertical as $t \to 5$.

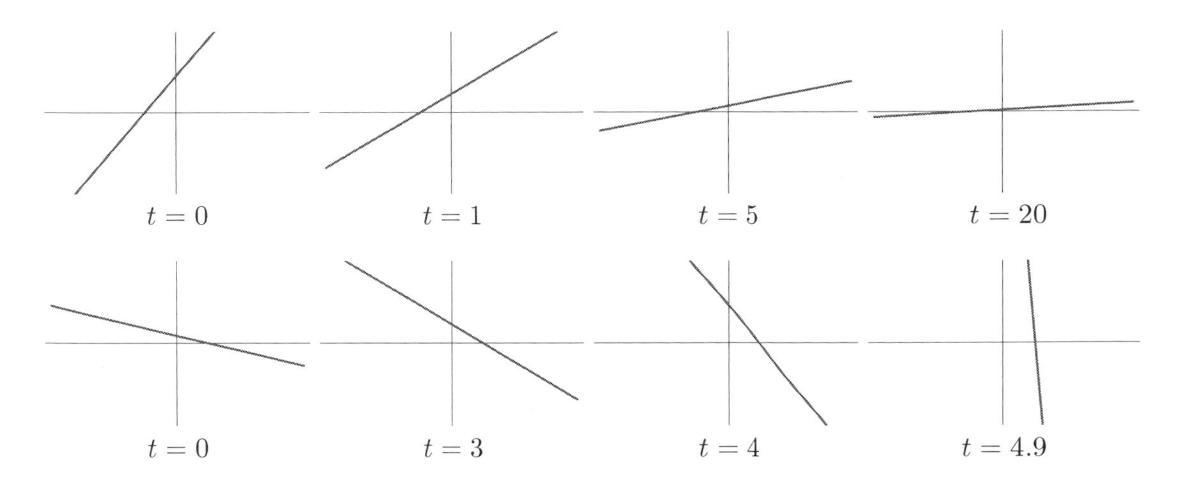

Figure 2.11. Two solutions to $u_t + u u_x = 0$.

Remark: Although (2.36) remains a valid solution formula after the blow-up time, $t > 5$, this is *not* to be viewed as a part of the original solution. With the appearance of such a singularity, the physical solution has broken down, and we stop tracking it.

To solve the general initial value problem

$$u(0,x) = f(x), \tag{2.37}$$

we note that, at $t = 0$, the implicit solution formula (2.35) reduces to (2.37), and hence the function f coincides with the initial data. However, because our solution formula (2.35) is an implicit equation, it is not immediately evident

(*a*) whether it can be solved to give a well-defined function $u(t,x)$, and,

(*b*) even granted this, how to describe the resulting solution's qualitative features and dynamical behavior.

A more instructive approach is founded on the following geometrical construction. Through each point $(0,y)$ on the x–axis, draw the characteristic line

$$x = t\,f(y) + y \tag{2.38}$$

whose slope, namely $f(y) = u(0,y)$, equals the value of the initial data (2.37) at that point. According to the preceding discussion, the solution will have the same value on the entire characteristic line (2.38), and so

$$u(t, t\,f(y) + y) = f(y) \qquad \text{for all} \quad t. \tag{2.39}$$

For example, if $f(y) = y$, then $u(t,x) = y$ whenever $x = ty + y$; eliminating y, we find $u(t,x) = x/(t+1)$, which agrees with one of our straight line solutions (2.36).

Now, the problem with this construction is immediately apparent from Figure 2.12, which plots the characteristic lines associated with the initial data

$$u(0,x) = \tfrac{1}{2}\pi - \tan^{-1} x.$$

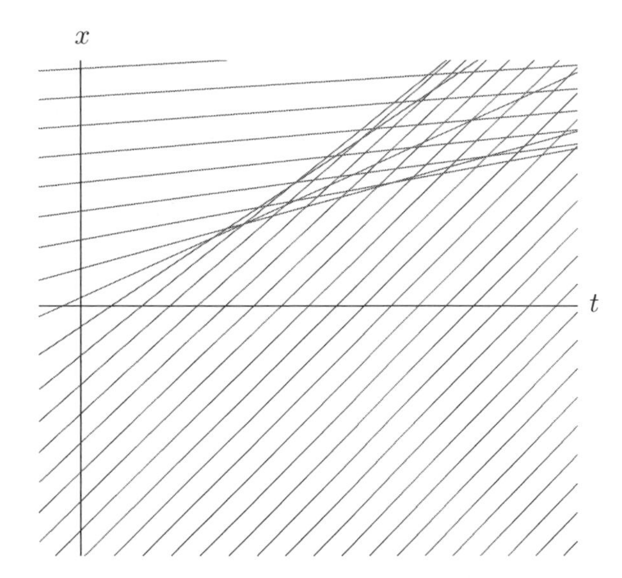

Figure 2.12. Characteristics lines for $u(0,x) = \frac{1}{2}\pi - \tan^{-1} x$.

Two characteristic lines that are not parallel must cross each other somewhere. The value of the solution is supposed to equal the slope of the characteristic line passing through the point. Hence, at a crossing point, the solution is required to assume two *different* values, one corresponding to each line. Something is clearly amiss, and we need to resolve this apparent paradox.

There are three principal scenarios. The first, trivial, situation occurs when all the characteristic lines are parallel, and so the difficulty does not arise. In this case, they all have the same slope, say c, which means that the solution has the same value on each one. Therefore, $u(t,x) \equiv c$ is a constant solution.

The next-simplest case occurs when the initial data is everywhere *nondecreasing*, so $f(x) \leq f(y)$ whenever $x \leq y$, which is assured if its derivative is never negative: $f'(x) \geq 0$. In this case, as sketched in Figure 2.13, the characteristic lines emanating from the x axis fan out into the right half-plane, and so never cross each other at any future time $t > 0$. Each point (t,x) with $t \geq 0$ lies on a unique characteristic line, and the value of the solution at (t,x) is equal to the slope of the line. We conclude that the solution $u(t,x)$ is well defined at all future times $t \geq 0$. Physically, such solutions represent *rarefaction waves*, which spread out as time progresses. A typical example, corresponding to initial data

$$u(0,x) = \tfrac{1}{2}\pi + \tan^{-1}(3x),$$

has its characteristic lines plotted in Figure 2.13, while Figure 2.14 graphs some representative solution profiles.

The more interesting case occurs when the initial data is a decreasing function, and so $f'(x) < 0$. Now, as in Figure 2.12, some of the characteristic lines starting at $t = 0$ will cross at some point in the future. If a point (t,x) lies on two or more distinct characteristic lines, the value of the solution $u(t,x)$, which should equal the characteristic slope, is no longer uniquely determined. Although, in a purely mathematical context, one might be tempted to allow such multiply valued solutions, from a physical standpoint this is unacceptable. The solution $u(t,x)$ is supposed to represent a measurable quantity, e.g., concentration,

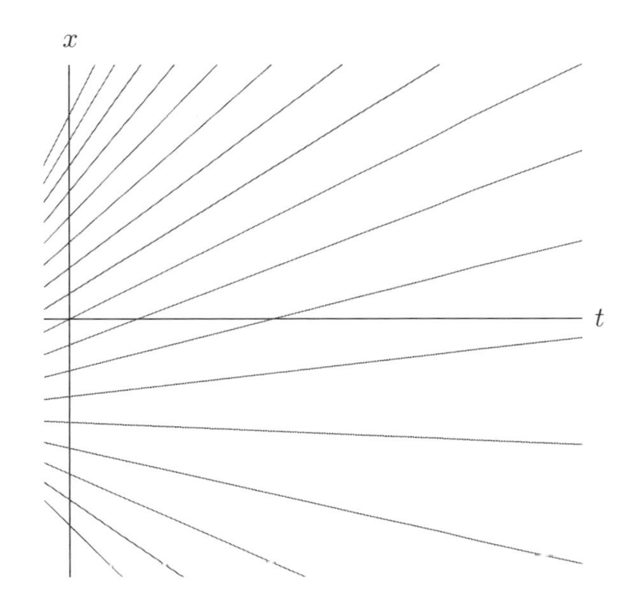

Figure 2.13. Characteristic lines for a rarefaction wave.

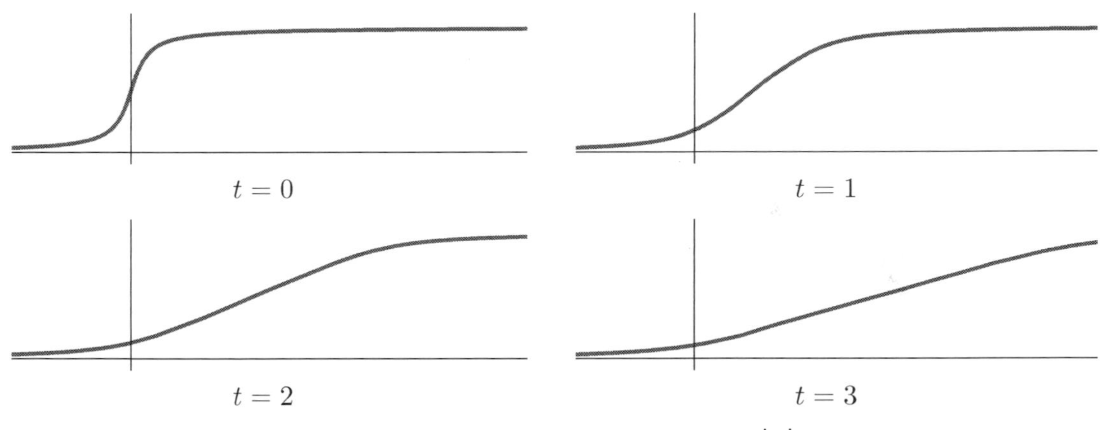

Figure 2.14. Rarefaction wave.

velocity, pressure, and must therefore assume a unique value at each point. In effect, the mathematical model has broken down and no longer conforms to physical reality.

However, before confronting this difficulty, let us first, from a purely theoretical standpoint, try to understand what happens if we mathematically continue the solution as a multiply valued function. For specificity, consider the initial data

$$u(0,x) = \tfrac{1}{2}\pi - \tan^{-1} x, \tag{2.40}$$

appearing in the first graph in Figure 2.15. The corresponding characteristic lines are displayed in Figure 2.12. Initially, they do not cross, and the solution remains a well-defined, single-valued function. However, after a while one reaches a critical time, $t_\star > 0$, when the first two characteristic lines cross each other. Subsequently, a wedge-shaped region appears in the (t,x)–plane, consisting of points that lie on the intersection of three

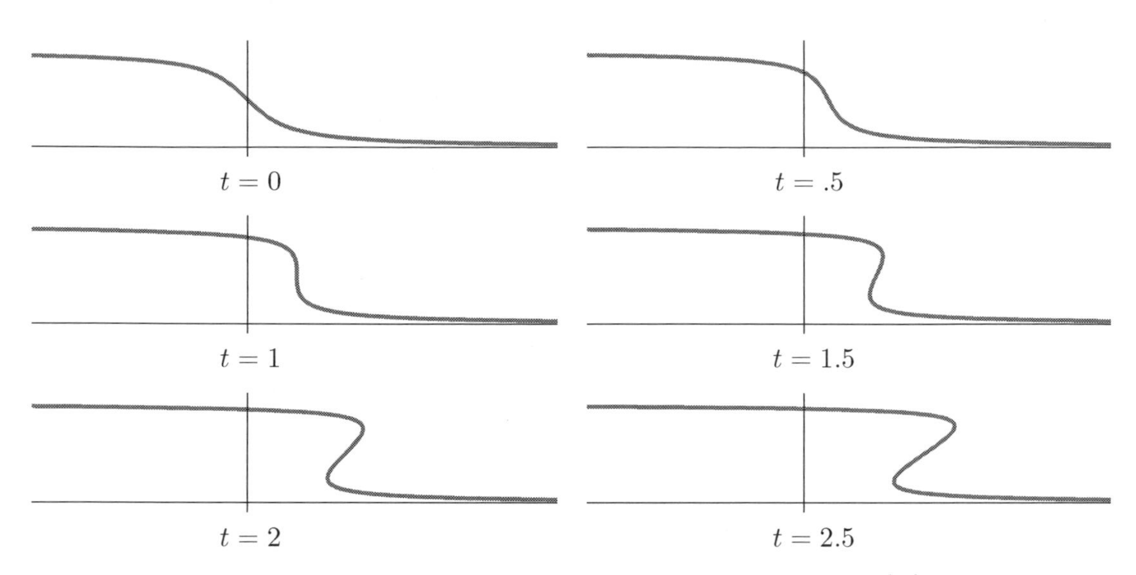

Figure 2.15. Multiply valued compression wave.

distinct characteristic lines with different slopes; at such points, the mathematical solution achieves three distinct values. Points outside the wedge lie on a single characteristic line, and the solution remains single-valued there. The boundary of the wedge consists of points where precisely two characteristic lines cross.

To fully appreciate what is going on, look now at the sequence of pictures of the multiply valued solution in Figure 2.15, plotted at six successive times. Since the initial data is positive, $f(x) > 0$, all the characteristic slopes are positive. As a consequence, every point on the solution curve moves to the right, at a speed equal to its height. Since the initial data is a decreasing function, points on the graph lying to the left will move faster than those to the right and eventually overtake them. At first, the solution merely steepens into a *compression wave*. At the critical time $t_\star$ when the first two characteristic lines cross, say at position $x_\star$, so that $(t_\star, x_\star)$ is the tip of the aforementioned wedge, the solution graph has become vertical:

$$\frac{\partial u}{\partial x}(t, x_\star) \longrightarrow \infty \qquad \text{as} \qquad t \longrightarrow t_\star,$$

and $u(t, x)$ is no longer a classical solution. Once this occurs, the solution graph ceases to be a single-valued function, and its overlapping lobes lie over the points (t, x) belonging to the wedge.

The critical time $t_\star$ can, in fact, be determined from the implicit solution formula (2.35). Indeed, if we differentiate with respect to x, we obtain

$$\frac{\partial u}{\partial x} = \frac{\partial}{\partial x} f(\xi) = f'(\xi)\,\frac{\partial \xi}{\partial x} = f'(\xi)\left(1 - t\,\frac{\partial u}{\partial x}\right), \qquad \text{where} \qquad \xi = x - t\,u.$$

Solving for

$$\frac{\partial u}{\partial x} = \frac{f'(\xi)}{1 + t\,f'(\xi)},$$

we see that the slope blows up:

$$\frac{\partial u}{\partial x} \longrightarrow \infty \qquad \text{as} \qquad t \longrightarrow -\frac{1}{f'(\xi)}.$$

In other words, if the initial data has negative slope at position x, so $f'(x) < 0$, then the solution along the characteristic line emanating from the point $(0, x)$ will fail to be smooth at the time $-1/f'(x)$. The earliest critical time is, thus,

$$t_\star := \min \left\{ -\frac{1}{f'(x)} \;\middle|\; f'(x) < 0 \right\}. \tag{2.41}$$

If x_0 is the value of x that produces the minimum $t_\star$, then the slope of the solution profile will first become infinite at the location where the characteristic starting at x_0 is at time $t_\star$, namely

$$x_\star = x_0 + f(x_0)\, t_\star. \tag{2.42}$$

For instance, for the particular initial configuration (2.40) represented in Figure 2.15,

$$f(x) = \frac{\pi}{2} - \tan^{-1} x, \qquad\qquad f'(x) = -\frac{1}{1+x^2},$$

and so the critical time is

$$t_\star = \min \{ 1 + x^2 \} = 1, \qquad \text{with} \qquad x_\star = f(0)\, t_\star = \tfrac{1}{2}\pi,$$

since the minimum value occurs at $x_0 = 0$.

Now, while mathematically plausible, such a multiply valued solution is physically untenable. So what really happens after the critical time $t_\star$? One needs to decide which (if any) of the possible solution values is physically appropriate. The mathematical model, in and of itself, is incapable of resolving this quandary. We must therefore revisit the underlying physics, and ask what sort of phenomenon we are trying to model.

Shock Dynamics

To be specific, let us regard the transport equation (2.31) as a model of compressible fluid flow in a single space variable, e.g., the motion of gas in a long pipe. If we push a piston into the pipe, then the gas will move ahead of it and thereby be compressed. However, if the piston moves too rapidly, then the gas piles up on top of itself, and a shock wave forms and propagates down the pipe. Mathematically, the shock is represented by a discontinuity where the solution abruptly changes value. The formulas (2.41) and (2.42) determine the time and position for the onset of the shock-wave discontinuity. Our goal now is to predict its subsequent behavior, and this will be based on use of a suitable physical conservation law. Indeed, one expects mass to be conserved – even through a shock discontinuity — since gas atoms can neither be created nor destroyed. And, as we will see, conservation of mass (almost) suffices to prescribe the subsequent motion of the shock wave.

Before investigating the implications of conservation of mass, let us first convince ourselves of its validity for the nonlinear transport model. (Just because a mathematical equation models a physical system does not automatically imply that it inherits any of its physical conservation laws.) If $u(t,x)$ represents density, then, at time t, the total mass lying in an interval $a \le x \le b$ is calculated by integration:

$$M_{a,b}(t) = \int_a^b u(t,x)\,dx. \tag{2.43}$$

Assuming that $u(t,x)$ is a classical solution to the nonlinear transport equation (2.31), we can determine the rate of change of mass on this interval by differentiation:

$$\begin{aligned}\frac{dM_{a,b}}{dt} &= \frac{d}{dt}\int_a^b u(t,x)\,dx = \int_a^b \frac{\partial u}{\partial t}(t,x)\,dx = -\int_a^b u(t,x)\,\frac{\partial u}{\partial x}(t,x)\,dx \\ &= -\int_a^b \frac{\partial}{\partial x}\left[\tfrac{1}{2}u(t,x)^2\right]dx = -\tfrac{1}{2}u(t,x)^2\,\Big|_{x=a}^{b} = \tfrac{1}{2}u(t,a)^2 - \tfrac{1}{2}u(t,b)^2.\end{aligned} \tag{2.44}$$

The final expression represents the net *mass flux* through the endpoints of the interval. Thus, the only way in which the mass on the interval $[a,b]$ changes is through its endpoints; inside, mass can be neither created nor destroyed, which is the precise meaning of the mass conservation law in continuum mechanics. In particular, if there is zero net mass flux, then the total mass is constant, and hence conserved. For example, if the initial data (2.37) has finite total mass,

$$\left|\int_{-\infty}^{\infty} f(x)\,dx\right| < \infty, \tag{2.45}$$

which requires that $f(x) \to 0$ reasonably rapidly as $|x| \to \infty$, then the total mass of the solution — at least up to the formation of a shock discontinuity — remains constant and equal to its initial value:

$$\int_{-\infty}^{\infty} u(t,x)\,dx = \int_{-\infty}^{\infty} u(0,x)\,dx = \int_{-\infty}^{\infty} f(x)\,dx. \tag{2.46}$$

Similarly, if $u(t,x)$ represents the traffic density on a highway at time t and position x, then the integrated conservation law (2.44) tells us that the rate of change in the number of vehicles on the stretch of road between a and b equals the number of vehicles entering at point a minus the number leaving at point b — which assumes that there are no other exits or entrances on this part of the highway. Thus, in the traffic model, (2.44) represents the conservation of vehicles.

The preceding calculation relied on the fact that the integrand can be written as an x derivative. This is a common feature of physical conservation laws in continuum mechanics, and motivates the following general definition.

Definition 2.7. A *conservation law*, in one space dimension, is an equation of the form

$$\frac{\partial T}{\partial t} + \frac{\partial X}{\partial x} = 0. \tag{2.47}$$

The function T is known as the *conserved density*, while X is the associated *flux*.

In the simplest situations, the conserved density $T(t,x,u)$ and flux $X(t,x,u)$ depend on the time t, the position x, and the solution $u(t,x)$ to the physical system. (Higher-order conservation laws, which also depend on derivatives of u, arise in the analysis of integrable partial differential equations; see Section 8.5 and [**36, 87**].) For example, the nonlinear transport equation (2.31) is itself a conservation law, since it can be written in the form

$$\frac{\partial u}{\partial t} + \frac{\partial}{\partial x}\left(\tfrac{1}{2}u^2\right) = 0, \tag{2.48}$$

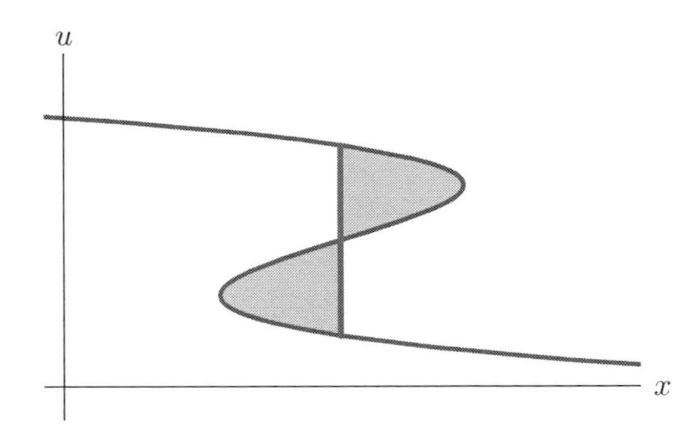

Figure 2.16. Equal Area Rule.

and so the conserved density is $T = u$ and the flux is $X = \frac{1}{2}u^2$. And indeed, it was this identity that made our computation (2.44) work. The general result, proved by an analogous computation, justifies calling (2.47) a conservation law.

Proposition 2.8. *Given a conservation law* (2.47), *then, on any closed interval* $a \leq x \leq b$,

$$\frac{d}{dt}\int_a^b T\,dx = -\left. X \right|_{x=a}^{b}. \tag{2.49}$$

Proof: The proof is an immediate consequence of the Fundamental Theorem of Calculus — assuming sufficient smoothness that allows one to bring the derivative inside the integral sign:

$$\frac{d}{dt}\int_a^b T\,dx = \int_a^b \frac{\partial T}{\partial t}\,dx = -\int_a^b \frac{\partial X}{\partial x}\,dx = -\left. X \right|_{x=a}^{b}. \qquad Q.E.D.$$

We will refer to (2.49) as the *integrated form* of the conservation law (2.47). It states that the rate of change of the total density, integrated over an interval, is equal to the amount of flux through its two endpoints. In particular, if there is no net flux into or out of the interval, then the integrated density is *conserved*, meaning that it remains constant over time. All physical conservation laws — mass, momentum, energy, and so on — for systems governed by partial differential equations are of this form or its multi-dimensional extensions, [**87**].

With this in hand, let us return to the physical context of the nonlinear transport equation. By definition, a *shock* is a discontinuity in the solution $u(t,x)$. We will make the physically plausible assumption that mass (or vehicle) conservation continues to hold even within the shock. Recall that the total mass, which at time t is the area† under the curve $u(t,x)$, must be conserved. This continues to hold even when the mathematical solution becomes multiply valued, in which case one employs a line integral $\int_C u\,dx$, where C represents the graph of the solution, to compute the mass/area. Thus, to construct a discontinuous shock solution with the *same* mass, one replaces part of the multiply valued

† We are implicitly assuming that the mass is finite, as in (2.45), although the overall construction does not rely on this restriction.

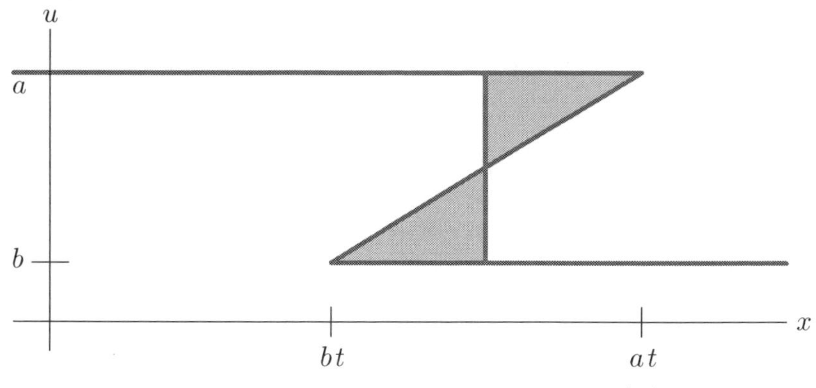

Figure 2.17. Multiply valued step wave.

graph by a vertical shock line in such a way that the resulting function is single-valued and has the same area under its graph. Referring to Figure 2.16, observe that the region under the shock graph is obtained from that under the multi-valued solution graph by deleting the upper shaded lobe and appending the lower shaded lobe. Thus the resulting area will be the same, provided the shock line is drawn so that the areas of the two shaded lobes are equal. This construction is known as the *Equal Area Rule*; it ensures that the total mass of the shock solution matches that of the multiply valued solution, which in turn is equal to the initial mass, as required by the physical conservation law.

Example 2.9. An illuminating special case occurs when the initial data has the form of a *step function* with a single discontinuity at the origin:

$$u(0,x) = \begin{cases} a, & x < 0, \\ b, & x > 0. \end{cases} \tag{2.50}$$

If $a > b$, then the initial data is already in the form of a shock wave. For $t > 0$, the mathematical solution constructed by continuing along the characteristic lines is multiply valued in the region $b\,t < x < a\,t$, where it assumes both values a and b; see Figure 2.17. Moreover, the initial vertical line of discontinuity has become a tilted line, because each point $(0,u)$ on it has moved along the associated characteristic a distance $u\,t$. The Equal Area Rule tells us to draw the shock line halfway along, at $x = \frac{1}{2}(a+b)\,t$, in order that the two triangles have the same area. We deduce that the shock moves with speed $c = \frac{1}{2}(a+b)$, equal to the average of the two speeds at the jump. The resulting shock-wave solution is

$$u(t,x) = \begin{cases} a, & x < c\,t, \\ b, & x > c\,t, \end{cases} \qquad \text{where} \qquad c = \frac{a+b}{2}. \tag{2.51}$$

A plot of its characteristic lines appears in Figure 2.18. Observe that colliding pairs of characteristic lines terminate at the shock line, whose slope is the average of their individual slopes.

The fact that the shock speed equals the *average* of the solution values on either side is, in fact, of general validity, and is known as the *Rankine–Hugoniot condition*, named after the nineteenth-century Scottish physicist William Rankine and French engineer Pierre Hugoniot, although historically these conditions first appeared in a 1849 paper by George Stokes, [**109**]. However, intimidated by criticism by his contemporary applied mathematicians Lords Kelvin and Rayleigh, Stokes thought he was mistaken, and even ended up

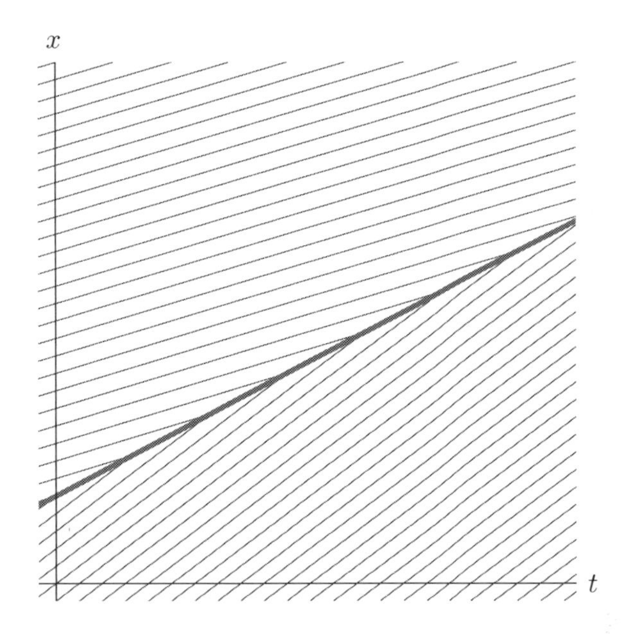

Figure 2.18. Characteristic lines for the step wave shock.

deleting the relevant part when his collected works were published in 1883, [**110**]. The missing section was restored in the 1966 reissue, [**111**].

Proposition 2.10. *Let $u(t,x)$ be a solution to the nonlinear transport equation that has a discontinuity at position $x = \sigma(t)$, with finite, unequal left- and right-hand limits*

$$u^-(t) = u\big(t,\sigma(t)^-\big) = \lim_{x \to \sigma(t)^-} u(t,x), \qquad u^+(t) = u\big(t,\sigma(t)^+\big) = \lim_{x \to \sigma(t)^+} u(t,x), \tag{2.52}$$

on either side of the shock discontinuity. Then, to maintain conservation of mass, the speed of the shock must equal the average of the solution values on either side:

$$\frac{d\sigma}{dt} = \frac{u^-(t) + u^+(t)}{2}. \tag{2.53}$$

Proof: Referring to Figure 2.19, consider a small time interval, from t to $t + \Delta t$, with $\Delta t > 0$. During this time, the shock moves from position $a = \sigma(t)$ to position $b = \sigma(t + \Delta t)$. The total mass contained in the interval $[a,b]$ at time t, before the shock has passed through, is

$$M(t) = \int_a^b u(t,x)\,dx \approx u^+(t)\,(b-a) = u^+(t)\,\big[\,\sigma(t+\Delta t) - \sigma(t)\,\big],$$

where we assume that $\Delta t \ll 1$ is very small, and so the integrand is well approximated by its limiting value (2.52). Similarly, after the shock has passed, the total mass remaining in the interval is

$$M(t+\Delta t) = \int_a^b u(t+\Delta t,x)\,dx \approx u^-(t+\Delta t)\,(b-a) = u^-(t+\Delta t)\,\big[\,\sigma(t+\Delta t) - \sigma(t)\,\big].$$

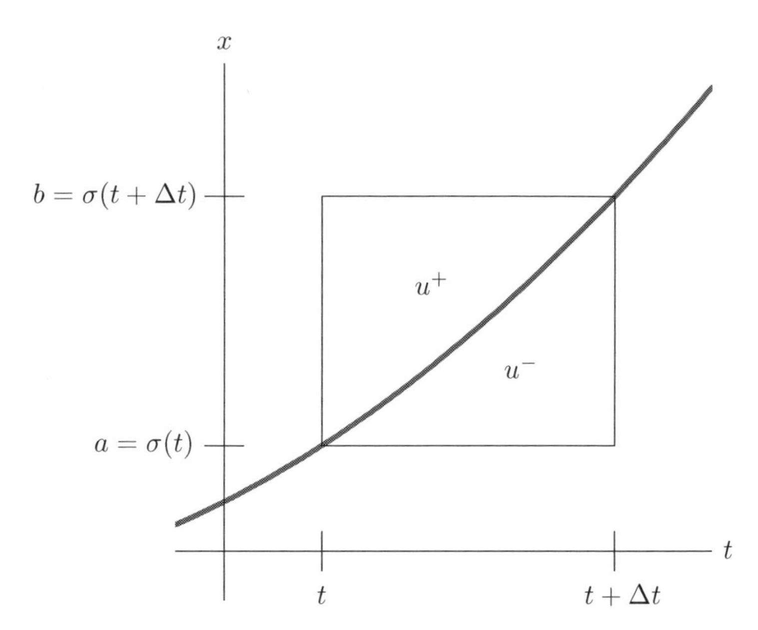

Figure 2.19. Conservation of mass near a shock.

Thus, the rate of change in mass across the shock at time t is given by

$$\begin{aligned}\frac{dM}{dt} &= \lim_{\Delta t \to 0} \frac{M(t+\Delta t) - M(t)}{\Delta t} \\ &= \lim_{\Delta t \to 0} \left[\, u^-(t+\Delta t) - u^+(t) \,\right] \frac{\sigma(t+\Delta t) - \sigma(t)}{\Delta t} = \left[\, u^-(t) - u^+(t) \,\right] \frac{d\sigma}{dt} \,.\end{aligned}$$

On the other hand, at any $t < \tau < t + \Delta t$, the mass flux into the interval $[a, b]$ through the endpoints is given by the right-hand side of (2.44):

$$\tfrac{1}{2}\left[\, u(\tau,a)^2 - u(\tau,b)^2 \,\right] \;\longrightarrow\; \tfrac{1}{2}\left[\, u^-(t)^2 - u^+(t)^2 \,\right], \qquad \text{since} \quad \tau \to t \quad \text{as} \quad \Delta t \to 0.$$

Conservation of mass requires that the rate of change in mass be equal to the mass flux:

$$\frac{dM}{dt} = \left[\, u^-(t) - u^+(t) \,\right] \frac{d\sigma}{dt} = \tfrac{1}{2}\left[\, u^-(t)^2 - u^+(t)^2 \,\right].$$

Solving for $d\sigma/dt$ establishes (2.53). *Q.E.D.*

Example 2.11. By way of contrast, let us investigate the case when the initial data is a step function (2.50), but with $a < b$, so the jump goes upwards. In this case, the characteristic lines diverge from the initial discontinuity, and the mathematical solution is not specified at all in the wedge-shaped region $at < x < bt$. Our task is to decide how to "fill in" the solution values between the two regions where the solution is well defined and constant.

One possible connection is by a straight line. Indeed, a simple modification of the rational solution (2.36) produces the *similarity solution*[†]

$$u(t, x) = \frac{x}{t} \,,$$

[†] See Section 8.2 for general techniques for constructing similarity (scale-invariant) solutions to partial differential equations.

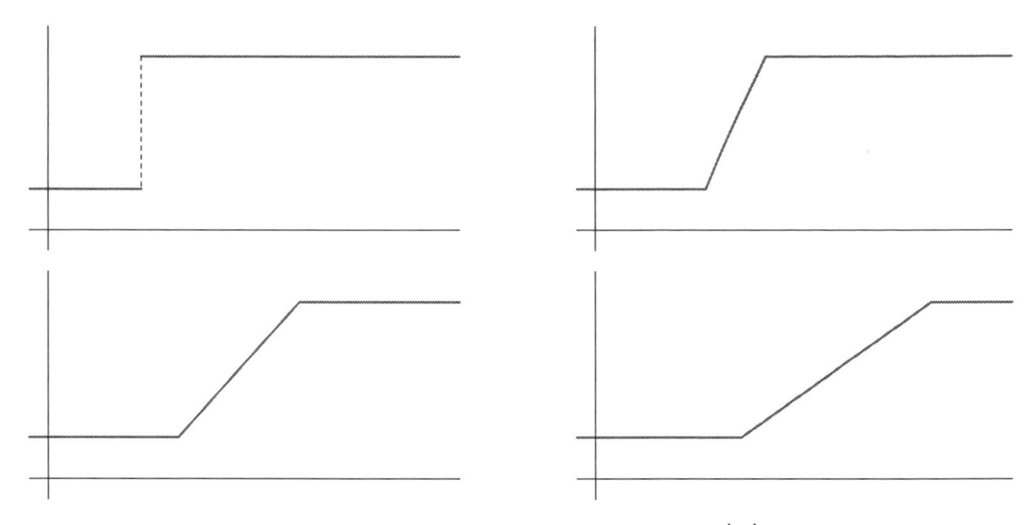

Figure 2.20. Rarefaction wave.

which not only solves the differential equation, but also has the required values $u(t, a\,t) = a$ and $u(t, b\,t) = b$ at the two edges of the wedge. This can be used to construct the piecewise affine *rarefaction wave*

$$u(t,x) = \begin{cases} a, & x \le a\,t, \\ x/t, & a\,t \le x \le b\,t, \\ b, & x \ge b\,t, \end{cases} \tag{2.54}$$

which is graphed at four representative times in Figure 2.20.

A second possibility would be to continue the discontinuity as a shock wave, whose speed is governed by the Rankine-Hugoniot condition, leading to a discontinuous solution having the same formula as (2.51). Which of the two competing solutions should we use? The first, (2.54), makes better physical sense; indeed, if we were to smooth out the discontinuity, then the resulting solutions would converge to the rarefaction wave and not the reverse shock wave; see Exercise 2.3.13. Moreover, the discontinuous solution (2.51) has characteristic lines emanating from the discontinuity, which means that the shock is creating new values for the solution as it moves along, and this can, in fact, be done in a variety of ways. In other words, the discontinuous solution violates *causality*, meaning that the solution profile at any given time uniquely prescribes its subsequent motion. Causality requires that, while characteristics may terminate at a shock discontinuity, they cannot begin there, because their slopes will not be uniquely prescribed by the shock profile, and hence the characteristics to the left of the shock must have larger slope (or speed), while those to the right must have smaller slope. Since the shock speed is the average of the two characteristic slopes, this requires the *Entropy Condition*

$$u^-(t) > \frac{d\sigma}{dt} = \frac{u^-(t) + u^+(t)}{2} > u^+(t). \tag{2.55}$$

With further analysis, it can be shown, [**57**], that the rarefaction wave (2.54) is the unique solution[†] to the initial value problem satisfying the entropy condition (2.55).

[†] Albeit not a classical solution, but rather a weak solution, as per Section 10.4.

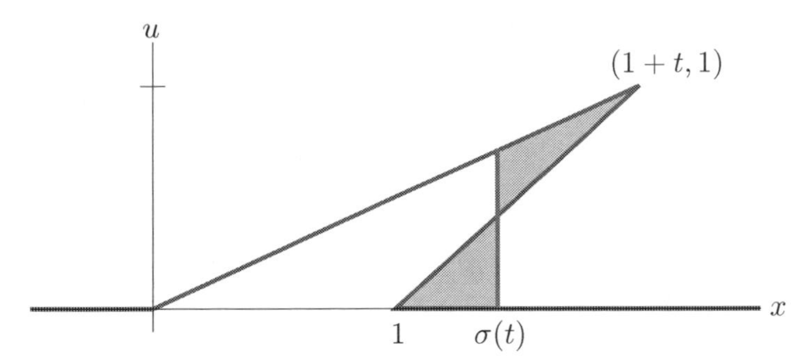

Figure 2.21. Equal Area Rule for the triangular wave.

These prototypical solutions epitomize the basic phenomena modeled by the nonlinear transport equation: *rarefaction waves*, which emanate from regions where the initial data satisfies $f'(x) > 0$, causing the solution to spread out as time progresses, and *compression waves*, emanting from regions where $f'(x) < 0$, causing the solution to progressively steepen and eventually break into a shock discontinuity. Anyone caught in a traffic jam recognizes the compression waves, where the vehicles are bunched together and almost stationary, while the interspersed rarefaction waves correspond to freely moving traffic. (An intelligent driver will take advantage of the rarefaction waves moving backwards through the jam to switch lanes!) The familiar, frustrating traffic jam phenomenon, even on accident- or construction-free stretches of highway, is, thus, an intrinsic effect of the nonlinear transport models that govern traffic flow, [**122**].

Example 2.12. *Triangular wave*: Suppose the initial data has the triangular profile

$$u(0,x) = f(x) = \begin{cases} x, & 0 \le x \le 1, \\ 0, & \text{otherwise}, \end{cases}$$

as in the first graph in Figure 2.22. The initial discontinuity at $x = 1$ will propagate as a shock wave, while the slanted line behaves as a rarefaction wave. To find the profile at time t, we first graph the multi-valued solution obtained by moving each point on the graph of f to the right an amount equal to t times its height. As noted above, this motion preserves straight lines. Thus, points on the x–axis remain fixed, and the diagonal line now goes from $(0,0)$ to $(1+t,1)$, which is where the uppermost point $(1,1)$ on the graph of f has moved to, and hence has slope $(1+t)^{-1}$, while the initial vertical shock line has become tilted, going from $(1,0)$ to $(0,1+t)$. We now need to find the position $\sigma(t)$ of the shock line in order to satisfy the Equal Area Rule, namely so that the areas of the two shaded regions in Figure 2.21 are identical. The reader is invited to determine this geometrically; instead, we invoke the Rankine–Hugoniot condition (2.53). At the shock line, $x = \sigma(t)$, the left- and right-hand limiting values are, respectively,

$$u^-(t) = u\big(t, \sigma(t)^-\big) = \frac{\sigma(t)}{1+t}, \qquad u^+(t) = u\big(t, \sigma(t)^+\big) = 0,$$

and hence (2.53) prescribes the shock speed to be

$$\frac{d\sigma}{dt} = \frac{1}{2}\left(\frac{\sigma(t)}{1+t} + 0 \right) = \frac{\sigma(t)}{2(1+t)}.$$

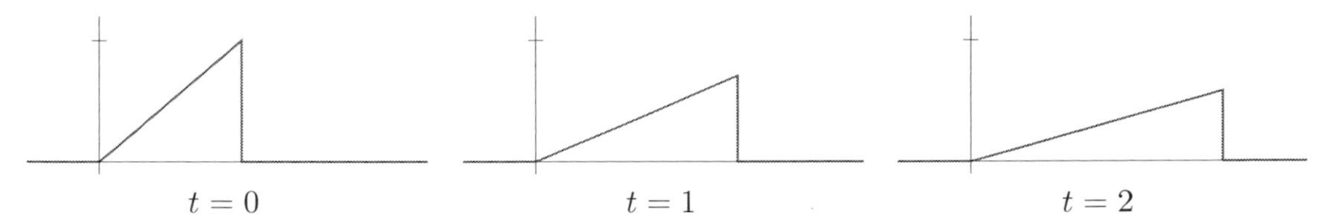

Figure 2.22. Triangular-wave solution.

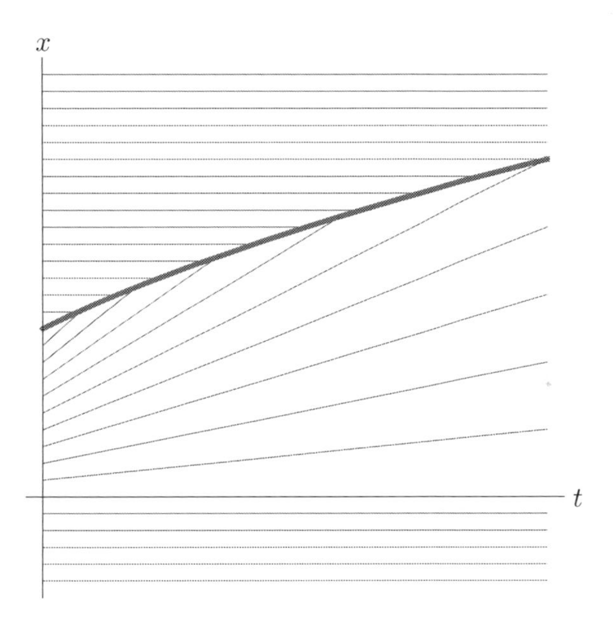

Figure 2.23. Characteristic lines for the triangular-wave shock.

The solution to the resulting separable ordinary differential equation is easily found. Since the shock starts out at $\sigma(0) = 1$, we deduce that

$$\sigma(t) = \sqrt{1+t}\,, \qquad \text{with} \qquad \frac{d\sigma}{dt} = \frac{1}{2\sqrt{1+t}}\,.$$

Further, the strength of the shock, namely its height, is

$$u^-(t) = \frac{\sigma(t)}{1+t} = \frac{1}{\sqrt{1+t}}\,.$$

We conclude that, as t increases, the solution remains a triangular wave, of steadily decreasing slope, while the shock moves off to $x = +\infty$ at a progressively slower speed and smaller height. Its position follows a parabolic trajectory in the (t, x)–plane. See Figure 2.22 for representative plots of the triangular wave solution, while Figure 2.23 illustrates the characteristic lines and shock wave trajectory.

In more general situations, continuing on after the initial shock formation, other characteristic lines may start to cross, thereby producing new shocks. The shocks themselves continue to propagate, often at different velocities. When a fast-moving shock catches up with a slow-moving shock, one must then decide how to merge the shocks so as to retain a physically meaningful solution. The Rankine–Hugoniot (Equal Area) and Entropy Conditions continue to uniquely specify the dynamics. However, at this point, the mathematical details have become too intricate for us to pursue any further, and we refer the interested reader to Whitham's book, [**122**]. See also [**57**] for a proof of the following existence theorem for shock-wave solutions to the nonlinear transport equation.

Theorem 2.13. *If the initial data* $u(0,x) = f(x)$ *is piecewise*† C^1 *with finitely many jump discontinuities, then, for* $t > 0$, *there exists a unique (weak) solution to the nonlinear transport equation* (2.31) *that also satisfies the Rankine–Hugoniot condition* (2.53) *and the entropy condition* (2.55).

Remark: Our derivation of the Rankine–Hugoniot shock speed condition (2.53) relied on the fact that we can write the original partial differential equation in the form of a conservation law. But there are, in fact, other ways to do this. For instance, multiplying the nonlinear transport equation (2.31) by u allows us write it in the alternative conservative form

$$u\,\frac{\partial u}{\partial t} + u^2\,\frac{\partial u}{\partial x} = \frac{\partial}{\partial t}\left(\tfrac{1}{2}u^2\right) + \frac{\partial}{\partial x}\left(\tfrac{1}{3}u^3\right) = 0. \tag{2.56}$$

In this formulation, the conserved density is $T = \frac{1}{2}u^2$, and the associated flux is $X = \frac{1}{3}u^3$. The integrated form (2.49) of the conservation law (2.56) is

$$\frac{d}{dt}\int_a^b \tfrac{1}{2}u(t,x)^2\,dx = \tfrac{1}{3}\left[\,u(t,a)^3 - u(t,b)^3\,\right]. \tag{2.57}$$

In some physical models, the integral on the left-hand side represents the energy within the interval $[a,b]$, and the conservation law tells us that energy can enter the interval as a flux only through its ends. If we assume that energy is conserved at a shock, then, repeating our previous argument, we are led to the alternative equation

$$\frac{d\sigma}{dt} = \frac{\frac{1}{3}\left[\,u^-(t)^3 - u^+(t)^3\,\right]}{\frac{1}{2}\left[\,u^-(t)^2 - u^+(t)^2\,\right]} = \frac{2}{3}\,\frac{u^-(t)^2 + u^-(t)\,u^+(t) + u^+(t)^2}{u^-(t) + u^+(t)} \tag{2.58}$$

for the shock speed. Thus, a shock that conserves energy moves at a different speed from one that conserves mass! The evolution of a shock wave depends not just on the underlying differential equation, but also on the physical assumptions governing the selection of a suitable conservation law.

More General Wave Speeds

Let us finish this section by considering a nonlinear transport equation

$$u_t + c(u)\,u_x = 0, \tag{2.59}$$

whose wave speed is a more general function of the disturbance u. (Further extensions, allowing c to depend also on t and x, are discussed in Exercise 2.3.20.) Most of the

† Meaning continuous everywhere, and continuously differentiable except at a discrete set of points; see Definition 3.7 below for the precise definition.

development is directly parallel to the special case (2.31) discussed above, and so the details are left for the reader to fill in, although the shock dynamics does require some care.

In this case, the *characteristic curve* equation is

$$\frac{dx}{dt} = c\big(u(t,x)\big). \tag{2.60}$$

As before, the solution u is constant on characteristics, and hence the characteristics are straight lines, now with slope $c(u)$. Thus, to solve the initial value problem

$$u(0,x) = f(x), \tag{2.61}$$

through each point $(0,y)$ on the x–axis, one draws the characteristic line of slope $c(u(0,y)) = c(f(y))$. Until the onset of a shock discontinuity, the solution maintains its initial value $u(0,y) = f(y)$ along the characteristic line.

A shock forms whenever two characteristic lines cross. As before, the mathematical equation no longer uniquely specifies the subsequent dynamics, and we need to appeal to an appropriate conservation law. We write the transport equation in the form

$$\frac{\partial u}{\partial t} + \frac{\partial}{\partial x} C(u) = 0, \qquad \text{where} \qquad C(u) = \int c(u)\,du \tag{2.62}$$

is any convenient anti-derivative of the wave speed. Thus, following the same computation as in (2.44), we discover that conservation of mass now takes the integrated form

$$\frac{d}{dt}\int_a^b u(t,x)\,dx = C(u(t,a)) - C(u(t,b)), \tag{2.63}$$

with $C(u)$ playing the role of the mass flux. Requiring the conservation of mass, i.e., of the area under the graph of the solution, means that the Equal Area Rule remains valid. However, the Rankine–Hugoniot shock-speed condition must be modified in accordance with the new dynamics. Mimicking the preceding argument, but with the modified mass flux, we find that the shock speed is now given by

$$\frac{d\sigma}{dt} = \frac{C(u^-(t)) - C(u^+(t))}{u^-(t) - u^+(t)}. \tag{2.64}$$

Note that if

$$c(u) = u, \qquad \text{then} \qquad C(u) = \int u\,du = \tfrac{1}{2}u^2,$$

and so (2.64) reduces to our earlier formula (2.53). Moreover, in the limit as the shock magnitude approaches zero, $u^-(t) - u^+(t) \to 0$, the right-hand side of (2.64) converges to the derivative $C'(u) = c(u)$ and hence recovers the wave speed, as it should.

Exercises

2.3.1. Discuss the behavior of the solution to the nonlinear transport equation (2.31) for the following initial data:

(a) $u(0,x) = \begin{cases} 2, & x < -1, \\ 1, & x > -1; \end{cases}$ (b) $u(0,x) = \begin{cases} -2, & x < -1, \\ 1, & x > -1; \end{cases}$ (c) $u(0,x) = \begin{cases} 1, & x < 1, \\ -2, & x > 1. \end{cases}$

2.3.2. Solve the following initial value problems: (a) $u_t + 3 u u_x = 0, \ u(0,x) = \begin{cases} 2, & x < 1, \\ 0, & x > 1; \end{cases}$

(b) $u_t - u u_x = 0, \ u(1,x) = \begin{cases} -1, & x < 0, \\ 3, & x > 0; \end{cases}$ (c) $u_t - 2 u u_x = 0, \ u(0,x) = \begin{cases} 1, & x < 1, \\ 0, & x > 1. \end{cases}$

2.3.3. Let $u(0,x) = (x^2+1)^{-1}$. Does the resulting solution to the nonlinear transport equation (2.31) produce a shock wave? If so, find the time of onset of the shock, and sketch a graph of the solution just before and soon after the shock wave. If not, explain what happens to the solution as t increases.

2.3.4. Solve Exercise 2.3.3 when $u(0,x) =$ (a) $-(x^2+1)^{-1}$, (b) $x(x^2+1)^{-1}$.

2.3.5. Consider the initial value problem $u_t - 2 u u_x = 0, \ u(0,x) = e^{-x^2}$. Does the resulting solution produce a shock wave? If so, find the time of onset of the shock and the position at which it first forms. If not, explain what happens to the solution as t increases.

2.3.6. (a) For what values of $\alpha, \beta, \gamma, \delta$ is $u(t,x) = \dfrac{\alpha x + \beta}{\gamma t + \delta}$ a solution to (2.31)?

(b) For what values of $\alpha, \beta, \gamma, \delta, \lambda, \mu$ is $u(t,x) = \dfrac{\lambda t + \alpha x + \beta}{\gamma t + \mu x + \delta}$ a solution to (2.31)?

2.3.7. A *triangular wave* is a shock-wave solution to the initial value problem for (2.31) that has initial data $u(0,x) = \begin{cases} m x, & 0 \le x \le \ell, \\ 0, & \text{otherwise.} \end{cases}$ Assuming $m > 0$, write down a formula for the triangular-wave solution at times $t > 0$. Discuss what happens to the triangular wave as time progresses.

2.3.8. Solve Exercise 2.3.7 when $m < 0$.

2.3.9. Solve (2.31) for $t > 0$ subject to the following initial conditions, and graph your solution at some representative times. In what sense does your solution conserve mass?

(a) $u(0,x) = \begin{cases} 1, & 0 < x < 1, \\ 0, & \text{otherwise,} \end{cases}$ (b) $u(0,x) = \begin{cases} x, & -1 < x < 1, \\ 0, & \text{otherwise,} \end{cases}$

(c) $u(0,x) = \begin{cases} -x, & -1 < x < 1, \\ 0, & \text{otherwise,} \end{cases}$ (d) $u(0,x) = \begin{cases} 1 - |\,x\,|, & -1 < x < 1, \\ 0, & \text{otherwise.} \end{cases}$

2.3.10. An *N–wave* is a solution to the nonlinear transport equation (2.31) that has initial conditions $u(0,x) = \begin{cases} m x, & -\ell \le x \le \ell, \\ 0, & \text{otherwise,} \end{cases}$ where $m > 0$. (a) Write down a formula for the N–wave solution at times $t > 0$. (b) What about when $m < 0$?

◇ 2.3.11. Suppose $u(t,x)$ and $\tilde{u}(t,x)$ are two solutions to the nonlinear transport equation (2.31) such that, for some $t_\star > 0$, they agree: $u(t_\star,x) = \tilde{u}(t_\star,x)$ for all x. Do the solutions necessarily have the same initial conditions: $u(0,x) = \tilde{u}(0,x)$? Use your answer to discuss the uniqueness of solutions to the nonlinear transport equation.

2.3.12. Suppose that $x_1 < x_2$ are such that the characteristic lines of (2.31) through $(0,x_1)$ and $(0,x_2)$ cross at a shock at $(t,\sigma(t))$ and, moreover, the left- and right-hand shock values (2.52) are $f(x_1) = u^-(t)$, $f(x_1) = u^+(t)$. Explain why the signed area of the region between the graph of $f(x)$ and the secant line connecting $(x_1, f(x_1))$ to $(x_2, f(x_2))$ is zero.

◇ 2.3.13. Consider the initial value problem $u^\varepsilon(0,x) = 2 + \tan^{-1}(x/\varepsilon)$ for the nonlinear transport equation (2.31). (a) Show that, as $\varepsilon \to 0^+$, the initial condition converges to a step function (2.51). What are the values of a, b? (b) Show that, moreover, the resulting solution $u^\varepsilon(0,x)$ to the nonlinear transport equation converges to the corresponding rarefaction wave (2.54) resulting from the limiting initial condition.

♢ 2.3.14. (a) Under what conditions can equation (2.35) be solved for a single-valued function $u(t,x)$? *Hint*: Use the Implicit Function Theorem. (*b*) Use implicit differentiation to prove that the resulting function $u(t,x)$ is a solution to the nonlinear transport equation.

2.3.15. For what values of $\alpha, \beta, \gamma, \delta, k$ is $u(t,x) = \left(\dfrac{\alpha x + \beta}{\gamma t + \delta}\right)^k$ a solution to the transport equation $u_t + u^2 u_x = 0$?

2.3.16. (a) Solve the initial value problem $u_t + u^2 u_x = 0$, $u(0,x) = f(x)$, by the method of characteristics. (*b*) Discuss the behavior of solutions and compare/contrast with (2.31).

2.3.17. (a) Determine the Rankine–Hugoniot condition, based on conservation of mass, for the speed of a shock for the equation $u_t + u^2 u_x = 0$. (*b*) Solve the initial value problem $u(0,x) = \begin{cases} a, & x < 0, \\ b, & x > 0, \end{cases}$ when (*i*) $|\,a\,| > |\,b\,|$, (*ii*) $|\,a\,| < |\,b\,|$. *Hint*: Use Exercise 2.3.15 to determine the shape of a rarefaction wave.

2.3.18. Solve Exercise 2.3.17 when the wave speed $c(u) =$ (*i*) $1 - 2u$, (*ii*) u^3, (*iii*) $\sin u$.

♢ 2.3.19. Justify the shock-speed formula (2.58).

♢ 2.3.20. Consider the general quasilinear first-order partial differential equation

$$\frac{\partial u}{\partial t} + c(t,x,u)\,\frac{\partial u}{\partial x} = h(t,x,u).$$

Let us define a *lifted characteristic curve* to be a solution $(t, x(t), u(t))$ to the system of ordinary differential equations $\dfrac{dx}{dt} = c(t,x,u)$, $\dfrac{du}{dt} = h(t,x,u)$. The corresponding *characteristic curve* $\big(t, x(t)\big)$ is obtained by projecting to the (t,x)–plane. Prove that if $u(t,x)$ is a solution to the partial differential equation, and $u(t_0, x_0) = u_0$, then the lifted characteristic curve passing through (t_0, x_0, u_0) lies on the graph of $u(t,x)$. Conclude that the graph of the solution to the initial value problem $u(t_0, x) = f(x)$ is the union of all lifted characteristic curves passing through the initial data points $\big(t_0, x_0, f(x_0)\big)$.

2.3.21. Let $a > 0$. (a) Apply the method of Exercise 2.3.20 to solve the initial value problem for the *damped transport equation*: $u_t + u\,u_x + a\,u = 0$, $u(0,x) = f(x)$.
(*b*) Does the damping eliminate shocks?

2.3.22. Apply the method of Exercise 2.3.20 to solve the initial value problem

$$u_t + t\,u_x = u^2, \qquad u(0,x) = \frac{1}{1+x^2}\,.$$

2.4 The Wave Equation: d'Alembert's Formula

Newton's Second Law states that force equals mass times acceleration. It forms the bedrock underlying the derivation of mathematical models describing all of classical dynamics. When applied to a one-dimensional medium, such as the transverse displacements of a violin string or the longitudinal motions of an elastic bar, the resulting model governing small vibrations is the second-order partial differential equation

$$\rho(x)\,\frac{\partial^2 u}{\partial t^2} = \frac{\partial}{\partial x}\left(\kappa(x)\,\frac{\partial u}{\partial x}\right). \tag{2.65}$$

Here $u(t,x)$ represents the displacement of the string or bar at time t and position x, while $\rho(x) > 0$ denotes its density and $\kappa(x) > 0$ its stiffness or tension, both of which are assumed not to vary with t. The right-hand side of the equation represents the restoring force due to a (small) displacement of the medium from its equilibrium, whereas the left-hand side is the product of mass per unit length and acceleration. A correct derivation of the model from first principles would require a significant detour, and we refer the reader to [**120**, **124**] for the details.

We will simplify the general model by assuming that the underlying medium is *uniform*, and so both its density ρ and stiffness κ are constant. Then (2.65) reduces to the one-dimensional *wave equation*

$$\frac{\partial^2 u}{\partial t^2} = c^2 \frac{\partial^2 u}{\partial x^2}, \qquad \text{where the constant} \qquad c = \sqrt{\frac{\kappa}{\rho}} > 0 \tag{2.66}$$

is known as the *wave speed*, for reasons that will soon become apparent.

In general, to uniquely specify the solution to any dynamical system arising from Newton's Second Law, including the wave equation (2.66) and the more general vibration equation (2.65), one must fix both its initial position and initial velocity. Thus, the initial conditions take the form

$$u(0,x) = f(x), \qquad \frac{\partial u}{\partial t}(0,x) = g(x), \tag{2.67}$$

where, for simplicity, we set the initial time $t_0 = 0$. (See also Exercise 2.4.6.) The *initial value problem* seeks the corresponding C^2 function $u(t,x)$ that solves the wave equation (2.66) and has the required initial values (2.67). In this section, we will learn how to solve the initial value problem on the entire line $-\infty < x < \infty$. The analysis of the wave equation on bounded intervals will be deferred until Chapters 4 and 7. The two- and three-dimensional versions of the wave equation are treated in Chapters 11 and 12, respectively.

d'Alembert's Solution

Let us now derive the explicit solution formula for the second-order wave equation (2.66) first found by d'Alembert. The starting point is to write the partial differential equation in the suggestive form

$$\Box u = (\partial_t^2 - c^2\,\partial_x^2)\,u = u_{tt} - c^2\,u_{xx} = 0. \tag{2.68}$$

Here

$$\Box = \partial_t^2 - c^2\,\partial_x^2$$

is a common mathematical notation for the *wave operator*, which is a linear second-order partial differential operator. In analogy with the elementary polynomial factorization

$$t^2 - c^2\,x^2 = (t - c\,x)(t + c\,x),$$

we can factor the wave operator into a product of two first-order partial differential operators:[†]

$$\Box = \partial_t^2 - c^2\,\partial_x^2 = (\partial_t - c\,\partial_x)\,(\partial_t + c\,\partial_x). \tag{2.69}$$

[†] The cross terms cancel, thanks to the equality of mixed partial derivatives: $\partial_t\partial_x u = \partial_x\partial_t u$. Constancy of the wave speed c is essential here.

Now, if the second factor annihilates the function $u(t,x)$, meaning

$$(\partial_t + c\,\partial_x)\,u = u_t + c\,u_x = 0, \tag{2.70}$$

then u is automatically a solution to the wave equation, since

$$\Box\, u = (\partial_t - c\,\partial_x)\,(\partial_t + c\,\partial_x)\,u = (\partial_t - c\,\partial_x)\,0 = 0.$$

We recognize (2.70) as the first-order transport equation (2.4) with constant wave speed c. Proposition 2.1 tells us that its solutions are traveling waves with wave speed c:

$$u(t,x) = p(\xi) = p(x - c\,t), \tag{2.71}$$

where p is an arbitrary function of the characteristic variable $\xi = x - c\,t$. As long as $p \in \mathrm{C}^2$ (i.e., is twice continuously differentiable), the resulting function $u(t,x)$ is a classical solution to the wave equation (2.66), as you can easily check.

Now, the factorization (2.69) can equally well be written in the reverse order:

$$\Box = \partial_t^2 - c^2\,\partial_x^2 = (\partial_t + c\,\partial_x)\,(\partial_t - c\,\partial_x). \tag{2.72}$$

The same argument tells us that any solution to the "backwards" transport equation

$$u_t - c\,u_x = 0, \tag{2.73}$$

with constant wave speed $-c$, also provides a solution to the wave equation. Again, by Proposition 2.1, with c replaced by $-c$, the general solution to (2.73) has the form

$$u(t,x) = q(\eta) = q(x + c\,t), \tag{2.74}$$

where q is an arbitrary function of the alternative characteristic variable $\eta = x + c\,t$. The solutions (2.74) represent traveling waves moving to the *left* with constant speed $c > 0$. Provided $q \in \mathrm{C}^2$, the functions (2.74) will provide a second family of solutions to the wave equation.

We conclude that, unlike first-order transport equations, the wave equation (2.68) is *bidirectional* in that it admits both left and right traveling-wave solutions. Moreover, by linearity the sum of any two solutions is again a solution, and so we can immediately construct solutions that are superpositions of left and right traveling waves. The remarkable fact is that *every* solution to the wave equation can be so represented.

Theorem 2.14. *Every solution to the wave equation (2.66) can be written as a superposition,*

$$u(t,x) = p(\xi) + q(\eta) = p(x - c\,t) + q(x + c\,t), \tag{2.75}$$

of right and left traveling waves. Here $p(\xi)$ and $q(\eta)$ are arbitrary C^2 functions, each depending on its respective characteristic variable

$$\xi = x - c\,t, \qquad\qquad \eta = x + c\,t. \tag{2.76}$$

Proof: As in our treatment of the transport equation, we will simplify the wave equation through an inspired change of variables. In this case, the new independent variables are the characteristic variables ξ, η defined by (2.76). We set

$$u(t,x) = v(x - c\,t, x + c\,t) = v(\xi,\eta), \qquad \text{whereby} \qquad v(\xi,\eta) = u\left(\frac{\eta - \xi}{2\,c}, \frac{\eta + \xi}{2}\right). \tag{2.77}$$

Then, employing the chain rule to compute the partial derivatives,

$$\frac{\partial u}{\partial t} = c\left(-\frac{\partial v}{\partial \xi} + \frac{\partial v}{\partial \eta}\right), \qquad \frac{\partial u}{\partial x} = \frac{\partial v}{\partial \xi} + \frac{\partial v}{\partial \eta}, \tag{2.78}$$

and, further,

$$\frac{\partial^2 u}{\partial t^2} = c^2\left(\frac{\partial^2 v}{\partial \xi^2} - 2\,\frac{\partial^2 v}{\partial \xi\,\partial \eta} + \frac{\partial^2 v}{\partial \eta^2}\right), \qquad \frac{\partial^2 u}{\partial x^2} = \frac{\partial^2 v}{\partial \xi^2} + 2\,\frac{\partial^2 v}{\partial \xi\,\partial \eta} + \frac{\partial^2 v}{\partial \eta^2}.$$

Therefore

$$\Box\, u = \frac{\partial^2 u}{\partial t^2} - c^2\,\frac{\partial^2 u}{\partial x^2} = -\,4\,c^2\,\frac{\partial^2 v}{\partial \xi\,\partial \eta}. \tag{2.79}$$

We conclude that $u(t,x)$ solves the wave equation $\Box\, u = 0$ if and only if $v(\xi,\eta)$ solves the second-order partial differential equation

$$\frac{\partial^2 v}{\partial \xi\,\partial \eta} = 0,$$

which we write in the form

$$\frac{\partial}{\partial \xi}\left(\frac{\partial v}{\partial \eta}\right) = \frac{\partial w}{\partial \xi} = 0, \qquad \text{where} \qquad w = \frac{\partial v}{\partial \eta}.$$

Thus, applying the methods of Section 2.1 (and making the appropriate assumptions on the domain of definition of w), we deduce that

$$w = \frac{\partial v}{\partial \eta} = r(\eta),$$

where r is an arbitrary function of the characteristic variable η. Integrating both sides of the latter partial differential equation with respect to η, we find

$$v(\xi,\eta) = p(\xi) + q(\eta), \qquad \text{where} \qquad q(\eta) = \int r(\eta)\, d\eta,$$

while $p(\xi)$ represents the η integration "constant". Replacing the characteristic variables by their formulas in terms of t and x completes the proof. *Q.E.D.*

Let us see how the solution formula (2.75) can be used to solve the initial value problem (2.67). Substituting into the initial conditions, we deduce that

$$u(0,x) = p(x) + q(x) = f(x), \qquad \frac{\partial u}{\partial t}(0,x) = -\,c\,p'(x) + c\,q'(x) = g(x). \tag{2.80}$$

To solve this pair of equations for the functions p and q, we differentiate the first,

$$p'(x) + q'(x) = f'(x),$$

and then subtract off the second equation divided by c; the result is

$$2\,p'(x) = f'(x) - \frac{1}{c}\,g(x).$$

Therefore,

$$p(x) = \frac{1}{2}\,f(x) - \frac{1}{2\,c}\int_0^x g(z)\,dz + a,$$

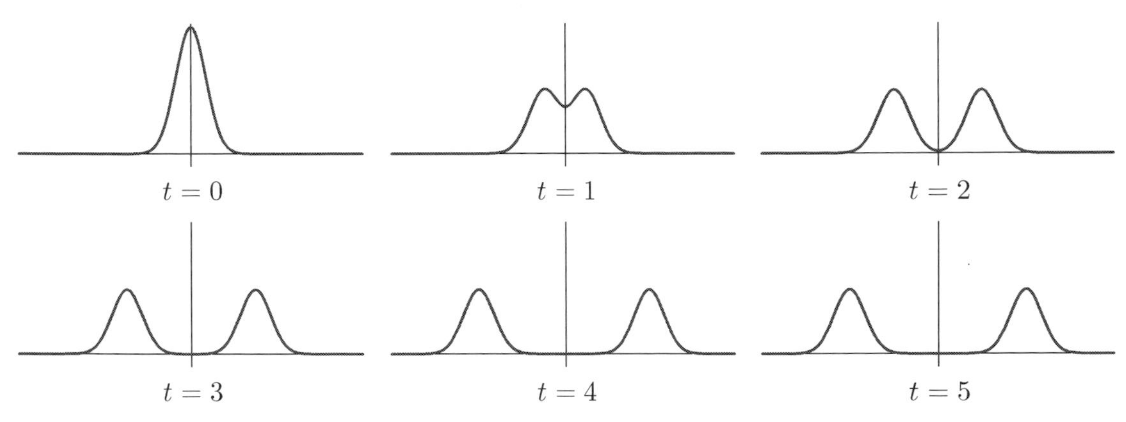

Figure 2.24. Splitting of waves.

where a is an integration constant. The first equation in (2.80) then yields

$$q(x) = f(x) - p(x) = \frac{1}{2}\, f(x) + \frac{1}{2\,c}\int_0^x g(z)\, dz - a.$$

Substituting these two expressions back into our solution formula (2.75), we obtain

$$\begin{aligned} u(t,x) = p(\xi) + q(\eta) &= \frac{f(\xi) + f(\eta)}{2} - \frac{1}{2\,c}\int_0^{\xi} g(z)\, dz + \frac{1}{2\,c}\int_0^{\eta} g(z)\, dz \\ &= \frac{f(\xi) + f(\eta)}{2} + \frac{1}{2\,c}\int_{\xi}^{\eta} g(z)\, dz, \end{aligned}$$

where ξ, η are the characteristic variables (2.76). In this manner, we have arrived at *d'Alembert's solution* to the initial value problem for the wave equation on the real line.

Theorem 2.15. *The solution to the initial value problem*

$$\frac{\partial^2 u}{\partial t^2} = c^2\, \frac{\partial^2 u}{\partial x^2}, \qquad u(0,x) = f(x), \qquad \frac{\partial u}{\partial t}\,(0,x) = g(x), \qquad -\infty < x < \infty, \tag{2.81}$$

is given by

$$u(t,x) = \frac{f(x - c\,t) + f(x + c\,t)}{2} + \frac{1}{2\,c}\int_{x - c\,t}^{x + c\,t} g(z)\, dz. \tag{2.82}$$

Remark: In order that (2.82) define a classical solution to the wave equation, we need $f \in \mathrm{C}^2$ and $g \in \mathrm{C}^1$. However, the formula itself makes sense for more general initial conditions. We will continue to treat the resulting functions as solutions, albeit nonclassical, since they fit under the more general rubric of "weak solution", to be developed in Section 10.4.

Example 2.16. Suppose there is no initial velocity, so $g(x) \equiv 0$, and hence the motion is purely the result of the initial displacement $u(0,x) = f(x)$. In this case, (2.82) reduces to

$$u(t,x) = \tfrac{1}{2}\, f(x - c\,t) + \tfrac{1}{2}\, f(x + c\,t). \tag{2.83}$$

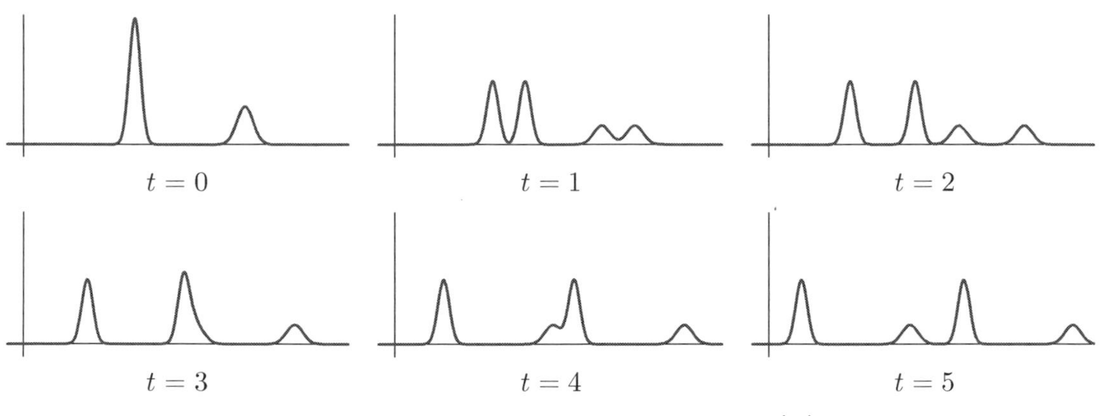

Figure 2.25. Interaction of waves.

The effect is that the initial displacement splits into two waves, one moving to the right and the other moving to the left, each of constant speed c, and each of exactly the same shape as $f(x)$, but only half as tall. For example, if the initial displacement is a localized pulse centered at the origin, say

$$u(0,x) = e^{-x^2}, \qquad \frac{\partial u}{\partial t}(0,x) = 0,$$

then the solution

$$u(t,x) = \tfrac{1}{2}\, e^{-(x-ct)^2} + \tfrac{1}{2}\, e^{-(x+ct)^2}$$

consists of two half size pulses running away from the origin with the same speed c, but in opposite directions. A graph of the solution at several successive times can be seen in Figure 2.24.

If we take two initially separated pulses, say

$$u(0,x) = e^{-x^2} + 2\, e^{-(x-1)^2}, \qquad \frac{\partial u}{\partial t}(0,x) = 0,$$

centered at $x = 0$ and $x = 1$, then the solution

$$u(t,x) = \tfrac{1}{2}\, e^{-(x-ct)^2} + e^{-(x-1-ct)^2} + \tfrac{1}{2}\, e^{-(x+ct)^2} + e^{-(x-1+ct)^2}$$

will consist of four pulses, two moving to the right and two to the left, all with the same speed. An important observation is that when a right-moving pulse collides with a left-moving pulse, they emerge from the collision unchanged, which is a consequence of the inherent linearity of the wave equation. In Figure 2.25, the first picture plots the initial displacement. In the second and third pictures, the two localized bumps have each split into two copies moving in opposite directions. In the fourth and fifth, the larger right-moving bump is in the process of interacting with the smaller left-moving bump. Finally, in the last picture the interaction is complete, and the individual pairs of left- and right-moving waves move off in tandem in opposing directions, experiencing no further collisions.

In general, if the initial displacement is localized, so that $|\, f(x) \,| \ll 1$ for $|\, x \,| \gg 0$, then, after a finite time, the left- and right-moving waves will separate, and the observer will see two half-size replicas running away, with speed c, in opposite directions. If the displacement

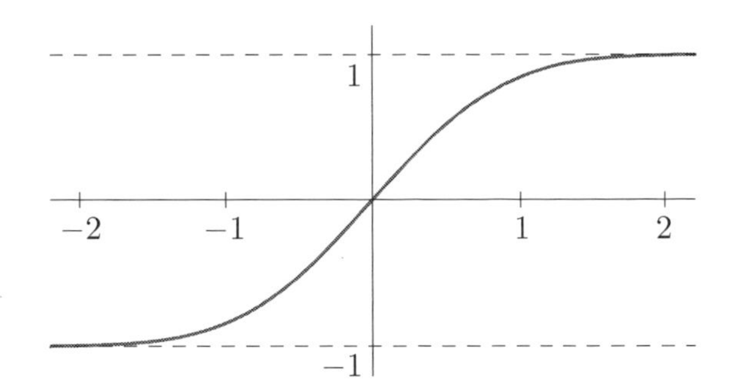

Figure 2.26. The error function $\operatorname{erf} x$.

is not localized, then the left and right traveling waves will never fully disengage, and one might be hard pressed to recognize that a complicated solution pattern is, in reality, just the superposition of two simple traveling waves. For example, consider the elementary trigonometric solution

$$\cos ct \ \cos x = \tfrac{1}{2}\cos(x - ct) + \tfrac{1}{2}\cos(x + ct). \tag{2.84}$$

In accordance with the left-hand expression, an observer will see a standing cosinusoidal wave that vibrates up and down with frequency c. However, the d'Alembert form of the solution on the right-hand side says that this is just the sum of left- and right-traveling cosine waves! The interactions of their peaks and troughs reproduce the standing wave. Thus, the same solution can be interpreted in two seemingly incompatible ways. And, in fact, this paradox lies at the heart of the perplexing wave-particle duality of quantum physics.

Example 2.17. By way of contrast, suppose there is no initial displacement, so $f(x) \equiv 0$, and the motion is purely the result of the initial velocity $u_t(0,x) = g(x)$. Physically, this models a violin string at rest being struck by a "hammer blow" at the initial time. In this case, the d'Alembert formula (2.82) reduces to

$$u(t,x) = \frac{1}{2\,c}\int_{x-ct}^{x+ct} g(z)\,dz. \tag{2.85}$$

For example, when $u(0,x) = 0$, $u_t(0,x) = e^{-x^2}$, the resulting solution (2.85) is

$$u(t,x) = \frac{1}{2\,c}\int_{x-ct}^{x+ct} e^{-x^2}\,dz = \frac{\sqrt{\pi}}{4\,c}\big[\operatorname{erf}(x+ct) - \operatorname{erf}(x-ct)\big], \tag{2.86}$$

where

$$\operatorname{erf} x = \frac{2}{\sqrt{\pi}}\int_0^x e^{-z^2}\,dz \tag{2.87}$$

is known as the *error function* due to its many applications throughout probability and statistics, [**39**]. The error function integral cannot be written in terms of elementary functions; nevertheless, its properties have been well studied and its values tabulated, [**86**]. A graph appears in Figure 2.26. The constant in front of the integral (2.87) has been chosen so that the error function has asymptotic values

$$\lim_{x\to\infty}\operatorname{erf} x = 1, \qquad \lim_{x\to-\infty}\operatorname{erf} x = -1, \tag{2.88}$$

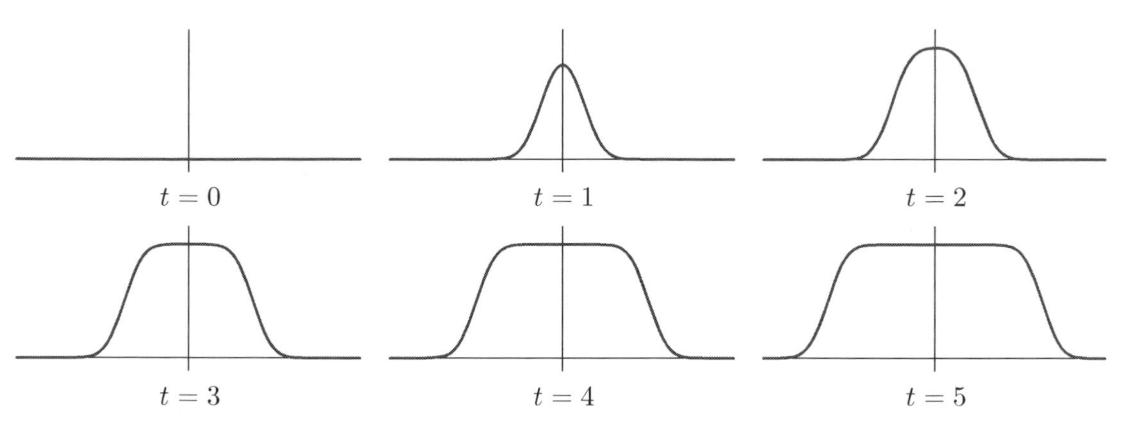

Figure 2.27. Error function solution to the wave equation.

which follow from a well-known integration formula to be derived in Exercise 2.4.21.

A graph of the solution (2.86) at successive times is displayed in Figure 2.27. The first graph shows the zero initial displacement. Gradually, the effect of the initial hammer blow is felt further and further away along the string, as the two wave fronts propagate away from the origin, both with speed c, but in opposite directions. Thus, unlike the case of a nonzero initial displacement in Figure 2.24, where the solution eventually returns to its equilibrium position $u = 0$ after the wave passes by, a nonzero initial velocity leaves the string permanently deformed.

In general, the lines of slope $\pm c$, where the respective characteristic variables are constant,

$$\xi = x - ct = a, \qquad \eta = x + ct = b, \tag{2.89}$$

are known as the *characteristics* of the wave equation. Thus, the second-order wave equation has *two* distinct characteristic lines passing through each point in the (t, x)–plane.

Remark: The characteristic lines are the one-dimensional counterparts of the light cone in Minkowski space-time, which plays a starring role in special relativity, [**70**, **75**]. See Section 12.5 for further details.

In Figure 2.28, we plot the two characteristics going through a point $(0, y)$ on the x axis. The wedge-shaped region $\{ y - ct \leq x \leq y + ct, \ t \geq 0 \}$ lying between them is known as the *domain of influence* of the point $(0, y)$, since, in general, the value of the initial data at a point will affect the subsequent solution values only in its domain of influence. Indeed, the effect of an initial displacement at the point y propagates along the two characteristic lines, while the effect of an initial velocity there will be felt at every point in the triangular wedge.

External Forcing and Resonance

When a homogeneous vibrating medium is subjected to external forcing, the wave equation acquires an additional, inhomogeneous term:

$$\frac{\partial^2 u}{\partial t^2} = c^2 \frac{\partial^2 u}{\partial x^2} + F(t, x), \tag{2.90}$$

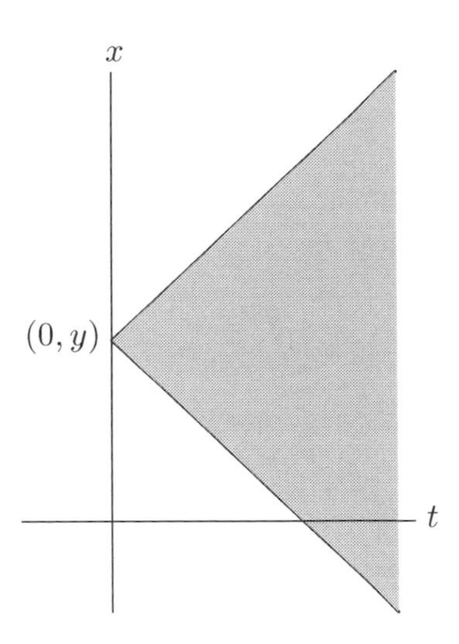

Figure 2.28. Characteristic lines and domain of influence.

in which $F(t, x)$ represents a force imposed at time t and spatial position x. With a bit more work, d'Alembert's solution technique can be readily adapted to incorporate the forcing term.

Let us, for simplicity, assume that the differential equation is supplemented by homogeneous initial conditions,

$$u(0, x) = 0, \qquad u_t(0, x) = 0, \tag{2.91}$$

meaning that there is no initial displacement or velocity. To solve the initial value problem (2.90–91), we switch to the same characteristic coordinates (2.76), setting

$$v(\xi, \eta) = u\left(\frac{\eta - \xi}{2c}, \frac{\eta + \xi}{2}\right).$$

Invoking the chain rule formulas (2.79), we find that the forced equation (2.90) becomes

$$\frac{\partial^2 v}{\partial \xi\, \partial \eta} = -\frac{1}{4c^2} F\left(\frac{\eta - \xi}{2c}, \frac{\eta + \xi}{2}\right). \tag{2.92}$$

Let us integrate both sides of the equation with respect to η, on the interval $\xi \leq \zeta \leq \eta$:

$$\frac{\partial v}{\partial \xi}(\xi, \eta) - \frac{\partial v}{\partial \xi}(\xi, \xi) = -\frac{1}{4c^2} \int_\xi^\eta F\left(\frac{\zeta - \xi}{2c}, \frac{\zeta + \xi}{2}\right) d\zeta. \tag{2.93}$$

But, recalling (2.78),

$$\frac{\partial v}{\partial \xi}(\xi, \eta) = \frac{1}{2c}\frac{\partial u}{\partial t}\left(\frac{\eta - \xi}{2c}, \frac{\eta + \xi}{2}\right) + \frac{1}{2}\frac{\partial u}{\partial x}\left(\frac{\eta - \xi}{2c}, \frac{\eta + \xi}{2}\right),$$

and so, in particular,

$$\frac{\partial v}{\partial \xi}(\xi, \xi) = \frac{1}{2c}\frac{\partial u}{\partial t}(0, \xi) + \frac{1}{2}\frac{\partial u}{\partial x}(0, \xi) = 0,$$

which vanishes owing to our choice of homogeneous initial conditions (2.91). Indeed, the initial velocity condition says that $u_t(0, x) = 0$, while differentiating the initial displacement condition $u(0,x) = 0$ with respect to x implies that $u_x(0,x) = 0$ for all x, including $x = \xi$. As a result, (2.93) simplifies to

$$\frac{\partial v}{\partial \xi}(\xi,\eta) = -\frac{1}{4c^2}\int_\xi^\eta F\left(\frac{\zeta-\xi}{2c}, \frac{\zeta+\xi}{2}\right) d\zeta.$$

We now integrate the latter equation with respect to ξ on the interval $\xi \le \chi \le \eta$, producing

$$-v(\xi,\eta) = v(\eta,\eta) - v(\xi,\eta) = -\frac{1}{4c^2}\int_\xi^\eta \int_\chi^\eta F\left(\frac{\zeta-\chi}{2c}, \frac{\zeta+\chi}{2}\right) d\zeta\, d\chi,$$

since $v(\eta,\eta) = u(0,\eta) = 0$, thanks again to the initial conditions. In this manner, we have produced an explicit formula for the solution to the characteristic variable version of the forced wave equation subject to the homogeneous initial conditions. Reverting to the original physical coordinates, the left-hand side of this equation becomes $-u(t,x)$. As for the double integral on the right-hand side, it takes place over the triangular region

$$T(\xi,\eta) = \{\, (\chi,\zeta) \mid \xi \le \chi \le \zeta \le \eta \,\}. \tag{2.94}$$

Let us introduce "physical" integration variables by setting

$$\chi = y - c\,s, \qquad \zeta = y + c\,s.$$

The defining inequalities of the triangle (2.94) become

$$x - ct \le y - c\,s \le y + c\,s \le x + ct,$$

and so, in the physical coordinates, the triangular integration domain assumes the form

$$D(t,x) = \{\, (s,y) \mid x - c\,(t-s) \le y \le x + c\,(t-s),\ 0 \le s \le t \,\}, \tag{2.95}$$

which is graphed in Figure 2.29. The change of variables formula for double integrals requires that we compute the Jacobian determinant

$$\det\begin{pmatrix} \partial\chi/\partial y & \partial\chi/\partial s \\ \partial\zeta/\partial y & \partial\zeta/\partial s \end{pmatrix} = \det\begin{pmatrix} 1 & -c \\ 1 & c \end{pmatrix} = 2c,$$

and so $d\chi\, d\zeta = 2c\, ds\, dy$. Therefore,

$$u(t,x) = \frac{1}{2c}\iint_{D(t,x)} F(s,y)\, ds\, dy = \frac{1}{2c}\int_0^t \int_{x-c\,(t-s)}^{x+c\,(t-s)} F(s,y)\, dy\, ds, \tag{2.96}$$

which gives the solution formula for the forced wave equation when subject to homogeneous initial conditions.

To solve the general initial value problem, we appeal to linear superposition, writing its solution as a sum of the solution (2.96) to the forced wave equation subject to homogeneous initial conditions plus the d'Alembert solution (2.82) to the unforced equation subject to inhomogeneous boundary conditions.

Theorem 2.18. *The solution to the general initial value problem*

$$u_{tt} = c^2 u_{xx} + F(t,x), \qquad u(0,x) = f(x), \qquad u_t(0,x) = g(x), \qquad -\infty < x < \infty, \qquad t > 0,$$

for the wave equation subject to an external forcing is given by

$$u(t,x) = \frac{f(x-ct) + f(x+ct)}{2} + \frac{1}{2c}\int_{x-ct}^{x+ct} g(y)\, dy + \frac{1}{2c}\int_0^t \int_{x-c\,(t-s)}^{x+c\,(t-s)} F(s,y)\, dy\, ds. \tag{2.97}$$

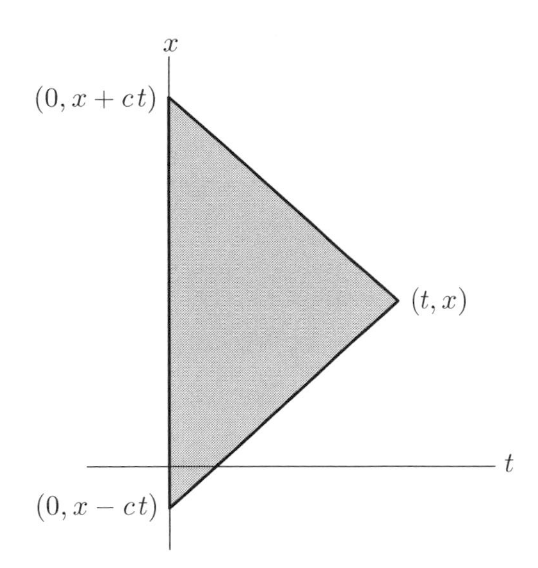

Figure 2.29. Domain of dependence.

Observe that the solution is a linear superposition of the respective effects of the initial displacement, the initial velocity, and the external forcing. The triangular integration region (2.95), lying between the x–axis and the characteristic lines going backwards from (t,x), is known as the *domain of dependence* of the point (t,x). This is because, for any $t>0$, the solution value $u(t,x)$ depends only on the values of the initial data and the forcing function at points lying within the domain of dependence $D(t,x)$. Indeed, the first term in the solution formula (2.97) requires only the initial displacement at the corners $(0,x+ct)$, $(0,x-ct)$; the second term requires only the initial velocity at points on the x–axis lying on the vertical side of $D(t,x)$; while the final term requires the value of the external force on the entire triangular region.

Example 2.19. Let us solve the initial value problem

$$u_{tt}=u_{xx}+\sin\omega t\,\sin x,\qquad u(0,x)=0,\qquad u_t(0,x)=0,$$

for the wave equation with unit wave speed subject to a sinusoidal forcing function whose amplitude varies periodically in time with frequency $\omega>0$. According to formula (2.96), the solution is

$$\begin{aligned}u(t,x)&=\frac{1}{2}\int_0^t\int_{x-t+s}^{x+t-s}\sin\omega s\sin y\,dy\,ds\\&=\frac{1}{2}\int_0^t\sin\omega s\,\big[\cos(x-t+s)-\cos(x+t-s)\big]\,ds\\&=\begin{cases}\dfrac{\sin\omega t-\omega\sin t}{1-\omega^2}\,\sin x, & 0<\omega\neq 1,\\[2ex] \dfrac{\sin t-t\cos t}{2}\,\sin x, & \omega=1.\end{cases}\end{aligned}$$

Notice that, when $\omega\neq 1$, the solution is bounded, being a combination of two vibrational modes: an externally induced mode at frequency ω along with an internal mode, at frequency 1. If $\omega=p/q\neq 1$ is a rational number, then the solution varies periodically in

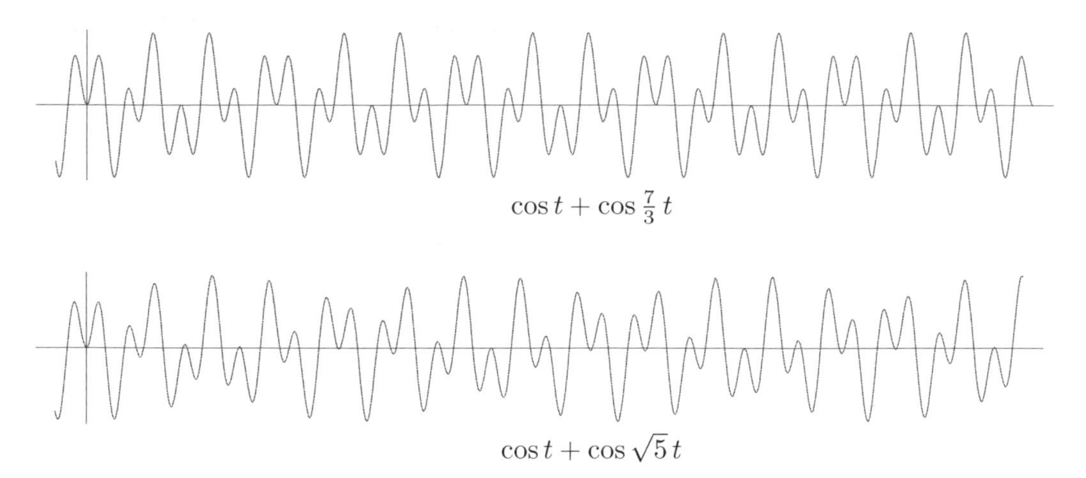

Figure 2.30. Periodic and quasiperiodic functions.

time. On the other hand, if ω is irrational, then the solution is only *quasiperiodic*, and never exactly repeats itself. Finally, if $\omega = 1$, the solution grows without limit as t increases, indicating that this is a *resonant frequency*. We will investigate external forcing and the mechanisms leading to resonance in dynamical partial differential equations in more detail in Chapters 4 and 6.

Example 2.20. To appreciate the difference between periodic and quasiperiodic vibrations, consider the elementary trigonometric function

$$u(t) = \cos t + \cos \omega t,$$

which is a linear combination of two simple periodic vibrations, of frequencies 1 and ω. If $\omega = p/q$ is a rational number, then $u(t)$ is a periodic function of period $2\pi q$, so $u(t+2\pi q) = u(t)$. However, if ω is an irrational number, then $u(t)$ is not periodic, and never repeats. You are encouraged to inspect the graphs in Figure 2.30. The first is periodic — can you spot where it begins to repeat? — whereas the second is only quasiperiodic. The only quasiperiodic functions we will encounter in this text are linear combinations of periodic trigonometric functions whose frequencies are *not* all rational multiples of each other. To the uninitiated, such quasiperiodic motions may appear to be random, even though they are built from a few simple periodic constituents. While ostensibly complicated, quasiperiodic motion is *not* true chaos, which is is an inherently nonlinear phenomenon, [**77**].

Exercises

2.4.1. Solve the initial value problem $u_{tt} = c^2 u_{xx}$, $u(0,x) = e^{-x^2}$, $u_t(0,x) = \sin x$.

2.4.2. (a) Solve the wave equation $u_{tt} = u_{xx}$ when the initial displacement is the box function $u(0,x) = \begin{cases} 1, & 1 < x < 2, \\ 0, & \text{otherwise,} \end{cases}$ while the initial velocity is 0.
(b) Sketch the resulting solution at several representative times.

2.4.3. Answer Exercise 2.4.2 when the initial velocity is the box function, while the initial displacement is zero.

2.4.4. Write the following solutions to the wave equation $u_{tt} = u_{xx}$ in d'Alembert form (2.82). *Hint*: What is the appropriate initial data?

(a) $\cos x \cos t$, (b) $\cos 2x \, \sin 2t$, (c) e^{x+t}, (d) $t^2 + x^2$, (e) $t^3 + 3tx^2$.

♡ 2.4.5. (a) Solve the *dam break problem*, that is, the wave equation when the initial displacement is a step function $\sigma(x) = \begin{cases} 1, & x > 0, \\ 0, & x < 0, \end{cases}$ and there is no initial velocity. (b) Analyze the case in which there is no initial displacement, while the initial velocity is a step function. (c) Are your solutions classical solutions? Explain your answer. (d) Prove that the step function is the limit, as $n \to \infty$, of the functions $f_n(x) = \dfrac{1}{\pi} \tan^{-1} n x + \dfrac{1}{2}$. (e) Show that, in both cases, the step function solution can be realized as the limit, as $n \to \infty$, of solutions to the initial value problems with the functions $f_n(x)$ as initial displacement or velocity.

◇ 2.4.6. Suppose $u(t,x)$ solves the initial value problem $u(0,x) = f(x)$, $u_t(0,x) = g(x)$, for the wave equation (2.66). Prove that the solution to the initial value problem $u(t_0,x) = f(x)$, $u_t(t_0,x) = g(x)$, is $u(t - t_0, x)$.

2.4.7. Find all resonant frequencies for the wave equation with wave speed c when subject to the external forcing function $F(t,x) = \sin \omega t \, \sin kx$ for fixed $\omega, k > 0$.

2.4.8. Consider the initial value problem $u_{tt} = 4u_{xx} + F(t,x)$, $u(0,x) = f(x)$, $u_t(0,x) = g(x)$. Determine (a) the domain of influence of the point $(0,2)$; (b) the domain of dependence of the point $(3,-1)$; (c) the domain of influence of the point $(3,-1)$.

2.4.9. (a) A solution to the wave equation $u_{tt} = 2u_{xx}$ is generated by a displacement concentrated at position $x_0 = 1$ and time $t_0 = 0$, but no initial velocity. At what time will an observer at position $x_1 = 5$ feel the effect of this displacement? Will the observer continue to feel an effect in the future? (b) Answer part (a) when there is an initial velocity concentrated at position $x_0 = 1$ and time $t_0 = 0$, but no initial displacement.

2.4.10. Suppose $u(t,x)$ solves the initial value problem $u_{tt} = 4u_{xx} + \sin \omega t \, \cos x$, $u(0,x) = 0$, $u_t(0,x) = 0$. Is $h(t) = u(t,0)$ a periodic function?

♡ 2.4.11. (a) Write down an explicit formula for the solution to the initial value problem

$$\frac{\partial^2 u}{\partial t^2} - 4\frac{\partial^2 u}{\partial x^2} = 0, \qquad u(0,x) = \sin x, \qquad \frac{\partial u}{\partial t}(0,x) = \cos x, \qquad -\infty < x < \infty, \quad t \geq 0.$$

(b) *True or false*: The solution is a periodic function of t.
(c) Now solve the forced initial value problem

$$\frac{\partial^2 u}{\partial t^2} - 4\frac{\partial^2 u}{\partial x^2} = \cos 2t, \qquad u(0,x) = \sin x, \qquad \frac{\partial u}{\partial t}(0,x) = \cos x, \qquad -\infty < x < \infty, \quad t \geq 0.$$

(d) *True or false*: The forced equation exhibits resonance. Explain.
(e) Does the answer to part (d) change if the forcing function is $\sin 2t$?

2.4.12. Given a classical solution $u(t,x)$ of the wave equation, let $E = \frac{1}{2}(u_t^2 + c^2 u_x^2)$ be the associated *energy density* and $P = u_t u_x$ the *momentum density*.
(a) Show that both E and P are conserved densities for the wave equation.
(b) Show that $E(t,x)$ and $P(t,x)$ both satisfy the wave equation.

◇ 2.4.13. Let $u(t,x)$ be a classical solution to the wave equation $u_{tt} = c^2 u_{xx}$. The *total energy*

$$E(t) = \int_{-\infty}^{\infty} \frac{1}{2}\left[\left(\frac{\partial u}{\partial t}\right)^2 + c^2\left(\frac{\partial u}{\partial x}\right)^2\right] dx \tag{2.98}$$

represents the sum of kinetic and potential energies of the displacement $u(t,x)$ at time t. Suppose that $\nabla u \to \mathbf{0}$ sufficiently rapidly as $x \to \pm\infty$; more precisely, one can find $\alpha > \frac{1}{2}$ and $C(t) > 0$ such that $|\,u_t(t,x)\,|$, $|\,u_x(t,x)\,| \leq C(t)/|\,x\,|^\alpha$ for each fixed t and all sufficiently large $|\,x\,| \gg 0$. For such solutions, establish the *Law of Conservation of Energy* by showing that $E(t)$ is finite and constant. *Hint*: You do not need the formula for the solution.

♢ 2.4.14. (a) Use Exercise 2.4.13 to prove that the only classical solution to the initial-boundary value problem $u_{tt} = c^2 u_{xx}$, $u(0,x) = 0$, $u_t(0,x) = 0$, satisfying the indicated decay assumptions is the trivial solution $u(t,x) \equiv 0$. (b) Establish the following *Uniqueness Theorem* for the wave equation: there is at most one such solution to the initial-boundary value problem $u_{tt} = c^2 u_{xx}$, $u(0,x) = f(x)$, $u_t(0,x) = g(x)$.

2.4.15. The *telegrapher's equation* $u_{tt} + a\,u_t = c^2 u_{xx}$, with $a > 0$, models the vibration of a string under frictional damping. (a) Show that, under the decay assumptions of Exercise 2.4.13, the wave energy (2.98) of a classical solution is a nonincreasing function of t. (b) Prove uniqueness of such solutions to the initial value problem for the telegrapher's equation.

2.4.16. What happens to the proof of Theorem 2.14 if $c = 0$?

2.4.17. (a) Explain why the d'Alembert factorization method doesn't work when the wave speed $c(x)$ depends on the spatial variable x.
(b) Does it work when $c(t)$ depends only on the time t?

2.4.18. The *Poisson–Darboux equation* is $\dfrac{\partial^2 u}{\partial t^2} - \dfrac{\partial^2 u}{\partial x^2} - \dfrac{2}{x}\dfrac{\partial u}{\partial x} = 0$. Solve the initial value problem $u(0,x) = 0$, $u_t(0,x) = g(x)$, where $g(x) = g(-x)$ is an even function. *Hint*: Set $w = x\,u$.

♡ 2.4.19. (a) Solve the initial value problem $u_{tt} - 2u_{tx} - 3u_{xx} = 0$, $u(0,x) = x^2$, $u_t(0,x) = e^x$. *Hint*: Factor the associated linear differential operator. (b) Determine the domain of influence of a point $(0,x)$. (c) Determine the domain of dependence of a point (t,x) with $t > 0$.

♢ 2.4.20. (a) Use polar coordinates to prove that, for any $a > 0$,

$$\iint_{\mathbb{R}^2} e^{-a(x^2+y^2)}\,dx\,dy = \frac{\pi}{a}. \tag{2.99}$$

(b) Explain why

$$\int_{-\infty}^{\infty} e^{-a x^2}\,dx = \sqrt{\frac{\pi}{a}}. \tag{2.100}$$

♢ 2.4.21. Use Exercise 2.4.20 to prove the error function formulae (2.88).

Chapter 3
Fourier Series

Just before 1800, the French mathematician/physicist/engineer Jean Baptiste Joseph Fourier made an astonishing discovery, [**42**]. Through his deep analytical investigations into the partial differential equations modeling heat propagation in bodies, Fourier was led to claim that "every" function could be represented as an infinite series of elementary trigonometric functions: sines and cosines. For example, consider the sound produced by a musical instrument, e.g., piano, violin, trumpet, or drum. Decomposing the signal into its trigonometric constituents reveals the fundamental frequencies (tones, overtones, etc.) that combine to produce the instrument's distinctive timbre. This Fourier decomposition lies at the heart of modern electronic music; a synthesizer combines pure sine and cosine tones to reproduce the diverse sounds of instruments, both natural and artificial, according to Fourier's general prescription.

Fourier's claim was so remarkable and counterintuitive that most of the leading mathematicians of the time did not believe him. Nevertheless, it was not long before scientists came to appreciate the power and far-ranging applicability of Fourier's method, thereby opening up vast new realms of mathematics, physics, engineering, and beyond. Indeed, Fourier's discovery easily ranks in the "top ten" mathematical advances of all time, a list that would also include Newton's invention of the calculus, and Gauss and Riemann's differential geometry, which, 70 years later, became the foundation of Einstein's general relativity. Fourier analysis is an essential component of much of modern applied (and pure) mathematics. It forms an exceptionally powerful analytic tool for solving a broad range of linear partial differential equations. Applications in physics, engineering, biology, finance, etc., are almost too numerous to catalogue: typing the word "Fourier" in the subject index of a modern science library will dramatically demonstrate just how ubiquitous these methods are. Fourier analysis lies at the heart of signal processing, including audio, speech, images, videos, seismic data, radio transmissions, and so on. Many modern technological advances, including television, music CDs and DVDs, cell phones, movies, computer graphics, image processing, and fingerprint analysis and storage, are, in one way or another, founded on the many ramifications of Fourier theory. In your career as a mathematician, scientist, or engineer, you will find that Fourier theory, like calculus and linear algebra, is one of the most basic weapons in your mathematical arsenal. Mastery of the subject is essential.

Furthermore, a surprisingly large fraction of modern mathematics rests on subsequent attempts to place Fourier series on a firm mathematical foundation. Thus, many of modern analysis' most basic concepts, including the definition of a function, the ε–δ definition of limit and continuity, convergence properties in function space, the modern theory of integration and measure, generalized functions such as the delta function, and many others, all owe a profound debt to the prolonged struggle to establish a rigorous framework for Fourier analysis. Even more remarkably, modern set theory, and, thus, the foundations of modern mathematics and logic, can be traced directly back to the nineteenth-century German mathematician Georg Cantor's attempts to understand the sets on which Fourier series converge!

We begin our development of Fourier methods by explaining why Fourier series naturally appear when we try to solve the one-dimensional heat equation. The reader uninterested in such motivations can safely omit this initial section, since the same material reappears in Chapter 4, where we apply Fourier methods to solve several important linear partial differential equations. Beginning in Section 3.2, we shall introduce the most basic computational techniques for Fourier series. The final section is an abbreviated introduction to the analytic background required to develop a rigorous foundation for Fourier series methods. While this section is a bit more mathematically sophisticated than what has appeared so far, the student is strongly encouraged to delve into it to gain additional insight and see further developments, including some of direct importance in applications.

3.1 Eigensolutions of Linear Evolution Equations

Following our studies of first-order partial differential equations in Chapter 2, the next important example to merit investigation is the second-order linear equation

$$\frac{\partial u}{\partial t} = \frac{\partial^2 u}{\partial x^2}, \tag{3.1}$$

known as the *heat equation*, since it models (among other diffusion processes) heat flow in a one-dimensional medium, e.g., a metal bar. For simplicity, we have set the physical parameters equal to 1 in order to focus on the solution techniques. A more complete discussion, including a brief derivation from physical principles, will appear in Chapter 4. Unlike the wave equation considered in Chapter 2, there is no comparably elementary formula for the general solution to the heat equation. Instead, we will write solutions as infinite series in certain simple, explicit solutions. This solution method, pioneered by Fourier, will lead us immediately to the definition of a Fourier series. The remainder of this chapter will be devoted to developing the basic properties and calculus of Fourier series. Once we have mastered these essential mathematical techniques, we will start applying them to partial differential equations in Chapter 4.

Let us begin by writing the heat equation (3.1) in a more abstract, but suggestive, linear *evolutionary form*

$$\frac{\partial u}{\partial t} = L[u], \tag{3.2}$$

in which

$$L[u] = \frac{\partial^2 u}{\partial x^2} \tag{3.3}$$

is a linear second-order differential operator. Recall, (1.11), that linearity imposes two requirements on the operator L:

$$L[u + v] = L[u] + L[v], \qquad L[cu] = c\,L[u], \tag{3.4}$$

for any functions[†] u, v and any constant c. Moreover, since L involves differentiation only with respect to x, it also satisfies

$$L[\,c(t)\,u\,] = c(t)\,L[\,u\,] \tag{3.5}$$

for any function $c(t)$ that does not depend on x.

Of course, there are many other possible linear differential operators, and so our abstract linear evolution equation (3.2) can represent a wide range of linear partial differential equations. For example, if

$$L[\,u\,] = -\,c(x)\,\frac{\partial u}{\partial x}\,, \tag{3.6}$$

where $c(x)$ is a function representing the wave speed in a nonuniform medium, then (3.2) becomes the transport equation

$$\frac{\partial u}{\partial t} = -\,c(x)\,\frac{\partial u}{\partial x} \tag{3.7}$$

that we studied in Chapter 2. If

$$L[\,u\,] = \frac{1}{\sigma(x)}\,\frac{\partial}{\partial x}\left(\kappa(x)\,\frac{\partial u}{\partial x}\right), \tag{3.8}$$

where $\sigma(x) > 0$ represents *heat capacity* and $\kappa(x) > 0$ *thermal conductivity*, then (3.2) becomes the *generalized heat equation*

$$\frac{\partial u}{\partial t} = \frac{1}{\sigma(x)}\,\frac{\partial}{\partial x}\left(\kappa(x)\,\frac{\partial u}{\partial x}\right), \tag{3.9}$$

governing the diffusion of heat in a nonuniform bar. If

$$L[\,u\,] = \frac{\partial^2 u}{\partial x^2} - \gamma\,u, \tag{3.10}$$

where $\gamma > 0$ is a positive constant, then (3.2) becomes the *damped heat equation*

$$\frac{\partial u}{\partial t} = \frac{\partial^2 u}{\partial x^2} - \gamma\,u, \tag{3.11}$$

which models the temperature of a bar that is cooling off due to radiation of heat energy. We can even take u to be a function of more than one space variable, e.g., $u(t,x,y)$ or $u(t,x,y,z)$, in which case (3.2) includes higher-dimensional versions of the heat equation for plates and solid bodies, which we will study in due course. In all cases, the key requirements on the operator L are (a) linearity, and (b) only differentiation with respect to the spatial variables is allowed.

Fourier's inspired idea for solving such linear evolution equations is a direct adaptation of the eigensolution method for first-order linear systems of ordinary differential equations, [**20, 23, 89**], which we now recall. The starting point is the elementary scalar ordinary differential equation

$$\frac{du}{dt} = \lambda\,u. \tag{3.12}$$

[†] We assume throughout that the functions are sufficiently smooth so that the indicated derivatives are well defined.

The general solution is an exponential function

$$u(t) = c\,e^{\lambda t}, \tag{3.13}$$

whose coefficient c is an arbitrary constant. This elementary observation motivates the solution method for a first-order homogeneous linear system of ordinary differential equations

$$\frac{d\mathbf{u}}{dt} = A\mathbf{u}, \tag{3.14}$$

in which A is a constant $n \times n$ matrix. Working by analogy, we will seek solutions of exponential form

$$\mathbf{u}(t) = e^{\lambda t}\,\mathbf{v}, \tag{3.15}$$

where $\mathbf{v} \in \mathbb{R}^n$ is a constant vector. We substitute this *ansatz*† into the equation. First,

$$\frac{d\mathbf{u}}{dt} = \frac{d}{dt}\left(e^{\lambda t}\,\mathbf{v}\right) = \lambda\,e^{\lambda t}\,\mathbf{v}.$$

On the other hand, since $e^{\lambda t}$ is a scalar, it commutes with matrix multiplication, and so

$$A\mathbf{u} = A\,e^{\lambda t}\,\mathbf{v} = e^{\lambda t}A\mathbf{v}.$$

Therefore, $\mathbf{u}(t)$ will solve the system (3.14) if and only if $\mathbf{v}$ satisfies

$$A\mathbf{v} = \lambda\mathbf{v}. \tag{3.16}$$

We recognize this as the *eigenequation* that determines the eigenvalues of the matrix A. Namely, (3.16) has a nonzero solution $\mathbf{v} \neq \mathbf{0}$ if and only if λ is an *eigenvalue* and $\mathbf{v}$ a corresponding *eigenvector*. Each eigenvalue λ and eigenvector $\mathbf{v}$ produces a nonzero, exponentially varying *eigensolution* (3.15) to the linear system of ordinary differential equations.

Remark: Any nonzero scalar multiple of an eigenvector $\widehat{\mathbf{v}} = c\mathbf{v}$, for $c \neq 0$, is automatically another eigenvector for the same eigenvalue λ. However, the only effect is to multiply the eigensolution by the scalar c. Thus, to obtain a complete system of independent solutions, we need only the independent eigenvectors.

For simplicity — and also because *all* of the linear partial differential equations we will treat will have the analogous property — suppose that the $n \times n$ matrix A has a *complete* system of real eigenvalues $\lambda_1, \ldots, \lambda_n$ and corresponding real, linearly independent eigenvectors $\mathbf{v}_1, \ldots, \mathbf{v}_n$, which therefore form an *eigenvector basis* of the underlying space $\mathbb{R}^n$. (We allow the possibility of repeated eigenvalues, but require that all eigenvectors be independent to avoid superfluous solutions.) For example, according to Theorem B.26 (see also [**89**; Theorem 8.20]), all real, symmetric matrices, $A = A^T$, are complete. Complex eigenvalues lead to complex exponential solutions, whose real and imaginary parts can be used to construct the associated real solutions. Incomplete matrices, having an insufficient number of eigenvectors, are trickier, and the solution to the corresponding linear system

† The German word *ansatz* refers to the method of finding a solution to a complicated equation by postulating that it is of a special form. Usually, an ansatz will depend on one or more free parameters — in this case, the entries of the vector $\mathbf{v}$ along with the scalar λ — that, with some luck, can be adjusted to fulfill the requirements imposed by the equation. Thus, a reasonable English translation of "ansatz" is "inspired guess".

requires use of the Jordan canonical form, [**89**; Section 8.6]. Fortunately, we do not have to deal with the latter, technically annoying, cases here.

Using our completeness assumption, we can produce n independent real exponential eigensolutions

$$u_1(t) = e^{\lambda_1 t}\mathbf{v}_1, \qquad \ldots \qquad u_n(t) = e^{\lambda_n t}\mathbf{v}_n,$$

to the linear system (3.14). The Linear Superposition Principle of Theorem 1.4 tells us that, for any choice of scalars $c_1, \ldots, c_n$, the linear combination

$$c_1 u_1(t) + \cdots + c_n u_n(t) = c_1 e^{\lambda_1 t}\mathbf{v}_1 + \cdots + c_n e^{\lambda_n t}\mathbf{v}_n \tag{3.17}$$

is also a solution. The basic Existence and Uniqueness Theorems for first-order systems of ordinary differential equations, [**18, 23, 52**], imply that (3.17) forms the *general solution* to the original linear system, and so the eigensolutions form a basis for the solution space.

Let us now adapt this seminal idea to construct exponentially varying solutions to the heat equation (3.1) or, for that matter, any linear evolution equation in the form (3.2). To this end, we introduce an analogous exponential ansatz:

$$u(t,x) = e^{\lambda t}\, v(x), \tag{3.18}$$

in which we replace the vector $\mathbf{v}$ in (3.15) by a function $v(x)$. We substitute the expression (3.18) into the dynamical equations (3.2). First, the time derivative of such a function is

$$\frac{\partial u}{\partial t} = \frac{\partial}{\partial t}\left[\, e^{\lambda t}\, v(x)\,\right] = \lambda\, e^{\lambda t}\, v(x).$$

On the other hand, in view of (3.5),

$$L[\,u\,] = L[\, e^{\lambda t}\, v(x)\,] = e^{\lambda t}\, L[\,v\,].$$

Equating these two expressions and canceling the common exponential factor, we conclude that $v(x)$ must satisfy the *eigenequation*

$$L[\,v\,] = \lambda\, v \tag{3.19}$$

for the linear differential operator L, in which λ is the *eigenvalue*, while $v(x)$ is the corresponding *eigenfunction*. Each eigenvalue and eigenfunction pair will produce an exponentially varying *eigensolution* (3.18) to the partial differential equation (3.2). We will then appeal to Linear Superposition to combine the resulting eigensolutions to form additional solutions. The key complication is that partial differential equations admit an infinite number of independent eigensolutions, and thus one cannot hope to write the general solution as a finite linear combination thereof. Rather, one is led to try constructing solutions as *infinite series* in the eigensolutions. However, justifying such series solution formulas requires additional analytical skills and sophistication. Not every infinite series converges to a bona fide function. Moreover, a convergent series of differentiable functions need not converge to a differentiable function, and hence the series may not represent a (classical) solution to the partial differential equation. We are being reminded, yet again, that partial differential equations are much wilder creatures than their relatively tame cousins, ordinary differential equations.

Let us, for specificity, focus our attention on the heat equation, for which the linear operator L is given by (3.3). If $v(x)$ is a function of x alone, then

$$L[\,v\,] = v''(x).$$

Thus, our eigenequation (3.19) becomes

$$v'' = \lambda v. \tag{3.20}$$

This is a linear second-order ordinary differential equation for $v(x)$, and so has two linearly independent solutions. The explicit solution formulas depend on the sign of the eigenvalue λ, and can be found in any basic text on ordinary differential equations, e.g., [**20, 23**]. The following table summarizes the results for real eigenvalues λ; the case of complex λ is left as Exercise 3.1.3 for the reader. The resulting exponential eigensolutions are also referred to as *separable solutions* to indicate that they are the product of a function of t alone and a function of x alone. The general method of separation of variables will be one of our main tools for solving linear partial differential equations, to be developed in detail starting in Chapter 4.

Real Eigensolutions of the Heat Equation

λ	Eigenfunctions $v(x)$	Eigensolutions $u(t,x) = e^{\lambda t}\, v(x)$
$\lambda = -\,\omega^2 < 0$	$\cos\omega x,\ \sin\omega x$	$e^{-\,\omega^2 t}\cos\omega x,\ e^{-\,\omega^2 t}\sin\omega x$
$\lambda = 0$	$1,\ x$	$1,\ x$
$\lambda = \omega^2 > 0$	$e^{-\,\omega x},\ e^{\omega x}$	$e^{\omega^2 t - \omega x},\ e^{\omega^2 t + \omega x}$

Remark: Thus, in the absence of boundary conditions, each real number λ qualifies as an eigenvalue of the linear differential operator (3.3), possessing two linearly independent eigenfunctions, and thus two linearly independent eigensolutions to the heat equation. As with eigenvectors, any (nonzero) linear combination of eigenfunctions (eigensolutions) with the same eigenvalue is also an eigenfunction (eigensolution). Thus, the preceding table lists only independent eigenfunctions and eigensolutions.

As noted above, any *finite* linear combination of these basic eigensolutions is automatically a solution. Thus, for example,

$$u(t,x) = c_1 e^{-\,t}\cos x + c_2 e^{-\,4t}\sin 2x + c_3 x + c_4$$

is a solution to the heat equation for any choice of constants c_1, c_2, c_3, c_4, as you can easily check. But, since there are infinitely many independent eigensolutions, we cannot expect to be able to represent *every* solution to the heat equation as a finite linear combination of eigensolutions. And so, we must learn how to deal with *infinite series* of eigensolutions.

Remark: Eigensolutions in the first class, where $\lambda < 0$, are exponentially decaying, which is in accord with our physical intuition as to how the temperature of a body should behave. Those in the second class are constant in time — also physically reasonable. However, those in the third class, corresponding to positive eigenvalues $\lambda > 0$, are exponentially growing in time. In the absence of external heat sources, physical bodies should approach some sort of thermal equilibrium, and certainly not exhibit an exponentially growing temperature! However, notice that the latter eigensolutions (as well as the solution x) are not bounded in space, and so include an infinite amount of heat energy being supplied to the system from infinity. As we will soon come to appreciate, physically relevant boundary conditions — posed either on a bounded interval or by specifying the asymptotics of the solutions at large distances — will separate out the physically reasonable solutions from the mathematically valid but physically irrelevant ones.

The Heated Ring

So far, we have not paid any attention to boundary conditions. As noted above, these will eliminate nonphysical eigensolutions and thereby reduce the collection to a manageable, albeit still infinite, number. In this subsection, we will discuss a particularly important case, which, following Fourier's line of reasoning, leads us directly into the heart of Fourier series.

Consider the heat equation on the interval $-\pi \leq x \leq \pi$, subject to the *periodic boundary conditions*

$$\frac{\partial u}{\partial t} = \frac{\partial^2 u}{\partial x^2}, \qquad u(t,-\pi) = u(t,\pi), \qquad \frac{\partial u}{\partial x}(t,-\pi) = \frac{\partial u}{\partial x}(t,\pi). \tag{3.21}$$

The physical problem being modeled is the thermodynamic behavior of an insulated circular ring, in which x represents the angular coordinate. The boundary conditions ensure that the temperature remains continuously differentiable at the junction point where the angle switches over from $-\pi$ to π. Given the ring's initial temperature distribution

$$u(0,x) = f(x), \qquad -\pi \leq x \leq \pi, \tag{3.22}$$

our task is to determine the temperature of the ring $u(t,x)$ at each subsequent time $t > 0$.

Let us find out which of the preceding eigensolutions respect the boundary conditions. Substituting our exponential ansatz (3.18) into the differential equation and boundary conditions (3.21), we find that the eigenfunction $v(x)$ must satisfy the periodic boundary value problem

$$v'' = \lambda v, \qquad v(-\pi) = v(\pi), \qquad v'(-\pi) = v'(\pi). \tag{3.23}$$

Our task is to find those values of λ for which (3.23) *has a nonzero solution* $v(x) \not\equiv 0$. These are the eigenvalues and eigenfunctions.

As noted above, there are three cases, depending on the sign of λ. First, suppose $\lambda = \omega^2 > 0$. Then the general solution to the ordinary differential equation is

$$v(x) = a e^{\omega x} + b e^{-\omega x},$$

where a, b are arbitrary constants. Substituting into the boundary conditions, we find that a, b must satisfy the pair of linear equations

$$a e^{-\omega \pi} + b e^{\omega \pi} = a e^{\omega \pi} + b e^{-\omega \pi}, \qquad a\omega e^{-\omega \pi} - b\omega e^{\omega \pi} = a\omega e^{\omega \pi} - b\omega e^{-\omega \pi}.$$

Since $\omega \neq 0$, the first equation implies that $a = b$, while the second requires $a = -b$. So, the only way to satisfy both boundary conditions is to take $a = b = 0$, and so $v(x) \equiv 0$ is a trivial solution. We conclude that there are no positive eigenvalues.

Second, if $\lambda = 0$, then the ordinary differential equation reduces to $v'' = 0$, with solution

$$v(x) = a + bx.$$

Substituting into the boundary conditions requires

$$a - b\pi = a + b\pi, \qquad b = b.$$

The first equation implies that $b = 0$, but this is the only condition. Therefore, any constant function, $v(x) \equiv a$, solves the boundary value problem, and hence $\lambda = 0$ is an eigenvalue. We take $v_0(x) \equiv 1$ as the unique independent eigenfunction, bearing in mind that any constant multiple of an eigenfunction is automatically also an eigenfunction. We will call 1 a *null eigenfunction*, indicating that it is associated with the zero eigenvalue $\lambda = 0$. The corresponding eigensolution (3.18) is $u(t,x) = e^{0\,t} v_0(x) = 1$, a constant solution to the heat equation.

Finally, we must deal with the case $\lambda = -\omega^2 < 0$. Now, the general solution to the differential equation in (3.23) is a trigonometric function:

$$v(x) = a\cos\omega x + b\sin\omega x. \tag{3.24}$$

Since

$$v'(x) = -a\omega\sin\omega x + b\omega\cos\omega x,$$

when we substitute into the boundary conditions, we obtain

$$\begin{aligned} a\cos\omega\pi - b\sin\omega\pi &= a\cos\omega\pi + b\sin\omega\pi, \\ a\sin\omega\pi + b\cos\omega\pi &= -a\sin\omega\pi + b\cos\omega\pi, \end{aligned}$$

where we canceled out a common factor of ω in the second equation. These simplify to

$$2b\sin\omega\pi = 0, \qquad 2a\sin\omega\pi = 0.$$

If $\sin\omega\pi \neq 0$, then $a = b = 0$, and so we have only the trivial solution $v(x) \equiv 0$. Thus, to obtain a nonzero eigenfunction, we must have

$$\sin\omega\pi = 0,$$

which requires that $\omega = 1, 2, 3, \ldots$ be a positive integer. For such $\omega_k = k$, *every* solution

$$v(x) = a\cos kx + b\sin kx, \qquad k = 1, 2, 3, \ldots\,,$$

satisfies both boundary conditions, and hence (unless identically zero) qualifies as an eigenfunction of the boundary value problem. Thus, the eigenvalue $\lambda_k = -k^2$ admits a two-dimensional space of eigenfunctions, with basis $v_k(x) = \cos kx$ and $\widetilde{v}_k(x) = \sin kx$.

Consequently, the basic trigonometric functions

$$1, \qquad \cos x, \qquad \sin x, \qquad \cos 2x, \qquad \sin 2x, \qquad \cos 3x, \qquad \ldots \tag{3.25}$$

form a system of independent eigenfunctions for the periodic boundary value problem (3.23). The corresponding exponentially varying eigensolutions are

$$u_k(x) = e^{-k^2 t}\cos kx, \qquad \widetilde{u}_k(x) = e^{-k^2 t}\sin kx, \qquad k = 0, 1, 2, 3, \ldots, \tag{3.26}$$

each of which, by design, is a solution to the heat equation (3.21) and satisfies the periodic boundary conditions. Note that we subsumed the case $\lambda_0 = 0$ into (3.26), keeping in mind that, when $k = 0$, the sine function is trivial, and hence $\widetilde{u}_0(x) \equiv 0$ is not needed. So the null eigenvalue $\lambda_0 = 0$ provides (up to a constant multiple) only one eigensolution, whereas the strictly negative eigenvalues $\lambda_k = -k^2 < 0$ each provide two independent eigensolutions.

Remark: For completeness, one should also consider the possibility of complex eigenvalues. If $\lambda = \omega^2 \neq 0$, where ω is now allowed to be complex, then all solutions to the differential equation (3.23) are of the form

$$v(x) = a e^{\omega x} + b e^{-\omega x}.$$

The periodic boundary conditions require

$$a e^{-\omega \pi} + b e^{\omega \pi} = a e^{\omega \pi} + b e^{-\omega \pi}, \qquad a\omega e^{-\omega \pi} - b\omega e^{\omega \pi} = a\omega e^{\omega \pi} - b\omega e^{-\omega \pi}.$$

If $e^{\omega \pi} \neq e^{-\omega \pi}$, or, equivalently, $e^{2\omega \pi} \neq 1$, then the first condition implies $a = b$, but then the second implies $a = b = 0$, and so $\lambda = \omega^2$ is not an eigenvalue. Thus, the eigenvalues only occur when $e^{2\omega \pi} = 1$. This implies $\omega = k\,\mathrm{i}$, where k is an integer, and so $\lambda = -k^2$, leading back to the known trigonometric solutions. Later, in Section 9.5, we will learn that the "self-adjoint" structure of the underlying boundary value problem implies, a priori, that all its eigenvalues are necessarily real and nonpositive. So a good part of the preceding analysis was, in fact, superfluous.

We conclude that there is an infinite number of independent eigensolutions (3.26) to the periodic heat equation (3.21). Linear Superposition, as described in Theorem 1.4, tells us that any *finite* linear combination of the eigensolutions is automatically a solution to the periodic heat equation. However, only solutions whose initial data $u(0,x) = f(x)$ happens to be a finite linear combination of the trigonometric eigenfunctions (a trigonometric polynomial) can be so represented. Fourier's brilliant idea was to propose taking infinite "linear combinations" of the eigensolutions in an attempt to solve the general initial value problem. Thus, we try representing a general solution to the periodic heat equation as an infinite series of the form[†]

$$u(t,x) = \frac{a_0}{2} + \sum_{k=1}^{\infty} \left[a_k\, e^{-k^2 t} \cos k x + b_k\, e^{-k^2 t} \sin k x \right]. \tag{3.27}$$

The coefficients $a_0, a_1, a_2, \ldots, b_1, b_2, \ldots$, are constants, to be fixed by the initial condition. Indeed, substituting our proposed solution formula (3.27) into (3.22), we obtain

$$f(x) = u(0,x) = \frac{a_0}{2} + \sum_{k=1}^{\infty} \left[a_k \cos k x + b_k \sin k x \right]. \tag{3.28}$$

Thus, we must represent the initial temperature distribution $f(x)$ as an infinite *Fourier series* in the elementary trigonometric eigenfunctions. Once we have prescribed the *Fourier coefficients* $a_0, a_1, a_2, \ldots, b_1, b_2, \ldots$, we expect that the corresponding eigensolution series (3.27) will provide an explicit formula for the solution to the periodic initial-boundary value problem for the heat equation.

However, infinite series are much more delicate than finite sums, and so this formal construction requires some serious mathematical analysis to place it on a rigorous foundation. The key questions are:

- When does an infinite trigonometric Fourier series converge?
- What kinds of functions $f(x)$ can be represented by a convergent Fourier series?

[†] For technical reasons, one takes the basic null eigenfunction to be $\frac{1}{2}$ instead of 1. The reason for this choice will be revealed in the following section.

- Given such a function, how do we determine its Fourier coefficients a_k, b_k?
- Are we allowed to differentiate a Fourier series?
- Does the result actually form a solution to the initial-boundary value problem for the heat equation?

These are the basic issues in Fourier analysis, which must be properly addressed before we can make any serious progress towards actually solving the heat equation. Thus, we will leave partial differential equations aside for the time being, and start a detailed investigation into the mathematics of Fourier series.

Exercises

3.1.1. For each of the following differential operators, (*i*) prove linearity; (*ii*) prove (3.5); (*iii*) write down the corresponding linear evolution equation (3.2):
(a) $\dfrac{\partial}{\partial x}$, (b) $\dfrac{\partial}{\partial x}+1$, (c) $\dfrac{\partial^2}{\partial x^2}+3\,\dfrac{\partial}{\partial x}$, (d) $\dfrac{\partial}{\partial x}\,e^x\,\dfrac{\partial}{\partial x}$, (e) $\dfrac{\partial^2}{\partial x^2}+\dfrac{\partial^2}{\partial y^2}$.

3.1.2. Find all separable eigensolutions to the heat equation $u_t = u_{xx}$ on the interval $0 \le x \le \pi$ subject to (a) homogeneous Dirichlet boundary conditions $u(t,0)=0$, $u(t,\pi)=0$; (b) mixed boundary conditions $u(t,0)=0$, $u_x(t,\pi)=0$; (c) Neumann boundary conditions $u_x(t,0)=0$, $u_x(t,\pi)=0$.

♢ 3.1.3. Complete the table of eigensolutions to the heat equation, in the absence of boundary conditions, by allowing the eigenvalue λ to be complex.

3.1.4. Find all separable eigensolutions to the following partial differential equations:
(a) $u_t = u_x$, (b) $u_t = u_x - u$, (c) $u_t = x\,u_x$.

3.1.5. (a) Find the real eigensolutions to the damped heat equation $u_t = u_{xx} - u$. (b) Which solutions satisfy the periodic boundary conditions $u(t,-\pi)=u(t,\pi)$, $u_x(t,-\pi)=u_x(t,\pi)$?

3.1.6. Answer Exercise 3.1.5 for the diffusive transport equation $u_t + c\,u_x = u_{xx}$ modeling the combined diffusion and transport of a solute in a uniform flow with constant wave speed c.

♡ 3.1.7. (a) Find the real eigensolutions to the diffusion equation $u_t = (x^2\,u_x)_x$ modeling diffusion in an inhomogeneous medium on the half-line $x > 0$.
(b) Which solutions satisfy the Dirichlet boundary conditions $u(t,1)=u(t,2)=0$?

3.2 Fourier Series

The preceding section served to motivate the development of Fourier series as a tool for solving partial differential equations. Our immediate goal is to represent a given function $f(x)$ as a convergent series in the elementary trigonometric functions:

$$f(x) = \frac{a_0}{2} + \sum_{k=1}^{\infty}\left[\,a_k\cos k\,x + b_k\sin k\,x\,\right]. \tag{3.29}$$

The first order of business is to determine the formulae for the Fourier coefficients a_k, b_k; only then will we deal with convergence issues.

The key that unlocks the Fourier treasure chest is orthogonality. Recall that two vectors in Euclidean space are called *orthogonal* if they meet at a right angle. More explicitly, $\mathbf{v}, \mathbf{w}$ are orthogonal if and only if their dot product is zero: $\mathbf{v} \cdot \mathbf{w} = 0$. Orthogonality, and particularly orthogonal bases, has profound consequences that underpin many modern computational algorithms. See Section B.4 for the basics, and [**89**] for full details on finite-dimensional developments. In infinite-dimensional function space, were it not for orthogonality, Fourier theory would be vastly more complicated, if not completely impractical for applications.

The starting point is the introduction of a suitable inner product on function space, to assume the role played by the dot product in the finite-dimensional context. For classical Fourier series, we use the rescaled L^2 *inner product*

$$\langle f , g \rangle = \frac{1}{\pi} \int_{-\pi}^{\pi} f(x)\, g(x)\, dx \tag{3.30}$$

on the space of continuous functions defined on the interval† $[-\pi, \pi]$. It is not hard to show that (3.30) satisfies the basic inner product axioms listed in Definition B.10. The associated norm is

$$\| f \| = \sqrt{\langle f , f \rangle} = \sqrt{\frac{1}{\pi} \int_{-\pi}^{\pi} f(x)^2\, dx}\,. \tag{3.31}$$

Lemma 3.1. *Under the rescaled* L^2 *inner product* (3.30), *the trigonometric functions* $1, \cos x, \sin x, \cos 2x, \sin 2x, \ \ldots\ ,$ *satisfy the following orthogonality relations*:

$$\begin{aligned} \langle \cos kx , \cos lx \rangle = \langle \sin kx , \sin lx \rangle &= 0, && \text{for} \quad k \neq l, \\ \langle \cos kx , \sin lx \rangle &= 0, && \text{for all} \quad k, l, \\ \| 1 \| = \sqrt{2}, \qquad \| \cos kx \| = \| \sin kx \| &= 1, && \text{for} \quad k \neq 0, \end{aligned} \tag{3.32}$$

where k and l indicate nonnegative integers.

Proof: The formulas follow immediately from the elementary integration identities

$$\int_{-\pi}^{\pi} \cos kx \ \cos lx \, dx = \begin{cases} 0, & k \neq l, \\ 2\pi, & k = l = 0, \\ \pi, & k = l \neq 0, \end{cases} \qquad \begin{aligned} &\int_{-\pi}^{\pi} \sin kx \ \sin lx \, dx = \begin{cases} 0, & k \neq l, \\ \pi, & k = l \neq 0, \end{cases} \\ &\int_{-\pi}^{\pi} \cos kx \ \sin lx \, dx = 0, \end{aligned} \tag{3.33}$$

which are valid for all nonnegative integers $k, l \geq 0$. *Q.E.D.*

Lemma 3.1 implies that the elementary trigonometric functions form an *orthogonal system*, meaning that any distinct pair are orthogonal under the chosen inner product. If we were to replace the constant function 1 by $\frac{1}{\sqrt{2}}$, then the resulting functions would form an *orthonormal system* meaning that, in addition, they all have norm 1. However, the extra $\sqrt{2}$ is utterly annoying, and best omitted.

† We have chosen to use the interval $[-\pi, \pi]$ for convenience. A common alternative is to develop Fourier series on the interval $[0, 2\pi]$. In fact, since the basic trigonometric functions are 2π–periodic, any interval of length 2π will serve equally well. Adapting Fourier series to other intervals will be discussed in Section 3.4.

Remark: As with all essential mathematical facts, the orthogonality of the trigonometric functions is not an accident, but indicates that something deeper is going on. Indeed, orthogonality is a consequence of the fact that the trigonometric functions are the eigenfunctions for the "self-adjoint" boundary value problem (3.23), which is the function space counterpart to the orthogonality of eigenvectors of symmetric matrices, cf. Theorem B.26. The general framework will be developed in detail in Section 9.5, and then applied to the more complicated systems of eigenfunctions we will encounter when dealing with higher-dimensional partial differential equations.

If we ignore convergence issues, then the trigonometric orthogonality relations serve to prescribe the Fourier coefficients: Taking the inner product of both sides of (3.29) with $\cos l x$ for $l > 0$, and invoking linearity of the inner product, yields

$$\begin{aligned}\langle f , \cos l x \rangle &= \frac{a_0}{2} \langle 1 , \cos l x \rangle + \sum_{k=1}^{\infty} \left[a_k \langle \cos k x , \cos l x \rangle + b_k \langle \sin k x , \cos l x \rangle \right] \\ &= a_l \langle \cos l x , \cos l x \rangle = a_l,\end{aligned}$$

since, by the orthogonality relations (3.32), all terms but the lth vanish. This serves to prescribe the Fourier coefficient a_l. A similar manipulation with $\sin l x$ fixes $b_l = \langle f , \sin l x \rangle$, while taking the inner product with the constant function 1 gives

$$\langle f , 1 \rangle = \frac{a_0}{2} \langle 1 , 1 \rangle + \sum_{k=1}^{\infty} \left[a_k \langle \cos k x , 1 \rangle + b_k \langle \sin k x , 1 \rangle \right] = \frac{a_0}{2} \| 1 \|^2 = a_0,$$

which agrees with the preceding formula for a_l when $l = 0$, and explains why we include the extra factor $\frac{1}{2}$ in the constant term. Thus, *if the Fourier series converges to the function $f(x)$, then its coefficients are prescribed by taking inner products with the basic trigonometric functions.*

Definition 3.2. The *Fourier series* of a function $f(x)$ defined on $-\pi \le x \le \pi$ is

$$f(x) \sim \frac{a_0}{2} + \sum_{k=1}^{\infty} \left[a_k \cos k x + b_k \sin k x \right], \tag{3.34}$$

whose coefficients are given by the inner product formulae

$$\begin{aligned} a_k &= \langle f , \cos k x \rangle = \frac{1}{\pi} \int_{-\pi}^{\pi} f(x) \cos k x \, dx, \qquad & k &= 0, 1, 2, 3, \ldots, \\ b_k &= \langle f , \sin k x \rangle = \frac{1}{\pi} \int_{-\pi}^{\pi} f(x) \sin k x \, dx, \qquad & k &= 1, 2, 3, \ldots. \end{aligned} \tag{3.35}$$

The function $f(x)$ cannot be completely arbitrary, since, at the very least, the integrals in the coefficient formulae must be well defined and finite. Even if the coefficients (3.35) are finite, there is no guarantee that the resulting infinite series converges, and, even if it converges, no guarantee that it converges to the original function $f(x)$. For these reasons, we will tend to use the $\sim$ symbol instead of an equal sign when writing down a Fourier series. Before tackling these critical issues, let us work through an elementary example.

Example 3.3. Consider the function $f(x) = x$. We may compute its Fourier coefficients directly, employing integration by parts to evaluate the integrals:

$$a_0 = \frac{1}{\pi}\int_{-\pi}^{\pi} x\,dx = 0, \qquad a_k = \frac{1}{\pi}\int_{-\pi}^{\pi} x\cos kx\,dx = \frac{1}{\pi}\left[\frac{x\sin kx}{k} + \frac{\cos kx}{k^2}\right]\Bigg|_{x=-\pi}^{\pi} = 0,$$

$$b_k = \frac{1}{\pi}\int_{-\pi}^{\pi} x\sin kx\,dx = \frac{1}{\pi}\left[-\frac{x\cos kx}{k} + \frac{\sin kx}{k^2}\right]\Bigg|_{x=-\pi}^{\pi} = \frac{2}{k}\,(-1)^{k+1}. \tag{3.36}$$

The resulting Fourier series is

$$x \sim 2\left(\sin x - \frac{\sin 2x}{2} + \frac{\sin 3x}{3} - \frac{\sin 4x}{4} + \cdots\right). \tag{3.37}$$

Establishing convergence of this infinite series is far from elementary. Standard calculus criteria, including the ratio and root tests, are inconclusive. Even if we know that the series converges (which it does — for all x), it is certainly not obvious what function it converges to. Indeed, it *cannot* converge to the function $f(x) = x$ everywhere! For instance, if $x = \pi$, then every term in the Fourier series is zero, and so it converges to 0 — which is not the same as $f(\pi) = \pi$.

Recall that the convergence of an infinite series is predicated on the convergence of its sequence of *partial sums*, which, in this case, are

$$s_n(x) = \frac{a_0}{2} + \sum_{k=1}^{n}\left[\,a_k\cos kx + b_k\sin kx\,\right]. \tag{3.38}$$

By definition, the Fourier series *converges* at a point x if and only if its partial sums have a limit:

$$\lim_{n\to\infty} s_n(x) = \widetilde{f}(x), \tag{3.39}$$

which may or may not equal the value of the original function $f(x)$. Thus, a key requirement is to find conditions on the function $f(x)$ that guarantee that the Fourier series converges, and, even more importantly, that the limiting sum reproduces the original function: $\widetilde{f}(x) = f(x)$. This will all be done in detail below.

Remark: A finite Fourier sum, of the form (3.38), is also known as a *trigonometric polynomial*. This is because, by trigonometric identities, it can be re-expressed as a polynomial $P(\cos x, \sin x)$ in the cosine and sine functions; vice versa, every such polynomial can be uniquely written as such a sum; see [**89**] for details.

The passage from trigonometric polynomials to Fourier series might be viewed as analogous to the passage from polynomials to power series. Recall that the *Taylor series* of an infinitely differentiable function $f(x)$ at the point $x = 0$ is

$$f(x) \sim c_0 + c_1 x + \cdots + c_n x^n + \cdots = \sum_{k=0}^{\infty} c_k x^k,$$

where, according to Taylor's formula, the coefficients $c_k = \dfrac{f^{(k)}(0)}{k!}$ are expressed in terms of its derivatives at the origin, *not* by an inner product. The partial sums

$$s_n(x) = c_0 + c_1 x + \cdots + c_n x^n = \sum_{k=0}^{n} c_k x^k$$

of a power series are ordinary polynomials, and the same basic convergence issues arise.

Although superficially similar, in actuality the two theories are profoundly different. Indeed, while the theory of power series was well established in the early days of the calculus, there remain, to this day, unresolved foundational issues in Fourier theory. A power series in a real variable x either converges everywhere, or on an interval centered at 0, or nowhere except at 0. On the other hand, a Fourier series can converge on quite bizarre sets. Secondly, when a power series converges, it converges to an analytic function, whose derivatives are represented by the differentiated power series. Fourier series may converge, not only to continuous functions, but also to a wide variety of discontinuous functions and even more general objects. Therefore, term-wise differentiation of a Fourier series is a nontrivial issue.

Once one appreciates how radically different the two subjects are, one begins to understand why Fourier's astonishing claims were initially widely disbelieved. Before that time, all functions were taken to be analytic. The fact that Fourier series might converge to a nonanalytic, even discontinuous function was extremely disconcerting, resulting in a profound re-evaluation of the foundations of function theory and the calculus, culminating in the modern definitions of function and convergence that you now learn in your first courses in analysis, [**8, 96, 97**]. Only through the combined efforts of many of the leading mathematicians of the nineteenth century was a rigorous theory of Fourier series firmly established. Section 3.5 contains the most important details, while more comprehensive treatments can be found in the advanced texts [**37, 68, 128**].

Exercises

3.2.1. Find the Fourier series of the following functions: (*a*) $\operatorname{sign} x$, (*b*) $|\,x\,|$, (*c*) $3x-1$, (*d*) x^2, (*e*) $\sin^3 x$, (*f*) $\sin x \cos x$, (*g*) $|\sin x\,|$, (*h*) $x \cos x$.

3.2.2. Find the Fourier series of the following functions:
(*a*) $\begin{cases} 1, & |\,x\,| < \frac{1}{2}\pi, \\ 0, & \text{otherwise}, \end{cases}$ (*b*) $\begin{cases} 1, & \frac{1}{2}\pi < |\,x\,| < \pi, \\ 0, & \text{otherwise}, \end{cases}$ (*c*) $\begin{cases} 1, & \frac{1}{2}\pi < x < \pi, \\ 0, & \text{otherwise}, \end{cases}$
(*d*) $\begin{cases} x, & |\,x\,| < \frac{1}{2}\pi, \\ 0, & \text{otherwise}, \end{cases}$ (*e*) $\begin{cases} \cos x, & |\,x\,| < \frac{1}{2}\pi, \\ 0, & \text{otherwise}. \end{cases}$

3.2.3. Find the Fourier series of $\sin^2 x$ and $\cos^2 x$ without directly calculating the Fourier coefficients. *Hint*: Use some standard trigonometric identities.

♢ 3.2.4. Let $g(x) = \frac{1}{2}p_0 + \sum_{k=1}^{n} (p_k \cos k\,x + q_k \sin k\,x)$ be a trigonometric polynomial. Explain why its Fourier coefficients are $a_k = p_k$ and $b_k = q_k$ for $k \le n$, while $a_k = b_k = 0$ for $k > n$.

3.2.5. *True or false*: (*a*) The Fourier series for the function $2f(x)$ is obtained by multiplying each term in the Fourier series for $f(x)$ by 2. (*b*) The Fourier series for the function $f(2x)$ is obtained by replacing x by $2x$ in the Fourier series for $f(x)$. (*c*) The Fourier coefficients of $f(x) + g(x)$ can be found by adding the corresponding Fourier coefficients of $f(x)$ and $g(x)$. (*d*) The Fourier coefficients of $f(x)\,g(x)$ can be found by multiplying the corresponding Fourier coefficients of $f(x)$ and $g(x)$.

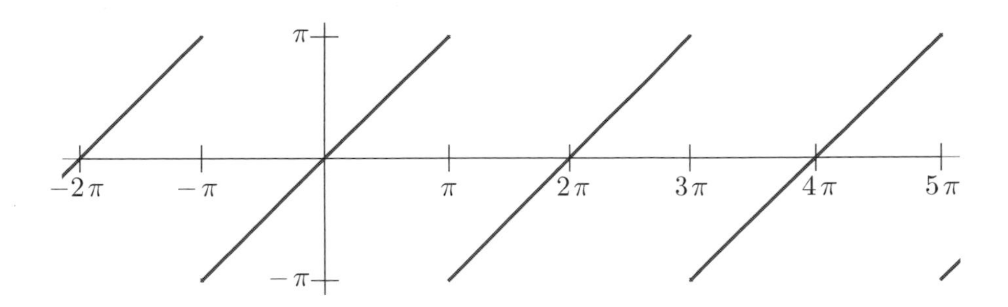

Figure 3.1. 2π–periodic extension of x.

Periodic Extensions

The trigonometric constituents (3.25) of a Fourier series are all periodic functions of period 2π. Therefore, if the series converges, the limiting function $\widetilde{f}(x)$ must also be periodic of period 2π:

$$\widetilde{f}(x+2\pi) = \widetilde{f}(x) \qquad \text{for all} \qquad x \in \mathbb{R}.$$

A Fourier series can converge only to a 2π–periodic function. So it was unreasonable to expect the Fourier series (3.37) to converge to the aperiodic function $f(x) = x$ everywhere. Rather, it should converge to its "periodic extension", which we now define.

Lemma 3.4. *If $f(x)$ is any function defined for $-\pi < x \le \pi$, then there is a unique 2π–periodic function $\widetilde{f}$, known as the* 2π–periodic extension *of f, that satisfies $\widetilde{f}(x) = f(x)$ for all $-\pi < x \le \pi$.*

Proof: Pictorially, the graph of the periodic extension of a function $f(x)$ is obtained by repeatedly copying the part of its graph between $-\pi$ and π to adjacent intervals of length 2π; Figure 3.1 shows a simple example. More formally, given $x \in \mathbb{R}$, there is a unique integer m such that $(2m-1)\pi < x \le (2m+1)\pi$. Periodicity of $\widetilde{f}$ leads us to define

$$\widetilde{f}(x) = \widetilde{f}(x - 2m\pi) = f(x - 2m\pi). \tag{3.40}$$

In particular, if $-\pi < x \le \pi$, then $m = 0$, and hence $\widetilde{f}(x) = f(x)$ for such x. The proof that the resulting function $\widetilde{f}$ is 2π–periodic is left as Exercise 3.2.8. *Q.E.D.*

Remark: The construction of the periodic extension in Lemma 3.4 uses the value $f(\pi)$ at the right endpoint and requires $\widetilde{f}(-\pi) = \widetilde{f}(\pi) = f(\pi)$. One could, alternatively, require $\widetilde{f}(\pi) = \widetilde{f}(-\pi) = f(-\pi)$, which, if $f(-\pi) \ne f(\pi)$, leads to a slightly different 2π–periodic extension of the function. There is no a priori reason to prefer one over the other. In fact, as we shall discover, the preferred Fourier periodic extension $\widetilde{f}(x)$ takes the average of the two values:

$$\widetilde{f}(\pi) = \widetilde{f}(-\pi) = \tfrac{1}{2}\left[\, f(\pi) + f(-\pi)\,\right], \tag{3.41}$$

which then fixes its values at the odd multiples of π.

Example 3.5. The 2π–periodic extension of $f(x) = x$ is the "sawtooth" function $\widetilde{f}(x)$ graphed in Figure 3.1. It agrees with x between $-\pi$ and π. Since $f(\pi) = \pi$, $f(-\pi) = -\pi$, the Fourier extension (3.41) sets $\widetilde{f}(k\pi) = 0$ for any odd integer k. Explicitly,

$$\widetilde{f}(x) = \begin{cases} x - 2m\pi, & (2m-1)\pi < x < (2m+1)\pi, \\ 0, & x = (2m-1)\pi, \end{cases} \qquad \text{where } m \text{ is any integer.}$$

With this convention, it can be proved that the Fourier series (3.37) converges everywhere to the 2π–periodic extension $\widetilde{f}(x)$. In particular,

$$2 \sum_{k=1}^{\infty} (-1)^{k+1} \frac{\sin kx}{k} = \begin{cases} x, & -\pi < x < \pi, \\ 0, & x = \pm\pi. \end{cases} \tag{3.42}$$

Even this very simple example has remarkable and nontrivial consequences. For instance, if we substitute $x = \frac{1}{2}\pi$ in (3.42) and divide by 2, we obtain *Gregory's series*

$$\frac{\pi}{4} = 1 - \frac{1}{3} + \frac{1}{5} - \frac{1}{7} + \frac{1}{9} - \cdots . \tag{3.43}$$

While this striking formula predates Fourier theory — it was, in fact, first discovered by Leibniz — a direct proof is not easy.

Remark: While numerologically fascinating, Gregory's series is of scant practical use for actually computing π, since its rate of convergence is painfully slow. The reader may wish to try adding up terms to see how far out one needs to go to accurately compute even the first two decimal digits of π. Round-off errors will eventually interfere with any attempt to numerically compute the summation with any reasonable degree of accuracy.

Exercises

3.2.6. Graph the 2π–periodic extension of each of the following functions. Which extensions are continuous? Differentiable? (a) x^2, (b) $(x^2 - \pi^2)^2$, (c) e^x, (d) $e^{-|x|}$, (e) $\sinh x$, (f) $1 + \cos^2 x$; (g) $\sin \frac{1}{2}\pi x$, (h) $\dfrac{1}{x}$, (i) $\dfrac{1}{1+x^2}$.

3.2.7. Sketch a graph of the 2π–periodic extension of each of the functions in Exercise 3.2.2.

♢ 3.2.8. Complete the proof of Lemma 3.4 by showing that $\widetilde{f}(x)$ is 2π periodic.

♢ 3.2.9. Suppose $f(x)$ is periodic with period ℓ and integrable. Prove that, for any a,

(a) $\displaystyle\int_a^{a+\ell} f(x)\,dx = \int_0^{\ell} f(x)\,dx$, (b) $\displaystyle\int_0^{\ell} f(x+a)\,dx = \int_0^{\ell} f(x)\,dx$.

♡ 3.2.10. Let $f(x)$ be a sufficiently nice 2π–periodic function. (a) Prove that $f'(x)$ is 2π–periodic. (b) Show that if $f(x)$ has *mean zero*, so $\displaystyle\int_{-\pi}^{\pi} f(x)\,dx = 0$, then $g(x) = \displaystyle\int_0^x f(y)\,dy$ is 2π–periodic; (c) Does the result in part (b) rely on the fact that the lower limit in the integral for $g(x)$ is 0? (d) More generally, prove that if $f(x)$ has mean $m = \dfrac{1}{2\pi}\displaystyle\int_{-\pi}^{\pi} f(x)\,dx$, then the function $g(x) = \displaystyle\int_0^x f(y)\,dy - mx$ is 2π–periodic.

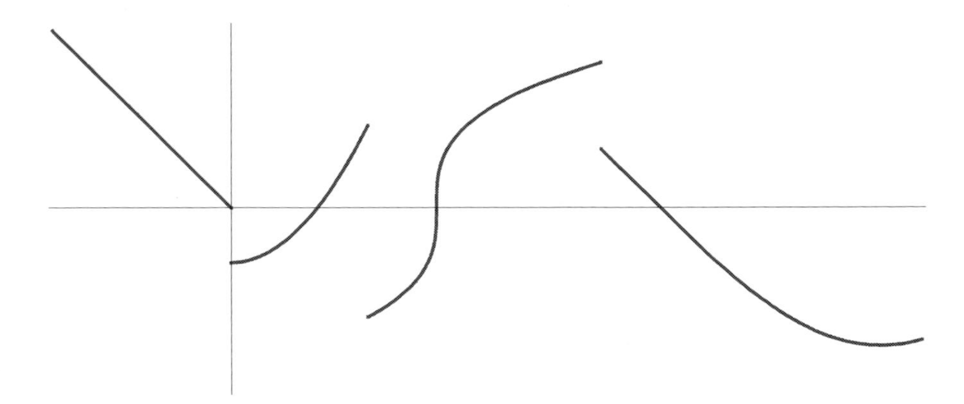

Figure 3.2. Piecewise continuous function.

♢ 3.2.11. Given a function $f(x)$ defined for $0 \leq x < \ell$, prove that there is a unique periodic function of period ℓ that agrees with f on the interval $[0, \ell)$. If $\ell = 2\pi$, is this the same periodic extension as we constructed in the text? Explain your answer. Try the case $f(x) = x$ as an illustrative example.

3.2.12. Use the method in Exercise 3.2.11 to construct and graph the 1–periodic extensions of the following functions: (*a*) x^2, (*b*) e^{-x}, (*c*) $\cos \pi x$, (*d*) $\begin{cases} 1, & |x| < \frac{1}{2}\pi, \\ 0, & \text{otherwise.} \end{cases}$

♠ 3.2.13. (*a*) How many terms in Gregory's series (3.43) are required to compute the first two decimal digits of π? (*b*) The first 10 decimal digits? *Hint*: Use the fact that it is an alternating series. (*c*) For part (*a*), try summing up the required number of terms on your computer, and check whether you obtain an accurate result.

Piecewise Continuous Functions

As we shall see, all continuously differentiable 2π–periodic functions can be represented as convergent Fourier series. More generally, we can allow functions that have simple discontinuities.

Definition 3.6. A function $f(x)$ is said to be *piecewise continuous* on an interval $[a, b]$ if it is defined and continuous except possibly at a finite number of points $a \leq x_1 < x_2 < \cdots < x_n \leq b$. Furthermore, at each point of discontinuity, we require that the left- and right-hand limits

$$f(x_k^-) = \lim_{x \to x_k^-} f(x), \qquad f(x_k^+) = \lim_{x \to x_k^+} f(x), \tag{3.44}$$

exist. (At the endpoints a, b, existence of only one of the limits, namely $f(a^+)$ and $f(b^-)$ is required.) Note that we do not require that $f(x)$ be defined at x_k. Even if $f(x_k)$ is defined, it does not necessarily equal either the left- or the right-hand limit.

A representative graph of a piecewise continuous function appears in Figure 3.2. The

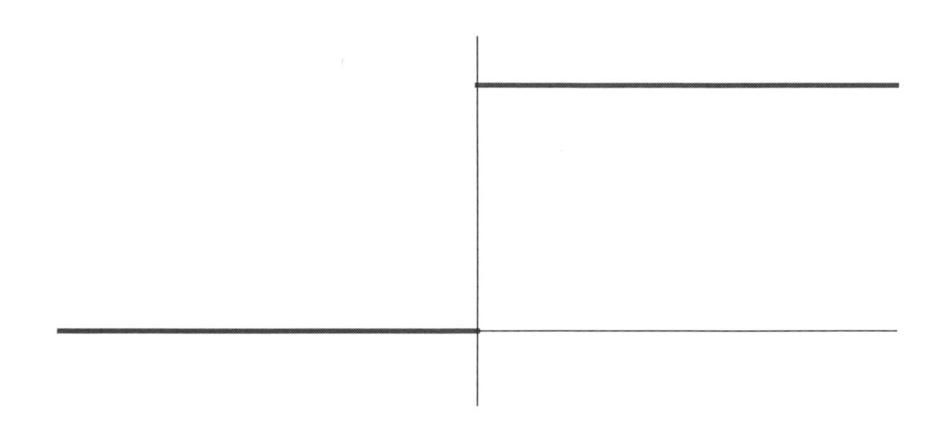

Figure 3.3. The unit step function.

points x_k are known as *jump discontinuities* of $f(x)$, and the difference

$$\beta_k = f(x_k^+) - f(x_k^-) = \lim_{x \to x_k^+} f(x) - \lim_{x \to x_k^-} f(x) \tag{3.45}$$

between the left- and right-hand limits is the *magnitude* of the jump. Note the value of the function at the discontinuity, namely $f(x_k)$ — which may not even be defined — plays no role in the specification of the jump magnitude. The jump magnitude is positive if the function jumps up (when moving from left to right) at x_k and negative if it jumps down. If the jump magnitude vanishes, $\beta_k = 0$, the left- and right-hand limits agree, and the discontinuity is *removable*, since redefining $f(x_k) = f(x_k^+) = f(x_k^-)$ makes $f(x)$ continuous at $x = x_k$. Since removable discontinuities have no effect in either the theory or applications, they can always be removed without penalty.

The simplest example of a piecewise continuous function is the *unit step function*

$$\sigma(x) = \begin{cases} 1, & x > 0, \\ 0, & x < 0, \end{cases} \tag{3.46}$$

graphed in Figure 3.3. It has a single jump discontinuity at $x = 0$ of magnitude 1:

$$\sigma(0^+) - \sigma(0^-) = 1 - 0 = 1,$$

and is continuous — indeed, locally constant — everywhere else. If we translate and scale the step function, we obtain a function

$$h(x) = \beta\,\sigma(x - \xi) = \begin{cases} \beta, & x > \xi, \\ 0, & x < \xi, \end{cases} \tag{3.47}$$

with a single jump discontinuity of magnitude β at the point $x = \xi$.

If $f(x)$ is any piecewise continuous function on $[-\pi, \pi]$, then its Fourier coefficients are well defined — the integrals (3.35) exist and are finite. Continuity, however, is not enough to ensure convergence of the associated Fourier series.

Definition 3.7. A function $f(x)$ is called *piecewise* C^1 on an interval $[a, b]$ if it is defined, continuous, and continuously differentiable except at a finite number of points

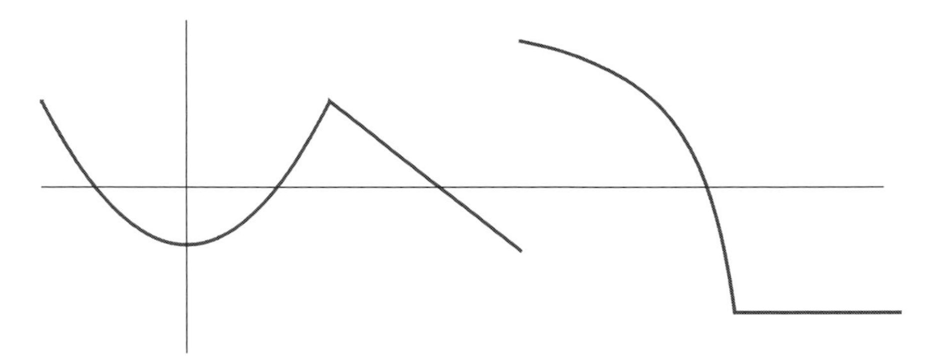

Figure 3.4. Piecewise C^1 function.

$a \le x_1 < x_2 < \cdots < x_n \le b$. At each exceptional point, the left- and right-hand limits† of both the function and its derivative exist:

$$f(x_k^-) = \lim_{x \to x_k^-} f(x), \qquad f(x_k^+) = \lim_{x \to x_k^+} f(x),$$
$$f'(x_k^-) = \lim_{x \to x_k^-} f'(x), \qquad f'(x_k^+) = \lim_{x \to x_k^+} f'(x).$$

See Figure 3.4 for a representative graph. For a piecewise C^1 function, an exceptional point x_k is either

- a *jump discontinuity* where the left- and right-hand derivatives exist, or
- a *corner*, meaning a point where f is continuous, so $f(x_k^-) = f(x_k^+)$, but has different left- and right-hand derivatives: $f'(x_k^-) \ne f'(x_k^+)$.

Thus, at each point, including jump discontinuities, the graph of $f(x)$ has well-defined right and left tangent lines. For example, the function $f(x) = |\,x\,|$ is piecewise C^1, since it is continuous everywhere and has a corner at $x = 0$, with $f'(0^+) = +1$, $f'(0^-) = -1$.

There is an analogous definition of piecewise C^n functions. One requires that the function have n continuous derivatives, except at a finite number of points. Moreover, at every point, the function must have well defined left- and right-hand limits of all its derivatives up to order n.

Finally, a function $f(x)$ defined for all $x \in \mathbb{R}$ is piecewise continuous (or C^1 or C^n) provided it is piecewise continuous (or C^1 or C^n) on any bounded interval. Thus, a piecewise continuous function on $\mathbb{R}$ can have an infinite number of discontinuities, but they are not allowed to accumulate at any finite limit point. In particular, a 2π-periodic function $\widetilde{f}(x)$ is piecewise continuous if and only if it is piecewise continuous on the interval $[-\pi, \pi]$.

Exercises

3.2.14. Find the discontinuities and the jump magnitudes for the following piecewise continuous functions:

† As before, at the endpoints we require only the appropriate one-sided limits, namely $f(a^+)$, $f'(a^+)$, and $f(b^-)$, $f'(b^-)$, to exist.

(a) $2\sigma(x) + \sigma(x+1) - 3\sigma(x-1)$, (b) $\operatorname{sign}(x^2 - 2x)$, (c) $\sigma(x^2 - 2x)$, (d) $|\,x^2 - 2x\,|$, (e) $\sqrt{|\,x-2\,|}$, (f) $\sigma(\sin x)$, (g) $\operatorname{sign}(\sin x)$, (h) $|\sin x\,|$, (i) $e^{\sigma(x)}$, (j) $\sigma(e^x)$, (k) $e^{|\,x-2\,|}$.

3.2.15. Graph the following piecewise continuous functions. List all discontinuities and jump magnitudes.

(a) $\begin{cases} e^x, & 1 < |\,x\,| < 2, \\ 0, & \text{otherwise}, \end{cases}$ (b) $\begin{cases} \sin x, & 0 < x < \frac{1}{2}\pi, \\ 0, & \text{otherwise}, \end{cases}$ (c) $\begin{cases} \dfrac{\sin x}{x}, & 0 < |\,x\,| < 2\pi, \\ 1, & x = 0, \\ 0, & \text{otherwise}, \end{cases}$

(d) $\begin{cases} x & |\,x\,| \le 1, \\ x^2, & |\,x\,| > 1, \end{cases}$ (e) $\begin{cases} x, & -1 < x < 0, \\ \sin x, & 0 < x < \pi, \\ 0, & \text{otherwise}, \end{cases}$ (f) $\begin{cases} -\dfrac{1}{x}, & |\,x\,| \ge 1, \\ \dfrac{2}{1+x^2}, & |\,x\,| < 1. \end{cases}$

3.2.16. Are the functions in Exercises 3.2.14 and 3.2.15 piecewise C^1? If so, list all corners.

3.2.17. Prove that the nth order *ramp function* $\rho_n(x-\xi) = \begin{cases} \dfrac{(x-\xi)^n}{n!}, & x > \xi, \\ 0, & x < \xi, \end{cases}$ is piecewise C^k for any $k \ge 0$.

3.2.18. Is $x^{1/3}$ piecewise continuous? piecewise C^1? piecewise C^2?

3.2.19. Answer Exercise 3.2.18 for

(a) $\sqrt{|\,x\,|}$, (b) $\dfrac{1}{x}$, (c) $e^{-1/|\,x\,|}$, (d) $x^3 \sin \dfrac{1}{x}$, (e) $|\,x\,|^3$, (f) $|\,x\,|^{3/2}$.

3.2.20. (a) Give an example of a function that is continuous but not piecewise C^1. (b) Give an example that is piecewise C^1 but not piecewise C^2.

3.2.21. (a) Prove that the sum $f + g$ of two piecewise continuous functions is piecewise continuous. (b) Where are the jump discontinuities of $f + g$? What are the jump magnitudes? (c) Check your result by summing the functions in parts (a) and (b) of Exercise 3.2.14.

3.2.22. Give an example of two piecewise continuous (but not continuous) functions f, g whose sum $f + g$ is continuous. Can you characterize all such pairs of functions?

◇ 3.2.23. (a) Prove that if $f(x)$ is piecewise continuous on $[-\pi, \pi]$, then its 2π–periodic extension is piecewise continuous on all of $\mathbb{R}$. Where are its jump discontinuities and what are their magnitudes? (b) Similarly, prove that if $f(x)$ is piecewise C^1, then its periodic extension is piecewise C^1. Where are the corners?

3.2.24. *True or false*: (a) If $f(x)$ is a piecewise continuous function, its absolute value $|\,f(x)\,|$ is piecewise continuous. If true, what are the jumps and their magnitudes?
(b) If $f(x)$ is piecewise C^1, then $|\,f(x)\,|$ is piecewise C^1. If true, what are the corners?

The Convergence Theorem

We are now able to state the fundamental convergence theorem for Fourier series. But we will postpone a discussion of its proof until the end of Section 3.5.

Theorem 3.8. *If $\widetilde{f}(x)$ is a 2π–periodic, piecewise C^1 function, then, at any $x \in \mathbb{R}$, its Fourier series converges to*

$$\begin{aligned} &\widetilde{f}(x), && \text{if } \widetilde{f} \text{ is continuous at } x, \\ &\tfrac{1}{2}\big[\,\widetilde{f}(x^+) + \widetilde{f}(x^-)\,\big], && \text{if } x \text{ is a jump discontinuity.} \end{aligned}$$

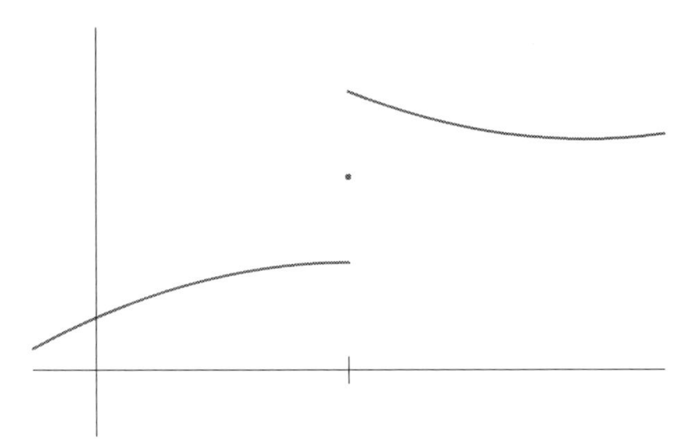

Figure 3.5. Splitting the difference.

Thus, the Fourier series converges, as expected, to $\widetilde{f}(x)$ at all points of continuity. At discontinuities, it apparently can't decide whether to converge to the left- or right-hand limit, and so ends up "splitting the difference" by converging to their average; see Figure 3.5. If we redefine $\widetilde{f}(x)$ at its jump discontinuities to have the average limiting value, so

$$\widetilde{f}(x) = \tfrac{1}{2}\left[\, \widetilde{f}(x^+) + \widetilde{f}(x^-) \,\right] \tag{3.48}$$

— an equation that automatically holds at all points of continuity — then Theorem 3.8 would say that the Fourier series converges to the 2π–periodic piecewise C^1 function $\widetilde{f}(x)$ everywhere.

Example 3.9. Let $\sigma(x)$ denote the unit step function (3.46). Its Fourier coefficients are easily computed:

$$\begin{aligned}
a_0 &= \frac{1}{\pi}\int_{-\pi}^{\pi} \sigma(x)\,dx = \frac{1}{\pi}\int_0^{\pi} dx = 1, \\
a_k &= \frac{1}{\pi}\int_{-\pi}^{\pi} \sigma(x)\cos kx\,dx = \frac{1}{\pi}\int_0^{\pi} \cos kx\,dx = 0, \\
b_k &= \frac{1}{\pi}\int_{-\pi}^{\pi} \sigma(x)\sin kx\,dx = \frac{1}{\pi}\int_0^{\pi} \sin kx\,dx = \begin{cases} \dfrac{2}{k\pi}, & k = 2l+1 \text{ odd}, \\ 0, & k = 2l \text{ even}. \end{cases}
\end{aligned}$$

Therefore, the Fourier series for the step function is

$$\sigma(x) \sim \frac{1}{2} + \frac{2}{\pi}\left(\sin x + \frac{\sin 3x}{3} + \frac{\sin 5x}{5} + \frac{\sin 7x}{7} + \cdots \right). \tag{3.49}$$

According to Theorem 3.8, the Fourier series will converge to its 2π–periodic extension,

$$\widetilde{\sigma}(x) = \begin{cases} 0, & (2m-1)\pi < x < 2m\pi, \\ 1, & 2m\pi < x < (2m+1)\pi, \\ \tfrac{1}{2}, & x = m\pi, \end{cases} \qquad \text{where } m \text{ is any integer,}$$

which is plotted in Figure 3.6. Observe that, in accordance with Theorem 3.8, $\widetilde{\sigma}(x)$ takes the midpoint value $\frac{1}{2}$ at the jump discontinuities $0, \pm\pi, \pm 2\pi, \ldots$.

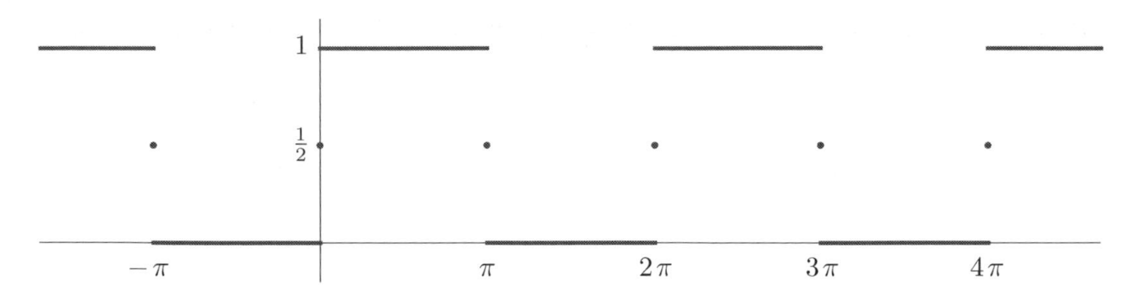

Figure 3.6. 2π–periodic step function.

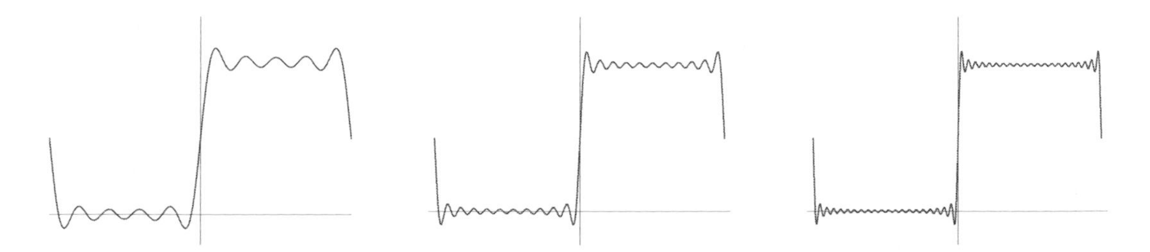

Figure 3.7. Gibbs phenomenon.

It is instructive to investigate the convergence of this particular Fourier series in some detail. Figure 3.7 displays a graph of the first few partial sums, taking, respectively, $n = 4, 10$, and 20 terms. The reader will notice that away from the discontinuities, the series indeed appears to be converging, albeit slowly. However, near the jumps there is a consistent overshoot of about 9% of the jump magnitude. The region where the overshoot occurs becomes narrower and narrower as the number of terms increases, but the actual amount of overshoot persists no matter how many terms are summed up. This was first noted by the American physicist Josiah Gibbs, and is now known as the *Gibbs phenomenon* in his honor. The Gibbs overshoot is a manifestation of the subtle nonuniform convergence of the Fourier series.

Exercises

3.2.25. (*a*) Sketch the 2π–periodic half-wave $f(x) = \begin{cases} \sin x, & 0 < x \le \pi, \\ 0, & -\pi \le x < 0. \end{cases}$ (*b*) Find its Fourier series. (*c*) Graph the first five Fourier sums and compare with the function. (*d*) Discuss convergence of the Fourier series.

3.2.26. Answer Exercise 3.2.25 for the cosine half-wave $f(x) = \begin{cases} \cos x, & 0 < x \le \pi, \\ 0, & -\pi \le x < 0. \end{cases}$

3.2.27. (*a*) Find the Fourier series for $f(x) = e^x$. (*b*) For which values of x does the Fourier series converge? (*c*) Graph the function it converges to.

♠ 3.2.28. (*a*) Use a graphing package to investigate the Gibbs phenomenon for the Fourier series (3.37) of the function x. Determine the amount of overshoot of the partial sums at the discontinuities. (*b*) How many terms do you need to approximate the function to within two decimal places at $x = 2.0$? At $x = 3.0$?

3.2.29. Use the Fourier series (3.49) for the step function to rederive Gregory's series (3.43).

♢ 3.2.30. Suppose a_k, b_k are the Fourier coefficients of the function $f(x)$. (a) To which function does the Fourier series $\dfrac{a_0}{2} + \sum_{k=1}^{\infty} \left[a_k \cos 2kx + b_k \sin 2kx \right]$ converge? *Hint*: The answer is *not* $f(2x)$. (b) Test your answer with the Fourier series (3.37) for $f(x) = x$.

Even and Odd Functions

We already noted that the Fourier cosine coefficients of the function $f(x) = x$ are all 0. This is not an accident, but, rather, a consequence of the fact that x is an odd function. Recall first the basic definition:

Definition 3.10. A function is called *even* if $f(-x) = f(x)$. A function is called *odd* if $f(-x) = -f(x)$.

For example, the functions 1, $\cos kx$, and x^2 are all even, whereas x, $\sin kx$, and $\operatorname{sign} x$ are odd. Note that an odd function necessarily has $f(0) = 0$. We require three elementary lemmas, whose proofs are left to the reader.

Lemma 3.11. *The sum, $f(x) + g(x)$, of two even functions is even; the sum of two odd functions is odd.*

Remark: *Every* function can be represented as the sum of an even and an odd function; see Exercise 3.2.32.

Lemma 3.12. *The product $f(x)\,g(x)$ of two even functions, or of two odd functions, is an even function. The product of an even and an odd function is odd.*

Lemma 3.13. *If $f(x)$ is odd and integrable on the symmetric interval $[-a, a]$, then $\int_{-a}^{a} f(x)\,dx = 0$. If $f(x)$ is even and integrable, then $\int_{-a}^{a} f(x)\,dx = 2\int_{0}^{a} f(x)\,dx$.*

The next result is an immediate consequence of applying Lemmas 3.12 and 3.13 to the Fourier integrals (3.35).

Proposition 3.14. *If $f(x)$ is even, then its Fourier sine coefficients all vanish, $b_k = 0$, and so $f(x)$ can be represented by a* Fourier cosine series

$$f(x) \sim \frac{a_0}{2} + \sum_{k=1}^{\infty} a_k \cos kx\,, \tag{3.50}$$

where

$$a_k = \frac{2}{\pi}\int_0^{\pi} f(x) \cos kx\,dx, \qquad k = 0, 1, 2, 3, \ldots\,. \tag{3.51}$$

If $f(x)$ is odd, then its Fourier cosine coefficients vanish, $a_k = 0$, and so $f(x)$ can be represented by a Fourier sine series

$$f(x) \sim \sum_{k=1}^{\infty} b_k \sin kx\,, \tag{3.52}$$

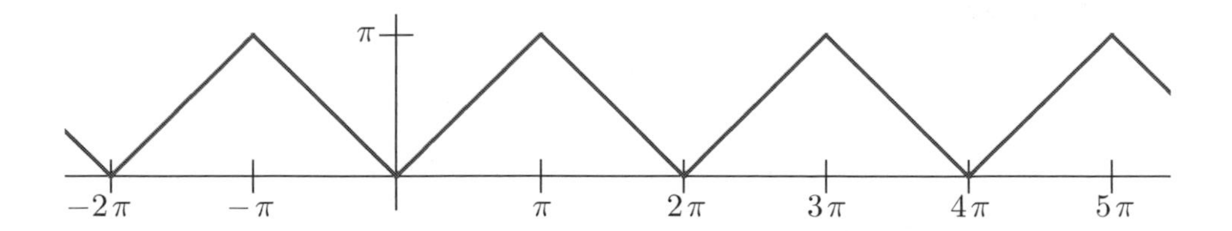

Figure 3.8. 2π–periodic extension of $|x|$.

where

$$b_k = \frac{2}{\pi}\int_0^{\pi} f(x)\sin k\,x\,dx, \qquad k = 1,2,3,\ldots. \tag{3.53}$$

Conversely, a convergent Fourier cosine series always represents an even function, while a convergent sine series always represents an odd function.

Example 3.15. The absolute value $f(x) = |x|$ is an even function, and hence has a Fourier cosine series. The coefficients are

$$a_0 = \frac{2}{\pi}\int_0^{\pi} x\,dx = \pi, \tag{3.54}$$

$$a_k = \frac{2}{\pi}\int_0^{\pi} x\cos k\,x\,dx = \frac{2}{\pi}\left[\frac{x\sin k\,x}{k} + \frac{\cos k\,x}{k^2}\right]_{x=0}^{\pi} = \begin{cases} 0, & 0 \neq k \text{ even}, \\ -\dfrac{4}{k^2\pi}, & k \text{ odd}. \end{cases}$$

Therefore

$$|x| \sim \frac{\pi}{2} - \frac{4}{\pi}\left(\cos x + \frac{\cos 3\,x}{9} + \frac{\cos 5\,x}{25} + \frac{\cos 7\,x}{49} + \cdots\right). \tag{3.55}$$

According to Theorem 3.8, this Fourier cosine series converges to the 2π–periodic extension of $|x|$, the "sawtooth function" graphed in Figure 3.8.

In particular, if we substitute $x = 0$, we obtain another interesting series:

$$\frac{\pi^2}{8} = 1 + \frac{1}{9} + \frac{1}{25} + \frac{1}{49} + \cdots = \sum_{j=0}^{\infty}\frac{1}{(2\,j+1)^2}. \tag{3.56}$$

It converges faster than Gregory's series (3.43), and, while far from optimal in this regard, can be used to compute reasonable approximations to π. One can further manipulate this result to compute the sum of the series

$$S = \sum_{k=1}^{\infty}\frac{1}{k^2} = 1 + \frac{1}{4} + \frac{1}{9} + \frac{1}{16} + \frac{1}{25} + \frac{1}{36} + \frac{1}{49} + \cdots.$$

We note that

$$\frac{S}{4} = \sum_{k=1}^{\infty}\frac{1}{4\,k^2} = \sum_{k=1}^{\infty}\frac{1}{(2\,k)^2} = \frac{1}{4} + \frac{1}{16} + \frac{1}{36} + \frac{1}{64} + \cdots.$$

Therefore, by (3.56),

$$\frac{3}{4}S = S - \frac{S}{4} = 1 + \frac{1}{9} + \frac{1}{25} + \frac{1}{49} + \cdots = \frac{\pi^2}{8},$$

from which we conclude that

$$S = \sum_{k=1}^{\infty} \frac{1}{k^2} = 1 + \frac{1}{4} + \frac{1}{9} + \frac{1}{16} + \frac{1}{25} + \cdots = \frac{\pi^2}{6}. \tag{3.57}$$

Remark: The most famous function in number theory — and the source of the most outstanding problem in mathematics, the *Riemann hypothesis* — is the *Riemann zeta function*

$$\zeta(s) = \sum_{k=1}^{\infty} \frac{1}{k^s}. \tag{3.58}$$

Formula (3.57) shows that $\zeta(2) = \frac{1}{6}\pi^2$. In fact, the value of the zeta function at any *even* positive integer $s = 2j$ is a rational polynomial in π, [**9**]. Because of its importance to the study of prime numbers, locating all the complex zeros of the zeta function will earn you \$1,000,000 — see `http://www.claymath.org` for details.

Any function $f(x)$ defined on $[0, \pi]$ has a unique even extension to $[-\pi, \pi]$, obtained by setting $f(-x) = f(x)$ for $-\pi \le x < 0$, and also a unique odd extension, where now $f(-x) = -f(x)$ and $f(0) = 0$. These in turn can be periodically extended to the entire real line. The *Fourier cosine series* of $f(x)$ is defined by the formulas (3.50–51), and represents the even, 2π–periodic extension. Similarly, the formulas (3.52–53) define the *Fourier sine series* of $f(x)$, representing its odd, 2π–periodic extension.

Example 3.16. Suppose $f(x) = \sin x$. Its Fourier cosine series has coefficients

$$a_k = \frac{2}{\pi}\int_0^{\pi} \sin x \cos kx \, dx = \begin{cases} \dfrac{4}{(1-k^2)\pi}, & k \text{ even}, \\ 0, & k \text{ odd}. \end{cases}$$

The resulting cosine series represents the even, 2π–periodic extension of $\sin x$, namely

$$|\sin x| \sim \frac{2}{\pi} - \frac{4}{\pi}\sum_{j=1}^{\infty} \frac{\cos 2jx}{4j^2 - 1}.$$

On the other hand, $f(x) = \sin x$ is already odd, and so its Fourier sine series coincides with its ordinary Fourier series, namely $\sin x$, all the other Fourier sine coefficients being zero; in other words, $b_1 = 1$, while $b_k = 0$ for $k > 1$.

Exercises

3.2.31. Are the following functions even, odd, or neither?
(a) x^2, (b) e^x, (c) $\sinh x$, (d) $\sin \pi x$, (e) $\dfrac{1}{x}$, (f) $\dfrac{1}{1+x^2}$, (g) $\tan^{-1} x$.

♢ 3.2.32. Prove that (a) the sum of two even functions is even; (b) the sum of two odd functions is odd; (c) *every* function is the sum of an even and an odd function.

♢ 3.2.33. Prove (a) Lemma 3.12; (b) Lemma 3.13.

3.2.34. If $f(x)$ is odd, is $f'(x)$ (*i*) even? (*ii*) odd? (*iii*) neither? (*iv*) could be either?

3.2.35. If $f'(x)$ is even, is $f(x)$ (*i*) even? (*ii*) odd? (*iii*) neither? (*iv*) could be either? How do you reconcile your answer with Exercise 3.2.34?

3.2.36. Answer Exercise 3.2.34 for $f''(x)$.

3.2.37. *True or false*: (*a*) If $f(x)$ is odd, its 2π–periodic extension is odd. (*b*) If the 2π–periodic extension of $f(x)$ is odd, then $f(x)$ is odd.

3.2.38. Let $\widetilde{f}(x)$ denote the odd, 2π–periodic Fourier extension of a function $f(x)$ defined on $[0,\pi]$. Explain why $\widetilde{f}(k\pi) = 0$ for any integer k.

3.2.39. Construct and graph the even and odd 2π–periodic extensions of the function $f(x) = 1 - x$. What are their Fourier series? Discuss convergence of each.

3.2.40. Find the Fourier series and discuss convergence for: (*a*) the box function $b(x) = \begin{cases} 1, & |x| < \frac{1}{2}\pi, \\ 0, & \frac{1}{2}\pi < |x| < \pi, \end{cases}$ (*b*) the hat function $h(x) = \begin{cases} 1 - |x|, & |x| < 1, \\ 0, & 1 < |x| < \pi. \end{cases}$

3.2.41. Find the Fourier sine and cosine series of the following functions. Then graph the function to which the series converges. (*a*) 1, (*b*) $\cos x$, (*c*) $\sin^3 x$, (*d*) $x(\pi - x)$.

3.2.42. Find the Fourier series of the hyperbolic functions $\cosh m x$ and $\sinh m x$.

3.2.43. Use the Fourier cosine series of the function $|\sin x|$ constructed in Example 3.16 to evaluate the sums $\sum_{k=1}^{\infty} (4k^2 - 1)^{-1}$ and $\sum_{k=1}^{\infty} (-1)^{k-1}(4k^2 - 1)^{-1}$.

3.2.44. *True or false*: The sum of the Fourier cosine series and the Fourier sine series of the function $f(x)$ is the Fourier series for $f(x)$. If false, what function is represented by the combined Fourier series?

3.2.45. (*a*) Show that if a function is periodic of period π, then its Fourier series contains only even terms, i.e., $a_k = b_k = 0$ whenever $k = 2j + 1$ is odd. (*b*) What if the period is $\frac{1}{2}\pi$?

3.2.46. Under what conditions on $f(x)$ does its Fourier sine series contain only even terms, i.e., its Fourier sine coefficients $b_k = 0$ whenever k is odd?

♠ 3.2.47. Graph the partial sums $s_3(x), s_5(x), s_{10}(x)$ of the Fourier series (3.55). Do you notice a Gibbs phenomenon? If so, what is the amount of overshoot? If not, explain why.

3.2.48. Explain why, in the case of the step function $\sigma(x)$, all its Fourier cosine coefficients vanish, $a_k = 0$, except for $a_0 = 1$.

♠ 3.2.49. How many terms do you need to sum in (3.56) to correctly approximate π to two decimal digits? To ten digits?

3.2.50. Prove that $\displaystyle \sum_{k=1}^{\infty} \frac{(-1)^{k-1}}{k^2} = 1 - \frac{1}{4} + \frac{1}{9} - \frac{1}{16} + \frac{1}{25} - \frac{1}{36} + \frac{1}{49} - \cdots = \frac{\pi^2}{12}$.

Complex Fourier Series

An alternative, and often more convenient, approach to Fourier series is to use complex exponentials instead of sines and cosines. Indeed, *Euler's formula*

$$e^{\,\mathrm{i}\,k x} = \cos k x + \mathrm{i}\,\sin k x, \qquad e^{-\,\mathrm{i}\,k x} = \cos k x - \mathrm{i}\,\sin k x, \tag{3.59}$$

shows how to write the trigonometric functions

$$\cos kx = \frac{e^{\,i\,kx} + e^{-\,i\,kx}}{2}, \qquad \sin kx = \frac{e^{\,i\,kx} - e^{-\,i\,kx}}{2\,i}, \tag{3.60}$$

in terms of complex exponentials, and so we can easily go back and forth between the two representations.

Like their trigonometric antecedents, complex exponentials are also endowed with an underlying orthogonality. But here, since we are dealing with the vector space of complex-valued functions on the interval $[-\pi,\pi]$, we need to use the rescaled L^2 *Hermitian inner product*

$$\langle f, g \rangle = \frac{1}{2\pi}\int_{-\pi}^{\pi} f(x)\,\overline{g(x)}\,dx, \tag{3.61}$$

in which the second function acquires a complex conjugate, as indicated by the overbar. This is needed to ensure that the associated L^2 *Hermitian norm*

$$\| f \| = \sqrt{\frac{1}{2\pi}\int_{-\pi}^{\pi} |\, f(x)\,|^2\,dx} \tag{3.62}$$

is real and positive for all nonzero complex functions: $\| f \| > 0$ when $f \not\equiv 0$. Orthonormality of the complex exponentials is proved by direct computation:

$$\begin{aligned} \langle e^{\,i\,kx}, e^{\,i\,lx} \rangle &= \frac{1}{2\pi}\int_{-\pi}^{\pi} e^{\,i\,(k-l)x}\,dx = \begin{cases} 1, & k = l, \\ 0, & k \neq l, \end{cases} \\ \| e^{\,i\,kx} \|^2 &= \frac{1}{2\pi}\int_{-\pi}^{\pi} |\, e^{\,i\,kx}\,|^2\,dx = 1. \end{aligned} \tag{3.63}$$

The *complex Fourier series* for a (piecewise continuous) real or complex function f is the doubly infinite series

$$f(x) \sim \sum_{k=-\infty}^{\infty} c_k\, e^{\,i\,kx} = \cdots + c_{-2}\,e^{-2\,i\,x} + c_{-1}\,e^{-\,i\,x} + c_0 + c_1\,e^{\,i\,x} + c_2\,e^{2\,i\,x} + \cdots. \tag{3.64}$$

The orthonormality formulae (3.63) imply that the *complex Fourier coefficients* are obtained by taking the inner products

$$c_k = \langle f, e^{\,i\,kx} \rangle = \frac{1}{2\pi}\int_{-\pi}^{\pi} f(x)\,e^{-\,i\,kx}\,dx. \tag{3.65}$$

Pay particular attention to the minus sign appearing in the integrated exponential, which happens because the second argument in the Hermitian inner product (3.61) requires a complex conjugate.

It must be emphasized that the real (3.34) and complex (3.64) Fourier formulae are just two different ways of writing the *same* series! Indeed, if we substitute Euler's formula (3.59) into (3.65) and compare the result with the real Fourier formulae (3.35), we find that the real and complex Fourier coefficients are related by

$$\begin{aligned} a_k &= c_k + c_{-k}, & c_k &= \tfrac{1}{2}(a_k - i\,b_k), \\ b_k &= i\,(c_k - c_{-k}), & c_{-k} &= \tfrac{1}{2}(a_k + i\,b_k), \end{aligned} \qquad k = 0, 1, 2, \ldots. \tag{3.66}$$

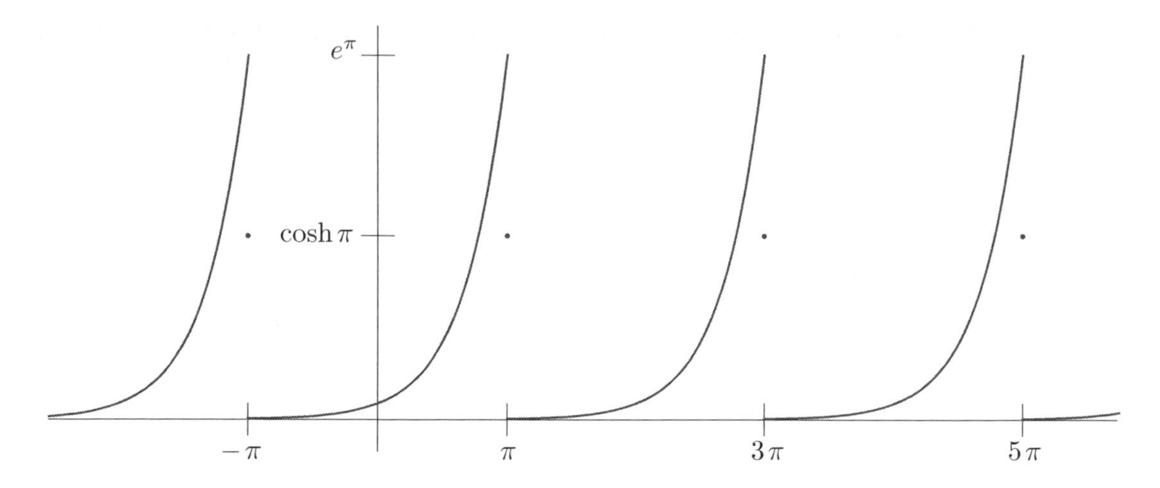

Figure 3.9. 2π–periodic extension of e^x.

Remark: We already see one advantage of the complex version. The constant function $1 = e^{0\,\mathrm{i}\,x}$ no longer plays an anomalous role — the annoying factor of $\frac{1}{2}$ in the real Fourier series (3.34) has mysteriously disappeared!

Example 3.17. For the unit step function $\sigma(x)$ considered in Example 3.9, the complex Fourier coefficients are

$$c_k = \frac{1}{2\pi}\int_{-\pi}^{\pi} \sigma(x)\, e^{-\mathrm{i}\,k\,x}\, dx = \frac{1}{2\pi}\int_{0}^{\pi} e^{-\mathrm{i}\,k\,x}\, dx = \begin{cases} \frac{1}{2}, & k = 0, \\ 0, & 0 \neq k \text{ even}, \\ \dfrac{1}{\mathrm{i}\,k\,\pi}, & k \text{ odd}. \end{cases}$$

Therefore, the step function has the complex Fourier series

$$\sigma(x) \sim \frac{1}{2} - \frac{\mathrm{i}}{\pi}\sum_{l=-\infty}^{\infty} \frac{e^{(2l+1)\,\mathrm{i}\,x}}{2l+1}\,. \tag{3.67}$$

You should convince yourself that this is *exactly the same series* as the real Fourier series (3.49). We are merely rewriting it using complex exponentials instead of real sines and cosines.

Example 3.18. Let us find the Fourier series for the exponential function $e^{a\,x}$. It is much easier to evaluate the integrals for the complex Fourier coefficients, and so

$$\begin{aligned} c_k = \langle\, e^{a\,x}, e^{\mathrm{i}\,k\,x}\,\rangle &= \frac{1}{2\pi}\int_{-\pi}^{\pi} e^{(a-\mathrm{i}\,k)\,x}\, dx = \left.\frac{e^{(a-\mathrm{i}\,k)\,x}}{2\pi(a-\mathrm{i}\,k)}\right|_{x=-\pi}^{\pi} \\ &= \frac{e^{(a-\mathrm{i}\,k)\,\pi} - e^{-(a-\mathrm{i}\,k)\,\pi}}{2\pi(a-\mathrm{i}\,k)} = (-1)^k\,\frac{e^{a\,\pi} - e^{-a\,\pi}}{2\pi(a-\mathrm{i}\,k)} = \frac{(-1)^k(a+\mathrm{i}\,k)\sinh a\,\pi}{\pi(a^2+k^2)}\,. \end{aligned}$$

Therefore, the desired Fourier series is

$$e^{a\,x} \sim \frac{\sinh a\,\pi}{\pi}\sum_{k=-\infty}^{\infty} \frac{(-1)^k(a+\mathrm{i}\,k)}{a^2+k^2}\, e^{\mathrm{i}\,k\,x}. \tag{3.68}$$

As an exercise, the reader should try writing this as a real Fourier series, either by breaking up the complex series into its real and imaginary parts, or by direct evaluation of the real coefficients via their integral formulae (3.35). According to Theorem 3.8 (which is equally valid for complex Fourier series), the Fourier series converges to the 2π–periodic extension of the exponential function, as graphed in Figure 3.9. In particular, its values at odd multiples of π is the average of the limiting values there, namely $\cosh a\pi = \frac{1}{2}(e^{a\pi} + e^{-a\pi})$.

Exercises

3.2.51. Find the complex Fourier series of the following functions: (a) $\sin x$, (b) $\sin^3 x$, (c) x, (d) $|x|$, (e) $|\sin x|$, (f) $\operatorname{sign} x$, (g) the ramp function $\rho(x) = \begin{cases} x, & x \geq 0, \\ 0, & x \leq 0. \end{cases}$

3.2.52. Let $-\pi < \xi < \pi$. Determine the complex Fourier series for the shifted step function $\sigma(x - \xi)$, and graph the function it converges to.

3.2.53. Let $a \in \mathbb{R}$. Find the real form of the Fourier series for the exponential function e^{ax}:
(a) by breaking up the complex series (3.68) into its real and imaginary parts;
(b) by direct evaluation of the real coefficients via their integral formulae (3.35).
Make sure that your results agree!

3.2.54. Prove that $\coth \pi = \dfrac{1}{\pi} + \dfrac{2}{\pi}\left(\dfrac{1}{1+1^2} + \dfrac{1}{1+2^2} + \dfrac{1}{1+3^2} + \cdots \right)$, where

$\coth x = \dfrac{\cosh x}{\sinh x} = \dfrac{e^x + e^{-x}}{e^x - e^{-x}}$ is the hyperbolic cotangent function.

3.2.55. (a) Find the complex Fourier series for $x\,e^{\,i x}$.
(b) Use your result to write down the real Fourier series for $x \cos x$ and $x \sin x$.

♦ 3.2.56. Prove that if $f(x) = \sum_{k=m}^{n} r_k\, e^{\,i k x}$ is a complex trigonometric polynomial, with $-\infty < m \leq n < \infty$, then its Fourier coefficients are $c_k = \begin{cases} r_k, & m \leq k \leq n, \\ 0, & \text{otherwise.} \end{cases}$

3.2.57. *True or false*: If the complex function $f(x) = g(x) + i\,h(x)$ has Fourier coefficients c_k, then $g(x) = \operatorname{Re} f(x)$ and $h(x) = \operatorname{Im} f(x)$ have, respectively, complex Fourier coefficients $\operatorname{Re} c_k$ and $\operatorname{Im} c_k$.

♦ 3.2.58. Let $f(x)$ be 2π–periodic. Explain how to construct the complex Fourier series for $f(x - a)$ from that of $f(x)$.

♦ 3.2.59. (a) Show that if c_k are the complex Fourier coefficients for $f(x)$, then the Fourier coefficients of $\widetilde{f}(x) = f(x)\,e^{\,i x}$ are $\widetilde{c}_k = c_{k-1}$. (b) Let m be an integer. Which function has complex Fourier coefficients $\widehat{c}_k = c_{k+m}$? (c) If a_k, b_k are the Fourier coefficients of the real function $f(x)$, what are the Fourier coefficients of $f(x)\cos x$ and $f(x)\sin x$?

♦ 3.2.60. Can you recognize whether a function is real by looking at its complex Fourier coefficients?

♦ 3.2.61. Can you characterize the complex Fourier coefficients of an even function? an odd function?

♦ 3.2.62. What does it mean for a doubly infinite series $\sum_{k=-\infty}^{\infty} c_k$ to converge? Be precise!

3.3 Differentiation and Integration

Under appropriate hypotheses, if a series of functions converges, then one will be able to integrate or differentiate it term by term, and the resulting series should converge to the integral or derivative of the original sum. For example, integration and differentiation of power series is always valid within the range of convergence, and is used extensively in the construction of series solutions of differential equations, series for integrals of non-elementary functions, and so on. (See Section 11.3 for further details.) The convergence of Fourier series is considerably more delicate, and so one must exercise due care when differentiating or integrating. Nevertheless, in favorable situations, both operations lead to valid results, and are quite useful for constructing Fourier series of more intricate functions.

Integration of Fourier Series

Integration is a smoothing operation — the integrated function is always nicer than the original. Therefore, we should anticipate being able to integrate Fourier series without difficulty. There is, however, one complication: the integral of a periodic function is not necessarily periodic. The simplest example is the constant function 1, which is certainly periodic, but its integral, namely x, is not. On the other hand, integrals of all the other periodic sine and cosine functions appearing in the Fourier series are periodic. Thus, only the constant term

$$\frac{a_0}{2} = \frac{1}{2\pi}\int_{-\pi}^{\pi} f(x)\,dx \tag{3.69}$$

might cause us difficulty when we try to integrate a Fourier series (3.34). Note that (3.69) is the *mean*, or *average*, of the function $f(x)$ over the interval $[-\pi,\pi]$, and so a function has no constant term in its Fourier series, i.e., $a_0 = 0$, if and only if it has *mean zero*. It is easily shown, cf. Exercise 3.2.10, that the mean-zero functions are precisely those that remain periodic upon integration. In particular, Lemma 3.13 implies that all odd functions automatically have mean zero, and hence have periodic integrals.

Lemma 3.19. *If $f(x)$ is 2π–periodic, then its integral $g(x) = \int_0^x f(y)\,dy$ is 2π–periodic if and only if $\int_{-\pi}^{\pi} f(x)\,dx = 0$, so that f has mean zero on the interval $[-\pi,\pi]$.*

In view of the elementary integration formulae

$$\int \cos kx\,dx = \frac{\sin kx}{k}, \qquad \int \sin kx\,dx = -\frac{\cos kx}{k}, \tag{3.70}$$

termwise integration of a Fourier series without constant term is straightforward.

Theorem 3.20. *If f is piecewise continuous and has mean zero on the interval $[-\pi,\pi]$, then its Fourier series*

$$f(x) \sim \sum_{k=1}^{\infty} [\, a_k \cos kx + b_k \sin kx \,]$$

can be integrated term by term, to produce the Fourier series

$$g(x) = \int_0^x f(y)\,dy \sim m + \sum_{k=1}^{\infty}\left[-\frac{b_k}{k}\cos kx + \frac{a_k}{k}\sin kx \right]. \tag{3.71}$$

The constant term

$$m = \frac{1}{2\pi} \int_{-\pi}^{\pi} g(x)\, dx \tag{3.72}$$

is the mean of the integrated function.

Example 3.21. The function $f(x) = x$ is odd, and so has mean zero: $\displaystyle\int_{-\pi}^{\pi} x\, dx = 0$. Let us integrate its Fourier series

$$x \sim 2 \sum_{k=1}^{\infty} \frac{(-1)^{k-1}}{k} \sin k\,x, \tag{3.73}$$

which we found in Example 3.3. The result is the Fourier series

$$\begin{aligned} \frac{1}{2}\,x^2 &\sim \frac{\pi^2}{6} - 2 \sum_{k=1}^{\infty} \frac{(-1)^{k-1}}{k^2} \cos k\,x \\ &= \frac{\pi^2}{6} - 2\left(\cos x - \frac{\cos 2\,x}{4} + \frac{\cos 3\,x}{9} - \frac{\cos 4\,x}{16} + \cdots\right), \end{aligned} \tag{3.74}$$

whose constant term is the mean of the left-hand side:

$$\frac{1}{2\pi} \int_{-\pi}^{\pi} \frac{x^2}{2}\, dx = \frac{\pi^2}{6}.$$

Let us revisit the derivation of the integrated Fourier series from a slightly different standpoint. If we were to integrate each trigonometric summand in a Fourier series (3.34) from 0 to x, we would obtain

$$\int_0^x \cos k\,y\, dy = \frac{\sin k\,x}{k}, \qquad \text{whereas} \qquad \int_0^x \sin k\,y\, dy = \frac{1}{k} - \frac{\cos k\,x}{k}.$$

The extra $1/k$ terms coming from the definite sine integrals did not appear explicitly in our previous expression for the integrated Fourier series, (3.71), and so must be hidden in the constant term m. We deduce that the mean value of the integrated function can be computed using the Fourier sine coefficients of f via the formula

$$\frac{1}{2\pi} \int_{-\pi}^{\pi} g(x)\, dx = m = \sum_{k=1}^{\infty} \frac{b_k}{k}. \tag{3.75}$$

For example, integrating both sides of the Fourier series (3.73) for $f(x) = x$ from 0 to x produces

$$\frac{x^2}{2} \sim 2 \sum_{k=1}^{\infty} \frac{(-1)^{k-1}}{k^2} (1 - \cos k\,x).$$

The constant terms sum to yield the mean value of the integrated function:

$$2\left(1 - \frac{1}{4} + \frac{1}{9} - \frac{1}{16} + \cdots\right) = 2 \sum_{k=1}^{\infty} \frac{(-1)^{k-1}}{k^2} = \frac{1}{2\pi} \int_{-\pi}^{\pi} \frac{x^2}{2}\, dx = \frac{\pi^2}{6}, \tag{3.76}$$

which reproduces a formula established in Exercise 3.2.50.

More generally, if $f(x)$ does not have mean zero, its Fourier series contains a nonzero constant term,

$$f(x) \sim \frac{a_0}{2} + \sum_{k=1}^{\infty} \left[\, a_k \cos k\,x + b_k \sin k\,x \,\right].$$

In this case, the result of integration will be

$$g(x) = \int_0^x f(y)\,dy \sim \frac{a_0}{2}\,x + m + \sum_{k=1}^{\infty} \left[-\frac{b_k}{k} \cos k\,x + \frac{a_k}{k} \sin k\,x \right], \tag{3.77}$$

where m is given in (3.75). The right-hand side is not, strictly speaking, a Fourier series. There are two ways to interpret this formula within the Fourier framework. We can write (3.77) as the Fourier series for the difference

$$g(x) - \frac{a_0}{2}\,x \sim m + \sum_{k=1}^{\infty} \left[-\frac{b_k}{k} \cos k\,x + \frac{a_k}{k} \sin k\,x \right], \tag{3.78}$$

which, by Exercise 3.2.10(d), is a 2π–periodic function. Alternatively, we can replace x by its Fourier series (3.37), and the result will be the Fourier series for the 2π–periodic extension of the integral $g(x) = \displaystyle\int_0^x f(y)\,dy$.

Differentiation of Fourier Series

Differentiation has the opposite effect — it makes a function worse. Therefore, to justify taking the derivative of a Fourier series, we need to know that the derived function remains reasonably nice. Since we need the derivative $f'(x)$ to be piecewise C^1 for the Convergence Theorem 3.8 to be applicable, we require that $f(x)$ itself be continuous and piecewise C^2.

Theorem 3.22. *If $f(x)$ has a piecewise C^2 and continuous 2π–periodic extension, then its Fourier series can be differentiated term by term, to produce the Fourier series for its derivative*

$$f'(x) \sim \sum_{k=1}^{\infty} \left[\, k\,b_k \cos k\,x - k\,a_k \sin k\,x \,\right] = \sum_{k=-\infty}^{\infty} \mathrm{i}\,k\,c_k\,e^{\,\mathrm{i}\,k\,x}. \tag{3.79}$$

Example 3.23. The derivative (6.31) of the absolute value function $f(x) = |\,x\,|$ is the sign function:

$$\frac{d}{dx}\,|\,x\,| = \operatorname{sign} x = \begin{cases} +1, & x > 0, \\ -1, & x < 0. \end{cases} \tag{3.80}$$

Therefore, if we differentiate its Fourier series (3.55), we obtain the Fourier series

$$\operatorname{sign} x \sim \frac{4}{\pi}\left(\sin x + \frac{\sin 3\,x}{3} + \frac{\sin 5\,x}{5} + \frac{\sin 7\,x}{7} + \cdots \right). \tag{3.81}$$

Note that $\operatorname{sign} x = \sigma(x) - \sigma(-\,x)$ is the difference of two step functions. Indeed, subtracting the step function Fourier series (3.49) at x from the same series at $-\,x$ reproduces (3.81).

Exercises

3.3.1. Starting with the Fourier series (3.49) for the step function $\sigma(x)$, use integration to:
(a) Find the Fourier series for the ramp function $\rho(x) = \begin{cases} x, & x > 0, \\ 0, & x < 0. \end{cases}$
(b) Then, find the Fourier series for the second-order ramp function $\rho_2(x) = \begin{cases} \frac{1}{2}x^2, & x > 0, \\ 0, & x < 0. \end{cases}$

3.3.2. Find the Fourier series for the function $f(x) = x^3$. If you differentiate your series, do you recover the Fourier series for $f'(x) = 3x^2$? If not, explain why not.

3.3.3. Answer Exercise 3.3.2 when $f(x) = x^4$.

3.3.4. Use Theorem 3.20 to construct the Fourier series for (a) x^3, (b) x^4.

3.3.5. Write down the identities obtained by substituting $x = 0, \frac{1}{2}\pi$, and $\frac{1}{3}\pi$ in the Fourier series (3.74).

♢ 3.3.6. Suppose $f(x)$ is a 2π–periodic function with complex Fourier coefficients c_k, and $g(x)$ is a 2π–periodic function with complex Fourier coefficients d_k. (a) Find the Fourier coefficients e_k of their *periodic convolution* $f(x) * g(x) = \int_{-\pi}^{\pi} f(x-y)\, g(y)\, dy$.
(b) Find the complex Fourier series for the periodic convolution of $\cos 3x$ and $\sin 2x$.
(c) Answer part (b) for the functions x and $\sin 2x$.

♢ 3.3.7. Suppose f is piecewise continuous on $[-\pi, \pi]$. Prove that the mean of the integrated function $g(x) = \int_0^x f(y)\, dy$ equals $\frac{1}{2}\int_{-\pi}^{\pi} \left(\operatorname{sign} x - \frac{x}{\pi}\right) f(x)\, dx$.

3.3.8. Suppose the 2π–periodic extension of $f(x)$ is continuous and piecewise C^1. Prove directly from the formulas (3.35) that the Fourier coefficients of its derivative $\widetilde{f}(x) = f'(x)$ are, respectively, $\widetilde{a}_k = k\, b_k$ and $\widetilde{b}_k = -k\, a_k$, where a_k, b_k are the Fourier coefficients of $f(x)$.

♢ 3.3.9. Explain how to integrate a complex Fourier series (3.64). Under what conditions is your formula valid?

♡ 3.3.10. The initial value problem $\dfrac{d^2u}{dt^2} + u = f(t)$, $u(0) = 0$, $\dfrac{du}{dt}(0) = 0$, describes the forced motion of an initially motionless unit mass attached to a unit spring.
(a) Solve the initial value problem when $f(t) = \cos kt$ and $f(t) = \sin kt$ for $k = 0, 1, \ldots$.
(b) Assuming that the forcing function $f(t)$ is 2π–periodic, write out its Fourier series, and then use your result from part (b) to write out a series for the solution $u(t)$.
(c) Under what conditions is the result a convergent Fourier series, and hence the solution $u(t)$ remains 2π–periodic?
(d) Explain why $f(t)$ induces a resonance of the mass-spring system if and only if its Fourier coefficients of order 1 are not both zero: $a_1^2 + b_1^2 \neq 0$.

3.4 Change of Scale

So far, we have dealt only with Fourier series on the standard interval of length 2π. We chose $[-\pi, \pi]$ for convenience, but all of the results and formulas are easily adapted to any other interval of the same length, e.g., $[0, 2\pi]$. However, since physical objects like bars and strings do not all come in this particular length, we need to understand how to adapt the formulas to more general intervals.

Any symmetric interval $[-\ell,\ell]$ of length 2ℓ can be rescaled (stretched) to the standard interval $[-\pi,\pi]$ through the linear change of variables

$$x = \frac{\ell}{\pi}\, y, \qquad \text{so that} \qquad -\pi \le y \le \pi \qquad \text{whenever} \qquad -\ell \le x \le \ell. \tag{3.82}$$

Given a function $f(x)$ defined on $[-\ell,\ell]$, the *rescaled function* $F(y) = f\left(\frac{\ell}{\pi}\, y\right)$ lives on $[-\pi,\pi]$. Let

$$F(y) \sim \frac{a_0}{2} + \sum_{k=1}^{\infty} \left[a_k \cos k y + b_k \sin k y \right]$$

be the standard Fourier series for $F(y)$, so that

$$a_k = \frac{1}{\pi}\int_{-\pi}^{\pi} F(y)\cos k y\, dy, \qquad\qquad b_k = \frac{1}{\pi}\int_{-\pi}^{\pi} F(y)\sin k y\, dy. \tag{3.83}$$

Then, reverting to the unscaled variable x, we deduce that

$$f(x) \sim \frac{a_0}{2} + \sum_{k=1}^{\infty} \left[a_k \cos\frac{k\pi x}{\ell} + b_k \sin\frac{k\pi x}{\ell} \right] \tag{3.84}$$

is the Fourier series of $f(x)$ on the interval $[-\ell,\ell]$. The Fourier coefficients a_k, b_k can, in fact, be computed directly without appealing to the rescaling. Indeed, replacing the integration variable in (3.83) by $y = \pi x/\ell$, and noting that $dy = (\pi/\ell)\, dx$, we deduce the rescaled formulae

$$a_k = \frac{1}{\ell}\int_{-\ell}^{\ell} f(x)\cos\frac{k\pi x}{\ell}\, dx, \qquad\qquad b_k = \frac{1}{\ell}\int_{-\ell}^{\ell} f(x)\sin\frac{k\pi x}{\ell}\, dx, \tag{3.85}$$

for the Fourier coefficients of $f(x)$ on the interval $[-\ell,\ell]$.

All of the convergence results, integration and differentiation formulae, etc., that are valid for the interval $[-\pi,\pi]$ carry over, essentially unchanged, to Fourier series on nonstandard intervals. In particular, adapting our basic convergence Theorem 3.8, we conclude that if $f(x)$ is piecewise C^1, then its rescaled Fourier series (3.84) converges to its 2ℓ periodic extension $\widetilde{f}(x)$, subject to the proviso that $\widetilde{f}(x)$ takes on the midpoint values at all jump discontinuities.

Example 3.24. Let us compute the Fourier series for the function $f(x) = x$ on the interval $-1 \le x \le 1$. Since f is odd, only the sine coefficients will be nonzero. We have

$$b_k = \int_{-1}^{1} x \sin k\pi x\, dx = \left[-\frac{x\cos k\pi x}{k\pi} + \frac{\sin k\pi x}{(k\pi)^2} \right]_{x=-1}^{1} = \frac{2(-1)^{k+1}}{k\pi}\,.$$

The resulting Fourier series is

$$x \sim \frac{2}{\pi}\left(\sin\pi x - \frac{\sin 2\pi x}{2} + \frac{\sin 3\pi x}{3} - \cdots \right).$$

The series converges to the 2–periodic extension of the function x, namely

$$\widetilde{f}(x) = \begin{cases} x - 2m, & 2m-1 < x < 2m+1, \\ 0, & x = m, \end{cases} \qquad \text{where } m \in \mathbb{Z} \text{ is an arbitrary integer,}$$

which is plotted in Figure 3.10.

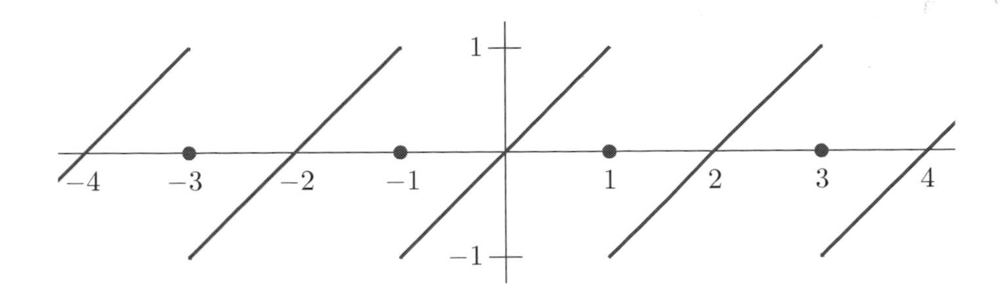

Figure 3.10. 2–periodic extension of x.

We can similarly reformulate complex Fourier series on the nonstandard interval $[-\ell,\ell]$. Using (3.82) to rescale the variables in (3.64), we obtain

$$f(x) \sim \sum_{k=-\infty}^{\infty} c_k\, e^{\,\mathrm{i}\,k\pi x/\ell}, \qquad \text{where} \qquad c_k = \frac{1}{2\ell}\int_{-\ell}^{\ell} f(x)\, e^{-\,\mathrm{i}\,k\pi x/\ell}\, dx. \tag{3.86}$$

Again, this is merely an alternative way of writing the real Fourier series (3.84).

When dealing with a more general interval $[a,b]$, there are two possible options. The first is to take a function $f(x)$ defined for $a \le x \le b$ and periodically extend it to a function $\widetilde{f}(x)$ that agrees with $f(x)$ on $[a,b]$ and has period $b-a$. One can then compute the Fourier series (3.84) for its periodic extension $\widetilde{f}(x)$ on the symmetric interval $[-\ell,\ell]$ of width $2\ell = b-a$; the resulting Fourier series will (under the appropriate hypotheses) converge to $\widetilde{f}(x)$ and hence agree with $f(x)$ on the original interval. An alternative approach is to translate the interval by an amount $\frac{1}{2}(a+b)$ so as to make it symmetric around the origin; this is accomplished by the change of variables $\widehat{x} = x - \frac{1}{2}(a+b)$, followed by an additional rescaling to convert the interval into $[-\pi,\pi]$. The two methods are essentially equivalent, and details are left to the reader.

Exercises

3.4.1. Let $f(x) = x^2$ for $0 \le x \le 1$. Find its (a) Fourier sine series; (b) Fourier cosine series.

3.4.2. Find the Fourier sine series and the Fourier cosine series of the following functions defined on the interval $[0,1]$; then graph the function to which the series converges:
(a) 1, (b) $\sin \pi x$, (c) $\sin^3 \pi x$, (d) $x(1-x)$.

3.4.3. Find the Fourier series for the following functions on the indicated intervals, and graph the function that the Fourier series converges to.
(a) $|x|$, $-3 \le x \le 3$, (b) x^2-4, $-2 \le x \le 2$, (c) e^x, $-10 \le x \le 10$,
(d) $\sin x$, $-1 \le x \le 1$, (e) $\sigma(x)$, $-2 \le x \le 2$.

3.4.4. For each of the functions in Exercise 3.4.3, write out the differentiated Fourier series, and determine whether it converges to the derivative of the original function.

3.4.5. Find the Fourier series for the integral of each of the functions in Exercise 3.4.3.

♢ 3.4.6. Write down formulas for the Fourier series of both even and odd functions on $[-\ell,\ell]$.

3.4.7. Let $f(x)$ be a continuous function on $[0, \ell]$.
(a) Under what conditions is its odd 2ℓ–periodic extension also continuous?
(b) Under what conditions is its odd extension also continuously differentiable?

3.4.8. (a) Write down the formulae for the Fourier series for a function $f(x)$ defined on the interval $0 \le x \le 2\pi$. (b) Use your formula in the case $f(x) = x$. Is the result the same as (3.37)? Explain, and, if different, discuss the connection between the two Fourier series.

3.4.9. Find the Fourier series for the function $f(x) = x$ on the interval $1 \le x \le 2$ using the two different methods described in the last paragraph of this subsection. Are your Fourier series the same? Explain. Graph the functions that the Fourier series converge to.

3.4.10. Answer Exercise 3.4.9 when $f(x) = \sin x$ on the interval $\pi \le x \le 2\pi$.

3.5 Convergence of Fourier Series

The goal of this final section is to establish some of the most basic convergence results for Fourier series. This is not a purely theoretical enterprise, since convergence considerations impinge directly upon applications. One particularly important consequence is the connection between the degree of smoothness of a function and the decay rate of its high-order Fourier coefficients — a result that is exploited in signal and image denoising and in the analytic properties of solutions to partial differential equations.

This section is written at a slightly more theoretically sophisticated level than what you have read so far. However, an appreciation of the full scope, and limitations, of Fourier analysis requires some familiarity with the underlying theory. Moreover, the required techniques and proofs serve as an excellent introduction to some of the most important tools of modern mathematical analysis, and the effort you expend to assimilate this material will be more than amply rewarded in both this book and your subsequent mathematical studies, be they applied or pure.

Unlike power series, which converge to analytic functions on the interval of convergence, and diverge elsewhere (the only tricky point being whether or not the series converges at the endpoints), the convergence of a Fourier series is a much subtler matter, and still not completely understood. A large part of the difficulty stems from the intricacies of convergence in infinite-dimensional function spaces. Let us therefore begin with a brief outline of the key issues.

We assume that you are familiar with the usual calculus definition of the limit of a sequence of real numbers: $\lim_{n\to\infty} a_n = a^\star$. In any finite-dimensional vector space, e.g., $\mathbb{R}^m$, there is essentially only one way for a sequence of vectors $\mathbf{v}^{(0)}, \mathbf{v}^{(1)}, \mathbf{v}^{(2)}, \ldots \in \mathbb{R}^m$ to converge, as guaranteed by any one of the following equivalent criteria:

- The vectors converge: $\mathbf{v}^{(n)} \to \mathbf{v}^\star \in \mathbb{R}^m$ as $n \to \infty$.
- The individual components of $\mathbf{v}^{(n)} = (v_1^{(n)}, \ldots, v_m^{(n)})$ converge, so $\lim_{n\to\infty} v_j^{(n)} = v_j^\star$ for all $j = 1, \ldots, m$.
- The norm of the difference goes to zero: $\| \mathbf{v}^{(n)} - \mathbf{v}^\star \| \to 0$ as $n \to \infty$.

The last requirement, known as *convergence in norm*, does not, in fact, depend on which norm is chosen. Indeed, on a finite-dimensional vector space, all norms are essentially equivalent, and if one norm goes to zero, so does any other norm, [**89**; Theorem 3.17].

On the other hand, the analogous convergence criteria are certainly *not the same* in infinite-dimensional spaces. There is, in fact, a bewildering variety of convergence mechanisms in function space, including pointwise convergence, uniform convergence, convergence in norm, weak convergence, and so on. Each plays a significant role in advanced mathematical analysis, and hence all are deserving of study. Here, though, we shall cover just the most basic aspects of convergence of the Fourier series and their applications to partial differential equations, leaving the complete development to a more specialized text, e.g., [**37**, **128**].

Pointwise and Uniform Convergence

The most familiar convergence mechanism for a sequence of functions $v_n(x)$ is *pointwise convergence*. This requires that the functions' values at each individual point converge in the usual sense:

$$\lim_{n \to \infty} v_n(x) = v_\star(x) \qquad \text{for all} \qquad x \in I, \tag{3.87}$$

where $I \subset \mathbb{R}$ denotes an interval contained in their common domain. Even more explicitly, pointwise convergence requires that, for every $\varepsilon > 0$ and every $x \in I$, there exist an integer N, depending on ε and x, such that

$$|\, v_n(x) - v_\star(x) \,| < \varepsilon \qquad \text{for all} \qquad n \geq N. \tag{3.88}$$

Pointwise convergence can be viewed as the function space version of the convergence of the components of a vector. We have already stated the Fundamental Theorem 3.8 regarding pointwise convergence of Fourier series; the proof will be deferred until the end of this section.

On the other hand, establishing uniform convergence of a Fourier series is not so difficult, and so we will begin there. The basic definition of uniform convergence looks very similar to that of pointwise convergence, with a subtle, but important, difference.

Definition 3.25. A sequence of functions $v_n(x)$ is said to converge *uniformly* to a function $v_\star(x)$ on a subset $I \subset \mathbb{R}$ if, for every $\varepsilon > 0$, there exists an integer N, depending solely on ε, such that

$$|\, v_n(x) - v_\star(x) \,| < \varepsilon \qquad \text{for all } x \in I \text{ and all } n \geq N. \tag{3.89}$$

Clearly, a uniformly convergent sequence of functions converges pointwise, but the converse does not hold. The key difference — and the reason for the term "uniform convergence" — is that the integer N depends only on ε and *not* on the point $x \in I$. According to (3.89), the sequence converges uniformly if and only if for every small ε, the graphs of the functions eventually lie inside a band of width 2ε centered on the graph of the limiting function, as in the first plot in Figure 3.11. The Gibbs phenomenon shown in Figure 3.7 is a prototypical example of nonuniform convergence: For a given $\varepsilon > 0$, the closer x is to the discontinuity, the larger n must be chosen so that the inequality in (3.89) holds. Hence, there is *no* uniform choice of N that makes the inequality (3.89) valid for *all* x and *all* $n \geq N$.

A key feature of uniform convergence is that it preserves continuity.

Theorem 3.26. *If each $v_n(x)$ is continuous and $v_n(x) \to v_\star(x)$ converges uniformly, then $v_\star(x)$ is also a continuous function.*

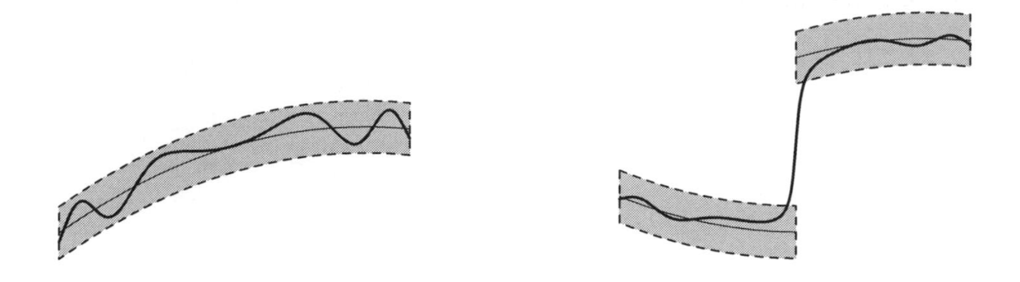

Figure 3.11. Uniform and nonuniform convergence of functions.

The proof is by contradiction. Intuitively, if $v_\star(x)$ were to have a discontinuity, then, as sketched in the second plot in Figure 3.11, a sufficiently small band around its graph would not connect together, and this prevents the connected graph of any continuous function, such as $v_n(x)$, from remaining entirely within the band. A detailed discussion of these issues, including the proofs of the basic theorems, can be found in any introductory real analysis text, [**8, 96, 97**].

Warning: A sequence of continuous functions can converge *nonuniformly* to a continuous function. For example, the sequence

$$v_n(x) = \frac{2nx}{1+n^2x^2}$$

converges pointwise to $v_\star(x) \equiv 0$ (why?) but not uniformly, since

$$\max |\, v_n(x) \,| = v_n\left(\tfrac{1}{n}\right) = 1,$$

which implies that (3.89) cannot hold when $\varepsilon < 1$.

The convergence (pointwise, uniform, etc.) of an infinite series $\sum_{k=1}^{\infty} u_k(x)$ is, by definition, dictated by the convergence of its sequence of *partial sums*

$$v_n(x) = \sum_{k=1}^{n} u_k(x). \tag{3.90}$$

The most useful test for uniform convergence of series of functions is known as the *Weierstrass M–test*, in honor of the nineteenth century German mathematician Karl Weierstrass, known as the "father of modern analysis".

Theorem 3.27. *Let $I \subset \mathbb{R}$. Suppose that, for each $k = 1, 2, 3, \ldots$, the function $u_k(x)$ is bounded:*

$$|\, u_k(x) \,| \leq m_k \qquad \text{for all} \qquad x \in I, \tag{3.91}$$

where $m_k \geq 0$ is a nonnegative constant. If the constant series

$$\sum_{k=1}^{\infty} m_k < \infty \tag{3.92}$$

converges, then the function series

$$\sum_{k=1}^{\infty} u_k(x) = f(x) \tag{3.93}$$

converges uniformly and absolutely[†] *to a function* $f(x)$ *for all* $x \in I$. *In particular, if the summands* $u_k(x)$ *are continuous, so is the sum* $f(x)$.

Warning: Failure of the M–test strongly indicates, but does not necessarily preclude, that a pointwise convergent series does not converge uniformly.

With some care, we can manipulate uniformly convergent series just like finite sums. Thus, if (3.93) is a uniformly convergent series, so is its term-wise product

$$\sum_{k=1}^{\infty} g(x)\, u_k(x) = g(x)\, f(x) \tag{3.94}$$

with any bounded function: $|\, g(x)\,| \leq C$ for all $x \in I$. We can integrate a uniformly convergent series term by term,[‡] and the resulting integrated series

$$\int_a^x \left(\sum_{k=1}^{\infty} u_k(y) \right) dy = \sum_{k=1}^{\infty} \int_a^x u_k(y)\, dy = \int_a^x f(y)\, dy \tag{3.95}$$

is uniformly convergent. Differentiation is also allowed — but only when the differentiated series converges uniformly.

Proposition 3.28. *Suppose the series* $\sum_{k=1}^{\infty} u_k(x) = f(x)$ *converges pointwise. If the differentiated series* $\sum_{k=1}^{\infty} u_k'(x) = g(x)$ *is uniformly convergent, then the original series is also uniformly convergent, and, moreover,* $f'(x) = g(x)$.

We are particularly interested in the convergence of a Fourier series, which, to facilitate the exposition, we take in its complex form

$$f(x) \sim \sum_{k=-\infty}^{\infty} c_k\, e^{\,i\,k\,x}. \tag{3.96}$$

Since x is real, $\left|\, e^{\,i\,k\,x} \,\right| \leq 1$, and hence the individual summands are bounded by

$$\left|\, c_k\, e^{\,i\,k\,x} \,\right| \leq |\, c_k \,| \qquad \text{for all } x.$$

Applying the Weierstrass M–test, we immediately deduce the basic result on uniform convergence of Fourier series.

[†] Recall that a series $\sum_{n=1}^{\infty} a_n = a^\star$ is said to converge absolutely if $\sum_{n=1}^{\infty} |\, a_n \,|$ converges.

[‡] Assuming that the individual functions are all integrable.

Theorem 3.29. *If the Fourier coefficients* c_k *of a function* $f(x)$ *satisfy*

$$\sum_{k=-\infty}^{\infty} |\, c_k \,| < \infty, \tag{3.97}$$

then the Fourier series (3.96) converges uniformly to a continuous function $\widetilde{f}(x)$ *that has the same Fourier coefficients:* $c_k = \langle\, f\,, e^{\,i\,k\,x}\,\rangle = \langle\, \widetilde{f}\,, e^{\,i\,k\,x}\,\rangle$.

Proof: Uniform convergence and continuity of the limiting function follow from Theorem 3.27. To show that the c_k actually are the Fourier coefficients of the sum, we multiply the Fourier series by $e^{-\,i\,k\,x}$ and integrate term by term from $-\pi$ to π. As in (3.94, 95), both operations are valid thanks to the uniform convergence of the series. *Q.E.D.*

Remark: As with the Weierstrass test, failure of condition (3.97) strongly indicates that the Fourier series does not converge uniformly, but does not completely rule it out; nor does it say anything about nonuniform convergence or lack thereof.

The one thing that Theorem 3.29 does not guarantee is that the original function $f(x)$ used to compute the Fourier coefficients c_k is the *same* as the function $\widetilde{f}(x)$ obtained by summing the resulting Fourier series! Indeed, this may very well not be the case. As we know, the function that the series converges to is necessarily 2π–periodic. Thus, at the very least, $\widetilde{f}(x)$ will be the 2π periodic extension of $f(x)$. But even this may not suffice. Indeed, two functions $f(x)$ and $\widehat{f}(x)$ that have the same values except at a finite set of points $x_1, \ldots, x_m$ have the same Fourier coefficients. (Why?) For example, the discontinuous function $f(x) = \begin{cases} 1, & x = 0, \\ 0, & \text{otherwise}, \end{cases}$ has all zero Fourier coefficients, and hence its Fourier series converges to the continuous zero function. More generally, two functions that agree everywhere outside a set of "measure zero" will have identical Fourier coefficients. In this way, a convergent Fourier series singles out a distinguished representative from a collection of essentially equivalent 2π–periodic functions.

Remark: The term "measure" refers to a rigorous generalization of the notion of the length of an interval to more general subsets $S \subset \mathbb{R}$. In particular, S has *measure zero* if it can be covered by a collection of intervals of arbitrarily small total length. For example, any set consisting of finitely many points, or even countably many points, e.g., the rational numbers, has measure zero; see Exercise 3.5.19. The proper development of the notion of measure, and the consequential Lebesgue theory of integration, is properly studied in a course in real analysis, [**96, 98**].

As a consequence of Theorem 3.26, a Fourier series cannot converge uniformly when discontinuities are present. However, it can be proved, [**128**], that even when the function is not everywhere continuous, its Fourier series is uniformly convergent on any closed subset of continuity.

Theorem 3.30. *Let* $f(x)$ *be* 2π*–periodic and piecewise* C^1. *If* $f(x)$ *is continuous on the open interval* $a < x < b$, *then its Fourier series converges uniformly to* $f(x)$ *on any closed subinterval* $a + \delta \leq x \leq b - \delta$ *for* $0 < \delta < \frac{1}{2}(b - a)$.

For example, the Fourier series (3.49) for the unit step function converges uniformly if we stay away from the discontinuities — for instance, by restriction to a subinterval of the form $[\,\delta, \pi - \delta\,]$ or $[\,-\pi + \delta, -\delta\,]$ for any $0 < \delta < \frac{1}{2}\pi$. This reconfirms our observation that the nonuniform Gibbs behavior becomes progressively more and more localized at the discontinuities.

Exercises

3.5.1. Consider the following sequence of planar vectors $\mathbf{v}^{(n)} = \left(1 - \frac{1}{n}, e^{-n}\right)$, $n = 1, 2, 3, \ldots$. Prove that $\mathbf{v}^{(n)}$ converges to $\mathbf{v}^\star = (1, 0)$ as $n \to \infty$ by showing that: (*a*) the individual components converge; (*b*) the Euclidean norms converge: $\| \mathbf{v}^{(n)} - \mathbf{v}^\star \|_2 \to 0$.

3.5.2. Which of the following sequences of vectors converge as $n \to \infty$? What is the limit? (*a*) $\left(\dfrac{1}{1+n^2}, \dfrac{n^2}{1+2n^2}\right)$, (*b*) $(\cos n, \sin n)$, (*c*) $\left(\dfrac{\cos n}{n}, \dfrac{\sin n}{n}\right)$, (*d*) $\left(\cos \dfrac{1}{n}, \sin \dfrac{1}{n}\right)$, (*e*) $\left(\dfrac{1}{n}\cos \dfrac{1}{n}, \dfrac{1}{n}\sin \dfrac{1}{n}\right)$, (*f*) $\left(e^{-n}, n\,e^{-n}, n^2 e^{-n}\right)$, (*g*) $\left(\dfrac{\log n}{n}, \dfrac{(\log n)^2}{n^2}, \dfrac{(\log n)^3}{n^3}\right)$, (*h*) $\left(\dfrac{1-n}{1+n}, \dfrac{1-n}{1+n^2}, \dfrac{1-n^2}{1+n^2}\right)$, (*i*) $\left(\left(1+\dfrac{1}{n}\right)^n, \left(1-\dfrac{1}{n}\right)^{-n}\right)$, (*j*) $\left(\dfrac{e^n - 1}{n}, \dfrac{\cos n - 1}{n^2}\right)$, (*k*) $\left(n\left(e^{1/n} - 1\right), n^2\left(\cos \dfrac{1}{n} - 1\right)\right)$.

3.5.3. Which of the following sequences of functions converge pointwise for $x \in \mathbb{R}$ as $n \to \infty$? What is the limit? (*a*) $1 - \dfrac{x^2}{n^2}$, (*b*) e^{-nx}, (*c*) e^{-nx^2}, (*d*) $|\,x - n\,|$, (*e*) $\dfrac{1}{1+(x-n)^2}$, (*f*) $\begin{cases} 1, & x < n, \\ 2, & x > n, \end{cases}$ (*g*) $\begin{cases} n^2, & \frac{1}{n} < x < \frac{2}{n}, \\ 0, & \text{otherwise}, \end{cases}$ (*h*) $\begin{cases} x, & |\,x\,| < n, \\ n x^{-2}, & |\,x\,| \geq n. \end{cases}$

3.5.4. Prove that the sequence $v_n(x) = \begin{cases} 1, & 0 < x < \frac{1}{n}, \\ 0, & \text{otherwise}, \end{cases}$ converges pointwise, but not uniformly, to the zero function.

3.5.5. Which of the following sequences of functions converge pointwise to the zero function for all $x \in \mathbb{R}$? Which converge uniformly? (*a*) $-\dfrac{x^2}{n^2}$, (*b*) $e^{-n|\,x\,|}$, (*c*) $x\,e^{-n|\,x\,|}$, (*d*) $\dfrac{1}{n(1+x^2)}$, (*e*) $\dfrac{1}{1+(x-n)^2}$, (*f*) $|\,x - n\,|$, (*g*) $\begin{cases} \frac{1}{n}, & 0 < |\,x\,| < n, \\ 0, & \text{otherwise}, \end{cases}$ (*h*) $\begin{cases} n, & 0 < |\,x\,| < \frac{1}{n}, \\ 0, & \text{otherwise}, \end{cases}$ (*i*) $\begin{cases} x/n, & |\,x\,| < 1, \\ 1/(nx), & |\,x\,| \geq 1. \end{cases}$

3.5.6. Does the sequence $v_n(x) = n\,x\,e^{-nx^2}$ converge pointwise to the zero function for $x \in \mathbb{R}$? Does it converge uniformly?

3.5.7. Answer Exercise 3.5.6 when (*a*) $v_n(x) = x\,e^{-nx^2}$, (*b*) $v_n(x) = \begin{cases} 1, & n < x < n+1, \\ 0, & \text{otherwise}, \end{cases}$ (*c*) $v_n(x) = \begin{cases} 1, & n < x < n + 1/n, \\ 0, & \text{otherwise}, \end{cases}$ (*d*) $v_n(x) = \begin{cases} 1/n, & n < x < 2n, \\ 0, & \text{otherwise}, \end{cases}$ (*e*) $v_n(x) = \begin{cases} 1/\sqrt{n}, & n < x < 2n, \\ 0, & \text{otherwise}, \end{cases}$ (*f*) $v_n(x) = \begin{cases} n^2 x^2 - 1, & -1/n < x < 1/n, \\ 0, & \text{otherwise}. \end{cases}$

3.5.8. (*a*) What is the limit of the functions $v_n(x) = \tan^{-1} n x$ as $n \to \infty$? (*b*) Is the convergence uniform on all of $\mathbb{R}$? (*c*) on the interval $[-1, 1]$? (*d*) on the subset $\{x \geq 1\}$?

3.5.9. *True or false*: If $p_n(x)$ is a sequence of polynomials that converge pointwise to a polynomial $p_\star(x)$, then the convergence is uniform.

3.5.10. Suppose $v_n(x)$ are continuous functions such that $v_n \to v_\star$ pointwise on all of $\mathbb{R}$. *True or false*: (a) $v_n - v_\star \to 0$ pointwise; (b) if $v_\star(x) \neq 0$ for all x, then $\dfrac{v_n}{v_\star} \to 1$ pointwise.

3.5.11. Which of the following series satisfy the M–test and hence converge uniformly on the interval $[0,1]$? (a) $\sum_{k=1}^{\infty} \dfrac{\cos k x}{k^2}$, (b) $\sum_{k=1}^{\infty} \dfrac{\sin k x}{k}$, (c) $\sum_{k=1}^{\infty} x^k$, (d) $\sum_{k=1}^{\infty} (x/2)^k$, (e) $\sum_{k=1}^{\infty} \dfrac{e^{kx}}{k^2}$, (f) $\sum_{k=1}^{\infty} \dfrac{e^{-kx}}{k^2}$, (g) $\sum_{k=1}^{\infty} \dfrac{e^{x/k}-1}{k}$.

3.5.12. Prove that the power series $\sum_{k=1}^{\infty} \dfrac{x^k}{k(k+1)}$ converges uniformly for $-1 \leq x \leq 1$.

♢ 3.5.13. (a) Prove the following result: Suppose $|\,g(x)\,| \leq M$ for all $x \in I$. If (3.93) is a uniformly convergent series on I, so is the term-wise product (3.94).
(b) Find a counterexample when $g(x)$ is not uniformly bounded.

♢ 3.5.14. Suppose each $u_k(x)$ is continuous, and the series $\sum_{k=1}^{\infty} u_k(x) = f(x)$ converges uniformly on the bounded interval $a \leq x \leq b$. Prove that the integrated series (3.95) is uniformly convergent.

♢ 3.5.15. Prove that if $\sum_{k=1}^{\infty} \sqrt{a_k^2 + b_k^2} < \infty$, then the real Fourier series (3.34) converges uniformly to a continuous 2π–periodic function.

3.5.16. Suppose $\sum_{k=1}^{\infty} |\,a_k\,| < \infty$ and $\sum_{k=1}^{\infty} |\,b_k\,| < \infty$. Does the conclusion of Exercise 3.5.15 still hold?

3.5.17. Explain why you only need check the inequalities (3.91) for all sufficiently large $k \gg 0$ in order to use the Weierstrass M–test.

3.5.18. Suppose we say that a sequence of vectors $\mathbf{v}^{(k)} \in \mathbb{R}^m$ converges *uniformly* to $\mathbf{v}^\star \in \mathbb{R}^m$ if, for every $\varepsilon > 0$, there is an N, depending only on ε, such that $|\,v_i^{(k)} - v_i^\star\,| < \varepsilon$, for all $k \geq N$ and all $i = 1, \ldots, m$. Prove that *every* convergent sequence of vectors converges uniformly.

♢ 3.5.19. (a) Let $S = \{x_1, x_2, x_3, \ldots\} \subset \mathbb{R}$ be a countable set. Prove that S has measure zero by showing that, for every $\varepsilon > 0$, there exists a collection of open intervals $I_1, I_2, I_3, \ldots \subset \mathbb{R}$, with respective lengths $\ell_1, \ell_2, \ell_3, \ldots$, such that $S \subset \bigcup I_j$, while the total length $\sum \ell_j = \varepsilon$.
(b) Explain why the set of rational numbers $\mathbb{Q} \subset \mathbb{R}$ is dense but nevertheless has measure zero.

Smoothness and Decay

The criterion (3.97), which guarantees uniform convergence of a Fourier series, requires, at the very least, that the Fourier coefficients go to zero: $c_k \to 0$ as $k \to \pm\infty$. And they cannot decay too slowly. For example, the individual summands of the infinite series

$$\sum_{0 \neq k=-\infty}^{\infty} \frac{1}{|\,k\,|^\alpha} \tag{3.98}$$

go to 0 as $k \to \infty$ whenever $\alpha > 0$, but the series converges only when $\alpha > 1$. (This is an immediate consequence of the standard integral convergence test, [**8, 97, 108**].) Thus, if we can bound the Fourier coefficients by

$$| c_k | \leq \frac{M}{| k |^\alpha} \qquad \text{for all} \qquad | k | \gg 0, \tag{3.99}$$

for some exponent $\alpha > 1$ and some positive constant $M > 0$, then the Weierstrass M–test will guarantee that the Fourier series converges uniformly to a continuous function.

An important consequence of the differentiation formula (3.79) for Fourier series is that one can detect the degree of smoothness of a function by seeing how rapidly its Fourier coefficients decay to zero. More rigorously:

Theorem 3.31. *Let $0 \leq n \in \mathbb{Z}$. If the Fourier coefficients of $f(x)$ satisfy*

$$\sum_{k=-\infty}^{\infty} | k |^n \, | c_k | < \infty, \tag{3.100}$$

then the Fourier series (3.64) converges uniformly to an n–times continuously differentiable function $\widetilde{f}(x) \in \mathrm{C}^n$, which is the 2π–periodic extension of $f(x)$. Furthermore, for any $0 < m \leq n$, the m–times differentiated Fourier series converges uniformly to the corresponding derivative $\widetilde{f}^{(m)}(x)$.

Proof: Iterating (3.79), the Fourier series for the nth derivative of a function is

$$f^{(n)}(x) \sim \sum_{k=-\infty}^{\infty} \mathrm{i}^n \, k^n \, c_k \, e^{\,\mathrm{i}\,k\,x}. \tag{3.101}$$

If (3.100) holds, the Weierstrass M–test implies the uniform convergence of the differentiated series (3.101) to a continuous 2π–periodic function. Proposition 3.28 guarantees that the limit is the nth derivative of the original Fourier series. *Q.E.D.*

This result enables us to quantify the rule of thumb that, the smaller the high-frequency Fourier coefficients, the smoother the function.

Corollary 3.32. *If the Fourier coefficients satisfy* (3.99) *for some $\alpha > n+1$, then the Fourier series converges uniformly to an n–times continuously differentiable 2π–periodic function.*

If the Fourier coefficients go to zero faster than any power of k, e.g., exponentially fast, then the function is infinitely differentiable. Analyticity is more delicate, and we refer the reader to [**128**] for details.

Example 3.33. The 2π–periodic extension of the function $| x |$ is continuous with piecewise continuous first derivative. Its Fourier coefficients (3.54) satisfy the estimate (3.99) for $\alpha = 2$, which is not quite fast enough to ensure a continuous second derivative. On the other hand, the Fourier coefficients (3.36) of the step function $\sigma(x)$ tend to zero only as $1/| k |$, so $\alpha = 1$, reflecting the fact that its periodic extension is piecewise continuous, but not continuous.

Exercises

3.5.20. (a) Prove that the complex Fourier series $f(x) = \sum_{k=1}^{\infty} \frac{1}{k^2} e^{ikx}$ converges uniformly on the interval $[-\pi,\pi]$. (b) Is the sum $f(x)$ continuous? Why or why not? (c) Is $f(x)$ continuously differentiable? Why or why not?

3.5.21. First, without explicitly evaluating them, how fast do you expect the Fourier coefficients of the following functions to go to zero as $k \to \infty$? Then prove your claim by evaluating the coefficients.
(a) $x - \pi$, (b) $|x|$, (c) x^2, (d) $x^4 - 2\pi^2 x^2$, (e) $\sin^2 x$, (f) $|\sin x|$.

3.5.22. Using the criteria of Theorem 3.31, determine how many continuous derivatives the functions represented by the following Fourier series have:
(a) $\sum_{k=-\infty}^{\infty} \frac{e^{ikx}}{1+k^4}$, (b) $\sum_{\substack{k=-\infty \\ k\neq 0}}^{\infty} \frac{e^{ikx}}{k^2+k^5}$, (c) $\sum_{k=-\infty}^{\infty} e^{ikx-k^2}$, (d) $\sum_{k=0}^{\infty} \frac{e^{ikx}}{k+1}$,
(e) $\sum_{k=-\infty}^{\infty} \frac{e^{ikx}}{|k|!}$, (f) $\sum_{k=1}^{\infty} \left(1 - \cos\frac{1}{k^2}\right) e^{ikx}$.

♣ 3.5.23. Discuss convergence of each of the following Fourier series. How smooth is the sum? Graph the partial sums to obtain a reasonable approximation to the graph of the summed series. How many summands are needed to obtain accuracy in the second decimal digit over the entire interval? Point out discontinuities, corners, and other features that you observe.
(a) $\sum_{k=0}^{\infty} e^{-k} \cos kx$, (b) $\sum_{k=0}^{\infty} \frac{\cos kx}{k+1}$, (c) $\sum_{k=1}^{\infty} \frac{\sin kx}{k^{3/2}}$, (d) $\sum_{k=1}^{\infty} \frac{\sin kx}{k^3+k}$.

3.5.24. Prove that if $|a_k|, |b_k| \leq M k^{-\alpha}$ for some $M > 0$ and $\alpha > n+1$, then the real Fourier series (3.34) converges uniformly to an n–times continuously differentiable 2π–periodic function $f \in \mathrm{C}^n$.

3.5.25. Give a simple explanation of why, if the Fourier coefficients $a_k = b_k = 0$ for all sufficiently large $k \gg 0$, then the Fourier series converges to an analytic function.

Hilbert Space

In order to make further progress, we must take a little detour. The proper setting for the rigorous theory of Fourier series turns out to be the most important function space in modern analysis and modern physics, known as *Hilbert space* in honor of the great late-nineteenth-/early-twentieth-century German mathematician David Hilbert. The precise definition of this infinite-dimensional inner product space is somewhat technical, but a rough version goes as follows:

Definition 3.34. A complex-valued function $f(x)$ is called *square-integrable* on the interval $[-\pi,\pi]$ if it has finite L^2 norm:

$$\| f \|^2 = \frac{1}{2\pi} \int_{-\pi}^{\pi} | f(x) |^2 \, dx < \infty. \tag{3.102}$$

The *Hilbert space* $\mathrm{L}^2 = \mathrm{L}^2[-\pi,\pi]$ is the vector space consisting of all complex-valued *square-integrable* functions.

The triangle inequality

$$\| f + g \| \le \| f \| + \| g \|$$

implies that if $f, g \in \mathrm{L}^2$, so $\| f \|, \| g \| < \infty$, then $\| f + g \| < \infty$, and so $f + g \in \mathrm{L}^2$. Moreover, for any complex constant c,

$$\| c f \| = | c | \, \| f \|,$$

and so $c f \in \mathrm{L}^2$ also. Thus, as claimed, Hilbert space is a complex vector space. The Cauchy–Schwarz inequality

$$| \langle f , g \rangle | \le \| f \| \, \| g \|$$

implies that the L^2 Hermitian inner product

$$\langle f , g \rangle = \frac{1}{2\pi} \int_{-\pi}^{\pi} f(x) \, \overline{g(x)} \, dx \tag{3.103}$$

of two square-integrable functions is well defined and finite. In particular, the Fourier coefficients of a function $f \in \mathrm{L}^2$ are specified by its inner products

$$c_k = \langle f , e^{\,\mathrm{i}\,k x} \rangle = \frac{1}{2\pi} \int_{-\pi}^{\pi} f(x) \, e^{-\,\mathrm{i}\,k x} \, dx$$

with the complex exponentials (which, by (3.63), are in L^2), and hence are all well defined and finite.

There are some interesting analytic subtleties that arise when one tries to prescribe precisely which functions are in the Hilbert space. Every piecewise continuous function belongs to L^2. But some functions with singularities are also members. For example, the power function $| x |^{-\alpha}$ belongs to L^2 for any $\alpha < \frac{1}{2}$, but not if $\alpha \ge \frac{1}{2}$.

Analysis relies on limiting procedures, and it is essential that Hilbert space be "complete" in the sense that appropriately convergent† sequences of functions have a limit. The completeness requirement is not elementary, and relies on the development of the more sophisticated Lebesgue theory of integration, which was formalized in the early part of the twentieth century by the French mathematician Henri Lebesgue. Any function which is square-integrable in the Lebesgue sense is admitted into L^2. This includes such non-piecewise-continuous functions as $\sin \dfrac{1}{x}$ and $x^{-1/3}$, as well as the strange function

$$r(x) = \begin{cases} 1 & \text{if } x \text{ is a rational number,} \\ 0 & \text{if } x \text{ is irrational.} \end{cases} \tag{3.104}$$

Thus, while well behaved in some respects, square-integrable functions can be quite wild in others.

Remark: The completeness of Hilbert space can be viewed as the infinite-dimensional analogue of the *completeness* of the real line $\mathbb{R}$, meaning that every convergent Cauchy sequence of real numbers has a limit in $\mathbb{R}$. On the other hand, the rational numbers $\mathbb{Q}$ are not complete — since a convergent sequence of rational numbers may well have an irrational limit — but form a *dense* subset of $\mathbb{R}$, because every real number can be arbitrarily closely

† The precise technical requirement is that every *Cauchy sequence* of functions $v_k \in \mathrm{L}^2$ converges to a function $v_\star \in \mathrm{L}^2$; see [**37, 96, 98**] and also Exercise 3.5.42 for details.

approximated by rational numbers, e.g., its truncated decimal expansions. Indeed, a fully rigorous definition of the real numbers $\mathbb{R}$ is somewhat delicate, [**96, 97**].

Similarly, the space of continuous functions $\mathrm{C}^0[-\pi,\pi]$ is not complete, in that (nonuniformly) convergent sequences of continuous functions are not, in general, continuous, but it does form a dense subspace of the Hilbert space $\mathrm{L}^2[-\pi,\pi]$, since every L^2 function can be arbitrarily closely approximated (in norm) by continuous functions, e.g., its approximating trigonometric polynomials. Thus, just as $\mathbb{R}$ can be viewed as the completion of $\mathbb{Q}$ under the Euclidean norm, so Hilbert space can be viewed as the completion of the space of continuous functions under the L^2 norm, and, just like that of $\mathbb{R}$, its fully rigorous definition is rather subtle.

A second complication is that (3.102) does not, strictly speaking, define a norm once we allow discontinuous functions into the fold. For example, the piecewise continuous function

$$f_0(x) = \begin{cases} 1, & x = 0, \\ 0, & x \neq 0, \end{cases} \tag{3.105}$$

has norm zero, $\| f_0 \| = 0$, even though it is not zero everywhere. Indeed, any function that is zero except on a set of measure zero also has norm zero, including the function (3.104). Therefore, in order to make (3.102) into a legitimate norm, we must agree to identify any two functions that have the same values except on a set of measure zero. Thus, the zero function 0 along with the preceding examples (3.104) and (3.105) are all viewed as defining the *same* element of Hilbert space. So, an element of Hilbert space is not, in fact, a function, but, rather, an equivalence class of functions all differing on a set of measure zero. All this may strike the applications-oriented reader as becoming much too abstract and arcane. In practice, you will not lose much by working with the elements of L^2 as if they were ordinary functions, and, even better, assuming that said "functions" are always piecewise continuous and square-integrable. Nevertheless, the full analytical power of Hilbert space theory is unleashed only by including completely general square-integrable functions.

After its invention by pure mathematicians around the turn of the twentieth century, physicists in the 1920s suddenly realized that Hilbert space was the ideal setting for the modern theory of quantum mechanics, [**66, 72, 115**]. A quantum-mechanical *wave function* is an element[‡] $\varphi \in \mathrm{L}^2$ that has unit norm: $\| \varphi \| = 1$. Thus, the set of wave functions is merely the "unit sphere" in Hilbert space. Quantum mechanics endows each physical wave function with a probabilistic interpretation. Suppose the wave function represents a single subatomic particle — photon, electron, etc. Then the squared modulus of the wave function, $|\varphi(x)|^2$, represents the probability density that quantifies the chance of the particle being located at position x. More precisely, the probability that the particle resides in a prescribed interval $[a,b] \subset [-\pi,\pi]$ is equal to $\sqrt{\dfrac{1}{2\pi}\displaystyle\int_a^b |\varphi(x)|^2\,dx}$. In particular, the wave function has unit norm,

$$\| \varphi \| = \sqrt{\frac{1}{2\pi}\int_{-\pi}^{\pi} |\varphi(x)|^2\,dx} = 1, \tag{3.106}$$

[‡] Here we are acting as if the physical universe were represented by the one-dimensional interval $[-\pi,\pi]$. The more apt context of three-dimensional physical space is developed analogously, replacing the single integral by a triple integral over all of $\mathbb{R}^3$. See also Section 7.4.

because the particle must certainly, i.e., with probability 1, be *somewhere*!

Convergence in Norm

We are now in a position to discuss convergence in norm of a Fourier series. We begin with the basic definition, which makes sense on any normed vector space.

Definition 3.35. Let V be a normed vector space. A sequence $s_1, s_2, s_3, \ldots \in V$ is said to *converge in norm* to $f \in V$ if $\| s_n - f \| \to 0$ as $n \to \infty$.

As we noted earlier, on finite-dimensional vector spaces such as $\mathbb{R}^m$, convergence in norm is equivalent to ordinary convergence. On the other hand, on infinite-dimensional function spaces, convergence in norm differs from pointwise convergence. For instance, it is possible to construct a sequence of functions that converges in norm to 0, but does not converge pointwise *anywhere*! (See Exercise 3.5.43.)

While our immediate interest is in the convergence of the Fourier series of a square-integrable function $f \in \mathrm{L}^2[-\pi,\pi]$, the methods we develop are of very general utility. Indeed, in later chapters we will require the analogous convergence results for other types of series solutions to partial differential equations, including multiple Fourier series as well as series involving Bessel functions, spherical harmonics, Laguerre polynomials, and so on. Since it distills the key issues down to their essence, the general, abstract version is, in fact, easier to digest, and, moreover, will be immediately applicable, not just to basic Fourier series, but to very general "eigenfunction series".

Let V be an infinite-dimensional inner product space, e.g., $\mathrm{L}^2[-\pi,\pi]$. Suppose $\varphi_1, \varphi_2, \varphi_3, \ldots$, are an *orthonormal* collection of elements of V, meaning that

$$\langle \varphi_j , \varphi_k \rangle = \begin{cases} 1 & j = k, \\ 0, & j \neq k. \end{cases} \tag{3.107}$$

A straightforward argument — see Exercise 3.5.33 — proves that the φ_k are linearly independent. Given $f \in V$, we form its *generalized Fourier series*

$$f \sim \sum_{k=1}^{\infty} c_k \varphi_k, \qquad \text{where} \qquad c_k = \langle f , \varphi_k \rangle. \tag{3.108}$$

The formula for the coefficient c_k is obtained by formally taking the inner product of the series with φ_k and invoking the orthonormality conditions (3.107). The two main examples are the real and complex L^2 spaces:

- V consists of real square-integrable functions defined on $[-\pi,\pi]$ under the rescaled L^2 inner product $\langle f , g \rangle = \dfrac{1}{\pi}\displaystyle\int_{-\pi}^{\pi} f(x)\, g(x)\, dx$. The orthonormal system $\{\varphi_k\}$ consists of the basic trigonometric functions, numbered as follows:

$$\varphi_1 = \frac{1}{\sqrt{2}}, \quad \varphi_2 = \cos x, \quad \varphi_3 = \sin x, \quad \varphi_4 = \cos 2x, \quad \varphi_5 = \sin 2x, \quad \varphi_6 = \cos 3x, \quad \ldots .$$

- V consists of complex square-integrable functions defined on $[-\pi,\pi]$ using the Hermitian inner product (3.103). The orthonormal system $\{\varphi_k\}$ consists of the complex exponentials, which we order as follows:

$$\varphi_1 = 1, \quad \varphi_2 = e^{\,\mathrm{i}\,x}, \quad \varphi_3 = e^{-\,\mathrm{i}\,x}, \quad \varphi_4 = e^{2\,\mathrm{i}\,x}, \quad \varphi_5 = e^{-2\,\mathrm{i}\,x}, \quad \varphi_6 = e^{3\,\mathrm{i}\,x}, \quad \ldots .$$

In each case, the generalized Fourier series (3.108) reduces to the ordinary Fourier series, with a minor change of indexing. Later, when we extend the separation of variables technique to partial differential equations in more than one space dimension, we will encounter a variety of other important examples, in which the φ_k are the eigenfunctions of a self-adjoint linear operator.

For the remainder of this section, to streamline the ensuing proofs, we will henceforth assume that V is a *real* inner product space. However, all results will be formulated so they are also valid for complex inner product spaces; the slightly more complicated proofs in the complex case are relegated to the exercises.

By definition, the generalized Fourier series (3.108) *converges in norm* to f if the sequence provided by its *partial sums*

$$s_n = \sum_{k=1}^{n} c_k \varphi_k \tag{3.109}$$

satisfies the criterion of Definition 3.35. Our first result states that the partial Fourier sum (3.109), with c_k given by the inner product formula in (3.108), is, in fact, the *best approximation* to $f \in V$ in the *least squares* sense, [**89**].

Theorem 3.36. *Let $V_n = \operatorname{span}\{\varphi_1, \varphi_2, \ldots, \varphi_n\} \subset V$ be the n-dimensional subspace spanned by the first n elements of the orthonormal system. Then the nth order Fourier partial sum $s_n \in V_n$ is the best least squares approximation to f that belongs to the subspace, meaning that it minimizes $\| f - p_n \|$ among all possible $p_n \in V_n$.*

Proof: Given any element

$$p_n = \sum_{k=1}^{n} d_k \varphi_k \in V_n,$$

we have, in view of the orthonormality relations (3.107),

$$\begin{aligned} \| p_n \|^2 &= \langle p_n, p_n \rangle \\ &= \left\langle \sum_{j=1}^{n} d_j \varphi_j, \sum_{k=1}^{n} d_k \varphi_k \right\rangle = \sum_{j,k=1}^{n} d_j d_k \langle \varphi_j, \varphi_k \rangle = \sum_{k=1}^{n} | d_k |^2, \end{aligned} \tag{3.110}$$

reproducing the formula (B.27) for the norm with respect to an orthonormal basis. Therefore, by the symmetry property of the real inner product,

$$\begin{aligned} \| f - p_n \|^2 &= \langle f - p_n, f - p_n \rangle = \| f \|^2 - 2 \langle f, p_n \rangle + \| p_n \|^2 \\ &= \| f \|^2 - 2 \sum_{k=1}^{n} d_k \langle f, \varphi_k \rangle + \| p_n \|^2 = \| f \|^2 - 2 \sum_{k=1}^{n} c_k d_k + \sum_{k=1}^{n} | d_k |^2 \\ &= \| f \|^2 - \sum_{k=1}^{n} | c_k |^2 + \sum_{k=1}^{n} | c_k - d_k |^2. \end{aligned}$$

The final equality results from adding and subtracting the squared norm of the partial sum (3.109),

$$\| s_n \|^2 = \sum_{k=1}^{n} | c_k |^2, \tag{3.111}$$

which is a particular case of (3.110). We conclude that

$$\| f - p_n \|^2 = \| f \|^2 - \| s_n \|^2 + \sum_{k=1}^{n} | c_k - d_k |^2. \tag{3.112}$$

The first and second terms on the right-hand side of (3.112) are uniquely determined by f and hence cannot be altered by the choice of $p_n \in V_n$, which affects only the final summation. Since the latter is a sum of nonnegative quantities, it is clearly minimized by setting all its summands to zero, i.e., setting $d_k = c_k$ for all $k = 1, \ldots, n$. We conclude that $\| f - p_n \|$ achieves its minimum value among all $p_n \in V_n$ if and only if $d_k = c_k$, which implies that $p_n = s_n$ is the Fourier partial sum (3.109). *Q.E.D.*

Example 3.37. Consider the ordinary real Fourier series. The subspace $\mathcal{T}^{(n)} \subset \mathrm{L}^2$ spanned by the trigonometric functions $\cos kx$, $\sin kx$, for $0 \le k \le n$, consists of all *trigonometric polynomials* (finite Fourier sums) of degree $\le n$:

$$p_n(x) = \frac{r_0}{2} + \sum_{k=1}^{n} \left[r_k \cos kx + s_k \sin kx \right]. \tag{3.113}$$

Theorem 3.36 implies that the nth Fourier partial sum (3.38) is distinguished as the one that best approximates $f(x)$ in the least squares sense, meaning that it minimizes the L^2 norm of the difference,

$$\| f - p_n \| = \sqrt{\frac{1}{\pi} \int_{-\pi}^{\pi} | f(x) - p_n(x) |^2 \, dx} \,, \tag{3.114}$$

among all such trigonometric polynomials (3.113).

Returning to the general framework, if we set $p_n = s_n$, so $d_k = c_k$, in (3.112), we conclude that the minimizing least squares error for the Fourier partial sum is

$$0 \le \| f - s_n \|^2 = \| f \|^2 - \| s_n \|^2 = \| f \|^2 - \sum_{k=1}^{n} | c_k |^2. \tag{3.115}$$

We conclude that the general Fourier coefficients of the function f must satisfy the inequality

$$\sum_{k=1}^{n} | c_k |^2 \le \| f \|^2. \tag{3.116}$$

Let us see what happens in the limit as $n \to \infty$. Since we are summing a sequence of nonnegative numbers, with uniformly bounded partial sums, the limiting summation must exist and be subject to the same bound. We have thus established *Bessel's inequality*, a key step on the road to the general theory.

Theorem 3.38. *The sum of the squares of the general Fourier coefficients of $f \in V$ is bounded by*

$$\sum_{k=1}^{\infty} | c_k |^2 \le \| f \|^2. \tag{3.117}$$

Now, if a series, such as that on the left-hand side of Bessel's inequality (3.117), is to converge, the individual summands must go to zero. Thus, we immediately deduce:

Corollary 3.39. *The general Fourier coefficients of $f \in V$ satisfy $c_k \to 0$ as $k \to \infty$.*

In the case of the trigonometric Fourier series, Corollary 3.39 yields the following simplified form of what is known as the *Riemann–Lebesgue Lemma.*

Lemma 3.40. *If $f \in \mathrm{L}^2[-\pi,\pi]$ is square-integrable, then its Fourier coefficients satisfy*

$$\left.\begin{aligned} a_k &= \frac{1}{\pi}\int_{-\pi}^{\pi} f(x)\cos kx\,dx \\ b_k &= \frac{1}{\pi}\int_{-\pi}^{\pi} f(x)\sin kx\,dx \end{aligned}\right\} \longrightarrow 0 \qquad \text{as} \qquad k \longrightarrow \infty. \tag{3.118}$$

Remark: This result is equivalent to the decay of the complex Fourier coefficients

$$c_k = \frac{1}{2\pi}\int_{-\pi}^{\pi} f(x)\,e^{-\,\mathrm{i}\,kx}\,dx \longrightarrow 0 \qquad \text{as} \qquad |\,k\,| \longrightarrow \infty, \tag{3.119}$$

of any complex-valued square-integrable function.

Convergence of the sum (3.117) requires that the coefficients c_k not tend to zero too slowly. For instance, requiring the power bound (3.99) for some $\alpha > \frac{1}{2}$ suffices to ensure that $\displaystyle\sum_{k=-\infty}^{\infty} |\,c_k\,|^2 < \infty$. Thus, as we should have expected, convergence in norm of the Fourier series imposes less-restrictive requirements on the decay of the Fourier coefficients than uniform convergence — which needed $\alpha > 1$. Indeed, a Fourier series with slowly decaying coefficients may very well converge in norm to a discontinuous L^2 function, which is not possible under uniform convergence.

Completeness

Calculations in vector spaces rely on the specification of a basis, meaning a set of linearly independent elements that span the space. The choice of basis serves to introduce a system of local coordinates on the space, namely, the coefficients in the expression of an element as a linear combination of basis elements. Orthogonal and orthonormal bases are particularly handy, since the coordinates are immediately calculated by taking inner products, while general bases require solving linear systems. In finite-dimensional vector spaces, all bases contain the same number of elements, which, by definition, is the *dimension* of the space. A vector space is, therefore, infinite-dimensional if it contains an infinite number of linearly independent elements. However, the question when such a collection forms a basis for the space is considerably more delicate, and mere counting will no longer suffice. Indeed, omitting a finite number of elements from an infinite collection would still leave an infinite number, but the latter will certainly not span the space. Moreover, we cannot, in general, expect to write a general element of an infinite-dimensional space as a finite linear combination of basis elements, and so subtle questions of convergence of infinite series must also be addressed if we are to properly formulate the concept.

The definition of a basis of an infinite-dimensional vector space rests on the idea of completeness. We shall discuss completeness in the general abstract setting, but the key example is, of course, the Hilbert space $\mathrm{L}^2[-\pi,\pi]$ and the systems of trigonometric or complex exponential functions. For simplicity, we define completeness in terms of orthonormal systems here. (Similar arguments will clearly apply to orthogonal systems, but normality helps to streamline the presentation.)

Definition 3.41. An orthonormal system $\varphi_1, \varphi_2, \varphi_3, \ldots \in V$ is called *complete* if, for every $f \in V$, its generalized Fourier series (3.108) converges in norm to f:

$$\| f - s_n \| \longrightarrow 0, \quad \text{as} \quad n \to \infty, \qquad \text{where} \quad s_n = \sum_{k=1}^{n} c_k \varphi_k, \qquad c_k = \langle f, \varphi_k \rangle, \tag{3.120}$$

is the nth partial sum of the generalized Fourier series (3.108).

Thus, completeness requires that every element of V can be arbitrarily closely approximated (in norm) by a finite linear combination of the basis elements. A complete orthonormal system should be viewed as the infinite-dimensional version of an orthonormal basis of a finite-dimensional vector space. An orthogonal system is called *complete* whenever the corresponding orthonormal system obtained by dividing the elements by their norms is complete. Existence of a complete orthonormal system is directly tied to completeness of the underlying Hilbert space.

Determining whether a given orthonormal or orthogonal system of functions is complete is a difficult problem, and requires some detailed analysis of their properties. The key result for classical Fourier series is that the trigonometric functions, or, equivalently, the complex exponentials, form a complete system; an indication of its proof will appear below. A general characterization of complete orthonormal eigenfunction systems can be found in Section 9.4.

Theorem 3.42. *The trigonometric functions* $1, \cos kx, \sin kx, \quad k = 1, 2, 3, \ldots,$ *form a complete orthogonal system in* $\mathrm{L}^2 = \mathrm{L}^2[-\pi, \pi]$. *In other words, if* $s_n(x)$ *denotes the* n^{th} *partial sum of the Fourier series of the square-integrable function* $f(x) \in \mathrm{L}^2$, *then* $\lim_{n \to \infty} \| f - s_n \| = 0$.

To better comprehend completeness, let us describe some equivalent characterizations and consequences. One is the infinite-dimensional counterpart of formula (B.27) for the norm of a vector in terms of its coordinates with respect to an orthonormal basis.

Theorem 3.43. *The orthonormal system* $\varphi_1, \varphi_2, \varphi_3, \ldots \in V$ *is complete if and only if* Plancherel's formula

$$\| f \|^2 = \sum_{k=1}^{\infty} | c_k |^2 = \sum_{k=1}^{\infty} \langle f, \varphi_k \rangle^2 \tag{3.121}$$

holds for every $f \in V$.

Proof: Theorem 3.43, thus, states that the system of functions is complete if and only if the Bessel inequality (3.117) is, in fact, an equality. Indeed, letting $n \to \infty$ in (3.115), we find

$$\lim_{n \to \infty} \| f - s_n \|^2 = \| f \|^2 - \lim_{n \to \infty} \sum_{k=1}^{n} | c_k |^2 = \| f \|^2 - \sum_{k=1}^{\infty} | c_k |^2.$$

Therefore, the completeness condition (3.120) holds if and only if the right-hand side vanishes, which is the Plancherel identity (3.121). *Q.E.D.*

An analogous result holds for the inner product between two elements, which we state in its general complex form, although the proof given here is for the real version; in Exercise 3.5.35 the reader is asked to supply the slightly more intricate complex proof.

Corollary 3.44. *The Fourier coefficients* $c_k = \langle f, \varphi_k \rangle$, $d_k = \langle g, \varphi_k \rangle$, *of any* $f, g \in V$ *satisfy* Parseval's formula

$$\langle f, g \rangle = \sum_{k=1}^{\infty} c_k \overline{d}_k. \tag{3.122}$$

Proof: Since, for a real inner product,

$$\langle f, g \rangle = \tfrac{1}{4} \left(\| f + g \|^2 - \| f - g \|^2 \right), \tag{3.123}$$

Parseval's formula results from applying Plancherel's formula (3.121) to each term on the right-hand side:

$$\langle f, g \rangle = \frac{1}{4} \sum_{k=1}^{\infty} \left[(c_k + d_k)^2 - (c_k - d_k)^2 \right] = \sum_{k=1}^{\infty} c_k \, d_k,$$

which agrees with (3.122), since we are assuming that $d_k = \overline{d}_k$ are all real. *Q.E.D.*

Note that Plancherel's formula is a special case of Parseval's formula,[†] obtained by setting $f = g$. In the particular case of the complex exponential basis $e^{\,i\,k\,x}$ of $L^2[-\pi, \pi]$, the Plancherel and Parseval formulae take the form

$$\frac{1}{2\pi} \int_{-\pi}^{\pi} |\, f(x) \,|^2 \, dx = \sum_{k=-\infty}^{\infty} |\, c_k \,|^2, \qquad \frac{1}{2\pi} \int_{-\pi}^{\pi} f(x) \, \overline{g(x)} \, dx = \sum_{k=-\infty}^{\infty} c_k \, \overline{d_k}, \tag{3.124}$$

where $c_k = \langle f, e^{\,i\,k\,x} \rangle$, $d_k = \langle g, e^{\,i\,k\,x} \rangle$ are the ordinary Fourier coefficients of the complex-valued functions $f(x)$ and $g(x)$. In Exercise 3.5.38, you are asked to write the corresponding formulas for the real Fourier coefficients.

Completeness also tells us that a function is uniquely determined by its Fourier coefficients.

Proposition 3.45. *If the orthonormal system* $\varphi_1, \varphi_2, \ldots \in V$ *is complete, then the only element* $f \in V$ *with all zero Fourier coefficients,* $0 = c_1 = c_2 = \cdots$, *is the zero element:* $f = 0$. *More generally, two elements* $f, g \in V$ *have the same Fourier coefficients if and only if they are the same:* $f = g$.

Proof: The proof is an immediate consequence of Plancherel's formula. Indeed, if $c_k = 0$, then (3.121) implies that $\| f \| = 0$ and hence $f = 0$. The second statement follows by applying the first to their difference $f - g$. *Q.E.D.*

Another way of stating this result is that the only function that is orthogonal to every element of a complete orthonormal system is the zero function.[‡] In other words, a complete orthonormal system is maximal in the sense that no further orthonormal elements can be appended to it.

[†] Curiously, Marc-Antoine Parseval des Chênes' contribution slightly predates Fourier, whereas Michel Plancherel's appeared almost a century later.

[‡] Or, to be more technically accurate, any function that is zero outside a set of measure zero.

Let us now discuss the completeness of the Fourier trigonometric and complex exponential functions. We shall establish the completeness property only for sufficiently smooth functions, leaving the harder general proof to the references, [**37, 128**].

According to Theorem 3.30, if $f(x)$ is continuous, 2π periodic, and piecewise C^1, its Fourier series converges uniformly,

$$f(x) = \sum_{k=-\infty}^{\infty} c_k \, e^{\,\mathrm{i}\,k\,x} \qquad \text{for all} \qquad -\pi \leq x \leq \pi.$$

The same holds for its complex conjugate $\overline{f(x)}$. Therefore,

$$|\, f(x) \,|^2 = f(x)\,\overline{f(x)} = f(x) \sum_{k=-\infty}^{\infty} \overline{c}_k \, e^{-\,\mathrm{i}\,k\,x} = \sum_{k=-\infty}^{\infty} \overline{c}_k \, f(x) \, e^{-\,\mathrm{i}\,k\,x},$$

which also converges uniformly by (3.94). Formula (3.95) permits us to integrate both sides from $-\pi$ to π, yielding

$$\|\, f \,\|^2 = \frac{1}{2\pi} \int_{-\pi}^{\pi} |\, f(x) \,|^2 \, dx = \sum_{k=-\infty}^{\infty} \frac{\overline{c}_k}{2\pi} \int_{-\pi}^{\pi} f(x) \, e^{-\,\mathrm{i}\,k\,x} \, dx = \sum_{k=-\infty}^{\infty} c_k \, \overline{c}_k = \sum_{k=-\infty}^{\infty} |\, c_k \,|^2.$$

Therefore, Plancherel's formula (3.121) holds for any continuous, piecewise C^1 function.

With some additional technical work, this result is used to establish the validity of Plancherel's formula for all $f \in \mathrm{L}^2$, the key step being to suitably approximate f by such continuous, piecewise C^1 functions. With this in hand, completeness is an immediate consequence of Theorem 3.43. *Q.E.D.*

Pointwise Convergence

Let us finally return to the Pointwise Convergence Theorem 3.8 for the trigonometric Fourier series. The goal is to prove that, under the appropriate hypotheses on $f(x)$, namely 2π–periodic and piecewise C^1, the limit of its partial Fourier sums is

$$\lim_{n \to \infty} s_n(x) = \tfrac{1}{2} \left[\, f(x^+) + f(x^-) \,\right]. \tag{3.125}$$

We begin by substituting the formulae (3.65) for the complex Fourier coefficients into the formula (3.109) for the nth partial sum:

$$\begin{aligned} s_n(x) = \sum_{k=-n}^{n} c_k \, e^{\,\mathrm{i}\,k\,x} &= \sum_{k=-n}^{n} \left(\frac{1}{2\pi} \int_{-\pi}^{\pi} f(y) \, e^{-\,\mathrm{i}\,k\,y} \, dy \right) e^{\,\mathrm{i}\,k\,x} \\ &= \frac{1}{2\pi} \int_{-\pi}^{\pi} f(y) \left(\sum_{k=-n}^{n} e^{\,\mathrm{i}\,k(x-y)} \right) dy. \end{aligned} \tag{3.126}$$

To proceed further, we need to calculate the final summation

$$\sum_{k=-n}^{n} e^{\,\mathrm{i}\,k\,x} = e^{-\,\mathrm{i}\,n\,x} + \cdots + e^{-\,\mathrm{i}\,x} + 1 + e^{\,\mathrm{i}\,x} + \cdots + e^{\,\mathrm{i}\,n\,x}.$$

This, in fact, has the form of a geometric sum,

$$\sum_{k=0}^{m} a\,r^k = a + a\,r + a\,r^2 + \cdots + a\,r^m = a\left(\frac{r^{m+1}-1}{r-1}\right), \tag{3.127}$$

with $m+1 = 2n+1$ summands, initial term $a = e^{-\,i\,n x}$, and ratio $r = e^{\,i\,x}$. Therefore,

$$\begin{aligned}\sum_{k=-n}^{n} e^{\,i\,k x} &= e^{-\,i\,n x}\left(\frac{e^{\,i\,(2n+1)x}-1}{e^{\,i\,x}-1}\right) = \frac{e^{\,i\,(n+1)x}-e^{-\,i\,n x}}{e^{\,i\,x}-1} \\ &= \frac{e^{\,i\left(n+\frac12\right)x}-e^{-\,i\left(n+\frac12\right)x}}{e^{\,i\,x/2}-e^{-\,i\,x/2}} = \frac{\sin\left(n+\frac12\right)x}{\sin\frac12 x}.\end{aligned} \tag{3.128}$$

In this computation, to pass from the first to the second line, we multiplied numerator and denominator by $e^{-\,i\,x/2}$, after which we used the formula (3.60) for the sine function in terms of complex exponentials. Incidentally, (3.128) is equivalent to the intriguing trigonometric summation formula

$$1 + 2\big(\cos x + \cos 2x + \cos 3x + \cdots + \cos n x\big) = \frac{\sin\left(n+\frac12\right)x}{\sin\frac12 x}. \tag{3.129}$$

Therefore, substituting back into (3.126), we obtain

$$\begin{aligned} s_n(x) &= \frac{1}{2\pi}\int_{-\pi}^{\pi} f(y)\,\frac{\sin\left(n+\frac12\right)(x-y)}{\sin\frac12(x-y)}\,dy \\ &= \frac{1}{2\pi}\int_{-\pi-x}^{\pi-x} f(x+\widehat{y})\,\frac{\sin\left(n+\frac12\right)\widehat{y}}{\sin\frac12\widehat{y}}\,d\widehat{y} = \frac{1}{2\pi}\int_{-\pi}^{\pi} f(x+y)\,\frac{\sin\left(n+\frac12\right)y}{\sin\frac12 y}\,dy.\end{aligned}$$

The second equality is the result of changing the integration variable to $\widehat{y} = x - y$ and canceling the minus signs in the resulting trigonometric fraction, while the final equality follows since the integrand is 2π–periodic, and so its integrals over *any* interval of length 2π all have the same value; see Exercise 3.2.9.

Thus, to prove (3.125), it suffices to show that

$$\begin{aligned}\lim_{n\to\infty}\frac{1}{\pi}\int_{0}^{\pi} f(x+y)\,\frac{\sin\left(n+\frac12\right)y}{\sin\frac12 y}\,dy &= f(x^+), \\ \lim_{n\to\infty}\frac{1}{\pi}\int_{-\pi}^{0} f(x+y)\,\frac{\sin\left(n+\frac12\right)y}{\sin\frac12 y}\,dy &= f(x^-).\end{aligned} \tag{3.130}$$

The proofs of the two formulas are identical, and so we concentrate on establishing the first. Using the fact that the integrand is even, and then our summation formula (3.128) in reverse, yields

$$\frac{1}{\pi}\int_{0}^{\pi}\frac{\sin\left(n+\frac12\right)y}{\sin\frac12 y}\,dy = \frac{1}{2\pi}\int_{-\pi}^{\pi}\frac{\sin\left(n+\frac12\right)y}{\sin\frac12 y}\,dy = \frac{1}{2\pi}\int_{-\pi}^{\pi}\sum_{k=-n}^{n} e^{\,i\,k y}\,dy = 1,$$

because only the constant term has a nonzero integral. Multiplying this formula by $f(x^+)$ and then subtracting the result from the first formula in (3.130) leads to

$$\lim_{n\to\infty}\frac{1}{\pi}\int_{0}^{\pi}\frac{f(x+y)-f(x^+)}{\sin\frac12 y}\,\sin\left(n+\tfrac12\right)y\,dy = 0, \tag{3.131}$$

which we now proceed to prove.

We claim that, for each fixed value of x, the function

$$g(y) = \frac{f(x+y) - f(x^+)}{\sin \frac{1}{2} y}$$

is piecewise continuous for all $0 \le y \le \pi$. Owing to our hypotheses on $f(x)$, the only problematic point is at $y = 0$, but then, by l'Hôpital's Rule (for one-sided limits),

$$\lim_{y \to 0^+} g(y) = \lim_{y \to 0^+} \frac{f(x+y) - f(x^+)}{\sin \frac{1}{2} y} = \lim_{y \to 0^+} \frac{f'(x+y)}{\frac{1}{2} \cos \frac{1}{2} y} = 2\, f'(x^+).$$

Consequently, (3.131) will be established if we can show that

$$\lim_{n \to \infty} \frac{1}{\pi} \int_0^\pi g(y) \sin \left(n + \tfrac{1}{2}\right) y \; dy = 0 \tag{3.132}$$

whenever g is piecewise continuous. Were it not for the extra $\frac{1}{2}$, this would immediately follow from the simplified Riemann–Lebesgue Lemma 3.40. More honestly, we can invoke the addition formula for $\sin \left(n + \frac{1}{2}\right) y$ to write

$$\frac{1}{\pi} \int_0^\pi g(y) \sin \left(n + \tfrac{1}{2}\right) y \, dy = \frac{1}{\pi} \int_0^\pi \left(g(y) \sin \tfrac{1}{2} y \right) \cos n y \, dy + \frac{1}{\pi} \int_0^\pi \left(g(y) \cos \tfrac{1}{2} y \right) \sin n y \, dy.$$

The first integral is the nth Fourier cosine coefficient for the piecewise continuous function $g(y) \sin \frac{1}{2} y$, while the second integral is the nth Fourier sine coefficient for the piecewise continuous function $g(y) \cos \frac{1}{2} y$. Lemma 3.40 implies that both of these converge to zero as $n \to \infty$, and hence (3.132) holds. This completes the proof, thus establishing pointwise convergence of the Fourier series. *Q.E.D.*

Remark: An alternative approach to the last part of the proof is to use the general *Riemann–Lebesgue Lemma*, whose proof can be found in [**37, 128**].

Lemma 3.46. *Suppose $g(x)$ is piecewise continuous on $[a, b]$. Then*

$$\begin{aligned} 0 &= \lim_{\omega \to \infty} \int_a^b g(x)\, e^{\,\mathrm{i}\,\omega x} \, dx \\ &= \lim_{\omega \to \infty} \int_a^b g(x) \cos \omega x \, dx + \mathrm{i} \lim_{\omega \to \infty} \int_a^b g(x) \sin \omega x \, dx. \end{aligned} \tag{3.133}$$

Intuitively, the Riemann–Lebesgue Lemma says that, as the frequency ω gets larger and larger, the increasingly rapid oscillations of the integrand tend to cancel each other out.

Remark: While the Fourier series of a merely continuous function need not converge pointwise everywhere, a deep theorem, proved by the Swedish mathematician Lennart Carleson in 1966, [**28**], states that the set of points where it does not converge has measure zero, and hence the exceptional points form a very small subset.

Exercises

3.5.26. Which of the following sequences converge in norm to the zero function on $\mathbb{R}$?

(a) $v_n(x) = \dfrac{n\,x}{1+n^2x^2}$, (b) $v_n(x) = \begin{cases} 1, & n < x < n+1, \\ 0, & \text{otherwise}, \end{cases}$

(c) $v_n(x) = \begin{cases} 1, & n < x < n+1/n, \\ 0, & \text{otherwise}, \end{cases}$ (d) $v_n(x) = \begin{cases} 1/n, & n < x < 2n, \\ 0, & \text{otherwise}, \end{cases}$

(e) $v_n(x) = \begin{cases} 1/\sqrt{n}, & n < x < 2n, \\ 0, & \text{otherwise}, \end{cases}$ (f) $v_n(x) = \begin{cases} n^2x^2 - 1, & -1/n < x < 1/n, \\ 0, & \text{otherwise}. \end{cases}$

3.5.27. Discuss pointwise and L^2 convergence of the following sequences on the interval $[0,1]$:

(a) $1 - \dfrac{x^2}{n^2}$, (b) $\begin{cases} n, & 1/n^2 < x < 1/n, \\ x, & \text{otherwise}, \end{cases}$ (c) e^{-nx}, (d) $\sin n\,x$.

3.5.28. Prove, directly from the definition, the convergence in norm of the Fourier series (3.49) of the step function.

3.5.29. Let $f(x) \in \mathrm{L}^2[a,b]$ be square integrable. Which constant function $g(x) \equiv c$ best approximates f in the least squares sense?

3.5.30. Suppose the sequence $f_n(x)$ converges pointwise to a function $f_\star(x)$ on an interval $[a,b]$, and converges to $g_\star(x)$ in the L^2 norm on $[a,b]$. Is $f_\star(x) = g_\star(x)$ at every $a \le x \le b$?

3.5.31. Find a formula for the L^2 norm of the Fourier series in Exercises 3.5.20 and 3.5.22.

3.5.32. Under what conditions on the function $f(x)$ is the least squares error due to the nth order Fourier partial sum equal to zero: $\| f - s_n \| = 0$?

♢ 3.5.33. Let V be an inner product space. Prove that the elements of a (finite or infinite) orthonormal system $\varphi_1, \varphi_2, \ldots \in V$ are linearly independent, meaning that any *finite* linear combination vanishes, $c_1\varphi_1 + \cdots + c_n\varphi_n = 0$, if and only if the coefficients are all zero: $c_1 = \cdots = c_n = 0$.

♢ 3.5.34. Let V be a complex inner product space. Prove that, for all $f, g \in V$,

(a) $\| f + g \|^2 = \| f \|^2 + 2\,\mathrm{Re}\,\langle f\,, g \rangle + \| g \|^2$;

(b) $\langle f\,, g \rangle = \frac{1}{4}\left(\| f + g \|^2 - \| f - g \|^2 + \mathrm{i}\,\| f + \mathrm{i}\,g \|^2 - \mathrm{i}\,\| f - \mathrm{i}\,g \|^2 \right)$.

♢ 3.5.35. Let V be an infinite-dimensional complex inner product space, and $\varphi_k \in V$ a complete orthonormal system. Prove the corresponding *Plancherel* and *Parseval formulas*.
Hint: Use the identities in Exercise 3.5.34.

3.5.36. What does Plancherel's formula (3.121) tell us in a finite-dimensional vector space? What about Parseval's formula (3.122)?

3.5.37. Let $f(x) = x$, $g(x) = \operatorname{sign} x$. (a) Write out Plancherel's formula for the complex Fourier coefficients of f. (b) Write out Plancherel's formula for the complex Fourier coefficients of g. (c) Write out Parseval's formula for the complex Fourier coefficients of f, g.

♢ 3.5.38. (a) Prove the real version of the *Plancherel formula*

$$\frac{1}{\pi}\int_{-\pi}^{\pi} |\, f(x)\,|^2\, dx = \tfrac{1}{2}\, a_0^2 + \sum_{k=1}^{\infty} (a_k^2 + b_k^2) \qquad (3.134)$$

for the trigonometric Fourier coefficients of a real function $f(x)$.
(b) What is the real version of Parseval's formula?

3.5.39. Give an alternative proof of formula (3.129) that does not require complex functions by first multiplying through by $\sin \frac{1}{2} x$ and then invoking a suitable trigonometric identity for the product terms.

3.5.40. (a) Prove that the functions $\varphi_n(x) = \sin\left(n - \frac{1}{2}\right)x$, for $n = 1, 2, 3, \ldots$, form an orthogonal sequence on the interval $[0, \pi]$ relative to the L^2 inner product $\langle f, g \rangle = \int_0^\pi f(x)\, g(x)\, dx$. (b) Find the formula for the Fourier coefficients of a function $f(x)$ relative to the orthogonal sequence $\varphi_n(x)$. (c) State Bessel's inequality and Plancherel's formula in this case. Carefully state any hypotheses that might be required for the validity of your formulas.

♢ 3.5.41. Prove that a sequence of vectors $\mathbf{v}^{(n)} \in \mathbb{R}^m$ converges in the Euclidean norm, $\| \mathbf{v}^{(n)} - \mathbf{v}^\star \| \to 0$ as $n \to \infty$, if and only if their individual components converge: $v_i^{(n)} \to v_i^\star$ for $i = 1, \ldots, m$.

♢ 3.5.42. Let V be a normed vector space. A sequence $\mathbf{v}_n \in V$ is called a *Cauchy sequence* if for every $\varepsilon > 0$ there exists an N such that $\| \mathbf{v}_m - \mathbf{v}_n \| < \varepsilon$ whenever both $m, n \geq N$. Prove that a sequence that converges in norm, $\| \mathbf{v}_n - \mathbf{v}^\star \| \to 0$ as $n \to \infty$, is necessarily a Cauchy sequence. *Remark*: A normed vector space is called *complete* if *every* Cauchy sequence converges in norm. It can be proved, [**96, 98**], that any finite-dimensional normed vector space is complete, but this is not necessarily the case in infinite dimensions. For example, the vector spaces consisting of all trigonometric polynomials and of all polynomials are not complete in the L^2 norm. The most important example of a complete infinite-dimensional vector space is the Hilbert space L^2.

♢ 3.5.43. For each $n = 1, 2, \ldots$, define the function $f_n(x) = \begin{cases} 1, & \frac{k}{m} \leq x \leq \frac{k+1}{m}, \\ 0, & \text{otherwise}, \end{cases}$ where $n = \frac{1}{2} m(m+1) + k$ and $0 \leq k \leq m$. Show first that m, k are uniquely determined by n. Then prove that, on the interval $[0, 1]$, the sequence $f_n(x)$ converges in norm to 0 but does not converge pointwise *anywhere*!

♡ 3.5.44. Let $u(t, x)$ solve the initial value problem $\dfrac{\partial^2 u}{\partial t^2} = c^2 \dfrac{\partial^2 u}{\partial x^2}$, $u(0, x) = f(x)$, $\dfrac{\partial u}{\partial t}(0, x) = 0$, for $-\infty < x < \infty$, where $f(x) \to 0$ as $|x| \to \infty$. *True or false*: As $t \to \infty$, the solution $u(t, x)$ converges to an equilibrium solution (a) pointwise; (b) uniformly; (c) in norm.

♡ 3.5.45. Answer Exercise 3.5.44 for the initial conditions $u(0, x) = 0$, $\dfrac{\partial u}{\partial t}(0, x) = g(x)$, with $g(x) \to 0$ as $|x| \to \infty$.

Chapter 4
Separation of Variables

Three cardinal linear second-order partial differential equations have collectively driven the development of the entire subject. The first two we have already encountered: The *wave equation* describes vibrations and waves in continuous media, including sound waves, water waves, elastic waves, electromagnetic waves, and so on. The *heat equation* models diffusion processes, including thermal energy in solids, solutes in liquids, and biological populations. Third, and in many ways the most important of all, is the *Laplace equation* and its inhomogeneous counterpart, the *Poisson equation*, which govern equilibrium mechanics. The latter two equations arise in an astonishing variety of mathematical and physical contexts, ranging through elasticity and solid mechanics, fluid mechanics, electromagnetism, potential theory, thermomechanics, geometry, probability, number theory, and many other fields. The solutions to the Laplace equation are known as harmonic functions, and the discovery of their many remarkable properties forms one of the most celebrated chapters in the history of mathematics. All three equations, along with their multi-dimensional kin, will appear repeatedly throughout this text.

The aim of the current chapter is to develop the method of separation of variables for solving these key partial differential equations in their two-independent-variable incarnations. For the wave and heat equations, the variables are time, t, and a single space coordinate, x, leading to initial-boundary value problems modeling the dynamical behavior of a one-dimensional medium. For the Laplace and Poisson equations, both variables represent space coordinates, x and y, and the associated boundary value problems model the equilibrium configuration of a planar body, e.g., the deformations of a membrane. Separation of variables seeks special solutions that can be written as the product of functions of the individual variables, thereby reducing the partial differential equation to a pair of ordinary differential equations. More-general solutions can then be expressed as infinite series in the appropriate separable solutions. For the two-variable equations considered here, this results in a Fourier series representation of the solution. In the case of the wave equation, separation of variables serves to focus attention on the vibrational character of the solution, whereas the earlier d'Alembert approach emphasizes its particle-like aspects. Unfortunately, for the Laplace equation, separation of variables applies only to boundary value problems in very special geometries, e.g., rectangles and disks. Further development of the separation of variables method for solving partial differential equations in three or more variables can be found in Chapters 11 and 12.

In the final section, we take the opportunity to summarize the fundamental tripartite classification of planar second-order partial differential equations. Each of the three paradigmatic equations epitomizes one of the classes: *hyperbolic*, such as the wave equation; *parabolic*, such as the heat equation; and *elliptic*, such as the Laplace and Poisson equations. Each category enjoys its own distinctive properties and features, both analytic and numeric, and, in effect, forms a separate mathematical subdiscipline.

4.1 The Diffusion and Heat Equations

Let us begin with a brief physical derivation of the heat equation from first principles. We consider a *bar* — meaning a thin, heat-conducting body. "Thin" means that we can regard the bar as a one-dimensional continuum with no significant transverse temperature variation. We will assume that the bar is fully insulated along its length, and so heat can enter (or leave) only through its uninsulated endpoints. We use t to represent time, and $a \leq x \leq b$ to denote spatial position along the bar, which occupies the interval $[a, b]$. Our goal is to find the temperature $u(t, x)$ of the bar at position x and time t.

The dynamical equations governing the temperature are based on three fundamental physical principles. First is the Law of Conservation of Heat Energy. Recalling the general Definition 2.7, this particular conservation law takes the form

$$\frac{\partial \varepsilon}{\partial t} + \frac{\partial w}{\partial x} = 0, \tag{4.1}$$

in which $\varepsilon(t, x)$ represents the thermal *energy density* at time t and position x, while $w(t, x)$ denotes the *heat flux*, i.e., the rate of flow of thermal energy along the bar. Our sign convention is that $w(t, x) > 0$ at points where the energy flows in the direction of increasing x (left to right). The integrated form (2.49) of the conservation law, namely

$$\frac{d}{dt} \int_a^b \varepsilon(t, x)\, dx = w(t, a) - w(t, b), \tag{4.2}$$

states that the rate of change in the thermal energy within the bar is equal to the total heat flux passing through its uninsulated ends. The signs of the boundary terms confirm that heat flux *into* the bar results in an increase in temperature.

The second ingredient is a *constitutive assumption* concerning the bar's material properties. It has been observed that, under reasonable conditions, thermal energy is proportional to temperature:

$$\varepsilon(t, x) = \sigma(x)\, u(t, x). \tag{4.3}$$

The factor

$$\sigma(x) = \rho(x)\, \chi(x) > 0 \tag{4.4}$$

is the product of the *density* ρ of the material and its *specific heat capacity* χ, which is the amount of heat energy required to raise the temperature of a unit mass of the material by one degree. Note that we are assuming that the medium is not changing in time, and so physical quantities such as density and specific heat depend only on position x. We also assume, perhaps with less physical justification, that its material properties do not depend upon the temperature; otherwise, we would be forced to deal with a much thornier nonlinear diffusion equation, [**70, 99**].

The third physical principle relates heat flux and temperature. Physical experiments show that the thermal energy moves from hot to cold at a rate that is in direct proportion to the temperature gradient, which, in the one-dimensional case, means its derivative $\partial u/\partial x$. The resulting relation

$$w(t,x) = -\,\kappa(x)\,\frac{\partial u}{\partial x} \tag{4.5}$$

is known as *Fourier's Law of Cooling*. The proportionality factor $\kappa(x) > 0$ is the *thermal conductivity* of the bar at position x, and the minus sign reflects the everyday observation that heat energy moves from hot to cold. A good heat conductor, e.g., silver, will have high conductivity, while a poor conductor, e.g., glass, will have low conductivity.

Combining the three laws (4.1, 3, 5) produces the *linear diffusion equation*

$$\frac{\partial}{\partial t}\bigl(\,\sigma(x)\,u\,\bigr) = \frac{\partial}{\partial x}\left(\kappa(x)\,\frac{\partial u}{\partial x}\right), \qquad a < x < b, \tag{4.6}$$

governing the thermodynamics of a one-dimensional medium. It is also used to model a wide variety of diffusive processes, including chemical diffusion, diffusion of contaminants in liquids and gases, population dispersion, and the spread of infectious diseases. If there is an external heat source along the length of the bar, then the diffusion equation acquires an additional prescribed inhomogeneous term:

$$\frac{\partial}{\partial t}\bigl(\,\sigma(x)\,u\,\bigr) = \frac{\partial}{\partial x}\left(\kappa(x)\,\frac{\partial u}{\partial x}\right) + h(t,x), \qquad a < x < b. \tag{4.7}$$

In order to uniquely prescribe the solution $u(t,x)$, we need to specify an initial temperature distribution

$$u(t_0,x) = f(x), \qquad a \le x \le b. \tag{4.8}$$

In addition, we must impose a suitable boundary condition at each end of the bar. There are three common types. The first is a *Dirichlet boundary condition*, where the end is held at a prescribed temperature. For example,

$$u(t,a) = \alpha(t) \tag{4.9}$$

fixes the temperature (possibly time-varying) at the left end. Alternatively, the *Neumann boundary condition*

$$\frac{\partial u}{\partial x}\,(t,a) = \mu(t) \tag{4.10}$$

prescribes the heat flux $w(t,a) = -\,\kappa(a) u_x(t,a)$ there. In particular, a homogeneous Neumann condition, $u_x(t,a) \equiv 0$, models an insulated end that prevents thermal energy flowing in or out. The *Robin*† *boundary condition*,

$$\frac{\partial u}{\partial x}\,(t,a) + \beta(t)\,u(t,a) = \tau(t), \tag{4.11}$$

models the heat exchange resulting from the end of the bar being placed in a heat bath (thermal reservoir) at temperature $\tau(t)$.

Each end of the bar is required to satisfy one of these boundary conditions. For example, a bar with both ends having prescribed temperatures is governed by the pair of Dirichlet boundary conditions

$$u(t,a) = \alpha(t), \qquad u(t,b) = \beta(t), \tag{4.12}$$

† Since it is named after the nineteenth-century French analyst Victor Gustave Robin, the pronunciation should be with a French accent.

whereas a bar with two insulated ends requires two homogeneous Neumann boundary conditions

$$\frac{\partial u}{\partial x}(t,a) = 0, \qquad \frac{\partial u}{\partial x}(t,b) = 0. \tag{4.13}$$

Mixed boundary conditions, with one end at a fixed temperature and the other insulated, are similarly formulated, e.g.,

$$u(t,a) = \alpha(t), \qquad \frac{\partial u}{\partial x}(t,b) = 0. \tag{4.14}$$

Finally, the *periodic boundary conditions*

$$u(t,a) = u(t,b), \qquad \frac{\partial u}{\partial x}(t,a) = \frac{\partial u}{\partial x}(t,b), \tag{4.15}$$

correspond to a circular *ring* obtained by joining the two ends of the bar. As before, we are assuming that the heat is allowed to flow only around the ring — insulation prevents the radiation of heat from one side of the ring affecting the other side.

The Heat Equation

In this book, we will retain the term "heat equation" to refer to the case in which the bar is composed of a uniform material, and so its density ρ, conductivity κ, and specific heat χ are all positive constants. We also exclude external heat sources (other than at the endpoints), meaning that the bar remains insulated along its entire length. Under these assumptions, the general diffusion equation (4.6) reduces to the homogeneous *heat equation*

$$\frac{\partial u}{\partial t} = \gamma \, \frac{\partial^2 u}{\partial x^2} \tag{4.16}$$

for the temperature $u(t,x)$ at time t and position x. The constant

$$\gamma = \frac{\kappa}{\sigma} = \frac{\kappa}{\rho\chi} \tag{4.17}$$

is called the *thermal diffusivity*; it incorporates all of the bar's relevant physical properties. The solution $u(t,x)$ will be uniquely prescribed once we specify initial conditions (4.8) and a suitable boundary condition at both of its endpoints.

As we learned in Section 3.1, the separable solutions to the heat equation are based on the exponential ansatz[†]

$$u(t,x) = e^{-\lambda t} v(x), \tag{4.18}$$

where $v(x)$ depends only on the spatial variable. Functions of this form, which "separate" into a product of a function of t times a function of x, are known as *separable solutions*. Substituting (4.18) into (4.16) and canceling the common exponential factors, we find that $v(x)$ must solve the second-order linear ordinary differential equation

$$-\gamma \, \frac{d^2 v}{dx^2} = \lambda \, v.$$

[†] Anticipating the eventual signs of the eigenvalues, and to facilitate later discussions, we now include a minus sign in the exponential term.

Each nontrivial solution $v(x) \not\equiv 0$ is an *eigenfunction*, with associated *eigenvalue* λ, for the linear differential operator $L[v] = -\gamma v''(x)$. With the separable eigensolutions (4.18) in hand, we will then be able to reconstruct the desired solution $u(t,x)$ as a linear combination, or rather infinite series, thereof.

Let us concentrate on the simplest case: a uniform, insulated bar of length ℓ that is held at zero temperature at both ends. We specify its initial temperature $f(x)$ at time $t_0 = 0$, and so the relevant initial and boundary conditions are

$$\begin{aligned} u(t,0) = 0, \qquad u(t,\ell) = 0, \qquad & t \geq 0, \\ u(0,x) = f(x), \qquad & 0 \leq x \leq \ell. \end{aligned} \tag{4.19}$$

The eigensolutions (4.18) are found by solving the Dirichlet boundary value problem

$$\gamma \frac{d^2 v}{dx^2} + \lambda v = 0, \qquad v(0) = 0, \qquad v(\ell) = 0. \tag{4.20}$$

By direct calculation (as you are asked to do in Exercises 4.1.19–20), one finds that if λ is either complex, or real and nonpositive, then the only solution to the boundary value problem (4.20) is the trivial solution $v(x) \equiv 0$. This means that all the eigenvalues must necessarily be real and positive. In fact, the reality and positivity of the eigenvalues need not be explicitly checked. Rather, they follow from very general properties of positive definite boundary value problems, of which (4.20) is a particular case. See Section 9.5 for the underlying theory and Theorem 9.34 for the relevant result.

When $\lambda > 0$, the general solution to the differential equation is a trigonometric function

$$v(x) = a \cos \omega x + b \sin \omega x, \qquad \text{where} \qquad \omega = \sqrt{\lambda/\gamma},$$

and a and b are arbitrary constants. The first boundary condition requires $v(0) = a = 0$. This serves to eliminate the cosine term, and then the second boundary condition requires

$$v(\ell) = b \sin \omega \ell = 0.$$

Therefore, since we require $b \neq 0$ — otherwise, the solution is trivial and does not qualify as an eigenfunction — $\omega \ell$ must be an integer multiple of π, and so

$$\omega = \frac{\pi}{\ell}, \qquad \frac{2\pi}{\ell}, \qquad \frac{3\pi}{\ell}, \qquad \ldots .$$

We conclude that the eigenvalues and eigenfunctions of the boundary value problem (4.20) are

$$\lambda_n = \gamma \left(\frac{n\pi}{\ell} \right)^2, \qquad v_n(x) = \sin \frac{n\pi x}{\ell}, \qquad n = 1, 2, 3, \ldots . \tag{4.21}$$

The corresponding eigensolutions (4.18) are

$$u_n(t,x) = \exp\left(-\frac{\gamma n^2 \pi^2 t}{\ell^2} \right) \sin \frac{n\pi x}{\ell}, \qquad n = 1, 2, 3, \ldots . \tag{4.22}$$

Each represents a trigonometrically oscillating temperature profile that maintains its form while decaying to zero at an exponentially fast rate.

To solve the general initial value problem, we assemble the eigensolutions into an infinite series,

$$u(t,x) = \sum_{n=1}^{\infty} b_n u_n(t,x) = \sum_{n=1}^{\infty} b_n \exp\left(-\frac{\gamma n^2 \pi^2 t}{\ell^2} \right) \sin \frac{n\pi x}{\ell}, \tag{4.23}$$

whose coefficients b_n are to be fixed by the initial conditions. Indeed, assuming that the series converges, the initial temperature profile is

$$u(0,x) = \sum_{n=1}^{\infty} b_n \sin \frac{n\pi x}{\ell} = f(x). \tag{4.24}$$

This has the form of a Fourier sine series (3.52) on the interval $[0,\ell]$. Thus, the coefficients are determined by the Fourier formulae (3.53), and so

$$b_n = \frac{2}{\ell} \int_0^\ell f(x) \sin \frac{n\pi x}{\ell}\, dx, \qquad n = 1,2,3,\ldots\,. \tag{4.25}$$

The resulting formula (4.23) describes the Fourier sine series for the temperature $u(t,x)$ of the bar at each later time $t \geq 0$.

Example 4.1. Consider the initial temperature profile

$$u(0,x) = f(x) = \begin{cases} -x, & 0 \leq x \leq \frac{1}{5}, \\ x - \frac{2}{5}, & \frac{1}{5} \leq x \leq \frac{7}{10}, \\ 1 - x, & \frac{7}{10} \leq x \leq 1, \end{cases} \tag{4.26}$$

on a bar of length 1, plotted in the first graph in Figure 4.1. Using (4.25), the first few Fourier coefficients of $f(x)$ are computed (by either exact or numerical integration) to be

$$\begin{array}{lllll} b_1 \approx .0897, & b_2 \approx -.1927, & b_3 \approx -.0289, & b_4 = 0, & \\ b_5 \approx -.0162, & b_6 \approx .0132, & b_7 \approx .0104, & b_8 = 0, & \ldots\,. \end{array}$$

The resulting Fourier series solution to the heat equation is

$$\begin{aligned} u(t,x) &= \sum_{n=1}^{\infty} b_n u_n(t,x) = \sum_{n=1}^{\infty} b_n e^{-\gamma n^2\pi^2 t} \sin n\pi x \\ &\approx .0897\, e^{-\gamma\pi^2 t} \sin \pi x - .1927\, e^{-4\gamma\pi^2 t} \sin 2\pi x - .0289\, e^{-9\gamma\pi^2 t} \sin 3\pi x - \cdots\,. \end{aligned}$$

In Figure 4.1, the solution, for $\gamma = 1$, is plotted at some representative times. Observe that the corners in the initial profile are immediately smoothed out. As time progresses, the solution decays, at a fast exponential rate of $e^{-\pi^2 t} \approx e^{-9.87t}$, to a uniform, zero temperature, which is the equilibrium temperature distribution for the homogeneous Dirichlet boundary conditions. As the solution decays to thermal equilibrium, the higher Fourier modes rapidly disappear, and the solution assumes the progressively more symmetric shape of a single sine arc, of rapidly decreasing amplitude.

Smoothing and Long–Time Behavior

The fact that we can write the solution to an initial-boundary value problem in the form of an infinite series (4.23) is progress of a sort. However, because we are unable to sum the series in closed form, this "solution" is much less satisfying than a direct, explicit formula. Nevertheless, there are important qualitative and quantitative features of the solution that can be easily gleaned from such series expansions.

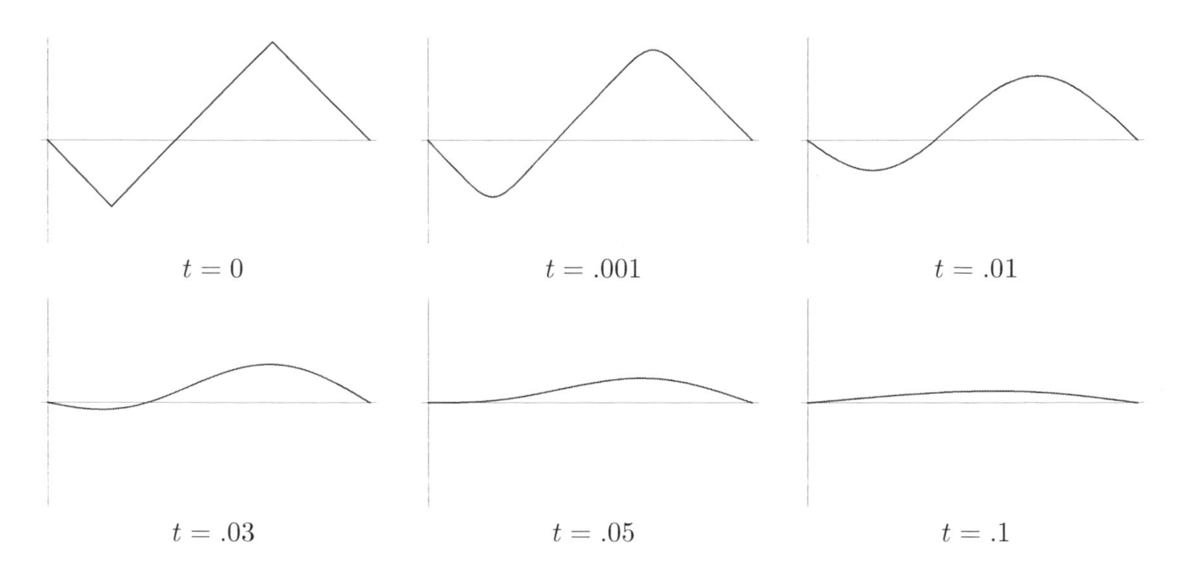

Figure 4.1. A solution to the heat equation.

If the initial data $f(x)$ is integrable (e.g., piecewise continuous), then its Fourier coefficients are uniformly bounded; indeed, for any $n \geq 1$,

$$| b_n | \leq \frac{2}{\ell} \int_0^\ell \left| f(x) \sin \frac{n \pi x}{\ell} \right| dx \leq \frac{2}{\ell} \int_0^\ell | f(x) | \, dx \equiv M. \tag{4.27}$$

This property holds even for quite irregular data. Under these conditions, each term in the series solution (4.23) is bounded by an exponentially decaying function

$$\left| b_n \exp\left(- \frac{\gamma n^2 \pi^2}{\ell^2} t \right) \sin \frac{n \pi x}{\ell} \right| \leq M \exp\left(- \frac{\gamma n^2 \pi^2}{\ell^2} t \right).$$

This means that, as soon as $t > 0$, most of the high-frequency terms, $n \gg 0$, will be extremely small. Only the first few terms will be at all noticeable, and so the solution essentially degenerates into a finite sum over the first few Fourier modes. As time increases, more and more of the Fourier modes will become negligible, and the sum further degenerates into fewer and fewer significant terms. Eventually, as $t \to \infty$, *all* of the Fourier modes will decay to zero. Therefore, the solution will converge exponentially fast to a zero temperature profile: $u(t, x) \to 0$ as $t \to \infty$, representing the bar in its final uniform thermal equilibrium. The fact that its equilibrium temperature is zero is the result of holding both ends of the bar fixed at zero temperature, whereby any initial thermal energy is eventually dissipated away through the ends. The small-scale temperature fluctuations tend to rapidly cancel out through diffusion of thermal energy, and the last term to disappear is the one with the slowest decay, namely

$$u(t, x) \approx b_1 \exp\left(- \frac{\gamma \pi^2}{\ell^2} t \right) \sin \frac{\pi x}{\ell}, \qquad \text{where} \qquad b_1 = \frac{2}{\ell} \int_0^\ell f(x) \sin \frac{\pi x}{\ell} \, dx. \tag{4.28}$$

For generic initial data, the coefficient $b_1 \neq 0$, and the solution approaches thermal equilibrium at an exponential rate prescribed by the smallest eigenvalue, $\lambda_1 = \gamma \pi^2 / \ell^2$, which is proportional to the thermal diffusivity divided by the square of the length of the bar. The

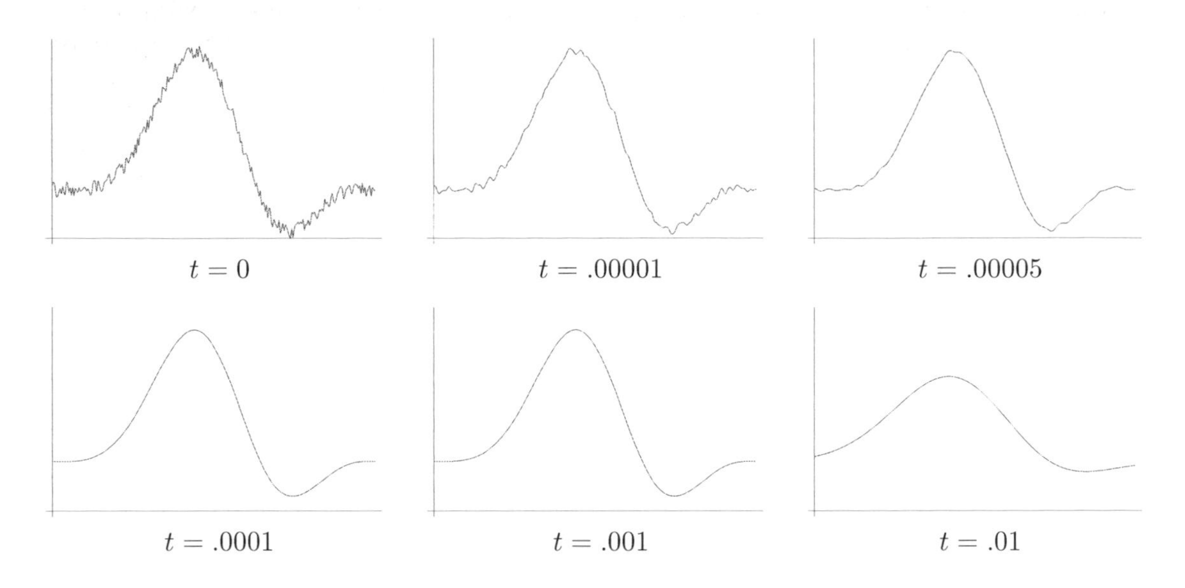

Figure 4.2. Denoising a signal with the heat equation.

longer the bar, or the smaller the diffusivity, the longer it takes for the effect of holding the ends at zero temperature to propagate along its entire length. Also, again provided $b_1 \neq 0$, the asymptotic shape of the temperature profile is a small, exponentially decaying sine arc, just as we observed in Example 4.1. In exceptional situations, namely when $b_1 = 0$, the solution decays even faster, at a rate equal to the eigenvalue $\lambda_k = \gamma\, k^2 \pi^2/\ell^2$ corresponding to the first nonzero term, $b_k \neq 0$, in the Fourier series; its asymptotic shape now oscillates k times over the interval.

Another, closely related, observation is that, for any fixed time $t > 0$ after the initial moment, the coefficients in the Fourier sine series (4.23) decay exponentially fast as $n \to \infty$. According to Corollary 3.32, this implies that the Fourier series converges to an infinitely differentiable function of x at each positive time t, *no matter how unsmooth the initial temperature profile*. We have discovered the basic smoothing property of heat flow, which we state for a general initial time t_0.

Theorem 4.2. *If $u(t,x)$ is a solution to the heat equation with piecewise continuous initial data $f(x) = u(t_0,x)$, or, more generally, initial data satisfying (4.27), then, for any $t > t_0$, the solution $u(t,x)$ is an infinitely differentiable function of x.*

In other words, the heat equation *instantaneously* smoothes out any discontinuities and corners in the initial temperature profile by fast damping of the high-frequency modes. The heat equation's effect on irregular initial data underlies its effectiveness for smoothing and denoising signals. We take the initial data $u(0,x) = f(x)$ to be a noisy signal, and then evolve the heat equation forward to a prescribed time $t^\star > 0$. The resulting function $g(x) = u(t^\star,x)$ will be a smoothed version of the original signal $f(x)$ in which most of the high-frequency noise has been eliminated. Of course, if we run the heat flow for too long, all of the low-frequency features will also be smoothed out and the result will be a uniform, constant signal. Thus, the choice of *stopping time* $t^\star$ is crucial to the success of this method. Figure 4.2 shows the effect of running the heat equation,† with $\gamma = 1$, on a signal that has been contaminated by random noise. Observe how quickly the noise is removed. By the final time, the overall smoothing effect of the heat flow has caused significant degradation (blurring) of the original signal. The heat equation approach to denoising has the advantage that no Fourier coefficients need be explicitly computed, nor does one need to reconstruct the smoothed signal. Basic numerical solution schemes for the heat equation are to be discussed in Chapter 5.

An important theoretical consequence of the smoothing property is that diffusion is a one-way process — one cannot run time backwards and accurately infer what a temperature distribution looked like in the past. In particular, if the initial data $u(0, x) = f(x)$ is not smooth, then the value of $u(t, x)$ for any $t < 0$ cannot be defined, because if $u(t_0, x)$ were defined and integrable at some $t_0 < 0$ then, by Theorem 4.2, $u(t, x)$ would be smooth at all subsequent times $t > t_0$, including $t = 0$, in contradiction to our assumption. Moreover, for most initial data, the Fourier coefficients in the solution formula (4.23) are, at any $t < 0$, exponentially *growing* as $n \to \infty$, indicating that high-frequency noise has completely overwhelmed the solution, thereby precluding any kind of convergence of the Fourier series.

Mathematically, we can reverse future and past by changing t to $-t$. In the differential equation, this merely reverses the sign of the time-derivative term; the x derivatives are unaffected. Thus, by the above reasoning, the *backwards heat equation*

$$\frac{\partial u}{\partial t} = -\gamma \frac{\partial^2 u}{\partial x^2}, \qquad \text{with a negative diffusion coefficient} \qquad -\gamma < 0, \tag{4.29}$$

is an *ill-posed problem* in the sense that small changes in the initial data — e.g., a small perturbation of a high-frequency mode — can produce arbitrarily large changes in the solution arbitrarily close to the initial time. In other words, the solution does not depend continuously on the initial data. Even worse, for nonsmooth initial data, the solution is not even well defined in forwards time $t > 0$ (although it is well-posed if we run t backwards). The same holds for more general diffusion processes, e.g., (4.6). If, as in all physically relevant cases, the coefficient of u_{xx} is everywhere positive, then the initial value problem is well-posed for $t > 0$, but ill-posed for $t < 0$. On the other hand, if the coefficient is everywhere negative, the reverse holds. A coefficient that changes signs would cause the differential equation to be ill-posed in both directions.

While theoretically undesirable, the unsmoothing effect of the backwards heat equation has potential benefits in certain contexts. For example, in image processing, diffusion will gradually blur an image by damping out the high-frequency modes. Image enhancement is the reverse process, and can be based on running the heat flow backwards in some stable manner. In forensics, determining the time of death based on the current temperature of a corpse also requires running the equations governing the dissipation of body heat backwards in time. One option would be to restrict the backwards evolution to the first few Fourier modes, which prevents the small-scale fluctuations from overwhelming the computation. Ill-posed problems also arise in the reconstruction of subterranean profiles from seismic data, a central problem of the oil and gas industry. These and other applications are driving contemporary research into how to cleverly circumvent the ill-posedness of backwards diffusion processes.

† To avoid artifacts at the ends of the interval, we are, in fact, using periodic boundary conditions in the plots. Away from the ends, running the equation with Dirichlet boundary conditions leads to almost identical results.

Remark: The irreversibility of the heat equation, along with the irreversibility of nonlinear transport in the presence of shock waves discussed in Section 2.3, highlight a crucial distinction between partial differential equations and ordinary differential equations. Ordinary differential equations are always reversible — the existence, uniqueness, and continuous dependence properties of solutions are all equally valid in reverse time (although their detailed qualitative and quantitative properties will, of course, depend upon whether time is running forwards or backwards). The irreversibility and ill-posedness of partial differential equations modeling thermodynamical, biological, and other diffusive processes in our universe may explain why Time's Arrow points exclusively to the future.

The Heated Ring Redux

Let us next consider the periodic boundary value problem modeling heat flow in an insulated circular ring. We fix the length of the ring to be $\ell = 2\pi$, with $-\pi \leq x \leq \pi$ representing the "angular" coordinate around the ring. For simplicity, we also choose units in which the thermal diffusivity is $\gamma = 1$. Thus, we seek to solve the heat equation

$$\frac{\partial u}{\partial t} = \frac{\partial^2 u}{\partial x^2}, \qquad -\pi < x < \pi, \qquad t > 0, \tag{4.30}$$

subject to periodic boundary conditions

$$u(t,-\pi) = u(t,\pi), \qquad \frac{\partial u}{\partial x}(t,-\pi) = \frac{\partial u}{\partial x}(t,\pi), \qquad t \geq 0, \tag{4.31}$$

that ensure continuity of the solution when the angular coordinate switches from $-\pi$ to π. The initial temperature distribution is

$$u(0,x) = f(x), \qquad -\pi < x \leq \pi. \tag{4.32}$$

The resulting temperature $u(t,x)$ will be a periodic function in x of period 2π.

Substituting the separable solution ansatz (3.15) into the heat equation and the boundary conditions results in the periodic eigenvalue problem

$$\frac{d^2 v}{dx^2} + \lambda v = 0, \qquad v(-\pi) = v(\pi), \qquad v'(-\pi) = v'(\pi). \tag{4.33}$$

As we already noted in Section 3.1, the eigenvalues of this particular boundary value problem are $\lambda_n = n^2$, where $n = 0, 1, 2, \ldots$ is a nonnegative integer; the corresponding eigenfunctions are the trigonometric functions

$$v_n(x) = \cos n x, \qquad \widetilde{v}_n(x) = \sin n x, \qquad n = 0, 1, 2, \ldots .$$

Note that $\lambda_0 = 0$ is a simple eigenvalue, with constant eigenfunction $\cos 0x = 1$ — the sine solution $\sin 0x \equiv 0$ is trivial — while the positive eigenvalues are, in fact, double, each possessing two linearly independent eigenfunctions. The corresponding eigensolutions to the heated ring equation (4.30–31) are

$$u_n(t,x) = e^{-n^2 t} \cos n x, \qquad \widetilde{u}_n(t,x) = e^{-n^2 t} \sin n x, \qquad n = 0, 1, 2, 3, \ldots .$$

The resulting infinite series solution is

$$u(t,x) = \tfrac{1}{2} a_0 + \sum_{n=1}^{\infty} \left(a_n e^{-n^2 t} \cos n x + b_n e^{-n^2 t} \sin n x \right), \qquad ⊕ \tag{4.34}$$

with as yet unspecified coefficients a_n, b_n. The initial conditions require

$$u(0,x) = \tfrac{1}{2}a_0 + \sum_{n=1}^{\infty} (a_n \cos nx + b_n \sin nx) = f(x), \tag{4.35}$$

which is precisely the complete Fourier series (3.34) of the initial temperature profile $f(x)$. Consequently,

$$a_n = \frac{1}{\pi}\int_{-\pi}^{\pi} f(x)\cos nx\, dx, \qquad b_n = \frac{1}{\pi}\int_{-\pi}^{\pi} f(x)\sin nx\, dx, \tag{4.36}$$

are its usual Fourier coefficients (3.35).

As in the Dirichlet problem, after the initial instant, the high-frequency terms in the series (4.34) become extremely small, since $e^{-n^2 t} \ll 1$ for $n \gg 0$. Therefore, as soon as $t > 0$, the solution instantaneously becomes smooth, and quickly degenerates into what is in essence a finite sum over the first few Fourier modes. Moreover, as $t \to \infty$, *all* of the Fourier modes will decay to zero with the exception of the constant mode, associated with the null eigenvalue $\lambda_0 = 0$. Consequently, the solution will converge, at an exponential rate, to a constant-temperature profile,

$$u(t,x) \longrightarrow \tfrac{1}{2}a_0 = \frac{1}{2\pi}\int_{-\pi}^{\pi} f(x)\, dx,$$

which equals the *average* of the initial temperature profile. In physical terms, since the insulation prevents any thermal energy from escaping the ring, it rapidly redistributes itself so that the ring achieves a uniform constant temperature — its eventual equilibrium state.

Prior to attaining equilibrium, only the very lowest frequency Fourier modes will still be noticeable, and so the solution will asymptotically look like

$$u(t,x) \approx \tfrac{1}{2}a_0 + e^{-t}(a_1 \cos x + b_1 \sin x) = \tfrac{1}{2}a_0 + r_1 e^{-t}\cos(x+\delta_1), \tag{4.37}$$

where

$$a_1 = r_1\cos\delta_1 = \frac{1}{2\pi}\int_{-\pi}^{\pi} f(x)\cos x\, dx, \qquad b_1 = r_1\sin\delta_1 = \frac{1}{2\pi}\int_{-\pi}^{\pi} f(x)\sin x\, dx.$$

Thus, for most initial data, the solution approaches thermal equilibrium at an exponential rate of e^{-t}. The exceptions are when $a_1 = b_1 = 0$, for which the rate of convergence is even faster, namely at a rate $e^{-k^2 t}$, where k is the smallest integer such that at least one of the kth order Fourier coefficients a_k, b_k is nonzero.

In fact, once we are convinced that the bar must tend to thermal equilibrium as $t \to \infty$, we can predict the final temperature without knowing the explicit solution formula. Our derivation in Section 4.1 implies that the heat equation has the form of a conservation law (4.1), with the conserved density being the temperature $u(t,x)$. As in (4.2), the integrated form of the conservation law reads

$$\begin{aligned}\frac{d}{dt}\int_{-\pi}^{\pi} u(t,x)\, dx &= \int_{-\pi}^{\pi} \frac{\partial u}{\partial t}(t,x)\, dx = \gamma\int_{-\pi}^{\pi} \frac{\partial^2 u}{\partial x^2}(t,x)\, dx \\ &= \gamma\left[\frac{\partial u}{\partial x}(t,\pi) - \frac{\partial u}{\partial x}(t,-\pi)\right] = 0,\end{aligned}$$

where the flux terms cancel thanks to the periodic boundary conditions (4.31). Physically, any flux out of one end of the circular bar is immediately fed into the other, abutting end, and so there is no net loss of thermal energy. We conclude that, for the periodic boundary value problem, the total *thermal energy*

$$E(t) = \int_{-\pi}^{\pi} u(t,x)\,dx = \text{constant} \tag{4.38}$$

remains constant for all time. (In contrast, the thermal energy does *not* remain constant for the Dirichlet boundary value problem, decaying steadily to 0 due to the out-flux of heat through the ends of the bar; see Exercise 4.1.13 for further details.)

Remark: More correctly, according to (4.3), the thermal energy is obtained by multiplying the temperature by the product, $\sigma = \rho\,\chi$, of the density and the specific heat of the body. For the heat equation, both are constant, and so the physical thermal energy equals $\sigma\,E(t)$. Mathematically, we can safely ignore this extra constant factor, or, equivalently, work in physical units in which $\sigma = 1$. This does not extend to nonuniform bodies, whose *thermal energy* is given by $E(t) = \displaystyle\int_{-\pi}^{\pi} \sigma(x)\,u(t,x)\,dx$, and whose constancy, under suitable boundary conditions, follows from the conservation-law form (4.6) of the linear diffusion equation.

In general, a system is in (static) *equilibrium* if it remains unaltered as time progresses. Thus, any equilibrium configuration has the form $u = u^\star(x)$, and hence satisfies $\partial u^\star/\partial t = 0$. If, in addition, $u^\star(x)$ is an equilibrium solution to the periodic heat equation (4.30–33), then it must satisfy

$$\frac{\partial u^\star}{\partial t} = 0 = \frac{\partial^2 u^\star}{\partial x^2}, \qquad u^\star(-\pi) = u^\star(\pi), \qquad \frac{\partial u^\star}{\partial x}(-\pi) = \frac{\partial u^\star}{\partial x}(\pi). \tag{4.39}$$

In other words, $u^\star$ is a solution to the periodic boundary value problem (4.33) for the null eigenvalue $\lambda = 0$. Thus, *the null eigenfunctions* (*including the zero solution*) *are all the possible equilibrium solutions*. In particular, for the periodic boundary value problem, the null eigenfunctions are constant, and therefore solutions to the periodic heat equation will tend to a constant equilibrium temperature.

Now, once we know that the solution tends to a constant, $u(t,x) \to a$ as $t \to \infty$, then its thermal energy tends to

$$E(t) = \int_{-\pi}^{\pi} u(t,x)\,dx \longrightarrow \int_{-\pi}^{\pi} a\,dx = 2\pi a \qquad \text{as} \qquad t \longrightarrow \infty.$$

On the other hand, as we just demonstrated, the thermal energy is constant, so

$$E(t) = E(0) = \int_{-\pi}^{\pi} u(0,x)\,dx = \int_{-\pi}^{\pi} f(x)\,dx.$$

Combining these two, we conclude that

$$\int_{-\pi}^{\pi} f(x)\,dx = 2\pi a, \quad \text{and so the equilibrium temperature} \quad a = \frac{1}{2\pi}\int_{-\pi}^{\pi} f(x)\,dx$$

equals the average initial temperature. This reconfirms our earlier result, but avoids having to know an explicit series solution formula. As a result, the latter method can be applied to a much wider range of situations.

Inhomogeneous Boundary Conditions

So far, we have concentrated our attention on homogeneous boundary conditions. There is a simple trick that will convert a boundary value problem with inhomogeneous but constant Dirichlet boundary conditions,

$$\frac{\partial u}{\partial t} = \gamma\, \frac{\partial^2 u}{\partial x^2}\,, \qquad u(t,0) = \alpha, \qquad u(t,\ell) = \beta, \qquad t \geq 0, \tag{4.40}$$

into a homogeneous Dirichlet problem. We begin by solving for the equilibrium temperature profile. As in (4.39), the equilibrium does not depend on t and hence satisfies the boundary value problem

$$\frac{\partial u^\star}{\partial t} = 0 = \gamma\, \frac{\partial^2 u^\star}{\partial x^2}\,, \qquad u^\star(0) = \alpha, \qquad u^\star(\ell) = \beta.$$

Solving the ordinary differential equation yields $u^\star(x) = a + b\,x$, where the constants a, b are fixed by the boundary conditions. We conclude that the equilibrium solution is a straight line connecting the boundary values:

$$u^\star(x) = \alpha + \frac{\beta - \alpha}{\ell}\, x. \tag{4.41}$$

The difference

$$\widetilde{u}(t,x) = u(t,x) - u^\star(x) = u(t,x) - \alpha - \frac{\beta - \alpha}{\ell}\, x \tag{4.42}$$

measures the deviation of the solution from equilibrium. It clearly satisfies the homogeneous boundary conditions at both ends:

$$\widetilde{u}(t,0) = 0 = \widetilde{u}(t,\ell).$$

Moreover, by linearity, since both $u(t,x)$ and $u^\star(x)$ are solutions to the heat equation, so is $\widetilde{u}(t,x)$. The initial data must be similarly adapted:

$$\widetilde{u}(0,x) = u(t,x) - u^\star(x) = f(x) - \alpha - \frac{\beta - \alpha}{\ell}\, x \equiv \widetilde{f}(x). \tag{4.43}$$

Solving the resulting homogeneous initial-boundary value problem, we write $\widetilde{u}(t,x)$ in Fourier series form (4.23), where the Fourier coefficients are specified by the modified initial data $\widetilde{f}(x)$ in (4.43). The solution to the inhomogeneous boundary value problem thus has the series form

$$u(t,x) = \alpha + \frac{\beta - \alpha}{\ell}\, x \; + \sum_{n=1}^{\infty} \widetilde{b}_n \exp\left(- \frac{\gamma\, n^2 \pi^2}{\ell^2}\, t \right) \sin \frac{n \pi x}{\ell}\,, \tag{4.44}$$

where

$$\widetilde{b}_n = \frac{2}{\ell} \int_0^\ell \widetilde{f}(x) \sin \frac{n \pi x}{\ell}\, dx, \qquad n = 1, 2, 3, \ldots\,. \tag{4.45}$$

Since $\widetilde{u}(t,0)$ decays to zero at an exponential rate as $t \to \infty$, the actual temperature profile (4.44) will asymptotically decay to the equilibrium profile,

$$u(t,x) \quad \longrightarrow \quad u^\star(x) = \alpha \; + \; \frac{\beta - \alpha}{\ell}\, x,$$

at the same exponentially fast rate, governed by the first eigenvalue $\lambda_1 = \pi^2/\ell^2$ — unless $\widetilde{b}_1 = 0$, in which case the decay rate is even faster.

This method does not work as well when the boundary conditions are time-dependent:

$$u(t,0)=\alpha(t), \qquad u(t,\ell)=\beta(t).$$

Attempting to mimic the preceding technique, we discover that the deviation[†]

$$\widetilde{u}(t,x)=u(t,x)-u^\star(t,x), \qquad \text{where} \qquad u^\star(t,x)=\alpha(t)+\frac{\beta(t)-\alpha(t)}{\ell}\,x, \tag{4.46}$$

satisfies the homogeneous boundary conditions, but now solves an inhomogeneous or forced version of the heat equation:

$$\frac{\partial \widetilde{u}}{\partial t}=\frac{\partial^2 \widetilde{u}}{\partial x^2}+h(t,x), \quad \text{where} \quad h(t,x)=-\frac{\partial u^\star}{\partial t}(t,x)=-\alpha'(t)-\frac{\beta'(t)-\alpha'(t)}{\ell}\,x. \tag{4.47}$$

Solution techniques for the latter partial differential equation will be discussed in Section 8.1 below.

Robin Boundary Conditions

Consider a bar of unit length and unit thermal diffusivity, insulated along its length, which has one of its ends held at 0° and the other put in a heat bath. The resulting thermodynamics are modeled by the heat equation subject to Dirichlet boundary conditions at $x=0$ and Robin boundary conditions at $x=1$:

$$\frac{\partial u}{\partial t}=\frac{\partial^2 u}{\partial x^2}, \qquad u(t,0)=0, \qquad \frac{\partial u}{\partial x}(t,1)+\beta\,u(t,1)=0, \tag{4.48}$$

where $\beta\neq 0$ is a constant[‡] that measures the rate of transfer of thermal energy, with $\beta>0$ when the bath is cold and so the energy is being extracted from the bar. As before, the general solution to the resulting initial-boundary value problem can be assembled from the separable eigensolutions based on our usual exponential ansatz $u(t,x)=e^{-\lambda t}\,v(x)$. Substituting this expression into (4.48), we find that the eigenfunction $v(x)$ must satisfy the boundary value problem

$$-\frac{d^2 v}{dx^2}=\lambda\,v, \qquad v(0)=0, \qquad v'(1)+\beta\,v(1)=0. \tag{4.49}$$

In order to find nontrivial solutions $v(x)\not\equiv 0$ to (4.49), let us first assume $\lambda=\omega^2>0$, where, without loss of generality, $\omega>0$. The solution to the ordinary differential equation that satisfies the Dirichlet boundary condition at $x=0$ is a constant multiple of $v(x)=\sin\omega\,x$. Substituting this function into the Robin boundary condition at $x=1$, we find

$$\omega\cos\omega+\beta\sin\omega=0, \qquad \text{or, equivalently,} \qquad \omega=-\beta\tan\omega. \tag{4.50}$$

It is not hard to see that there is an infinite number of real, positive solutions $0<\omega_1<\omega_2<\omega_3<\cdots\to\infty$ to the latter transcendental equation. Indeed, they can be characterized as the abscissas $\omega_n>0$ of the intersection points of the graphs of the two functions $f(\omega)=\omega$

[†] In this case, $u^\star(t,x)$ is not an equilibrium solution. Indeed, we do not expect the bar to go to equilibrium if the temperature of its endpoints is constantly changing.

[‡] The case $\beta=0$ reduces to the mixed boundary value problem, whose analysis is left to the reader.

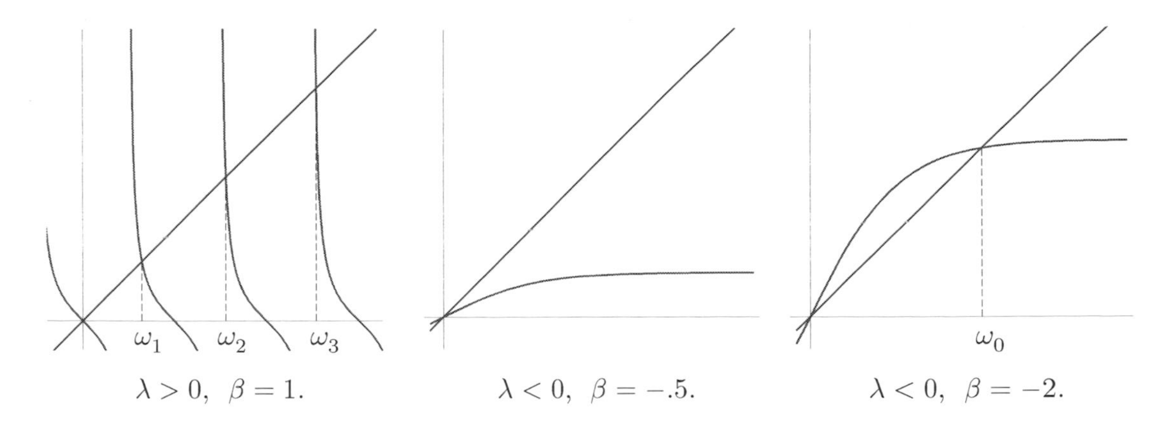

Figure 4.3. Eigenvalue equation for Robin boundary conditions.

and $g(\omega) = -\beta \tan \omega$, as shown in the first plot in Figure 4.3. Each root ω_n defines a positive eigenvalue $\lambda_n = \omega_n^2 > 0$ to the boundary value problem (4.49) and hence an exponentially decaying eigensolution

$$u_n(t,x) = e^{-\lambda_n t} \sin \omega_n x \tag{4.51}$$

to the Robin boundary value problem (4.48). While there is no explicit formula, numerical approximations to the eigenvalues are easily found via a numerical root finder, e.g., Newton's Method, [**24, 94**]. In particular, for $\beta = 1$, the first three eigenvalues are $\lambda_1 = \omega_1^2 \approx 4.1159$, $\lambda_2 = \omega_2^2 \approx 24.1393$, $\lambda_3 = \omega_3^2 \approx 63.6591$.

What about a zero eigenvalue? If $\lambda = 0$ in (4.49), then the solution to the ordinary differential equation that satisfies the Dirichlet boundary condition is a constant multiple of $v(x) = x$. This function satisfies the Robin boundary condition $v'(1) + \beta\, v(1) = 0$ if and only if $\beta = -1$. In this special configuration, the heat equation admits a time-independent eigensolution $u_0(t,x) = x$ with eigenvalue $\lambda_0 = 0$. Physically, the rate of transfer of thermal energy into the bar through its end in the heat bath is exactly enough to cancel the heat loss through the Dirichlet end, resulting in a steady-state solution. All other eigenmodes correspond to positive eigenvalues, and hence are exponentially decaying. The general solution decays to the steady state, which is a constant multiple of the null eigensolution: $u(t,x) \to c\,x$ as $t \to \infty$, at an exponential rate prescribed, generically, by the first positive eigenvalue $\lambda_1 > 0$.

However, in contrast to the more common types of boundary conditions (Dirichlet, Neumann, mixed, periodic), we cannot automatically rule out the existence of negative eigenvalues in the Robin case. Suppose $\lambda = -\omega^2 < 0$ with $\omega > 0$. Now the solution to (4.49) that satisfies the Dirichlet boundary condition at $x = 0$ is a constant multiple of the hyperbolic sine function $v(x) = \sinh \omega\, x$. Substituting this expression into the Robin boundary condition at $x = 1$ produces

$$\omega \cosh \omega + \beta \sinh \omega = 0, \qquad \text{or, equivalently,} \qquad \omega = -\,\beta \tanh \omega, \tag{4.52}$$

where

$$\tanh \omega = \frac{\sinh \omega}{\cosh \omega} = \frac{e^{\omega} - e^{-\omega}}{e^{\omega} + e^{-\omega}} \tag{4.53}$$

is the hyperbolic tangent. If $\beta > -1$, there are no solutions $\omega > 0$ to this transcendental equation, and in this case all the eigenvalues are strictly positive and all solutions to the heat equation are exponentially decaying. On the other hand, if $\beta < -1$, there is a single solution $\omega_0 > 0$, which produces a single negative eigenvalue $\lambda_0 = -\omega_0^2$. Representative graphs illustrating the two possibilities appear in Figure 4.3; in the first, the graph of $f(\omega) = \omega$ does not intersect the graph of $g(\omega) = \frac{1}{2}\tanh\omega$ when $\omega > 0$, whereas it intersects the graph of $\widehat{g}(\omega) = 2\tanh\omega$ at a single point, with abscissa $\omega_0 \approx 1.9150$, producing the negative eigenvalue $\lambda_0 \approx -\omega_0^2 \approx -3.6673$. Thus, when $\beta < -1$, there is, in addition to all the exponentially decaying eigenmodes associated with the positive eigenvalues, a single unstable exponentially *growing* eigenmode

$$u_0(t,x) = e^{\lambda_0 t} \sinh \omega_0 x. \tag{4.54}$$

Physically, $\beta < -1$ implies that thermal energy is entering the Robin end of the bar at a faster rate than can be removed through the Dirichlet end, and hence the bar experiences an exponential increase in its overall temperature.

Remark: Even though some Robin boundary conditions admit exponentially growing solutions, and hence lead to *unstable* dynamics, the initial-boundary value problem remains *well-posed* because the solution exists and is uniquely determined by the initial data, and, moreover, small changes in the initial conditions induce relatively small changes in the resulting solution on bounded time intervals.

The Root Cellar Problem

As a final example, we discuss a problem that involves analysis of the heat equation on a semi-infinite interval. The question is this: how deep should you dig a root cellar? In the prerefrigeration era, a root cellar was used to keep food cool in the summer, but not freeze in the winter. We assume that the temperature inside the Earth depends only on the depth and the time of year. Let $u(t,x)$ denote the deviation in the temperature from its annual mean at depth $x > 0$ and time t. We shall assume that the temperature at the Earth's surface, $x = 0$, fluctuates in a periodic manner; specifically, we set

$$u(t,0) = a \cos \omega t, \tag{4.55}$$

where the oscillatory frequency

$$\omega = \frac{2\pi}{365.25 \text{ days}} = 2.0 \times 10^{-7}\text{sec}^{-1} \tag{4.56}$$

refers to yearly temperature variations. In this model, we shall ignore daily temperature fluctuations, since their effect is not significant below a very thin surface layer. At large depths the temperature is assumed to be unvarying:

$$u(t,x) \longrightarrow 0 \qquad \text{as} \qquad x \longrightarrow \infty, \tag{4.57}$$

where 0 refers to the mean temperature.

Thus, we must solve the heat equation on a semi-infinite bar $0 < x < \infty$, with time-dependent boundary conditions (4.55, 57) at the ends. The analysis will be simplified a little if we replace the cosine by a complex exponential, and so we look for a complex solution with boundary conditions

$$u(t,0) = a\, e^{\,i\,\omega\, t}, \qquad\qquad \lim_{x\to\infty} u(t,x) = 0. \tag{4.58}$$

Let us try a separable solution of the form

$$u(t,x) = v(x)\, e^{\,\mathrm{i}\,\omega\, t}. \tag{4.59}$$

Substituting this expression into the heat equation $u_t = \gamma\, u_{xx}$ leads to

$$\mathrm{i}\,\omega\, v(x)\, e^{\,\mathrm{i}\,\omega\, t} = \gamma\, v''(x)\, e^{\,\mathrm{i}\,\omega\, t}.$$

Canceling the common exponential factors, we conclude that $v(x)$ should solve the boundary value problem

$$\gamma\, v''(x) = \mathrm{i}\,\omega\, v, \qquad v(0) = a, \qquad \lim_{x\to\infty} v(x) = 0.$$

The solutions to the ordinary differential equation are

$$v_1(x) = e^{\sqrt{\mathrm{i}\,\omega/\gamma}\, x} = e^{\sqrt{\omega/(2\gamma)}\,(1+\mathrm{i})\, x}, \qquad v_2(x) = e^{-\sqrt{\mathrm{i}\,\omega/\gamma}\, x} = e^{-\sqrt{\omega/(2\gamma)}\,(1+\mathrm{i})\, x}.$$

The first solution is exponentially growing as $x \to \infty$, and so not germane to our problem. The solution to the boundary value problem must therefore be a multiple of the exponentially decaying solution:

$$v(x) = a\, e^{-\sqrt{\omega/(2\gamma)}\,(1+\mathrm{i})\, x}.$$

Substituting back into (4.59), we find the (complex) solution to the root cellar problem to be

$$u(t,x) = a\, e^{-x\sqrt{\omega/(2\gamma)}}\; e^{\,\mathrm{i}\,(\omega\, t - \sqrt{\omega/(2\gamma)}\, x)}. \tag{4.60}$$

The corresponding real solution is obtained by taking the real part,

$$u(t,x) = a\, e^{-x\sqrt{\omega/(2\gamma)}}\, \cos\left(\omega\, t - \sqrt{\frac{\omega}{2\gamma}}\, x\right). \tag{4.61}$$

The first factor in (4.61) is exponentially decaying as a function of the depth. Thus, the further underground one is, the less noticeable is the effect of the surface temperature fluctuations. The second factor is periodic in time, with the same annual frequency ω. The interesting feature is that the temperature variations (4.61) are typically out of phase with respect to the surface temperature fluctuations, having an overall *phase lag* of

$$\delta = \sqrt{\frac{\omega}{2\gamma}}\, x$$

that depends linearly on the depth x. In particular, a cellar built at a depth where δ is an odd multiple of π will be completely out of phase, being hottest in the winter, and coldest in the summer. Thus, the (shallowest) ideal depth at which to build a root cellar would take $\delta = \pi$, corresponding to a depth of

$$x = \pi\sqrt{\frac{2\gamma}{\omega}}\,. \tag{4.62}$$

For typical soils in the Earth, $\gamma \approx 10^{-6}$ meters^2 sec^{-1}, and so, with ω given by (4.56), $x \approx 9.9$ meters. However, at this depth, the relative amplitude of the oscillations is

$$e^{-x\sqrt{\omega/2\gamma}} = e^{-\pi} = .04,$$

and hence there is only a 4% temperature fluctuation. In Minneapolis, the temperature varies, roughly, from $-40°$C to $+40°$C, and hence our 10-meter-deep root cellar would experience only a 3.2°C annual temperature deviation from the winter, when it is the warmest, to the summer, when it is the coldest. Building the cellar twice as deep would lead to a temperature fluctuation of .2%, now in phase with the surface variations, which means that the cellar would be, for all practical purposes, at constant temperature year round.

Exercises

4.1.1. Suppose the ends of a bar of length 1 and thermal diffusivity $\gamma = 1$ are held fixed at respective temperatures 0° and 10°. (*a*) Determine the equilibrium temperature profile. (*b*) Determine the rate at which the equilibrium temperature profile is approached. (*c*) What does the temperature profile look like as it nears equilibrium?

4.1.2. A uniform insulated bar 1 meter long is stored at room temperature of 20° Celsius. An experimenter places one end of the bar in boiling water and the other end in ice water.
(*a*) Set up an initial-boundary value problem that models the temperature in the bar.
(*b*) Find the equilibrium temperature distribution.
(*c*) Discuss how your answer depends on the material properties of the bar.

4.1.3. Consider the initial-boundary value problem

$$\frac{\partial u}{\partial t} = \frac{\partial^2 u}{\partial x^2}, \qquad \begin{array}{ll} u(t,0) = 0 = u(t,10), & t > 0, \\ u(0,x) = f(x), & 0 < x < 10, \end{array}$$

for the heat equation where the initial data has the following form:

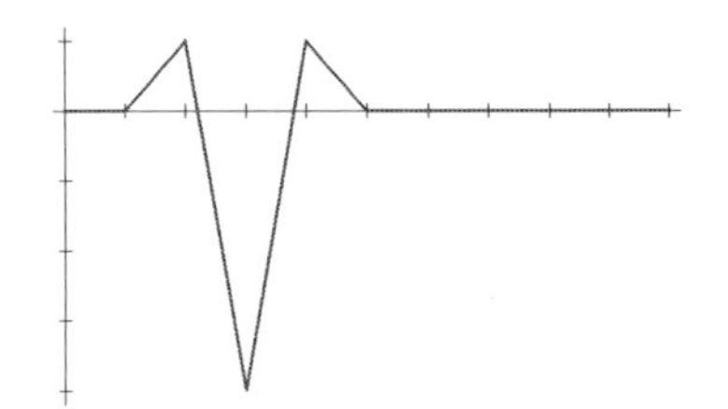

$$f(x) = \begin{cases} x - 1, & 1 \le x \le 2, \\ 11 - 5x, & 2 \le x \le 3, \\ 5x - 19, & 3 \le x \le 4, \\ 5 - x, & 4 \le x \le 5, \\ 0, & \text{otherwise.} \end{cases}$$

Discuss what happens to the solution as t increases. You do *not* need to write down an explicit formula, but for full credit you must explain (sketches can help) at least three or four interesting things that happen to the solution as time progresses.

4.1.4. Find a series solution to the initial-boundary value problem for the heat equation $u_t = u_{xx}$ for $0 < x < 1$ when one the end of the bar is held at 0° and the other is insulated. Discuss the asymptotic behavior of the solution as $t \to \infty$.

4.1.5. Answer Exercise 4.1.4 when both ends of the bar are insulated.

4.1.6. A metal bar, of length $\ell = 1$ meter and thermal diffusivity $\gamma = 2$, is taken out of a 100° oven and then fully insulated except for *one* end, which is fixed to a large ice cube at 0°. (*a*) Write down an initial-boundary value problem that describes the temperature $u(t,x)$ of the bar at all subsequent times. (*b*) Write a series formula for the temperature distribution $u(t,x)$ at time $t > 0$. (*c*) What is the equilibrium temperature distribution in the bar, i.e., for $t \gg 0$? How fast does the solution go to equilibrium? (*d*) Just before the temperature distribution reaches equilibrium, what does it look like? Sketch a picture and discuss.

4.1.7. A metal bar of length $\ell = 1$ and thermal diffusivity $\gamma = 1$ is fully insulated, including its ends. Suppose the initial temperature distribution is $u(0,x) = \begin{cases} x, & 0 \le x \le \frac{1}{2}, \\ 1 - x, & \frac{1}{2} \le x \le 1. \end{cases}$
(*a*) Use Fourier series to write down the temperature distribution at time $t > 0$.

(b) What is the equilibrium temperature distribution in the bar, i.e., for $t \gg 0$?
(c) How fast does the solution go to equilibrium? (d) Just before the temperature distribution reaches equilibrium, what does it look like? Sketch a picture and discuss.

4.1.8. (a) Find the series solution to the heat equation $u_t = u_{xx}$ on $-2 < x < 2$, $t > 0$, when subject to the Dirichlet boundary conditions $u(t,-2) = u(t,2) = 0$ and the initial condition $u(0,x) = \begin{cases} x, & |x| < 1, \\ 0, & \text{otherwise.} \end{cases}$ (b) Sketch a graph of the solution at some representative times. (c) At what rate does the temperature approach thermal equilibrium?

4.1.9. Solve the heat equation when the right-hand end of a bar of unit length is held at a fixed constant temperature α while the left-hand end is insulated. Discuss the asymptotic behavior of the solution.

4.1.10. For each of the following initial temperature distributions, (i) write out the Fourier series solution to the heated ring (4.30–32), and (ii) find the resulting equilibrium temperature as $t \to \infty$: (a) $\cos x$, (b) $\sin^3 x$, (c) $|x|$, (d) $\begin{cases} 1, & -\pi < x < 0, \\ 0, & 0 < x < \pi. \end{cases}$

♢ 4.1.11. Suppose that the temperature $u(t,x)$ of a homogeneous bar satisfies the heat equation. Show that the associated heat flux $w(t,x)$ is also a solution to the same heat equation.

♢ 4.1.12. Show that the time derivative $v = u_t$ of any solution to the heat equation is also a solution. If $u(t,x)$ satisfies the initial condition $u(0,x) = f(x)$, what initial condition does $v(t,x)$ inherit?

♢ 4.1.13. Explain why the thermal energy $E(t) = \displaystyle\int_0^\ell u(t,x)\,dx$ is not constant for the Dirichlet initial-boundary value problem for the heat equation on the interval $[0,\ell]$.

♢ 4.1.14. (a) Show that the thermal energy $E(t) = \displaystyle\int_0^\ell u(t,x)\,dx$ is constant for the Neumann boundary value problem on the interval $[0,\ell]$. (b) Use part (a) to prove that the constant equilibrium solution for the homogeneous Neumann boundary value problem is equal to the mean initial temperature $u(0,x)$.

4.1.15. Let $u(t,x)$ be any nonconstant solution to the periodic heat equation (4.30–31). Prove that the squared L^2 norm of the solution, $N(t) = \displaystyle\int_{-\pi}^{\pi} u(t,x)^2\,dx$, is a strictly decreasing function of t. *Remark*: Interestingly, comparing this result with formula (4.38), we find that, for the periodic boundary value problem, the integral of u is constant, but the integral of u^2 is strictly decreasing. How is this possible?

♡ 4.1.16. The *cable equation* $v_t = \gamma v_{xx} - \alpha v$, with $\gamma, \alpha > 0$, also known as the *lossy heat equation*, was derived by the nineteenth-century Scottish physicist William Thomson to model propagation of signals in a transatlantic cable. Later, in honor of his work on thermodynamics, including determining the value of absolute zero temperature, he was named Lord Kelvin by Queen Victoria. The cable equation was later used to model the electrical activity of neurons. (a) Show that the general solution to the cable equation is given by $v(t,x) = e^{-\alpha t} u(t,x)$, where $u(t,x)$ solves the heat equation $u_t = \gamma u_{xx}$.
(b) Find a Fourier series solution to the Dirichlet initial-boundary value problem
$$v_t = \gamma v_{xx} - \alpha v, \qquad v(0,x) = f(x), \qquad v(t,0) = 0 = v(t,1), \qquad 0 \le x \le 1, \qquad t > 0.$$
Does your solution approach an equilibrium value? If so, how fast?
(c) Answer part (b) for the Neumann problem
$$v_t = \gamma v_{xx} - \alpha v, \qquad v(0,x) = f(x), \qquad v_x(t,0) = 0 = v_x(t,1), \qquad 0 \le x \le 1, \qquad t > 0.$$

♢ 4.1.17. The *convection-diffusion equation* $u_t + c u_x = \gamma u_{xx}$ is a simple model for the diffusion of a pollutant in a fluid flow moving with constant speed c. Show that $v(t,x) = u(t, x + ct)$ solves the heat equation. What is the physical interpretation of this change of variables?

4.1.18. Combine Exercises 4.1.16–17 to solve the *lossy convection-diffusion equation*
$$u_t = \gamma u_{xx} + c u_x - \alpha u.$$

◇ 4.1.19. Let $\gamma > 0$ and $\lambda \leq 0$. (a) Find all solutions to the differential equation $\gamma\, v'' + \lambda\, v = 0$. (b) Prove that the only solution that satisfies the boundary conditions $v(0) = 0$, $v(\ell) = 0$, is the zero solution $v(x) \equiv 0$.

◇ 4.1.20. Answer Exercise 4.1.19 when λ is a non-real complex number.

4.2 The Wave Equation

Let us return to the one-dimensional *wave equation*

$$\frac{\partial^2 u}{\partial t^2} = c^2 \frac{\partial^2 u}{\partial x^2}, \tag{4.63}$$

with constant wave speed $c > 0$, used to model the vibrations of bars and strings. In Chapter 2, we learned how to explicitly solve the wave equation by the method of d'Alembert. Unfortunately, d'Alembert's approach does not extend to other equations of interest to us, and so alternative solution techniques, particularly those based on Fourier methods, are worth developing. Indeed, the resulting series solutions provide valuable insight into wave dynamics on bounded intervals.

Separation of Variables and Fourier Series Solutions

One of the oldest — and still one of the most widely used — techniques for constructing explicit analytic solutions to a wide range of linear partial differential equations is the method of *separation of variables*. We have, in fact, already employed a simplified version of the method when constructing each eigensolution to the heat equation as an exponential function of t times a function of x. In general, the separation of variables method seeks solutions to the partial differential equation that can be written as the product of functions of the individual independent variables. For the wave equation, we seek solutions

$$u(t,x) = w(t)\, v(x) \tag{4.64}$$

that can be written as the product of a function of t alone and a function of x alone. When the method succeeds (which is not guaranteed in advance), both factors are found as solutions to certain ordinary differential equations.

Let us see whether such an expression can possibly solve the wave equation. First of all,

$$\frac{\partial^2 u}{\partial t^2} = w''(t)\, v(x), \qquad\qquad \frac{\partial^2 u}{\partial x^2} = w(t)\, v''(x),$$

where the primes indicate ordinary derivatives. Substituting these expressions into the wave equation (4.63), we obtain

$$w''(t)\, v(x) = c^2\, w(t)\, v''(x).$$

Dividing both sides by $w(t)\, v(x)$ (which we assume is not identically zero, since otherwise, the solution would be trivial) yields

$$\frac{w''(t)}{w(t)} = c^2 \frac{v''(x)}{v(x)},$$

which effectively "separates" the t and x variables on each side of the equation, whence the name "separation of variables".

Now, how could a function of t alone be equal to a function of x alone? A moment's reflection should convince the reader that this can happen if and only if the two functions are constant,[†] so

$$\frac{w''(t)}{w(t)} = c^2\,\frac{v''(x)}{v(x)} = \lambda, \tag{4.65}$$

where we use λ to indicate the common *separation constant*. Thus, the individual factors $w(t)$ and $v(x)$ must satisfy ordinary differential equations

$$\frac{d^2 w}{dt^2} - \lambda\, w = 0, \qquad\qquad \frac{d^2 v}{dx^2} - \frac{\lambda}{c^2}\, v = 0,$$

as promised. We already know how to solve both of these ordinary differential equations by elementary techniques. There are three different cases, depending on the sign of the separation constant λ. As a result, each value of λ leads to four independent separable solutions to the wave equation, as listed in the accompanying table.

Separable Solutions to the Wave Equation

λ	$w(t)$	$v(x)$	$u(t,x) = w(t)\,v(x)$
$\lambda = -\omega^2 < 0$	$\cos\omega t,\ \sin\omega t$	$\cos\dfrac{\omega x}{c},\ \sin\dfrac{\omega x}{c}$	$\cos\omega t\,\cos\dfrac{\omega x}{c},\ \cos\omega t\,\sin\dfrac{\omega x}{c},$ $\sin\omega t\,\cos\dfrac{\omega x}{c},\ \sin\omega t\,\sin\dfrac{\omega x}{c}$
$\lambda = 0$	$1,\ t$	$1,\ x$	$1,\ x,\ t,\ tx$
$\lambda = \omega^2 > 0$	$e^{-\omega t},\ e^{\omega t}$	$e^{-\omega x/c},\ e^{\omega x/c}$	$e^{-\omega(t+x/c)},\ e^{\omega(t-x/c)},$ $e^{-\omega(t-x/c)},\ e^{\omega(t+x/c)}$

So far, we have not taken the boundary conditions into account. Consider first the case of a string of length ℓ with two fixed ends, and thus subject to homogeneous Dirichlet boundary conditions

$$u(t,0) = 0 = u(t,\ell).$$

Substituting the separable ansatz (4.65), we find that $v(x)$ must satisfy

$$\frac{d^2 v}{dx^2} - \frac{\lambda}{c^2}\, v = 0, \qquad\qquad v(0) = 0 = v(\ell). \tag{4.66}$$

The complete system of (nontrivial) solutions to this boundary value problem were found in (4.21):

$$v_n(x) = \sin\frac{n\pi x}{\ell}, \qquad \lambda_n = -\left(\frac{n\pi c}{\ell}\right)^2, \qquad n = 1, 2, 3, \ldots .$$

[†] *Technical detail*: one should assume that the underlying domain is connected for this to be valid as stated. In practice, this technicality can be safely ignored.

Hence, according to the table, the corresponding separable solutions are

$$u_n(t,x) = \cos\frac{n\pi c\,t}{\ell}\,\sin\frac{n\pi x}{\ell}, \qquad \widetilde{u}_n(t,x) = \sin\frac{n\pi c\,t}{\ell}\,\sin\frac{n\pi x}{\ell}. \tag{4.67}$$

We will now employ these solutions to construct a candidate series solution to the wave equation subject to the prescribed boundary conditions:

$$u(t,x) = \sum_{n=1}^{\infty}\left[b_n\cos\frac{n\pi c\,t}{\ell}\,\sin\frac{n\pi x}{\ell} + d_n\sin\frac{n\pi c\,t}{\ell}\,\sin\frac{n\pi x}{\ell}\right]. \tag{4.68}$$

The solution is thus a linear combination of the natural Fourier modes vibrating with frequencies

$$\omega_n = \frac{n\pi c}{\ell} = \frac{n\pi}{\ell}\sqrt{\frac{\kappa}{\rho}}, \qquad n = 1,2,3,\ldots\,, \tag{4.69}$$

where the second expression follows from (2.66). Observe that, the longer the length ℓ of the string, or the higher its density ρ, the slower the vibrations, whereas increasing its stiffness or tension κ speeds them up — in exact accordance with our physical intuition.

The Fourier coefficients b_n and d_n in (4.68) will be uniquely determined by the initial conditions

$$u(0,x) = f(x), \qquad \frac{\partial u}{\partial t}(0,x) = g(x), \qquad 0 < x < \ell.$$

Differentiating the series term by term, we discover that we must represent the initial displacement and velocity as Fourier sine series

$$u(0,x) = \sum_{n=1}^{\infty} b_n\sin\frac{n\pi x}{\ell} = f(x), \qquad \frac{\partial u}{\partial t}(0,x) = \sum_{n=1}^{\infty} d_n\frac{n\pi c}{\ell}\,\sin\frac{n\pi x}{\ell} = g(x).$$

Therefore,

$$b_n = \frac{2}{\ell}\int_0^{\ell} f(x)\,\sin\frac{n\pi x}{\ell}\,dx, \qquad n = 1,2,3,\ldots\,, \tag{4.70}$$

are the Fourier sine coefficients (3.85) of the initial displacement $f(x)$, while

$$d_n = \frac{2}{n\pi c}\int_0^{\ell} g(x)\,\sin\frac{n\pi x}{\ell}\,dx, \qquad n = 1,2,3,\ldots\,. \tag{4.71}$$

are rescaled versions of the Fourier sine coefficients of the initial velocity $g(x)$.

Example 4.3. A string of unit length fixed at both ends is held taut at its center and then released. Our task is to describe the ensuing vibrations. Let us assume that the physical units are chosen so that $c^2 = 1$, and so we are asked to solve the initial-boundary value problem

$$u_{tt} = u_{xx}, \qquad u(0,x) = f(x), \qquad u_t(0,x) = 0, \qquad u(t,0) = u(t,1) = 0. \tag{4.72}$$

To be specific, we assume that the center of the string has been moved by half a unit, and so the initial displacement is

$$f(x) = \begin{cases} x, & 0 \le x \le \frac{1}{2}, \\ 1-x, & \frac{1}{2} \le x \le 1. \end{cases}$$

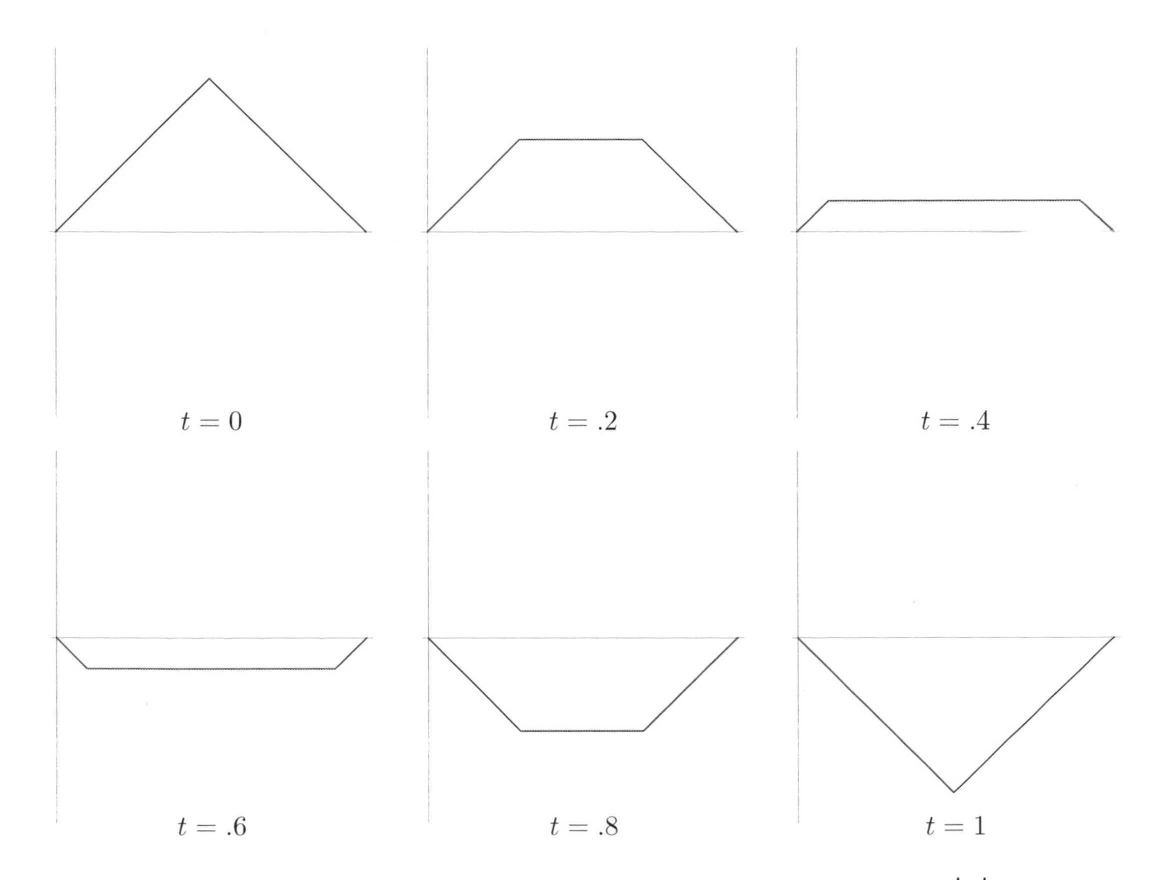

Figure 4.4. Plucked string solution of the wave equation.

The vibrational frequencies $\omega_n = n\pi$ are the integral multiples of π, and so the natural modes of vibration are

$$\cos n\pi t \ \sin n\pi x \qquad \text{and} \qquad \sin n\pi t \ \sin n\pi x \qquad \text{for} \qquad n = 1, 2, \ldots .$$

Consequently, the general solution to the boundary value problem is

$$u(t,x) = \sum_{n=1}^{\infty} \big(b_n \cos n\pi t \ \sin n\pi x + d_n \sin n\pi t \ \sin n\pi x \big),$$

where

$$b_n = 2\int_0^1 f(x) \sin n\pi x \, dx = \begin{cases} 4\displaystyle\int_0^{1/2} x \sin n\pi x \, dx = \dfrac{4(-1)^k}{(2k+1)^2\pi^2}, & n = 2k+1, \\ 0, & n = 2k, \end{cases}$$

while $d_n = 0$. Therefore, the solution is the Fourier sine series

$$u(t,x) = \frac{4}{\pi^2} \sum_{k=0}^{\infty} (-1)^k \frac{\cos(2k+1)\pi t \ \sin(2k+1)\pi x}{(2k+1)^2}, \tag{4.73}$$

whose profile is depicted in Figure 4.4. At time $t = 1$, the original displacement is reproduced exactly, but upside down. The subsequent dynamics proceeds as before, but in mirror-image form. The original displacement reappears at time $t = 2$, after which time the motion is periodically repeated. Interestingly, at times $t_k = .5, 1.5, 2.5, \ldots$, the displacement is identically zero, $u(t_k, x) \equiv 0$, although the velocity is not, $u_t(t_k, x) \not\equiv 0$. The solution appears to be piecewise affine, i.e., its graph is a collection of straight lines. This can, in fact, be proved as a consequence of the d'Alembert formula; see Exercise 4.2.13. Observe that, unlike the heat equation, the wave equation does *not* smooth out discontinuities and corners in the initial data. And, although we will loosely refer to such piecewise C^2 functions as "solutions", they are not, in fact, classical solutions. (Their status as weak solutions, though, can be established using the methods of Section 10.4.)

While the series form (4.68) of the solution is perhaps less satisfying than a d'Alembert-style formula, we can still use it to deduce important qualitative properties. First of all, since each term is periodic in t with period $2\ell/c$, the entire solution is time periodic with that period: $u(t + 2\ell/c, x) = u(t, x)$. In fact, after half a period, the solution reduces to

$$u\left(\frac{\ell}{c}, x\right) = \sum_{n=1}^{\infty} (-1)^n b_n \sin\frac{n\pi x}{\ell} = -\sum_{n=1}^{\infty} b_n \sin\frac{n\pi(\ell - x)}{\ell} = -u(0, \ell - x) = -f(\ell - x).$$

In general,

$$u\left(t + \frac{\ell}{c},\, x\right) = -u(t, \ell - x), \qquad u\left(t + \frac{2\ell}{c},\, x\right) = u(t, x). \tag{4.74}$$

Therefore, the initial wave form is reproduced, first as an upside down mirror image of itself at time $t = \ell/c$, and then in its original form at time $t = 2\ell/c$. This has the important consequence that vibrations of (homogeneous) one-dimensional media are inherently periodic, because the fundamental frequencies (4.69) are all integer multiples of the lowest one: $\omega_n = n\omega_1$.

Remark: The immediately preceding remark has important musical consequences. To the human ear, sonic vibrations that are integral multiples of a single frequency, and thus periodic in time, sound harmonious, whereas those with irrationally related frequencies, and hence experiencing aperiodic vibrations, sound dissonant. This is why most tonal instruments rely on vibrations in one dimension, be it a violin or piano string, a column of air in a wind instrument (flute, clarinet, trumpet, or saxophone), a xylophone bar, or a triangle. On the other hand, most percussion instruments rely on the vibrations of two-dimensional media, e.g., drums and cymbals, or three-dimensional solid bodies, e.g., blocks. As we shall see in Chapters 11 and 12, the frequency ratios of the latter are irrationally related, and hence their motion is only quasiperiodic, as in Example 2.20. For some reason, our appreciation of music is psychologically attuned to the differences between rationally related/periodic and irrationally related/quasiperiodic vibrations, [**105**].

Consider next a string with both ends left free, and so subject to the Neumann boundary conditions

$$\frac{\partial u}{\partial x}(t, 0) = 0 = \frac{\partial u}{\partial x}(t, \ell). \tag{4.75}$$

The solutions of (4.66) satisfying $v'(0) = 0 = v'(\ell)$ are now

$$v_n(x) = \cos\frac{n\pi x}{\ell} \qquad \text{with} \qquad \omega_n = \frac{n\pi c}{\ell}, \qquad n = 0, 1, 2, 3, \ldots .$$

The resulting solution takes the form of a Fourier cosine series

$$u(t, x) = a_0 + c_0 t + \sum_{n=1}^{\infty} \left(a_n \cos\frac{n\pi c t}{\ell} \cos\frac{n\pi x}{\ell} + c_n \sin\frac{n\pi c t}{\ell} \cos\frac{n\pi x}{\ell} \right). \tag{4.76}$$

The first two terms come from the null eigenfunction $v_0(x) = 1$ with $\omega_0 = 0$. The string vibrates with the same fundamental frequencies (4.69) as in the fixed-end case, but there is now an additional *unstable mode* $c_0 t$ that is no longer periodic, but grows linearly in time. In general, the presence of null eigenfunctions implies that the wave equation admits unstable modes.

Substituting (4.76) into the initial conditions

$$u(0,x) = f(x), \qquad \frac{\partial u}{\partial t}(0,x) = g(x), \qquad 0 < x < \ell,$$

we find that the Fourier coefficients are prescribed, as before, by the initial displacement and velocity:

$$a_n = \frac{2}{\ell}\int_0^\ell f(x)\cos\frac{n\pi x}{\ell}\,dx, \qquad c_n = \frac{2}{n\pi c}\int_0^\ell g(x)\cos\frac{n\pi x}{\ell}\,dx, \qquad n = 1,2,3,\ldots.$$

The order-zero coefficients[†]

$$a_0 = \frac{1}{\ell}\int_0^\ell f(x)\,dx, \qquad c_0 = \frac{1}{\ell}\int_0^\ell g(x)\,dx,$$

are equal to the average initial displacement and average initial velocity of the string. In particular, when $c_0 = 0$, there is no net initial velocity, and the unstable mode is not excited. In this case, the solution is time-periodic, oscillating around the position given by the average initial displacement. On the other hand, if $c_0 \neq 0$, the string will move off with constant average speed c_0, all the while vibrating at the same fundamental frequencies.

Similar considerations apply to the periodic boundary value problem for the wave equation on a circular ring. The details are left as Exercise 4.2.6 for the reader.

Exercises

4.2.1. In music, an octave corresponds to doubling the frequency of the sound waves. On my piano, the middle C string has length .7 meter, while the string for the C an octave higher has length .6 meter. Assuming that they have the same density, how much tighter does the shorter string need to be tuned?

4.2.2. How much longer would a piano string have to be to make the same sound when it is pulled twice as tight?

4.2.3. Write down the solutions to the following initial-boundary value problems for the wave equation in the form of a Fourier series:
(a) $u_{tt} = u_{xx}$, $u(t,0) = u(t,\pi) = 0$, $u(0,x) = 1$, $u_t(0,x) = 0$;
(b) $u_{tt} = 2u_{xx}$, $u(t,0) = u(t,\pi) = 0$, $u(0,x) = 0$, $u_t(0,x) = 1$;
(c) $u_{tt} = 3u_{xx}$, $u(t,0) = u(t,\pi) = 0$, $u(0,x) = \sin^3 x$, $u_t(0,x) = 0$;
(d) $u_{tt} = 4u_{xx}$, $u(t,0) = u(t,1) = 0$, $u(0,x) = x$, $u_t(0,x) = -x$;
(e) $u_{tt} = u_{xx}$, $u(t,0) = u_x(t,1) = 0$, $u(0,x) = 1$, $u_t(0,x) = 0$;
(f) $u_{tt} = 2u_{xx}$, $u_x(t,0) = u_x(t,2\pi) = 0$, $u(0,x) = -1$, $u_t(0,x) = 1$;
(g) $u_{tt} = u_{xx}$, $u_x(t,0) = u_x(t,1) = 0$, $u(0,x) = x(1-x)$, $u_t(0,x) = 0$.

[†] Note that we have not included the usual $\frac{1}{2}$ factor in the constant terms in the Fourier series (4.76).

4.2.4. Find all separable solutions to the wave equation $u_{tt} = u_{xx}$ on the interval $0 \le x \le \pi$ subject to (a) mixed boundary conditions $u(t,0) = 0$, $u_x(t,\pi) = 0$;
(b) Neumann boundary conditions $u_x(t,0) = 0$, $u_x(t,\pi) = 0$.

4.2.5. (a) Under what conditions is the solution to the Neumann boundary value problem (4.75) a periodic function of t? What is the period? (b) Establish explicit periodicity formulas of the form (4.74). (c) Under what conditions is the velocity $\partial u/\partial t$ periodic in t?

♡ 4.2.6. (a) Formulate the periodic initial-boundary value problem for the wave equation on the interval $-\pi \le x \le \pi$, modeling the vibrations of a circular ring. (b) Write out a formula for the solution to your problem in the form of a Fourier series. (c) Is the solution a periodic function of t? If so, what is the period? (d) Suppose the initial displacement coincides with that in Figure 4.6, while the initial velocity is zero. Describe what happens to the solution as time evolves.

4.2.7. Show that the time derivative, $v = \partial u/\partial t$, of any solution to the wave equation is also a solution. If you know the initial conditions of u, what initial conditions does v satisfy?

4.2.8. Find all the separable real solutions to the wave equation subject to a restoring force: $u_{tt} = u_{xx} - u$. Discuss their long-term behavior.

♡ 4.2.9. Let $a, c > 0$ be positive constants. The *telegrapher's equation* $u_{tt} + a\,u_t = c^2 u_{xx}$ represents a damped version of the wave equation. Consider the Dirichlet boundary value problem $u(t,0) = u(t,1) = 0$, on the interval $0 \le x \le 1$, with initial conditions $u(0,x) = f(x)$, $u_t(0,x) = 0$. (a) Find all separable solutions to the telegrapher's equation that satisfy the boundary conditions. (b) Write down a series solution for the initial boundary value problem. (c) Discuss the long term behavior of your solution. (d) State a criterion that distinguishes overdamped from underdamped versions of the equation.

4.2.10. The fourth-order partial differential equation $u_{tt} = -u_{xxxx}$ is a simple model for a vibrating elastic beam. (a) Find all separable real solutions to the beam equation.
(b) Show that any C^4 (complex) solution to the Schrödinger equation $i\,u_t = u_{xx}$ solves the beam equation.

4.2.11. The initial-boundary value problem

$$u_{tt} = -u_{xxxx}, \qquad \begin{array}{ll} u(t,0) = u_{xx}(t,0) = u(t,1) = u_{xx}(t,1) = 0, & 0 < x < 1, \\ u(0,x) = f(x), \qquad u_t(0,x) = 0, & t > 0, \end{array}$$

models the vibrations of an elastic beam of unit length with simply supported ends, subject to a nonzero initial displacement $f(x)$ and zero initial velocity. (a) What are the vibrational frequencies for the beam? (b) Write down the solution to the initial-boundary value problem as a Fourier series. (c) Does the beam vibrate periodically
(*i*) for all initial conditions? (*ii*) for some initial conditions? (*iii*) for no initial conditions?

4.2.12. *Multiple choice*: The initial-boundary value problem

$$u_{tt} = u_{xxxx}, \qquad \begin{array}{ll} u(t,0) = u_{xx}(t,0) = u(t,1) = u_{xx}(t,1) = 0, & 0 < x < 1, \\ u(0,x) = f(x), \qquad u_t(0,x) = g(x), & t > 0, \end{array}$$

is well-posed for (a) $t > 0$; (b) $t < 0$; (c) all t; (d) no t. Explain your answer.

The d'Alembert Formula for Bounded Intervals

In Theorem 2.15, we derived the explicit d'Alembert formula

$$u(t,x) = \frac{f(x-ct) + f(x+ct)}{2} + \frac{1}{2c}\int_{x-ct}^{x+ct} g(z)\,dz, \tag{4.77}$$

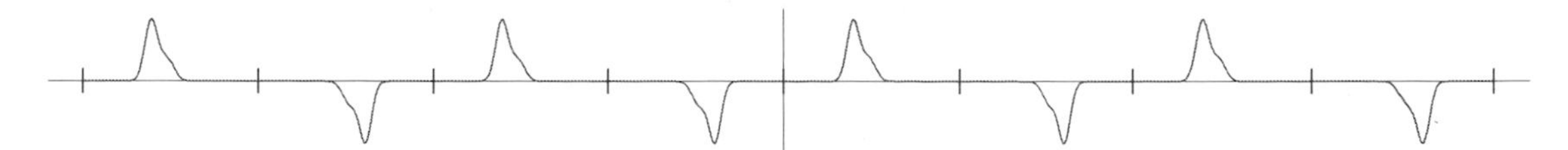

Figure 4.5. Odd periodic extension of a concentrated pulse.

for solving the basic initial value problem for the wave equation on an infinite interval:

$$\frac{\partial^2 u}{\partial t^2} = c^2 \frac{\partial^2 u}{\partial x^2}, \qquad u(0,x) = f(x), \qquad \frac{\partial u}{\partial t}(0,x) = g(x), \qquad -\infty < x < \infty.$$

In this section we explain how to adapt the formula in order to solve initial-boundary value problems on bounded intervals, thereby effectively summing the Fourier series solution.

The easiest case to deal with is the periodic problem on $0 \le x \le \ell$, with boundary conditions

$$u(t,0) = u(t,\ell), \qquad u_x(t,0) = u_x(t,\ell). \tag{4.78}$$

If we extend the initial displacement $f(x)$ and velocity $g(x)$ to be periodic functions of period ℓ, so $f(x+\ell) = f(x)$ and $g(x+\ell) = g(x)$ for all $x \in \mathbb{R}$, then the resulting d'Alembert solution (4.77) will also be periodic in x, so $u(t,x+\ell) = u(t,x)$. In particular, it satisfies the boundary conditions (4.78) and so coincides with the desired solution. Details are to be supplied in Exercises 4.2.27–28.

Next, suppose we have fixed (Dirichlet) boundary conditions

$$u(t,0) = 0, \qquad u(t,\ell) = 0. \tag{4.79}$$

The resulting solution can be written as a Fourier sine series (4.68), and hence is both odd and 2ℓ–periodic in x. Therefore, to write the solution in d'Alembert form (4.77), we extend the initial displacement $f(x)$ and velocity $g(x)$ to be odd, periodic functions of period 2ℓ:

$$f(-x) = -f(x), \qquad f(x+2\ell) = f(x), \qquad g(-x) = -g(x), \qquad g(x+2\ell) = g(x).$$

This will ensure that the d'Alembert solution also remains odd and periodic. As a result, it satisfies the homogeneous Dirichlet boundary conditions (4.79) for all t, cf. Exercise 4.2.31. Keep in mind that, while the solution $u(t,x)$ is defined for all x, the only physically relevant values occur on the interval $0 \le x \le \ell$. Nevertheless, the effects of displacements in the unphysical regime will eventually be felt as the propagating waves pass through the physical interval.

For example, consider an initial displacement that is concentrated near $x = \xi$ for some $0 < \xi < \ell$. Its odd 2ℓ–periodic extension consists of two sets of replicas: those of the same form occurring at positions $\xi \pm 2\ell$, $\xi \pm 4\ell$, $\ldots$, and their upside-down mirror images at the intermediate positions $-\xi$, $-\xi \pm 2\ell$, $-\xi \pm 4\ell$, $\ldots$; Figure 4.5 shows a representative example. The resulting solution begins with each of the pulses, both positive and negative, splitting into two half-size replicas that propagate with speed c in opposite directions. When a left and right moving pulse meet, they emerge from the interaction unaltered. The process repeats periodically, with an infinite row of half-size pulses moving to the right kaleidoscopically interacting with an infinite row moving to the left.

However, only the part of this solution that lies on $0 \le x \le \ell$ is actually observed on the physical string. The effect is as if one were watching the full solution as it passes by a window of length ℓ. Such observers will interpret what they see a bit differently. To

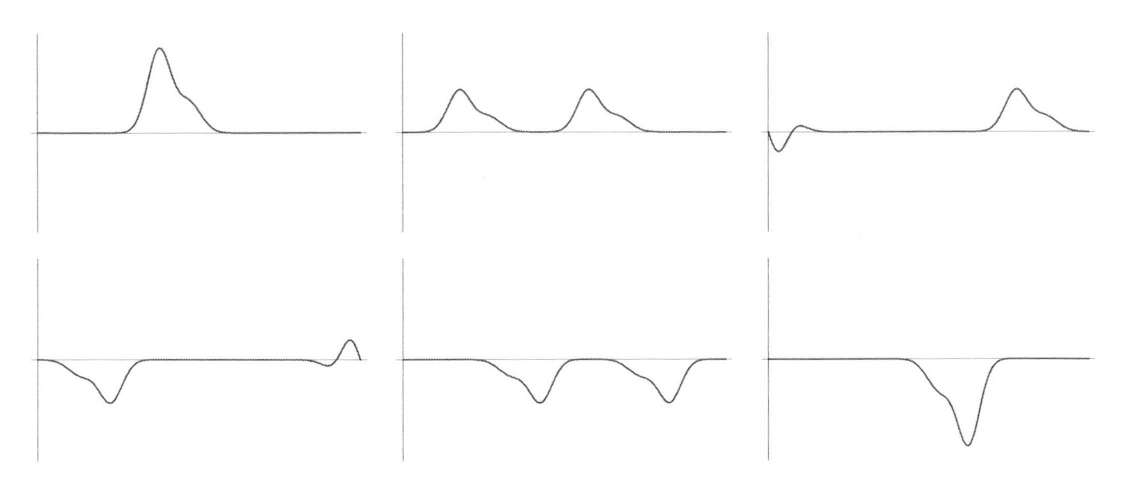

Figure 4.6. Solution to wave equation with fixed ends.

wit, the original pulse starting at position $0 < \xi < \ell$ splits up into two half-size replicas that move off in opposite directions. As each half-size pulse reaches an end of the string, it meets a mirror-image pulse that has been propagating in the opposite direction from the nonphysical regime. The pulse is reflected at the end of the interval and becomes an upside-down mirror image moving in the opposite direction. The original positive pulse has moved off the end of the string just as its mirror image has moved into the physical regime. (A common physical realization is a pulse propagating down a jump rope that is held fixed at its end; the reflected pulse returns upside down.) A similar reflection occurs as the other half-size pulse hits the other end of the physical interval, after which the solution consists of two upside-down half-size pulses moving back towards each other. At time $t = \ell/c$ they recombine at the point $\ell - \xi$ to instantaneously form a full-sized, but upside-down mirror image of the original disturbance — in accordance with (4.74). The recombined pulse in turn splits apart into two upside-down half-size pulses that, when each collides with the end, reflect and return to their original upright form. At time $t = 2\ell/c$, the pulses recombine to exactly reproduce the original displacement. The process then repeats, and the solution is periodic in time with period $2\ell/c$.

In Figure 4.6, the first picture displays the initial displacement. In the second, it has split into left- and right-moving half-size clones. In the third picture, the left-moving bump is in the process of colliding with the left end of the string. In the fourth picture, it has emerged from the collision, and is now upside down, reflected, and moving to the right. Meanwhile, the right-moving pulse is starting to collide with the right end. In the fifth picture, both pulses have completed their collisions and are now moving back towards each other, where, in the last picture, they recombine into an upside-down mirror image of the original pulse. The process then repeats itself, in mirror image, finally recombining to the original pulse, at which point the entire process starts over.

The Neumann (free) boundary value problem

$$\frac{\partial u}{\partial x}(t,0) = 0, \qquad \frac{\partial u}{\partial x}(t,\ell) = 0, \tag{4.80}$$

is handled similarly. Since the solution has the form of a Fourier cosine series in x, we extend the initial conditions to be *even* 2ℓ–periodic functions

$$f(-x) = f(x), \qquad f(x+2\ell) = f(x), \qquad g(-x) = g(x), \qquad g(x+2\ell) = g(x).$$

The resulting d'Alembert solution (4.77) is also even and 2ℓ–periodic in x, and hence satisfies the boundary conditions, cf. Exercise 4.2.31(b). In this case, when a pulse hits one of the ends, its reflection remains upright, but becomes a mirror image of the original; a familiar physical illustration is a water wave that reflects off a solid wall. Further details are left to the reader in Exercise 4.2.22

In summary, we have now studied two very different ways to solve the one-dimensional wave equation. The first, based on the d'Alembert formula, emphasizes their particle-like aspects, where individual wave packets collide with each other, or reflect at the boundary, all the while maintaining their overall form, while the second, based on Fourier analysis, emphasizes the vibrational or wave-like character of the solutions. Some solutions look like vibrating waves, while others appear much more like interacting particles. But, like the proverbial blind men describing an elephant, these are merely two facets of the *same* solution. The Fourier series formula shows how every particle-like solution can be decomposed into its constituent vibrational modes, while the d'Alembert formula demonstrates how vibrating solutions combine into moving wave packets.

The coexistence of particle and wave features is reminiscent of the long-running historical debate over the nature of light. Newton and his disciples proposed a particle-based theory, anticipating the modern concept of photons. However, until the beginning of the twentieth century, most physicists advocated a wave-like or vibrational viewpoint. Einstein's explanation of the photoelectric effect served to resurrect the particle interpretation. Only with the establishment of quantum mechanics was the debate resolved — light, and, indeed, all subatomic particles manifest *both* particle and wave features, depending upon the experiment and the physical situation. But a theoretical basis for the perplexing wave-particle duality could have been found already in Fourier's and d'Alembert's competing solution formulae for the classical wave equation!

Exercises

♢ 4.2.13. (a) Solve the initial-boundary value problem from Example 4.3 using the d'Alembert method. (b) Verify that your solution coincides with the Fourier series solution derived above. (c) Justify our earlier observation that, at each time t, the solution $u(t,x)$ is a piecewise affine function of x.

4.2.14. Sketch the solution of the wave equation $u_{tt} = u_{xx}$ and describe its behavior when the initial displacement is the box function $u(0,x) = \begin{cases} 1, & 1 < x < 2, \\ 0, & \text{otherwise}, \end{cases}$ while the initial velocity is 0 in each of the following scenarios: (a) on the entire line $-\infty < x < \infty$; (b) on the half-line $0 \le x < \infty$, with homogeneous Dirichlet boundary condition at the end; (c) on the half-line $0 \le x < \infty$, with homogeneous Neumann boundary condition at the end; (d) on the bounded interval $0 \le x \le 5$ with homogeneous Dirichlet boundary conditions; (e) on the bounded interval $0 \le x \le 5$ with homogeneous Neumann boundary conditions.

4.2.15. Answer Exercise 4.2.14 when the initial velocity is the box function, while the initial displacement is zero.

4.2.16. Consider the initial-boundary value problem

$$\frac{\partial^2 u}{\partial t^2} = \frac{\partial^2 u}{\partial x^2}, \qquad \begin{array}{ll} u(t,0) = 0 = u(t,10), & t > 0, \\ u(0,x) = f(x), \quad u_t(0,x) = 0, & 0 < x < 10, \end{array}$$

for the wave equation, where the initial data has the following form:

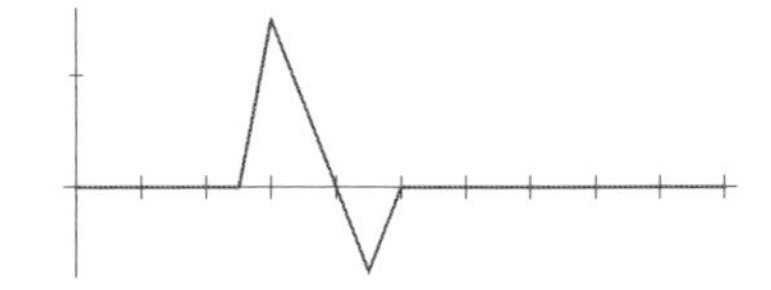

$$f(x) = \begin{cases} 3x - 7.5, & 2.5 \le x \le 3, \\ 6 - 1.5x, & 3 \le x \le 4.5, \\ 1.5x - 7.5, & 4.5 \le x \le 5, \\ 0, & \text{otherwise.} \end{cases}$$

Discuss what happens to the solution. You do *not* need to write down an explicit formula for the solution, but for full credit you must explain (sketches can help) at least three or four interesting things that happen to the solution as time progresses.

4.2.17. Repeat Exercise 4.2.16 for the Neumann boundary conditions.

4.2.18. Suppose the initial displacement of a string of length ℓ looks like the graph to the right. Assuming that the ends of the string are held fixed, graph the string's profile at times $t = \ell/c$ and $2\ell/c$.

♣ 4.2.19. Consider the wave equation $u_{tt} = u_{xx}$ on the interval $0 \le x \le 1$, with homogeneous Dirichlet boundary conditions at both ends. (a) Use the d'Alembert formula to explicitly solve the initial value problem $u(0,x) = x - x^2$, $u_t(0,x) = 0$. (b) Graph the solution profile at some representative times, and discuss what you observe. (c) Find the Fourier series at each t of your solution and compare the two. (d) How many terms do you need to sum to obtain a reasonable approximation to the exact solution?

♣ 4.2.20. Solve Exercise 4.2.19 for the initial conditions $u(0,x) = 0$, $u_t(0,x) = x^2 - x$.

♣ 4.2.21. Solve (*i*) Exercise 4.2.19, (*ii*) Exercise 4.2.20, when the solution is subject to homogeneous Neumann boundary conditions.

◇ 4.2.22. Under what conditions is the solution to the Neumann boundary value problem for the wave equation on a bounded interval $[0,\ell]$ periodic in time? What is the period?

4.2.23. Discuss and sketch the behavior of the solution to the Neumann boundary value problem $u_{tt} = 4u_{xx}$, $0 < x < 1$, $u_x(t,0) = 0 = u_x(t,1)$, $u(0,x) = f(x)$, $u_t(0,x) = g(x)$, for

(a) a localized initial displacement: $f(x) = \begin{cases} 1, & .2 < x < .3, \\ 0, & \text{otherwise.} \end{cases}$ $\quad g(x) = 0$;

(b) a localized initial velocity: $f(x) = 0$, $g(x) = \begin{cases} 1, & .2 < x < .3, \\ 0, & \text{otherwise.} \end{cases}$

4.2.24. (a) Explain how to solve the Neumann initial-boundary value problem

$$\frac{\partial^2 u}{\partial t^2} = \frac{\partial^2 u}{\partial x^2}, \qquad \frac{\partial u}{\partial x}(t,0) = 0 = \frac{\partial u}{\partial x}(t,1), \qquad u(0,x) = f(x), \qquad \frac{\partial u}{\partial t}(0,x) = g(x),$$

on the interval $0 \le x \le 1$.

(b) Let $f(x) = \begin{cases} x - \frac{1}{4}, & \frac{1}{4} \le x \le \frac{1}{2}, \\ \frac{3}{4} - x, & \frac{1}{2} \le x \le \frac{3}{4}, \\ 0, & \text{otherwise,} \end{cases}$ and $g(x) = 0$. Sketch the graph of the solution at a few representative times, and discuss what is happening. Is the solution periodic in time? If so, what is the period?

(c) Do the same when $f(x) = 0$ and $g(x) = x$.

4.2.25. (a) Write down a formula for the solution $u(t,x)$ to the initial-boundary value problem

$$\frac{\partial^2 u}{\partial t^2} - 4\,\frac{\partial^2 u}{\partial x^2} = 0, \quad u(0,x) = \sin x, \quad \frac{\partial u}{\partial t}\,(0,x) = \frac{\partial u}{\partial x}\,(t,0) = \frac{\partial u}{\partial x}\,(t,\pi) = 0, \quad 0 < x < \pi, \quad t > 0.$$

(b) Find $u\left(\frac{\pi}{2},\frac{\pi}{2}\right)$. (c) Prove that $h(t) = u\left(t,\frac{\pi}{2}\right)$ is a periodic function of t and find its period. (d) Does $\dfrac{\partial u}{\partial x}$ have any discontinuities? If so, discuss their behavior.

4.2.26. Answer Exercise 4.2.25 for the mixed boundary conditions $u(t,0) = 0 = u_x(t,\pi)$.

♡ 4.2.27. (a) Explain how to use d'Alembert's formula (4.77) to solve the periodic initial-boundary value problem for the wave equation given in Exercise 4.2.6.
(b) Do the d'Alembert and Fourier series formulae represent the same solution? If so, can you justify it? If not, explain why they are different.

♢ 4.2.28. Show that the solution $u(t,x)$ to the wave equation on an interval $[0,\ell]$, subject to periodic boundary conditions $u(t,0) = u(t,\ell)$, $u_x(t,0) = u_x(t,\ell)$, is a periodic function of t if and only if there is no net initial velocity: $\int_0^\ell g(x)\,dx = 0$.

4.2.29. (a) Explain how to solve the wave equation on a half-line $x > 0$ when subject to Dirichlet boundary conditions $u(t,0) = 0$. (b) Assuming $c = 1$, find the solution satisfying $u(0,x) = (x-2)\,e^{-5\,(x-2.2)^2}$, $u_t(0,x) = 0$. (c) Sketch a picture of your solution at some representative times, and discuss what is happening.

4.2.30. Solve Exercise 4.2.29 for homogeneous Neumann boundary conditions at $x = 0$.

♢ 4.2.31. (a) Given that $f(x)$ is odd and 2ℓ–periodic, explain why $f(0) = 0 = f(\ell)$.
(b) Given that $f(x)$ is even and 2ℓ–periodic, explain why $f'(0) = 0 = f'(\ell)$.

♢ 4.2.32. (a) Prove that if $f(-x) = -f(x)$, $f(x+2\ell) = f(x)$, for all x, then $u(t,x) = \frac{1}{2}\,[\,f(x-ct) + f(x+ct)\,]$ satisfies the Dirichlet boundary conditions (4.79).
(b) Prove that if $g(-x) = -g(x)$, $g(x+2\ell) = g(x)$ for all x, then $u(t,x) = \dfrac{1}{2c}\displaystyle\int_{x-ct}^{x+ct} g(z)\,dz$ also satisfies the Dirichlet boundary conditions.

4.2.33. If both $u(0,x) = f(x)$ and $u_t(0,x) = g(x)$ are even functions, show that the solution $u(t,x)$ of the wave equation is even in x for all t.

4.2.34. (a) Prove that the solution $u(t,x)$ to the wave equation for $x \in \mathbb{R}$ is an even function of t if and only if its initial velocity, at $t = 0$, is zero.
(b) Under what conditions is $u(t,x)$ an odd function of t?

♢ 4.2.35. Let $u(t,x)$ be a classical solution to the wave equation $u_{tt} = c^2 u_{xx}$ on the interval $0 < x < \ell$, satisfying homogeneous Dirichlet boundary conditions. The *total energy* of u at time t is

$$E(t) = \int_0^\ell \frac{1}{2}\left[\left(\frac{\partial u}{\partial t}\right)^2 + c^2\left(\frac{\partial u}{\partial x}\right)^2\right] dx. \tag{4.81}$$

Establish the *Law of Conservation of Energy* by showing that $E(t) = E(0)$ is a constant function.

♢ 4.2.36. (a) Use Exercise 4.2.35 to prove that the only C^2 solution to the initial-boundary value problem $v_{tt} = c^2 v_{xx}$, $v(t,0) = v(t,\ell) = 0$, $v(0,x) = 0$, $v_t(0,x) = 0$, is the trivial solution $v(t,x) \equiv 0$. (b) Establish the following *Uniqueness Theorem* for the wave equation: given $f(x), g(x) \in \mathrm{C}^2$, there is at most one C^2 solution $u(t,x)$ to the initial-boundary value problem $u_{tt} = c^2 u_{xx}$, $u(t,0) = u(t,\ell) = 0$, $u(0,x) = f(x)$, $u_t(0,x) = g(x)$.

4.2.37. Referring back to Exercises 4.2.35 and 4.2.36: (a) Does conservation of energy hold for solutions to the homogeneous Neumann initial-boundary value problem?
(b) Can you establish a uniqueness theorem for the Neumann problem?

4.2.38. Explain how to solve the Dirichlet initial-boundary value problem

$$u_{tt} = c^2 u_{xx} + F(t,x), \qquad u(0,x) = f(x), \qquad u_t(0,x) = g(x), \qquad u(t,0) = u(t,\ell) = 0,$$

for the wave equation subject to an external forcing on the interval $[0,\ell]$.

4.3 The Planar Laplace and Poisson Equations

The two-dimensional *Laplace equation* is the second-order linear partial differential equation

$$\frac{\partial^2 u}{\partial x^2} + \frac{\partial^2 u}{\partial y^2} = 0, \tag{4.82}$$

named in honor of the influential eighteenth-century French mathematician Pierre–Simon Laplace. It, along with its higher-dimensional versions, is arguably the most important differential equation in all of mathematics. A real-valued solution $u(x,y)$ to the Laplace equation is known as a *harmonic function*. The space of harmonic functions can thus be identified as the kernel of the second-order linear partial differential operator

$$\Delta = \frac{\partial^2}{\partial x^2} + \frac{\partial^2}{\partial y^2}, \tag{4.83}$$

known as the *Laplace operator*, or *Laplacian* for short. The inhomogeneous or forced version, namely

$$-\Delta[u] = -\frac{\partial^2 u}{\partial x^2} - \frac{\partial^2 u}{\partial y^2} = f(x,y), \tag{4.84}$$

is known as *Poisson's equation*, named after Siméon–Denis Poisson, who was taught by Laplace. The mathematical and physical reasons for including the minus sign will gradually become clear.

Besides their theoretical importance, the Laplace and Poisson equations arise as the basic equilibrium equations in a remarkable variety of physical systems. For example, we may interpret $u(x,y)$ as the displacement of a *membrane*, e.g., a drum skin; the inhomogeneity $f(x,y)$ in the Poisson equation represents an external forcing over the surface of the membrane. Another example is in the thermal equilibrium of flat plates; here $u(x,y)$ represents the temperature and $f(x,y)$ an external heat source. In fluid mechanics, $u(x,y)$ represents the potential function whose gradient $\mathbf{v} = \nabla u$ is the velocity vector field of a steady planar fluid flow. Similar considerations apply to two-dimensional electrostatic and gravitational potentials. The dynamical counterparts to the Laplace equation are the two-dimensional versions of the heat and wave equations, to be analyzed in Chapter 11.

Since both the Laplace and Poisson equations describe equilibrium configurations, they almost always appear the context of boundary value problems. We seek a solution $u(x,y)$ to the partial differential equation defined at points (x,y) belonging to a bounded, open domain $\Omega \subset \mathbb{R}^2$. The solution is required to satisfy suitable conditions on the boundary of the domain, denoted by $\partial\Omega$, which will consist of one or more simple closed curves, as illustrated in Figure 4.7. As in one-dimensional boundary value problems, there are several especially important types of boundary conditions.

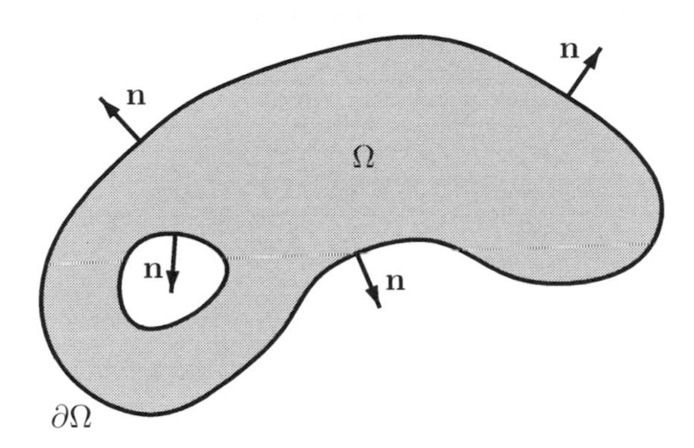

Figure 4.7. A planar domain with outward unit normals on its boundary.

The first are the *fixed* or *Dirichlet boundary conditions*, which specify the value of the function u on the boundary:

$$u(x,y) = h(x,y) \qquad \text{for} \qquad (x,y) \in \partial\Omega. \tag{4.85}$$

Under mild regularity conditions on the domain Ω, the boundary values h, and the forcing function f, the Dirichlet conditions (4.85) serve to uniquely specify the solution $u(x,y)$ to the Laplace or the Poisson equation. Physically, in the case of a free or forced membrane, the Dirichlet boundary conditions correspond to gluing the edge of the membrane to a wire at height $h(x,y)$ over each boundary point $(x,y) \in \partial\Omega$, as illustrated in Figure 4.8. A physical realization can be easily obtained by dipping the wire in a soap solution; the resulting soap film spanning the wire forms a *minimal surface*, which, if the wire is reasonably close to planar shape,[†] is the solution to the Dirichlet problem prescribed by the wire. Similarly, in the modeling of thermal equilibrium, a Dirichlet boundary condition represents the imposition of a prescribed temperature distribution, represented by the function h, along the boundary of the plate.

The second important class consists of the *Neumann boundary conditions*

$$\frac{\partial u}{\partial \mathbf{n}} = \nabla u \cdot \mathbf{n} = k(x,y) \qquad \text{on} \qquad \partial\Omega, \tag{4.86}$$

in which the *normal derivative* of the solution u on the boundary is prescribed. In general, $\mathbf{n}$ denotes the *unit outwards normal* to the boundary $\partial\Omega$, i.e., the vector of unit length, $\|\,\mathbf{n}\,\| = 1$, that is orthogonal to the tangent to the boundary and points *away* from the domain; see Figure 4.7. For example, in thermomechanics, a Neumann boundary condition specifies the heat flux out of a plate through its boundary. The "no-flux" or homogeneous Neumann boundary conditions, where $k(x,y) \equiv 0$, correspond to a fully insulated boundary. In the case of a membrane, homogeneous Neumann boundary conditions correspond to a free, unattached edge of a drum. In fluid mechanics, the Neumann conditions prescribe the fluid flux through the boundary; in particular, homogeneous Neumann boundary conditions

[†] More generally, the minimal surface formed by the soap film solves the vastly more complicated nonlinear *minimal surface equation* $(1+u_x^2)u_{xx} - 2u_x u_y u_{xy} + (1+u_y^2)u_{yy} = 0$, which, for surfaces with small variation, i.e., with $\|\,\nabla u\,\| \ll 1$, can be approximated by the Laplace equation.

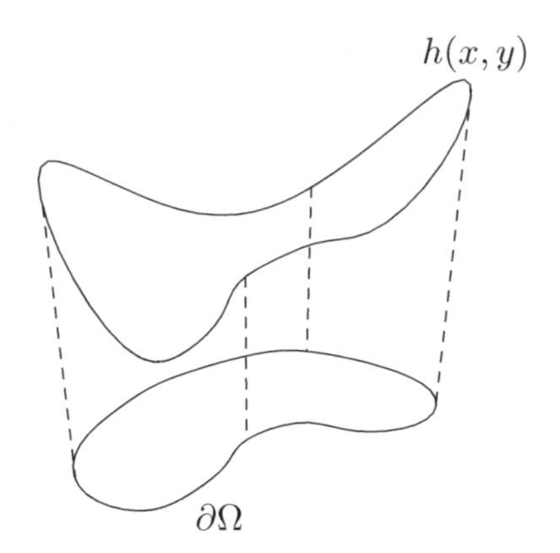

Figure 4.8. Dirichlet boundary conditions.

correspond to a solid boundary that the fluid cannot penetrate. More generally, the *Robin boundary conditions*

$$\frac{\partial u}{\partial \mathbf{n}} + \beta(x,y)\,u = k(x,y) \qquad \text{on} \qquad \partial\Omega,$$

also known as *impedance boundary conditions* due to their applications in electromagnetism, are used to model insulated plates in heat baths, or membranes attached to springs.

Finally, one can mix the previous kinds of boundary conditions, imposing, say, Dirichlet conditions on part of the boundary and Neumann conditions on the complementary part. A typical *mixed boundary value problem* has the form

$$-\Delta u = f \quad \text{in} \quad \Omega, \qquad u = h \quad \text{on} \quad D, \qquad \frac{\partial u}{\partial \mathbf{n}} = k \quad \text{on} \quad N, \tag{4.87}$$

with the boundary $\partial\Omega = D \cup N$ being the disjoint union of a "Dirichlet segment", denoted by D, and a "Neumann segment" N. For example, if u represents the equilibrium temperature in a plate, then the Dirichlet segment of the boundary is where the temperature is fixed, while the Neumann segment is insulated, or, more generally, has prescribed heat flux. Similarly, when modeling the displacement of a membrane, the Dirichlet segment is where the edge of the drum is attached to a support, while the homogeneous Neumann segment is left hanging free.

Exercises

4.3.1. (a) Solve the boundary value problem $\Delta u = 1$ for $x^2 + y^2 < 1$ and $u(x,y) = 0$ for $x^2 + y^2 = 1$ directly. *Hint*: The solution is a simple polynomial.
(*b*) Graph your solution, interpreting it as the equilibrium displacement of a circular drum under a constant gravitational force.

4.3.2. Set up the boundary value problem corresponding to the equilibrium of a circular membrane subject to a constant downwards gravitational force, half of whose boundary is glued to a flat semicircular wire, while the other half is unattached.

4.3.3. Set up the boundary value problem corresponding to the thermal equilibrium of a rectangular plate that is insulated on two of its sides, has 0° at its top edge and 100° at the bottom edge. Where do you expect the maximum temperature to be located? What is its value? Can you find a formula for the temperature inside the plate? *Hint*: The solution is constant along horizontal lines.

4.3.4. Set up the boundary value problem corresponding to the thermal equilibrium of an insulated semi-circular plate with unit diameter, whose curved edge is kept at 0° and whose straight edge is at 50°.

4.3.5. Explain why the solution to the homogeneous Neumann boundary value problem for the Laplace equation is *not* unique.

4.3.6. Write down the Dirichlet boundary value problem for the Laplace equation on the unit square $0 \le x, y \le 1$ that is satisfied by $u(x,y) = 1 + x\,y$.

4.3.7. Write down the Neumann boundary value problem for the Poisson equation on the unit disk $x^2 + y^2 \le 1$ that is satisfied by $u(x,y) = x^3 + x\,y^2$.

♢ 4.3.8. Suppose $u(x,y)$ is a solution to the Laplace equation.
(a) Show that any translate $U(x,y) = u(x-a, y-b)$, where $a, b \in \mathbb{R}$, is also a solution.
(b) Show that the rotated function $U(x,y) = u(x\cos\theta + y\sin\theta, -x\sin\theta + y\cos\theta)$, where $-\pi < \theta \le \pi$, is also a solution.

♢ 4.3.9. (a) Show that if $u(x,y)$ solves the Laplace equation, then so does the rescaled function $U(x,y) = c\,u(\alpha\,x, \alpha\,y)$ for any constants c, α.
(b) Discuss the effect of scaling on the Dirichlet boundary value problem.
(c) What happens if we use different scaling factors in x and y?

Separation of Variables

Our first approach to solving the Laplace equation

$$\Delta u = \frac{\partial^2 u}{\partial x^2} + \frac{\partial^2 u}{\partial y^2} = 0 \tag{4.88}$$

will be based on the method of *separation of variables*. As in (4.64), we seek solutions that can be written as a product

$$u(x,y) = v(x)\,w(y) \tag{4.89}$$

of a function of x alone times a function of y alone. We compute

$$\frac{\partial^2 u}{\partial x^2} = v''(x)\,w(y), \qquad\qquad \frac{\partial^2 u}{\partial y^2} = v(x)\,w''(y),$$

and so

$$\Delta u = \frac{\partial^2 u}{\partial x^2} + \frac{\partial^2 u}{\partial y^2} = v''(x)\,w(y) + v(x)\,w''(y) = 0.$$

We then *separate the variables* by placing all the terms involving x on one side of the equation and all the terms involving y on the other; this is accomplished by dividing by $v(x)\,w(y)$ and then writing the resulting equation in the separated form

$$\frac{v''(x)}{v(x)} = -\,\frac{w''(y)}{w(y)} = \lambda. \tag{4.90}$$

As we argued in (4.65), the only way a function of x alone can be equal to a function of y alone is if both functions are equal to a common *separation constant* λ. Thus, the factors $v(x)$ and $w(y)$ must satisfy the elementary ordinary differential equations

$$v'' - \lambda v = 0, \qquad\qquad w'' + \lambda w = 0.$$

As before, the solution formulas depend on the sign of the separation constant λ. We list the resulting collection of separable harmonic functions in the following table:

Separable Solutions to Laplace's Equation

λ	$v(x)$	$w(y)$	$u(x,y) = v(x)\,w(y)$
$\lambda = -\omega^2 < 0$	$\cos\omega x,\ \sin\omega x$	$e^{-\omega y},\ e^{\omega y},$	$e^{\omega y}\cos\omega x,\quad e^{\omega y}\sin\omega x,$ $e^{-\omega y}\cos\omega x,\ e^{-\omega y}\sin\omega x$
$\lambda = 0$	$1,\ x$	$1,\ y$	$1,\ x,\ y,\ xy$
$\lambda = \omega^2 > 0$	$e^{-\omega x},\ e^{\omega x}$	$\cos\omega y,\ \sin\omega y$	$e^{\omega x}\cos\omega y,\quad e^{\omega x}\sin\omega y,$ $e^{-\omega x}\cos\omega y,\ e^{-\omega x}\sin\omega y$

Since Laplace's equation is a homogeneous linear system, any linear combination of solutions is also a solution. So, we can build more general solutions as finite linear combinations, or, provided we pay proper attention to convergence issues, infinite series in the separable solutions. Our goal is to solve boundary value problems, and so we must ensure that the resulting combination satisfies the boundary conditions. But this is not such an easy task, unless the underlying domain has a rather special geometry.

In fact, the only bounded domains on which we can explicitly solve boundary value problems using the preceding separable solutions are rectangles. So, we will concentrate on boundary value problems for Laplace's equation

$$\Delta u = 0 \qquad \text{on a rectangle} \qquad R = \{\,0 < x < a,\quad 0 < y < b\,\}. \tag{4.91}$$

To make progress, we will allow nonzero boundary values on only one of the four sides of the rectangle. To illustrate, we will focus on the following Dirichlet boundary conditions:

$$u(x,0) = f(x), \qquad u(x,b) = 0, \qquad u(0,y) = 0, \qquad u(a,y) = 0. \tag{4.92}$$

Once we know how to solve this type of problem, we can employ linear superposition to solve the general Dirichlet boundary value problem on a rectangle; see Exercise 4.3.12 for details. Other boundary conditions can be treated in a similar fashion — with the proviso that the condition on each side of the rectangle is either entirely Dirichlet or entirely Neumann or, more generally, entirely Robin with constant transfer coefficient.

To solve the boundary value problem (4.91–92), the first step is to narrow down the separable solutions to only those that respect the three homogeneous boundary conditions. The separable function $u(x,y) = v(x)\,w(y)$ will vanish on the top, right, and left sides of the rectangle, provided

$$v(0) = v(a) = 0 \qquad \text{and} \qquad w(b) = 0.$$

Referring to the preceding table, the first condition $v(0) = 0$ requires

$$v(x) = \begin{cases} \sin \omega x, & \lambda = -\omega^2 < 0, \\ x, & \lambda = 0, \\ \sinh \omega x, & \lambda = \omega^2 > 0, \end{cases}$$

where $\sinh z = \frac{1}{2}(e^z - e^{-z})$ is the usual hyperbolic sine function. However, the second and third cases cannot satisfy the second boundary condition $v(a) = 0$, and so we discard them. The first case leads to the condition

$$v(a) = \sin \omega a = 0, \qquad \text{and hence} \qquad \omega a = \pi,\ 2\pi,\ 3\pi,\ \ldots .$$

The corresponding separation constants and solutions (up to constant multiple) are

$$\lambda_n = -\omega^2 = -\frac{n^2\pi^2}{a^2}, \qquad v_n(x) = \sin \frac{n\pi x}{a}, \qquad n = 1, 2, 3, \ldots . \tag{4.93}$$

Note: So far, we have merely recomputed the known eigenvalues and eigenfunctions of the familiar boundary value problem $v'' - \lambda v = 0$, $v(0) = v(a) = 0$.

Next, since $\lambda = -\omega^2 < 0$, we have $w(y) = c_1 e^{\omega y} + c_2 e^{-\omega y}$ for constants c_1, c_2. The third boundary condition $w(b) = 0$ then requires that, up to constant multiple,

$$w_n(y) = \sinh \omega\,(b - y) = \sinh \frac{n\pi\,(b - y)}{a}. \tag{4.94}$$

We conclude that the harmonic functions

$$u_n(x, y) = \sin \frac{n\pi x}{a} \sinh \frac{n\pi\,(b - y)}{a}, \qquad n = 1, 2, 3, \ldots, \tag{4.95}$$

provide a complete list of separable solutions that satisfy the three homogeneous boundary conditions. It remains to analyze the inhomogeneous boundary condition along the bottom edge of the rectangle. To this end, let us try a linear superposition of the relevant separable solutions in the form of an infinite series

$$u(x, y) = \sum_{n=1}^{\infty} c_n u_n(x, y) = \sum_{n=1}^{\infty} c_n \sin \frac{n\pi x}{a} \sinh \frac{n\pi\,(b - y)}{a},$$

whose coefficients $c_1, c_2, \ldots$ are to be prescribed by the remaining boundary condition. At the bottom edge, $y = 0$, we find

$$u(x, 0) = \sum_{n=1}^{\infty} c_n \sinh \frac{n\pi b}{a} \sin \frac{n\pi x}{a} = f(x), \qquad 0 \le x \le a, \tag{4.96}$$

which takes the form of a Fourier sine series for the function $f(x)$. Let

$$b_n = \frac{2}{a} \int_0^a f(x) \sin \frac{n\pi x}{a}\, dx \tag{4.97}$$

be its Fourier sine coefficients, whence $c_n = b_n / \sinh(n\pi b/a)$. We thus anticipate that the solution to the boundary value problem can be expressed as the infinite series

$$u(x, y) = \sum_{n=1}^{\infty} \frac{b_n \sin \dfrac{n\pi x}{a} \sinh \dfrac{n\pi\,(b - y)}{a}}{\sinh \dfrac{n\pi b}{a}}. \tag{4.98}$$

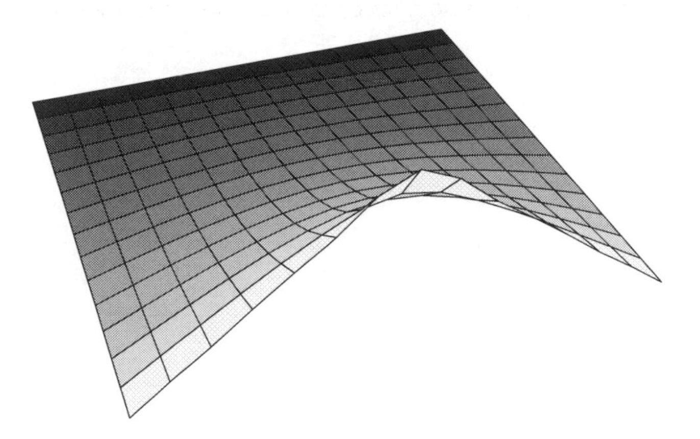

Figure 4.9. Square membrane on a wire.

Does this series actually converge to the solution to the boundary value problem? Fourier analysis says that, under very mild conditions on the boundary function $f(x)$, the answer is yes. Suppose that its Fourier coefficients are uniformly bounded,

$$|\, b_n \,| \leq M \qquad \text{for all} \qquad n \geq 1, \tag{4.99}$$

which, according to (4.27), is true whenever $f(x)$ is piecewise continuous or, more generally, integrable: $\displaystyle\int_0^a |\, f(x) \,|\, dx < \infty$. In this case, as you are asked to prove in Exercise 4.3.20, the coefficients of the Fourier sine series (4.98) go to zero exponentially fast:

$$\frac{b_n \sinh \dfrac{n\pi(b-y)}{a}}{\sinh \dfrac{n\pi b}{a}} \longrightarrow 0 \qquad \text{as} \quad n \longrightarrow \infty \qquad \text{for all} \quad 0 < y \leq b, \tag{4.100}$$

and so, at each point inside the rectangle, the series can be well approximated by partial summation. Theorem 3.31 tells us that, for each $0 < y \leq b$, the solution $u(x,y)$ is an infinitely differentiable function of x. Moreover, by term-wise differentiation of the series with respect to y and use of Proposition 3.28, we also establish that the solution is infinitely differentiable with respect to y; see Exercise 4.3.21. (In fact, as we shall see, solutions to the Laplace equation are *always analytic* functions inside their domain of definition — even when their boundary values are rather rough.) Since the individual terms all satisfy the Laplace equation, we conclude that the series (4.98) is indeed a classical solution to the boundary value problem.

Example 4.4. A membrane is stretched over a wire in the shape of a unit square with one side bent in half, as graphed in Figure 4.9. The precise boundary conditions are

$$u(x,y) = \begin{cases} x, & 0 \leq x \leq \frac{1}{2}, \quad y = 0, \\ 1 - x, & \frac{1}{2} \leq x \leq 1, \quad y = 0, \\ 0, & 0 \leq x \leq 1, \quad y = 1, \\ 0, & x = 0, \quad 0 \leq y \leq 1, \\ 0, & x = 1, \quad 0 \leq y \leq 1. \end{cases}$$

The Fourier sine series of the inhomogeneous boundary function is readily computed:

$$f(x) = \begin{cases} x, & 0 \le x \le \frac{1}{2}, \\ 1 - x, & \frac{1}{2} \le x \le 1, \end{cases}$$
$$= \frac{4}{\pi^2}\left(\sin \pi x - \frac{\sin 3\pi x}{9} + \frac{\sin 5\pi x}{25} - \cdots \right) = \frac{4}{\pi^2} \sum_{j=0}^{\infty} (-1)^j \frac{\sin(2j+1)\pi x}{(2j+1)^2}.$$

Specializing (4.98) to $a = b = 1$, we conclude that the solution to the boundary value problem can be expressed as a Fourier series

$$u(x,y) = \frac{4}{\pi^2} \sum_{j=0}^{\infty} (-1)^j \frac{\sin(2j+1)\pi x \, \sinh(2j+1)\pi(1-y)}{(2j+1)^2 \sinh(2j+1)\pi}.$$

In Figure 4.9 we plot the sum of the first 10 terms in the series. This gives a reasonably good approximation to the actual solution, except when we are very close to the raised corner of the boundary wire — which is the point of maximal displacement of the membrane.

Exercises

4.3.10. Solve the following boundary value problems for Laplace's equation on the square $\Omega = \{ 0 \le x \le \pi, \quad 0 \le y \le \pi \}$.
(a) $u(x,0) = \sin^3 x, \quad u(x,\pi) = 0, \quad u(0,y) = 0, \quad u(\pi,y) = 0.$
(b) $u(x,0) = 0, \quad u(x,\pi) = 0, \quad u(0,y) = \sin y, \quad u(\pi,y) = 0.$
(c) $u(x,0) = 0, \quad u(x,\pi) = 1, \quad u(0,y) = 0, \quad u(\pi,y) = 0.$
(d) $u(x,0) = 0, \quad u(x,\pi) = 0, \quad u(0,y) = 0, \quad u(\pi,y) = y(\pi - y).$

♢ 4.3.11. (a) Explain how to use linear superposition to solve the boundary value problem
$$\Delta u = 0, \qquad u(x,0) = f(x), \qquad u(x,b) = g(x), \qquad u(0,y) = h(y), \qquad u(a,y) = k(y),$$
on the rectangle $R = \{ 0 < x < a, \ 0 < y < b \}$, by splitting it into four separate boundary value problems for which each of the solutions vanishes on three sides of the rectangle.
(b) Write down a series formula for the resulting solution.

4.3.12. Solve the following Dirichlet problems for Laplace's equation on the unit square $S = \{ 0 < x, y < 1 \}$. *Hint*: Use superposition as in Exercise 4.3.11.
(a) $u(x,0) = \sin \pi x, \quad u(x,1) = 0, \quad u(0,y) = \sin \pi y, \quad u(1,y) = 0;$
(b) $u(x,0) = 1, \quad u(x,1) = 0, \quad u(0,y) = 1, \quad u(1,y) = 0;$
(c) $u(x,0) = 1, \quad u(x,1) = 1, \quad u(0,y) = 0, \quad u(1,y) = 0;$
(d) $u(x,0) = x, \quad u(x,1) = 1 - x, \quad u(0,y) = y, \quad u(1,y) = 1 - y.$

4.3.13. Solve the following mixed boundary value problems for Laplace's equation $\Delta u = 0$ on the square $S = \{ 0 < x, y < \pi \}$.
(a) $u(x,0) = \sin \frac{1}{2} x, \quad u_y(x,\pi) = 0, \quad u(0,y) = 0, \quad u_x(\pi,y) = 0;$
(b) $u(x,0) = \sin \frac{1}{2} x, \quad u_y(x,\pi) = 0, \quad u_x(0,y) = 0, \quad u_x(\pi,y) = 0;$
(c) $u(x,0) = x, \quad u(x,\pi) = 0, \quad u_x(0,y) = 0, \quad u_x(\pi,y) = 0;$
(d) $u(x,0) = x, \quad u(x,\pi) = 0, \quad u(0,y) = 0, \quad u_x(\pi,y) = 0.$

4.3.14. Find the solution to the boundary value problem
$$\Delta u = 0, \qquad \begin{matrix} u_y(x,0) = u_y(x,2) = 0, & 0 < x < 1, \\ u(0,y) = 2\cos \pi y - 1, \quad u(1,y) = 0, & 0 < y < 2. \end{matrix}$$

4.3.15. Find the solution to the boundary value problem

$$\Delta u = 0, \qquad \begin{array}{c} u(x,0) = 2\cos 7\pi x - 4, \quad u(x,1) = 5\cos 3\pi x, \\ u_x(0,y) = u_x(1,y) = 0, \end{array} \qquad 0 < x, y < 1.$$

4.3.16. Let $u(x,y)$ be the solution to the boundary value problem

$$\Delta u = 0,\ u(x,-1) = f(x),\ u(x,1) = 0,\ u(-1,y) = 0,\ u(1,y) = 0,\ -1 < x < 1,\ -1 < y < 1.$$

(a) *True or false*: If $f(-x) = -f(x)$ is odd, then $u(0,y) = 0$ for all $-1 \le y \le 1$.
(b) *True or false*: If $f(0) = 0$, then $u(0,y) = 0$ for all $-1 \le y \le 1$.
(c) Under what conditions on $f(x)$ is $u(x,0) = 0$ for all $-1 \le x \le 1$?

4.3.17. Use separation of variables to solve the following boundary value problem:

$$u_{xx} + 2u_y + u_{yy} = 0, \quad u(x,0) = 0, \quad u(x,1) = f(x), \quad u(0,y) = 0, \quad u(1,y) = 0.$$

4.3.18. Use separation of variables to solve the Helmholtz boundary value problem $\Delta u = u$, $u(x,0) = 0$, $u(x,1) = f(x)$, $u(0,y) = 0$, $u(1,y) = 0$, on the unit square $0 < x, y < 1$.

♢ 4.3.19. Provide the details for the derivation of (4.94).

♢ 4.3.20. Justify the statement that if $|\,b_n\,| \le M$ are uniformly bounded, then the coefficients given in (4.100) go to zero exponentially fast as $n \to \infty$ for any $0 < y \le b$.

♢ 4.3.21. Let $u(x,y)$ denote the solution to the boundary value problem (4.91–92). (a) Write down the Fourier sine series for $\partial u/\partial y$. (b) Prove that $\partial u/\partial y$ is an infinitely differentiable function of x. (c) Justify the same result for the functions $\partial^k u/\partial y^k$ for each $k \ge 0$. *Hint*: Don't forget that $u(x,y)$ solves the Laplace equation.

Polar Coordinates

The method of separation of variables can be successfully exploited in certain other very special geometries. One particularly important case is a circular disk. To be specific, let us take the disk to have radius 1 and be centered at the origin. Consider the Dirichlet boundary value problem

$$\Delta u = 0, \qquad x^2 + y^2 < 1, \qquad \text{and} \qquad u = h, \qquad x^2 + y^2 = 1, \tag{4.101}$$

so that the function $u(x,y)$ satisfies the Laplace equation on the unit disk and the specified Dirichlet boundary conditions on the unit circle. For example, $u(x,y)$ might represent the displacement of a circular drum that is attached to a wire of height

$$h(x,y) = h(\cos\theta, \sin\theta) \equiv h(\theta), \qquad -\pi < \theta \le \pi, \tag{4.102}$$

at each point $(x,y) = (\cos\theta, \sin\theta)$ on its edge.

The rectangular separable solutions are not particularly helpful in this situation, and so we look for solutions that are better adapted to a circular geometry. This inspires us to adopt *polar coordinates*

$$x = r\cos\theta, \qquad y = r\sin\theta, \qquad \text{or} \qquad r = \sqrt{x^2 + y^2}\,, \qquad \theta = \tan^{-1}\frac{y}{x}\,, \tag{4.103}$$

and write the solution $u(r,\theta)$ as a function thereof.

Warning: We will often retain the same symbol, e.g., u, when rewriting a function in a different coordinate system. This is the convention of tensor analysis, physics, and differential geometry, [**3**], that treats the function (scalar field) as an intrinsic object, which is concretely realized through its formula in any chosen coordinate system. For instance, if $u(x,y) = x^2 + 2y$ in rectangular coordinates, then its expression in polar coordinates is $u(r,\theta) = (r\cos\theta)^2 + 2r\sin\theta$, *not* $r^2 + 2\theta$. This convention avoids the inconvenience of having to devise new symbols when changing coordinates.

We need to relate derivatives with respect to x and y to those with respect to r and θ. Performing a standard multivariate chain rule computation based on (4.103), we obtain

$$\begin{aligned} \frac{\partial}{\partial r} &= \cos\theta\,\frac{\partial}{\partial x} + \sin\theta\,\frac{\partial}{\partial y}, \\ \frac{\partial}{\partial \theta} &= -r\sin\theta\,\frac{\partial}{\partial x} + r\cos\theta\,\frac{\partial}{\partial y}, \end{aligned} \qquad \text{so} \qquad \begin{aligned} \frac{\partial}{\partial x} &= \cos\theta\,\frac{\partial}{\partial r} - \frac{\sin\theta}{r}\,\frac{\partial}{\partial \theta}, \\ \frac{\partial}{\partial y} &= \sin\theta\,\frac{\partial}{\partial r} + \frac{\cos\theta}{r}\,\frac{\partial}{\partial \theta}. \end{aligned} \tag{4.104}$$

Applying the squares of the latter differential operators to $u(r,\theta)$, we find, after a calculation in which many of the terms cancel, the *polar coordinate form of the Laplace equation*:

$$\Delta u = \frac{\partial^2 u}{\partial x^2} + \frac{\partial^2 u}{\partial y^2} = \frac{\partial^2 u}{\partial r^2} + \frac{1}{r}\frac{\partial u}{\partial r} + \frac{1}{r^2}\frac{\partial^2 u}{\partial \theta^2} = 0. \tag{4.105}$$

The boundary conditions are imposed on the unit circle $r = 1$, and so, by (4.102), take the form

$$u(1,\theta) = h(\theta). \tag{4.106}$$

Keep in mind that, in order to be single-valued functions of x, y, the solution $u(r,\theta)$ and its boundary values $h(\theta)$ must both be 2π–periodic functions of the angular coordinate:

$$u(r,\theta + 2\pi) = u(r,\theta), \qquad h(\theta + 2\pi) = h(\theta). \tag{4.107}$$

Polar *separation of variables* is based on the ansatz

$$u(r,\theta) = v(r)\,w(\theta), \tag{4.108}$$

which assumes that the solution is a product of functions of the individual variables. Substituting (4.108) into the polar form (4.105) of Laplace's equation yields

$$v''(r)\,w(\theta) + \frac{1}{r}\,v'(r)\,w(\theta) + \frac{1}{r^2}\,v(r)\,w''(\theta) = 0.$$

We now separate variables by moving all the terms involving r onto one side of the equation and all the terms involving θ onto the other. This is accomplished by first multiplying the equation by $r^2/\big(v(r)\,w(\theta)\big)$ and then moving the final term to the right-hand side:

$$\frac{r^2\,v''(r) + r\,v'(r)}{v(r)} = -\,\frac{w''(\theta)}{w(\theta)} = \lambda.$$

As in the rectangular case, a function of r can equal a function of θ if and only if both are equal to a common separation constant, which we call λ. The partial differential equation thus splits into a pair of ordinary differential equations

$$r^2\,v'' + r\,v' - \lambda\,v = 0, \qquad w'' + \lambda\,w = 0, \tag{4.109}$$

that will prescribe the separable solution (4.108). Observe that both have the form of an eigenfunction equation in which the separation constant λ plays the role of the eigenvalue. We are, as always, interested only in nonzero solutions.

We have already solved the eigenvalue problem for $w(\theta)$. According to (4.107), $w(\theta + 2\pi) = w(\theta)$ must be a 2π–periodic function. Therefore, by our earlier discussion, this periodic boundary value problem has the nonzero eigenfunctions

$$1, \qquad \sin n\theta, \qquad \cos n\theta, \qquad n = 1, 2, \ldots\,, \tag{4.110}$$

corresponding to the eigenvalues (separation constants)

$$\lambda = n^2, \qquad n = 0, 1, 2, \ldots .$$

With the value of λ fixed, the linear ordinary differential equation for the radial component,

$$r^2 v'' + r v' - n^2 v = 0, \tag{4.111}$$

does not have constant coefficients. But, fortunately, it has the form of a second-order *Euler ordinary differential equation*, [**23**, **89**], and hence can be readily solved by substituting the power ansatz $v(r) = r^k$. (See also Exercise 4.3.23.) Note that

$$v'(r) = k r^{k-1}, \qquad v''(r) = k(k-1)\, r^{k-2},$$

and hence, by substituting into the differential equation,

$$r^2 v'' + r v' - n^2 v = \left[\, k(k-1) + k - n^2 \,\right] r^k = (k^2 - n^2) r^k.$$

Thus, r^k is a solution if and only if

$$k^2 - n^2 = 0, \qquad \text{and hence} \qquad k = \pm\, n.$$

For $n \neq 0$, we have found the two linearly independent solutions:

$$v_1(r) = r^n, \qquad v_2(r) = r^{-n}, \qquad n = 1, 2, \ldots . \tag{4.112}$$

When $n = 0$, the power ansatz yields only the constant solution. But in this case, the equation $r^2 v'' + r v' = 0$ is effectively of first order and linear in v', and hence readily integrated. This provides the two independent solutions

$$v_1(r) = 1, \qquad v_2(r) = \log r, \qquad n = 0. \tag{4.113}$$

Combining (4.110) and (4.112–113), we produce the complete list of separable polar coordinate solutions to the Laplace equation:

$$\begin{matrix} 1, & r^n \cos n\theta, & r^n \sin n\theta, \\ \log r, & r^{-n} \cos n\theta, & r^{-n} \sin n\theta, \end{matrix} \qquad n = 1, 2, 3, \ldots . \tag{4.114}$$

Now, the solutions in the top row of (4.114) are continuous (in fact analytic) at the origin, where $r = 0$, whereas the solutions in the bottom row have singularities as $r \to 0$. The latter are not of use in the present situation, since we require that the solution remain bounded and smooth — even at the center of the disk. Thus, we should use only the nonsingular solutions to concoct a candidate series solution

$$u(r, \theta) = \frac{a_0}{2} + \sum_{n=1}^{\infty} \left(a_n r^n \cos n\theta + b_n r^n \sin n\theta \right). \tag{4.115}$$

The coefficients a_n, b_n will be prescribed by the boundary conditions (4.106). Substituting $r = 1$, we obtain

$$u(1,\theta) = \frac{a_0}{2} + \sum_{n=1}^{\infty} \bigl(a_n \cos n\theta + b_n \sin n\theta \bigr) = h(\theta).$$

We recognize this as a standard Fourier series (3.29) (with θ replacing x) for the 2π periodic function $h(\theta)$. Therefore,

$$a_n = \frac{1}{\pi}\int_{-\pi}^{\pi} h(\theta)\cos n\theta \, d\theta, \qquad b_n = \frac{1}{\pi}\int_{-\pi}^{\pi} h(\theta)\sin n\theta \, d\theta, \tag{4.116}$$

are precisely its Fourier coefficients, cf. (3.35). In this manner, we have produced a series solution (4.115) to the boundary value problem (4.105–106).

Remark: Introducing the complex variable

$$z = x + \mathrm{i}\,y = r\,e^{\mathrm{i}\,\theta} = r\cos\theta + \mathrm{i}\,r\sin\theta \tag{4.117}$$

allows us to write

$$z^n = r^n\,e^{\mathrm{i}\,n\theta} = r^n\cos n\theta + \mathrm{i}\,r^n\sin n\theta. \tag{4.118}$$

Therefore, the nonsingular separable solutions are the *harmonic polynomials*

$$r^n\cos n\theta = \operatorname{Re} z^n, \qquad r^n \sin n\theta = \operatorname{Im} z^n. \tag{4.119}$$

The first few are listed in the following table:

n	$\operatorname{Re} z^n$	$\operatorname{Im} z^n$
0	1	0
1	x	y
2	$x^2 - y^2$	$2xy$
3	$x^3 - 3xy^2$	$3x^2y - y^3$
4	$x^4 - 4x^2y^2 + y^4$	$4x^3y - 4xy^3$

Their general expression is obtained using the Binomial Formula:

$$\begin{aligned} z^n &= (x + \mathrm{i}\,y)^n \\ &= x^n + n x^{n-1}(\mathrm{i}\,y) + \binom{n}{2} x^{n-2}(\mathrm{i}\,y)^2 + \binom{n}{3} x^{n-3}(\mathrm{i}\,y)^3 + \cdots + (\mathrm{i}\,y)^n \\ &= x^n + \mathrm{i}\,n x^{n-1} y - \binom{n}{2} x^{n-2}y^2 - \mathrm{i}\binom{n}{3} x^{n-3}y^3 + \cdots, \end{aligned}$$

where

$$\binom{n}{k} = \frac{n!}{k!\,(n-k)!} \tag{4.120}$$

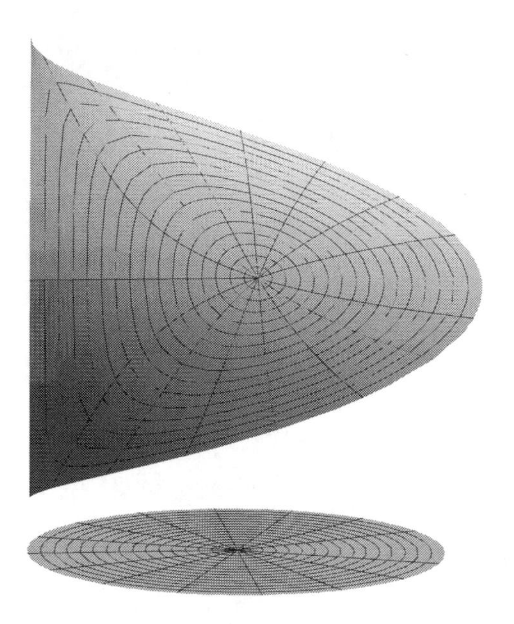

Figure 4.10. Membrane attached to a helical wire.

are the usual *binomial coefficients*. Separating the real and imaginary terms, we produce the explicit formulae

$$\begin{aligned} r^n \cos n\theta &= \operatorname{Re} z^n = x^n - \binom{n}{2} x^{n-2} y^2 + \binom{n}{4} x^{n-4} y^4 + \cdots, \\ r^n \sin n\theta &= \operatorname{Im} z^n = n x^{n-1} y - \binom{n}{3} x^{n-3} y^3 + \binom{n}{5} x^{n-5} y^5 + \cdots, \end{aligned} \tag{4.121}$$

for the two independent harmonic polynomials of degree n.

Example 4.5. Consider the Dirichlet boundary value problem on the unit disk with

$$u(1,\theta) = \theta \qquad \text{for} \qquad -\pi < \theta < \pi. \tag{4.122}$$

The boundary data can be interpreted as a wire in the shape of a single turn of a spiral helix sitting over the unit circle. The wire has a single jump discontinuity, of magnitude 2π, at the boundary point $(-1,0)$. The required Fourier series

$$h(\theta) = \theta \;\sim\; 2\left(\sin\theta - \frac{\sin 2\theta}{2} + \frac{\sin 3\theta}{3} - \frac{\sin 4\theta}{4} + \cdots \right)$$

was already computed in Example 3.3. Therefore, invoking our solution formula (4.115–116), we have

$$u(r,\theta) = 2\left(r\sin\theta - \frac{r^2\sin 2\theta}{2} + \frac{r^3\sin 3\theta}{3} - \frac{r^4\sin 4\theta}{4} + \cdots \right) \tag{4.123}$$

is the desired solution, which is plotted in Figure 4.10. In fact, this series can be explicitly summed. In view of (4.119) and the usual formula (A.13) for the complex logarithm, we have

$$u = 2\,\operatorname{Im}\left(z - \frac{z^2}{2} + \frac{z^3}{3} - \frac{z^4}{4} + \cdots \right) = 2\,\operatorname{Im}\,\log(1+z) = 2\psi, \tag{4.124}$$

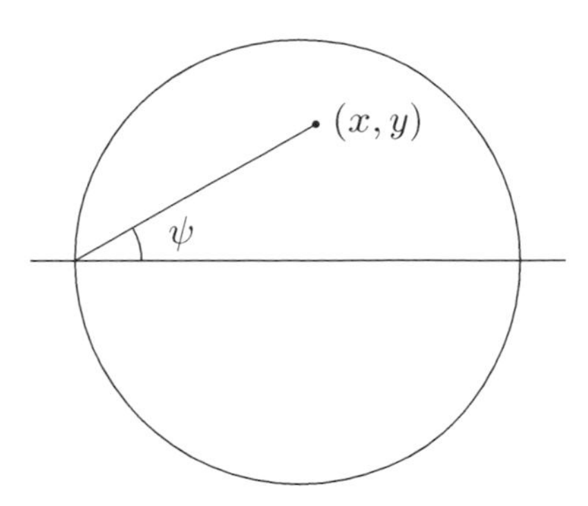

Figure 4.11. Geometric construction of the solution.

where

$$\psi = \tan^{-1} \frac{y}{1+x}$$

is the angle that the line passing through the two points (x, y) and $(-1, 0)$ makes with the x-axis, as sketched in Figure 4.11. You should try to convince yourself that, on the unit circle, $2\psi = \theta$ has the correct boundary values. Observe that, even though the boundary values are discontinuous, the solution is an analytic function inside the disk.

In fact, unlike the rectangular series (4.98), the general polar series solution formula (4.115) can, in fact, be summed in closed form! If we substitute the explicit Fourier formulae (4.116) into (4.115) — remembering to change the integration variable to, say, ϕ to avoid a notational conflict — we obtain

$$\begin{aligned} u(r,\theta) &= \frac{a_0}{2} + \sum_{n=1}^{\infty} \left(a_n r^n \cos n\theta + b_n r^n \sin n\theta \right) \\ &= \frac{1}{2\pi} \int_{-\pi}^{\pi} h(\phi)\, d\phi + \sum_{n=1}^{\infty} \left[\frac{r^n \cos n\theta}{\pi} \int_{-\pi}^{\pi} h(\phi) \cos n\phi\, d\phi + \frac{r^n \sin n\theta}{\pi} \int_{-\pi}^{\pi} h(\phi) \sin n\phi\, d\phi \right] \\ &= \frac{1}{\pi} \int_{-\pi}^{\pi} h(\phi) \left[\frac{1}{2} + \sum_{n=1}^{\infty} r^n \left(\cos n\theta \cos n\phi + \sin n\theta \sin n\phi \right) \right] d\phi \\ &= \frac{1}{\pi} \int_{-\pi}^{\pi} h(\phi) \left[\frac{1}{2} + \sum_{n=1}^{\infty} r^n \cos n(\theta - \phi) \right] d\phi. \end{aligned} \tag{4.125}$$

We next show how to sum the final series. Using (4.118), we can write it as the real part of a geometric series:

$$\begin{aligned} \frac{1}{2} + \sum_{n=1}^{\infty} r^n \cos n\theta &= \operatorname{Re}\left(\frac{1}{2} + \sum_{n=1}^{\infty} z^n \right) = \operatorname{Re}\left(\frac{1}{2} + \frac{z}{1-z} \right) = \operatorname{Re}\left(\frac{1+z}{2(1-z)} \right) \\ &= \operatorname{Re}\left(\frac{(1+z)(1-\overline{z})}{2\,|\,1-z\,|^2} \right) = \frac{\operatorname{Re}\,(1+z-\overline{z}-|\,z\,|^2)}{2\,|\,1-z\,|^2} = \frac{1-|\,z\,|^2}{2\,|\,1-z\,|^2} = \frac{1-r^2}{2(1+r^2-2r\cos\theta)}, \end{aligned}$$

which is known as the *Poisson kernel*. Substituting back into (4.125) establishes the important *Poisson Integral Formula* for the solution to the boundary value problem.

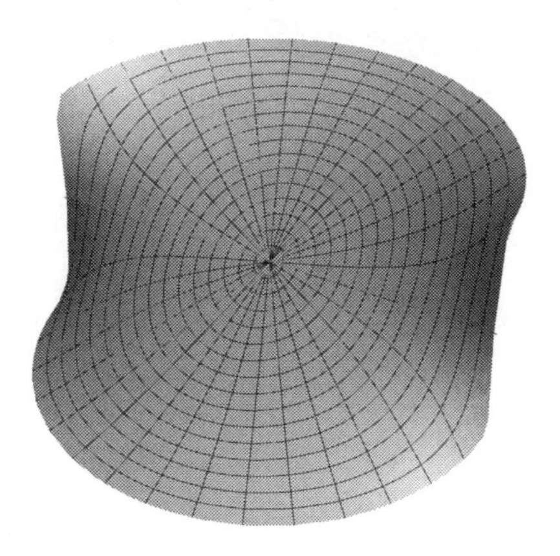

Figure 4.12. Equilibrium temperature of a disk.

Theorem 4.6. *The solution to the Laplace equation in the unit disk subject to Dirichlet boundary conditions $u(1,\theta) = h(\theta)$ is*

$$u(r,\theta) = \frac{1}{2\pi}\int_{-\pi}^{\pi} h(\phi)\,\frac{1-r^2}{1+r^2-2\,r\cos(\theta-\phi)}\,d\phi. \tag{4.126}$$

Example 4.7. A uniform metal disk of unit radius has half of its circular boundary held at 1°, while the other half is held at 0°. Our task is to find the equilibrium temperature $u(x,y)$. In other words, we seek the solution to the Dirichlet boundary value problem

$$\Delta u = 0, \qquad x^2+y^2 < 1, \qquad u(x,y) = \begin{cases} 1, & x^2+y^2 = 1, \quad y > 0, \\ 0, & x^2+y^2 = 1, \quad y < 0. \end{cases} \tag{4.127}$$

In polar coordinates, the boundary data is a (periodic) step function

$$h(\theta) = \begin{cases} 1, & 0 < \theta < \pi, \\ 0, & -\pi < \theta < 0. \end{cases}$$

Therefore, according to the Poisson formula (4.126), the solution is given by[†]

$$u(r,\theta) = \frac{1}{2\pi}\int_{0}^{\pi} \frac{1-r^2}{1+r^2-2\,r\cos(\theta-\phi)}\,d\phi = \begin{cases} 1 - \dfrac{1}{\pi}\tan^{-1}\left(\dfrac{1-r^2}{2\,r\sin\theta}\right), & 0 < \theta < \pi, \\ \dfrac{1}{2}, & \theta = 0, \pm\pi, \\ -\dfrac{1}{\pi}\tan^{-1}\left(\dfrac{1-r^2}{2\,r\sin\theta}\right), & -\pi < \theta < 0, \end{cases} \tag{4.128}$$

† The detailed derivation of the final expressions is left to the reader as Exercise 4.3.40.

where we use the principal branch $-\frac{1}{2}\pi < \tan^{-1} t < \frac{1}{2}\pi$ of the inverse tangent. Reverting to rectangular coordinates, we find that the equilibrium temperature has the explicit formula

$$u(x,y) = \begin{cases} 1 - \dfrac{1}{\pi}\tan^{-1}\left(\dfrac{1-x^2-y^2}{2y}\right), & x^2+y^2<1, \quad y>0, \\ \dfrac{1}{2}, & x^2+y^2<1, \quad y=0, \\ -\dfrac{1}{\pi}\tan^{-1}\left(\dfrac{1-x^2-y^2}{2y}\right), & x^2+y^2<1, \quad y<0. \end{cases} \tag{4.129}$$

The result is depicted in Figure 4.12.

Averaging, the Maximum Principle, and Analyticity

Let us investigate some important consequences of the Poisson integral formula (4.126). First, setting $r = 0$ yields

$$u(0,\theta) = \frac{1}{2\pi}\int_{-\pi}^{\pi} h(\phi)\,d\phi. \tag{4.130}$$

The left-hand side is the value of u at the origin — the center of the disk — and so independent of θ; the right-hand side is the *average* of its boundary values around the unit circle. This formula is a particular instance of an important general fact.

Theorem 4.8. *Let $u(x,y)$ be harmonic inside a disk of radius a centered at a point (x_0,y_0) with piecewise continuous (or, more generally, integrable) boundary values on the circle $C = \{(x-x_0)^2+(y-y_0)^2 = a^2\}$. Then its value at the center of the disk is equal to the average of its values on the boundary circle*:

$$u(x_0,y_0) = \frac{1}{2\pi a}\oint_C u\,ds = \frac{1}{2\pi}\int_{-\pi}^{\pi} u(x_0 + a\cos\theta, y_0 + a\sin\theta)\,d\theta. \tag{4.131}$$

Proof: One approach is to use the scaling and translation symmetries of the Laplace equation, cf. Exercises 4.3.8–9, to map the disk of radius a centered at (x_0,y_0) to the unit disk centered at the origin, and then invoke (4.130). Specifically, we set

$$U(x,y) = u(x_0 + ax, y_0 + ay). \tag{4.132}$$

An easy chain rule computation proves that $U(x,y)$ also satisfies the Laplace equation on the unit disk $x^2+y^2<1$, with boundary values

$$h(\theta) = U(\cos\theta,\sin\theta) = u(x_0 + a\cos\theta, y_0 + a\sin\theta).$$

Therefore, by (4.130),

$$U(0,0) = \frac{1}{2\pi}\int_{-\pi}^{\pi} h(\theta)\,d\theta = \frac{1}{2\pi}\int_{-\pi}^{\pi} U(\cos\theta,\sin\theta)\,d\theta.$$

Replacing U by its formula (4.132) produces the desired result for solutions defined by the Poisson integral formula. However, it is not a priori clear that all solutions, i.e., all harmonic functions, are necessarily of this form. This will follow eventually from the Uniqueness Theorem 4.10; however, its proof relies on formula (4.131), leading to a circular argument.

A better proof, which does not rely on the solution formula (4.115, 116), is the following. Given the harmonic function $u(x,y)$, consider the scalar function

$$g(a) = \frac{1}{2\pi}\int_{-\pi}^{\pi} u(x_0 + a\cos\theta, y_0 + a\sin\theta)\, d\theta,$$

which is well defined for $a > 0$ sufficiently small. Since $u \in \mathrm{C}^2$, we can calculate the derivative of g as follows:

$$\begin{aligned} g'(a) &= \frac{1}{2\pi}\int_{-\pi}^{\pi}\left[\cos\theta\,\frac{\partial u}{\partial x}(x_0{+}a\cos\theta, y_0{+}a\sin\theta) + \sin\theta\,\frac{\partial u}{\partial y}(x_0{+}a\cos\theta, y_0{+}a\sin\theta)\right] d\theta \\ &= \frac{1}{2\pi a}\oint_C \frac{\partial u}{\partial \mathbf{n}}\, ds, \end{aligned}$$

where $\mathbf{n} = (\cos\theta, \sin\theta)$ defines the unit normal to C at the point $(x_0 + a\cos\theta, y_0 + a\sin\theta)$ and $ds = a\, d\theta$ is the arc length element. Letting $D = \{\,(x - x_0)^2 + (y - y_0)^2 \le a^2\,\}$ denote the disk of radius a, so $C = \partial D$ is its boundary, the divergence identity (6.89), which is an easy consequence of Green's Theorem, implies that the latter integral equals

$$\oint_C \frac{\partial u}{\partial \mathbf{n}}\, ds = \iint_D \Delta u\, dx\, dy = 0$$

because u is harmonic. Thus, $g'(a) = 0$ for all $a > 0$ sufficiently small, which implies $g(a) = c$ is constant. But $g(a)$ represents the average of $u(x,y)$ on the circle C of radius a and hence, as $a \to 0$, the average $g(a) \to u(x_0, y_0)$. We conclude that $g(a) = u(x_0, y_0)$ for all $a > 0$ such that $u(x,y)$ is harmonic in the disk of radius a, which establishes (4.131) for all such harmonic functions. *Q.E.D.*

An important consequence of the integral formula (4.131) is the *Strong Maximum Principle* for harmonic functions.

Theorem 4.9. *Let u be a nonconstant harmonic function defined on a bounded domain Ω and continuous on $\partial\Omega$. Then u achieves its maximum and minimum values only at boundary points of the domain. In other words, if*

$$m = \min\{\, u(x,y) \mid (x,y) \in \partial\Omega\,\}, \qquad M = \max\{\, u(x,y) \mid (x,y) \in \partial\Omega\,\},$$

are, respectively, its maximum and minimum values on the boundary, then

$$m < u(x,y) < M \qquad \text{at all interior points} \qquad (x,y) \in \Omega.$$

Proof: Let $M^\star \ge M$ be the maximum value of u on all of $\overline{\Omega} = \Omega \cup \partial\Omega$, and assume $u(x_0, y_0) = M^\star$ at some interior point $(x_0, y_0) \in \Omega$. Theorem 4.8 implies that $u(x_0, y_0)$ equals its average over any circle C centered at (x_0, y_0) that bounds a closed disk contained in Ω. Since u is continuous and $\le M^\star$ on C, its average must be strictly less than $M^\star$ — except in the trivial case in which it is constant and equal to $M^\star$ on all of C. Thus, our assumption implies that $u(x,y) = M^\star = u(x_0, y_0)$ for all (x,y) belonging to any circle $C \subset \Omega$ centered at (x_0, y_0). Since Ω is connected, this allows us to conclude[†] that $u(x,y) = M^\star$ is constant throughout Ω, in contradiction to our original assumption.

A similar argument works for the minimum; alternatively, one can interchange maximum and minimum by replacing u by $-u$. *Q.E.D.*

[†] You are asked to supply the details in Exercise 4.3.42.

Physically, if we interpret $u(x,y)$ as the vertical displacement of a membrane stretched over a wire, then Theorem 4.9 says that, in the absence of external forcing, the membrane cannot have any internal bumps — its highest and lowest points are necessarily on the boundary of the domain. This reconfirms our physical intuition: the restoring force exerted by the stretched membrane will serve to flatten any bump, and hence a membrane with a local maximum or minimum cannot be in equilibrium. A similar interpretation holds for heat conduction. A body in thermal equilibrium will achieve its maximum and minimum temperature only at boundary points. Indeed, thermal energy would flow away from any internal maximum, or towards any local minimum, and so if the body contained a local maximum or minimum in its interior, it could not remain in thermal equilibrium.

The Maximum Principle immediately implies the uniqueness of solutions to the Dirichlet boundary value problem for both the Laplace and Poisson equations:

Theorem 4.10. *If u and $\widetilde{u}$ both satisfy the same Poisson equation $-\Delta u = f = -\Delta \widetilde{u}$ within a bounded domain Ω, and $u = \widetilde{u}$ on $\partial\Omega$, then $u \equiv \widetilde{u}$ throughout Ω.*

Proof: By linearity, the difference $v = u - \widetilde{u}$ satisfies the homogeneous boundary value problem $\Delta v = 0$ in Ω and $v = 0$ on $\partial\Omega$. Our assumption implies that the maximum and minimum boundary values of v are both $0 = m = M$. Theorem 4.9 implies that $v(x,y) \equiv 0$ at all $(x,y) \in \Omega$, and hence $u \equiv \widetilde{u}$ everywhere in Ω. *Q.E.D.*

Finally, let us discuss the analyticity of harmonic functions. In view of (4.119), the nth order term in the polar series solution (4.115), namely,

$$a_n r^n \cos n\theta + b_n r^n \sin n\theta = a_n \operatorname{Re} z^n + b_n \operatorname{Im} z^n = \operatorname{Re}\left[\,(a_n - \mathrm{i}\, b_n)\, z^n\,\right],$$

is, in fact, a homogeneous polynomial in (x,y) of degree n. This means that, when written in rectangular coordinates x and y, (4.115) is, in fact, a *power series* for the harmonic function $u(x,y)$. It is well known, [**8**, **23**, **97**], that any convergent power series converges to an analytic function — in this case $u(x,y)$. Moreover, the power series must, in fact, be the *Taylor series* for $u(x,y)$ based at the origin, and so its coefficients are multiples of the derivatives of u at $x = y = 0$. Details are worked out in Exercise 4.3.49.

We can adapt this argument to prove analyticity of *all* solutions to the Laplace equation. Note especially the contrast with the wave equation, which has many non-analytic solutions.

Theorem 4.11. *A harmonic function is analytic at every point in the interior of its domain of definition.*

Proof: Let $u(x,y)$ be a solution to the Laplace equation on the open domain $\Omega \subset \mathbb{R}^2$. Let $\mathbf{x}_0 = (x_0, y_0) \in \Omega$, and choose $a > 0$ such that the closed disk of radius a centered at $\mathbf{x}_0$ is entirely contained within Ω:

$$D_a(\mathbf{x}_0) = \{\, \| \mathbf{x} - \mathbf{x}_0 \| \le a \,\} \subset \Omega,$$

where $\|\cdot\|$ is the usual Euclidean norm. Then the function $U(x,y)$ defined by (4.132) is harmonic on the unit disk, with well-defined boundary values. Thus, by the preceding remarks, $U(x,y)$ is analytic at every point inside the unit disk, and hence so is

$$u(x,y) = U\left(\frac{x - x_0}{a}, \frac{y - y_0}{a}\right)$$

at every point (x,y) in the interior of the disk $D_a(\mathbf{x}_0)$. Since $\mathbf{x}_0 \in \Omega$ was arbitrary, this establishes the analyticity of u throughout the domain. *Q.E.D.*

This concludes our discussion of the method of separation of variables for the planar Laplace equation and some of its important consequences. The method can be used in a few other special coordinate systems. See [**78, 79**] for a complete account, including the fascinating connections with the underlying symmetry properties of the equation.

Exercises

4.3.22. Solve the following Euler differential equations by use of the power ansatz:
(a) $x^2 u'' + 5 x u' - 5 u = 0$, (b) $2x^2 u'' - x u' - 2u = 0$, (c) $x^2 u'' - u = 0$,
(d) $x^2 u'' + x u' - 3u = 0$, (e) $3x^2 u'' - 5 x u' - 3u = 0$, (f) $\dfrac{d^2 u}{dx^2} + \dfrac{2}{x}\dfrac{du}{dx} = 0$.

♢ 4.3.23. (*i*) Show that if $u(x)$ solves the *Euler differential equation*

$$a x^2 \frac{d^2 u}{dx^2} + b x \frac{du}{dx} + c u = 0, \tag{4.133}$$

then $v(y) = u(e^y)$ solves a linear constant-coefficient differential equation.
(*ii*) Use this technique to solve the Euler differential equations in Exercise 4.3.22.

4.3.24. (a) Use the method in Exercise 4.3.23 to solve an Euler equation whose characteristic equation has a double root $r_1 = r_2 = r$. (*b*) Solve the specific equations

$$(i)\ x^2 u'' - x u' + u = 0, \qquad (ii)\ \frac{d^2 u}{dx^2} + \frac{1}{x}\frac{du}{dx} = 0.$$

4.3.25. Solve the following boundary value problems:
(a) $\Delta u = 0$, $x^2 + y^2 < 1$, $u = x^3$, $x^2 + y^2 = 1$;
(b) $\Delta u = 0$, $x^2 + y^2 < 2$, $u = \log(x^2 + y^2)$, $x^2 + y^2 = 1$;
(c) $\Delta u = 0$, $x^2 + y^2 < 4$, $u = x^4$, $x^2 + y^2 = 4$;
(d) $\Delta u = 0$, $x^2 + y^2 < 1$, $\dfrac{\partial u}{\partial \mathbf{n}} = x$, $x^2 + y^2 = 1$.

4.3.26. Let $u(x, y)$ be the solution to the boundary value problem $u_{xx} + u_{yy} = 0$, $x^2 + y^2 < 1$, $u(x, y) = x^2$, $x^2 + y^2 = 1$. Find $u(0, 0)$.

♡ 4.3.27. (a) Find the equilibrium temperature on a disk of radius 1 when half the boundary is held at 1° and the other half is held at -1°. (*b*) Find the equilibrium temperature on a half-disk of radius 1 when the temperature is held to 1° on the curved edge and 0° on the straight edge. (*c*) Find the equilibrium temperature on a half disk of radius 1 when the temperature is held to 0° on the curved edge and 1° on the straight edge.

4.3.28. Find the solution to Laplace's equation $u_{xx} + u_{yy} = 0$ on the semi-disk $x^2 + y^2 < 1$, $y > 0$, that satisfies the boundary conditions $u(x, 0) = 0$ for $-1 < x < 1$ and $u(x, y) = y^3$ for $x^2 + y^2 = 1$, $y > 0$.

4.3.29. Find the equilibrium temperature on a half-disk of radius 1 when the temperature is held to 1° on the curved edge, while the straight edge is insulated.

4.3.30. Solve the Dirichlet boundary value problem for the Laplace equation on the pie wedge $W = \{0 < \theta < \frac{1}{4}\pi,\ 0 < r < 1\}$, when the nonzero boundary data $u(1, \theta) = h(\theta)$ appears only on the curved portion of its boundary.

4.3.31. Find a harmonic function $u(x, y)$ defined on the annulus $\frac{1}{2} < r < 1$ subject to the constant Dirichlet boundary conditions $u = a$ on $r = \frac{1}{2}$ and $u = b$ on $r = 1$.

4.3.32. Boiling water flows continually through a long circular metal pipe of inner radius 1 cm and outer radius 1.2 cm placed in an ice water bath. *True or false*: The temperature at the midpoint, at radius 1.1 cm, is 50°. If false, what is the temperature at this point?

4.3.33. Write out the series solution to the boundary value problem $u(1,\theta)=0,\ u(2,\theta)=h(\theta)$, for the Laplace equation on an annulus $1<r<2$. *Hint*: Use all of the separable solutions listed in (4.114).

4.3.34. Solve the following boundary value problems for the Laplace equation on the annulus $1<r<2$: (a) $u(1,\theta)=0,\ u(2,\theta)=1$, (b) $u(1,\theta)=0,\ u(2,\theta)=\cos\theta$, (c) $u(1,\theta)=\sin 2\theta,\ u(2,\theta)=\cos 2\theta$, (d) $u_r(1,\theta)=0,\ u(2,\theta)=1$, (e) $u_r(1,\theta)=0,\ u(2,\theta)=\sin 2\theta$, (f) $u_r(1,\theta)=0,\ u_r(2,\theta)=1$, (g) $u_r(1,\theta)=2,\ u_r(2,\theta)=1$.

4.3.35. Solve the following boundary value problems for the Laplace equation on the semi-annular domain $D=\{1<x^2+y^2<2,\ y>0\}$:
(a) $u(x,y)=0,\quad x^2+y^2=1,\quad u(x,y)=1,\quad x^2+y^2=2,\quad u(x,0)=0;$
(b) $u(x,y)=0,\quad x^2+y^2=1$ or $2,\quad u(x,0)=0,\quad x>0,\quad u(x,0)=1,\quad x<0.$

4.3.36. Solve the following boundary value problem:
$$(x^2+y^2)(u_{xx}+u_{yy})+2xu_x+2yu_y=0,\ x^2+y^2<1,\ u(x,y)=1+3x,\ x^2+y^2=1.$$

♢ 4.3.37. Justify the chain rule computation (4.104). Then justify formula (4.105) for the Laplacian in polar coordinates.

4.3.38. Suppose $\int_{-\pi}^{\pi}|h(\theta)|\,d\theta<\infty$. Prove that (4.115) converges uniformly to the solution to the boundary value problem (4.101) on any smaller disk $D_{r_\star}=\{r\le r_\star<1\}\subsetneq D_1$.

4.3.39. Prove directly that (4.124) satisfies the boundary conditions (4.122).

♢ 4.3.40. Justify the integration formula in (4.128).

4.3.41. Provide a complete proof that (4.129) is indeed the solution to the boundary value problem (4.127).

♢ 4.3.42. Complete the proof of Theorem 4.9 by showing that $u(x,y)=M^\star$ for all $(x,y)\in\Omega$. *Hint*: Join (x_0,y_0) to (x,y) by a curve $C\subset\Omega$ of finite length, and use the preceding part of the proof to inductively deduce the existence of a finite sequence of points $(x_i,y_i)\in C$, $i=0,\ldots,n$, with $(x_n,y_n)=(x,y)$, and such that $u(x_i,y_i)=M^\star$.

♢ 4.3.43. Derive the analogue of the Poisson integral formula for the solution to the Neumann boundary value problem $\Delta u=0,\ x^2+y^2<1,\ \partial u/\partial\mathbf{n}=h,\ x^2+y^2=1$, on the unit disk. Pay careful attention to the existence and uniqueness of solutions in your formulation.

4.3.44. Give an example of a solution to Poisson's equation on the unit disk that achieves its maximum at an interior point. Interpret your construction physically.

4.3.45. Let $p(x,y)$ be a polynomial (not necessarily harmonic). Suppose $u(x,y)$ is harmonic and equals $p(x,y)$ on the unit circle $x^2+y^2=1$. Prove that $u(x,y)$ is a harmonic polynomial.

4.3.46. Write down an integral formula for the solution to the Dirichlet boundary value problem on a disk of radius $R>0$, namely, $\Delta u=0,\ x^2+y^2<R^2,\ u=h,\ x^2+y^2=R^2$.

4.3.47. State and prove a one-dimensional version of Theorem 4.8. Does the analogue of Theorem 4.9 hold?

4.3.48. A unit area square plate has 100° temperature on its top edge and 0° on its three other edges. *True or false*: The temperature at the center equals the average edge temperature.

♢ 4.3.49. Let $u(x,y)$ be a harmonic function on the unit disk with boundary values $h(\theta)$ when $r=1$. Using the fact that (4.115) is the Taylor series for $u(x,y)$ at the origin: (a) Find integral formulas for its partial derivatives $u_x(0,0),\ u_y(0,0)$, involving the boundary values $h(\theta)$. (b) Generalize part (a) to the second-order derivatives $u_{xx}(0,0),\ u_{xy}(0,0),\ u_{yy}(0,0)$.

4.3.50. Prove that if $u(x,y)$ is a bounded harmonic function defined on all of $\mathbb{R}^2$, then u is constant. *Hint*: First generalize Exercise 4.3.49(a) to find the value of its gradient, $\nabla u(x_0,y_0)$, in terms of the values of u on a circle of radius a centered at (x_0,y_0). Then see what happens when the radius of the circle goes to ∞.

4.4 Classification of Linear Partial Differential Equations

We have, at last, been introduced to the three paradigmatic linear second-order partial differential equations for functions of two variables. The homogeneous versions are

(a) The wave equation:	$u_{tt} - c^2\, u_{xx} = 0,$	*hyperbolic,*
(b) The heat equation:	$u_t - \gamma\, u_{xx} = 0,$	*parabolic,*
(c) Laplace's equation:	$u_{xx} + u_{yy} = 0,$	*elliptic.*

The last column indicates the equation's *type*, in accordance with the standard taxonomy of partial differential equations; an explanation will appear momentarily. The wave, heat, and Laplace equations are the prototypical representatives of these three fundamental genres. Each genre has its own distinctive analytic features, physical manifestations, and even numerical solution schemes. Equations governing vibrations, such as the wave equation, are typically hyperbolic. Equations modeling diffusion, such as the heat equation, are parabolic. Hyperbolic and parabolic equations both typically represent dynamical processes, and so one of the independent variables is identified as time. On the other hand, equations modeling equilibrium phenomena, including the Laplace and Poisson equations, are usually elliptic, and involve only spatial variables. Elliptic partial differential equations are associated with boundary value problems, whereas parabolic and hyperbolic equations require initial and initial-boundary value problems.

The classification theory of real linear second-order partial differential equations for a scalar-valued function $u(t,x)$ depending on two variables[†] proceeds as follows. The most general such equation has the form

$$L[u] = A\,u_{tt} + B\,u_{tx} + C\,u_{xx} + D\,u_t + E\,u_x + F\,u = G, \tag{4.134}$$

where the coefficients A,B,C,D,E,F are all allowed to be functions of (t,x), as is the inhomogeneity or forcing function $G(t,x)$. The equation is *homogeneous* if and only if $G \equiv 0$. We assume that at least one of the leading coefficients A,B,C is not identically zero, since otherwise, the equation degenerates to a first-order equation.

The key quantity that determines the *type* of such a partial differential equation is its *discriminant*

$$\Delta = B^2 - 4\,A\,C. \tag{4.135}$$

This should (and for good reason) remind the reader of the discriminant of the quadratic equation

$$Q(x,y) = A\,x^2 + B\,x\,y + C\,y^2 + D\,x + E\,y + F = 0, \tag{4.136}$$

whose solutions trace out a plane curve — a conic section. In the nondegenerate cases, the discriminant (4.135) fixes its geometric type:

[†] For equilibrium equations, we identify t with the space variable y.

- a hyperbola when $\Delta > 0$,
- a parabola when $\Delta = 0$,
- an ellipse when $\Delta < 0$.

This motivates the choice of terminology used to classify second-order partial differential equations.

Definition 4.12. At a point (t, x), the linear second-order partial differential equation (4.134) is called

- *hyperbolic* if $\Delta(t,x) > 0$,
- *parabolic* if $\Delta(t,x) = 0$, but $A^2 + B^2 + C^2 \neq 0$,
- *elliptic* if $\Delta(t,x) < 0$,
- *singular* if $A = B = C = 0$.

In particular:

- The wave equation $u_{tt} - u_{xx} = 0$ has discriminant $\Delta = 4$, and is hyperbolic.
- The heat equation $u_{xx} - u_t = 0$ has discriminant $\Delta = 0$, and is parabolic.
- The Poisson equation $u_{tt} + u_{xx} = -f$ has discriminant $\Delta = -4$, and is elliptic.

Example 4.13. When the coefficients A, B, C vary, the type of the partial differential equation may not remain fixed over the entire domain. Equations that change type are less common, as well as being much harder to analyze and solve, both analytically and numerically. One example arising in the theory of supersonic aerodynamics, [**44**], is the *Tricomi equation*

$$x \frac{\partial^2 u}{\partial t^2} - \frac{\partial^2 u}{\partial x^2} = 0. \tag{4.137}$$

Comparing with (4.134), we find that

$$A = x, \quad B = 0, \quad C = -1, \qquad \text{while} \qquad D = E = F = G = 0.$$

The discriminant in this particular case is

$$\Delta = B^2 - 4AC = 4x,$$

and hence the equation is hyperbolic when $x > 0$, elliptic when $x < 0$, and parabolic on the transition line $x = 0$. In the physical model, the hyperbolic region corresponds to subsonic flow, while the supersonic regions are of elliptic type. The transitional parabolic boundary represents the shock line between the sub- and super-sonic regions — the familiar sonic boom as an airplane crosses the sound barrier.

While this tripartite classification into hyperbolic, parabolic, and elliptic equations initially appears in the bivariate context, the terminology, underlying properties, and associated physical models carry over to second-order partial differential equations in higher dimensions. Most of the partial differential equations arising in applications fall into one of these three categories, and it is fair to say that the field of partial differential equations splits into three distinct subfields. Or rather four subfields, the last containing all the equations, including higher-order equations, that do not fit into the preceding categorization. (One important example appears in Section 8.5.)

Remark: The classification into hyperbolic, parabolic, elliptic, and singular types carries over as stated to *quasilinear* second-order equations, whose coefficients $A, \ldots, G$ are allowed to depend on u and its first-order derivatives, u_t, u_x. Here the type of the equation can vary with both the point in the domain and the particular solution being considered. Even more generally, for a *fully nonlinear* second-order partial differential equation

$$H(t, x, u, u_t, u_x, u_{tt}, u_{tx}, u_{xx}) = 0, \tag{4.138}$$

one defines its *discriminant* to be

$$\Delta = \left(\frac{\partial H}{\partial u_{tx}}\right)^2 - 4\,\frac{\partial H}{\partial u_{tt}}\,\frac{\partial H}{\partial u_{xx}}\,. \tag{4.139}$$

Its sign determines the type of the equation as above — again depending on the point in the domain and the solution under consideration.

Exercises

4.4.1. Plot the following conic sections and classify their type:
(a) $x^2 + 3y^2 = 1$, (b) $xy + x + y = 4$, (c) $x^2 - xy + y^2 = x - 2y$,
(d) $x^2 + 2xy + y^2 + y = 1$, (e) $x^2 - 2y^2 = 6x + 8y + 1$.

4.4.2. Determine the type of the following partial differential equations:
(a) $u_{tt} + 3u_{xx} = 0$, (b) $u_{tx} + u_t + u_x = u$, (c) $u_{tt} + u_t + u_x = 0$,
(d) $u_{tt} - u_{tx} + u_{xx} = u$, (e) $u_{tt} + 4u_{tx} + 4u_{xx} = u_t$, (f) $u_{tx} + u_{xx} = 0$.

4.4.3. Consider the partial differential equation $x\,u_{tt} + (t + x)\,u_{xx} = 0$. At what points of the plane is the equation elliptic? hyperbolic? parabolic? degenerate?

4.4.4. Answer Exercise 4.4.3 for the equations
(a) $x^2\,u_{xx} + x\,u_x + u_{yy} = 0$, (b) $\partial_x(x\,u_x) = \partial_y(y\,u_y)$, (c) $u_t = \partial_x[(x + t)u_x]$,
(d) $\nabla \cdot (c(x, y)\nabla u) = u$, where $c(x, y)$ is a given function.

4.4.5. Steady flow of air past an airplane is modeled by the partial differential equation $(m^2 - 1)u_{xx} + u_{yy} = 0$, in which x is the flight direction, y the transverse direction, and $m \geq 0$ is the *Mach number* — the ratio of the airplane's speed to the speed of sound. Show that the equation is hyperbolic for subsonic flight, but elliptic for supersonic flight.

4.4.6. Show that the second-order partial differential equation

$$-\frac{\partial}{\partial x}\left(p(x, y)\,\frac{\partial u}{\partial x}\right) - \frac{\partial}{\partial y}\left(q(x, y)\,\frac{\partial u}{\partial y}\right) + r(x, y)\,u = f(x, y)$$

is elliptic if and only if $p(x, y)$ and $q(x, y)$ are nonzero and have the same sign.

◇ 4.4.7. *True or false*: The type of a linear second-order partial differential equation is not affected by a change of independent variables: $\tau = \varphi(t, x)$, $\xi = \psi(t, x)$.

4.4.8. Let $v(t, x) = a(t, x)\,u(t, x) + b(t, x)$, where a, b are fixed functions with $a \neq 0$. Suppose u is a solution to a second-order linear partial differential equation. Prove that v also solves a linear partial differential equation of the same type.

◇ 4.4.9. *True or false*: The polar coordinate form (4.105) of the Laplace equation is elliptic.

4.4.10. Rewrite the Laplace equation $u_{xx} + u_{yy} = 0$ in terms of *parabolic coordinates* ξ, η, as defined by the equations $x = \xi^2 - \eta^2$, $y = 2\xi\eta$. Is the resulting equation elliptic?

♢ 4.4.11. Prove that the complex change of variables $x = x$, $t = \mathrm{i}\, y$, maps the Laplace equation $u_{xx} + u_{yy} = 0$ to the wave equation $u_{tt} = u_{xx}$. Explain why the type of a partial differential equation is *not* necessarily preserved under a complex change of variables.

♡ 4.4.12. Suppose, against all advice, we pose the elliptic Laplace equation as an initial value problem, namely

$$u_{tt} = -\,u_{xx} \qquad \text{for} \qquad 0 < x < 1, \qquad t > 0,$$
$$u(0,x) = f(x), \qquad u_t(0,x) = 0, \qquad 0 \le x \le 1, \qquad u(t,0) = 0 = u(t,1), \qquad t \ge 0.$$

(a) Prove that for any positive integer $n > 0$, the function $u_n(t,x) = \dfrac{\sin n\pi t \ \cosh n\pi x}{n}$ satisfies the initial value problem. Determine the initial condition $u_n(0,x) = f_n(x)$.
(b) Prove that, as $n \to \infty$, the initial condition $f_n(x) \to 0$ becomes vanishingly small, whereas, at any $t > 0$, the solution value $u_n\left(t, \frac{1}{2}\right) \to \infty$.
(c) Explain why this represents an ill-posed problem.

4.4.13. The *minimal surface equation* $(1+u_x^2)u_{xx} - 2\,u_x u_y u_{xy} + (1+u_y^2)u_{yy} = 0$ is (a) hyperbolic, (b) parabolic, (c) elliptic, (d) singular, (e) of variable type depending on the point in the domain, or (f) of variable type depending on the solution and the point in the domain.

Characteristics and the Cauchy Problem

In Chapter 2, we discovered that the characteristic curves guide the behavior of solutions to first-order partial differential equations. Characteristics play a similarly fundamental role in the analysis of more general hyperbolic partial differential equations and systems. In particular, they provide a mechanism for distinguishing among the various classes of second-order partial differential equations.

As above, we will focus our attention on partial differential equations involving two independent variables. The starting point is the general initial value problem, also known as the Cauchy problem, in honor of the prolific nineteenth-century French mathematician Augustin–Louis Cauchy, justly famous for his wide-ranging contributions throughout mathematics and its applications, including the Cauchy–Schwarz inequality, many of the fundamental concepts in complex analysis, as well as the foundations of elasticity and materials science. The general *Cauchy problem* specifies appropriate initial data along a smooth curve[†] $\Gamma \subset \mathbb{R}^2$ and seeks a solution to the partial differential equation that assumes the given initial data on Γ. In all our examples, the curve in question has been a straight line, e.g., the x–axis, but one could easily envisage more general situations. If the partial differential equation has order n, then the *Cauchy data* consists of the values of the dependent variable u along with all its partial differential equations up to order $n-1$ on the curve Γ. For most curves, there is a unique solution $u(t,x)$ to the partial differential equation that achieves the specified values along Γ. More rigorously, if we are in the analytic category, meaning that the partial differential equation, the curve, and the Cauchy data are all specified by analytic functions, then the fundamental *Cauchy–Kovalevskaya Theorem* guarantees the existence of an analytic solution $u(t,x)$ to the Cauchy problem near any point on the initial curve. The statement of proof of this important theorem, due to Cauchy and, in general form, the influential nineteenth-century Russian mathematician Sofia Kovalevskaya, relies on the construction of convergent power series for the desired

[†] More generally, for partial differential equations in $m > 2$ independent variables, the curve is replaced by a hypersurface $S \subset \mathbb{R}^m$ of dimension $m-1$.

solution and would take us too far afield. We refer the interested reader to [**35, 44**]. The exceptional curves, for which the Cauchy–Kovalevskaya Existence Theorem does not apply, are called the characteristics of the underlying partial differential equations.

More prosaically, a curve Γ will be called *non-characteristic* for the given partial differential equation if one can determine the values of *all* the derivatives of u along Γ from the specified Cauchy data. Indeed, the determination of the values of the higher-order derivatives along the curve is a necessary preliminary step towards establishing the Cauchy–Kovalevskaya existence result. As we will now show, this requirement serves to distinguish the characteristic and non-characteristic curves for the examples we have already encountered, and hence to lead to their characterization in much more general contexts.

To illustrate the preceding requirement, let us begin with a first-order linear partial differential equation of the form

$$\frac{\partial u}{\partial t} + c(t,x)\,\frac{\partial u}{\partial x} = f(t,x). \tag{4.140}$$

Let $\Gamma \subset \mathbb{R}^2$ be a smooth curve parametrized[‡] by $\mathbf{x}(s) = \bigl(t(s), x(s)\bigr)^T$, where smoothness necessitates that its tangent vector not vanish: $\mathbf{x}'(s) = (dt/ds, dx/ds)^T \neq \mathbf{0}$. Since the equation is of order $n = 1$, the Cauchy data requires specifying the values of the dependent variable u only along Γ — in other words, the function

$$h(s) = u\bigl(t(s), x(s)\bigr). \tag{4.141}$$

The curve will be non-characteristic if we can then determine the values of the derivatives of u along Γ, starting with

$$\frac{\partial u}{\partial t}\bigl(t(s), x(s)\bigr), \qquad \frac{\partial u}{\partial x}\bigl(t(s), x(s)\bigr). \tag{4.142}$$

To this end, let us differentiate the Cauchy data (4.141): applying the chain rule, we obtain

$$h'(s) = \frac{d}{ds}\,u\bigl(t(s), x(s)\bigr) = \frac{\partial u}{\partial t}\bigl(t(s), x(s)\bigr)\,\frac{dt}{ds} + \frac{\partial u}{\partial x}\bigl(t(s), x(s)\bigr)\,\frac{dx}{ds}\,. \tag{4.143}$$

On the other hand, we are assuming that $u(t,x)$ solves the partial differential equation (4.140) at all points in its domain of definition. In particular, at points on the curve Γ, the partial differential equation requires

$$\frac{\partial u}{\partial t}\bigl(t(s), x(s)\bigr) + c\bigl(t(s), x(s)\bigr)\,\frac{\partial u}{\partial x}\bigl(t(s), x(s)\bigr) = f\bigl(t(s), x(s)\bigr). \tag{4.144}$$

We can regard (4.143–144) as a pair of inhomogeneous linear algebraic equations, which can be uniquely solved for the as yet unknown quantities (4.142), *unless* the determinant of their coefficient matrix vanishes:

$$\det\begin{pmatrix} 1 & c\bigl(t(s), x(s)\bigr) \\ dt/ds & dx/ds \end{pmatrix} = \frac{dx}{ds} - c\bigl(t(s), x(s)\bigr)\,\frac{dt}{ds} = 0. \tag{4.145}$$

This condition serves to define a *characteristic curve* for the first-order partial differential equation (4.140). In particular, if the curve is parametrized by $s = t$, i.e., can be identified with the graph of a function $x = g(t)$, then the characteristic condition (4.145) reduces to

$$\frac{dx}{dt} = c(t,x), \tag{4.146}$$

[‡] The parameter s could be the arc length, but this is not required. See also Exercise 4.4.20.

thus reproducing our original definition of characteristic curve, as in (2.18) and, more generally, Exercise 2.2.26. On the other hand, if the determinant (4.145) is nonzero, then one can solve (4.143–144) for the values of the first-order derivatives (4.142) along Γ. Further differentiation of these conditions proves that one can, in fact, determine the values of all the higher-order derivatives of the solution u along the curve, which is hence non-characteristic.

Next, consider a nonsingular linear second-order partial differential equation of the form (4.134). Since the equation has order $n = 2$, the Cauchy data along a curve Γ parametrized as above consists of the values of the function and its first derivatives:

$$u\big(t(s), x(s)\big), \qquad \frac{\partial u}{\partial t}\big(t(s), x(s)\big), \qquad \frac{\partial u}{\partial x}\big(t(s), x(s)\big). \tag{4.147}$$

However, the latter cannot be specified independently. Indeed, given the value of the dependent variable, $h(s) = u\big(t(s), x(s)\big)$, along Γ, its derivative

$$h'(s) = \frac{d}{ds}\, u\big(t(s), x(s)\big) = \frac{\partial u}{\partial t}\big(t(s), x(s)\big)\,\frac{dt}{ds} + \frac{\partial u}{\partial x}\big(t(s), x(s)\big)\,\frac{dx}{ds} \tag{4.148}$$

prescribes a particular combination of the two first-order derivatives. Thus, once the value of one derivative of u on Γ is known, the other is automatically fixed by the relation (4.148). For example, if $dx/ds \neq 0$, we can use (4.148) to determine $u_x\big(t(s), x(s)\big)$, knowing $u\big(t(s), x(s)\big)$ and $u_t\big(t(s), x(s)\big)$. Similarly, if we differentiate the values of the first-order derivatives with respect to the curve parameter, we can determine two combinations of second-order derivatives along the curve Γ:

$$\begin{aligned} \frac{d}{ds}\,\frac{\partial u}{\partial t}\big(t(s), x(s)\big) &= \frac{\partial^2 u}{\partial t^2}\big(t(s), x(s)\big)\,\frac{dt}{ds} + \frac{\partial^2 u}{\partial t\,\partial x}\big(t(s), x(s)\big)\,\frac{dx}{ds}\,, \\ \frac{d}{ds}\,\frac{\partial u}{\partial x}\big(t(s), x(s)\big) &= \frac{\partial^2 u}{\partial t\,\partial x}\big(t(s), x(s)\big)\,\frac{dt}{ds} + \frac{\partial^2 u}{\partial x^2}\big(t(s), x(s)\big)\,\frac{dx}{ds}\,. \end{aligned} \tag{4.149}$$

On the other hand, the partial differential equation (4.134) induces yet a third relation among the second-order partial derivatives u_{tt}, u_{tx}, u_{xx}. These three linear equations can be uniquely solved for values of these derivatives on Γ if and only if the determinant of their coefficient matrix is nonzero:

$$\det\begin{pmatrix} A(t,x) & B(t,x) & C(t,x) \\ dt/ds & dx/ds & 0 \\ 0 & dt/ds & dx/ds \end{pmatrix} = A(t,x)\left(\frac{dx}{ds}\right)^2 - B(t,x)\,\frac{dt}{ds}\,\frac{dx}{ds} + C(t,x)\left(\frac{dt}{ds}\right)^2 = 0. \tag{4.150}$$

We conclude that a smooth curve $\mathbf{x}(s) = \big(t(s), x(s)\big)^T \subset \mathbb{R}^2$ is a *characteristic curve* for the nonsingular linear second-order partial differential equation (4.134) whenever its tangent vector $\mathbf{x}'(s) = (dt/ds, dx/ds)^T \neq \mathbf{0}$ satisfies the quadratic *characteristic equation* (4.150). Conversely, if the curve is non-characteristic, meaning that its tangent does not satisfy (4.150) anywhere, then one can, with some further work, determine all the higher-order derivatives of the solution $u(t, x)$ along Γ, and then, at least in the analytic category, prove existence of a solution to the Cauchy problem, [**35**].

According to Exercise 4.4.20, the status of a curve as characteristic or not does not depend on the choice of parametrization. In particular, if the curve is given by the graph of the function $x = x(t)$, which we parametrize by $s = t$, then the characteristic equation

(4.150) takes the form of a quadratically nonlinear first-order ordinary differential equation

$$A(t,x)\left(\frac{dx}{dt}\right)^2 - B(t,x)\,\frac{dx}{dt} + C(t,x) = 0, \tag{4.151}$$

whose solutions are characteristic curves of the second-order partial differential equation.

Warning: If $A(t,x) = 0$, then the partial differential equation admits characteristic curves with vertical tangents that cannot be parametrized by $s = t$. For example, if $A(t,x) \equiv 0$, then the vertical lines e.g., $t =$ constant, $x = s$, are characteristic, satisfying (4.150), but do not appear as solutions to (4.151).

For example, consider the hyperbolic wave equation

$$u_{tt} - c^2\,u_{xx} = 0.$$

According to (4.151), any characteristic curve that is given by the graph of $x(t)$ must solve

$$\left(\frac{dx}{dt}\right)^2 - c^2 = 0, \qquad \text{which implies that} \qquad \frac{dx}{dt} = \pm\,c.$$

Thus, in accordance with our previous analysis, the characteristic curves are the straight lines of slope $\pm\,c$, and there are two characteristic curves passing through each point of the (t,x) plane. On the other hand, the elliptic Laplace equation

$$u_{tt} + u_{xx} = 0$$

has no (real) characteristic curves, since the characteristic equation (4.150) reduces to

$$\left(\frac{dx}{ds}\right)^2 + \left(\frac{dt}{ds}\right)^2 = 0,$$

and t_s and x_s are not allowed to vanish simultaneously. Finally, for the parabolic heat equation

$$u_{xx} - u_t = 0,$$

the characteristic curve equation (4.150) is simply

$$\left(\frac{dt}{ds}\right)^2 = 0$$

(since the first-derivative term plays no role), and so there is only one characteristic curve passing through each point, namely the vertical line $t =$ constant. Observe that the standard initial value problem $u(0,x) = f(x)$ for the heat equation takes place on a characteristic curve — the x–axis — but does not take the form of a Cauchy problem, which would also require specifying the first-order derivatives $u_t(0,x), u_x(0,x)$ there. And indeed, the standard initial value problem is not well-posed near the characteristic x–axis for negative $t < 0$.

In general, the number of real solutions to the nondegenerate quadratic characteristic curve equation (4.150) depends on its discriminant $\Delta = B^2 - 4AC$: In the hyperbolic case, $\Delta > 0$, and there are two real characteristic curves passing through each point; in the parabolic case, $\Delta = 0$, and there is just one real characteristic curve passing through each point; in the elliptic case, $\Delta < 0$, and there are no real characteristic curves. In this manner, elliptic, parabolic, and hyperbolic partial differential equations are distinguished by the number of (real) characteristic curves passing through a point — namely, zero, one, and two, respectively. First-order partial differential equations are also viewed as *hyperbolic*, since they always admit real characteristic curves.

With further analysis, [**35**, **70**, **122**], it can be shown that, as with the wave equation, signals and disturbances propagate along characteristic curves. Thus, hyperbolic equations share many qualitative properties with the wave equation, with signals moving in two different directions. For example, light rays move along characteristic curves, and are thereby subject to the optical phenomena of refraction and focusing. Similarly, since the characteristic curves for the parabolic heat equation are the vertical lines, this indicates that the effect of a disturbance at a point $(t, x) = (t_0, x_0)$ is simultaneously felt along the entire contemporaneous vertical line $t = t_0$. This has the implication that disturbances in the heat equation propagate at infinite speed — a counterintuitive fact that will be further expounded on in Section 8.1. Elliptic equations have no characteristics, and as a consequence, do not support propagating signals; indeed, the effect of a localized disturbance is immediately felt throughout the domain. For example, even when an external force is concentrated near a single point, it displaces the entire membrane.

Exercises

4.4.14. Find and graph the real characteristic curves for each of the partial differential equations in Exercise 4.4.2.

4.4.15. Graph the characteristic curves for the Tricomi equation (4.137) in its hyperbolic region. What happens to the characteristics as one approaches the parabolic transition boundary?

4.4.16. *True or false*: The characteristic curves of the *Helmholtz equation* $u_{xx} + u_{yy} - u = 0$ are circles.

4.4.17. (*a*) At what points of the plane is the partial differential equation $x\,u_{xx} + y\,u_{yy} = 0$ elliptic? parabolic? hyperbolic? (*b*) How many characteristics are there through the point $(1, -1)$? (*c*) Find them explicitly.

4.4.18. Consider the partial differential equation $u_{xx} + y\,u_{xy} = y^2$.
(*a*) On which regions of the (x, y)–plane is the equation elliptic? parabolic? hyperbolic?
(*b*) Find the characteristics in the hyperbolic region.
(*c*) Find the general solution in the hyperbolic region. *Hint*: Use characteristic coordinates.

4.4.19. Find a partial differential equation whose characteristic curves are:
(*a*) the lines $x - y = a$, $x + 2y = b$, where $a, b \in \mathbb{R}$ are arbitrary constants;
(*b*) the exponential curves $y = ce^x$ for $c \in \mathbb{R}$;
(*c*) the concentric circles $x^2 + y^2 = a$ for $a \geq 0$, and the rays $y = bx$.

◇ 4.4.20. Prove that any reparametrization of a characteristic curve for a given second-order linear partial differential equation is also a characteristic curve.

4.4.21. *True or false*: You can uniquely recover a second-order partial differential equation by knowing all its characteristic curves.

◇ 4.4.22. Prove that any invertible change of variables, as in Exercise 4.4.7, maps the characteristic curves of the original linear partial differential equation to the characteristic curves of the transformed equation. Thus, characteristic curves are intrinsic: they do not depend on the parametrization, nor on the coordinates used to represent the partial differential equation.

Chapter 5
Finite Differences

As one quickly learns, the differential equations that can be solved by explicit analytic formulas are few and far between. Consequently, the development of accurate numerical approximation schemes is an essential tool for extracting quantitative information as well as achieving a qualitative understanding of the possible behaviors of solutions to the vast majority of partial differential equations. (On the other hand, the successful design of numerical algorithms necessitates a fairly deep understanding of their basic analytic properties, and so exclusive reliance on numerics is not an option.) Even in cases, such as the heat and wave equations, in which explicit solution formulas (either in closed form or infinite series) exist, numerical methods can still be profitably employed. Indeed, one can accurately test a proposed numerical algorithm by running it on a known solution. As we will see, the lessons learned in the design and testing of numerical algorithms on simpler "solved" examples are of inestimable value when confronting more challenging problems.

Many of the basic numerical solution schemes for partial differential equations can be fit into two broad themes. The first, to be presented in the present chapter, is that of *finite difference methods*, obtained by replacing the derivatives in the equation by appropriate numerical differentiation formulae. We thus start with a brief discussion of some elementary finite difference formulas used to numerically approximate first- and second-order derivatives of functions. We then establish and analyze some of the most basic finite difference schemes for the heat equation, first-order transport equations, the second-order wave equation, and the Laplace and Poisson equations. As we will learn, not all finite difference schemes produce accurate numerical approximations, and one must confront issues of stability and convergence in order to distinguish reliable from worthless methods. In fact, inspired by Fourier analysis, the key numerical stability criterion is a consequence of the scheme's handling of complex exponentials.

The second category of numerical solution techniques comprises the *finite element methods*, which will be the topic of Chapter 10. These two chapters should be regarded as but a preliminary excursion into this vast and active area of contemporary research. More sophisticated variations and extensions, as well as other classes of numerical integration schemes, e.g., spectral, pseudo-spectral, multigrid, multipole, probabilistic (Monte Carlo, etc.), geometric, symplectic, and many more, can be found in specialized numerical analysis texts, including [**6**, **51**, **60**, **80**, **94**], and research papers. Also, the journal *Acta Numerica* is an excellent source of survey papers on state-of-the-art numerical methods for a broad range of disciplines.

5.1 Finite Difference Approximations

In general, a *finite difference* approximation to the value of some derivative of a scalar function $u(x)$ at a point x_0 in its domain, say $u'(x_0)$ or $u''(x_0)$, relies on a suitable combination of sampled function values at nearby points. The underlying formalism used to construct these approximation formulas is known as the *calculus of finite differences*. Its development has a long and influential history, dating back to Newton.

We begin with the first-order derivative. The simplest finite difference approximation is the ordinary *difference quotient*

$$\frac{u(x+h)-u(x)}{h} \approx u'(x), \tag{5.1}$$

which appears in the original calculus definition of the derivative. Indeed, if u is differentiable at x, then $u'(x)$ is, by definition, the limit, as $h \to 0$ of the finite difference quotients. Geometrically, the difference quotient measures the slope of the secant line through the two points $(x, u(x))$ and $(x+h, u(x+h))$ on its graph. For small enough h, this should be a reasonably good approximation to the slope of the tangent line, $u'(x)$, as illustrated in the first picture in Figure 5.1. Throughout our discussion, h, the *step size*, which may be either positive or negative, is assumed to be small: $|\,h\,| \ll 1$. When $h > 0$, (5.1) is referred to as a *forward difference*, while $h < 0$ yields a *backward difference*.

How close an approximation is the difference quotient? To answer this question, we assume that $u(x)$ is at least twice continuously differentiable, and examine its first-order Taylor expansion

$$u(x+h) = u(x) + u'(x)\,h + \tfrac{1}{2}\,u''(\xi)\,h^2 \tag{5.2}$$

at the point x. We have used Lagrange's formula for the remainder term, [**8**, **97**], in which ξ, which depends on both x and h, is a point lying between x and $x+h$. Rearranging (5.2), we obtain

$$\frac{u(x+h)-u(x)}{h} - u'(x) = \tfrac{1}{2}\,u''(\xi)\,h.$$

Thus, the *error* in the finite difference approximation (5.1) can be bounded by a multiple of the step size:

$$\left|\,\frac{u(x+h)-u(x)}{h} - u'(x)\,\right| \le C\,|\,h\,|,$$

where $C = \max \tfrac{1}{2}\,|\,u''(\xi)\,|$ depends on the magnitude of the second derivative of the function over the interval in question. Since the error is proportional to the first power of h, we say that the finite difference quotient (5.1) is a *first-order* approximation to the derivative $u'(x)$. When the precise formula for the error is not so important, we will write

$$u'(x) = \frac{u(x+h)-u(x)}{h} + \mathrm{O}(h). \tag{5.3}$$

The "big Oh" notation $\mathrm{O}(h)$ refers to a term that is proportional to h, or, more precisely, whose absolute value is bounded by a constant multiple of $|\,h\,|$ as $h \to 0$.

Example 5.1. Let $u(x) = \sin x$. Let us try to approximate

$$u'(1) = \cos 1 = .5403023\ldots$$

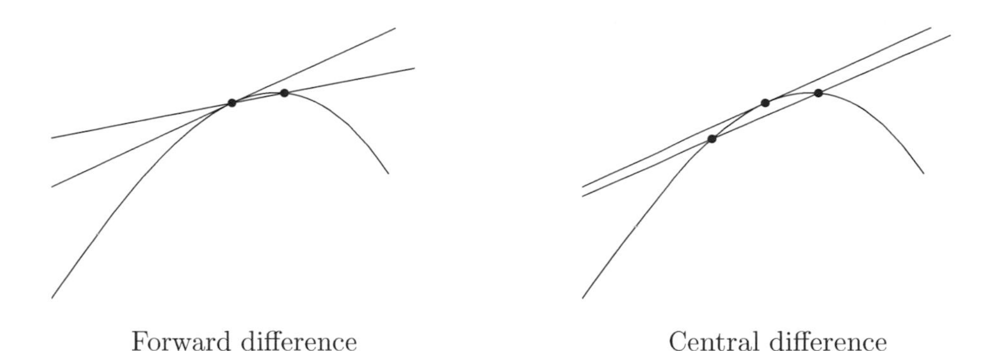

Figure 5.1. Finite difference approximations.

by computing finite difference quotients

$$\cos 1 \approx \frac{\sin(1+h) - \sin 1}{h}.$$

The result for smaller and smaller (positive) values of h is listed in the following table.

h	.1	.01	.001	.0001
approximation	.497364	.536086	.539881	.540260
error	−.042939	−.004216	−.000421	−.000042

We observe that reducing the step size by a factor of $\frac{1}{10}$ reduces the size of the error by approximately the same factor. Thus, to obtain 10 decimal digits of accuracy, we anticipate needing a step size of about $h = 10^{-11}$. The fact that the error is more or less proportional to the step size confirms that we are dealing with a first-order numerical approximation.

To approximate higher-order derivatives, we need to evaluate the function at more than two points. In general, an approximation to the nth order derivative $u^{(n)}(x)$ requires at least $n+1$ distinct sample points. For simplicity, we restrict our attention to equally spaced sample points, although the methods introduced can be readily extended to more general configurations.

For example, let us try to approximate $u''(x)$ by sampling u at the particular points x, $x+h$, and $x-h$. Which combination of the function values $u(x-h), u(x), u(x+h)$ should be used? The answer is found by consideration of the relevant Taylor expansions[†]

$$\begin{aligned} u(x+h) &= u(x) + u'(x)\,h + u''(x)\,\frac{h^2}{2} + u'''(x)\,\frac{h^3}{6} + \mathrm{O}(h^4), \\ u(x-h) &= u(x) - u'(x)\,h + u''(x)\,\frac{h^2}{2} - u'''(x)\,\frac{h^3}{6} + \mathrm{O}(h^4), \end{aligned} \tag{5.4}$$

where the error terms are proportional to h^4. Adding the two formulas together yields

$$u(x+h) + u(x-h) = 2\,u(x) + u''(x)\,h^2 + \mathrm{O}(h^4).$$

[†] Throughout, the function $u(x)$ is assumed to be sufficiently smooth so that any derivatives that appear are well defined and the expansion formula is valid.

Dividing by h^2 and rearranging terms, we arrive at the *centered finite difference approximation* to the second derivative of a function:

$$u''(x) = \frac{u(x+h) - 2\,u(x) + u(x-h)}{h^2} + \mathrm{O}(h^2). \tag{5.5}$$

Since the error is proportional to h^2, this forms a second-order approximation.

Example 5.2. Let $u(x) = e^{x^2}$, with $u''(x) = (4\,x^2 + 2)\,e^{x^2}$. Let us approximate

$$u''(1) = 6\,e = 16.30969097\ldots$$

using the finite difference quotient (5.5):

$$u''(1) = 6\,e \approx \frac{e^{(1+h)^2} - 2\,e + e^{(1-h)^2}}{h^2}.$$

The results are listed in the following table.

h	.1	.01	.001	.0001
approximation	16.48289823	16.31141265	16.30970819	16.30969115
error	.17320726	.00172168	.00001722	.00000018

Each reduction in step size by a factor of $\frac{1}{10}$ reduces the size of the error by a factor of about $\frac{1}{100}$, thereby gaining two new decimal digits of accuracy, which confirms that the centered finite difference approximation is of second order.

However, this prediction is not completely borne out in practice. If we take $h = .00001$ then the formula produces the approximation 16.3097002570, with an error of .0000092863 — which is *less* accurate than the approximation with $h = .0001$. The problem is that round-off errors due to the finite precision of numbers stored in the computer (in the preceding computation we used single-precision floating-point arithmetic) have now begun to affect the computation. This highlights the inherent difficulty with numerical differentiation: Finite difference formulae inevitably require dividing very small quantities, and so round-off inaccuracies may produce noticeable numerical errors. Thus, while they typically produce reasonably good approximations to the derivatives for moderately small step sizes, achieving high accuracy requires switching to higher-precision computer arithmetic. Indeed, a similar comment applies to the previous computation in Example 5.1. Our expectations about the error were not, in fact, fully justified, as you may have discovered had you tried an extremely small step size.

Another way to improve the order of accuracy of finite difference approximations is to employ more sample points. For instance, if the first-order approximation (5.3) to $u'(x)$ based on the two points x and $x+h$ is not sufficiently accurate, one can try combining the function values at three points, say x, $x+h$, and $x-h$. To find the appropriate combination of function values $u(x-h), u(x), u(x+h)$, we return to the Taylor expansions (5.4). To solve for $u'(x)$, we subtract the two formulas, and so

$$u(x+h) - u(x-h) = 2\,u'(x)\,h + \mathrm{O}(h^3).$$

Rearranging the terms, we are led to the well-known *centered difference formula*

$$u'(x) = \frac{u(x+h) - u(x-h)}{2\,h} + \mathrm{O}(h^2), \tag{5.6}$$

which is a second-order approximation to the first derivative. Geometrically, the centered difference quotient represents the slope of the secant line passing through the two points $(x-h, u(x-h))$ and $(x+h, u(x+h))$ on the graph of u, which are centered symmetrically about the point x. Figure 5.1 illustrates the two approximations, and the advantage of the centered difference version is graphically evident. Higher-order approximations can be found by evaluating the function at yet more sample points, say, $x+2h$, $x-2h$, etc.

Example 5.3. Return to the function $u(x) = \sin x$ considered in Example 5.1. The centered difference approximation to its derivative $u'(1) = \cos 1 = .5403023\ \ldots$ is

$$\cos 1 \approx \frac{\sin(1+h) - \sin(1-h)}{2h}.$$

The results are tabulated as follows:

h	.1	.01	.001	.0001
approximation	.53940225217	.54029330087	.54030221582	.54030230497
error	−.00090005370	−.00000900499	−.00000009005	−.00000000090

As advertised, the results are much more accurate than the one-sided finite difference approximation used in Example 5.1 at the same step size. Since it is a second-order approximation, each reduction in the step size by a factor of $\frac{1}{10}$ results in two more decimal places of accuracy — up until the point where the effects of round-off error kick in.

Many additional finite difference approximations can be constructed by similar manipulations of Taylor expansions, but these few very basic formulas, along with a couple that are derived in the exercises, will suffice for our purposes. (For a thorough treatment of the calculus of finite differences, the reader can consult [**74**].) In the following sections, we will employ the finite difference formulas to devise numerical solution schemes for a variety of partial differential equations. Applications to the numerical integration of ordinary differential equations can be found, for example, in [**24, 60, 63**].

Exercises

♣ 5.1.1. Use the finite difference formula (5.3) with step sizes $h = .1, .01$, and $.001$ to approximate the derivative $u'(1)$ of the following functions $u(x)$. Discuss the accuracy of your approximation. (a) x^4, (b) $\dfrac{1}{1+x^2}$, (c) $\log x$, (d) $\cos x$, (e) $\tan^{-1} x$.

♣ 5.1.2. Repeat Exercise 5.1.1 using the centered difference formula (5.6). Compare your approximations with those in the previous exercise — are the values in accordance with the claimed orders of accuracy?

♣ 5.1.3. Approximate the second derivative $u''(1)$ of the functions in Exercise 5.1.1 using the finite difference formula (5.5) with $h = .1, .01$, and $.001$. Discuss the accuracy of your approximations.

5.1.4. Construct finite difference approximations to the first and second derivatives of a function $u(x)$ using its values at the points $x-k, x, x+h$, where $h, k \ll 1$ are of comparable size, but not necessarily equal. What can you say about the error in the approximation?

♠ 5.1.5. In this exercise, you are asked to derive some basic *one-sided finite difference formulas*, which are used for approximating derivatives of functions at or near the boundary of their domain. (a) Construct a finite difference formula that approximates the derivative $u'(x)$ using the values of $u(x)$ at the points $x, x+h$, and $x+2h$. What is the order of your formula? (b) Find a finite difference formula for $u''(x)$ that involves the same three function values. What is its order? (c) Test your formulas by computing approximations to the first and second derivatives of $u(x) = e^{x^2}$ at $x = 1$ using step sizes $h = .1, .01$, and $.001$. What is the error in your numerical approximations? Are the errors compatible with the theoretical orders of the finite difference formulas? Discuss why or why not. (d) Answer part (c) at the point $x = 0$.

♣ 5.1.6. (a) Using the function values $u(x)$, $u(x+h)$, $u(x+3h)$, construct a numerical approximation to the derivative $u'(x)$. (b) What is the order of accuracy of your approximation? (c) Test your approximation on the function $u(x) = \cos x$ at $x = 1$ using the step sizes $h = .1, .01$, and $.001$. Are the errors consistent with your answer in part (b)?

♣ 5.1.7. Answer Exercise 5.1.6 for the second derivative $u''(x)$.

5.1.8. (a) Find the order of the five-point centered finite difference approximation

$$u'(x) \approx \frac{-u(x+2h) + 8u(x+h) - 8u(x-h) + u(x-2h)}{12h}.$$

(b) Test your result on the function $(1+x^2)^{-1}$ at $x = 1$ using the values $h = .1, .01, .001$.

5.1.9. (a) Using the formula in Exercise 5.1.8 as a guide, find five-point finite difference formulas to approximate (i) $u''(x)$, (ii) $u'''(x)$, (iii) $u^{(iv)}(x)$. What is the order of accuracy? (b) Test your formulas on the function $(1+x^2)^{-1}$ at $x = 1$ using the values $h = .1, .01, .001$.

5.2 Numerical Algorithms for the Heat Equation

Consider the heat equation

$$\frac{\partial u}{\partial t} = \gamma \frac{\partial^2 u}{\partial x^2}, \qquad 0 < x < \ell, \qquad t > 0, \tag{5.7}$$

on an interval of length ℓ, with constant thermal diffusivity $\gamma > 0$. We impose time-dependent Dirichlet boundary conditions

$$u(t,0) = \alpha(t), \qquad u(t,\ell) = \beta(t), \qquad t > 0, \tag{5.8}$$

fixing the temperature at the ends of the interval, along with the initial conditions

$$u(0,x) = f(x), \qquad 0 \le x \le \ell, \tag{5.9}$$

specifying the initial temperature distribution. In order to effect a numerical approximation to the solution to this initial-boundary value problem, we begin by introducing a *rectangular mesh* consisting of *nodes* $(t_j, x_m) \in \mathbb{R}^2$ with

$$0 = t_0 < t_1 < t_2 < \cdots \qquad \text{and} \qquad 0 = x_0 < x_1 < \cdots < x_n = \ell.$$

For simplicity, we maintain a uniform mesh spacing in both directions, with

$$\Delta t = t_{j+1} - t_j, \qquad \Delta x = x_{m+1} - x_m = \frac{\ell}{n},$$

representing, respectively, the time step size and the spatial mesh size. It will be essential that we do *not* a priori require that the two be the same. We shall use the notation

$$u_{j,m} \approx u(t_j, x_m), \qquad \text{where} \qquad t_j = j\,\Delta t, \qquad x_m = m\,\Delta x, \tag{5.10}$$

to denote the numerical approximation to the solution value at the indicated node.

As a first attempt at designing a numerical solution scheme, we shall employ the simplest finite difference approximations to the derivatives appearing in the equation. The second-order space derivative is approximated by the centered difference formula (5.5), and hence

$$\begin{aligned} \frac{\partial^2 u}{\partial x^2}(t_j, x_m) &\approx \frac{u(t_j, x_{m+1}) - 2\,u(t_j, x_m) + u(t_j, x_{m-1})}{(\Delta x)^2} + \mathrm{O}\big((\Delta x)^2\big) \\ &\approx \frac{u_{j,m+1} - 2\,u_{j,m} + u_{j,m-1}}{(\Delta x)^2} + \mathrm{O}\big((\Delta x)^2\big), \end{aligned} \tag{5.11}$$

where the error in the approximation is proportional to $(\Delta x)^2$. Similarly, the one-sided finite difference approximation (5.3) is used to approximate the time derivative, and so

$$\frac{\partial u}{\partial t}(t_j, x_m) \approx \frac{u(t_{j+1}, x_m) - u(t_j, x_m)}{\Delta t} + \mathrm{O}(\Delta t) \approx \frac{u_{j+1,m} - u_{j,m}}{\Delta t} + \mathrm{O}(\Delta t), \tag{5.12}$$

where the error is proportional to Δt. In general, one should try to ensure that the approximations have similar orders of accuracy, which leads us to require

$$\Delta t \approx (\Delta x)^2. \tag{5.13}$$

Assuming $\Delta x < 1$, this implies that the time steps must be *much* smaller than the space mesh size.

Remark: At this stage, the reader might be tempted to replace (5.12) by the second-order central difference approximation (5.6). However, this introduces significant complications, and the resulting numerical scheme is not practical; see Exercise 5.2.10.

Replacing the derivatives in the heat equation (5.14) by their finite difference approximations (5.11, 12) and rearranging terms, we end up with the linear system

$$u_{j+1,m} = \mu\, u_{j,m+1} + (1 - 2\,\mu) u_{j,m} + \mu\, u_{j,m-1}, \qquad \begin{array}{l} j = 0, 1, 2, \ldots\,, \\ m = 1, \ldots, n-1, \end{array} \tag{5.14}$$

in which

$$\mu = \frac{\gamma\,\Delta t}{(\Delta x)^2}\,. \tag{5.15}$$

The resulting scheme is of iterative form, whereby the solution values $u_{j+1,m} \approx u(t_{j+1}, x_m)$ at time t_{j+1} are successively calculated, via (5.14), from those at the preceding time t_j.

The initial condition (5.9) indicates that we should initialize our numerical data by sampling the initial temperature at the nodes:

$$u_{0,m} = f_m = f(x_m), \qquad m = 1, \ldots, n-1. \tag{5.16}$$

Similarly, the boundary conditions (5.8) require that

$$u_{j,0} = \alpha_j = \alpha(t_j), \qquad u_{j,n} = \beta_j = \beta(t_j), \qquad j = 0, 1, 2, \ldots\,. \tag{5.17}$$

For consistency, we should assume that the initial and boundary conditions agree at the corners of the domain:

$$f_0 = f(0) = u(0,0) = \alpha(0) = \alpha_0, \qquad f_n = f(\ell) = u(0,\ell) = \beta(0) = \beta_0.$$

The three equations (5.14, 16, 17) completely prescribe the numerical approximation scheme for the solution to the initial-boundary value problem (5.7–9).

Let us rewrite the preceding equations in a more transparent vectorial form. First, let

$$\mathbf{u}^{(j)} = \big(u_{j,1}, u_{j,2}, \ldots, u_{j,n-1}\big)^T \approx \big(u(t_j,x_1), u(t_j,x_2), \ldots, u(t_j,x_{n-1})\big)^T \tag{5.18}$$

be the vector whose entries are the numerical approximations to the solution values at time t_j at the *interior* nodes. We omit the boundary nodes (t_j, x_0), (t_j, x_n), since those values are fixed by the boundary conditions (5.17). Then (5.14) takes the form

$$\mathbf{u}^{(j+1)} = A\,\mathbf{u}^{(j)} + \mathbf{b}^{(j)}, \tag{5.19}$$

where

$$A = \begin{pmatrix} 1-2\mu & \mu & & & & \\ \mu & 1-2\mu & \mu & & & \\ & \mu & 1-2\mu & \mu & & \\ & & \mu & \ddots & \ddots & \\ & & & \ddots & \ddots & \mu \\ & & & & \mu & 1-2\mu \end{pmatrix}, \qquad \mathbf{b}^{(j)} = \begin{pmatrix} \mu\,\alpha_j \\ 0 \\ 0 \\ \vdots \\ 0 \\ \mu\,\beta_j \end{pmatrix}. \tag{5.20}$$

The $(n-1)\times(n-1)$ coefficient matrix A is symmetric and tridiagonal, and only its nonzero entries are displayed. The contributions (5.17) of the boundary nodes appear in the vector $\mathbf{b}^{(j)} \in \mathbb{R}^{n-1}$. This numerical method is known as an *explicit scheme*, since each iterate is computed directly from its predecessor without having to solve any auxiliary equations — unlike the implicit schemes to be discussed next.

Example 5.4. Let us fix the diffusivity $\gamma = 1$ and the interval length $\ell = 1$. For illustrative purposes, we take a spatial step size of $\Delta x = .1$. We work with the initial data

$$u(0,x) = f(x) = \begin{cases} -x, & 0 \le x \le \frac{1}{5}, \\ x - \frac{2}{5}, & \frac{1}{5} \le x \le \frac{7}{10}, \\ 1 - x, & \frac{7}{10} \le x \le 1, \end{cases}$$

used earlier in Example 4.1. In Figure 5.2 we compare the numerical solutions resulting from two (slightly) different time step sizes. The first row uses $\Delta t = (\Delta x)^2 = .01$ and plots the solution at the indicated times. The numerical solution is already showing signs of instability (the final plot does not even fit in the window), and indeed, soon thereafter, it becomes completely wild. The second row takes $\Delta t = .005$. Even though we are employing a rather coarse mesh, the numerical solution is not too far away from the true solution to the initial value problem, which can be seen in Figure 4.1.

Stability Analysis

In light of the preceding calculation, we need to understand why our numerical scheme sometimes gives reasonable answers but sometimes utterly fails. To this end, we investigate

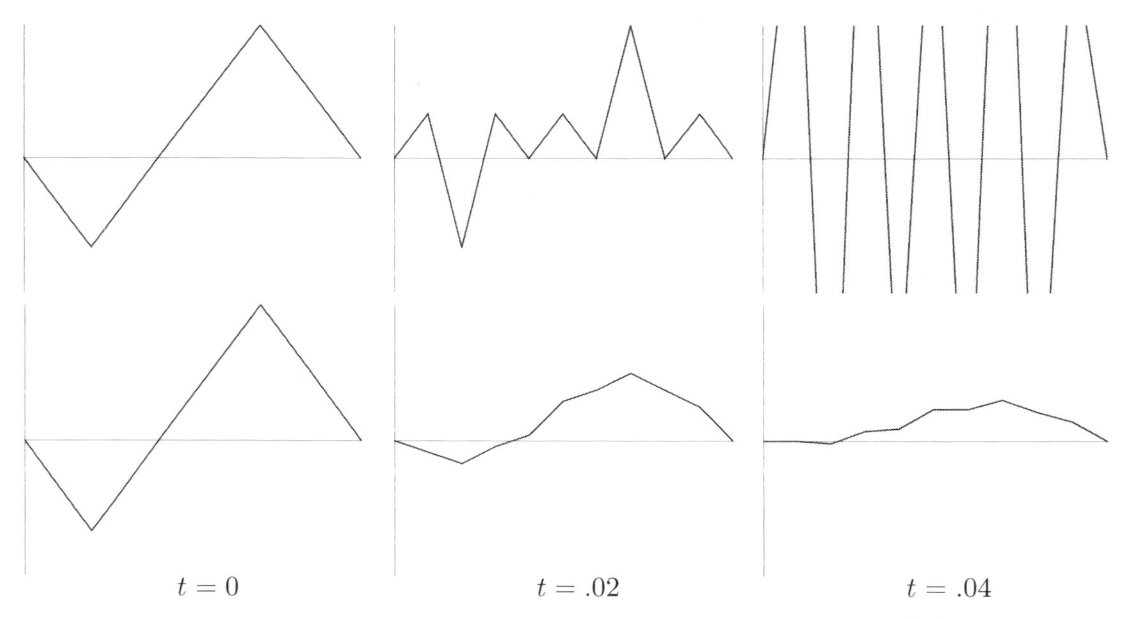

Figure 5.2. Numerical solutions for the heat equation based on the explicit scheme.

the effect of the numerical scheme on simple functions. As we know, the general solution to the heat equation can be decomposed into a sum over the various Fourier modes. Thus, we can concentrate on understanding what the numerical scheme does to an individual complex exponential,† bearing in mind that we can then reconstruct its effect on more general initial data by taking suitable linear combinations of exponentials.

To this end, suppose that, at time $t = t_j$, the solution is a sampled exponential

$$u(t_j, x) = e^{\,\mathrm{i}\,k\,x}, \qquad \text{and so} \qquad u_{j,m} = u(t_j, x_m) = e^{\,\mathrm{i}\,k\,x_m}, \tag{5.21}$$

where k is a real parameter. Substituting the latter values into our numerical equations (5.14), we find that the updated value at time t_{j+1} is also a sampled exponential:

$$\begin{aligned} u_{j+1,m} &= \mu\, u_{j,m+1} + (1 - 2\,\mu) u_{j,m} + \mu\, u_{j,m-1} \\ &= \mu\, e^{\,\mathrm{i}\,k\,x_{m+1}} + (1 - 2\,\mu) e^{\,\mathrm{i}\,k\,x_m} + \mu\, e^{\,\mathrm{i}\,k\,x_{m-1}} \\ &= \mu\, e^{\,\mathrm{i}\,k(x_m + \Delta x)} + (1 - 2\,\mu) e^{\,\mathrm{i}\,k\,x_m} + \mu\, e^{\,\mathrm{i}\,k(x_m - \Delta x)} \\ &= \lambda\, e^{\,\mathrm{i}\,k\,x_m}, \end{aligned} \tag{5.22}$$

where

$$\begin{aligned} \lambda = \lambda(k) &= \mu\, e^{\,\mathrm{i}\,k\,\Delta x} + (1 - 2\,\mu) + \mu\, e^{-\,\mathrm{i}\,k\,\Delta x} \\ &= 1 - 2\,\mu \big[\, 1 - \cos(k\,\Delta x) \,\big] = 1 - 4\,\mu \sin^2\!\big(\tfrac{1}{2}\,k\,\Delta x\big). \end{aligned} \tag{5.23}$$

Thus, the effect of a single step is to multiply the complex exponential (5.21) by the *magnification factor* λ:

$$u(t_{j+1}, x) = \lambda\, e^{\,\mathrm{i}\,k\,x}. \tag{5.24}$$

† As usual, complex exponentials are easier to work with than real trigonometric functions.

In other words, $e^{\,\mathrm{i}\,k\,x}$ plays the role of an *eigenfunction*, with the magnification factor $\lambda(k)$ the corresponding *eigenvalue*, of the linear operator governing each step of the numerical scheme. Continuing in this fashion, we find that the effect of p further iterations of the scheme is to multiply the exponential by the pth power of the magnification factor:

$$u(t_{j+p}, x) = \lambda^p \, e^{\,\mathrm{i}\,k\,x}. \tag{5.25}$$

As a result, the stability is governed by the size of the magnification factor: If $|\,\lambda\,| > 1$, then λ^p grows exponentially, and so the numerical solutions (5.25) become unbounded as $p \to \infty$, which is clearly incompatible with the analytical behavior of solutions to the heat equation. Therefore, an evident necessary condition for the stability of our numerical scheme is that its magnification factor satisfy

$$|\,\lambda\,| \le 1. \tag{5.26}$$

This method of stability analysis was developed by the mid-twentieth-century Hungarian/American mathematician — and father of the electronic computer — John von Neumann. The *stability criterion* (5.26) effectively distinguishes the stable, and hence valid, numerical algorithms from the unstable, and hence ineffectual, schemes. For the particular case (5.23), the von Neumann stability criterion (5.26) requires

$$-1 \le 1 - 4\,\mu \sin^2\!\left(\tfrac{1}{2}\,k\,\Delta x\right) \le 1, \qquad \text{or, equivalently,} \qquad 0 \le \mu \sin^2\!\left(\tfrac{1}{2}\,k\,\Delta x\right) \le \tfrac{1}{2}.$$

Since this is required to hold for all possible k, we must have

$$0 \le \mu = \frac{\gamma\,\Delta t}{(\Delta x)^2} \le \frac{1}{2}, \qquad \text{and hence} \qquad \Delta t \le \frac{(\Delta x)^2}{2\,\gamma}, \tag{5.27}$$

since $\gamma > 0$. Thus, once the space mesh size is fixed, stability of the numerical scheme places a restriction on the allowable time step size. For instance, if $\gamma = 1$, and the space mesh size $\Delta x = .01$, then we must adopt a minuscule time step size $\Delta t \le .00005$. It would take an exorbitant number of time steps to compute the value of the solution at even moderate times, e.g., $t = 1$. Moreover, the accumulation of round-off errors might then cause a significant reduction in the overall accuracy of the final solution values. Since not all choices of space and time steps lead to a convergent scheme, the explicit scheme (5.14) is called *conditionally stable*.

Implicit and Crank–Nicolson Methods

An unconditionally stable method — one that does not restrict the time step — can be constructed by replacing the forward difference formula (5.12) used to approximate the time derivative by the backwards difference formula

$$\frac{\partial u}{\partial t}\,(t_j, x_m) \approx \frac{u(t_j, x_m) - u(t_{j-1}, x_m)}{\Delta t} + \mathrm{O}\big((\Delta t)^2\big). \tag{5.28}$$

Substituting (5.28) and the same centered difference approximation (5.11) for u_{xx} into the heat equation, and then replacing j by $j+1$, leads to the iterative system

$$-\,\mu\,u_{j+1,m+1} + (1 + 2\,\mu)u_{j+1,m} - \mu\,u_{j+1,m-1} = u_{j,m}, \qquad \begin{aligned} j &= 0, 1, 2, \ldots\,, \\ m &= 1, \ldots, n-1, \end{aligned} \tag{5.29}$$

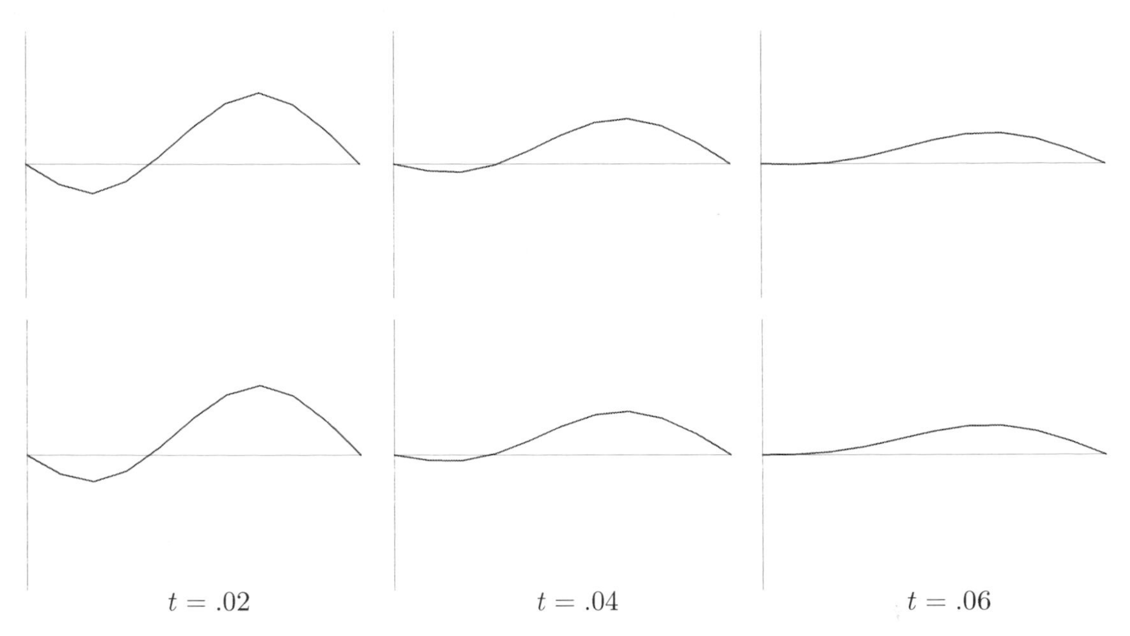

Figure 5.3. Numerical solutions for the heat equation based on the implicit scheme.

where the parameter $\mu = \gamma\,\Delta t/(\Delta x)^2$ is as before. The initial and boundary conditions have the same form (5.16, 17). The latter system can be written in the matrix form

$$\widehat{A}\,\mathbf{u}^{(j+1)} = \mathbf{u}^{(j)} + \mathbf{b}^{(j+1)}, \tag{5.30}$$

where $\widehat{A}$ is obtained from the matrix A in (5.20) by replacing μ by $-\mu$. This serves to define an *implicit scheme*, since we have to solve a linear system of algebraic equations at each step in order to compute the next iterate $\mathbf{u}^{(j+1)}$. However, since the coefficient matrix $\widehat{A}$ is tridiagonal, the solution can be computed extremely rapidly, [**89**], and so its calculation is not an impediment to the practical implementation of this implicit scheme.

Example 5.5. Consider the same initial-boundary value problem considered in Example 5.4. In Figure 5.3, we plot the numerical solutions obtained using the implicit scheme. The initial data is not displayed, but we graph the numerical solutions at times $t = .02, .04, .06$ with a mesh size of $\Delta x = .1$. In the top row, we use a time step of $\Delta t = .01$, while in the bottom row $\Delta t = .005$. In contrast to the explicit scheme, there is very little difference between the two — indeed, both come much closer to the actual solution than the explicit scheme. In fact, even significantly larger time steps yield reasonable numerical approximations to the solution.

Let us apply the von Neumann analysis to investigate the stability of the implicit scheme. Again, we need only look at the effect of the scheme on a complex exponential. Substituting (5.21, 24) into (5.29) and canceling the common exponential factor leads to the equation

$$\lambda\,(-\mu\,e^{\,\mathrm{i}\,k\,\Delta x} + 1 + 2\,\mu - \mu\,e^{-\,\mathrm{i}\,k\,\Delta x}) = 1.$$

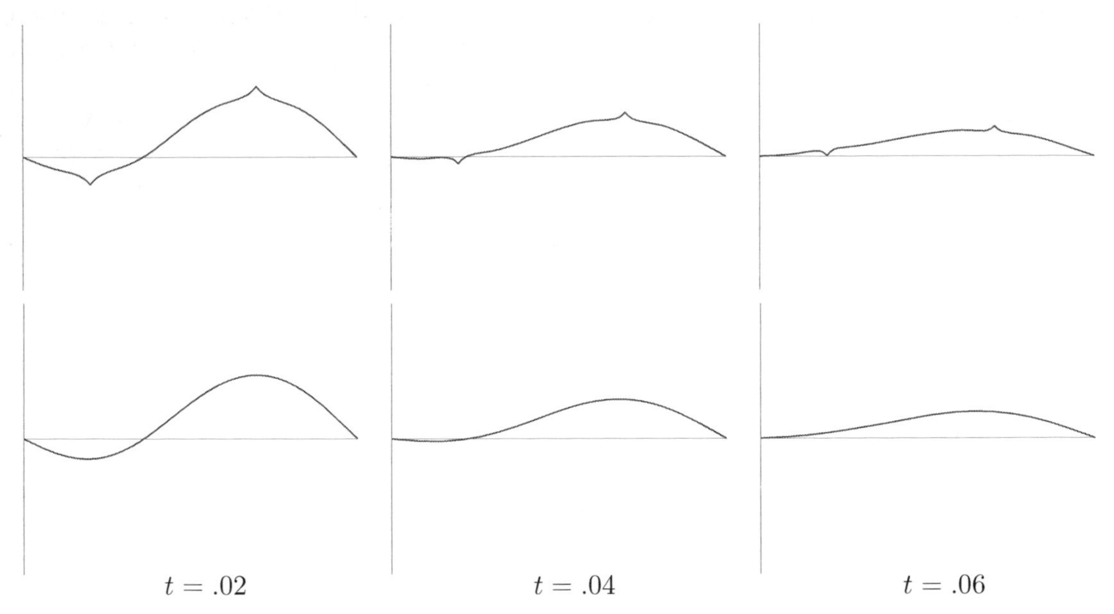

Figure 5.4. Numerical Solutions for the heat equation based on the Crank–Nicolson scheme.

We solve for the magnification factor

$$\lambda = \frac{1}{1 + 2\mu\big(1 - \cos(k\,\Delta x)\big)} = \frac{1}{1 + 4\mu\sin^2\big(\frac{1}{2}k\,\Delta x\big)}\,. \tag{5.31}$$

Since $\mu > 0$, the magnification factor is *always* less than 1 in absolute value, and so the stability criterion (5.26) is satisfied *for any choice of step sizes*. We conclude that the implicit scheme (5.14) is *unconditionally stable*.

Another popular numerical scheme for solving the heat equation is the *Crank–Nicolson method*, due to the British numerical analysts John Crank and Phyllis Nicolson:

$$u_{j+1,m} - u_{j,m} = \tfrac{1}{2}\mu\,(u_{j+1,m+1} - 2\,u_{j+1,m} + u_{j+1,m-1} + u_{j,m+1} - 2\,u_{j,m} + u_{j,m-1}), \tag{5.32}$$

which can be obtained by averaging the explicit and implicit schemes (5.14) and (5.29). We can write (5.32) in vectorial form

$$\widehat{B}\,\mathbf{u}^{(j+1)} = B\,\mathbf{u}^{(j)} + \tfrac{1}{2}\big(\mathbf{b}^{(j)} + \mathbf{b}^{(j+1)}\big),$$

where

$$\widehat{B} = \begin{pmatrix} 1+\mu & -\frac{1}{2}\mu & & \\ -\frac{1}{2}\mu & 1+\mu & -\frac{1}{2}\mu & \\ & -\frac{1}{2}\mu & \ddots & \ddots \\ & & \ddots & \ddots \end{pmatrix}, \qquad B = \begin{pmatrix} 1-\mu & \frac{1}{2}\mu & & \\ \frac{1}{2}\mu & 1-\mu & \frac{1}{2}\mu & \\ & \frac{1}{2}\mu & \ddots & \ddots \\ & & \ddots & \ddots \end{pmatrix}, \tag{5.33}$$

are both tridiagonal.

Applying the von Neumann analysis as before, we deduce that the magnification factor has the form

$$\lambda = \frac{1 - 2\mu\sin^2\big(\frac{1}{2}k\,\Delta x\big)}{1 + 2\mu\sin^2\big(\frac{1}{2}k\,\Delta x\big)}\,. \tag{5.34}$$

Since $\mu > 0$, we see that $|\lambda| \leq 1$ for all choices of step size, and so the Crank–Nicolson scheme is also unconditionally stable. A detailed analysis based on a Taylor expansion of the solution reveals that the errors are of order $(\Delta t)^2$ and $(\Delta x)^2$, and so it is reasonable to choose the time step to have the same order of magnitude as the space step: $\Delta t \approx \Delta x$. This gives the Crank–Nicolson scheme a significant advantage over the previous two methods, in that one can get away with far fewer time steps. However, applying it to the initial value problem considered above reveals a subtle weakness. The top row in Figure 5.4 has space and time step sizes $\Delta t = \Delta x = .01$, and does a reasonable job of approximating the solution except near the corners, where an annoying and incorrect local oscillation persists as the solution decays. The bottom row uses $\Delta t = \Delta x = .001$, and performs much better, although a similar oscillatory error can be observed at much smaller times. Indeed, unlike the implicit scheme, the Crank–Nicolson method fails to rapidly damp out the high-frequency Fourier modes associated with small-scale features such as discontinuities and corners in the initial data, although it performs quite well in smooth regimes. Thus, when dealing with irregular initial data, a good strategy is to first run the implicit scheme until the small-scale noise is dissipated away, and then switch to Crank–Nicolson with a much larger time step to determine the later large scale dynamics.

Finally, we remark that the finite difference schemes developed above for the heat equation can all be readily adapted to more general parabolic partial differential equations. The stability criteria and observed behaviors are fairly similar, and a couple of illustrative examples can be found in the exercises.

Exercises

5.2.1. Suppose we seek to approximate the solution to the initial-boundary value problem

$$u_t = 5\,u_{xx}, \qquad u(t,0) = u(t,3) = 0, \qquad u(0,x) = x(x-1)(x-3), \qquad 0 \leq x \leq 3,$$

by employing the explicit scheme (5.14). (a) Given the spatial mesh size $\Delta x = .1$, what range of time steps Δt can be used to produce an accurate numerical approximation? (*b*) Test your prediction by implementing the scheme using one value of Δt in the allowed range and one value outside.

5.2.2. Solve the following initial-boundary value problem

$$u_t = u_{xx}, \qquad u(t,0) = u(t,1) = 0, \qquad u(0,x) = f(x), \qquad 0 \leq x \leq 1,$$

with initial data $f(x) = \begin{cases} 2\left|x - \frac{1}{6}\right| - \frac{1}{3}, & 0 \leq x \leq \frac{1}{3}, \\ 0, & \frac{1}{3} \leq x \leq \frac{2}{3}, \\ \frac{1}{2} - 3\left|x - \frac{5}{6}\right|, & \frac{2}{3} \leq x \leq 1, \end{cases}$ using

(*i*) the explicit scheme (5.14); (*ii*) the implicit scheme (5.29); and (*iii*) the Crank–Nicolson scheme (5.32). Use space step sizes $\Delta x = .1$ and .05, and suitably chosen time steps Δt. Discuss which features of the solution can be observed in your numerical approximations.

5.2.3. Repeat Exercise 5.2.2 for the initial-boundary value problem $u_t = 3u_{xx}$, $u(0,x) = 0$, $u(t,-1) = 1$, $u(t,1) = -1$, using space step sizes $\Delta x = .2$ and .1.

5.2.4. (a) Solve the initial-boundary value problem

$$u_t = u_{xx}, \qquad u(t,-1) = u(t,1) = 0, \qquad u(0,x) = |\,x\,|^{1/2} - x^2, \qquad -1 \leq x \leq 1,$$

using (*i*) the explicit scheme (5.14); (*ii*) the implicit scheme (5.29); (*iii*) the Crank–Nicolson scheme (5.32). Use $\Delta x = .1$ and an appropriate time step Δt. Compare your numerical solutions at times $t = 0, .01, , .02, .05, .1, .3, .5, 1.0$, and discuss your findings. (*b*) Repeat part (a) for the implicit and Crank-Nicolson schemes with $\Delta x = .01$. Why aren't you being asked to implement the explicit scheme?

5.2.5. Use the implicit scheme with spatial mesh sizes $\Delta x = .1$ and $.05$ and appropriately chosen values of the time step Δt to investigate the solution to the periodically forced boundary value problem $u_t = u_{xx}$, $u(0,x) = 0$, $u(t,0) = \sin 5\pi t$, $u(t,1) = \cos 5\pi t$. Is your solution periodic in time?

♡ 5.2.6. (a) How would you modify (*i*) the explicit scheme; (*ii*) the implicit scheme; to deal with Neumann boundary conditions? *Hint*: Use the one-sided finite difference formulae found in Exercise 5.1.5 to approximate the derivatives at the boundary.
(*b*) Test your proposals on the boundary value problem

$$u_t = u_{xx}, \qquad u(0,x) = \tfrac{1}{2} + \cos 2\pi x - \tfrac{1}{2}\cos 3\pi x, \qquad u_x(t,0) = 0 = u_x(t,1),$$

using space step sizes $\Delta x = .1$ and $.01$ and appropriate time steps. Compare your numerical solution with the exact solution at times $t = .01, .03, .05$, and explain any discrepancies.

5.2.7. (*a*) Design an explicit numerical scheme for approximating the solution to the initial-boundary value problem

$$u_t = \gamma\, u_{xx} + s(x), \qquad u(t,0) = u(t,1) = 0, \qquad u(0,x) = f(x), \qquad 0 \le x \le 1,$$

for the heat equation with a *source term* $s(x)$. (*b*) Test your scheme when

$$\gamma = \tfrac{1}{6}, \qquad s(x) = x(1-x)(10-22x), \qquad f(x) = \begin{cases} 2\left|x-\tfrac{1}{6}\right| - \tfrac{1}{3}, & 0 \le x \le \tfrac{1}{3}, \\ 0, & \tfrac{1}{3} \le x \le \tfrac{2}{3}, \\ \tfrac{1}{2} - 3\left|x - \tfrac{5}{6}\right|, & \tfrac{2}{3} \le x \le 1, \end{cases}$$

using space step sizes $\Delta x = .1$ and $.05$, and a suitably chosen time step Δt. Are your two numerical solutions close? (*c*) What is the long-term behavior of the solution? Can you find a formula for its eventual profile? (*d*) Design an implicit scheme for the same problem. Does this affect the behavior of your numerical solution? What are the advantages of the implicit scheme?

5.2.8. Consider the initial-boundary value problem for the *lossy diffusion equation*

$$\frac{\partial u}{\partial t} = \frac{\partial^2 u}{\partial x^2} - \alpha\, u, \qquad u(t,0) = u(t,1) = 0, \qquad u(0,x) = f(x), \qquad \begin{array}{c} t \ge 0, \\ 0 \le x \le 1, \end{array}$$

where $\alpha > 0$ is a positive constant. (*a*) Devise an explicit finite difference method for computing a numerical approximation to the solution. (*b*) For what mesh sizes would you expect your method to provide a good approximation to the solution?
(*c*) Discuss the case when $\alpha < 0$.

5.2.9. Consider the initial-boundary value problem for the *diffusive transport equation*

$$\frac{\partial u}{\partial t} = \frac{\partial^2 u}{\partial x^2} + 2\,\frac{\partial u}{\partial x}, \qquad u(t,0) = u(t,1) = 0, \qquad u(0,x) = x(1-x), \qquad \begin{array}{c} t \ge 0, \\ 0 \le x \le 1. \end{array}$$

(*a*) Devise an explicit finite difference scheme for computing numerical approximations to the solution. *Hint*: Make sure your approximations are of comparable order. (*b*) For what range of time step sizes would you expect your method to provide a decent approximation to the solution? (*c*) Test your answer in part (*b*) for the spatial step size $\Delta x = .1$.

◇ 5.2.10. (*a*) Show that using the centered difference approximation (5.6) to approximate the time derivative leads to *Richardson's method* for numerically solving the heat equation:

$$u_{j+1,m} = u_{j-1,m} + 2\mu\,(u_{j,m+1} - 2u_{j,m} + u_{j,m-1}), \qquad \begin{array}{c} j = 1, 2, \ldots\,, \\ m = 1, \ldots, n-1, \end{array}$$

where $\mu = \gamma\,\Delta t/(\Delta x)^2$ is as in (5.15). (*b*) Discuss how to start Richardson's method. (*c*) Discuss the stability of Richardson's method. (*d*) Test Richardson's method on the initial-boundary value problem in Exercise 5.2.2. Does your numerical solution conform with your expectations from part (*b*)?

5.3 Numerical Algorithms for First–Order Partial Differential Equations

Let us next apply the method of finite differences to construct some basic numerical methods for first-order partial differential equations. As noted in Section 4.4, first-order partial differential equations are prototypes for hyperbolic equations, and so many of the lessons learned here carry over to the general hyperbolic regime, including the second-order wave equation, which we analyze in detail in the following section.

Consider the initial value problem for the elementary transport equation

$$\frac{\partial u}{\partial t} + c\,\frac{\partial u}{\partial x} = 0, \qquad u(0,x) = f(x), \qquad -\infty < x < \infty, \tag{5.35}$$

with constant wave speed c. Of course, as we learned in Section 2.2, the solution is a simple traveling wave

$$u(t,x) = f(x - c\,t) \tag{5.36}$$

that is constant along the characteristic lines of slope c in the (t,x)–plane. Although the analytical solution is completely elementary, there will be valuable lessons to be learned from our attempt to reproduce it by numerical approximation. Indeed, each of the numerical schemes developed below has an evident adaptation to transport equations with variable wave speeds $c(t,x)$, and even to nonlinear transport equations whose wave speed depends on the solution u, and so admit shock-wave solutions.

As before, we restrict our attention to a rectangular mesh (t_j, x_m) with uniform time step size $\Delta t = t_{j+1} - t_j$ and space mesh size $\Delta x = x_{m+1} - x_m$. We use $u_{j,m} \approx u(t_j, x_m)$ to denote our numerical approximation to the solution $u(t,x)$ at the indicated node. The simplest numerical scheme is obtained by replacing the time and space derivatives by their first-order finite difference approximations (5.1):

$$\frac{\partial u}{\partial t}(t_j, x_m) \approx \frac{u_{j+1,m} - u_{j,m}}{\Delta t} + \mathrm{O}(\Delta t), \qquad \frac{\partial u}{\partial x}(t_j, x_m) \approx \frac{u_{j,m+1} - u_{j,m}}{\Delta x} + \mathrm{O}(\Delta x). \tag{5.37}$$

Substituting these expressions into the transport equation (5.35) leads to the explicit numerical scheme

$$u_{j+1,m} = -\,\sigma\, u_{j,m+1} + (\sigma + 1) u_{j,m}, \tag{5.38}$$

in which the parameter

$$\sigma = \frac{c\,\Delta t}{\Delta x} \tag{5.39}$$

depends on the wave speed and the ratio of time to space step sizes. Since we are employing first-order approximations to both derivatives, we should choose the step sizes to be comparable: $\Delta t \approx \Delta x$. When working on a bounded interval, say $0 \le x \le \ell$, we will need to specify a value for the numerical solution at the right end, e.g., setting $u_{j,n} = 0$, which corresponds to imposing the boundary condition $u(t,\ell) = 0$.

In Figure 5.5, we plot the numerical solutions, at times $t = .1, .2, .3$, arising from the following initial condition:

$$u(0,x) = f(x) = .4\, e^{-300(x-.5)^2} + .1\, e^{-300(x-.65)^2}. \tag{5.40}$$

We use step sizes $\Delta t = \Delta x = .005$, and try four different values of the wave speed. The cases $c = .5$ and $c = -1.5$ clearly exhibit some form of numerical instability. The numerical

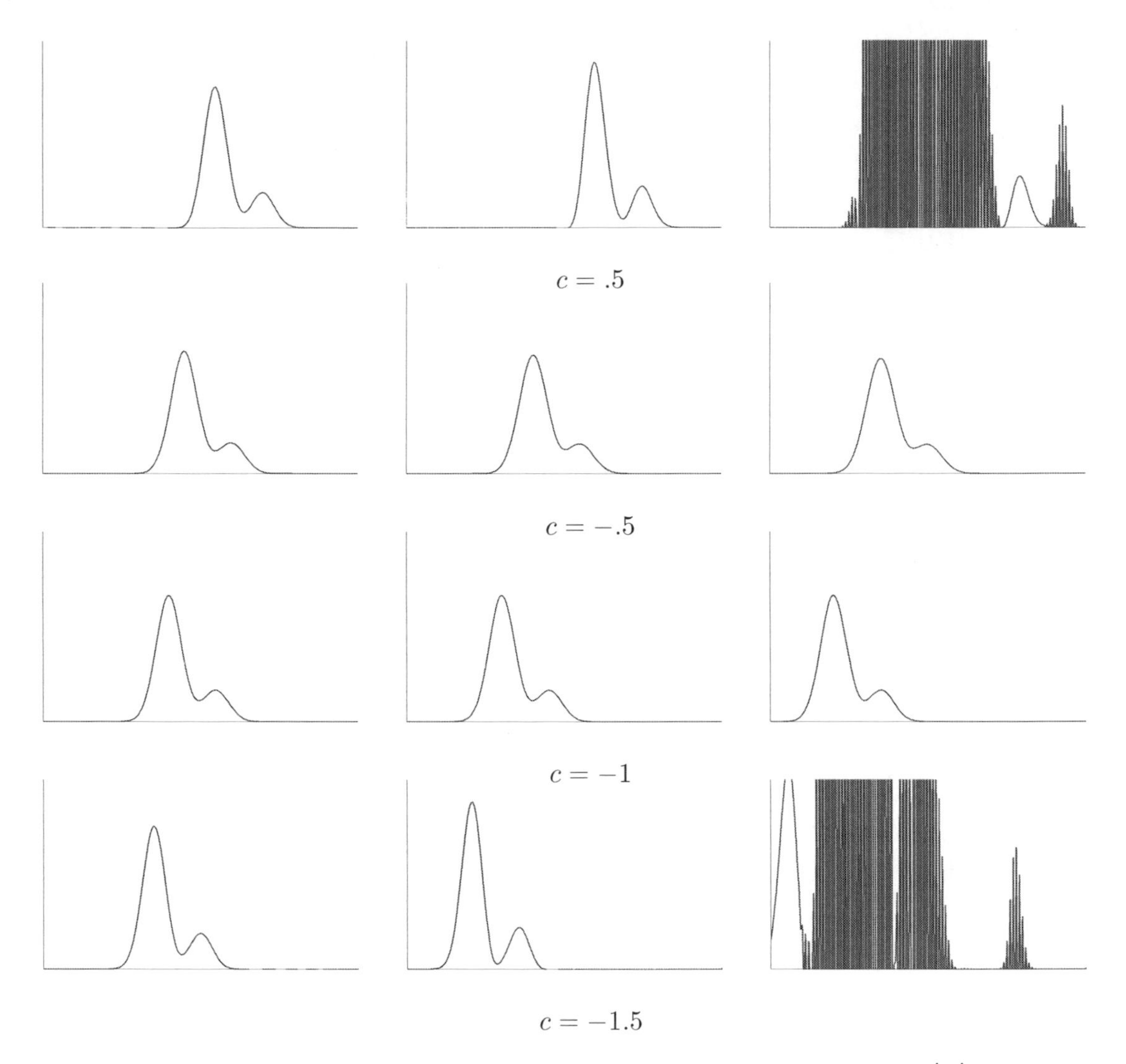

Figure 5.5. Numerical solutions to the transport equation.

solution when $c = -.5$ is a bit more reasonable, although one can already observe some degradation due to the relatively low accuracy of the scheme. This can be alleviated by employing a smaller step size. The case $c = -1$ looks exceptionally good, and you are asked to provide an explanation in Exercise 5.3.6.

The CFL Condition

There are two ways to understand the observed numerical instability. First, we recall that the exact solution (5.36) is constant along the characteristic lines $x = ct + \xi$, and hence the value of $u(t,x)$ depends only on the initial value $f(\xi)$ at the point $\xi = x - ct$. On the other hand, at time $t = t_j$, the numerical solution $u_{j,m} \approx u(t_j, x_m)$ computed using (5.38) depends on the values of $u_{j-1,m}$ and $u_{j-1,m+1}$. The latter two values have

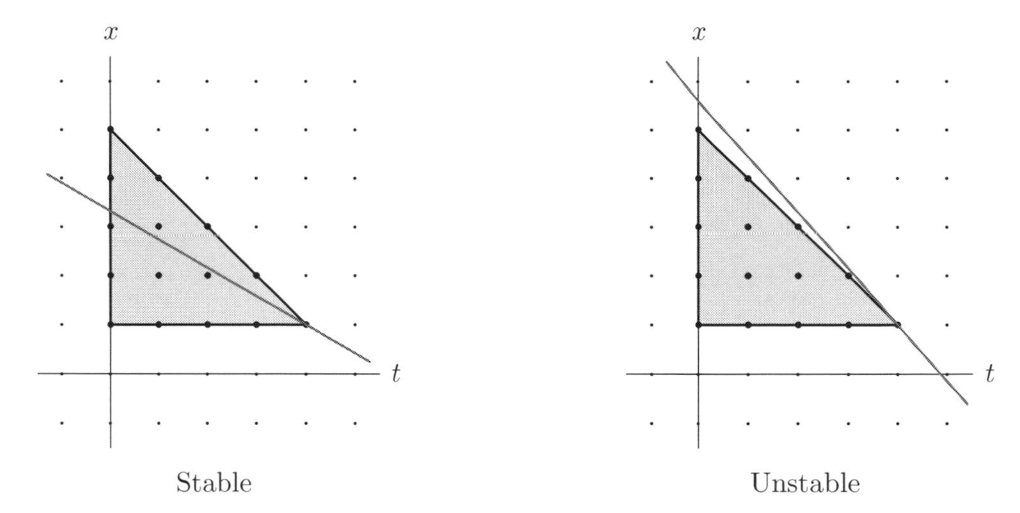

Figure 5.6. The CFL condition.

been computed from the previous approximations $u_{j-2,m}, u_{j-2,m+1}, u_{j-2,m+2}$. And so on. Going all the way back to the initial time $t_0 = 0$, we find that $u_{j,m}$ depends on the initial values $u_{0,m} = f(x_m), \ldots, u_{0,m+j} = f(x_m + j\,\Delta x)$ at the nodes lying in the interval $x_m \leq x \leq x_m + j\,\Delta x$. On the other hand, the actual solution $u(t_j, x_m)$ depends only on the value of $f(\xi)$, where

$$\xi = x_m - c\,t_j = x_m - c\,j\,\Delta t.$$

Thus, if ξ lies outside the interval $[x_m, x_m + j\,\Delta x]$, then varying the initial condition near the point $x = \xi$ will change the actual solution value $u(t_j, x_m)$ without altering its numerical approximation $u_{j,m}$ at all! So the numerical scheme cannot possibly provide an accurate approximation to the solution value. As a result, we must require

$$x_m \leq \xi = x_m - c\,j\,\Delta t \leq x_m + j\,\Delta x, \qquad \text{and hence} \qquad 0 \leq -c\,\Delta t \leq \Delta x,$$

which we rewrite as

$$0 \geq \sigma = \frac{c\,\Delta t}{\Delta x} \geq -1, \qquad \text{or, equivalently,} \qquad -\frac{\Delta x}{\Delta t} \leq c \leq 0. \tag{5.41}$$

This is the simplest manifestation of what is known as the *Courant–Friedrichs–Lewy condition*, or *CFL condition* for short, which was established in the groundbreaking 1928 paper [**33**] by three of the pioneers in the development of numerical methods for partial differential equations: the German (soon to be American) applied mathematicians Richard Courant, Kurt Friedrichs, and Hans Lewy. Note that the CFL condition requires that the wave speed be *negative*, and the time step size not too large. Thus, for allowable wave speeds, the finite difference scheme (5.38) is conditionally stable.

The CFL condition can be recast in a more geometrically transparent manner as follows. For the finite difference scheme (5.38), the *numerical domain of dependence* of a point (t_j, x_m) is the triangle

$$T_{(t_j,x_m)} = \left\{ (t,x) \;\middle|\; 0 \leq t \leq t_j,\ x_m \leq x \leq x_m + t_j - t \right\}. \tag{5.42}$$

The reason for this nomenclature is that, as we have just seen, the numerical approximation to the solution at the node (t_j, x_m) depends on the computed values at the nodes lying within its numerical domain of dependence; see Figure 5.6. The CFL condition (5.41) requires that, for all $0 \le t \le t_j$, the characteristic passing through the point (t_j, x_m) lie entirely within the numerical domain of dependence (5.42). If the characteristic ventures outside the domain, then the scheme will be numerically unstable. With this geometric reformulation, the CFL criterion can be applied to both linear and nonlinear transport equations that have nonuniform wave speeds.

The CFL criterion (5.41) is reconfirmed by a von Neumann stability analysis. As before, we test the numerical scheme on an exponential function. Substituting

$$u_{j,m} = e^{\,\mathrm{i}\,k x_m}, \qquad u_{j+1,m} = \lambda\, e^{\,\mathrm{i}\,k x_m}, \tag{5.43}$$

into (5.38) leads to

$$\lambda\, e^{\,\mathrm{i}\,k x_m} = -\,\sigma\, e^{\,\mathrm{i}\,k x_{m+1}} + (\sigma+1) e^{\,\mathrm{i}\,k x_m} = \bigl(-\,\sigma\, e^{\,\mathrm{i}\,k \Delta x} + \sigma + 1\bigr) e^{\,\mathrm{i}\,k x_m}.$$

The resulting (complex) magnification factor

$$\lambda = 1 + \sigma\bigl(1 - e^{\,\mathrm{i}\,k\Delta x}\bigr) = \bigl(1 + \sigma - \sigma\cos(k\,\Delta x)\bigr) - \mathrm{i}\,\sigma\sin(k\,\Delta x)$$

satisfies the stability criterion $|\,\lambda\,| \le 1$ if and only if

$$\begin{aligned} |\,\lambda\,|^2 &= \bigl(1 + \sigma - \sigma\cos(k\,\Delta x)\bigr)^2 + \bigl(\sigma\sin(k\,\Delta x)\bigr)^2 \\ &= 1 + 2\,\sigma(\sigma+1)\bigl(1 - \cos(k\,\Delta x)\bigr) = 1 + 4\,\sigma(\sigma+1)\sin^2\bigl(\tfrac{1}{2}\,k\,\Delta x\bigr) \le 1 \end{aligned}$$

for all k. Thus, stability requires that $\sigma(\sigma+1) \le 0$, and thus $-1 \le \sigma \le 0$, in complete accord with the CFL condition (5.41).

Upwind and Lax–Wendroff Schemes

To obtain a finite difference scheme that can be used for positive wave speeds, we replace the forward finite difference approximation to $\partial u/\partial x$ by the corresponding backwards difference quotient, namely, (5.1) with $h = -\,\Delta x$, leading to the alternative first-order numerical scheme

$$u_{j+1,m} = -\,(\sigma - 1)\,u_{j,m} + \sigma\,u_{j,m-1}, \tag{5.44}$$

where $\sigma = c\,\Delta t/\Delta x$ is as before. A similar analysis, left to the reader, produces the corresponding CFL stability criterion

$$0 \le \sigma = \frac{c\,\Delta t}{\Delta x} \le 1,$$

and so this scheme can be applied for suitable positive wave speeds.

In this manner, we have produced one numerical scheme that works for negative wave speeds, and an alternative scheme for positive speeds. The question arises — particularly when one is dealing with equations with variable wave speeds — whether one can devise a scheme that is (conditionally) stable for *both* positive and negative wave speeds. One might be tempted to use the centered difference approximation (5.6):

$$\frac{\partial u}{\partial x}\,(t_j, x_m) \approx \frac{u_{j,m+1} - u_{j,m-1}}{\Delta x} + \mathrm{O}\bigl((\Delta x)^2\bigr). \tag{5.45}$$

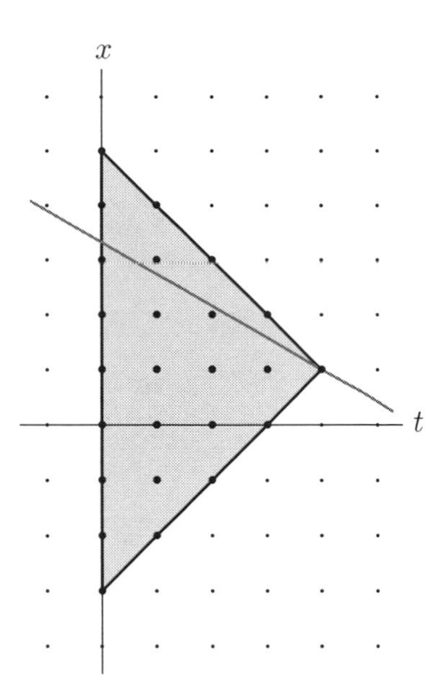

Figure 5.7. The CFL condition for the centered difference scheme.

Substituting (5.45) and the previous approximation to the time derivative (5.37) into (5.35) leads to the numerical scheme

$$u_{j+1,m} = -\tfrac{1}{2}\sigma\, u_{j,m+1} + u_{j,m} + \tfrac{1}{2}\sigma\, u_{j,m-1}, \tag{5.46}$$

where, as usual, $\sigma = c\,\Delta t/\Delta x$. In this case, the *numerical domain of dependence* of the node (t_j, x_m) consists of the nodes in the triangle

$$\widetilde{T}_{(t_j,x_m)} = \left\{\, (t,x) \;\middle|\; 0 \le t \le t_j,\ x_m - t_j + t \le x \le x_m + t_j - t \,\right\}. \tag{5.47}$$

The CFL condition requires that, for $0 \le t \le t_j$, the characteristic going through (t_j, x_m) lie within this triangle, as in Figure 5.7, which imposes the condition

$$|\,\sigma\,| = \left|\frac{c\,\Delta t}{\Delta x}\right| \le 1, \qquad \text{or, equivalently,} \qquad |\,c\,| \le \frac{\Delta x}{\Delta t}\,. \tag{5.48}$$

Unfortunately, although it satisfies the CFL condition over this range of wave speeds, the centered difference scheme is, in fact, always *unstable*! For instance, the instability of the numerical solution to the preceding initial value problem (5.40) for $c = 1$ can be observed in Figure 5.8. This is confirmed by applying a von Neumann analysis: substitute (5.43) into (5.46), and cancel the common exponential factors. Provided $\sigma \ne 0$, which means that $c \ne 0$, the resulting magnification factor

$$\lambda = 1 - \,\mathrm{i}\,\sigma \sin(k\,\Delta x)$$

satisfies $|\,\lambda\,| > 1$ for all k with $\sin(k\,\Delta x) \ne 0$. Thus, for $c \ne 0$, the centered difference scheme (5.46) is unstable for all (nonzero) wave speeds!

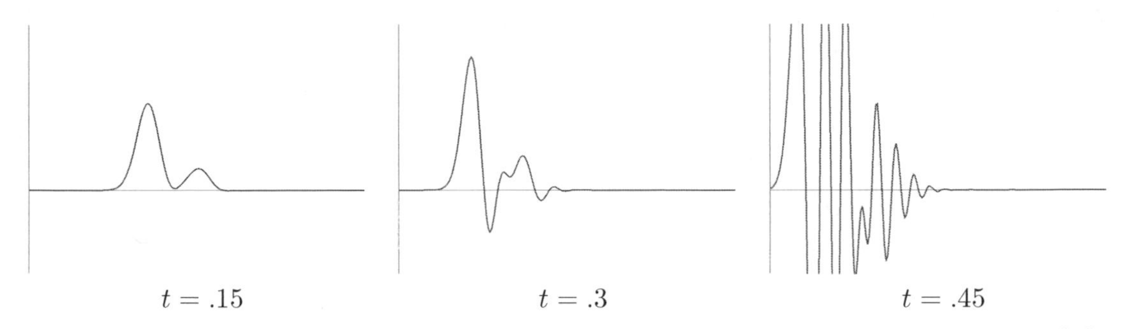

Figure 5.8. Centered difference numerical solution to the transport equation.

One possible means of overcoming the sign restriction on the wave speed is to use the forward difference scheme (5.38) when the wave speed is negative and the backwards scheme (5.44) when it is positive. The resulting scheme, valid for varying wave speeds $c(t,x)$, takes the form

$$u_{j+1,m} = \begin{cases} -\sigma_{j,m} u_{j,m+1} + (\sigma_{j,m}+1) u_{j,m}, & c_{j,m} \le 0, \\ -(\sigma_{j,m}-1) u_{j,m} + \sigma_{j,m} u_{j,m-1}, & c_{j,m} > 0, \end{cases} \tag{5.49}$$

where

$$\sigma_{j,m} = c_{j,m} \frac{\Delta t}{\Delta x}, \qquad c_{j,m} = c(t_j, x_m). \tag{5.50}$$

This is referred to as an *upwind scheme*, since the second node always lies "upwind" — that is, away from the direction of motion — from the reference point (t_j, x_m). The upwind scheme works reasonably well over short time intervals, assuming that the space step size is sufficiently small and the time step satisfies the CFL condition $\Delta x/\Delta t \le |\, c_{j,m} \,|$ at each node, cf. (5.41). However, over longer time intervals, as we already observed in Figure 5.5, the simple upwind scheme tends to produce a noticeable damping of waves or, alternatively, require an unacceptably small step size. One way of overcoming this defect is to use the popular *Lax–Wendroff scheme*, which is based on second-order approximations to the derivatives. In the case of constant wave speed, the iterative step takes the form

$$u_{j+1,m} = \tfrac{1}{2}\sigma(\sigma-1)\, u_{j,m+1} - (\sigma^2-1)\, u_{j,m} + \tfrac{1}{2}\sigma(\sigma+1)\, u_{j,m-1}. \tag{5.51}$$

The stability analysis of the Lax–Wendroff scheme is relegated to the exercises. Extensions to variable wave speeds are more subtle, and we refer the reader to [**80**] for a detailed derivation.

Exercises

5.3.1. Solve the initial value problem $u_t = 3u_x$, $u(0,x) = 1/(1+x^2)$, on the interval $[-10, 10]$ using an upwind scheme with space step size $\Delta x = .1$. Decide on an appropriate time step size, and graph your solution at times $t = .5, 1, 1.5$. Discuss what you observe.

5.3.2. Solve Exercise 5.3.1 for the nonuniform transport equations

(a) $u_t + 4(1+x^2)^{-1} u_x = 0$, (b) $u_t = \left(3 - 2e^{-x^2/4}\right) u_x$,

(c) $u_t + 7x(1+x^2)^{-1} u_x = 0$, (d) $u_t + \left(2\tan^{-1} \frac{1}{2} x\right) u_x = 0$.

5.3.3. Consider the initial value problem

$$u_t + \frac{3x}{x^2+1} u_x = 0, \qquad u(0,x) = \left(1 - \tfrac{1}{2}x^2\right)e^{-x^2/3}.$$

On the interval $[-5,5]$, using space step size $\Delta x = .1$ and time step size $\Delta t = .025$, apply (a) the forward scheme (5.38) (suitably modified for variable wave speed), (b) the backward scheme (5.44) (suitably modified for variable wave speed), and (c) the upwind scheme (5.49). Graph the resulting numerical solutions at times $t = .5, 1, 1.5$, and discuss what you observe in each case. Which of the schemes are stable?

5.3.4. Use the centered difference scheme (5.46) to solve the initial value problem in Exercise 5.3.1. Do you observe any instabilities in your numerical solution?

5.3.5. Use the Lax–Wendroff scheme (5.51) to solve the initial value problem in Exercise 5.3.1. Discuss the accuracy of your solution in comparison with the upwind scheme.

♢ 5.3.6. Can you explain why, in Figure 5.5, the numerical solution in the case $c = -1$ is significantly better than for $c = -.5$, or, indeed, for any other c in the stable range.

5.3.7. Nonlinear transport equations are often solved numerically by writing them in the form of a conservation law, and then applying the finite difference formulas directly to the conserved density and flux. (a) Devise an upwind scheme for numerically solving our favorite nonlinear transport equation, $u_t + \frac{1}{2}(u^2)_x = 0$.
(b) Test your scheme on the initial value problem $u(0,x) = e^{-x^2}$.

5.3.8. (a) Design a stable numerical solution scheme for the damped transport equation $u_t + \frac{3}{4} u_x + u = 0$. (b) Test your scheme on the initial value problem with $u(0,x) = e^{-x^2}$.

♢ 5.3.9. Analyze the stability of the numerical scheme (5.44) by applying (a) the CFL condition; (b) a von Neumann analysis. Are your conclusions the same?

♢ 5.3.10. For what choices of step size $\Delta t, \Delta x$ is the Lax–Wendroff scheme (5.51) stable?

5.4 Numerical Algorithms for the Wave Equation

Let us now develop some basic numerical solution techniques for the second-order wave equation. As above, although we are in possession of the explicit d'Alembert solution formula (2.82), the lessons learned in designing viable schemes here will carry over to more complicated situations, including inhomogeneous media and higher-dimensional problems, for which analytic solution formulas may no longer be readily available.

Consider the wave equation

$$\frac{\partial^2 u}{\partial t^2} = c^2 \frac{\partial^2 u}{\partial x^2}, \qquad 0 < x < \ell, \qquad t \geq 0, \tag{5.52}$$

on a bounded interval of length ℓ with constant wave speed $c > 0$. For specificity, we impose (possibly time-dependent) Dirichlet boundary conditions

$$u(t,0) = \alpha(t), \qquad u(t,\ell) = \beta(t), \qquad t \geq 0, \tag{5.53}$$

along with the usual initial conditions

$$u(0,x) = f(x), \qquad \frac{\partial u}{\partial t}(0,x) = g(x), \qquad 0 \le x \le \ell. \tag{5.54}$$

As usual, we adopt a uniformly spaced mesh

$$t_j = j\,\Delta t, \qquad x_m = m\,\Delta x, \qquad \text{where} \qquad \Delta x = \frac{\ell}{n}.$$

Discretization is implemented by replacing the second-order derivatives in the wave equation by their standard finite difference approximations (5.5):

$$\begin{aligned} \frac{\partial^2 u}{\partial t^2}(t_j, x_m) &\approx \frac{u(t_{j+1}, x_m) - 2\,u(t_j, x_m) + u(t_{j-1}, x_m)}{(\Delta t)^2} + \mathrm{O}\big((\Delta t)^2\big), \\ \frac{\partial^2 u}{\partial x^2}(t_j, x_m) &\approx \frac{u(t_j, x_{m+1}) - 2\,u(t_j, x_m) + u(t_j, x_{m-1})}{(\Delta x)^2} + \mathrm{O}\big((\Delta x)^2\big). \end{aligned} \tag{5.55}$$

Since the error terms are both of second order, we anticipate being able to choose the space and time step sizes to have comparable magnitudes: $\Delta t \approx \Delta x$. Substituting the finite difference formulas (5.55) into the partial differential equation (5.52) and rearranging terms, we are led to the iterative system

$$u_{j+1,m} = \sigma^2\, u_{j,m+1} + 2\,(1-\sigma^2)\, u_{j,m} + \sigma^2\, u_{j,m-1} - u_{j-1,m}, \qquad \begin{array}{l} j = 1, 2, \ldots\,, \\ m = 1, \ldots, n-1, \end{array} \tag{5.56}$$

for the numerical approximations $u_{j,m} \approx u(t_j, x_m)$ to the solution values at the nodes. The parameter

$$\sigma = \frac{c\,\Delta t}{\Delta x} > 0 \tag{5.57}$$

depends on the wave speed and the ratio of space and time step sizes. The boundary conditions (5.53) require that

$$u_{j,0} = \alpha_j = \alpha(t_j), \qquad u_{j,n} = \beta_j = \beta(t_j), \qquad j = 0, 1, 2, \ldots\,. \tag{5.58}$$

This allows us to rewrite the iterative system in vectorial form

$$\mathbf{u}^{(j+1)} = B\,\mathbf{u}^{(j)} - \mathbf{u}^{(j-1)} + \mathbf{b}^{(j)}, \tag{5.59}$$

where

$$B = \begin{pmatrix} 2\,(1-\sigma^2) & \sigma^2 & & & \\ \sigma^2 & 2\,(1-\sigma^2) & \sigma^2 & & \\ & \sigma^2 & \ddots & \ddots & \\ & & \ddots & \ddots & \sigma^2 \\ & & & \sigma^2 & 2\,(1-\sigma^2) \end{pmatrix}, \quad \mathbf{u}^{(j)} = \begin{pmatrix} u_{j,1} \\ u_{j,2} \\ \vdots \\ u_{j,n-2} \\ u_{j,n-1} \end{pmatrix}, \quad \mathbf{b}^{(j)} = \begin{pmatrix} \sigma^2\,\alpha_j \\ 0 \\ \vdots \\ 0 \\ \sigma^2\,\beta_j \end{pmatrix}. \tag{5.60}$$

The entries of $\mathbf{u}^{(j)} \in \mathbb{R}^{n-1}$ are, as in (5.18), the numerical approximations to the solution values at the *interior* nodes. Note that (5.59) describes a *second-order iterative scheme*, since computing the subsequent iterate $\mathbf{u}^{(j+1)}$ requires knowing the values of the preceding two: $\mathbf{u}^{(j)}$ and $\mathbf{u}^{(j-1)}$.

The one subtlety is how to get the method started. We know $\mathbf{u}^{(0)}$, since its entries $u_{0,m} = f_m = f(x_m)$ are determined by the initial position. However, we also need $\mathbf{u}^{(1)}$

in order to launch the iteration and compute $\mathbf{u}^{(2)}, \mathbf{u}^{(3)}, \ldots$. Its entries $u_{1,m} \approx u(\Delta t, x_m)$ approximate the solution at time $t_1 = \Delta t$, whereas the initial velocity $u_t(0,x) = g(x)$ prescribes the derivatives $u_t(0,x_m) = g_m = g(x_m)$ at the initial time $t_0 = 0$. To resolve this difficulty, a first thought might be to use the finite difference approximation

$$g_m = \frac{\partial u}{\partial t}(0,x_m) \approx \frac{u(\Delta t, x_m) - u(0,x_m)}{\Delta t} \approx \frac{u_{1,m} - f_m}{\Delta t} \tag{5.61}$$

to compute the required values $u_{1,m} = f_m + g_m \Delta t$. However, the approximation (5.61) is accurate only to order Δt, whereas the rest of the scheme has errors proportional to $(\Delta t)^2$. The effect would be to introduce an unacceptably large error at the initial step, and the resulting solution would fail to conform to the desired order of accuracy.

To construct an initial approximation to $\mathbf{u}^{(1)}$ with error on the order of $(\Delta t)^2$, we need to analyze the error in the approximation (5.61) in more depth. Note that, by Taylor's Theorem,

$$\begin{aligned}\frac{u(\Delta t, x_m) - u(0,x_m)}{\Delta t} &= \frac{\partial u}{\partial t}(0,x_m) + \frac{1}{2}\frac{\partial^2 u}{\partial t^2}(0,x_m)\Delta t + \mathrm{O}\big((\Delta t)^2\big) \\ &= \frac{\partial u}{\partial t}(0,x_m) + \frac{c^2}{2}\frac{\partial^2 u}{\partial x^2}(0,x_m)\,\Delta t + \mathrm{O}\big((\Delta t)^2\big),\end{aligned}$$

since $u(t,x)$ solves the wave equation. Therefore,

$$\begin{aligned}u_{1,m} = u(\Delta t, x_m) &\approx u(0,x_m) + \frac{\partial u}{\partial t}(0,x_m)\Delta t + \frac{c^2}{2}\frac{\partial^2 u}{\partial x^2}(0,x_m)(\Delta t)^2 \\ &= f(x_m) + g(x_m)\,\Delta t + \frac{c^2}{2} f''(x_m)(\Delta t)^2 \\ &\approx f_m + g_m\,\Delta t + \frac{c^2(f_{m+1} - 2f_m + f_{m-1})(\Delta t)^2}{2(\Delta x)^2},\end{aligned}$$

where the last line, which employs the finite difference approximation (5.5) to the second derivative, can be used if the explicit formula for $f''(x)$ is either not known or too complicated to evaluate directly. Therefore, we initiate the scheme by setting

$$u_{1,m} = \tfrac{1}{2}\sigma^2 f_{m+1} + (1-\sigma^2) f_m + \tfrac{1}{2}\sigma^2 f_{m-1} + g_m\,\Delta t, \tag{5.62}$$

or, in vectorial form,

$$\mathbf{u}^{(0)} = \mathbf{f}, \qquad \mathbf{u}^{(1)} = \tfrac{1}{2} B\,\mathbf{u}^{(0)} + \mathbf{g}\,\Delta t + \tfrac{1}{2}\mathbf{b}^{(0)}, \tag{5.63}$$

where $\mathbf{f} = \big(f_1, f_2, \ldots, f_{n-1}\big)^T$, $\mathbf{g} = \big(g_1, g_2, \ldots, g_{n-1}\big)^T$, are the sampled values of the initial data. This serves to maintain the desired second-order accuracy of the scheme.

Example 5.6. Consider the particular initial value problem

$$u_{tt} = u_{xx}, \qquad \begin{aligned} u(0,x) = e^{-400\,(x-.3)^2}, \qquad u_t(0,x) = 0, \qquad & 0 \le x \le 1, \\ u(t,0) = u(t,1) = 0, \qquad & t \ge 0, \end{aligned}$$

subject to homogeneous Dirichlet boundary conditions on the interval $[0,1]$. The initial data is a fairly concentrated hump centered at $x = .3$. As time progresses, we expect the initial hump to split into two half-sized humps, which then collide with the ends of the interval, reversing direction and orientation.

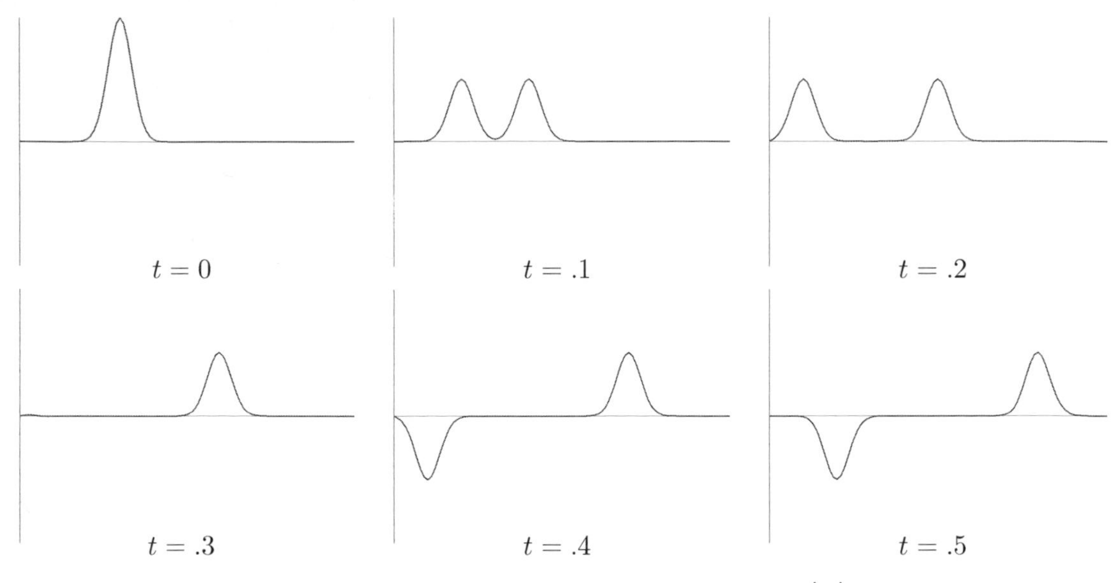

Figure 5.9. Numerically stable waves.

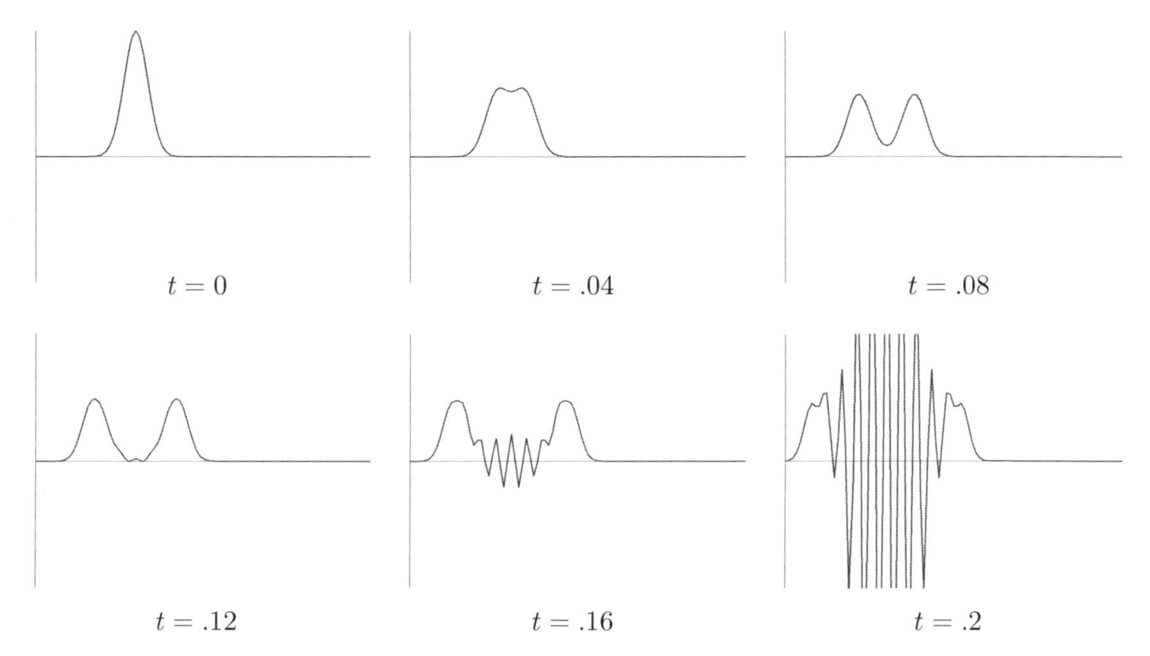

Figure 5.10. Numerically unstable waves.

For our numerical approximation, let us use a space discretization consisting of 90 equally spaced points, and so $\Delta x = \frac{1}{90} = .0111\ldots$. If we choose a time step of $\Delta t = .01$, whereby $\sigma = .9$, then we obtain a reasonably accurate solution over a fairly long time range, as plotted in Figure 5.9. On the other hand, if we double the time step, setting $\Delta t = .02$, so $\sigma = 1.8$, then, as shown in Figure 5.10, we induce an instability that eventually

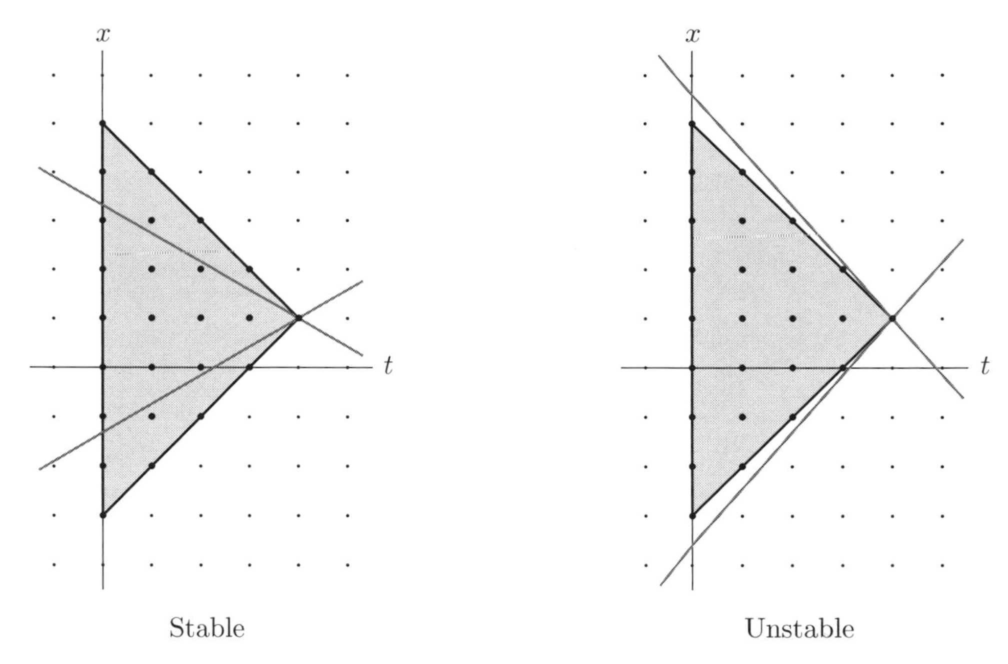

Figure 5.11. The CFL condition for the wave equation.

overwhelms the numerical solution. Thus, the preceding numerical scheme appears to be only conditionally stable.

Stability analysis proceeds along the same lines as in the first-order case. The CFL condition requires that the characteristics emanating from a node (t_j, x_m) remain, for times $0 \le t \le t_j$, in its numerical domain of dependence, which, for our particular numerical scheme, is the same triangle

$$\widetilde{T}_{(t_j,x_m)} = \left\{ (t,x) \;\middle|\; 0 \le t \le t_j,\; x_m - t_j + t \le x \le x_m + t_j - t \right\},$$

now plotted in Figure 5.11. Since the characteristics are the lines of slope $\pm c$, the CFL condition is the same as in (5.48):

$$\sigma = \frac{c\,\Delta t}{\Delta x} \le 1, \qquad \text{or, equivalently,} \qquad 0 < c \le \frac{\Delta x}{\Delta t}. \tag{5.64}$$

The resulting stability criterion explains the observed difference between the numerically stable and unstable cases.

However, as we noted above, the CFL condition is, in general, only necessary for stability of the numerical scheme; sufficiency requires that we perform a von Neumann stability analysis. To this end, we specialize the calculation to a single complex exponential $e^{\,\mathrm{i}\,k\,x}$. After one time step, the scheme will have the effect of multiplying it by the *magnification factor* $\lambda = \lambda(k)$, after another time step by λ^2, and so on. To determine λ, we substitute the relevant sampled exponential values

$$u_{j-1,m} = e^{\,\mathrm{i}\,k\,x_m}, \qquad u_{j,m} = \lambda\, e^{\,\mathrm{i}\,k\,x_m}, \qquad u_{j+1,m} = \lambda^2\, e^{\,\mathrm{i}\,k\,x_m}, \tag{5.65}$$

into the scheme (5.56). After canceling the common exponential, we find that the magnification factor satisfies the following quadratic equation:

$$\lambda^2 = \bigl(2 - 4\sigma^2 \sin^2\bigl(\tfrac{1}{2}\,k\,\Delta x\bigr)\bigr)\lambda - 1,$$

whence

$$\lambda = \alpha \pm \sqrt{\alpha^2 - 1}\,, \qquad \text{where} \qquad \alpha = 1 - 2\,\sigma^2 \sin^2\left(\tfrac{1}{2}\,k\,\Delta x\right). \tag{5.66}$$

Thus, there are *two* different magnification factors associated with each complex exponential — which is, in fact, a consequence of the scheme being of second order. Stability requires that *both* be ≤ 1 in modulus. Now, if the CFL condition (5.64) holds, then $|\,\alpha\,| \leq 1$, which implies that both magnification factors (5.66) are complex numbers of modulus $|\,\lambda\,| = 1$, and thus the numerical scheme satisfies the stability criterion (5.26). On the other hand, if $\sigma > 1$, then $\alpha < -1$ over a range of values of k, which implies that the two magnification factors (5.66) are both real and one of them is < -1, thus violating the stability criterion. Consequently, the CFL condition (5.64) does indeed distinguish between the stable and unstable finite difference schemes for the wave equation.

Exercises

5.4.1. Suppose you are asked to numerically approximate the solution to the initial-boundary value problem

$$u_{tt} = 64\,u_{xx}, \quad u(t,0) = u(t,3) = 0, \quad u(0,x) = \begin{cases} 1 - 2\,|\,x - 1\,|, & \frac{1}{2} \leq x \leq \frac{3}{2}, \\ 0, & \text{otherwise}, \end{cases} \quad u_t(0,x) = 0,$$

on the interval $0 \leq x \leq 3$, using (5.56) with space step size $\Delta x = .1$. (*a*) What range of time steps Δt are allowed? (*b*) Test your answer by implementing the numerical solution for one value of Δt in the allowable range and one value outside. Discuss what you observe in your numerical solutions. (*c*) In the stable range, compare your numerical solution with that obtained using the smaller step size $\Delta x = .01$ and a suitable time step Δt.

5.4.2. Solve Exercise 5.4.1 for the boundary value problem

$$u_{tt} = 64\,u_{xx}, \quad u(t,0) = 0 = u(t,3), \quad u(0,x) = 0, \quad u_t(0,x) = \begin{cases} 1 - 2\,|\,x - 1\,|, & \frac{1}{2} \leq x \leq \frac{3}{2}, \\ 0, & \text{otherwise}. \end{cases}$$

5.4.3. Solve the following initial-boundary value problem

$$u_{tt} = 9\,u_{xx}, \quad u(t,0) = u(t,1) = 0, \quad u(0,x) = \tfrac{1}{2} + \left|\,x - \tfrac{1}{4}\,\right| - \left|\,2x - \tfrac{3}{4}\,\right|, \quad u_t(0,x) = 0,$$

on the interval $0 \leq x \leq 1$, using the numerical scheme (5.56) with space step sizes $\Delta x = .1, .01$ and $.001$ and suitably chosen time steps. Discuss which features of the solution can be observed in your numerical approximations.

5.4.4. (*a*) Use a numerical integrator with space step size $\Delta x = .05$ to solve the periodically forced boundary value problem

$$u_{tt} = u_{xx}, \qquad u(0,x) = u_t(0,x) = 0, \qquad u(t,0) = \sin t, \qquad u(t,1) = 0.$$

Is your solution periodic? (*b*) Repeat the computation using the alternative boundary condition $u(t,0) = \sin \pi t$. Discuss any observed differences between the two problems.

5.4.5. (*a*) Design an explicit numerical scheme for solving the initial-boundary value problem

$$u_{tt} = c^2 u_{xx} + F(t,x), \quad u(t,0) = u(t,1) = 0, \quad u(0,x) = f(x), \quad u_t(0,x) = g(x), \quad 0 \leq x \leq 1,$$

for the wave equation with an external *forcing term* $F(t,x)$. Clearly state any stability conditions that need to be imposed on the time and space step sizes.
(*b*) Test your scheme on the particular case $c = \frac{1}{4}$, $F(t,x) = 3\,\mathrm{sign}\left(x - \frac{1}{2}\right) \sin \pi t$, $f(x) \equiv g(x) \equiv 0$, using space step sizes $\Delta x = .05$ and $.01$, and suitably chosen time steps.

5.4.6. Let $\beta > 0$. (a) Design a finite difference scheme for approximating the solution to the initial-boundary value problem

$$u_{tt} + \beta\,u_t = c^2 u_{xx}, \qquad u(t,0) = u(t,1) = 0, \qquad u(0,x) = f(x), \qquad u_t(0,x) = g(x),$$

for the damped wave equation on the interval $0 \le x \le 1$. (b) Discuss the stability of your scheme. What choice of step sizes will ensure stability? (c) Test your scheme with $c = 1$, $\beta = 1$, using the initial data $f(x) = e^{-(x-.7)^2}$, $g(x) = 0$.

5.5 Finite Difference Algorithms for the Laplace and Poisson Equations

Finally, let us discuss the implementation of finite diffference numerical schemes for elliptic boundary value problems. We concentrate on the simplest cases: the two-dimensional Laplace and Poisson equations. The basic issues are already apparent in this particular context, and extensions to more general equations, higher dimensions, and higher-order schemes are all reasonably straightforward. In Chapter 10, we will present a competitor — the renowned finite element method — which, while relying on more sophisticated mathematical machinery, enjoys several advantages, including more immediate adaptability to variable mesh sizes and more sophisticated geometries.

For specificity, we concentrate on the Dirichlet boundary value problem

$$\begin{aligned} -\Delta u = -u_{xx} - u_{yy} &= f(x,y), & (x,y) &\in \Omega, \\ u(x,y) &= g(x,y), & \text{for} \quad (x,y) &\in \partial\Omega, \end{aligned} \tag{5.67}$$

on a bounded planar domain $\Omega \subset \mathbb{R}^2$. The first step is to discretize the domain Ω by constructing a rectangular mesh. Thus, the finite difference method is particularly suited to domains whose boundary lines up with the coordinate axes; otherwise, the mesh nodes do not, generally, lie exactly on $\partial\Omega$, making the approximation of the boundary data more challenging — although not insurmountable.

For simplicity, let us study the case in which

$$\Omega = \{ a < x < b, \ c < y < d \}$$

is a rectangle. We introduce a regular rectanglar mesh, with x and y spacings given, respectively, by

$$\Delta x = \frac{b-a}{m}, \qquad \Delta y = \frac{c-d}{n},$$

for positive integers m, n. Thus, the interior of the rectangle contains $(m-1)(n-1)$ *interior nodes*

$$(x_i, y_j) = (a + i\,\Delta x, c + j\,\Delta y) \qquad \text{for} \qquad 0 < i < m, \quad 0 < j < n.$$

In addition, the $2m + 2n$ *boundary nodes* $(x_0, y_j) = (a, y_j)$, $(x_m, y_j) = (b, y_j)$, $(x_i, y_0) = (x_i, c)$, $(x_i, y_n) = (x_i, d)$, lie on the boundary of the rectangle.

At each interior node, we employ the centered difference formula (5.5) to approximate the relevant second-order derivatives:

$$\begin{aligned} \frac{\partial^2 u}{\partial x^2}(x_i, y_j) &= \frac{u(x_{i+1}, y_j) - 2\,u(x_i, y_j) + u(x_{i-1}, y_j)}{(\Delta x)^2} + \mathrm{O}\big((\Delta x)^2 \big), \\ \frac{\partial^2 u}{\partial y^2}(x_i, y_j) &= \frac{u(x_i, y_{j+1}) - 2\,u(x_i, y_j) + u(x_i, y_{j-1})}{(\Delta y)^2} + \mathrm{O}\big((\Delta y)^2 \big). \end{aligned} \tag{5.68}$$

Substituting these finite difference formulae into the Poisson equation produces the linear system

$$-\frac{u_{i+1,j}-2u_{i,j}+u_{i-1,j}}{(\Delta x)^2}-\frac{u_{i,j+1}-2u_{i,j}+u_{i,j-1}}{(\Delta y)^2}=f_{i,j},\qquad \begin{aligned} i&=1,\ldots,m-1,\\ j&=1,\ldots,n-1,\end{aligned} \tag{5.69}$$

in which $u_{i,j}$ denotes our numerical approximation to the solution values $u(x_i, y_j)$ at the nodes, while $f_{i,j} = f(x_i, y_j)$. If we set

$$\rho=\frac{\Delta x}{\Delta y}, \tag{5.70}$$

then (5.69) can be rewritten in the form

$$\begin{aligned} 2(1+\rho^2)u_{i,j}-(u_{i-1,j}+u_{i+1,j})-\rho^2(u_{i,j-1}+u_{i,j+1})&=(\Delta x)^2 f_{i,j},\\ i=1,\ldots,m-1,\quad j&=1,\ldots,n-1.\end{aligned} \tag{5.71}$$

Since both finite difference approximations (5.68) are of second order, one should choose Δx and Δy to be of comparable size, thus keeping ρ around 1.

The linear system (5.71) forms the finite difference approximation to the Poisson equation at the interior nodes. It is supplemented by the discretized Dirichlet boundary conditions

$$\begin{aligned} u_{i,0}&=g_{i,0}, & u_{i,n}&=g_{i,n}, & i&=0,\ldots,m,\\ u_{0,j}&=g_{0,j}, & u_{m,j}&=g_{m,j}, & j&=0,\ldots,n.\end{aligned} \tag{5.72}$$

These boundary values can be substituted directly into the system, making (5.71) a system of $(m-1)(n-1)$ linear equations involving the $(m-1)(n-1)$ unknowns $u_{i,j}$ for $1 \le i \le m-1$, $1 \le j \le n-1$. We impose some convenient ordering for these entries, e.g., from left to right and then bottom to top, forming the column vector of unknowns

$$\begin{aligned} \mathbf{w}&=(w_1,w_2,\ \ldots\ ,w_{(m-1)(n-1)})^T\\ &=(u_{1,1},u_{2,1},\ldots,u_{m-1,1},u_{1,2},u_{2,2},\ldots,u_{m-1,2},u_{1,3},\ldots,u_{m-1,n-1})^T.\end{aligned} \tag{5.73}$$

The combined linear system (5.71–72) can then be rewritten in matrix form

$$A\mathbf{w}=\widehat{\mathbf{f}}, \tag{5.74}$$

where the right-hand side is obtained by combining the column vector $\mathbf{f} = (\ \ldots\ f_{i,j}\ \ldots\ )^T$ with the boundary data provided by (5.72) according to where they appear in the system. The implementation will become clearer once we work through a small-scale example.

Example 5.7. To better understand how the process works, let us look at the case in which $\Omega = \{0 < x < 1,\ 0 < y < 1\}$ is the unit square. In order to write everything in full detail, we start with a very coarse mesh with $\Delta x = \Delta y = \frac{1}{4}$; see Figure 5.12. Thus $m = n = 4$, resulting in a total of nine interior nodes. In this case, $\rho = 1$, and hence the

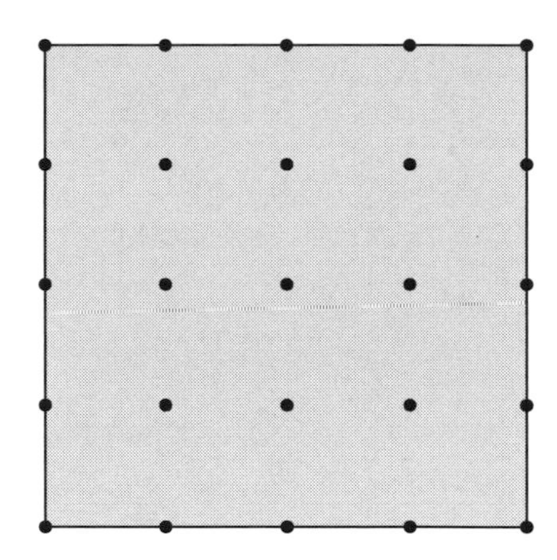

Figure 5.12. Square mesh with $\Delta x = \Delta y = \frac{1}{4}$.

finite difference system (5.71) consists of the following nine equations:

$$
\begin{aligned}
-u_{1,0} - u_{0,1} + 4u_{1,1} - u_{2,1} - u_{1,2} &= \tfrac{1}{16} f_{1,1}, \\
-u_{2,0} - u_{1,1} + 4u_{2,1} - u_{3,1} - u_{2,2} &= \tfrac{1}{16} f_{2,1}, \\
-u_{3,0} - u_{2,1} + 4u_{3,1} - u_{4,1} - u_{3,2} &= \tfrac{1}{16} f_{3,1}, \\
-u_{1,1} - u_{0,2} + 4u_{1,2} - u_{2,2} - u_{1,3} &= \tfrac{1}{16} f_{1,2}, \\
-u_{2,1} - u_{1,2} + 4u_{2,2} - u_{3,2} - u_{2,3} &= \tfrac{1}{16} f_{2,2}, \\
-u_{3,1} - u_{2,2} + 4u_{3,2} - u_{4,2} - u_{3,3} &= \tfrac{1}{16} f_{3,2}, \\
-u_{1,2} - u_{0,3} + 4u_{1,3} - u_{2,3} - u_{1,4} &= \tfrac{1}{16} f_{1,3}, \\
-u_{2,2} - u_{1,3} + 4u_{2,3} - u_{3,3} - u_{2,4} &= \tfrac{1}{16} f_{2,3}, \\
-u_{3,2} - u_{2,3} + 4u_{3,3} - u_{4,3} - u_{3,4} &= \tfrac{1}{16} f_{3,3}.
\end{aligned}
\tag{5.75}
$$

(Note that the values at the four corner nodes, $u_{0,0}, u_{4,0}, u_{0,4}, u_{4,4}$, do not appear.) The boundary data imposes the additional conditions (5.72), namely

$$
\begin{array}{llllll}
u_{0,1} = g_{0,1}, & u_{0,2} = g_{0,2}, & u_{0,3} = g_{0,3}, & u_{1,0} = g_{1,0}, & u_{2,0} = g_{2,0}, & u_{3,0} = g_{3,0}, \\
u_{4,1} = g_{4,1}, & u_{4,2} = g_{4,2}, & u_{4,3} = g_{4,3}, & u_{1,4} = g_{1,4}, & u_{2,4} = g_{2,4}, & u_{3,4} = g_{3,4}.
\end{array}
$$

The system (5.75) can be written in matrix form $A\mathbf{w} = \widehat{\mathbf{f}}$, where

$$
A = \begin{pmatrix}
4 & -1 & 0 & -1 & 0 & 0 & 0 & 0 & 0 \\
-1 & 4 & -1 & 0 & -1 & 0 & 0 & 0 & 0 \\
0 & -1 & 4 & 0 & 0 & -1 & 0 & 0 & 0 \\
-1 & 0 & 0 & 4 & -1 & 0 & -1 & 0 & 0 \\
0 & -1 & 0 & -1 & 4 & -1 & 0 & -1 & 0 \\
0 & 0 & -1 & 0 & -1 & 4 & 0 & 0 & -1 \\
0 & 0 & 0 & -1 & 0 & 0 & 4 & -1 & 0 \\
0 & 0 & 0 & 0 & -1 & 0 & -1 & 4 & -1 \\
0 & 0 & 0 & 0 & 0 & -1 & 0 & -1 & 4
\end{pmatrix},
\tag{5.76}
$$

and

$$\mathbf{w} = \begin{pmatrix} w_1 \\ w_2 \\ w_3 \\ w_4 \\ w_5 \\ w_6 \\ w_7 \\ w_8 \\ w_9 \end{pmatrix} = \begin{pmatrix} u_{1,1} \\ u_{2,1} \\ u_{3,1} \\ u_{1,2} \\ u_{2,2} \\ u_{3,2} \\ u_{1,3} \\ u_{2,3} \\ u_{3,3} \end{pmatrix}, \qquad \widehat{\mathbf{f}} = \begin{pmatrix} \frac{1}{16} f_{1,1} + g_{1,0} + g_{0,1} \\ \frac{1}{16} f_{2,1} + g_{2,0} \\ \frac{1}{16} f_{3,1} + g_{3,0} + g_{4,1} \\ \frac{1}{16} f_{1,2} + g_{0,2} \\ \frac{1}{16} f_{2,2} \\ \frac{1}{16} f_{3,2} + g_{4,2} \\ \frac{1}{16} f_{1,3} + g_{0,3} + g_{1,4} \\ \frac{1}{16} f_{2,3} + g_{2,4} \\ \frac{1}{16} f_{3,3} + g_{4,3} + g_{3,4} \end{pmatrix}.$$

Note that the known boundary values, namely $u_{i,j} = g_{i,j}$ when i or j equals 0 or 4, have been incorporated into the right-hand side $\widehat{\mathbf{f}}$ of the finite difference linear system (5.74). The resulting linear system is easily solved by Gaussian Elimination, [**89**]. Finer meshes lead to correspondingly larger linear systems, all endowed with a common overall structure, as discussed below.

For example, the function

$$u(x,y) = y \sin(\pi x)$$

solves the particular boundary value problem

$$-\Delta u = \pi^2 y \sin(\pi x), \quad u(x,0) = u(0,y) = u(1,y) = 0, \quad u(x,1) = \sin(\pi x), \quad 0 < x, y < 1.$$

Setting up and solving the linear system (5.75) produces the finite difference solution values

$$\begin{aligned} u_{1,1} &= .1831, & u_{1,2} &= .2589, & u_{1,3} &= .1831, \\ u_{2,1} &= .3643, & u_{2,2} &= .5152, & u_{2,3} &= .3643, \\ u_{3,1} &= .5409, & u_{3,2} &= .7649, & u_{3,3} &= .5409, \end{aligned}$$

leading to the numerical approximation plotted in the first graph[†] of Figure 5.13. The maximal error between the numerical and exact solution values is .01520, which occurs at the center of the square. In the second and third graphs, the mesh spacing is successively reduced by half, so there are, respectively, $m = n = 8$ and 16 nodes in each coordinate direction. The corresponding maximal numerical errors at the nodes are .004123 and .001035. Observe that halving the step size reduces the error by a factor of $\frac{1}{4}$, which is consistent with the numerical scheme being of second order.

Remark: The preceding test is a particular instance of the method of *manufactured solutions*, in which one starts with a preselected function that almost certainly is not a solution to the exact problem at hand. Nevertheless, substituting this function into the differential equation and the relevant initial and/or boundary conditions leads to an inhomogeneous problem of the same character as the original. After running the numerical scheme on the modified problem, one can test for accuracy by comparing the numerical output with the preselected function.

[†] We are using flat triangles to interpolate the nodal data. Smoother interpolation schemes, e.g., splines, [**102**], will produce a more realistic reproduction of the analytic solution graph.

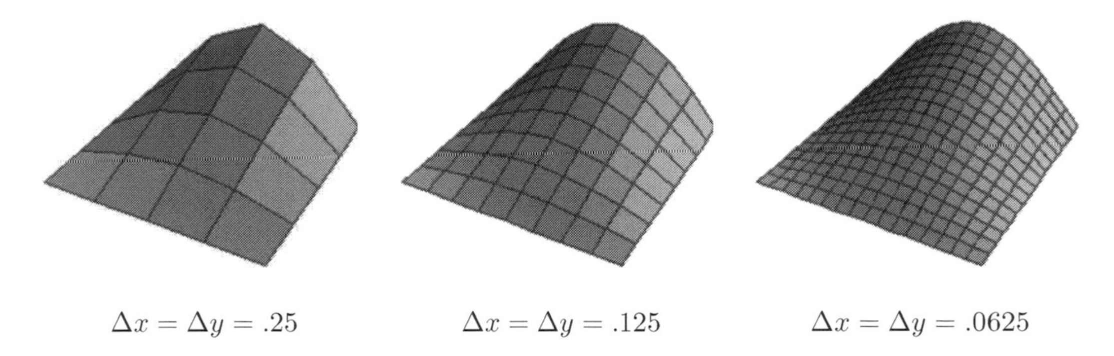

Figure 5.13. Finite difference solutions to a Poisson boundary value problem.

Solution Strategies

The linear algebraic system resulting from a finite difference discretization can be rather large, and it behooves us to devise efficient solution strategies. The general finite difference coefficient matrix A has a very structured form, which can already be inferred from the very simple case (5.76). When the underlying domain is a rectangle, it assumes a *block tridiagonal form*

$$A = \begin{pmatrix} B_\rho & -\rho^2\,\mathrm{I} & & & & \\ -\rho^2\,\mathrm{I} & B_\rho & -\rho^2\,\mathrm{I} & & & \\ & -\rho^2\,\mathrm{I} & B_\rho & -\rho^2\,\mathrm{I} & & \\ & & \ddots & \ddots & \ddots & \\ & & & -\rho^2\,\mathrm{I} & B_\rho & -\rho^2\,\mathrm{I} \\ & & & & -\rho^2\,\mathrm{I} & B_\rho \end{pmatrix}, \tag{5.77}$$

where I is the $(m-1)\times(m-1)$ identity matrix, while

$$B_\rho = \begin{pmatrix} 2(1+\rho^2) & -\rho^2 & & & & & \\ -\rho^2 & 2(1+\rho^2) & -\rho^2 & & & & \\ & -\rho^2 & 2(1+\rho^2) & -\rho^2 & & & \\ & & -\rho^2 & 2(1+\rho^2) & -\rho^2 & & \\ & & & \ddots & \ddots & \ddots & \\ & & & & -\rho^2 & 2(1+\rho^2) & -\rho^2 \\ & & & & & -\rho^2 & 2(1+\rho^2) \end{pmatrix} \tag{5.78}$$

is itself an $(m-1)\times(m-1)$ tridiagonal matrix. (Here and below, all entries not explicitly indicated are zero.) There are $n-1$ blocks in both the row and column directions.

When the finite difference linear system is of moderate size, it can be efficiently solved by Gaussian Elimination, which effectively factorizes $A = LU$ into a product of lower and upper triangular matrices. (This follows since A is symmetric and nonsingular, as guaranteed by Theorem 5.8 below.) In the present case, the factors are *block bidiagonal* matrices:

$$L = \begin{pmatrix} \mathrm{I} & & & & & \\ L_1 & \mathrm{I} & & & & \\ & L_2 & \mathrm{I} & & & \\ & & \ddots & \ddots & & \\ & & & L_{n-3} & \mathrm{I} & \\ & & & & L_{n-2} & \mathrm{I} \end{pmatrix},$$
$$U = \begin{pmatrix} U_1 & -\rho^2\,\mathrm{I} & & & & \\ & U_2 & -\rho^2\,\mathrm{I} & & & \\ & & U_3 & -\rho^2\,\mathrm{I} & & \\ & & & \ddots & \ddots & \\ & & & & U_{n-2} & -\rho^2\,\mathrm{I} \\ & & & & & U_{n-1} \end{pmatrix}, \tag{5.79}$$

where the individual blocks are again of size $(m-1)\times(m-1)$. Indeed, multiplying out the matrix product LU and equating the result to (5.77) leads to the iterative matrix system

$$U_1 = B_\rho, \qquad L_j = -\rho^2 U_j^{-1}, \qquad U_{j+1} = B_\rho + \rho^2 L_j, \qquad j = 1, \ldots, n-2, \tag{5.80}$$

which produces the individual blocks.

With the LU factors in place, we can apply Forward and Back Substitution to solve the block tridiagonal linear system $A\mathbf{w} = \widehat{\mathbf{f}}$ by solving the block lower and upper triangular systems

$$L\,\mathbf{z} = \widehat{\mathbf{f}}, \qquad U\,\mathbf{w} = \mathbf{z}. \tag{5.81}$$

In view of the forms (5.79) of L and U, if we write

$$\mathbf{w} = \begin{pmatrix} \mathbf{w}^{(1)} \\ \mathbf{w}^{(2)} \\ \vdots \\ \mathbf{w}^{(n-1)} \end{pmatrix}, \qquad \mathbf{z} = \begin{pmatrix} \mathbf{z}^{(1)} \\ \mathbf{z}^{(2)} \\ \vdots \\ \mathbf{z}^{(n-1)} \end{pmatrix}, \qquad \widehat{\mathbf{f}} = \begin{pmatrix} \widehat{\mathbf{f}}^{(1)} \\ \widehat{\mathbf{f}}^{(2)} \\ \vdots \\ \widehat{\mathbf{f}}^{(n-1)} \end{pmatrix},$$

so that each $\mathbf{w}^{(j)}, \mathbf{z}^{(j)}, \widehat{\mathbf{f}}^{(j)}$, is a vector with $m-1$ entries, then we must successively solve

$$\begin{aligned} &\mathbf{z}^{(1)} = \widehat{\mathbf{f}}^{(1)}, && \mathbf{z}^{(j+1)} = \widehat{\mathbf{f}}^{(j+1)} - L_j \mathbf{z}^{(j)}, && j = 1, 2, \ldots, n-2, \\ &\mathbf{w}^{(n-1)} = \mathbf{z}^{(n-1)}, && U_j \mathbf{w}^{(k)} = \mathbf{z}^{(k)} - \rho^2\,\mathbf{w}^{(k+1)}, && k = n-2, n-3, \ldots, 1, \end{aligned} \tag{5.82}$$

in the prescribed order. In view of the identification of L_j with $-\rho^2$ times the inverse of U_j, the last set of equations in (5.82) is perhaps better written as

$$\mathbf{w}^{(k)} = L_j\big(\mathbf{w}^{(k+1)} - \rho^{-2}\mathbf{z}^{(k)}\big), \qquad k = n-2, n-3, \ldots, 1. \tag{5.83}$$

As the number of nodes becomes large, the preceding elimination/factorization approach to solving the linear system becomes increasingly inefficient, and one often switches to an iterative solution method such as Gauss–Seidel, Jacobi, or, even better, Successive Over–Relaxation (SOR); indeed, SOR was originally designed to speed up the solution of the large-scale linear systems arising from the numerical solution of elliptic partial differential equations. Detailed discussions of iterative matrix methods can be found in

[**89**; Chapter 10] and [**118**]. For the SOR method, a good choice for the relaxation parameter is

$$\omega = \frac{4}{2 + \sqrt{4 - \cos^2(\pi/m) - \cos^2(\pi/n)}}. \tag{5.84}$$

Iterative solution methods are even more attractive in dealing with irregular domains, whose finite difference coefficient matrix, while still sparse, is less structured than in the rectangular case, and hence less amenable to fast Gaussian Elimination algorithms.

Finally, let us address the question of unique solvability of the finite difference linear system obtained by discretization of the Poisson equation on a bounded domain subject to Dirichlet boundary conditions. As in the Uniqueness Theorem 4.10 for the original boundary value, this will follow from an easily established Maximum Principle for the discrete system that directly mimics the Laplace equation maximum principle of Theorem 4.9.

Theorem 5.8. *Let Ω be a bounded domain. Then the finite difference linear system* (5.74) *has a unique solution.*

Proof: The result will follow if we can prove that the only solution to the corresponding homogeneous linear system $A\mathbf{w} = \mathbf{0}$ is the trivial solution $\mathbf{w} = \mathbf{0}$. The homogeneous system corresponds to discretizing the Laplace equation subject to zero Dirichlet boundary conditions.

Now, in view of (5.71), each equation in the homogeneous linear system can be written in the form

$$u_{i,j} = \frac{u_{i-1,j} + u_{i+1,j} + \rho^2 u_{i,j-1} + \rho^2 u_{i,j+1}}{2(1+\rho^2)}. \tag{5.85}$$

If $\rho = 1$, then (5.85) says that the value of $u_{i,j}$ at the node (x_i, y_j) is equal to the *average* of the values at the four neighboring nodes. For general ρ, it says that $u_{i,j}$ is a *weighted average* of the four neighboring values. In either case, the value of $u_{i,j}$ must lie *strictly* between the maximum and minimum values of $u_{i-1,j}, u_{i+1,j}, u_{i,j-1}$ and $u_{i,j+1}$ — unless all these values are the same, in which case $u_{i,j}$ also has the same value. This observation suffices to establish a *Maximum Principle* for the finite difference system for the Laplace equation — namely, that its solution cannot achieve a local maximum or minimum at an interior node.

Now suppose that the homogeneous finite difference system $A\mathbf{w} = \mathbf{0}$ for the domain has a nontrivial solution $\mathbf{w} \neq \mathbf{0}$. Let $u_{i,j} = w_k$ be the maximal entry of this purported solution. The Maximum Principle requires that all four of its neighboring values must have the *same* maximal value. But then the same argument applies to the neighbors of those entries, to their neighbors, and so on. Eventually one of the neighbors is at a boundary node, but, since we are dealing with the homogeneous Dirichlet boundary value problem, its value is zero. This immediately implies that all the entries of $\mathbf{w}$ must be zero, which is a contradiction. *Q.E.D.*

Rigorously establishing convergence of the finite difference solution to the analytic solution to the boundary value problem as the step size goes to zero will not be discussed here, and we refer the reader to [**6**, **80**] for precise results and proofs.

Exercises

♠ 5.5.1. Solve the Dirichlet problem $\Delta u = 0$, $u(x,0) = \sin^3 x$, $u(x,\pi) = 0$, $u(0,y) = 0$, $u(\pi,y) = 0$, numerically using a finite difference scheme. Compare your approximation with the solution you obtained in Exercise 4.3.10(a).

♠ 5.5.2. Solve the Dirichlet problem $\Delta u = 0$, $u(x,0) = x$, $u(x,1) = 1 - x$, $u(0,y) = y$, $u(1,y) = 1 - y$, numerically via finite differences. Compare your approximation with the solution you obtained in Exercise 4.3.12(*d*).

♠ 5.5.3. Consider the Dirichlet boundary value problem $\Delta u = 0$ $u(x,0) = \sin x$, $u(x,\pi) = 0$, $u(0,y) = 0$, $u(\pi,y) = 0$, on the square $\{0 < x, y < \pi\}$. (a) Find the exact solution. (*b*) Set up and solve the finite difference equations based on a square mesh with $m = n = 2$ squares on each side of the full square. How close is this value to the exact solution at the center of the square: $u\left(\frac{1}{2}\pi, \frac{1}{2}\pi\right)$? (*c*) Repeat part (*b*) for $m = n = 4$ squares per side. Is the value of your approximation at the center of the unit square closer to the true solution? (*d*) Use a computer to find a finite difference approximation to $u\left(\frac{1}{2}\pi, \frac{1}{2}\pi\right)$ using $m = n = 8$ and 16 squares per side. Is your approximation converging to the exact solution as the mesh becomes finer and finer? Is the convergence rate consistent with the order of the finite difference approximation?

♠ 5.5.4. (a) Use finite differences to approximate a solution to the Helmholtz boundary value problem $\Delta u = u$, $u(x,0) = u(x,1) = u(0,y) = 0$, $u(1,y) = 1$, on the unit square $0 < x, y < 1$. (*b*) Use separation of variables to construct a series solution. Do your analytic and numerical solutions match? Explain any discrepancies.

♠ 5.5.5. A drum is in the shape of an L, as in the accompanying figure, whose short sides all have length 1. (a) Use a finite difference scheme with mesh spacing $\Delta x = \Delta y = .1$ to find and graph the equilibrium configuration when the drum is subject to a unit upwards force while all its sides are fixed to the (x,y)–plane. What is the maximal deflection, and at which point(s) does it occur? (*b*) Check the accuracy of your answer in part (a) by reducing the step size by half: $\Delta x = \Delta y = .05$.

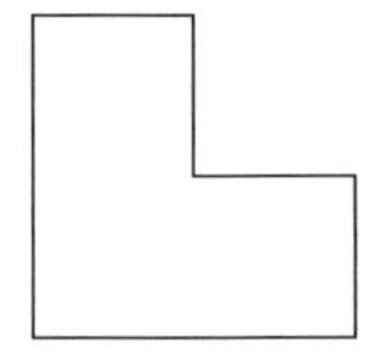

♣ 5.5.6. A metal plate has the shape of a 3 cm square with a 1 cm square hole cut out of the middle. The plate is heated by making the inner edge have temperature 100° while keeping the outer edge at 0°. (a) Find the (approximate) equilibrium temperature using finite differences with a mesh width of $\Delta x = \Delta y = .5$ cm. Plot your approximate solution using a three-dimensional graphics program. (*b*) Let C denote the square contour lying midway between the inner and outer square boundaries of the plate. Using your finite difference approximation, determine at what point(s) on C the temperature is (*i*) minimized; (*ii*) maximimized; (*iii*) equal to the average of the two boundary temperatures. (*c*) Repeat part (a) using a smaller mesh width of $\Delta x = \Delta y = .2$. How much does this affect your answers in part (*b*)?

♣ 5.5.7. Answer Exercise 5.5.6 when the plate is additionally subjected to a constant heat source

$$f(x,y) = 600x + 800y - 2400.$$

♠ 5.5.8. (a) Explain how to adapt the finite difference method to a mixed boundary value problem on a rectangle with inhomogeneous Neumann conditions. *Hint*: Use a one-sided difference formula of the appropriate order to approximate the normal derivative at the boundary. (*b*) Apply your method to the problem

$$\Delta u = 0, \qquad u(x,0) = 0, \qquad u(x,1) = 0, \qquad \frac{\partial u}{\partial x}(0,y) = y(1-y), \qquad u(1,y) = 0,$$

using mesh sizes $\Delta x = \Delta y = .1, .01$, and $.001$. Compare your answers. (*c*) Solve the boundary value problem via separation of variables, and compare the value of the solution and the numerical approximations at the center of the square.

Chapter 6
Generalized Functions and Green's Functions

Boundary value problems, involving both ordinary and partial differential equations, can be profitably viewed as the infinite-dimensional function space versions of finite-dimensional systems of linear algebraic equations. As a result, linear algebra not only provides us with important insights into their underlying mathematical structure, but also motivates both analytical and numerical solution techniques. In the present chapter, we develop the method of Green's functions, pioneered by the early-nineteenth-century self-taught English mathematician (and miller!) George Green, whose famous Theorem you already encountered in multivariable calculus. We begin with the simpler case of ordinary differential equations, and then move on to solving the two-dimensional Poisson equation, where the Green's function provides a powerful alternative to the method of separation of variables.

For inhomogeneous linear systems, the basic Superposition Principle says that the response to a combination of external forces is the self-same combination of responses to the individual forces. In a finite-dimensional system, any forcing function can be decomposed into a linear combination of unit impulse forces, each applied to a single component of the system, and so the full solution can be obtained by combining the solutions to the individual impulse problems. This simple idea will be adapted to boundary value problems governed by differential equations, where the response of the system to a concentrated impulse force is known as the Green's function. With the Green's function in hand, the solution to the inhomogeneous system with a general forcing function can be reconstructed by superimposing the effects of suitably scaled impulses. Understanding this construction will become increasingly important as we progress to partial differential equations, where direct analytic solution techniques are far harder to come by.

The obstruction blocking a direct implementation of this idea is that there is no ordinary function that represents an idealized concentrated impulse! Indeed, while this approach was pioneered by Green and Cauchy in the early 1800s, and then developed into an effective computational tool by Heaviside in the 1880s, it took another 60 years before mathematicians were able to develop a completely rigorous theory of *generalized functions*, also known as *distributions*. In the language of generalized functions, a unit impulse is represented by a *delta function*.† While we do not have the analytic tools to completely develop the mathematical theory of generalized functions in its full, rigorous glory, we will spend the first section learning the basic concepts and developing the practical computational skills, including Fourier methods, required for applications. The second

† *Warning*: We follow common practice and refer to the "delta distribution" as a function, even though, as we will see, it is most definitely not a function in the usual sense.

section will discuss the method of Green's functions in the context of one-dimensional boundary value problems governed by ordinary differential equations. In the final section, we develop the Green's function method for solving basic boundary value problems for the two-dimensional Poisson equation, which epitomizes the class of planar elliptic boundary value problems.

6.1 Generalized Functions

Our goal is to solve inhomogeneous linear boundary value problems by first determining the effect of a concentrated impulse force. The response to a general forcing function is then found by linear superposition. But before diving in, let us first review the relevant constructions in the case of linear systems of algebraic equations.

Consider a system of n linear equations in n unknowns† $\mathbf{u} = (u_1, u_2, \ldots, u_n)^T$, written in matrix form

$$A\mathbf{u} = \mathbf{f}. \tag{6.1}$$

Here A is a fixed $n \times n$ matrix, assumed to be nonsingular, which ensures the existence of a unique solution $\mathbf{u}$ for any choice of right-hand side $\mathbf{f} = (f_1, f_2, \ldots, f_n)^T \in \mathbb{R}^n$. We regard the linear system (6.1) as representing the equilibrium equations of some physical system, e.g., a system of masses interconnected by springs. In this context, the right hand side $\mathbf{f}$ represents an external forcing, so that its ith entry, f_i, represents the amount of force exerted on the ith mass, while the ith entry of the solution vector, u_i, represents the ith mass' induced displacement.

Let

$$\mathbf{e}_1 = \begin{pmatrix} 1 \\ 0 \\ 0 \\ \vdots \\ 0 \\ 0 \end{pmatrix}, \qquad \mathbf{e}_2 = \begin{pmatrix} 0 \\ 1 \\ 0 \\ \vdots \\ 0 \\ 0 \end{pmatrix}, \qquad \cdots \qquad \mathbf{e}_n = \begin{pmatrix} 0 \\ 0 \\ 0 \\ \vdots \\ 0 \\ 1 \end{pmatrix}, \tag{6.2}$$

denote the *standard basis vectors* of $\mathbb{R}^n$, so that $\mathbf{e}_j$ has a single 1 in its jth entry and all other entries 0. We interpret each $\mathbf{e}_j$ as a concentrated *unit impulse force* that is applied solely to the jth mass in our physical system. Let $\mathbf{u}_j = (u_{j,1}, \ldots, u_{j,n})^T$ be the induced response of the system, that is, the solution to

$$A\mathbf{u}_j = \mathbf{e}_j. \tag{6.3}$$

Let us suppose that we have calculated the response vectors $\mathbf{u}_1, \ldots, \mathbf{u}_n$ to each such impulse force. We can express any other force vector as a linear combination,

$$\mathbf{f} = \begin{pmatrix} f_1 \\ f_2 \\ \vdots \\ f_n \end{pmatrix} = f_1\mathbf{e}_1 + f_2\mathbf{e}_2 + \cdots + f_n\mathbf{e}_n, \tag{6.4}$$

† All vectors are column vectors, but we sometimes write the transpose, which is a row vector, to save space.

of the impulse forces. The Superposition Principle of Theorem 1.7 then implies that the solution to the inhomogeneous system (6.1) is the selfsame linear combination of the individual impulse responses:

$$\mathbf{u} = f_1 \mathbf{u}_1 + f_2 \mathbf{u}_2 + \cdots + f_n \mathbf{u}_n. \tag{6.5}$$

Thus, knowing how the linear system responds to each impulse force allows us to immediately calculate its response to a general external force.

Remark: The alert reader will recognize that $\mathbf{u}_1, \ldots, \mathbf{u}_n$ are the columns of the inverse matrix, A^{-1}, and so formula (6.5) is, in fact, reconstructing the solution to the linear system (6.1) by inverting its coefficient matrix: $\mathbf{u} = A^{-1}\mathbf{f}$. Thus, this observation is merely a restatement of a standard linear algebraic system solution technique.

The Delta Function

The aim of this chapter is to adapt the preceding algebraic solution technique to boundary value problems. Suppose we want to solve a linear boundary value problem governed by an ordinary differential equation on an interval $a < x < b$, the boundary conditions being imposed at the endpoints. The key issue is how to characterize an impulse force that is concentrated at a single point.

In general, a *unit impulse* at position $a < \xi < b$ will be described by something called the *delta function*, and denoted by $\delta_\xi(x)$. Since the impulse is supposed to be concentrated solely at $x = \xi$, our first requirement is

$$\delta_\xi(x) = 0 \qquad \text{for} \qquad x \neq \xi. \tag{6.6}$$

Moreover, since the delta function represents a *unit* impulse, we want the total amount of force to be equal to one. Since we are dealing with a continuum, the total force is represented by an integral over the entire interval, and so we also require that the delta function satisfy

$$\int_a^b \delta_\xi(x)\,dx = 1, \qquad \text{provided} \qquad a < \xi < b. \tag{6.7}$$

Alas, there is no bona fide function that enjoys both of the required properties! Indeed, according to the basic facts of Riemann (or even Lebesgue) integration, two functions that are the same everywhere except at a single point have exactly the same integral, [**96, 98**]. Thus, since δ_ξ is zero except at one point, its integral should be 0, not 1. The mathematical conclusion is that the two requirements, (6.6–7) are inconsistent!

This unfortunate fact stopped mathematicians dead in their tracks. It took the imagination of a British engineer, Oliver Heaviside, who was not deterred by the lack of rigorous justification, to start utilizing delta functions in practical applications — with remarkable effect. Despite his success, Heaviside was ridiculed by the mathematicians of his day, and eventually succumbed to mental illness. But, some thirty years later, the great British theoretical physicist Paul Dirac resurrected the delta function for quantum-mechanical applications, and this finally made the mathematicians sit up and take notice. (Indeed, the term "Dirac delta function" is quite common, even though Heaviside should rightly have priority.) In 1944, the French mathematician Laurent Schwartz finally established a rigorous theory of *distributions* that incorporated such useful but nonstandard objects, [**103**]. Thus, to be more accurate, we should really refer to the *delta distribution*; however, we will retain the more common, intuitive designation "delta function" throughout. It is beyond the scope of this introductory text to develop a fully rigorous theory of distributions. Rather, in the spirit of Heaviside, we shall concentrate on learning, through practice with computations and applications, how to make effective use of these exotic mathematical creatures.

There are two possible ways to introduce the delta distribution. Both are important and worth understanding.

Method #1. Limits: The first approach is to regard the delta function $\delta_\xi(x)$ as a limit of a sequence of ordinary smooth functions† $g_n(x)$. These will represent progressively more and more concentrated unit forces, which, in the limit, converge to the desired unit impulse concentrated at a single point, $x = \xi$. Thus, we require

$$\lim_{n \to \infty} g_n(x) = 0, \qquad x \neq \xi, \tag{6.8}$$

while the total amount of force remains fixed at

$$\int_a^b g_n(x)\, dx = 1 \qquad \text{for all } n. \tag{6.9}$$

On a formal level, the limit "function"

$$\delta_\xi(x) = \lim_{n \to \infty} g_n(x)$$

will satisfy the key properties (6.6–7).

An explicit example of such a sequence is provided by the rational functions

$$g_n(x) = \frac{n}{\pi(1 + n^2 x^2)}. \tag{6.10}$$

These functions satisfy

$$\lim_{n \to \infty} g_n(x) = \begin{cases} 0, & x \neq 0, \\ \infty, & x = 0, \end{cases} \tag{6.11}$$

while‡

$$\int_{-\infty}^{\infty} g_n(x)\, dx = \frac{1}{\pi} \tan^{-1} nx \,\bigg|_{x=-\infty}^{\infty} = 1. \tag{6.12}$$

Therefore, formally, we identify the limiting function

$$\lim_{n \to \infty} g_n(x) = \delta(x) = \delta_0(x) \tag{6.13}$$

with the unit-impulse delta function concentrated at $x = 0$. As sketched in Figure 6.1, as n gets larger and larger, each successive function $g_n(x)$ forms a more and more concentrated spike, while maintaining a unit total area under its graph. Thus, the limiting delta function can be thought of as an infinitely tall spike of zero width, entirely concentrated at the origin.

† To keep the notation compact, we suppress the dependence of the functions g_n on the point ξ where the limiting delta function is concentrated.

‡ For the moment, it will be slightly simpler to consider the entire real line $-\infty < x < \infty$. Exercise 6.1.8 discusses how to adapt the construction to a finite interval.

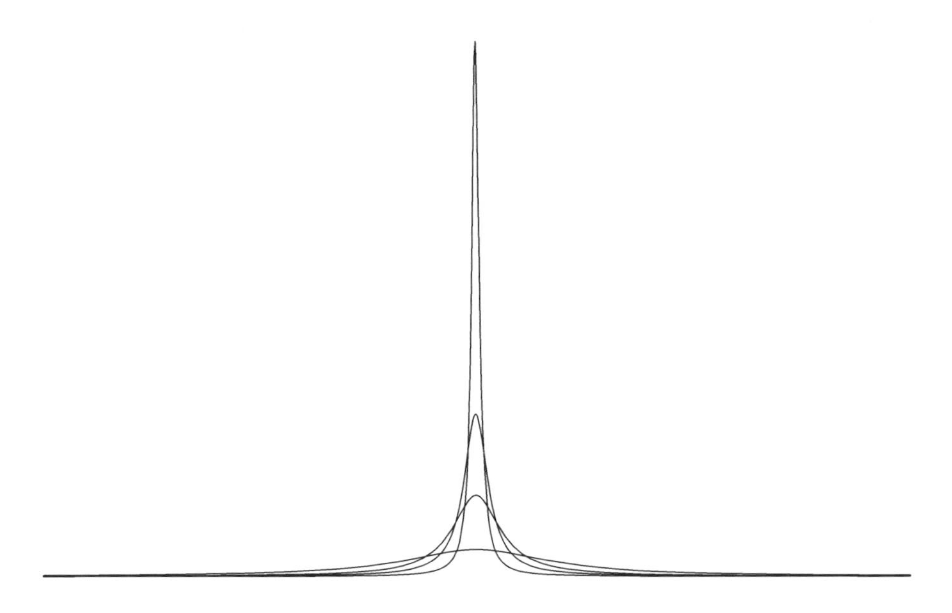

Figure 6.1. Delta function as limit.

Remark: There are many other possible choices for the limiting functions $g_n(x)$. See Exercise 6.1.7 for another important example.

Remark: This construction of the delta function highlights the perils of interchanging limits and integrals without rigorous justification. In any standard theory of integration (Riemann, Lebesgue, etc.), the limit of the functions g_n would be indistinguishable from the zero function, so the limit of their integrals (6.12) would *not* equal the integral of their limit:

$$1 = \lim_{n\to\infty} \int_{-\infty}^{\infty} g_n(x)\,dx \neq \int_{-\infty}^{\infty} \lim_{n\to\infty} g_n(x)\,dx = 0.$$

The delta function is, in a sense, a means of sidestepping this analytic inconvenience. The full ramifications and theoretical constructions underlying such limits must, however, be deferred to a rigorous course in real analysis, [**96, 98**].

Once we have defined the basic delta function $\delta(x) = \delta_0(x)$ concentrated at the origin, we can obtain the delta function concentrated at any other position ξ by a simple translation:

$$\delta_\xi(x) = \delta(x-\xi). \tag{6.14}$$

Thus, $\delta_\xi(x)$ can be realized as the limit, as $n \to \infty$, of the translated functions

$$\widehat{g}_n(x) = g_n(x-\xi) = \frac{n}{\pi\left[\,1 + n^2(x-\xi)^2\,\right]}\,. \tag{6.15}$$

Method #2. Duality: The second approach is a bit more abstract, but much closer in spirit to the proper rigorous formulation of the theory of distributions like the delta function. The critical property is that if $u(x)$ is any continuous function, then

$$\int_a^b \delta_\xi(x)\,u(x)\,dx = u(\xi), \qquad \text{for} \qquad a < \xi < b. \tag{6.16}$$

Indeed, since $\delta_\xi(x) = 0$ for $x \neq \xi$, the integrand depends only on the value of u at the point $x = \xi$, and so

$$\int_a^b \delta_\xi(x)\, u(x)\, dx = \int_a^b \delta_\xi(x)\, u(\xi)\, dx = u(\xi) \int_a^b \delta_\xi(x)\, dx = u(\xi).$$

Equation (6.16) serves to define a linear functional† $L_\xi \colon \mathrm{C}^0[a,b] \to \mathbb{R}$ that maps a continuous function $u \in \mathrm{C}^0[a,b]$ to its value at the point $x = \xi$:

$$L_\xi[u] = u(\xi). \tag{6.17}$$

The basic linearity requirements (1.11) are immediately established:

$$L_\xi[u+v] = u(\xi) + v(\xi) = L_\xi[u] + L_\xi[v], \qquad L_\xi[cu] = cu(\xi) = c\, L_\xi[u],$$

for any functions $u(x), v(x)$. In the dual approach to generalized functions, the delta function is, in fact, *defined* as this particular linear functional (6.17). The function $u(x)$ is sometimes referred to as a *test function*, since it serves to "test" the form of the linear functional L_ξ.

Remark: If the impulse point ξ lies outside the integration domain, then

$$\int_a^b \delta_\xi(x)\, u(x)\, dx = 0 \qquad \text{whenever} \qquad \xi < a \qquad \text{or} \qquad \xi > b, \tag{6.18}$$

because the integrand is identically zero on the entire interval. For technical reasons, we will not attempt to define the integral (6.18) if the impulse point $\xi = a$ or $\xi = b$ lies on the boundary of the interval of integration.

The interpretation of the linear functional L_ξ as representing a kind of function $\delta_\xi(x)$ is based on the following line of thought. According to Corollary B.34, every scalar-valued linear function $L\colon \mathbb{R}^n \to \mathbb{R}$ on the finite-dimensional vector space $\mathbb{R}^n$ is obtained by taking the dot product with a fixed element $\mathbf{a} \in \mathbb{R}^n$, so

$$L[\mathbf{u}] = \mathbf{a} \cdot \mathbf{u}.$$

In this sense, linear functions on $\mathbb{R}^n$ are the "same" as vectors. Similarly, on the infinite-dimensional function space $\mathrm{C}^0[a,b]$, the L^2 inner product

$$L_g[u] = \langle\, g\,, u\,\rangle = \int_a^b g(x)\, u(x)\, dx, \tag{6.19}$$

taken with a fixed continuous function $g \in \mathrm{C}^0[a,b]$, defines a real-valued linear functional $L_g\colon \mathrm{C}^0[a,b] \to \mathbb{R}$. However, unlike the finite-dimensional situation, *not* every real-valued linear functional is of this form! In particular, there is no bona fide function $\delta_\xi(x)$ such that the identity

$$L_\xi[u] = \langle\, \delta_\xi\,, u\,\rangle = \int_a^b \delta_\xi(x)\, u(x)\, dx = u(\xi) \tag{6.20}$$

holds for every continuous function $u(x)$. The bottom line is that every (continuous) function defines a linear functional, but not every linear functional arises in this manner.

† The term "functional" is used to refer to a linear function whose domain is a function space, thus avoiding confusion with the functions it acts on.

But the dual interpretation of generalized functions acts as if this were true. *Generalized functions are, in actuality, real-valued linear functionals on function space, but intuitively interpreted as a kind of function via the* L^2 *inner product*. Although this identification is not to be taken too literally, one can, with some care, manipulate generalized functions as if they were actual functions, but always keeping in mind that a rigorous justification of such computations must ultimately rely on their innate characterization as linear functionals.

The two approaches — limits and duality — are completely compatible. Indeed, one can recover the dual formula (6.20) as the limit

$$u(\xi) = \lim_{n \to \infty} \langle\, g_n\,, u \,\rangle = \lim_{n \to \infty} \int_a^b g_n(x)\, u(x)\, dx = \int_a^b \delta_\xi(x)\, u(x)\, dx = \langle\, \delta_\xi\,, u \,\rangle \tag{6.21}$$

of the inner products of the function u with the approximating concentrated impulse functions $g_n(x)$ satisfying (6.8–9). In this manner, the limiting linear functional represents the delta function:

$$u(\xi) = L_\xi[\,u\,] = \lim_{n \to \infty} L_n[\,u\,], \qquad \text{where} \qquad L_n[\,u\,] = \int_0^\ell g_n(x)\, u(x)\, dx.$$

The choice of interpretation of the generalized delta function is, at least on an operational level, a matter of taste. For the beginner, the limit version is perhaps easier to digest initially. However, the dual, linear functional interpretation has stronger connections with the rigorous theory and, even in applications, offers some significant advantages.

Although the delta function might strike you as somewhat bizarre, its utility throughout modern applied mathematics and mathematical physics more than justifies including it in your analytical toolbox. While probably not yet comfortable with either definition, you are advised to press on and familiarize yourself with its basic properties. With a little care, you usually won't go far wrong by treating it as if it were a genuine function. After you gain more practical experience, you can, if desired, return to contemplate just exactly what kind of creature the delta function really is.

Calculus of Generalized Functions

In order to make use of the delta function, we need to understand how it behaves under the basic operations of linear algebra and calculus. First, we can take linear combinations of delta functions. For example,

$$h(x) = 2\,\delta(x) - 3\,\delta(x-1) = 2\,\delta_0(x) - 3\,\delta_1(x)$$

represents a combination of an impulse of magnitude 2 concentrated at $x = 0$ and one of magnitude -3 concentrated at $x = 1$. In the dual interpretation, h defines the linear functional

$$L_h[\,u\,] = \langle\, h\,, u \,\rangle = \langle\, 2\,\delta_0 - 3\,\delta_1\,, u \,\rangle = 2\,\langle\, \delta_0\,, u \,\rangle - 3\,\langle\, \delta_1\,, u \,\rangle = 2\,u(0) - 3\,u(1),$$

or, more explicitly, provided $a < 0$ and $b > 1$,

$$\begin{aligned} L_h[\,u\,] = \int_a^b h(x)\, u(x)\, dx &= \int_a^b \big[\, 2\,\delta(x) - 3\,\delta(x-1) \,\big] u(x)\, dx \\ &= 2 \int_a^b \delta(x)\, u(x)\, dx - 3 \int_a^b \delta(x-1)\, u(x)\, dx = 2\,u(0) - 3\,u(1). \end{aligned}$$

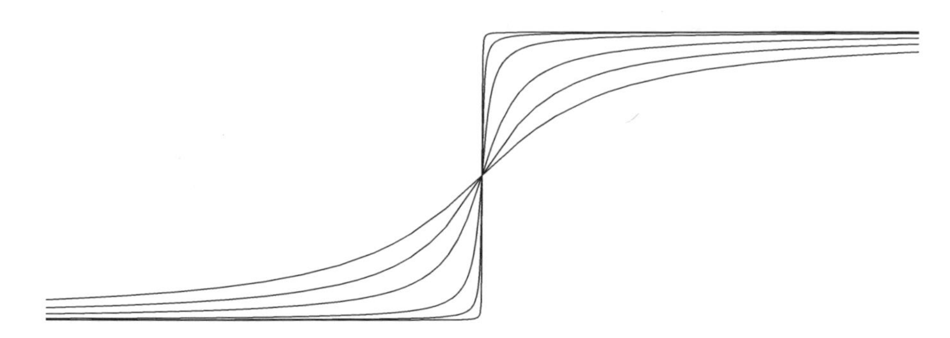

Figure 6.2. Step function as limit.

Next, since $\delta_\xi(x) = 0$ for any $x \neq \xi$, multiplying the delta function by an ordinary function is the same as multiplying by a constant:

$$g(x)\,\delta_\xi(x) = g(\xi)\,\delta_\xi(x), \tag{6.22}$$

provided $g(x)$ is continuous at $x = \xi$. For example, $x\,\delta(x) \equiv 0$ is the same as the constant zero function.

Warning: Since they are inherently *linear* functionals, it is *not* permissible to multiply delta functions together, or to apply more complicated *nonlinear* operations to them. Expressions like $\delta(x)^2$, $1/\delta(x)$, $e^{\delta(x)}$, etc., are *not* well defined in the theory of generalized functions — although this makes their application to nonlinear differential equations problematic.

The integral of the delta function is the *unit step function*:

$$\int_a^x \delta_\xi(t)\,dt = \sigma_\xi(x) = \sigma(x-\xi) = \begin{cases} 0, & x < \xi, \\ 1, & x > \xi, \end{cases} \qquad \text{provided} \qquad a < \xi. \tag{6.23}$$

Unlike the delta function, the step function $\sigma_\xi(x)$ is an ordinary function. It is continuous — indeed constant — except at $x = \xi$. The value of the step function at the discontinuity $x = \xi$ is left unspecified, although a wise choice — compatible with Fourier theory — is to set $\sigma_\xi(y) = \frac{1}{2}$, the average of its left- and right-hand limits.

We note that the integration formula (6.23) is compatible with our characterization of the delta function as the limit of highly concentrated forces. Integrating the approximating functions (6.10), we obtain

$$f_n(x) = \int_{-\infty}^x g_n(t)\,dt = \frac{1}{\pi}\,\tan^{-1} n\,x + \frac{1}{2}\,.$$

Since

$$\lim_{y\to\infty} \tan^{-1} y = \tfrac{1}{2}\pi, \qquad \text{while} \qquad \lim_{y\to-\infty} \tan^{-1} y = -\tfrac{1}{2}\pi,$$

these functions converge (nonuniformly) to the step function:

$$\lim_{n\to\infty} f_n(x) = \sigma(x) = \begin{cases} 0, & x < 0, \\ \frac{1}{2}, & x = 0, \\ 1, & x > 0. \end{cases} \tag{6.24}$$

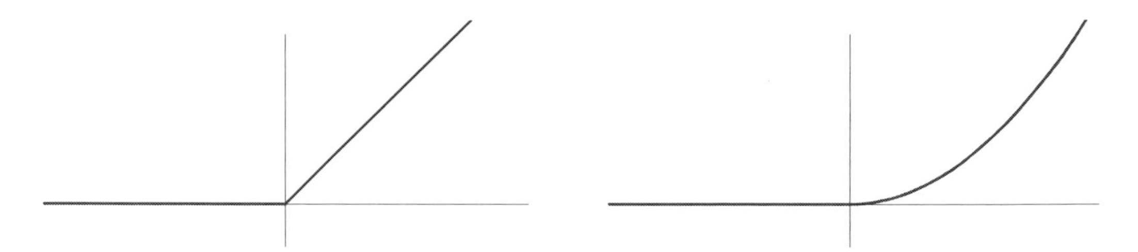

Figure 6.3. First and second-order ramp functions.

A graphical illustration of this limiting process appears in Figure 6.2.

The integral of the discontinuous step function (6.23) is the continuous *ramp function*

$$\int_a^x \sigma_\xi(t)\,dt = \rho_\xi(x) = \rho(x-\xi) = \begin{cases} 0, & x<\xi, \\ x-\xi, & x>\xi, \end{cases} \qquad \text{provided} \qquad a<\xi, \tag{6.25}$$

which is graphed in Figure 6.3. Note that $\rho_\xi(x)$ has a corner at $x=\xi$, and so is not differentiable there; indeed, its derivative $\rho'(x-\xi)=\sigma(x-\xi)$ has a jump discontinuity. We can continue to integrate; the $(n+1)^{\text{st}}$ integral of the delta function is the nth *order ramp function*

$$\rho_{n,\xi}(x) = \rho_n(x-\xi) = \begin{cases} 0, & x<\xi, \\ \dfrac{(x-\xi)^n}{n!}, & x>\xi. \end{cases} \tag{6.26}$$

Note that $\rho_{n,\xi}\in \mathrm{C}^{n-1}$ has only $n-1$ continuous derivatives.

What about differentiation? Motivated by the Fundamental Theorem of Calculus, we shall use formula (6.23) to identify the derivative of the step function with the delta function

$$\frac{d\sigma}{dx} = \delta. \tag{6.27}$$

This fact is highly significant. In elementary calculus, one is not allowed to differentiate a discontinuous function. Here, we discover that the derivative can be defined, not as an ordinary function, but rather as a generalized delta function!

In general, the derivative of a piecewise C^1 function with jump discontinuities is a generalized function that includes a delta function concentrated at each discontinuity, whose magnitude equals the jump magnitude. More explicitly, suppose that $f(x)$ is differentiable, in the usual calculus sense, everywhere except at a point ξ, where it has a jump discontinuity of magnitude β. Using the step function (3.47), we can re-express

$$f(x) = g(x) + \beta\,\sigma(x-\xi), \tag{6.28}$$

where $g(x)$ is continuous everywhere, with a removable discontinuity at $x=\xi$, and differentiable except possibly at the jump. Differentiating (6.28), we find that

$$f'(x) = g'(x) + \beta\,\delta(x-\xi) \tag{6.29}$$

has a delta spike of magnitude β at the discontinuity. Thus, the derivatives of f and g coincide everywhere except at the discontinuity.

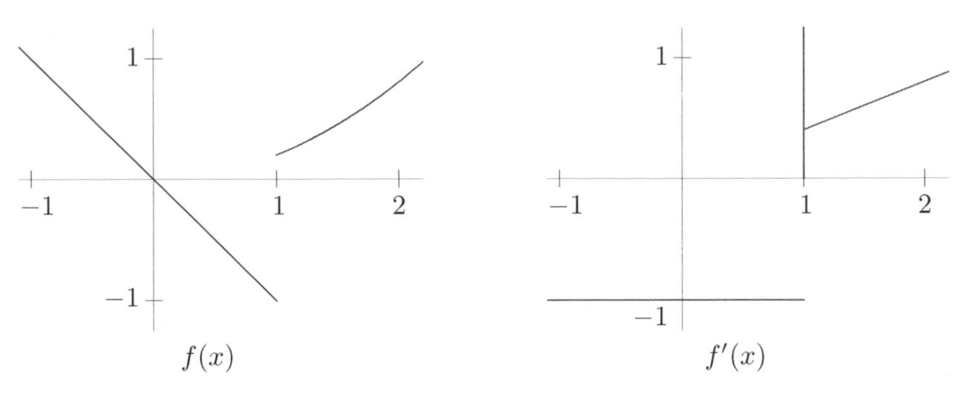

Figure 6.4. The derivative of the discontinuous function in Example 6.1.

Example 6.1. Consider the function

$$f(x) = \begin{cases} -x, & x < 1, \\ \frac{1}{5}x^2, & x > 1, \end{cases} \tag{6.30}$$

which we graph in Figure 6.4. We note that f has a single jump discontinuity at $x = 1$ of magnitude

$$f(1^+) - f(1^-) = \tfrac{1}{5} - (-1) = \tfrac{6}{5}.$$

This means that

$$f(x) = g(x) + \tfrac{6}{5}\,\sigma(x-1), \qquad \text{where} \qquad g(x) = \begin{cases} -x, & x < 1, \\ \frac{1}{5}x^2 - \frac{6}{5}, & x > 1, \end{cases}$$

is continuous everywhere, since its right- and left-hand limits at the original discontinuity are equal: $g(1^+) = g(1^-) = -1$. Therefore,

$$f'(x) = g'(x) + \tfrac{6}{5}\,\delta(x-1), \qquad \text{where} \qquad g'(x) = \begin{cases} -1, & x < 1, \\ \frac{2}{5}x, & x > 1, \end{cases}$$

while $g'(1)$ and $f'(1)$ are not defined. In Figure 6.4, the delta spike in the derivative of f is symbolized by a vertical line, although this pictorial device fails to indicate its magnitude of $\frac{6}{5}$.

Note that in this particular example, $g'(x)$ can be found by directly differentiating the formula for $f(x)$. Indeed, in general, once we determine the magnitude and location of the jump discontinuities of $f(x)$, we can compute its derivative without introducing the auxiliary function $g(x)$.

Example 6.2. As a second, more streamlined, example, consider the function

$$f(x) = \begin{cases} -x, & x < 0, \\ x^2 - 1, & 0 < x < 1, \\ 2\,e^{-x}, & x > 1, \end{cases}$$

which is plotted in Figure 6.5. This function has jump discontinuities of magnitude -1 at $x = 0$, and of magnitude $2/e$ at $x = 1$. Therefore, in light of the preceding remark,

$$f'(x) = -\,\delta(x) + \frac{2}{e}\,\delta(x-1) + \begin{cases} -1, & x < 0, \\ 2\,x, & 0 < x < 1, \\ -2\,e^{-x}, & x > 1, \end{cases}$$

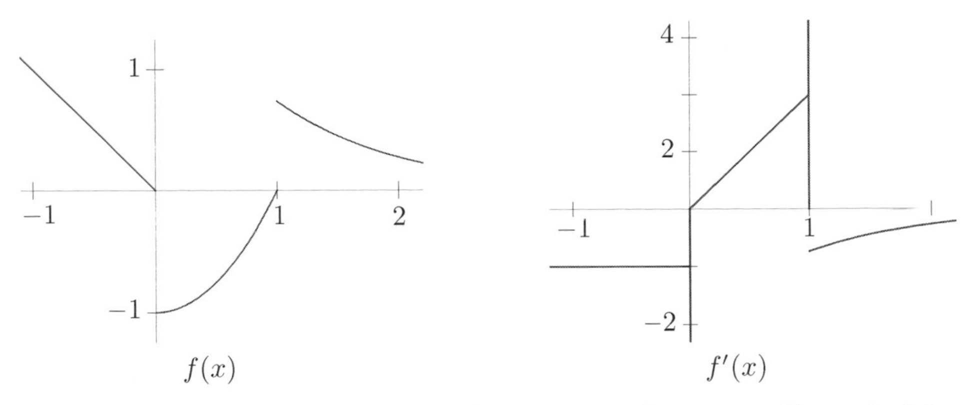

Figure 6.5. The derivative of the discontinuous function in Example 6.2.

where the final terms are obtained by directly differentiating $f(x)$.

Example 6.3. The derivative of the absolute value function

$$a(x) = |\, x \,| = \begin{cases} x, & x > 0, \\ -x, & x < 0, \end{cases}$$

is the *sign function*

$$a'(x) = \operatorname{sign} x = \begin{cases} +1, & x > 0, \\ -1, & x < 0. \end{cases} \tag{6.31}$$

Note that there is no delta function in $a'(x)$ because $a(x)$ is continuous everywhere. Since $\operatorname{sign} x$ has a jump of magnitude 2 at the origin and is otherwise constant, its derivative is twice the delta function:

$$a''(x) = \frac{d}{dx} \operatorname{sign} x = 2\,\delta(x).$$

Example 6.4. We are even allowed to differentiate the delta function. Its first derivative $\delta'(x)$ can be interpreted in two ways. First, as the limit of the derivatives of the approximating functions (6.10):

$$\frac{d\delta}{dx} = \lim_{n \to \infty} \frac{dg_n}{dx} = \lim_{n \to \infty} \frac{-2n^3 x}{\pi(1 + n^2 x^2)^2}\,. \tag{6.32}$$

The graphs of these rational functions take the form of more and more concentrated spiked "doublets", as illustrated in Figure 6.6. To determine the effect of the derivative on a test function $u(x)$, we compute the limiting integral

$$\begin{aligned} \langle\, \delta'\,, u \,\rangle &= \int_{-\infty}^{\infty} \delta'(x)\, u(x)\, dx = \lim_{n \to \infty} \int_{-\infty}^{\infty} g_n'(x)\, u(x)\, dx \\ &= -\lim_{n \to \infty} \int_{-\infty}^{\infty} g_n(x)\, u'(x)\, dx = -\int_{-\infty}^{\infty} \delta(x)\, u'(x)\, dx = -\,u'(0). \end{aligned} \tag{6.33}$$

The middle step is the result of an integration by parts, noting that the boundary terms at $\pm\infty$ vanish, provided that $u(x)$ is continuously differentiable and bounded as $|\, x \,| \to \infty$. Pay attention to the minus sign in the final answer.

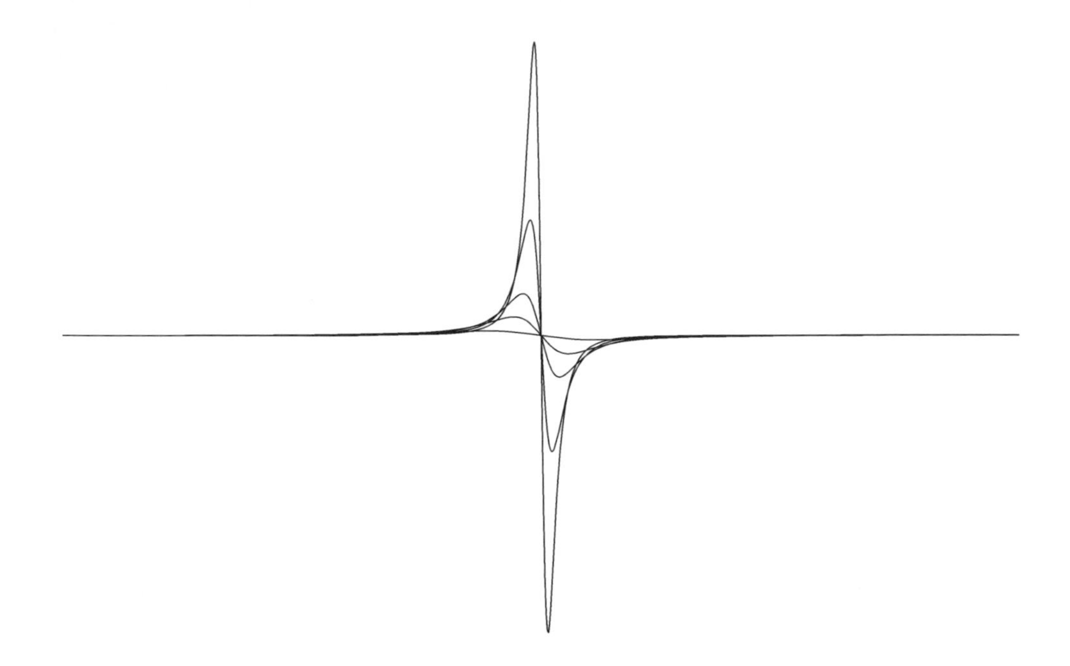

Figure 6.6. Derivative of delta function as limit of doublets.

In the dual interpretation, the generalized function $\delta'(x)$ corresponds to the linear functional

$$L'[u] = -\,u'(0) = \langle\, \delta'\,, u\,\rangle = \int_a^b \delta'(x)\,u(x)\,dx, \qquad \text{where} \qquad a < 0 < b, \tag{6.34}$$

which maps a continuously differentiable function $u(x)$ to *minus* its derivative at the origin. We note that (6.34) is compatible with a formal integration by parts:

$$\int_a^b \delta'(x)\,u(x)\,dx = \delta(x)\,u(x)\,\Big|_{x=a}^{b} - \int_a^b \delta(x)\,u'(x)\,dx = -\,u'(0).$$

The boundary terms at $x = a$ and $x = b$ automatically vanish, since $\delta(x) = 0$ for $x \neq 0$.

Remark: While we can test the delta function with any continuous function, we are permitted to test its derivative only on continuously differentiable functions. To avoid keeping track of such technicalities, one often restricts to only infinitely differentiable test functions.

Warning: The functions $\widetilde{g}_n(x) = g_n(x) + g_n'(x)$, cf. (6.10, 32), satisfy $\lim_{n\to\infty} \widetilde{g}_n(x) = 0$ for all $x \neq 0$, while $\displaystyle\int_{-\infty}^{\infty} \widetilde{g}_n(x)\,dx = 1$. However, $\lim_{n\to\infty} \widetilde{g}_n = \lim_{n\to\infty} g_n + \lim_{n\to\infty} g_n' = \delta + \delta'$. Thus, our original conditions (6.8–9) are *not* in fact sufficient to characterize whether a sequence of functions has the delta function as a limit. To be absolutely sure, one must, in fact, verify the more comprehensive limiting formula (6.21).

Exercises

6.1.1. Evaluate the following integrals: (a) $\int_{-\pi}^{\pi} \delta(x)\,\cos x\,dx$, (b) $\int_{1}^{2} \delta(x)\,(x-2)\,dx$,
(c) $\int_{0}^{3} \delta_1(x)\,e^x\,dx$, (d) $\int_{1}^{e} \delta(x-2)\,\log x\,dx$, (e) $\int_{0}^{1} \delta\bigl(x-\frac{1}{3}\bigr)\,x^2\,dx$, (f) $\int_{-1}^{1} \dfrac{\delta(x+2)\,dx}{1+x^2}$.

6.1.2. Simplify the following generalized functions; then write out how they act on a suitable test function $u(x)$: (a) $e^x\,\delta(x)$, (b) $x\,\delta(x-1)$, (c) $3\,\delta_1(x) - 3\,x\,\delta_{-1}(x)$,
(d) $\dfrac{\delta(x-1)}{x+1}$, (e) $(\cos x)\bigl[\,\delta(x)+\delta(x-\pi)+\delta(x+\pi)\bigr]$, (f) $\dfrac{\delta_1(x)-\delta_2(x)}{x^2+1}$.

6.1.3. Define the generalized function $\varphi(x) = \delta(x+1) - \delta(x-1)$:
(a) as a limit of ordinary functions; (b) using duality.

6.1.4. Find and sketch a graph of the derivative (in the context of generalized functions) of the following functions:

(a) $f(x) = \begin{cases} x^2, & 0 < x < 3, \\ x, & -1 < x < 0, \\ 0, & \text{otherwise}, \end{cases}$ (b) $g(x) = \begin{cases} \sin|\,x\,|, & |\,x\,| < \frac{1}{2}\pi, \\ 0, & \text{otherwise}, \end{cases}$

(c) $h(x) = \begin{cases} \sin \pi x, & x > 1, \\ 1 - x^2, & -1 < x < 1, \\ e^x, & x < -1, \end{cases}$ (d) $k(x) = \begin{cases} \sin x, & x < -\pi, \\ x^2 - \pi^2, & -\pi < x < 0, \\ e^{-x}, & x > 0. \end{cases}$

6.1.5. Find the first and second derivatives of the functions (a) $f(x) = \begin{cases} x+1, & -1 < x < 0, \\ 1-x, & 0 < x < 1, \\ 0, & \text{otherwise}, \end{cases}$

(b) $k(x) = \begin{cases} |\,x\,|, & -2 < x < 2, \\ 0, & \text{otherwise}, \end{cases}$ (c) $s(x) = \begin{cases} 1 + \cos \pi x, & -1 < x < 1, \\ 0, & \text{otherwise}. \end{cases}$

6.1.6. Find the first and second derivatives of $f(x) =$ (a) $e^{-|\,x\,|}$, (b) $2\,|\,x\,| - |\,x-1\,|$,
(c) $|\,x^2 + x\,|$, (d) $x\,\operatorname{sign}(x^2 - 4)$, (e) $\sin|\,x\,|$, (f) $|\sin x\,|$, (g) $\operatorname{sign}(\sin x)$.

♢ 6.1.7. Explain why the Gaussian functions $g_n(x) = \dfrac{n}{\sqrt{\pi}}\, e^{-n^2 x^2}$ have the delta function $\delta(x)$ as their limit as $n \to \infty$.

♢ 6.1.8. In this exercise, we realize the delta function $\delta_\xi(x)$ as a limit of functions on a finite interval $[\,a,b\,]$. Let $a < \xi < b$.

(a) Prove that the functions $\widetilde{g}_n(x) = \dfrac{g_n(x-\xi)}{M_n}$, where $g_n(x)$ is given by (6.10) and $M_n = \int_a^b g_n(x-\xi)\,dx$, satisfy (6.8–9), and hence $\lim_{n\to\infty} \widetilde{g}_n(x) = \delta_\xi(x)$.

(b) One can, alternatively, relax the second condition (6.9) to $\lim_{n\to\infty} \int_a^b g_n(x-\xi)\,dx = 1$. Show that, under this relaxed definition, $\lim_{n\to\infty} g_n(x-\xi) = \delta_\xi(x)$.

♡ 6.1.9. For each positive integer n, let $g_n(x) = \begin{cases} \frac{1}{2}n, & |\,x\,| < 1/n, \\ 0, & \text{otherwise}. \end{cases}$ (a) Sketch a graph of $g_n(x)$. (b) Show that $\lim_{n\to\infty} g_n(x) = \delta(x)$. (c) Evaluate $f_n(x) = \int_{-\infty}^{x} g_n(y)\,dy$ and sketch a graph. Does the sequence $f_n(x)$ converge to the step function $\sigma(x)$ as $n \to \infty$? (d) Find the derivative $h_n(x) = g_n'(x)$. (e) Does the sequence $h_n(x)$ converge to $\delta'(x)$ as $n \to \infty$?

♡ 6.1.10. Answer Exercise 6.1.9 for the *hat functions* $g_n(x) = \begin{cases} n - n^2\,|\,x\,|, & |\,x\,| < 1/n, \\ 0, & \text{otherwise}. \end{cases}$

6.1.11. Justify the formula $x\,\delta(x) = 0$ using (a) limits, (b) duality.

♢ 6.1.12. (a) Justify the formula $\delta(2x) = \frac{1}{2}\,\delta(x)$ by (i) limits, (ii) duality. (b) Find a similar formula for $\delta(a\,x)$ when $a > 0$. (c) What about when $a < 0$?

6.1.13. (a) Prove that $\sigma(\lambda x) = \sigma(x)$ for any $\lambda > 0$. (b) What about if $\lambda < 0$? (c) Use parts (a,b) to deduce that $\delta(\lambda x) = \dfrac{1}{|\,\lambda\,|}\,\delta(x)$ for any $\lambda \neq 0$.

6.1.14. Let $g(x)$ be a continuously differentiable function with $g'(x) \neq 0$ for all $x \in \mathbb{R}$. Does the composition $\delta(g(x))$ make sense as a distribution? If so, can you identify it?

6.1.15. Let $\xi < a$. Sketch the graphs of (a) $s(x) = \displaystyle\int_a^x \delta_\xi(z)\,dz$, (b) $r(x) = \displaystyle\int_a^x \sigma_\xi(z)\,dz$.

6.1.16. Justify the formula $\displaystyle\lim_{n\to\infty} n\left[\,\delta\big(x - \tfrac{1}{n}\big) - \delta\big(x + \tfrac{1}{n}\big)\,\right] = -2\,\delta'(x)$.

6.1.17. Define the generalized function $\delta''(x)$:
(a) as a limit of ordinary functions; (b) using duality.

6.1.18. Let $\delta_\xi^{(k)}(x)$ denote the kth derivative of the delta function $\delta_\xi(x)$. Justify the formula $\langle\, \delta_\xi^{(k)}\,, u\,\rangle = (-1)^k\, u^{(k)}(\xi)$ whenever $u \in \mathrm{C}^k$ is k–times continuously differentiable.

6.1.19. According to (6.22), $x\,\delta(x) = 0$. On the other hand, by Leibniz' rule, $(x\,\delta(x))' = \delta(x) + x\,\delta'(x)$ is apparently not zero. Can you explain this paradox?

6.1.20. If $f \in \mathrm{C}^1$, should $(f\,\delta)' = f\,\delta'$ or $f'\,\delta + f\,\delta'$?

♢ 6.1.21. (a) Use duality to justify the formula $f(x)\,\delta'(x) = f(0)\,\delta'(x) - f'(0)\,\delta(x)$ when $f \in \mathrm{C}^1$. (b) Find a similar formula for $f(x)\,\delta^{(n)}(x)$ as the product of a sufficiently smooth function and the nth derivative of the delta function.

6.1.22. Use Exercise 6.1.21 to simplify the following generalized functions; then write out how they act on a suitable test function $u(x)$:
(a) $\varphi(x) = (x-2)\,\delta'(x)$, (b) $\psi(x) = (1+\sin x)\big[\,\delta(x) + \delta'(x)\,\big]$,
(c) $\chi(x) = x^2\big[\,\delta(x-1) - \delta'(x-2)\,\big]$, (d) $\omega(x) = e^x\,\delta''(x+1)$.

♢ 6.1.23. Prove that if $f(x)$ is a continuous function, and $\displaystyle\int_a^b f(x)\,dx = 0$ for every interval $[a,b]$, then $f(x) \equiv 0$ everywhere.

♢ 6.1.24. Write out a rigorous proof that there is no continuous function $\delta_\xi(x)$ such that the inner product identity (6.20) holds for every continuous function $u(x)$.

♢ 6.1.25. *True or false*: The sequence (6.24) converges uniformly.

6.1.26. *True or false*: $\|\,\delta\,\| = 1$.

The Fourier Series of the Delta Function

Let us next investigate the capability of Fourier series to represent generalized functions. We begin with the delta function $\delta(x)$, based at the origin. Using the characterizing properties (6.16), its real Fourier coefficients are

$$a_k = \frac{1}{\pi}\int_{-\pi}^{\pi} \delta(x)\cos k\,x\,dx = \frac{1}{\pi}\cos k\,0 = \frac{1}{\pi}\,, \qquad b_k = \frac{1}{\pi}\int_{-\pi}^{\pi} \delta(x)\sin k\,x\,dx = \frac{1}{\pi}\sin k\,0 = 0. \tag{6.35}$$

Therefore, at least on a formal level, its Fourier series is

$$\delta(x) \sim \frac{1}{2\pi} + \frac{1}{\pi}\left(\cos x + \cos 2x + \cos 3x + \cdots\right). \tag{6.36}$$

Since $\delta(x) = \delta(-x)$ is an even function (why?), it should come as no surprise that it has a cosine series. Alternatively, we can rewrite the series in complex form

$$\delta(x) \sim \frac{1}{2\pi} \sum_{k=-\infty}^{\infty} e^{\,i\,k\,x} = \frac{1}{2\pi}\left(\cdots + e^{-2\,i\,x} + e^{-\,i\,x} + 1 + e^{\,i\,x} + e^{2\,i\,x} + \cdots\right), \tag{6.37}$$

where the complex Fourier coefficients are computed† as

$$c_k = \frac{1}{2\pi}\int_{-\pi}^{\pi} \delta(x)\, e^{-\,i\,k\,x}\, dx = \frac{1}{2\pi}.$$

Remark: Although we stated that the Fourier series (6.36) represents the delta function, this is not entirely correct. Remember that a Fourier series converges to the 2π–periodic extension of the original function. Therefore, (6.37) actually represents the periodic extension of the delta function, sometimes called the *Dirac comb*,

$$\widetilde{\delta}(x) = \cdots + \delta(x+4\pi) + \delta(x+2\pi) + \delta(x) + \delta(x-2\pi) + \delta(x-4\pi) + \delta(x-6\pi) + \cdots, \tag{6.38}$$

consisting of a periodic array of unit impulses concentrated at all integer multiples of 2π.

Let us investigate in what sense (if any) the Fourier series (6.36) or, equivalently, (6.37), represents the delta function. The first observation is that, because its summands do not tend to zero, the series certainly doesn't converge in the usual, calculus, sense. Nevertheless, in a "weak" sense, the series can be regarded as converging to the (periodic extension of the) delta function.

To understand the convergence mechanism, we recall that we already established a formula (3.129) for the partial sums:

$$s_n(x) = \frac{1}{2\pi}\sum_{k=-n}^{n} e^{\,i\,k\,x} = \frac{1}{2\pi} + \frac{1}{\pi}\sum_{k=1}^{n} \cos k\,x = \frac{1}{2\pi}\,\frac{\sin\left(n+\frac{1}{2}\right)x}{\sin \frac{1}{2}x}. \tag{6.39}$$

Graphs of some of the partial sums on the interval $[-\pi, \pi]$ are displayed in Figure 6.7. Note that, as n increases, the spike at $x = 0$ becomes progressively taller and thinner, converging to an infinitely tall delta spike. (We had to truncate the last two graphs; the spike extends beyond the top.) Indeed, by l'Hôpital's Rule,

$$\lim_{x\to 0} \frac{1}{2\pi}\,\frac{\sin\left(n+\frac{1}{2}\right)x}{\sin\frac{1}{2}x} = \lim_{x\to 0}\frac{1}{2\pi}\,\frac{\left(n+\frac{1}{2}\right)\cos\left(n+\frac{1}{2}\right)x}{\frac{1}{2}\cos\frac{1}{2}x} = \frac{n+\frac{1}{2}}{\pi} \longrightarrow \infty \quad \text{as} \quad n\to\infty.$$

(An elementary proof of this formula is to note that, at $x = 0$, every term in the original sum (6.36) is equal to 1.) Furthermore, the integrals remain fixed,

$$\frac{1}{2\pi}\int_{-\pi}^{\pi} s_n(x)\,dx = \frac{1}{2\pi}\int_{-\pi}^{\pi}\frac{\sin\left(n+\frac{1}{2}\right)x}{\sin\frac{1}{2}x}\,dx = \frac{1}{2\pi}\int_{-\pi}^{\pi}\sum_{k=-n}^{n} e^{\,i\,k\,x}\,dx = 1, \tag{6.40}$$

† Or we could use (3.66).

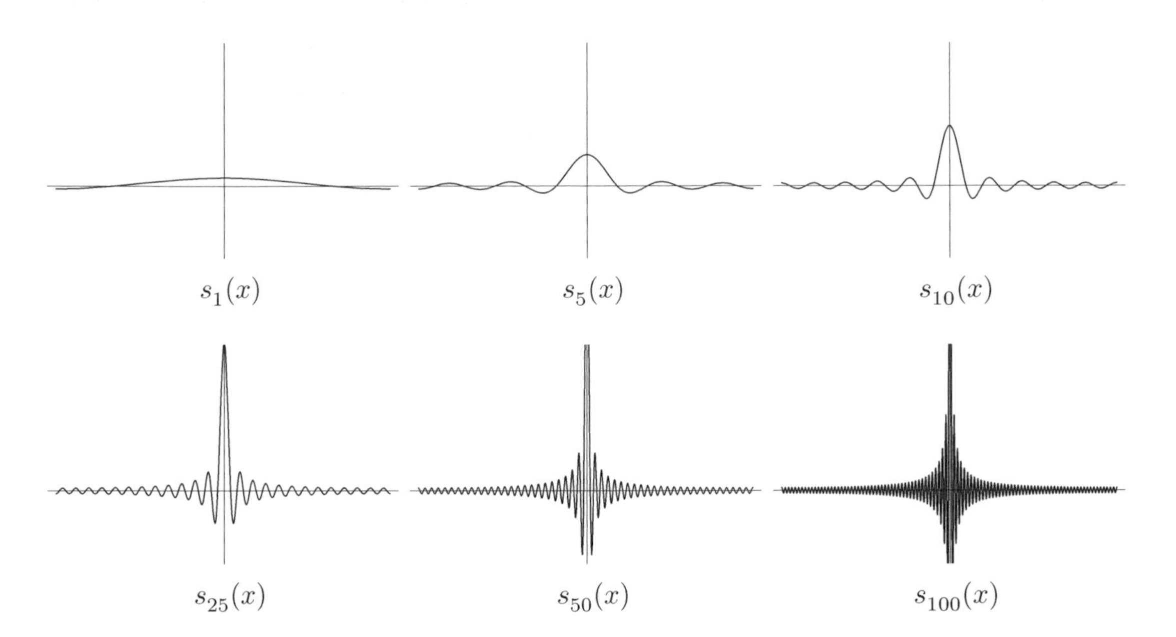

Figure 6.7. Partial Fourier sums approximating the delta function.

as required for convergence to the delta function. However, away from the spike, the partial sums do *not* go to zero! Rather, they oscillate ever more rapidly, while maintaining a fixed overall amplitude of

$$\frac{1}{2\pi} \csc \tfrac{1}{2} x = \frac{1}{2\pi \sin \tfrac{1}{2} x}. \tag{6.41}$$

As n increases, the amplitude function (6.41) can be seen, as in Figure 6.7, as the envelope of the increasingly rapid oscillations. So, roughly speaking, the convergence $s_n(x) \to \delta(x)$ means that the "infinitely fast" oscillations are somehow canceling each other out, and the net effect is zero away from the spike at $x = 0$. So the convergence of the Fourier sums to $\delta(x)$ is much more subtle than in the original limiting definition (6.10).

The technical term is *weak convergence*, which plays a very important role in advanced mathematical analysis, signal processing, composite materials, and elsewhere.

Definition 6.5. A sequence of functions $f_n(x)$ is said to *converge weakly* to $f_\star(x)$ on an interval $[a,b]$ if their L^2 inner products with every continuous *test function* $u(x) \in C^0[a,b]$ converge:

$$\int_a^b f_n(x)\, u(x)\, dx \longrightarrow \int_a^b f_\star(x)\, u(x)\, dx \qquad \text{as} \qquad n \longrightarrow \infty. \tag{6.42}$$

Weak convergence is often indicated by a half-pointed arrow: $f_n \rightharpoonup g$.

Remark: On unbounded intervals, one usually restricts the test functions to have *compact support*, meaning that $u(x) = 0$ for all sufficiently large $|\,x\,| \gg 0$. One can also restrict to smooth test functions only, e.g., require that $u \in C^\infty[a,b]$.

Example 6.6. Let us show that the trigonometric functions $f_n(x) = \cos nx$ converge weakly to the zero function:

$$\cos nx \rightharpoonup 0 \quad \text{as} \quad n \longrightarrow \infty \quad \text{on the interval} \quad [-\pi, \pi].$$

(Actually, this holds on any interval; see Exercise 6.1.38.) According to the definition, we need to prove that

$$\lim_{n \to \infty} \int_{-\pi}^{\pi} u(x) \cos nx \, dx = 0$$

for any continuous function $u \in \mathrm{C}^0[-\pi, \pi]$. But this is just a restatement of the Riemann–Lebesgue Lemma 3.40, which says that the high-frequency Fourier coefficients of a continuous (indeed, even square-integrable) function $u(x)$ go to zero. The same remark establishes the weak convergence $\sin nx \rightharpoonup 0$.

Observe that the functions $\cos nx$ fail to converge pointwise to 0 at *any* value of x. Indeed, if x is an integer multiple of 2π, then $\cos nx = 1$ for all n. If x is any other rational multiple of π, the values of $\cos nx$ periodically cycle through a finite number of different values, and never go to 0, while if x is an irrational multiple of π, they oscillate aperiodically between -1 and $+1$. The functions also fail to converge in norm to 0, since their (unscaled) L^2 norms remain fixed at

$$\| \cos nx \| = \sqrt{\int_{-\pi}^{\pi} \cos^2 nx \, dx} = \sqrt{\pi} \qquad \text{for all} \qquad n > 0.$$

The cancellation of oscillations in the high-frequency limit is a characteristic feature of weak convergence.

Let us now explain why, although the Fourier series (6.36) does not converge to the delta function either pointwise or in norm (indeed, $\| \delta \|$ is not even defined!), it does converge weakly on $[-\pi, \pi]$. More specifically, we need to prove that the partial sums $s_n \rightharpoonup \delta$, meaning that

$$\lim_{n \to \infty} \int_{-\pi}^{\pi} s_n(x)\, u(x)\, dx = \int_{-\pi}^{\pi} \delta(x)\, u(x)\, dx = u(0) \tag{6.43}$$

for every sufficiently nice function u, or, equivalently,

$$\lim_{n \to \infty} \frac{1}{2\pi} \int_{-\pi}^{\pi} u(x) \frac{\sin\left(n + \frac{1}{2}\right) x}{\sin \frac{1}{2} x} \, dx = u(0). \tag{6.44}$$

But this is a restatement of a special case of the identities (3.130) used in the proof of the Pointwise Convergence Theorem 3.8 for the Fourier series of a (piecewise) C^1 function. Indeed, summing the two identities in (3.130) and then setting $x = 0$ reproduces (6.44), since, by continuity, $u(0) = \frac{1}{2}\left[u(0^+) + u(0^-) \right]$. In other words, the pointwise convergence of the Fourier series of a C^1 function is equivalent to the weak convergence[†] of the Fourier series of the delta function!

[†] Definition 6.5 only requires continuity of the test functions, whereas in (6.44) they need to be C^1, so the notion of weak convergence here is slightly slightly more refined. One often restricts further to allow only C^∞ test functions.

Example 6.7. If we differentiate the Fourier series

$$x \sim 2 \sum_{k=1}^{\infty} \frac{(-1)^{k-1}}{k} \sin kx = 2 \left(\sin x - \frac{\sin 2x}{2} + \frac{\sin 3x}{3} - \frac{\sin 4x}{4} + \cdots \right),$$

we obtain an apparent contradiction:

$$1 \sim 2 \sum_{k=1}^{\infty} (-1)^{k+1} \cos kx = 2\cos x - 2\cos 2x + 2\cos 3x - 2\cos 4x + \cdots . \tag{6.45}$$

But the Fourier series for 1 consists of just a single constant term! (Why?)

The resolution of this paradox is not difficult. The Fourier series (3.37) does *not* converge to x, but rather to its 2π–periodic extension $\widetilde{f}(x)$, which has jump discontinuities of magnitude 2π at odd multiples of π; see Figure 3.1. Thus, Theorem 3.22 is *not* directly applicable. Nevertheless, we can assign a consistent interpretation to the differentiated series. The derivative $\widetilde{f}'(x)$ of the periodic extension is *not* equal to the constant function 1, but rather has an additional delta function concentrated at each jump discontinuity:

$$\widetilde{f}'(x) = 1 - 2\pi \sum_{j=-\infty}^{\infty} \delta\big(x - (2j+1)\pi\big) = 1 - 2\pi\, \widetilde{\delta}(x - \pi),$$

where $\widetilde{\delta}$ denotes the 2π–periodic extension of the delta function, cf. (6.38). The differentiated Fourier series (6.45) does, in fact, represent $\widetilde{f}'(x)$. Indeed, the Fourier coefficients of $\widetilde{\delta}(x-\pi)$ are

$$a_k = \frac{1}{\pi} \int_0^{2\pi} \delta(x-\pi) \cos kx \, dx = \frac{1}{\pi} \cos k\pi = \frac{(-1)^k}{\pi},$$

$$b_k = \frac{1}{\pi} \int_0^{2\pi} \delta(x-\pi) \sin kx \, dx = \frac{1}{\pi} \sin k\pi = 0.$$

Observe that we changed the interval of integration to $[0, 2\pi]$ to avoid placing the delta function singularities at the endpoints. Thus,

$$\delta(x-\pi) \sim \frac{1}{2\pi} + \frac{1}{\pi} \big(-\cos x + \cos 2x - \cos 3x + \cdots \big), \tag{6.46}$$

which serves to resolve the contradiction.

Example 6.8. Let us differentiate the Fourier series

$$\sigma(x) \sim \frac{1}{2} + \frac{2}{\pi} \left(\sin x + \frac{\sin 3x}{3} + \frac{\sin 5x}{5} + \frac{\sin 7x}{7} + \cdots \right)$$

for the unit step function we found in Example 3.9 and see whether we end up with the Fourier series (6.36) for the delta function. We compute

$$\frac{d\sigma}{dx} \sim \frac{2}{\pi} \big(\cos x + \cos 3x + \cos 5x + \cos 7x + \cdots \big), \tag{6.47}$$

which does *not* agree with (6.36) — half the terms are missing! The explanation is similar to the preceding example: the 2π–periodic extension $\widetilde{\sigma}(x)$ of the step function has two jump discontinuities, of magnitudes $+1$ at even multiples of π and -1 at odd multiples; see Figure 3.6. Therefore, its derivative

$$\frac{d\widetilde{\sigma}}{dx} = \widetilde{\delta}(x) - \widetilde{\delta}(x-\pi)$$

is the difference of the 2π–periodic extension of the delta function at 0, with Fourier series (6.36), minus the 2π–periodic extension of the delta function at π, with Fourier series (6.46), which produces (6.47).

It is a remarkable, profound fact that Fourier analysis is entirely compatible with the calculus of generalized functions, [**68**]. For instance, term-wise differentiation of the Fourier series for a piecewise C^1 function leads to the Fourier series for the differentiated function that incorporates delta functions of the appropriate magnitude at each jump discontinuity. This fact further reassures us that the rather mysterious construction of delta functions and their generalizations is indeed the right way to extend calculus to functions that do not possess derivatives in the ordinary sense.

Exercises

6.1.27. Determine the real and complex Fourier series for $\delta(x-\xi)$, where $-\pi < \xi < \pi$. What periodic generalized function(s) do they represent?

6.1.28. Determine the Fourier sine series and the Fourier cosine series for $\delta(x-\xi)$, where $0 < \xi < \pi$. Which periodic generalized functions do they represent?

♡ 6.1.29. Let $n > 0$ be a positive integer. (a) For integers $0 \le j < n$, find the complex Fourier series of the 2π–periodically extended delta functions $\widetilde{\delta}_j(x) = \widetilde{\delta}(x - 2j\pi/n)$. (b) Prove that their Fourier coefficients satisfy the periodicity condition $c_k = c_l$ whenever $k \equiv l \mod n$. (c) Conversely, given complex Fourier coefficients that satisfy the periodicity condition $c_k = c_l$ whenever $k \equiv l \mod n$, prove that the corresponding Fourier series represents a linear combination of the preceding periodically extended delta functions $\widetilde{\delta}_0(x), \ldots, \widetilde{\delta}_{n-1}(x)$. *Hint*: Use Example B.22. (d) Prove that a complex Fourier series represents a 2π–periodic function that is constant on the subintervals $2\pi j/n < x < 2\pi(j+1)/n$, for $j \in \mathbb{Z}$, if and only if its Fourier coefficients satisfy the conditions

$$k\, c_k = l\, c_l, \qquad k \equiv l \not\equiv 0 \mod n, \qquad\qquad c_k = 0, \qquad 0 \ne k \equiv 0 \mod n.$$

♣ 6.1.30. (a) Find the complex Fourier series for the derivative of the delta function $\delta'(x)$ by direct evaluation of the coefficient formulas. (b) Verify that your series can be obtained by term-by-term differentiation of the series for $\delta(x)$. (c) Write a formula for the nth partial sum of your series. (d) Use a computer graphics package to investigate the convergence of the series.

6.1.31. What is the Fourier series for the generalized function $g(x) = x\,\delta(x)$? Can you obtain this result through multiplication of the individual Fourier series (3.37), (6.37)?

6.1.32. Apply the method of Exercise 3.2.59 to find the complex Fourier series for the function $f(x) = \delta(x)\, e^{\mathrm{i}\,x}$. Which Fourier series do you get? Can you explain what is going on?

6.1.33. In Exercise 6.1.12 we established the identity $\delta(x) = 2\,\delta(2x)$. Does this hold on the level of Fourier series? Can you explain why or why not?

6.1.34. How should one interpret the formula (6.38) for the periodic extension of the delta function (a) as a limit? (b) as a linear functional?

6.1.35. Write down the complex Fourier series for e^x. Differentiate term by term. Do you get the same series? Explain your answer.

6.1.36. *True or false*: If you integrate the Fourier series for the delta function $\delta(x)$ term by term, you obtain the Fourier series for the step function $\sigma(x)$.

6.1.37. Find the Fourier series for the function $\delta(x)$ on the interval $-1 \le x \le 1$. Which (generalized) function does the Fourier series represent?

♢ 6.1.38.. Prove that $\cos nx \rightharpoonup 0$ (weakly) as $n \to \infty$ on any bounded interval $[a, b]$.

♢ 6.1.39. Prove that if $u_n \to u$ in norm, then $u_n \rightharpoonup u$ weakly.

6.1.40. *True or false*: (a) If $u_n \to u$ uniformly on $[a, b]$, then $u_n \rightharpoonup u$ weakly.
(b) If $u_n(x) \to u(x)$ pointwise, then $u_n \rightharpoonup u$ weakly.

6.1.41. Prove that the sequence $f_n(x) = \cos^2 nx$ converges weakly on $[-\pi, \pi]$. What is the limiting function?

6.1.42. Answer Exercise 6.1.41 when $f_n(x) = \cos^3 nx$.

6.1.43. Discuss the weak convergence of the Fourier series for the derivative $\delta'(x)$ of the delta function.

6.2 Green's Functions for One–Dimensional Boundary Value Problems

We will now put the delta function to work by developing a general method for solving inhomogeneous linear boundary value problems. The key idea, motivated by the linear algebra technique outlined at the beginning of the previous section, is to first solve the system when subject to a unit delta function impulse, which produces the Green's function. We then apply linear superposition to write down the solution for a general forcing inhomogeneity. The Green's function approach has wide applicability, but will be developed here in the context of a few basic examples.

Example 6.9. The boundary value problem

$$-c u'' = f(x), \qquad u(0) = 0 = u(1), \tag{6.48}$$

models the longitudinal deformation $u(x)$ of a homogeneous elastic bar of unit length and constant stiffness c that is fixed at both ends while subject to an external force $f(x)$. The associated *Green's function* refers to the family of solutions

$$u(x) = G_\xi(x) = G(x; \xi)$$

induced by unit-impulse forces concentrated at a single point $0 < \xi < 1$:

$$-c u'' = \delta(x - \xi), \qquad u(0) = 0 = u(1). \tag{6.49}$$

The solution to the differential equation can be straightforwardly obtained by direct integration. First, by (6.23),

$$u'(x) = -\frac{\sigma(x-\xi)}{c} + a,$$

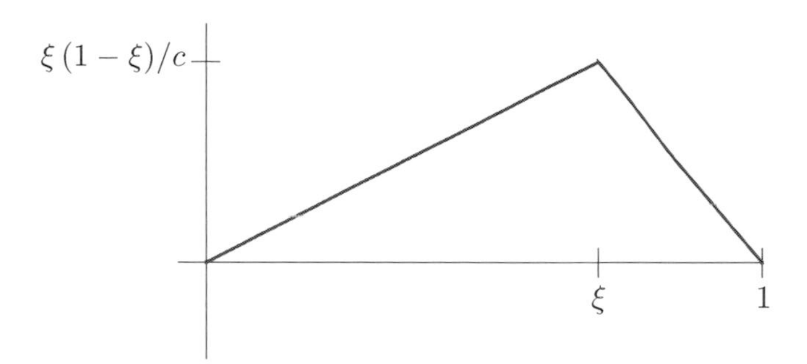

Figure 6.8. Green's function for a bar with fixed ends.

where a is a constant of integration. A second integration leads to

$$u(x) = -\,\frac{\rho(x-\xi)}{c} + a\,x + b, \tag{6.50}$$

where ρ is the ramp function (6.25). The integration constants a, b are fixed by the boundary conditions; since $0 < \xi < 1$, we have

$$u(0) = b = 0, \qquad u(1) = -\,\frac{1-\xi}{c} + a + b = 0, \qquad \text{and so} \qquad a = \frac{1-\xi}{c}, \qquad b = 0.$$

We deduce that the Green's function for the problem is

$$G(x;\xi) = \frac{(1-\xi)\,x - \rho(x-\xi)}{c} = \begin{cases} (1-\xi)\,x/c, & x \le \xi, \\ \xi\,(1-x)/c, & x \ge \xi. \end{cases} \tag{6.51}$$

As sketched in Figure 6.8, for each fixed ξ, the function $G_\xi(x) = G(x;\xi)$ depends continuously on x; its graph consists of two connected straight line segments, with a corner at the point of application of the unit impulse force.

Once we have determined the Green's function, we are able to solve the general inhomogeneous boundary value problem (6.48) by linear superposition. We first express the forcing function $f(x)$ as a linear combination of impulses concentrated at various points along the bar. Since there is a continuum of possible positions $0 < \xi < 1$ at which impulse forces may be applied, we will use an integral to sum them, thereby writing the external force as

$$f(x) = \int_0^1 \delta(x-\xi)\,f(\xi)\,d\xi. \tag{6.52}$$

We can interpret (6.52) as the (continuous) superposition of an infinite collection of impulses, namely $f(\xi)\,\delta(x-\xi)$, of magnitude $f(\xi)$ and concentrated at position ξ.

The Superposition Principle states that linear combinations of inhomogeneities produce the selfsame linear combinations of solutions. Again, we adapt this principle to the continuum by replacing the sums by integrals. Thus, the solution to the boundary value problem will be the linear superposition

$$u(x) = \int_0^1 G(x;\xi)\,f(\xi)\,d\xi \tag{6.53}$$

of the Green's function solutions to the individual unit-impulse problems.

For the particular boundary value problem (6.48), we use the formula (6.51) for the Green's function. Breaking the resulting integral (6.53) into two parts, over the subintervals $0 \le \xi \le x$ and $x \le \xi \le 1$, we arrive at the explicit solution formula

$$u(x) = \frac{1}{c}\int_0^x (1-x)\,\xi\, f(\xi)\, d\xi \;+\; \frac{1}{c}\int_x^1 x\,(1-\xi)\, f(\xi)\, d\xi. \tag{6.54}$$

For example, under a constant unit force f, (6.54) yields the solution

$$u(x) = \frac{f}{c}\int_0^x (1-x)\,\xi\, d\xi \;+\; \frac{f}{c}\int_x^1 x\,(1-\xi)\, d\xi = \frac{f}{2\,c}\,(1-x)\,x^2 + \frac{f}{2\,c}\,x\,(1-x)^2 = \frac{f}{2\,c}\,(x-x^2).$$

Let us, finally, convince ourselves that the superposition formula (6.54) indeed gives the correct answer. First,

$$\begin{aligned} c\,\frac{du}{dx} &= (1-x)\,x\,f(x) + \int_0^x \big(-\xi\, f(\xi)\big)\, d\xi - x\,(1-x)\,f(x) + \int_x^1 (1-\xi)\,f(\xi)\, d\xi \\ &= -\int_0^1 \xi\, f(\xi)\, d\xi + \int_x^1 f(\xi)\, d\xi. \end{aligned}$$

Differentiating again with respect to x, we see that the first term is constant, and so $-\,c\,\dfrac{d^2u}{dx^2} = f(x)$, as claimed.

Remark: In computing the derivatives of u, we made use of the calculus formula

$$\frac{d}{dx}\int_{\alpha(x)}^{\beta(x)} F(x,\xi)\, d\xi = F(x,\beta(x))\,\frac{d\beta}{dx} - F(x,\alpha(x))\,\frac{d\alpha}{dx} + \int_{\alpha(x)}^{\beta(x)} \frac{\partial F}{\partial x}(x,\xi)\, d\xi \tag{6.55}$$

for the derivative of an integral with variable limits — which is a straightforward consequence of the Fundamental Theorem of Calculus and the chain rule, [**8**, **108**]. As always, one must exercise due care when interchanging differentiation and integration.

We note the following basic properties, which serve to uniquely characterize the Green's function. First, since the delta forcing vanishes except at the point $x = \xi$, the Green's function satisfies the homogeneous differential equation[†]

$$-\,c\,\frac{\partial^2 G}{\partial x^2}\,(x;\xi) = 0 \qquad \text{for all} \qquad x \neq \xi. \tag{6.56}$$

Second, by construction, it must satisfy the boundary conditions

$$G(0;\xi) = 0 = G(1;\xi).$$

Third, for each fixed ξ, $G(x;\xi)$ is a continuous function of x, but its derivative $\partial G/\partial x$ has a jump discontinuity of magnitude $-1/c$ at the impulse point $x = \xi$. As a result, the second derivative $\partial^2 G/\partial x^2$ has a delta function discontinuity there, and hence solves the original impulse boundary value problem (6.49).

Finally, we cannot help but notice that the Green's function (6.51) is a symmetric function of its two arguments: $G(x;\xi) = G(\xi;x)$. Symmetry has the interesting physical consequence that the displacement of the bar at position x due to an impulse force

[†] Since $G(x;\xi)$ is a function of two variables, we switch to partial derivative notation to indicate its derivatives.

concentrated at position ξ is exactly the same as the displacement of the bar at ξ due to an impulse of the same magnitude being applied at x. This turns out to be a rather general, although perhaps unanticipated, phenomenon. Symmetry of the Green's function is a consequence of the underlying symmetry, or, more accurately, "self-adjointness", of the boundary value problem, a topic that will be developed in detail in Section 9.2.

Example 6.10. Let $\omega^2 > 0$ be a fixed positive constant. Let us solve the inhomogeneous boundary value problem

$$-u'' + \omega^2 u = f(x), \qquad u(0) = u(1) = 0, \tag{6.57}$$

by constructing its Green's function. To this end, we first analyze the effect of a delta function inhomogeneity

$$-u'' + \omega^2 u = \delta(x - \xi), \qquad u(0) = u(1) = 0. \tag{6.58}$$

Rather than try to integrate this differential equation directly, let us appeal to the defining properties of the Green's function. The general solution to the homogeneous equation is a linear combination of the two basic exponentials $e^{\omega x}$ and $e^{-\omega x}$, or better, the hyperbolic functions

$$\cosh \omega x = \frac{e^{\omega x} + e^{-\omega x}}{2}, \qquad \sinh \omega x = \frac{e^{\omega x} - e^{-\omega x}}{2}. \tag{6.59}$$

The solutions satisfying the first boundary condition are multiples of $\sinh \omega x$, while those satisfying the second boundary condition are multiples of $\sinh \omega (1 - x)$. Therefore, the solution to (6.58) has the form

$$G(x;\xi) = \begin{cases} a \sinh \omega x, & x \le \xi, \\ b \sinh \omega (1 - x), & x \ge \xi. \end{cases}$$

Continuity of $G(x;\xi)$ at $x = \xi$ requires

$$a \sinh \omega \xi = b \sinh \omega (1 - \xi). \tag{6.60}$$

At $x = \xi$, the derivative $\partial G/\partial x$ must have a jump discontinuity of magnitude -1 in order that the second derivative term in (6.58) match the delta function. (The $\omega^2 u$ term clearly cannot produce the required singularity.) Since

$$\frac{\partial G}{\partial x}(x;\xi) = \begin{cases} a\,\omega \cosh \omega x, & x < \xi, \\ -b\,\omega \cosh \omega (1 - x), & x > \xi, \end{cases}$$

the jump condition requires

$$a\,\omega \cosh \omega \xi - 1 = -b\,\omega \cosh \omega (1 - \xi). \tag{6.61}$$

Multiplying (6.60) by $\omega \cosh \omega (1 - \xi)$ and (6.61) by $\sinh \omega (1 - \xi)$, and then adding the results together, we obtain

$$\sinh \omega (1 - \xi) = a\,\omega \big[\sinh \omega \xi \, \cosh \omega (1 - \xi) + \cosh \omega \xi \, \sinh \omega (1 - \xi) \big] = a\,\omega \sinh \omega, \tag{6.62}$$

where we made use of the addition formula for the hyperbolic sine:

$$\sinh(\alpha + \beta) = \sinh \alpha \, \cosh \beta + \cosh \alpha \, \sinh \beta, \tag{6.63}$$

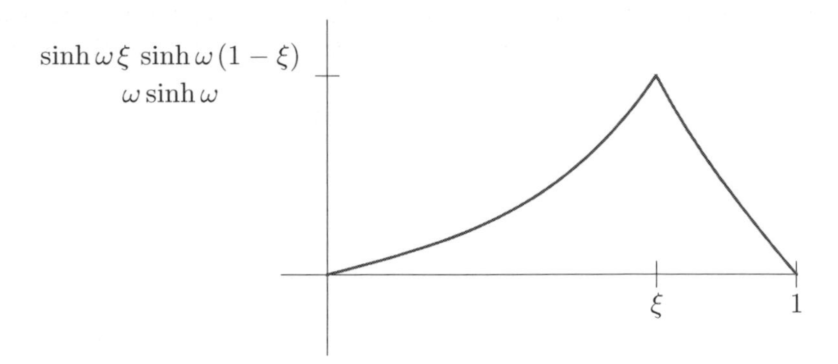

Figure 6.9. Green's function for the boundary value problem (6.57).

which you are asked to prove in Exercise 6.2.13. Therefore, solving (6.61–62) for

$$a = \frac{\sinh \omega\,(1-\xi)}{\omega \sinh \omega}, \qquad\qquad b = \frac{\sinh \omega\,\xi}{\omega \sinh \omega},$$

produces the explicit formula

$$G(x;\xi) = \begin{cases} \dfrac{\sinh \omega\, x \,\sinh \omega\,(1-\xi)}{\omega \sinh \omega}, & x \le \xi, \\[2ex] \dfrac{\sinh \omega\,(1-x) \,\sinh \omega\,\xi}{\omega \sinh \omega}, & x \ge \xi. \end{cases} \tag{6.64}$$

A representative graph appears in Figure 6.9. As before, a corner, indicating a discontinuity in the first derivative, appears at the point $x = \xi$ where the impulse force is applied. Moreover, as in the previous example, $G(x;\xi) = G(\xi;x)$ is a symmetric function.

The general solution to the inhomogeneous boundary value problem (6.57) is then given by the superposition formula (6.53); explicitly,

$$\begin{aligned} u(x) &= \int_0^1 G(x;\xi) f(\xi)\,d\xi \\ &= \int_0^x \frac{\sinh \omega\,(1-x) \sinh \omega\,\xi}{\omega \sinh \omega} f(\xi)\,d\xi \;+\; \int_x^1 \frac{\sinh \omega\,x \,\sinh \omega\,(1-\xi)}{\omega \sinh \omega} f(\xi)\,d\xi. \end{aligned} \tag{6.65}$$

For example, under a constant unit force $f(x) \equiv 1$, the solution is

$$\begin{aligned} u(x) &= \int_0^x \frac{\sinh \omega\,(1-x) \sinh \omega\,y}{\omega \sinh \omega}\,d\xi \;+\; \int_x^1 \frac{\sinh \omega\,x \,\sinh \omega\,(1-\xi)}{\omega \sinh \omega}\,d\xi \\ &= \frac{\sinh \omega\,(1-x)\big(\cosh \omega\,x - 1\big)}{\omega^2 \sinh \omega} + \frac{\sinh \omega\,x\big(\cosh \omega\,(1-x) - 1\big)}{\omega^2 \sinh \omega} \\ &= \frac{1}{\omega^2} - \frac{\sinh \omega\,x + \sinh \omega\,(1-x)}{\omega^2 \sinh \omega}. \end{aligned}$$

For comparative purposes, the reader may wish to rederive this particular solution by a direct calculation, without appealing to the Green's function.

Example 6.11. Finally, consider the Neumann boundary value problem

$$-\,c\,u'' = f(x), \qquad\qquad u'(0) = 0 = u'(1), \tag{6.66}$$

modeling the equilibrium deformation of a homogeneous bar with two free ends when subject to an external force $f(x)$. The *Green's function* should satisfy the particular case

$$-c u'' = \delta(x-\xi), \qquad u'(0) = 0 = u'(1),$$

when the forcing function is a concentrated impulse. As in Example 6.9, the general solution to the latter differential equation is

$$u(x) = -\frac{\rho(x-\xi)}{c} + a x + b,$$

where a, b are integration constants, and ρ is the ramp function (6.25). However, the Neumann boundary conditions require that

$$u'(0) = a = 0, \qquad u'(1) = -\frac{1}{c} + a = 0,$$

which cannot both be satisfied. We conclude that there is no Green's function in this case.

The difficulty is that the Neumann boundary value problem (6.66) does not have a unique solution, and hence cannot admit a Green's function solution formula (6.53). Indeed, integrating twice, we find that the general solution to the differential equation is

$$u(x) = a x + b - \frac{1}{c}\int_0^x \int_0^y f(z)\,dz\,dy,$$

where a, b are integration constants. Since

$$u'(x) = a - \frac{1}{c}\int_0^x f(z)\,dz,$$

the boundary conditions require that

$$u'(0) = a = 0, \qquad u'(1) = a - \frac{1}{c}\int_0^1 f(z)\,dz = 0.$$

These equations are compatible if and only if

$$\int_0^1 f(z)\,dz = 0. \tag{6.67}$$

Thus, the Neumann boundary value problem admits a solution if and only if there is no net force on the bar. Indeed, physically, if (6.67) does not hold, then, because its ends are not attached to any support, the bar cannot stay in equilibrium, but will move off in the direction of the net force. On the other hand, if (6.67) holds, then the solution

$$u(x) = b - \frac{1}{c}\int_0^x \int_0^y f(z)\,dz\,dy$$

is *not unique*, since b is not constrained by the boundary conditions, and so can assume any constant value. Physically, this means that any equilibrium configuration of the bar can be freely translated to assume another valid equilibrium.

Remark: The constraint (6.67) is a manifestation of the *Fredholm Alternative*, to be developed in detail in Section 9.1.

Let us summarize the fundamental properties that serve to completely characterize the Green's function of boundary value problems governed by second-order linear ordinary differential equations

$$p(x)\,\frac{d^2u}{dx^2} + q(x)\,\frac{du}{dx} + r(x)\,u(x) = f(x), \tag{6.68}$$

combined with a pair of homogeneous boundary conditions at the ends of the interval $[a,b]$. We assume that the coefficient functions are continuous, $p,q,r,f \in \mathrm{C}^0[a,b]$, and that $p(x) \neq 0$ for all $a \leq x \leq b$.

Basic Properties of the Green's Function $G(x;\xi)$

(*i*) Solves the homogeneous differential equation at all points $x \neq \xi$.
(*ii*) Satisfies the homogeneous boundary conditions.
(*iii*) Is a continuous function of its arguments.
(*iv*) For each fixed ξ, its derivative $\partial G/\partial x$ is piecewise C^1, with a single jump discontinuity of magnitude $1/p(\xi)$ at the impulse point $x=\xi$.

With the Green's function in hand, we deduce that the solution to the general boundary value problem (6.68) subject to the appropriate homogeneous boundary conditions is expressed by the *Green's Function Superposition Formula*

$$u(x) = \int_a^b G(x;\xi)\,f(\xi)\,d\xi. \tag{6.69}$$

The symmetry of the Green's function is more subtle, for it relies on the self-adjointness of the boundary value problem, an issue to be addressed in detail in Chapter 9. In the present situation, self-adjointness requires that $q(x) = p'(x)$, in which case $G(\xi;x) = G(x;\xi)$ will be symmetric in its arguments.

Finally, as we saw in Example 6.11, not every such boundary value problem admits a solution, and one expects to find a Green's function only in cases in which the solution exists and is unique.

Theorem 6.12. *The following are equivalent*:

- *The only solution to the homogeneous boundary value problem is the zero function.*
- *The inhomogeneous boundary value problem has a unique solution for every choice of forcing function.*
- *The boundary value problem admits a Green's function.*

Exercises

6.2.1. Let $c > 0$. Find the Green's function for the boundary value problem $-c\,u'' = f(x)$, $u(0) = 0$, $u'(1) = 0$, which models the displacement of a uniform bar of unit length with one fixed and one free end under an external force. Then use superposition to write down a formula for the solution. Verify that your integral formula is correct by direct differentiation and substitution into the differential equation and boundary conditions.

6.2.2. A uniform bar of length $\ell = 4$ has constant stiffness $c = 2$. Find the Green's function for the case that (*a*) both ends are fixed; (*b*) one end is fixed and the other is free. (*c*) Why is there no Green's function when both ends are free?

6.2.3. A point 2 cm along a 10 cm bar experiences a displacement of 1 mm under a concentrated force of 2 newtons applied at the midpoint of the bar. How far does the midpoint deflect when a concentrated force of 1 newton is applied at the point 2 cm along the bar?

♡ 6.2.4. The boundary value problem $-\dfrac{d}{dx}\left(c(x)\dfrac{du}{dx}\right) = f(x)$, $u(0) = u(1) = 0$, models the displacement $u(x)$ of a nonuniform elastic bar with stiffness $c(x) = \dfrac{1}{1+x^2}$ for $0 \le x \le 1$. (*a*) Find the displacement when the bar is subjected to a constant external force, $f \equiv 1$. (*b*) Find the Green's function for the boundary value problem. (*c*) Use the resulting superposition formula to check your solution to part (*a*). (*d*) Which point $0 < \xi < 1$ on the bar is the "weakest", i.e., the bar experiences the largest displacement under a unit impulse concentrated at that point?

6.2.5. Answer Exercise 6.2.4 when $c(x) = 1 + x$.

♡ 6.2.6. Consider the boundary value problem $-u'' = f(x)$, $u(0) = 0$, $u(1) = 2u'(1)$. (*a*) Find the Green's function. (*b*) Which of the fundamental properties does your Green's function satisfy? (*c*) Write down an explicit integral formula for the solution to the boundary value problem, and prove its validity by a direct computation. (*d*) Explain why the related boundary value problem $-u'' = f$, $u(0) = 0$, $u(1) = u'(1)$, does not have a Green's function.

♡ 6.2.7. For n a positive integer, set $f_n(x) = \begin{cases} \frac{1}{2}n, & |x-\xi| < \frac{1}{n}, \\ 0, & \text{otherwise.} \end{cases}$

(*a*) Find the solution $u_n(x)$ to the boundary value problem $-u'' = f_n(x)$, $u(0) = u(1) = 0$, assuming $0 < \xi - \frac{1}{n} < \xi + \frac{1}{n} < 1$. (*b*) Prove that $\lim\limits_{n\to\infty} u_n(x) = G(x;\xi)$ converges to the Green's function (6.51). Why should this be the case? (*c*) Reconfirm the result in part (*b*) by graphing $u_5(x), u_{15}(x), u_{25}(x)$, along with $G(x;\xi)$ when $\xi = .3$.

6.2.8. Solve the boundary value problem $-4u'' + 9u = 0$, $u(0) = 0$, $u(2) = 1$. Is your solution unique?

6.2.9. *True or false*: The Neumann boundary value problem $-u'' + u = 1$, $u'(0) = u'(1) = 0$, has a unique solution.

6.2.10. Use the Green's function (6.64) to solve the boundary value problem (6.57) when the forcing function is $f(x) = \begin{cases} 1, & 0 \le x < \frac{1}{2}, \\ -1, & \frac{1}{2} < x \le 1. \end{cases}$

6.2.11. Let $\omega > 0$. (*a*) Find the Green's function for the mixed boundary value problem

$$-u'' + \omega^2 u = f(x), \quad u(0) = 0, \quad u'(1) = 0.$$

(*b*) Use your Green's function to find the solution when $f(x) = \begin{cases} 1, & 0 \le x < \frac{1}{2}, \\ -1, & \frac{1}{2} < x \le 1. \end{cases}$

6.2.12. Suppose $\omega > 0$. Does the Neumann boundary value problem $-u'' + \omega^2 u = f(x)$, $u'(0) = u'(1) = 0$ admit a Green's function? If not, explain why not. If so, find it, and then write down an integral formula for the solution of the boundary value problem.

◇ 6.2.13. (*a*) Prove the addition formula (6.63) for the hyperbolic sine function. (*b*) Find the corresponding addition formula for the hyperbolic cosine.

◇ 6.2.14. Prove the differentiation formula (6.55).

6.3 Green's Functions for the Planar Poisson Equation

Now we develop the Green's function approach to solving boundary value problems involving the two-dimensional Poisson equation (4.84). As before, the Green's function is characterized as the solution to the homogeneous boundary value problem in which the inhomogeneity is a concentrated unit impulse — a delta function. The solution to the general forced boundary value problem is then obtained via linear superposition, that is, as a convolution integral with the Green's function.

However, before proceeding, we need to quickly review some basic facts concerning vector calculus in the plane. The student may wish to consult a standard multivariable calculus text, e.g., [**8**, **108**], for additional details.

Calculus in the Plane

Let $\mathbf{x} = (x, y)$ denote the usual Cartesian coordinates on $\mathbb{R}^2$. The term *scalar field* is synonymous with a real-valued function $u(x, y)$, defined on a domain $\Omega \subset \mathbb{R}^2$. A vector-valued function

$$\mathbf{v}(\mathbf{x}) = \mathbf{v}(x, y) = \begin{pmatrix} v_1(x, y) \\ v_2(x, y) \end{pmatrix} \tag{6.70}$$

is known as a (planar) *vector field*. A vector field assigns a vector $\mathbf{v}(x, y) \in \mathbb{R}^2$ to each point $(x, y) \in \Omega$ in its domain of definition, and hence defines a function $\mathbf{v}\colon \Omega \to \mathbb{R}^2$. Physical examples include velocity vector fields of fluid flows, heat flux fields in thermodynamics, and gravitational and electrostatic force fields.

The *gradient* operator ∇ maps a scalar field $u(x, y)$ to the vector field

$$\nabla u = \begin{pmatrix} \partial u / \partial x \\ \partial u / \partial y \end{pmatrix}. \tag{6.71}$$

The scalar field u is often referred to as a *potential function* for its gradient vector field $\mathbf{v} = \nabla u$. On a connected domain Ω, the potential, when it exists, is uniquely determined up to addition of a constant.

The *divergence* of the planar vector field $\mathbf{v} = (v_1, v_2)^T$ is the scalar field

$$\nabla \cdot \mathbf{v} = \operatorname{div} \mathbf{v} = \frac{\partial v_1}{\partial x} + \frac{\partial v_2}{\partial y}. \tag{6.72}$$

Its *curl* is defined as

$$\nabla \times \mathbf{v} = \operatorname{curl} \mathbf{v} = \frac{\partial v_2}{\partial x} - \frac{\partial v_1}{\partial y}. \tag{6.73}$$

Notice that the curl of a planar vector field is a scalar field. (In contrast, in three dimensions, the curl of a vector field is another vector field.) Given a smooth potential $u \in \mathrm{C}^2$, the curl of its gradient vector field automatically vanishes:

$$\nabla \times \nabla u = \frac{\partial}{\partial x}\frac{\partial u}{\partial y} - \frac{\partial}{\partial y}\frac{\partial u}{\partial x} \equiv 0,$$

by the equality of mixed partials. Thus, a necessary condition for a vector field $\mathbf{v}$ to admit a potential is that it be *irrotational*, meaning $\nabla \times \mathbf{v} = 0$; this condition is sufficient if

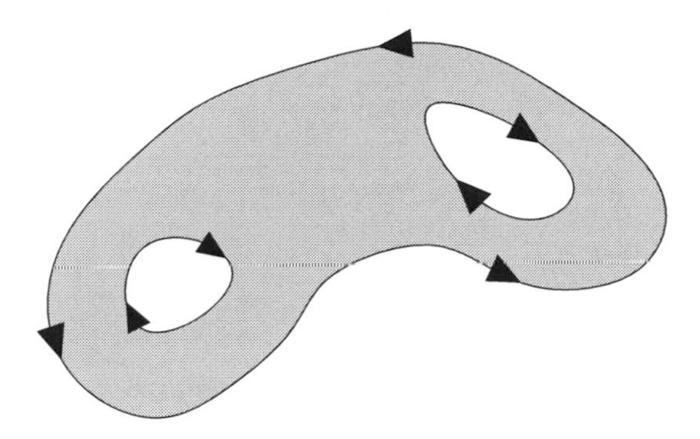

Figure 6.10. Orientation of the boundary of a planar domain.

the underlying domain Ω is *simply connected*, i.e., has no holes. On the other hand, the divergence of a gradient vector field coincides with the Laplacian of the potential function:

$$\nabla \cdot \nabla u = \Delta u = \frac{\partial^2 u}{\partial x^2} + \frac{\partial^2 u}{\partial y^2}. \tag{6.74}$$

A vector field is *incompressible* if it has zero divergence: $\nabla \cdot \mathbf{v} = 0$; for the velocity vector field of a steady-state fluid flow, incompressibility means that the fluid does not change volume. (Water is, for all practical purposes, an incompressible fluid.) Therefore, an irrotational vector field with potential u is also incompressible if and only if the potential solves the Laplace equation $\Delta u = 0$.

Remark: Because of formula (6.74), the Laplacian operator is also sometimes written as $\Delta = \nabla^2$. The factorization of the Laplacian into the product of the divergence and the gradient operators is, in fact, of great importance, and underlies its "self-adjointness", a fundamental property whose ramifications will be explored in depth in Chapter 9.

Let $\Omega \subset \mathbb{R}^2$ be a bounded domain whose boundary $\partial\Omega$ consists of one or more piecewise smooth closed curves. We orient the boundary so that the domain is always on one's left as one goes around the boundary curve(s). Figure 6.10 sketches a domain with two holes; its three boundary curves are oriented according to the directions of the arrows. Note that the outer boundary curve is traversed in a counterclockwise direction, while the two inner boundary curves are oriented clockwise.

Green's Theorem, first formulated by George Green to use in his seminal study of partial differential equations and potential theory, relates certain double integrals over a domain to line integrals around its boundary. It should be viewed as the extension of the Fundamental Theorem of Calculus to double integrals.

Theorem 6.13. *Let $\mathbf{v}(\mathbf{x})$ be a smooth*† *vector field defined on a bounded domain $\Omega \subset \mathbb{R}^2$. Then the line integral of $\mathbf{v}$ around the boundary $\partial\Omega$ equals the double integral of its curl over the domain*:

$$\iint_\Omega \nabla \times \mathbf{v} \; dx\, dy = \oint_{\partial\Omega} \mathbf{v} \cdot d\mathbf{x}, \tag{6.75}$$

† To be precise, we require $\mathbf{v}$ to be continuously differentiable within the domain, and continuous up to the boundary, so $\mathbf{v} \in \mathrm{C}^0(\overline{\Omega}) \cap \mathrm{C}^1(\Omega)$, where $\overline{\Omega} = \Omega \cup \partial\Omega$ denotes the closure of the domain Ω.

or, in full detail,

$$\iint_\Omega \left(\frac{\partial v_2}{\partial x} - \frac{\partial v_1}{\partial y} \right) dx\,dy = \oint_{\partial\Omega} v_1\,dx + v_2\,dy\,. \tag{6.76}$$

Example 6.14. Let us apply Green's Theorem 6.13 to the particular vector field $\mathbf{v} = (\,y, 0\,)^T$. Since $\nabla \times \mathbf{v} \equiv -1$, we obtain

$$\oint_{\partial\Omega} y\,dx = \iint_\Omega (-1)\,dx\,dy = -\text{ area } \Omega. \tag{6.77}$$

This means that we can determine the area of a planar domain by computing the negative of the indicated line integral around its boundary.

For later purposes, we rewrite the basic Green identity (6.75) in an equivalent "divergence form". Given a planar vector field $\mathbf{v} = (\,v_1, v_2\,)^T$, let

$$\mathbf{v}^\perp = \begin{pmatrix} -v_2 \\ v_1 \end{pmatrix} \tag{6.78}$$

denote the "perpendicular" vector field. We note that its curl

$$\nabla \times \mathbf{v}^\perp = \frac{\partial v_1}{\partial x} + \frac{\partial v_2}{\partial y} = \nabla \cdot \mathbf{v} \tag{6.79}$$

coincides with the divergence of the original vector field.

When we replace $\mathbf{v}$ in Green's identity (6.75) by $\mathbf{v}^\perp$, the result is

$$\iint_\Omega \nabla \cdot \mathbf{v}\; dx\,dy = \iint_\Omega \nabla \times \mathbf{v}^\perp\, dx\,dy = \oint_{\partial\Omega} \mathbf{v}^\perp \cdot\, d\mathbf{x} = \oint_{\partial\Omega} \mathbf{v} \cdot \mathbf{n}\; ds,$$

where $\mathbf{n}$ denotes the *unit outwards normal* to the boundary of our domain, while ds denotes the arc-length element along the boundary curve. This yields the *divergence form* of Green's Theorem:

$$\iint_\Omega \nabla \cdot \mathbf{v}\; dx\,dy = \oint_{\partial\Omega} \mathbf{v} \cdot \mathbf{n}\; ds. \tag{6.80}$$

Physically, if $\mathbf{v}$ represents the velocity vector field of a steady-state fluid flow, then the line integral in (6.80) represents the net fluid flux out of the region Ω. As a result, the divergence $\nabla \cdot \mathbf{v}$ represents the local change in area of the fluid at each point, which serves to justify our earlier statement on incompressibility.

Consider next the product vector field $u\,\mathbf{v}$ obtained by multiplying a vector field $\mathbf{v}$ by a scalar field u. An elementary computation proves that its divergence is

$$\nabla \cdot (u\,\mathbf{v}) = u\,\nabla \cdot \mathbf{v} + \nabla u \cdot \mathbf{v}. \tag{6.81}$$

Replacing $\mathbf{v}$ by $u\,\mathbf{v}$ in the divergence formula (6.80), we deduce what is usually referred to as *Green's formula*

$$\iint_\Omega (\,u\,\nabla \cdot \mathbf{v} + \nabla u \cdot \mathbf{v}\,)\,dx\,dy = \oint_{\partial\Omega} u\,(\mathbf{v} \cdot \mathbf{n})\,ds, \tag{6.82}$$

which is valid for arbitrary bounded domains Ω, and arbitrary C^1 scalar and vector fields defined thereon. Rearranging the terms produces

$$\iint_\Omega \nabla u \cdot \mathbf{v}\; dx\,dy = \oint_{\partial\Omega} u\,(\mathbf{v} \cdot \mathbf{n})\,ds - \iint_\Omega u\,\nabla \cdot \mathbf{v}\,dx\,dy. \tag{6.83}$$

We will view this identity as an *integration by parts* formula for double integrals. Indeed, comparing with the one-dimensional integration by parts formula

$$\int_a^b u'(x)\, v(x)\, dx = u(x)\, v(x) \Big|_{x=a}^{b} - \int_a^b u(x)\, v'(x)\, dx, \tag{6.84}$$

we observe that the single integrals have become double integrals; the derivatives are vector derivatives (gradient and divergence), while the boundary contributions at the endpoints of the interval are replaced by a line integral around the entire boundary of the two-dimensional domain.

A useful special case of (6.82) is that in which $\mathbf{v} = \nabla v$ is the gradient of a scalar field v. Then, in view of (6.74), Green's formula (6.82) becomes

$$\iint_\Omega \big(u\, \Delta v + \nabla u \cdot \nabla v \big)\, dx\, dy = \oint_{\partial\Omega} u\, \frac{\partial v}{\partial \mathbf{n}}\, ds, \tag{6.85}$$

where $\partial v/\partial \mathbf{n} = \nabla v \cdot \mathbf{n}$ is the *normal derivative* of the scalar field v on the boundary of the domain. In particular, setting $v = u$, we deduce

$$\iint_\Omega \big(u\, \Delta u + \| \nabla u \|^2 \big)\, dx\, dy = \oint_{\partial\Omega} u\, \frac{\partial u}{\partial \mathbf{n}}\, ds. \tag{6.86}$$

As an application, we establish a basic uniqueness theorem for solutions to the boundary value problems for the Poisson equation:

Theorem 6.15. *Suppose $\widetilde{u}$ and u both satisfy the same inhomogeneous Dirichlet or mixed boundary value problem for the Poisson equation on a connected, bounded domain Ω. Then $\widetilde{u} = u$. On the other hand, if $\widetilde{u}$ and u satisfy the same Neumann boundary value problem, then $\widetilde{u} = u + c$ for some constant c.*

Proof: Since, by assumption, $-\Delta \widetilde{u} = f = -\Delta u$, the difference $v = \widetilde{u} - u$ satisfies the Laplace equation $\Delta v = 0$ in Ω, and satisfies the homogeneous boundary conditions. Therefore, applying (6.86) to v, we find

$$\iint_\Omega \| \nabla v \|^2\, dx\, dy = \oint_{\partial\Omega} v\, \frac{\partial v}{\partial \mathbf{n}}\, ds = 0,$$

since, at every point on the boundary, either $v = 0$ or $\partial v/\partial \mathbf{n} = 0$. Since the integrand is continuous and everywhere nonnegative, we immediately conclude that $\| \nabla v \|^2 = 0$, and hence $\nabla v = \mathbf{0}$ throughout Ω. On a connected domain, the only functions annihilated by the gradient operator are the constants:

Lemma 6.16. *If $v(x, y)$ is a C^1 function defined on a connected domain $\Omega \subset \mathbb{R}^2$, then $\nabla v \equiv 0$ if and only if $v(x, y) \equiv c$ is a constant.*

Proof: Let $\mathbf{a}, \mathbf{b}$ be any two points in Ω. Then, by connectivity, we can find a curve C connecting them. The Fundamental Theorem for line integrals, [**8**, **108**], states that

$$\int_C \nabla v \cdot d\mathbf{x} = v(\mathbf{b}) - v(\mathbf{a}).$$

Thus, if $\nabla v \equiv 0$, then $v(\mathbf{b}) = v(\mathbf{a})$ for all $\mathbf{a}, \mathbf{b} \in \Omega$, which implies that v must be constant. *Q.E.D.*

Returning to our proof, we conclude that $\widetilde{u} = u + v = u + c$, which proves the result in the Neumann case. In the Dirichlet or mixed problems, there is at least one point on the boundary where $v = 0$, and hence the only possible constant is $v = c = 0$, proving that $\widetilde{u} = u$. *Q.E.D.*

Thus, the Dirichlet and mixed boundary value problems admit at most one solution, while the Neumann boundary value problem has either no solutions or infinitely many solutions. Proof of existence of solutions is more challenging, and will be left to a more advanced text, e.g., [**35, 44, 61, 70**].

If we subtract from formula (6.85) the formula

$$\iint_{\Omega} \left(v\,\Delta u + \nabla u \cdot \nabla v \right) dx\,dy = \oint_{\partial\Omega} v\,\frac{\partial u}{\partial \mathbf{n}}\,ds, \tag{6.87}$$

obtained by interchanging u and v, we obtain the identity

$$\iint_{\Omega} \left(u\,\Delta v - v\,\Delta u \right) dx\,dy = \oint_{\partial\Omega} \left(u\,\frac{\partial v}{\partial \mathbf{n}} - v\,\frac{\partial u}{\partial \mathbf{n}} \right) ds, \tag{6.88}$$

which will play a major role in our analysis of the Poisson equation. Setting $v = 1$ in (6.87) yields

$$\iint_{\Omega} \Delta u\,dx\,dy = \oint_{\partial\Omega} \frac{\partial u}{\partial \mathbf{n}}\,ds. \tag{6.89}$$

Suppose u solves the Neumann boundary value problem

$$-\Delta u = f, \quad \text{in} \quad \Omega \qquad\qquad \frac{\partial u}{\partial \mathbf{n}} = h \quad \text{on} \quad \partial\Omega.$$

Then (6.89) requires that

$$\iint_{\Omega} f\,dx\,dy + \oint_{\partial\Omega} h\,ds = 0, \tag{6.90}$$

which thus forms a necessary condition for the existence of a solution u to the inhomogeneous Neumann boundary value problem. Physically, if u represents the equilibrium temperature of a plate, then the integrals in (6.89) measure the net gain or loss in heat energy due to, respectively, the external heat source and the heat flux through the boundary. Equation (6.90) is telling us that, for the plate to remain in thermal equilibrium, there can be no net change in its total heat energy.

The Two–Dimensional Delta Function

Now let us return to the business at hand — solving the Poisson equation on a bounded domain $\Omega \subset \mathbb{R}^2$. We will subject the solution to either homogeneous Dirichlet boundary conditions or homogeneous mixed boundary conditions. (As we just noted, the Neumann boundary value problem does not admit a unique solution, and hence does not possess a Green's function.) The Green's function for the boundary value problem arises when the forcing function is a unit impulse concentrated at a single point in the domain.

Thus, our first task is to establish the proper form for a unit impulse in our two-dimensional context. The *delta function* concentrated at a point $\boldsymbol{\xi} = (\xi, \eta) \in \mathbb{R}^2$ is denoted by

$$\delta_{(\xi,\eta)}(x, y) = \delta_{\boldsymbol{\xi}}(\mathbf{x}) = \delta(\mathbf{x} - \boldsymbol{\xi}) = \delta(x - \xi, y - \eta), \tag{6.91}$$

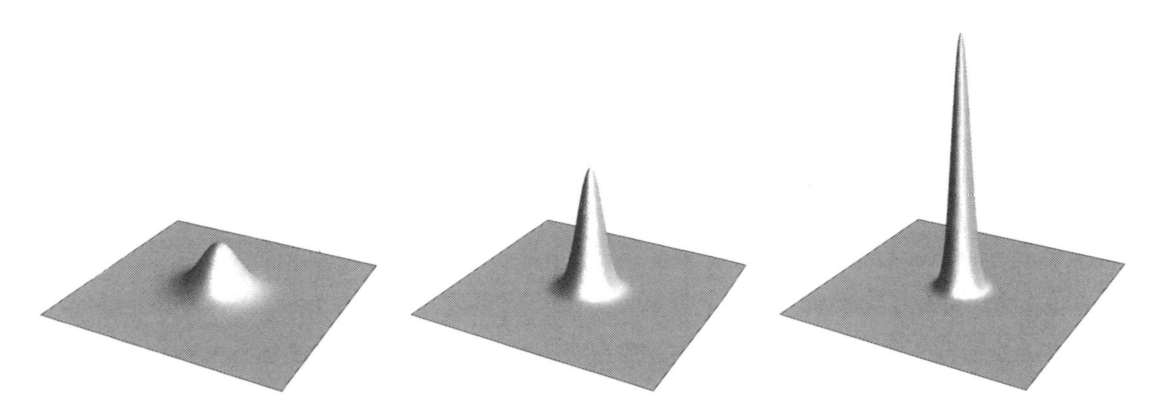

Figure 6.11. Gaussian functions converging to the delta function.

and is designed so that

$$\delta_{\boldsymbol{\xi}}(\mathbf{x}) = 0, \qquad \mathbf{x} \neq \boldsymbol{\xi}, \qquad \iint_{\Omega} \delta_{(\xi,\eta)}(x,y)\, dx\, dy = 1, \qquad \boldsymbol{\xi} \in \Omega. \tag{6.92}$$

In particular, $\delta(x,y) = \delta_{\mathbf{0}}(x,y)$ represents the delta function at the origin. As in the one-dimensional version, there is no ordinary function that satisfies both criteria; rather, $\delta(x,y)$ is to be viewed as the limit of a sequence of more and more highly concentrated functions $g_n(x,y)$, with

$$\lim_{n \to \infty} g_n(x,y) = 0, \quad \text{for} \quad (x,y) \neq (0,0), \qquad \text{while} \qquad \iint_{\mathbb{R}^2} g_n(x,y)\, dx\, dy = 1.$$

A good example of a suitable sequence is provided by the *radial Gaussian functions*

$$g_n(x,y) = \frac{n}{\pi}\, e^{-n\,(x^2+y^2)}. \tag{6.93}$$

As plotted in Figure 6.11, as $n \to \infty$, the Gaussian profiles become more and more concentrated near the origin, while maintaining a unit volume underneath their graphs. The fact that their integral over $\mathbb{R}^2$ equals 1 is a consequence of (2.99).

Alternatively, one can assign the delta function a dual interpretation as the linear functional

$$L_{(\xi,\eta)}[u] = L_{\boldsymbol{\xi}}[u] = u(\boldsymbol{\xi}) = u(\xi,\eta), \tag{6.94}$$

which assigns to each continuous function $u \in \mathrm{C}^0(\overline{\Omega})$ its value at the point $\boldsymbol{\xi} = (\xi,\eta) \in \Omega$. Then, using the L^2 inner product

$$\langle u, v \rangle = \iint_{\Omega} u(x,y)\, v(x,y)\, dx\, dy \tag{6.95}$$

between scalar fields $u, v \in \mathrm{C}^0(\overline{\Omega})$, we formally identify the linear functional $L_{(\xi,\eta)}$ with the delta "function" by the integral formula

$$\langle \delta_{(\xi,\eta)}, u \rangle = \iint_{\Omega} \delta_{(\xi,\eta)}(x,y)\, u(x,y)\, dx\, dy = \begin{cases} u(\xi,\eta), & (\xi,\eta) \in \Omega, \\ 0, & (\xi,\eta) \in \mathbb{R}^2 \setminus \overline{\Omega}, \end{cases} \tag{6.96}$$

for any $u \in \mathrm{C}^0(\overline{\Omega})$. As in the one-dimensional version, we will avoid defining the integral when the delta function is concentrated at a boundary point of the domain.

Since double integrals can be evaluated as repeated one-dimensional integrals, we can conveniently view

$$\delta_{(\xi,\eta)}(x,y) = \delta_\xi(x)\,\delta_\eta(y) = \delta(x-\xi)\,\delta(y-\eta) \tag{6.97}$$

as the product† of a pair of one-dimensional delta functions. Indeed, if the impulse point

$$(\xi,\eta) \in R = \big\{\, a < x < b,\ c < y < d \,\big\} \subset \Omega$$

is contained in a rectangle that lies within the domain, then

$$\begin{aligned}\iint_\Omega \delta_{(\xi,\eta)}(x,y)\,u(x,y)\,dx\,dy &= \iint_R \delta_{(\xi,\eta)}(x,y)\,u(x,y)\,dx\,dy \\ = \int_a^b \left(\int_c^d \delta(x-\xi)\,\delta(y-\eta)\,u(x,y)\,dy \right) dx &= \int_a^b \delta(x-\xi)\,u(x,\eta)\,dx = u(\xi,\eta).\end{aligned}$$

The Green's Function

As in the one-dimensional context, the Green's function is defined as the solution to the inhomogeneous differential equation when subject to a concentrated unit delta impulse at a prescribed point $\boldsymbol{\xi} = (\xi,\eta) \in \Omega$ inside the domain. In the current situation, the Poisson equation takes the form

$$-\Delta u = \delta_{\boldsymbol{\xi}}, \qquad \text{or, explicitly,} \qquad -\frac{\partial^2 u}{\partial x^2} - \frac{\partial^2 u}{\partial y^2} = \delta(x-\xi)\,\delta(y-\eta). \tag{6.98}$$

The function $u(x,y)$ is also subject to some homogeneous boundary conditions, e.g., the Dirichlet conditions $u = 0$ on $\partial\Omega$. The resulting solution is called the *Green's function* for the boundary value problem, and written

$$G_{\boldsymbol{\xi}}(\mathbf{x}) = G(\mathbf{x};\boldsymbol{\xi}) = G(x,y;\xi,\eta). \tag{6.99}$$

Once we know the Green's function, the solution to the general Poisson boundary value problem

$$-\Delta u = f \quad \text{in} \quad \Omega, \qquad u = 0 \quad \text{on} \quad \partial\Omega \tag{6.100}$$

is reconstructed as follows. We regard the forcing function

$$f(x,y) = \iint_\Omega \delta(x-\xi)\,\delta(y-\eta) f(\xi,\eta)\,d\xi\,d\eta$$

as a superposition of delta impulses, whose strength equals the value of f at the impulse point. Linearity implies that the solution to the boundary value problem is the corresponding superposition of Green's function responses to each of the constituent impulses. The net result is the fundamental *superposition formula*

$$u(x,y) = \iint_\Omega G(x,y;\xi,\eta)\,f(\xi,\eta)\,d\xi\,d\eta \tag{6.101}$$

† This is an exception to our earlier injunction not to multiply delta functions. Multiplication is allowed when they depend on *different* variables.

for the solution to the boundary value problem. Indeed,

$$
\begin{aligned}
-\Delta u(x,y) &= \iint_\Omega -\Delta G(x,y;\xi,\eta)\, f(\xi,\eta)\, d\xi\, d\eta \\
&= \iint_\Omega \delta(x-\xi, y-\eta)\, f(\xi,\eta)\, d\xi\, d\eta = f(x,y),
\end{aligned}
$$

while the fact that $G(x,y;\xi,\eta) = 0$ for all $(x,y) \in \partial\Omega$ implies that $u(x,y) = 0$ on the boundary.

The Green's function inevitably turns out to be symmetric under interchange of its arguments:

$$G(\xi,\eta;x,y) = G(x,y;\xi,\eta). \tag{6.102}$$

As in the one-dimensional case, symmetry is a consequence of the self-adjointness of the boundary value problem, and will be explained in full in Chapter 9. Symmetry has the following intriguing physical interpretation: Let $\mathbf{x}, \boldsymbol{\xi} \in \Omega$ be any two points in the domain. We apply a concentrated unit force to the membrane at the first point and measure its deflection at the second; the result is exactly the same as if we applied the impulse at the second point and measured the deflection at the first. (Deflections at other points in the domain will typically have no obvious relation with one another.) Similarly, in electrostatics, the solution $u(x,y)$ is interpreted as the electrostatic potential for a system of charges in equilibrium. A delta function corresponds to a point charge, e.g., an electron. The symmetry property says that the electrostatic potential at $\mathbf{x}$ due to a point charge placed at position $\boldsymbol{\xi}$ is exactly the same as the potential at $\boldsymbol{\xi}$ due to a point charge at $\mathbf{x}$. The reader may wish to meditate on the physical plausibility of these striking facts.

Unfortunately, most Green's functions cannot be written down in closed form. One important exception occurs when the domain is the entire plane: $\Omega = \mathbb{R}^2$. The solution to the Poisson equation (6.98) is the *free-space Green's function* $G_0(x,y;\xi,\eta) = G_0(\mathbf{x};\boldsymbol{\xi})$, which measures the effect of a unit impulse, concentrated at $\boldsymbol{\xi}$, throughout two-dimensional space, e.g., the gravitational potential due to a point mass or the electrostatic potential due to a point charge. To motivate the construction, let us appeal to physical intuition. First, since the concentrated impulse is zero when $\mathbf{x} \neq \boldsymbol{\xi}$, the function must solve the homogeneous Laplace equation

$$-\Delta G_0 = 0 \qquad \text{for all} \qquad \mathbf{x} \neq \boldsymbol{\xi}. \tag{6.103}$$

Second, since the Poisson equation is modeling a homogeneous, uniform medium, in the absence of boundary conditions the effect of a unit impulse should depend only on the distance from its source. Therefore, we expect G_0 to be a function of the radial variable alone:

$$G_0(x,y;\xi,\eta) = v(r), \qquad \text{where} \qquad r = \|\,\mathbf{x} - \boldsymbol{\xi}\,\| = \sqrt{(x-\xi)^2 + (y-\eta)^2}\,.$$

According to (4.113), the only radially symmetric solutions to the Laplace equation are

$$v(r) = a + b \log r, \tag{6.104}$$

where a and b are constants. The constant term a has zero derivative, and so cannot contribute to the delta function singularity. Therefore, we expect the required solution to be a multiple of the logarithmic term. To determine the multiple, consider a closed disk of radius $\varepsilon > 0$ centered at $\boldsymbol{\xi}$,

$$D_\varepsilon = \{\, 0 \leq r \leq \varepsilon \,\} = \{\, \|\,\mathbf{x} - \boldsymbol{\xi}\,\| \leq \varepsilon \,\},$$

with circular boundary

$$C_\varepsilon = \partial D_\varepsilon = \{ r = \| \mathbf{x} - \boldsymbol{\xi} \| = \varepsilon \} = \{ (\xi + \varepsilon \cos\theta, \eta + \varepsilon \sin\theta) \mid -\pi \le \theta \le \pi \}.$$

Then, by (6.74) and the divergence form (6.80) of Green's Theorem,

$$\begin{aligned} 1 &= \iint_{D_\varepsilon} \delta(x,y)\, dx\, dy = -b \iint_{D_\varepsilon} \Delta(\log r)\, dx\, dy = -b \iint_{D_\varepsilon} \nabla \cdot \nabla(\log r)\, dx\, dy \\ &= -b \oint_{C_\varepsilon} \frac{\partial(\log r)}{\partial \mathbf{n}}\, ds = -b \oint_{C_\varepsilon} \frac{\partial(\log r)}{\partial r}\, ds = -b \oint_{C_\varepsilon} \frac{1}{r}\, ds = -b \int_{-\pi}^{\pi} d\theta = -2\pi b, \end{aligned} \tag{6.105}$$

and hence $b = -1/(2\pi)$. We conclude that the free-space Green's function should have the logarithmic form

$$G_0(x,y;\xi,\eta) = -\frac{1}{2\pi} \log r = -\frac{1}{2\pi} \log \| \mathbf{x} - \boldsymbol{\xi} \| = -\frac{1}{4\pi} \log\big[(x-\xi)^2 + (y-\eta)^2 \big]. \tag{6.106}$$

A fully rigorous, albeit more difficult, justification of (6.106) comes from the following important result, known as *Green's representation formula*.

Theorem 6.17. *Let $\Omega \subset \mathbb{R}^2$ be a bounded domain, with piecewise C^1 boundary $\partial\Omega$. Suppose $u \in \mathrm{C}^2(\Omega) \cap \mathrm{C}^1(\overline{\Omega})$. Then, for any $(x,y) \in \Omega$,*

$$\begin{aligned} u(x,y) = &- \iint_\Omega G_0(x,y;\xi,\eta)\, \Delta u(\xi,\eta)\, d\xi\, d\eta \\ &+ \oint_{\partial\Omega} \left(G_0(x,y;\xi,\eta) \frac{\partial u}{\partial \mathbf{n}}(\xi,\eta) - \frac{\partial G_0}{\partial \mathbf{n}}(x,y;\xi,\eta)\, u(\xi,\eta) \right) ds, \end{aligned} \tag{6.107}$$

where the Laplacian and the normal derivatives on the boundary are all taken with respect to the integration variables $\boldsymbol{\xi} = (\xi,\eta)$.

In particular, if both u and $\partial u/\partial \mathbf{n}$ vanish on $\partial\Omega$, then (6.107) reduces to

$$u(x,y) = - \iint_{\mathbb{R}^2} G_0(x,y;\xi,\eta)\, \Delta u(\xi,\eta)\, d\xi\, d\eta.$$

Invoking the definition of the delta function on the left-hand side and formally applying the Green identity (6.88) to the right-hand side produces

$$\iint_{\mathbb{R}^2} \delta(x-\xi)\, \delta(y-\eta)\, u(\xi,\eta)\, d\xi\, d\eta = \iint_{\mathbb{R}^2} -\Delta G_0(x,y;\xi,\eta)\, u(\xi,\eta)\, d\xi\, d\eta. \tag{6.108}$$

It is in this dual sense that we justify the desired formula

$$-\Delta G_0(\mathbf{x};\boldsymbol{\xi}) = \frac{1}{2\pi} \Delta\big(\log \| \mathbf{x} - \boldsymbol{\xi} \| \big) = \delta(\mathbf{x} - \boldsymbol{\xi}). \tag{6.109}$$

Proof of Theorem 6.17: We first note that, even though $G_0(\mathbf{x},\boldsymbol{\xi})$ has a logarithmic singularity at $\mathbf{x} = \boldsymbol{\xi}$, the double integral in (6.107) is finite. Indeed, after introducing polar coordinates $\xi = x + r\cos\theta$, $\eta = y + r\sin\theta$, and recalling $d\xi\, d\eta = r\, dr\, d\theta$, we see that it equals

$$\frac{1}{2\pi} \iint (r \log r)\, \Delta u\, dr\, d\theta.$$

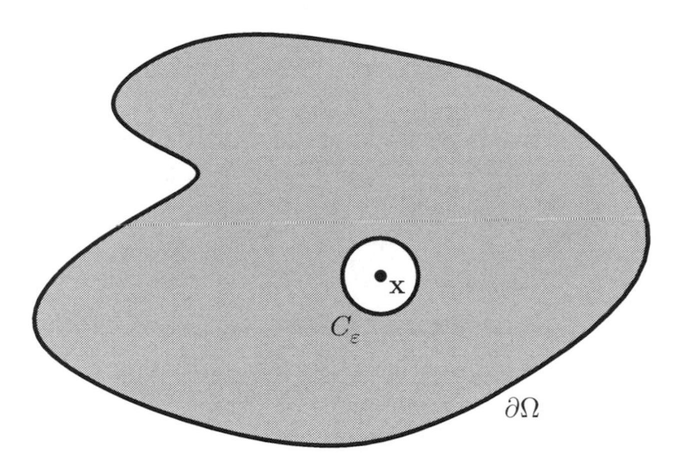

Figure 6.12. Domain $\Omega_\varepsilon = \Omega \setminus D_\varepsilon(\mathbf{x})$.

The product $r \log r$ is everywhere continuous — even at $r = 0$ — and so, provided Δu is well behaved, e.g., continuous, the integral is finite. There is, of course, no problem with the line integral in (6.107), since the contour does not go through the singularity.

Let us now avoid dealing directly with the singularity by working on a subdomain

$$\Omega_\varepsilon = \Omega \setminus D_\varepsilon(\mathbf{x}) = \{\, \boldsymbol{\xi} \in \Omega \mid \| \mathbf{x} - \boldsymbol{\xi} \| > \varepsilon \,\}$$

obtained by cutting out a small disk

$$D_\varepsilon(\mathbf{x}) = \{\, \boldsymbol{\xi} \mid \| \mathbf{x} - \boldsymbol{\xi} \| \leq \varepsilon \,\}$$

of radius $\varepsilon > 0$ centered at $\mathbf{x}$. We choose ε sufficiently small in order that $D_\varepsilon(\mathbf{x}) \subset \Omega$, and hence

$$\partial\Omega_\varepsilon = \partial\Omega \cup C_\varepsilon, \qquad \text{where} \qquad C_\varepsilon = \{\, \| \mathbf{x} - \boldsymbol{\xi} \| = \varepsilon \,\}$$

is the circular boundary of the disk. The subdomain Ω_ε is represented by the shaded region in Figure 6.12. Since the double integral is well defined, we can approximate it by integrating over Ω_ε:

$$\iint_\Omega G_0(x,y;\xi,\eta)\, \Delta u(\xi,\eta)\, d\xi\, d\eta = \lim_{\varepsilon \to 0} \iint_{\Omega_\varepsilon} G_0(x,y;\xi,\eta)\, \Delta u(\xi,\eta)\, d\xi\, d\eta. \tag{6.110}$$

Since G_0 has no singularities in Ω_ε, we are able to apply the Green formula (6.85) and then (6.103) to evaluate

$$\begin{aligned} &\iint_{\Omega_\varepsilon} G_0(x,y;\xi,\eta)\, \Delta u(\xi,\eta)\, d\xi\, d\eta \\ &\quad = \oint_{\partial\Omega} \left(G_0(x,y;\xi,\eta)\, \frac{\partial u}{\partial \mathbf{n}}(\xi,\eta) - \frac{\partial G_0}{\partial \mathbf{n}}(x,y;\xi,\eta)\, u(\xi,\eta) \right) ds \qquad (6.111) \\ &\qquad\quad - \oint_{C_\varepsilon} \left(G_0(x,y;\xi,\eta)\, \frac{\partial u}{\partial \mathbf{n}}(\xi,\eta) - \frac{\partial G_0}{\partial \mathbf{n}}(x,y;\xi,\eta)\, u(\xi,\eta) \right) ds, \end{aligned}$$

where the line integral around C_ε is taken in the usual counterclockwise direction — the opposite orientation to that induced by its status as part of the boundary of Ω_ε. Now, on the circle C_ε,

$$G_0(x,y;\xi,\eta) = -\left.\frac{\log r}{2\pi}\right|_{r=\varepsilon} = -\frac{\log\varepsilon}{2\pi}, \tag{6.112}$$

while, in view of Exercise 6.3.1,

$$\frac{\partial G_0}{\partial \mathbf{n}}(x,y;\xi,\eta) = -\left.\frac{1}{2\pi}\frac{\partial(\log r)}{\partial r}\right|_{r=\varepsilon} = -\frac{1}{2\pi\varepsilon}. \tag{6.113}$$

Therefore,

$$\oint_{C_\varepsilon}\frac{\partial G_0}{\partial \mathbf{n}}(x,y;\xi,\eta)\,u(\xi,\eta)\,ds = -\frac{1}{2\pi\varepsilon}\oint_{C_\varepsilon} u(\xi,\eta)\,ds,$$

which we recognize as minus the *average* of u on the circle of radius ε. As $\varepsilon \to 0$, the circles shrink down to their common center, and so, by continuity, the averages tend to the value $u(x,y)$ at the center; thus,

$$\lim_{\varepsilon\to 0}\oint_{C_\varepsilon}\frac{\partial G_0}{\partial \mathbf{n}}(x,y;\xi,\eta)\,u(\xi,\eta)\,ds = -\,u(x,y). \tag{6.114}$$

On the other hand, using (6.112), and then (6.89) on the disk D_ε, we have

$$\begin{aligned}\oint_{C_\varepsilon} G_0(x,y;\xi,\eta)\,\frac{\partial u}{\partial \mathbf{n}}(\xi,\eta)\,ds &= -\frac{\log\varepsilon}{2\pi}\oint_{C_\varepsilon}\frac{\partial u}{\partial \mathbf{n}}(\xi,\eta)\,ds \\ &= -\frac{\log\varepsilon}{2\pi}\iint_{D_\varepsilon}\Delta u(\xi,\eta)\,d\xi\,d\eta = -\,(\varepsilon^2\log\varepsilon)\,\overline{\Delta u}_\varepsilon,\end{aligned}$$

where

$$\overline{\Delta u}_\varepsilon = \frac{1}{2\pi\varepsilon^2}\iint_{D_\varepsilon}\Delta u(\xi,\eta)\,d\xi\,d\eta$$

is the *average* of Δu over the disk D_ε. As above, as $\varepsilon \to 0$, the averages over the disks converge to the value at their common center, $\overline{\Delta u}_\varepsilon \to \Delta u(x,y)$, and hence

$$\lim_{\varepsilon\to 0}\oint_{C_\varepsilon} G_0(x,y;\xi,\eta)\,\frac{\partial u}{\partial \mathbf{n}}(\xi,\eta)\,ds = \lim_{\varepsilon\to 0}\,(-\,\varepsilon^2\log\varepsilon)\,\overline{\Delta u}_\varepsilon = 0. \tag{6.115}$$

In view of (6.110, 114, 115), the $\varepsilon \to 0$ limit of (6.111) is exactly the Green representation formula (6.107). *Q.E.D.*

As noted above, the free space Green's function (6.106) represents the gravitational potential in empty two-dimensional space due to a unit point mass, or, equivalently, the two-dimensional electrostatic potential due to a unit point charge sitting at position $\boldsymbol{\xi}$. The corresponding gravitational or electrostatic force field is obtained by taking its gradient:

$$\mathbf{F} = \nabla G_0 = -\frac{\mathbf{x}-\boldsymbol{\xi}}{2\pi\,\|\,\mathbf{x}-\boldsymbol{\xi}\,\|^2}.$$

Its magnitude

$$\|\,\mathbf{F}\,\| = \frac{1}{2\pi\,\|\,\mathbf{x}-\boldsymbol{\xi}\,\|}$$

is inversely proportional to the distance from the mass or charge, which is the two-dimensional form of Newton's and Coulomb's three-dimensional inverse square laws.

The gravitational potential due to a two-dimensional mass, e.g., a flat plate, in the shape of a domain $\Omega \subset \mathbb{R}^2$ is obtained by superimposing delta function sources with strengths equal to the density of the material at each point. The result is the potential function

$$u(x,y) = -\frac{1}{4\pi}\iint_\Omega \rho(\xi,\eta)\,\log\left[\,(x-\xi)^2+(y-\eta)^2\,\right]d\xi\,d\eta, \tag{6.116}$$

in which $\rho(\xi,\eta)$ denotes the density at position $(\xi,\eta)\in\Omega$.

Example 6.18. The gravitational potential due to a circular disk $D=\{x^2+y^2\le 1\}$ of unit radius and unit density $\rho\equiv 1$ is

$$u(x,y) = -\frac{1}{4\pi}\iint_D \log\left[\,(x-\xi)^2+(y-\eta)^2\,\right]d\xi\,d\eta. \tag{6.117}$$

A direct evaluation of this double integral is not so easy. However, we can write down the potential in closed form by recalling that it solves the Poisson equation

$$-\Delta u = \begin{cases} 1, & \|\,\mathbf{x}\,\|<1, \\ 0, & \|\,\mathbf{x}\,\|>1. \end{cases} \tag{6.118}$$

Moreover, u is clearly radially symmetric, and hence a function of r alone. Thus, in the polar coordinate expression (4.105) for the Laplacian, the θ derivative terms vanish, and so (6.118) reduces to

$$\frac{d^2u}{dr^2}+\frac{1}{r}\,\frac{du}{dr} = \begin{cases} -1, & r<1, \\ 0, & r>1, \end{cases}$$

which is effectively a first-order linear ordinary differential equation for du/dr. Solving separately on the two subintervals produces

$$u(r) = \begin{cases} a+b\log r-\frac14 r^2, & r<1, \\ c+d\log r, & r>1, \end{cases}$$

where a,b,c,d are constants. Continuity of $u(r)$ and $u'(r)$ at $r=1$ implies $c=a-\frac14$, $d=b-\frac12$. Moreover, the potential for a non-concentrated mass cannot have a singularity at the origin, and so $b=0$. Direct evaluation of (6.117) at $x=y=0$, using polar coordinates, proves that $a=\frac14$. We conclude that the gravitational potential (6.117) due to a uniform disk of unit radius, and hence total mass (area) π, is, explicitly,

$$u(x,y) = \begin{cases} \frac14(1-r^2)=\frac14(1-x^2-y^2), & x^2+y^2\le 1, \\ -\frac12\log r = -\frac14\log(x^2+y^2), & x^2+y^2\ge 1. \end{cases} \tag{6.119}$$

Observe that, outside the disk, the potential is exactly the same as the logarithmic potential due to a point mass of magnitude π located at the origin. Consequently, the gravitational force field outside a uniform disk is the same as if all its mass were concentrated at the origin.

With the free-space logarithmic potential in hand, let us return to the question of finding the Green's function for a boundary value problem on a bounded domain $\Omega\subset\mathbb{R}^2$. Since the logarithmic potential (6.106) is a particular solution to the Poisson equation (6.98), the general solution, according to Theorem 1.6, is given by $u=G_0+z$, where z is an arbitrary solution to the homogeneous equation $\Delta z=0$, i.e., an arbitrary harmonic function. Thus, constructing the Green's function has been reduced to the problem of finding the harmonic function z such that $G=G_0+z$ satisfies the desired homogeneous boundary conditions. Let us explicitly formulate this result for the (inhomogeneous) Dirichlet problem.

Theorem 6.19. *The Green's function for the Dirichlet boundary value problem for the Poisson equation on a bounded domain* $\Omega \subset \mathbb{R}^2$ *has the form*

$$G(x,y;\xi,\eta) = G_0(x,y;\xi,\eta) + z(x,y;\xi,\eta), \tag{6.120}$$

where the first term is the logarithmic potential (6.106), *while, for each* $(\xi,\eta) \in \Omega$, *the second term is the harmonic function that solves the boundary value problem*

$$\begin{aligned} \Delta z &= 0 \qquad \text{on} \qquad \Omega, \\ z(x,y;\xi,\eta) &= \frac{1}{4\pi} \log\big[\,(x-\xi)^2 + (y-\eta)^2\,\big] \qquad \text{for} \qquad (x,y) \in \partial\Omega. \end{aligned} \tag{6.121}$$

If $u(x,y)$ *is a solution to the inhomogeneous Dirichlet problem*

$$-\Delta u = f, \qquad \mathbf{x} \in \Omega, \qquad\qquad u = h, \qquad \mathbf{x} \in \partial\Omega, \tag{6.122}$$

then

$$u(x,y) = \iint_\Omega G(x,y;\xi,\eta)\, f(\xi,\eta)\, d\xi\, d\eta - \oint_{\partial\Omega} \frac{\partial G}{\partial \mathbf{n}}(x,y;\xi,\eta)\, h(\xi,\eta)\, ds, \tag{6.123}$$

where the normal derivative of G *is taken with respect to* $(\xi,\eta) \in \partial\Omega$.

Proof: To show that (6.120) is the Green's function, we note that

$$-\Delta G = -\Delta G_0 - \Delta z = \delta_{(\xi,\eta)} \qquad \text{in} \qquad \Omega, \tag{6.124}$$

while

$$G(x,y;\xi,\eta) = G_0(x,y;\xi,\eta) + z(x,y;\xi,\eta) = 0 \qquad \text{on} \qquad \partial\Omega. \tag{6.125}$$

Next, to establish the solution formula (6.123), since both z and u are C^2, we can use (6.88) (with $v = z$, keeping in mind that $\Delta z = 0$) to establish

$$\begin{aligned} 0 = &- \iint_\Omega z(x,y;\xi,\eta)\, \Delta u(\xi,\eta)\, d\xi\, d\eta \\ &+ \oint_{\partial\Omega} \left(z(x,y;\xi,\eta)\, \frac{\partial u}{\partial \mathbf{n}}(\xi,\eta) - \frac{\partial z}{\partial \mathbf{n}}(x,y;\xi,\eta)\, u(\xi,\eta) \right) ds. \end{aligned}$$

Adding this to Green's representation formula (6.107), and using (6.125), we deduce that

$$u(x,y) = -\iint_\Omega G(x,y;\xi,\eta)\, \Delta u(\xi,\eta)\, d\xi\, d\eta - \oint_{\partial\Omega} \frac{\partial G(x,y;\xi,\eta)}{\partial \mathbf{n}}\, u(\xi,\eta)\, ds,$$

which, given (6.122), produces (6.123). *Q.E.D.*

The one subtle issue left unresolved is the existence of the solution. Read properly, Theorem 6.19 states that *if* a classical solution exists, *then* it is necessarily given by the Green's function formula (6.123). Proving existence of the solution — and also the existence of the Green's function, or equivalently, the solution z to (6.121) — requires further in-depth analysis, lying beyond the scope of this text. In particular, to guarantee existence, the underlying domain must have a reasonably nice boundary, e.g., a piecewise smooth curve without sharp cusps. Interestingly, lack of regularity at sharp cusps in the boundary underlies the electromagnetic phenomenon known as St. Elmo's fire, cf. [**121**]. Extensions to irregular domains, e.g., those with fractal boundaries, is an active area of contemporary research. Moreover, unlike one-dimensional boundary value problems, mere continuity of the forcing function f is not quite sufficient to ensure the existence of a classical solution to the Poisson boundary value problem; differentiability does suffice, although this assumption can be weakened. We refer to [**61, 70**], for a development of the Perron method based on approximating the solution by a sequence of *subsolutions*, which, by definition, solve the differential inequality $-\Delta u \leq f$. An alternative proof, using the direct method of the calculus of variations, can be found in [**35**]. The latter proof relies on the characterization of the solution by a minimization principle, which we discuss in some detail in Chapter 9.

Exercises

♢ 6.3.1. Let C_R be a circle of radius R centered at the origin and $\mathbf{n}$ its unit outward normal. Let $f(r,\theta)$ be a function expressed in polar coordinates. Prove that $\partial f/\partial \mathbf{n} = \partial f/\partial r$ on C_R.

6.3.2. Let $f(x) > 0$ be a continuous, positive function on the interval $a \leq x \leq b$. Let Ω be the domain lying between the graph of $f(x)$ on the interval $[a,b]$ and the x–axis. Explain why (6.77) reduces to the usual calculus formula for the area under the graph of f.

6.3.3. Explain what happens to the conclusion of Lemma 6.16 if Ω is not a connected domain.

6.3.4. Can you find constants c_n such that the functions $g_n(x,y) = c_n[1+n^2(x^2+y^2)]^{-1}$ converge to the two-dimensional delta function: $g_n(x,y) \to \delta(x,y)$ as $n \to \infty$?

♢ 6.3.5. Explain why the two-dimensional delta function satisfies the scaling law

$$\delta(\beta\, x, \beta\, y) = \frac{1}{\beta^2}\,\delta(x,y), \qquad \text{for} \qquad \beta \neq 0.$$

♢ 6.3.6. Write out a polar coordinate formula, in terms of $\delta(r-r_0)$ and $\delta(\theta-\theta_0)$, for the two-dimensional delta function $\delta(x-x_0, y-y_0) = \delta(x-x_0)\,\delta(y-y_0)$.

6.3.7. *True or false*: $\delta(\mathbf{x}) = \delta(\|\,\mathbf{x}\,\|)$.

♢ 6.3.8. Suppose that $\xi = f(x,y)$, $\eta = g(x,y)$ defines a one-to-one C^1 map from a domain $D \subset \mathbb{R}^2$ to the domain $\Omega = \{\,(\xi,\eta) = (f(x,y), g(x,y)) \mid (x,y) \in D\,\} \subset \mathbb{R}^2$, and has nonzero Jacobian determinant: $J(x,y) = f_x g_y - f_y g_x \neq 0$ for all $(x,y) \in D$. Suppose further that $(0,0) = (f(x_0,y_0), g(x_0,y_0)) \in \Omega$ for $(x_0,y_0) \in D$. Prove the following formula governing the effect of the map on the two-dimensional delta function:

$$\delta(f(x,y), g(x,y)) = \frac{\delta(x-x_0, y-y_0)}{|\,J(x_0,y_0)\,|}\,. \tag{6.126}$$

6.3.9. Suppose $f(x,y) = \begin{cases} 1, & 3x-2y>1, \\ 0, & 3x-2y<1. \end{cases}$ Compute its partial derivatives $\dfrac{\partial f}{\partial x}$ and $\dfrac{\partial f}{\partial y}$ in the sense of generalized functions.

6.3.10. Find a series solution to the rectangular boundary value problem (4.91–92) when the boundary data $f(x) = \delta(x-\xi)$ is a delta function at a point $0 < \xi < a$. Is your solution infinitely differentiable inside the rectangle?

6.3.11. Answer Exercise 6.3.10 when $f(x) = \delta'(x-\xi)$ is the derivative of the delta function.

6.3.12. A 1 meter square plate is subject to the Neumann boundary conditions $\partial u/\partial \mathbf{n} = 1$ on its entire boundary. What is the equilibrium temperature? Explain.

♢ 6.3.13. A *conservation law* for an equilibrium system in two dimensions is, by definition, a divergence expression

$$\frac{\partial X}{\partial x} + \frac{\partial Y}{\partial y} = 0 \tag{6.127}$$

that vanishes for all solutions.

(a) Given a conservation law prescribed by $\mathbf{v} = (X, Y)$ defined on a simply connected domain D, show that the line integral $\int_C \mathbf{v} \cdot \mathbf{n}\, ds = \int_C X\, dy - Y\, dx$ is path-independent, meaning that its value depends only on the endpoints of the curve C.

(b) Show that the Laplace equation can be written as a conservation law, and write down the corresponding path-independent line integral.

Note: Path-independent integrals are of importance in the study of cracks, dislocations, and other material singularities, [**49**].

♢ 6.3.14. In two-dimensional dynamics, a *conservation law* is an equation of the form

$$\frac{\partial T}{\partial t} + \frac{\partial X}{\partial x} + \frac{\partial Y}{\partial y} = 0, \tag{6.128}$$

in which T is the *conserved density*, while $\mathbf{v} = (X, Y)$ represents the associated *flux*.

(a) Prove that, on a bounded domain $\Omega \subset \mathbb{R}^2$, the rate of change of the integral $\iint_\Omega T\, dx\, dy$ of the conserved density depends only on the flux through the boundary $\partial\Omega$.

(b) Write the partial differential equation $u_t + u u_x + u u_y = 0$ as a conservation law. What is the integrated version?

The Method of Images

The preceding analysis exposes the underlying form of the Green's function, but we are still left with the determination of the harmonic component $z(x, y)$ required to match the logarithmic potential boundary values, cf. (6.121). We will discuss two principal analytic techniques employed to produce explicit formulas. The first is an adaptation of the method of separation of variables, which leads to infinite series expressions. We will not dwell on this approach here, although a couple of the exercises ask the reader to work through some of the details; see also the discussion leading up to (9.110). The second is the *Method of Images*, which will be developed in this section. Another approach is based on the theory of *conformal mapping*; it can be found in books on complex analysis, including [**53, 98**]. While the first two methods are limited to a fairly small class of domains, they extend to higher-dimensional problems, as well as to certain other types of elliptic boundary value problems, whereas conformal mapping is, unfortunately, restricted to two-dimensional problems involving the Laplace and Poisson equations.

We already know that the singular part of the Green's function for the two-dimensional Poisson equation is provided by a logarithmic potential. The problem, then, is to construct the harmonic part, called $z(x, y)$ in (6.120), so that the sum has the correct homogeneous boundary values, or, equivalently, so that $z(x, y)$ has the same boundary values as the logarithmic potential. In certain cases, $z(x, y)$ can be thought of as the potential induced by one or more hypothetical electric charges (or, equivalently, gravitational point masses) that are located *outside* the domain Ω, arranged in such a manner that their combined electrostatic potential happens to coincide with the logarithmic potential on the boundary of the domain. The goal, then, is to place image charges of suitable strengths in the appropriate positions.

Here, we will only consider the case of a single image charge, located at a position $\boldsymbol{\eta} \notin \Omega$. We scale the logarithmic potential (6.106) by the charge strength, and, for added

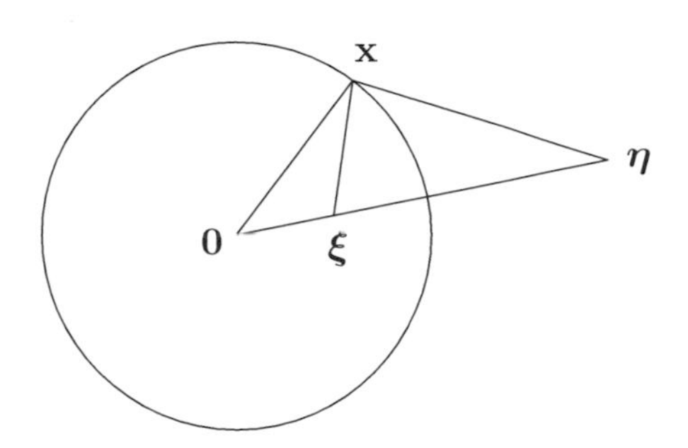

Figure 6.13. Method of Images for the unit disk.

flexibility, include an additional constant — the charge's potential baseline:

$$z(x,y) = a \log \| \mathbf{x} - \boldsymbol{\eta} \| + b, \qquad \boldsymbol{\eta} \in \mathbb{R}^2 \setminus \overline{\Omega}.$$

The function $z(x,y)$ is harmonic inside Ω, since the logarithmic potential is harmonic everywhere except at the external singularity $\boldsymbol{\eta}$. For the Dirichlet boundary value problem, then, for each point $\boldsymbol{\xi} \in \Omega$, we must find a corresponding image point $\boldsymbol{\eta} \in \mathbb{R}^2 \setminus \overline{\Omega}$ and constants $a, b \in \mathbb{R}$ such that[†]

$$\log \| \mathbf{x} - \boldsymbol{\xi} \| = a \log \| \mathbf{x} - \boldsymbol{\eta} \| + b \qquad \text{for all} \qquad \mathbf{x} \in \partial\Omega,$$

or, equivalently,

$$\| \mathbf{x} - \boldsymbol{\xi} \| = \lambda \| \mathbf{x} - \boldsymbol{\eta} \|^a \qquad \text{for all} \qquad \mathbf{x} \in \partial\Omega, \tag{6.129}$$

where $\lambda = e^b$. For each fixed $\boldsymbol{\xi}, \boldsymbol{\eta}, \lambda, a$, the equation in (6.129) will, typically, implicitly prescribe a plane curve, but it is not clear that one can always arrange that these curves all coincide with the boundary of our domain.

To make further progress, we appeal to a geometric construction based on similar triangles. Let us select $\boldsymbol{\eta} = c\boldsymbol{\xi}$ to be a point lying on the ray through $\boldsymbol{\xi}$. Its location is chosen so that the triangle with vertices $\mathbf{0}, \mathbf{x}, \boldsymbol{\eta}$ is similar to the triangle with vertices $\mathbf{0}, \boldsymbol{\xi}, \mathbf{x}$, noting that they have the same angle at the common vertex $\mathbf{0}$ — see Figure 6.13. Similarity requires that the triangles' corresponding sides have a common ratio, and so

$$\frac{\| \boldsymbol{\xi} \|}{\| \mathbf{x} \|} = \frac{\| \mathbf{x} \|}{\| \boldsymbol{\eta} \|} = \frac{\| \mathbf{x} - \boldsymbol{\xi} \|}{\| \mathbf{x} - \boldsymbol{\eta} \|} = \lambda. \tag{6.130}$$

The last equality implies that (6.129) holds with $a = 1$. Consequently, if we choose

$$\| \boldsymbol{\eta} \| = \frac{1}{\| \boldsymbol{\xi} \|}, \qquad \text{so that} \qquad \boldsymbol{\eta} = \frac{\boldsymbol{\xi}}{\| \boldsymbol{\xi} \|^2}, \tag{6.131}$$

then

$$\| \mathbf{x} \|^2 = \| \boldsymbol{\xi} \| \, \| \boldsymbol{\eta} \| = 1.$$

[†] To simplify the formulas, we have omitted the $1/(2\pi)$ factor, which can easily be reinstated at the end of the analysis.

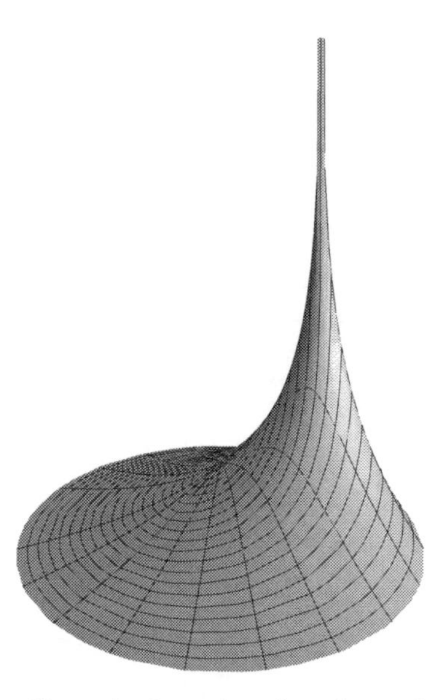

Figure 6.14. Green's function for the unit disk.

Thus $\mathbf{x}$ lies on the unit circle, and, as a result, $\lambda = \|\boldsymbol{\xi}\| = 1/\|\boldsymbol{\eta}\|$. The map taking a point $\boldsymbol{\xi}$ inside the disk to its image point $\boldsymbol{\eta}$ defined by (6.131) is known as *inversion* with respect to the unit circle.

We have now demonstrated that the potentials

$$\frac{1}{2\pi}\log\|\mathbf{x}-\boldsymbol{\xi}\| = \frac{1}{2\pi}\log\big(\|\boldsymbol{\xi}\|\,\|\mathbf{x}-\boldsymbol{\eta}\|\big) = \frac{1}{2\pi}\log\frac{\|\,\|\boldsymbol{\xi}\|^2\mathbf{x}-\boldsymbol{\xi}\,\|}{\|\boldsymbol{\xi}\|}, \qquad \|\mathbf{x}\|=1, \tag{6.132}$$

have the same boundary values on the unit circle. Consequently, their difference

$$G(\mathbf{x};\boldsymbol{\xi}) = -\frac{1}{2\pi}\log\|\mathbf{x}-\boldsymbol{\xi}\| + \frac{1}{2\pi}\log\frac{\|\,\|\boldsymbol{\xi}\|^2\mathbf{x}-\boldsymbol{\xi}\,\|}{\|\boldsymbol{\xi}\|} = \frac{1}{2\pi}\log\frac{\|\,\|\boldsymbol{\xi}\|^2\mathbf{x}-\boldsymbol{\xi}\,\|}{\|\boldsymbol{\xi}\|\,\|\mathbf{x}-\boldsymbol{\xi}\|} \tag{6.133}$$

has the required properties for the Green's function for the Dirichlet problem on the unit disk. Writing this in terms of polar coordinates

$$\mathbf{x} = (r\cos\theta, r\sin\theta), \qquad \boldsymbol{\xi} = (\rho\cos\phi, \rho\sin\phi),$$

and applying the Law of Cosines to the triangles in Figure 6.13 produces the explicit formula

$$G(r,\theta;\rho,\phi) = \frac{1}{4\pi}\log\left(\frac{1+r^2\rho^2-2r\rho\cos(\theta-\phi)}{r^2+\rho^2-2r\rho\cos(\theta-\phi)}\right). \tag{6.134}$$

In Figure 6.14 we sketch the Green's function for the Dirichlet boundary value problem corresponding to a unit impulse being applied at a point halfway between the center and the edge of the disk. We also require its radial derivative

$$\frac{\partial G}{\partial r}(r,\theta;\rho,\phi) = -\frac{1}{2\pi}\,\frac{1-r^2}{1+r^2-2r\cos(\theta-\phi)}, \tag{6.135}$$

which coincides with its normal derivative on the unit circle. Thus, specializing (6.123), we arrive at a solution to the general Dirichlet boundary value problem for the Poisson equation in the unit disk.

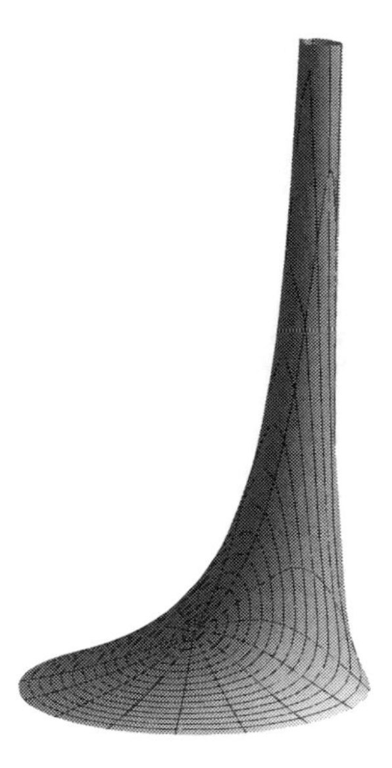

Figure 6.15. The Poisson kernel.

Theorem 6.20. *The solution to the inhomogeneous Dirichlet boundary value problem*

$$-\Delta u = f, \quad \text{for} \quad r = \| \mathbf{x} \| < 1, \qquad u = h, \quad \text{for} \quad r = 1,$$

is, when expressed in polar coordinates,

$$u(r,\theta) = \frac{1}{4\pi} \int_{-\pi}^{\pi} \int_0^1 f(\rho,\phi) \log\left(\frac{1 + r^2\rho^2 - 2\, r\rho \cos(\theta - \phi)}{r^2 + \rho^2 - 2\, r\rho \cos(\theta - \phi)} \right) \rho\, d\rho\, d\phi \\ + \frac{1}{2\pi} \int_{-\pi}^{\pi} h(\phi)\, \frac{1 - r^2}{1 + r^2 - 2\, r \cos(\theta - \phi)}\, d\phi. \tag{6.136}$$

When $f \equiv 0$, formula (6.136) recovers the Poisson integral formula (4.126) for the solution to the Dirichlet boundary value problem for the Laplace equation. In particular, the boundary data $h(\theta) = \delta(\theta - \phi)$, corresponding to a concentrated unit heat source applied to a single point on the boundary, produces the *Poisson kernel*

$$u(r,\theta) = \frac{1 - r^2}{2\pi\big(1 + r^2 - 2\, r \cos(\theta - \phi)\big)}. \tag{6.137}$$

The reader may enjoy verifying that this function indeed solves the Laplace equation and has the correct boundary values in the limit as $r \to 1$.

Exercises

6.3.15. A circular disk of radius 1 is subject to a heat source of unit magnitude on the subdisk $r \leq \frac{1}{2}$. Its boundary is kept at 0°.

(*a*) Write down an integral formula for the equilibrium temperature.

(*b*) Use radial symmetry to find an explicit formula for the equilibrium temperature.

6.3.16. A circular disk of radius 1 meter is subject to a unit concentrated heat source at its center and has completely insulated boundary. What is the equilibrium temperature?

♡ 6.3.17. (a) For $n > 0$, find the solution to the boundary value problem

$$-\Delta u = \frac{n}{\pi}\, e^{-n\,(x^2+y^2)}, \qquad x^2+y^2<1, \qquad u(x,y)=0, \qquad x^2+y^2=1.$$

(b) Discuss what happens in the limit as $n \to \infty$.

♡ 6.3.18. (a) Use the Method of Images to construct the Green's function for a half-plane $\{ y > 0 \}$ that is subject to homogeneous Dirichlet boundary conditions. *Hint*: The image point is obtained by reflection. (b) Use your Green's function to solve the boundary value problem

$$-\Delta u = \frac{1}{1+y}, \qquad y>0, \qquad u(x,0)=0.$$

6.3.19. Construct the Green's function for the half-disk $\Omega = \{ x^2+y^2<1,\ y>0 \}$ when subject to homogeneous Dirichlet boundary conditions. *Hint*: Use three image points.

6.3.20. Prove directly that the Poisson kernel (6.137) solves the Laplace equation for all $r < 1$.

♡ 6.3.21. Provide the details for the following alternative method for solving the homogeneous Dirichlet boundary value problem for the Poisson equation on the unit square:

$$-u_{xx}-u_{yy}=f(x,y), \quad u(x,0)=0, \quad u(x,1)=0, \quad u(0,y)=0, \quad u(1,y)=0, \quad 0<x,\ y<1.$$

(a) Write both $u(x,y)$ and $f(x,y)$ as Fourier sine series in y whose coefficients depend on x. (b) Substitute these series into the differential equation, and equate Fourier coefficients to obtain an infinite system of ordinary boundary value problems for the x-dependent Fourier coefficients of u. (c) Use the Green's functions for each boundary value problem to write out the solution and hence a series for the solution to the original boundary value problem. (d) Implement this method for the following forcing functions:
(i) $f(x,y)=\sin \pi y$, (ii) $f(x,y)=\sin \pi x\, \sin 2\pi y$, (iii) $f(x,y)=1$.

◇ 6.3.22. Use the method of Exercise 6.3.21 to find a series representation for the Green's function of a unit square subject to Dirichlet boundary conditions.

6.3.23. Write out the details of how to derive (6.134) from (6.133).

6.3.24. *True or false*: If the gravitational potential at a point **a** is greater than its value at the point **b**, then the magnitude of the gravitational force at **a** is greater than its value at **b**.

♠ 6.3.25. (a) Write down integral formulas for the gravitational potential and force due to a square plate $S = \{ -1 \le x, y \le 1 \}$ of unit density $\rho = 1$. (b) Use numerical integration to calculate the gravitational force at the points $(2,0)$ and $\left(\sqrt{2}, \sqrt{2}\right)$. Before starting, try to predict which point experiences the stronger force, and then check your prediction.

♠ 6.3.26. An equilateral triangular plate with unit area exerts a gravitational force on an observer sitting a unit distance away from its center. Is the force greater if the observer is located opposite a vertex of the triangle or opposite a side? Is the force greater than or less than that exerted by a circular plate of the same area? Use numerical integration to evaluate the double integrals.

6.3.27. Consider the wave equation $u_{tt} = c^2 u_{xx}$ on the line $-\infty < x < \infty$. Use the d'Alembert formula (2.82) to solve the initial value problem $u(0,x) = \delta(x-a)$, $u_t(0,x)=0$. Can you realize your solution as the limit of classical solutions?

◇ 6.3.28. Consider the wave equation $u_{tt} = c^2 u_{xx}$ on the line $-\infty < x < \infty$. Use the d'Alembert formula (2.82) to solve the initial value problem $u(0,x)=0$, $u_t(0,x)=\delta(x-a)$, modeling the effect of striking the string with a highly concentrated blow at the point $x=a$. Graph the solution at several times. Discuss the behavior of any discontinuities in the solution. In particular, show that $u(t,x) \neq 0$ on the domain of influence of the point $(0,a)$.

6.3.29. (a) Write down the solution $u(t,x)$ to the wave equation $u_{tt} = 4u_{xx}$ on the real line with initial data $u(0,x) = \begin{cases} 1 - |\,x\,|, & |\,x\,| \le 1, \\ 0, & \text{otherwise}, \end{cases}$ $\dfrac{\partial u}{\partial t}(0,x) = 0$. (b) Explain why $u(t,x)$ is not a classical solution to the wave equation. (c) Determine the derivatives $\partial^2 u/\partial t^2$ and $\partial^2 u/\partial x^2$ in the sense of distributions (generalized functions) and use this to justify the fact that $u(t,x)$ solves the wave equation in a distributional sense.

♡ 6.3.30. A piano string of length $\ell = 3$ and wave speed $c = 2$ with both ends fixed is hit by a hammer $\frac{1}{3}$ of the way along. The initial-boundary value problem that governs the resulting vibrations of the string is

$$\frac{\partial^2 u}{\partial t^2} = 4\,\frac{\partial^2 u}{\partial x^2}, \qquad u(t,0) = 0 = u(t,3), \qquad u(0,x) = 0, \qquad \frac{\partial u}{\partial t}(0,x) = \delta(x-1).$$

(a) What are the fundamental frequencies of vibration?
(b) Write down the solution to the initial-boundary value problem in Fourier series form.
(c) Write down the Fourier series for the velocity $\partial u/\partial t$ of your solution.
(d) Write down the d'Alembert formula for the solution, and sketch a picture of the string at four or five representative times.
(e) *True or false*: The solution is periodic in time. If true, what is the period? If false, explain what happens as t increases.

6.3.31. (a) Write down a Fourier series for the solution to the initial-boundary value problem

$$\frac{\partial^2 u}{\partial t^2} = \frac{\partial^2 u}{\partial x^2}, \qquad u(t,-1) = 0 = u(t,1), \qquad u(0,x) = \delta(x), \qquad \frac{\partial u}{\partial t}(0,x) = 0.$$

(b) Write down an analytic formula for the solution, i.e., sum your series.
(c) In what sense does the series solution in part (a) converge to the true solution? Do the partial sums provide a good approximation to the actual solution?

6.3.32. Answer Exercise 6.3.31 for

$$\frac{\partial^2 u}{\partial t^2} = \frac{\partial^2 u}{\partial x^2}, \qquad u(t,-1) = 0 = u(t,1), \qquad u(0,x) = 0, \qquad \frac{\partial u}{\partial t}(0,x) = \delta(x).$$

Chapter 7
Fourier Transforms

Fourier series and their ilk are designed to solve boundary value problems on bounded intervals. The extension of the Fourier calculus to the entire real line leads naturally to the *Fourier transform*, a powerful mathematical tool for the analysis of aperiodic functions. The Fourier transform is of fundamental importance in a remarkably broad range of applications, including both ordinary and partial differential equations, probability, quantum mechanics, signal and image processing, and control theory, to name but a few.

In this chapter, we motivate the construction by investigating how (rescaled) Fourier series behave as the length of the interval goes to infinity. The resulting Fourier transform maps a function defined on physical space to a function defined on the space of frequencies, whose values quantify the "amount" of each periodic frequency contained in the original function. The inverse Fourier transform then reconstructs the original function from its transformed frequency components. The integrals defining the Fourier transform and its inverse are, remarkably, almost identical, and this symmetry is often exploited, for example in assembling tables of Fourier transforms.

One of the most important properties of the Fourier transform is that it converts calculus — differentiation and integration — into algebra — multiplication and division. This underlies its application to linear ordinary differential equations and, in the following chapters, partial differential equations. In engineering applications, the Fourier transform is sometimes overshadowed by the Laplace transform, which is a particular subcase. The Fourier transform is used to analyze boundary value problems on the entire line. The Laplace transform is better suited to solving initial value problems, [**23**], but will not be developed in this text.

The Fourier transform is, like Fourier series, completely compatible with the calculus of generalized functions, [**68**]. The final section contains a brief introduction to the analytic foundations of the subject, including the basics of Hilbert space. However, a full, rigorous development requires more powerful analytical tools, including the Lebesgue integral and complex analysis, and the interested reader is therefore referred to more advanced texts, including [**37, 68, 98, 117**].

7.1 The Fourier Transform

We begin by motivating the Fourier transform as a limiting case of Fourier series. Although the rigorous details are subtle, the underlying idea can be straightforwardly explained. Let $f(x)$ be a function defined for all $-\infty < x < \infty$. The goal is to construct a Fourier expansion for $f(x)$ in terms of basic trigonometric functions. One evident approach is to construct its Fourier series on progressively longer and longer intervals, and then take the limit as their lengths go to infinity. This limiting process converts the Fourier sums into integrals, and the resulting representation of a function is renamed the Fourier transform. Since we are dealing with an infinite interval, there are no longer any periodicity requirements on the function $f(x)$. Moreover, the frequencies represented in the Fourier transform are no longer constrained by the length of the interval, and so we are effectively decomposing a quite general aperiodic function into a continuous superposition of trigonometric functions of all possible frequencies.

Let us present the details in a more concrete form. The computations will be significantly simpler if we work with the complex version of the Fourier series from the outset. Our starting point is the rescaled Fourier series (3.86) on a symmetric interval $[-\ell,\ell]$ of length 2ℓ, which we rewrite in the adapted form

$$f(x) \sim \sum_{\nu=-\infty}^{\infty} \sqrt{\frac{\pi}{2}}\, \frac{\widehat{f}_\ell(k_\nu)}{\ell}\, e^{\,\mathrm{i}\, k_\nu x}. \tag{7.1}$$

The sum is over the discrete collection of frequencies

$$k_\nu = \frac{\pi\nu}{\ell}, \qquad \nu = 0, \pm 1, \pm 2, \ldots\,, \tag{7.2}$$

corresponding to those trigonometric functions that have period 2ℓ. For reasons that will soon become apparent, the Fourier coefficients of f are now denoted as

$$c_\nu = \frac{1}{2\ell}\int_{-\ell}^{\ell} f(x)\, e^{-\mathrm{i}\, k_\nu x}\, dx = \sqrt{\frac{\pi}{2}}\, \frac{\widehat{f}_\ell(k_\nu)}{\ell}, \tag{7.3}$$

so that

$$\widehat{f}_\ell(k_\nu) = \frac{1}{\sqrt{2\pi}}\int_{-\ell}^{\ell} f(x)\, e^{-\mathrm{i}\, k_\nu x}\, dx. \tag{7.4}$$

This reformulation of the basic Fourier series formula allows us to easily pass to the limit as the interval's length $\ell \to \infty$.

On an interval of length 2ℓ, the frequencies (7.2) required to represent a function in Fourier series form are equally distributed, with interfrequency spacing

$$\Delta k = k_{\nu+1} - k_\nu = \frac{\pi}{\ell}\,. \tag{7.5}$$

As $\ell \to \infty$, the spacing $\Delta k \to 0$, and so the relevant frequencies become more and more densely packed in the line $-\infty < k < \infty$. In the limit, we thus anticipate that *all* possible frequencies will be represented. Indeed, letting $k_\nu = k$ be arbitrary in (7.4), and sending $\ell \to \infty$, results in the infinite integral

$$\widehat{f}(k) = \frac{1}{\sqrt{2\pi}}\int_{-\infty}^{\infty} f(x)\, e^{-\mathrm{i}\, k x}\, dx, \tag{7.6}$$

known as the *Fourier transform* of the function $f(x)$. If $f(x)$ is a sufficiently nice function, e.g., piecewise continuous and decaying to 0 reasonably quickly as $|\,x\,| \to \infty$, its Fourier transform $\widehat{f}(k)$ is defined for all possible frequencies $k \in \mathbb{R}$. The preceding formula will sometimes conveniently be abbreviated as

$$\widehat{f}(k) = \mathcal{F}[\,f(x)\,], \tag{7.7}$$

where $\mathcal{F}$ is the *Fourier transform operator*, which maps each (sufficiently nice) function of the spatial variable x to a function of the frequency variable k.

To reconstruct the function from its Fourier transform, we apply a similar limiting procedure to the Fourier series (7.1), which we first rewrite in a more suggestive form,

$$f(x) \sim \frac{1}{\sqrt{2\pi}} \sum_{\nu=-\infty}^{\infty} \widehat{f}_\ell(k_\nu)\, e^{\,\mathrm{i}\, k_\nu x}\, \Delta k, \tag{7.8}$$

using (7.5). For each fixed value of x, the right-hand side has the form of a Riemann sum approximating the integral

$$\frac{1}{\sqrt{2\pi}} \int_{-\infty}^{\infty} \widehat{f}_\ell(k)\, e^{\,\mathrm{i}\, k x}\, dk.$$

As $\ell \to \infty$, the functions (7.4) converge to the Fourier transform: $\widehat{f}_\ell(k) \to \widehat{f}(k)$; moreover, the interfrequency spacing $\Delta k = \pi/\ell \to 0$, and so one expects the Riemann sums to converge to the limiting integral

$$f(x) \sim \frac{1}{\sqrt{2\pi}} \int_{-\infty}^{\infty} \widehat{f}(k)\, e^{\,\mathrm{i}\, k x}\, dk. \tag{7.9}$$

The resulting formula serves to define the *inverse Fourier transform*, which is used to recover the original signal from its Fourier transform. In this manner, the Fourier series has become a Fourier integral that reconstructs the function $f(x)$ as a (continuous) superposition of complex exponentials $e^{\,\mathrm{i}\, k x}$ of *all* possible frequencies, with $\widehat{f}(k)/\sqrt{2\pi}$ quantifying the amount contributed by the complex exponential of frequency k. In abbreviated form, formula (7.9) can be written

$$f(x) = \mathcal{F}^{-1}[\,\widehat{f}(k)\,], \tag{7.10}$$

thus defining the inverse of the Fourier transform operator (7.7).

It is worth pointing out that both the Fourier transform (7.7) and its inverse (7.10) define linear operators on function space. This means that the Fourier transform of the sum of two functions is the sum of their individual transforms, while multiplying a function by a constant multiplies its Fourier transform by the same factor:

$$\begin{aligned} \mathcal{F}[\,f(x) + g(x)\,] &= \mathcal{F}[\,f(x)\,] + \mathcal{F}[\,g(x)\,] = \widehat{f}(k) + \widehat{g}(k), \\ \mathcal{F}[\,c\,f(x)\,] &= c\,\mathcal{F}[\,f(x)\,] = c\,\widehat{f}(k). \end{aligned} \tag{7.11}$$

A similar statement holds for the inverse Fourier transform $\mathcal{F}^{-1}$.

Recapitulating, by letting the length of the interval go to ∞, the discrete Fourier series has become a continuous Fourier integral, while the Fourier coefficients, which were defined only at a discrete collection of possible frequencies, have become a complete function $\widehat{f}(k)$ defined on all of frequency space. The reconstruction of $f(x)$ from its Fourier transform $\widehat{f}(k)$ via (7.9) can be rigorously justified under suitable hypotheses. For example, if $f(x)$ is piecewise C^1 on all of $\mathbb{R}$ and decays reasonably rapidly, $f(x) \to 0$ as $|\,x\,| \to \infty$, so that its Fourier integral (7.6) converges absolutely, then it can be proved, [**37, 117**], that the inverse Fourier integral (7.9) will converge to $f(x)$ at all points of continuity, and to the midpoint $\frac{1}{2}(f(x^-) + f(x^+))$ at jump discontinuities — just like a Fourier series. In particular, its Fourier transform $\widehat{f}(k) \to 0$ must also decay as $|\,k\,| \to \infty$, implying that (as with Fourier series) the very high frequency modes make negligible contributions to the

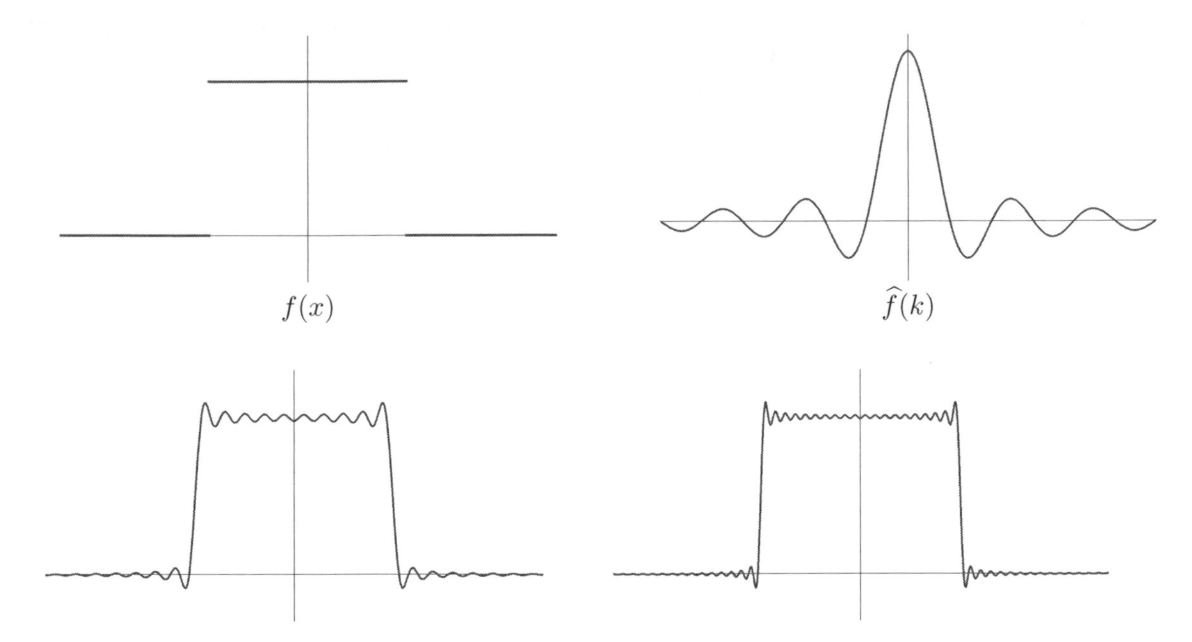

Figure 7.1. Fourier transform of a rectangular pulse.

reconstruction of such a signal. A more precise result will be formulated in Theorem 7.15 below.

Example 7.1. The Fourier transform of the *rectangular pulse*[†]

$$f(x) = \sigma(x+a) - \sigma(x-a) = \begin{cases} 1, & -a < x < a, \\ 0, & |\,x\,| > a, \end{cases} \tag{7.12}$$

of width $2a$, is easily computed:

$$\widehat{f}(k) = \frac{1}{\sqrt{2\pi}} \int_{-a}^{a} e^{-\,\mathrm{i}\,k\,x}\,dx = \frac{e^{\,\mathrm{i}\,k\,a} - e^{-\,\mathrm{i}\,k\,a}}{\sqrt{2\pi}\,\mathrm{i}\,k} = \sqrt{\frac{2}{\pi}}\,\frac{\sin a\,k}{k}\,. \tag{7.13}$$

On the other hand, the reconstruction of the pulse via the inverse transform (7.9) tells us that

$$\frac{1}{\pi} \int_{-\infty}^{\infty} \frac{e^{\,\mathrm{i}\,k\,x} \sin a\,k}{k}\,dk = f(x) = \begin{cases} 1, & -a < x < a, \\ \frac{1}{2}, & x = \pm\,a, \\ 0, & |\,x\,| > a. \end{cases} \tag{7.14}$$

Note the convergence to the middle of the jump discontinuities at $x = \pm\,a$. The real part of this complex integral produces a striking trigonometric integral identity:

$$\frac{1}{\pi} \int_{-\infty}^{\infty} \frac{\cos x\,k\,\sin a\,k}{k}\,dk = \begin{cases} 1, & -a < x < a, \\ \frac{1}{2}, & x = \pm\,a, \\ 0, & |\,x\,| > a. \end{cases} \tag{7.15}$$

Just as many Fourier series yield nontrivial summation formulas, the reconstruction of a function from its Fourier transform often leads to nontrivial integration formulas. One

[†] $\sigma(x)$ is the unit step function (3.46).

cannot compute the integral (7.14) by the Fundamental Theorem of Calculus, since there is no elementary function whose derivative equals the integrand.[†] In Figure 7.1 we display the box function with $a = 1$, its Fourier transform, along with a reconstruction obtained by numerically integrating (7.15). Since we are dealing with an infinite integral, we must break off the numerical integrator by restricting it to a finite interval. The first graph in the second row is obtained by integrating from $-5 \le k \le 5$, while the second is from $-10 \le k \le 10$. The nonuniform convergence of the integral leads to the appearance of a Gibbs phenomenon at the two discontinuities, similar to what we observed in the nonuniform convergence of a Fourier series.

On the other hand, the identity resulting from the imaginary part,

$$\frac{1}{\pi}\int_{-\infty}^{\infty}\frac{\sin k x\,\sin a k}{k}\,dk = 0,$$

is, on the surface, not surprising, because the integrand is odd. However, it is far from obvious that either integral converges; indeed, the amplitude of the oscillatory integrand decays like $1/|\,k\,|$, but the latter function does not have a convergent integral, and so the usual comparison test for infinite integrals, [**8**, **97**], fails to apply. Their convergence is marginal at best, and the trigonometric oscillations somehow manage to ameliorate the slow rate of decay of $1/k$.

Example 7.2. Consider an exponentially decaying right-handed pulse[‡]

$$f_r(x) = \begin{cases} e^{-a x}, & x > 0, \\ 0, & x < 0, \end{cases} \tag{7.16}$$

where $a > 0$. We compute its Fourier transform directly from the definition:

$$\widehat{f}_r(k) = \frac{1}{\sqrt{2\pi}}\int_0^{\infty} e^{-a x}\,e^{-\,\mathrm{i}\,k x}\,dx = -\,\frac{1}{\sqrt{2\pi}}\,\frac{e^{-(a+\mathrm{i}\,k)x}}{a+\mathrm{i}\,k}\,\bigg|_{x=0}^{\infty} = \frac{1}{\sqrt{2\pi}\,(a+\mathrm{i}\,k)}\,.$$

As in the preceding example, the inverse Fourier transform produces a nontrivial integral identity:

$$\frac{1}{2\pi}\int_{-\infty}^{\infty}\frac{e^{\,\mathrm{i}\,k x}}{a+\mathrm{i}\,k}\,dk = \begin{cases} e^{-a x}, & x > 0, \\ \frac{1}{2}, & x = 0, \\ 0, & x < 0. \end{cases} \tag{7.17}$$

Similarly, a pulse that decays to the left,

$$f_l(x) = \begin{cases} e^{a x}, & x < 0, \\ 0, & x > 0, \end{cases} \tag{7.18}$$

where $a > 0$ is still positive, has Fourier transform

$$\widehat{f}_l(k) = \frac{1}{\sqrt{2\pi}\,(a-\mathrm{i}\,k)}\,. \tag{7.19}$$

[†] One can use Euler's formula (3.59) to reduce (7.14) to a complex version of the *exponential integral* $\int (e^{\alpha k}/k)\,dk$, but it can be proved, [**25**], that neither integral can be written in terms of elementary functions.

[‡] Note that we cannot Fourier transform the entire exponential function $e^{-a x}$, because it does not go to zero at both $\pm\infty$, which is required for the integral (7.6) to converge.

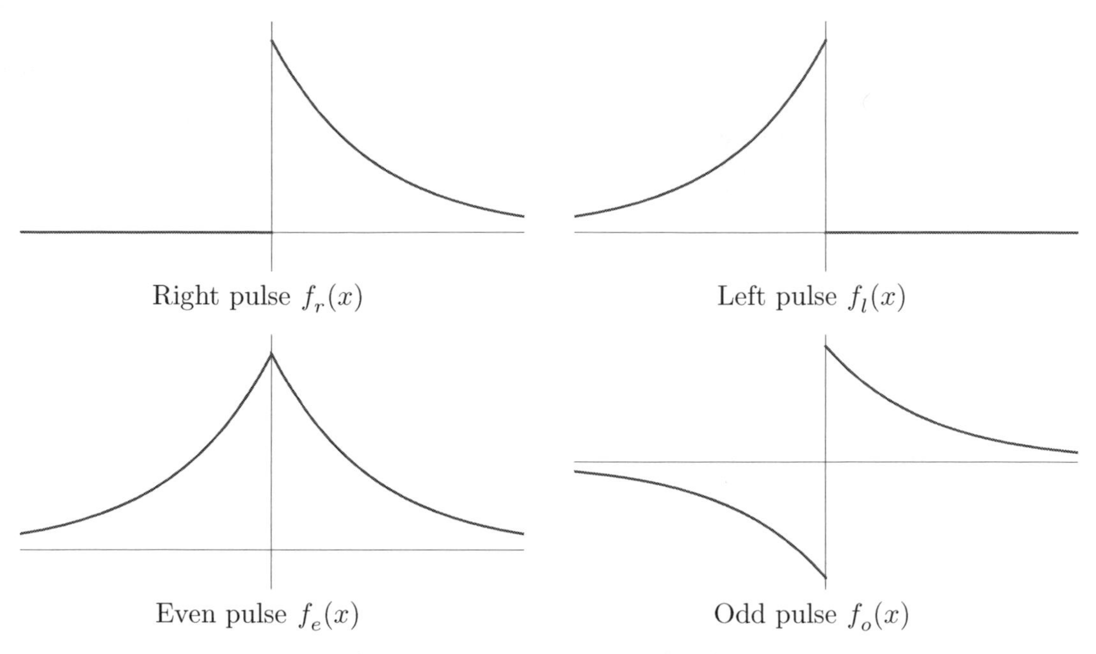

Figure 7.2. Exponential pulses.

This also follows from the general fact that the Fourier transform of $f(-x)$ is $\widehat{f}(-k)$; see Exercise 7.1.10. The even exponentially decaying pulse

$$f_e(x) = e^{-a|x|} \tag{7.20}$$

is merely the sum of left and right pulses: $f_e = f_r + f_l$. Thus, by linearity,

$$\widehat{f}_e(k) = \widehat{f}_r(k) + \widehat{f}_l(k) = \frac{1}{\sqrt{2\pi}\,(a + \mathrm{i}\,k)} + \frac{1}{\sqrt{2\pi}\,(a - \mathrm{i}\,k)} = \sqrt{\frac{2}{\pi}}\,\frac{a}{k^2 + a^2}\,. \tag{7.21}$$

The resulting Fourier transform is real and even because $f_e(x)$ is a real-valued even function; see Exercise 7.1.12. The inverse Fourier transform (7.9) produces another nontrivial integral identity:

$$e^{-a|x|} = \frac{1}{\pi}\int_{-\infty}^{\infty} \frac{a\, e^{\,\mathrm{i}\,k x}}{k^2 + a^2}\, dk = \frac{a}{\pi}\int_{-\infty}^{\infty} \frac{\cos k x}{k^2 + a^2}\, dk. \tag{7.22}$$

(The imaginary part of the integral vanishes, because its integrand is odd.) On the other hand, the odd exponentially decaying pulse,

$$f_o(x) = (\operatorname{sign} x)\, e^{-a|x|} = \begin{cases} e^{-a x}, & x > 0, \\ -\,e^{a x}, & x < 0, \end{cases} \tag{7.23}$$

is the difference of the right and left pulses, $f_o = f_r - f_l$, and has purely imaginary and odd Fourier transform

$$\widehat{f}_o(k) = \widehat{f}_r(k) - \widehat{f}_l(k) = \frac{1}{\sqrt{2\pi}\,(a + \mathrm{i}\,k)} - \frac{1}{\sqrt{2\pi}\,(a - \mathrm{i}\,k)} = -\,\mathrm{i}\,\sqrt{\frac{2}{\pi}}\,\frac{k}{k^2 + a^2}\,. \tag{7.24}$$

The inverse transform is

$$(\operatorname{sign} x)\, e^{-a|x|} = -\frac{\mathrm{i}}{\pi}\int_{-\infty}^{\infty}\frac{k\, e^{\mathrm{i}kx}}{k^2+a^2}\,dk = \frac{1}{\pi}\int_{-\infty}^{\infty}\frac{k\sin kx}{k^2+a^2}\,dk. \tag{7.25}$$

As a final example, consider the rational function

$$f(x) = \frac{1}{x^2+a^2}, \qquad \text{where} \qquad a>0. \tag{7.26}$$

Its Fourier transform requires integrating

$$\widehat{f}(k) = \frac{1}{\sqrt{2\pi}}\int_{-\infty}^{\infty}\frac{e^{-\mathrm{i}kx}}{x^2+a^2}\,dx. \tag{7.27}$$

The indefinite integral (anti-derivative) does not appear in basic integration tables, and, in fact, cannot be done in terms of elementary functions. However, we have just managed to evaluate this particular integral! Look at (7.22). If we change x to k and k to $-x$, then we exactly recover the integral (7.27) up to a factor of $a\sqrt{2/\pi}$. We conclude that the Fourier transform of (7.26) is

$$\widehat{f}(k) = \sqrt{\frac{\pi}{2}}\,\frac{e^{-a|k|}}{a}. \tag{7.28}$$

This last example is indicative of an important general fact. The reader has no doubt already noted the remarkable similarity between the Fourier transform (7.6) and its inverse (7.9). Indeed, the only difference is that the former has a minus sign in the exponential. This implies the following *Symmetry Principle* relating the direct and inverse Fourier transforms.

Theorem 7.3. *If the Fourier transform of the function $f(x)$ is $\widehat{f}(k)$, then the Fourier transform of $\widehat{f}(x)$ is $f(-k)$.*

The Symmetry Principle allows us to reduce the tabulation of Fourier transforms by half. For instance, referring back to Example 7.1, we deduce that the Fourier transform of the function

$$f(x) = \sqrt{\frac{2}{\pi}}\,\frac{\sin ax}{x}$$

is

$$\widehat{f}(k) = \sigma(-k+a) - \sigma(-k-a) = \sigma(k+a) - \sigma(k-a) = \begin{cases} 1, & -a<k<a, \\ \frac{1}{2}, & k=\pm a, \\ 0, & |k|>a. \end{cases} \tag{7.29}$$

Note that, by linearity, we can divide both $f(x)$ and $\widehat{f}(k)$ by $\sqrt{2/\pi}$ to deduce the Fourier transform of $\dfrac{\sin ax}{x}$.

Warning: Some authors omit the $\sqrt{2\pi}$ factor in the definition (7.6) of the Fourier transform $\widehat{f}(k)$. This alternative convention does have a slight advantage of eliminating many $\sqrt{2\pi}$ factors in the transformed expressions. However, this necessitates an extra such factor in the reconstruction formula (7.9), which is achieved by replacing $\sqrt{2\pi}$ by 2π. A significant disadvantage is that the resulting formulas for the Fourier transform and its inverse are less similar, and so the Symmetry Principle of Theorem 7.3 requires some modification. (On the other hand, convolution — to be discussed below — is a little easier without the extra factor.) Yet another, more recent, convention can be found in Exercise 7.1.18. When consulting any particular reference, the reader *always* needs to check which version of the Fourier transform is being used.

All of the functions in Example 7.2 required $a > 0$ for the Fourier integrals to converge. The functions that emerge in the limit as a goes to 0 are of special interest. Let us start with the odd exponential pulse (7.23). When $a \to 0$, the function $f_o(x)$ converges to the *sign function*

$$f(x) = \operatorname{sign} x = \sigma(x) - \sigma(-x) = \begin{cases} +1, & x > 0, \\ -1, & x < 0. \end{cases} \tag{7.30}$$

Taking the limit of the Fourier transform (7.24) leads to

$$\widehat{f}(k) = -\,\mathrm{i}\,\sqrt{\frac{2}{\pi}}\,\frac{1}{k}\,. \tag{7.31}$$

The nonintegrable singularity of $\widehat{f}(k)$ at $k = 0$ is indicative of the fact that the sign function does *not* decay as $|\,x\,| \to \infty$. In this case, neither the Fourier transform integral nor its inverse are well defined as standard (Riemann, or even Lebesgue) integrals. Nevertheless, it is possible to rigorously justify these results within the framework of generalized functions.

More interesting are the even pulse functions $f_e(x)$, which, in the limit $a \to 0$, become the constant function

$$f(x) \equiv 1. \tag{7.32}$$

The limit of the Fourier transform (7.21) is

$$\lim_{a \to 0} \sqrt{\frac{2}{\pi}}\,\frac{a}{k^2 + a^2} = \begin{cases} 0, & k \neq 0, \\ \infty, & k = 0. \end{cases} \tag{7.33}$$

This limiting behavior should remind the reader of our construction (6.10) of the delta function as the limit of the functions

$$\delta(x) = \lim_{n \to \infty} \frac{n}{\pi\,(1 + n^2 x^2)} = \lim_{a \to 0} \frac{a}{\pi\,(a^2 + x^2)}\,.$$

Comparing with (7.33), we conclude that the Fourier transform of the constant function (7.32) is a multiple of the delta function in the frequency variable:

$$\widehat{f}(k) = \sqrt{2\pi}\;\delta(k). \tag{7.34}$$

The direct transform integral

$$\delta(k) = \frac{1}{2\pi} \int_{-\infty}^{\infty} e^{-\,\mathrm{i}\,k x}\,dx \tag{7.35}$$

is, strictly speaking, not defined, because the infinite integrals of the oscillatory sine and cosine functions don't converge! However, this identity can be validly interpreted within the framework of weak convergence and generalized functions. On the other hand, the inverse transform formula (7.9) yields

$$\int_{-\infty}^{\infty} \delta(k)\,e^{\,\mathrm{i}\,k x}\,dk = e^{\,\mathrm{i}\,k\,0} = 1,$$

which is in accord with the basic definition (6.16) of the delta function. As in the preceding case, the delta function singularity at $k = 0$ manifests the lack of decay of the constant function.

Conversely, the delta function $\delta(x)$ has constant Fourier transform

$$\widehat{\delta}(k) = \frac{1}{\sqrt{2\pi}} \int_{-\infty}^{\infty} \delta(x)\, e^{-\,\mathrm{i}\,k\,x}\, dx = \frac{e^{-\,\mathrm{i}\,k\,0}}{\sqrt{2\pi}} \equiv \frac{1}{\sqrt{2\pi}}\,, \tag{7.36}$$

a result that also follows from the Symmetry Principle of Theorem 7.3. To determine the Fourier transform of a delta spike $\delta_\xi(x) = \delta(x - \xi)$ concentrated at position $x = \xi$, we compute

$$\widehat{\delta_\xi}(k) = \frac{1}{\sqrt{2\pi}} \int_{-\infty}^{\infty} \delta(x-\xi)\, e^{-\,\mathrm{i}\,k\,x}\, dx = \frac{e^{-\,\mathrm{i}\,k\,\xi}}{\sqrt{2\pi}}\,. \tag{7.37}$$

The result is a pure exponential in frequency space. Applying the inverse Fourier transform (7.9) leads, at least on a formal level, to the remarkable identity

$$\delta_\xi(x) = \delta(x-\xi) = \frac{1}{2\pi} \int_{-\infty}^{\infty} e^{\,\mathrm{i}\,k(x-\xi)}\, dk = \frac{1}{2\pi}\, \langle\, e^{\,\mathrm{i}\,k\,x}\,, e^{\,\mathrm{i}\,k\,\xi}\,\rangle\,, \tag{7.38}$$

where $\langle\,\cdot\,,\cdot\,\rangle$ denotes the L^2 Hermitian inner product of complex-valued functions of $k \in \mathbb{R}$. Since the delta function vanishes for $x \neq \xi$, this identity is telling us that complex exponentials of differing frequencies are mutually orthogonal. However, as with (7.35), this makes sense only within the language of generalized functions. On the other hand, multiplying both sides of (7.38) by $f(\xi)$ and then integrating with respect to ξ produces

$$f(x) = \frac{1}{2\pi} \int_{-\infty}^{\infty} \int_{-\infty}^{\infty} f(\xi)\, e^{\,\mathrm{i}\,k(x-\xi)}\, dx\, dk. \tag{7.39}$$

This *is* a perfectly valid formula, being a restatement (or, rather, combination) of the basic formulas (7.6) and (7.9) connecting the direct and inverse Fourier transforms of the function $f(x)$.

Conversely, the Symmetry Principle tells us that the Fourier transform of a pure exponential $e^{\,\mathrm{i}\,\kappa\,x}$ will be a shifted delta spike $\sqrt{2\pi}\,\delta(k-\kappa)$, concentrated at frequency $k = \kappa$. Both results are particular cases of the following *Shift Theorem*, whose proof is left as an exercise for the reader.

Theorem 7.4. *If $f(x)$ has Fourier transform $\widehat{f}(k)$, then the Fourier transform of the shifted function $f(x-\xi)$ is $e^{-\,\mathrm{i}\,k\,\xi}\,\widehat{f}(k)$. Similarly, the transform of the product function $e^{\,\mathrm{i}\,\kappa\,x} f(x)$, for real κ, is the shifted transform $\widehat{f}(k-\kappa)$.*

In a similar vein, the *Dilation Theorem* gives the effect of a scaling transformation on the Fourier transform. Again, the proof is left to the reader.

Theorem 7.5. *If $f(x)$ has Fourier transform $\widehat{f}(k)$, then the Fourier transform of the rescaled function $f(c\,x)$ for $0 \neq c \in \mathbb{R}$ is $\dfrac{1}{|\,c\,|}\,\widehat{f}\left(\dfrac{k}{c}\right)$.*

Concise Table of Fourier Transforms

$f(x)$	$\widehat{f}(k)$
1	$\sqrt{2\pi}\,\delta(k)$
$\delta(x)$	$\dfrac{1}{\sqrt{2\pi}}$
$\sigma(x)$	$\sqrt{\dfrac{\pi}{2}}\,\delta(k) - \dfrac{\mathrm{i}}{\sqrt{2\pi}\,k}$
$\operatorname{sign} x$	$-\,\mathrm{i}\,\sqrt{\dfrac{2}{\pi}}\,\dfrac{1}{k}$
$\sigma(x+a) - \sigma(x-a)$	$\sqrt{\dfrac{2}{\pi}}\,\dfrac{\sin a\,k}{k}$
$e^{-a\,x}\,\sigma(x)$	$\dfrac{1}{\sqrt{2\pi}\,(a+\mathrm{i}\,k)}$
$e^{a\,x}\,(1-\sigma(x))$	$\dfrac{1}{\sqrt{2\pi}\,(a-\mathrm{i}\,k)}$
$e^{-a\,\lvert x\rvert}$	$\sqrt{\dfrac{2}{\pi}}\,\dfrac{a}{k^2+a^2}$
$e^{-a\,x^2}$	$\dfrac{e^{-k^2/(4\,a)}}{\sqrt{2\,a}}$
$\tan^{-1} x$	$-\,\mathrm{i}\,\sqrt{\dfrac{\pi}{2}}\,\dfrac{e^{-\lvert k\rvert}}{k}$
$f(c\,x+d)$	$\dfrac{e^{\,\mathrm{i}\,k\,d/c}}{\lvert c\rvert}\,\widehat{f}\left(\dfrac{k}{c}\right)$
$\overline{f(x)}$	$\overline{\widehat{f}(-k)}$
$\widehat{f}(x)$	$f(-k)$
$f'(x)$	$\mathrm{i}\,k\,\widehat{f}(k)$
$x\,f(x)$	$\mathrm{i}\,\widehat{f}\,'(k)$
$f * g(x)$	$\sqrt{2\pi}\,\widehat{f}(k)\,\widehat{g}(k)$

Note: The parameters a, c, d are real, with $a > 0$ and $c \neq 0$.

Example 7.6. Let us determine the Fourier transform of the Gaussian function $g(x) = e^{-x^2}$. To evaluate its Fourier integral, we first complete the square in the exponent:

$$\widehat{g}(k) = \frac{1}{\sqrt{2\pi}} \int_{-\infty}^{\infty} e^{-x^2 - i\,k\,x}\,dx = \frac{1}{\sqrt{2\pi}} \int_{-\infty}^{\infty} e^{-(x+i\,k/2)^2 - k^2/4}\,dx$$
$$= \frac{e^{-k^2/4}}{\sqrt{2\pi}} \int_{-\infty}^{\infty} e^{-y^2}\,dy = \frac{e^{-k^2/4}}{\sqrt{2}}.$$

The next-to-last equality employed the change of variables† $y = x + \frac{1}{2}\,i\,k$, while the final step used formula (2.100).

More generally, to find the Fourier transform of $g_a(x) = e^{-a x^2}$, where $a > 0$, we invoke the Dilation Theorem 7.5 with $c = \sqrt{a}$ to deduce that $\widehat{g}_a(k) = e^{-k^2/(4a)}/\sqrt{2a}$.

Since the Fourier transform uniquely associates a function $\widehat{f}(k)$ on frequency space with each (reasonable) function $f(x)$ on physical space, one can characterize functions by their transforms. Many practical applications rely on tables (or, even better, computer algebra systems such as MATHEMATICA and MAPLE) that recognize a wide variety of transforms of basic functions of importance in applications. The accompanying table lists some of the most important examples of functions and their Fourier transforms, based on our convention (7.6). Keep in mind that, by applying the Symmetry Principle of Theorem 7.3, each entry can be used to deduce two different Fourier transforms. A more extensive collection of Fourier transforms can be found in [**82**].

Exercises

7.1.1. Find the Fourier transform of the following functions:
(a) $e^{-(x+4)^2}$, (b) $e^{-|x+1|}$, (c) $\begin{cases} x, & |x| < 1, \\ 0, & \text{otherwise}, \end{cases}$ (d) $\begin{cases} e^{-2x}, & x \geq 0, \\ e^{3x}, & x \leq 0, \end{cases}$
(e) $\begin{cases} e^{-|x|}, & |x| \geq 1, \\ e^{-1}, & |x| \leq 1, \end{cases}$ (f) $\begin{cases} e^{-x}\sin x, & x > 0, \\ 0, & x \leq 0, \end{cases}$ (g) $\begin{cases} 1 - |x|, & |x| \leq 1, \\ 0, & \text{otherwise}. \end{cases}$

7.1.2. Find the Inverse Fourier transform of the following functions: (a) e^{-k^2}, (b) $e^{-|k|}$,
(c) $\begin{cases} e^{-k}\sin k, & k \geq 0, \\ 0, & k \leq 0, \end{cases}$ (d) $\begin{cases} 1, & \alpha < k < \beta, \\ 0, & \text{otherwise}, \end{cases}$ (e) $\begin{cases} 1 - |k|, & |k| < 1, \\ 0, & \text{otherwise}. \end{cases}$

7.1.3. Find the inverse Fourier transform of the function $1/(k + c)$ when (a) $c = a$ is real; (b) $c = i\,b$ is purely imaginary; (c) $c = a + i\,b$ is an arbitrary complex number.

7.1.4. Find the inverse Fourier transform of $1/(k^2 - a^2)$, where $a > 0$ is real.
Hint: Use Exercise 7.1.3.

♢ 7.1.5. (a) Find the Fourier transform of $e^{i\,\omega x}$. (b) Use this to find the Fourier transforms of the basic trigonometric functions $\cos\omega x$ and $\sin\omega x$.

7.1.6. Write down two real integral identites that result from the inverse Fourier transform of (7.28).

† Since this represents a complex change of variables, a fully rigorous justification of this step requires the use of complex integration.

7.1.7. Write down two real integral identities that follow from (7.17).

7.1.8. (a) Find the Fourier transform of the hat function $f_n(x) = \begin{cases} n - n^2\,|\,x\,|, & |\,x\,| \le 1/n, \\ 0, & \text{otherwise.} \end{cases}$
(b) What is the limit, as $n \to \infty$, of $\widehat{f}_n(k)$?
(c) In what sense is the limit the Fourier transform of the limit of $f_n(x)$?

7.1.9. (a) Justify the linearity of the Fourier transform, as in (7.11).
(b) State and justify the linearity of the inverse Fourier transform.

♢ 7.1.10. If the Fourier transform of $f(x)$ is $\widehat{f}(k)$, prove that (a) the Fourier transform of $f(-x)$ is $\widehat{f}(-k)$; (b) the Fourier transform of the complex conjugate function $\overline{f(x)}$ is $\overline{\widehat{f}(-k)}$.

7.1.11. *True or false*: If the complex-valued function $f(x) = g(x) + \mathrm{i}\,h(x)$ has Fourier transform $\widehat{f}(k) = \widehat{g}(k) + \mathrm{i}\,\widehat{h}(k)$, then $g(x)$ has Fourier transform $\widehat{g}(k)$ and $h(x)$ has Fourier transform $\widehat{h}(k)$.

♢ 7.1.12. (a) Prove that the Fourier transform of an even function is even. (b) Prove that the Fourier transform of a real even function is real and even. (c) What can you say about the Fourier transform of an odd function? (d) Of a real odd function? (e) What about a general real function?

♢ 7.1.13. Prove the Shift Theorem 7.4.

♢ 7.1.14. Prove the Dilation Theorem 7.5.

7.1.15. Given that the Fourier transform of $f(x)$ is $\widehat{f}(k)$, find, from first principles, the Fourier transform of $g(x) = f(a\,x + b)$, where a and b are fixed real constants.

7.1.16. Let a be a real constant. Given the Fourier transform $\widehat{f}(k)$ of $f(x)$, find the Fourier transforms of (a) $f(x)\,e^{\,\mathrm{i}\,a\,x}$, (b) $f(x)\cos a\,x$, (c) $f(x)\sin a\,x$.

♢ 7.1.17. A common alternative convention for the Fourier transform is to define

$$\widehat{f}_1(k) = \int_{-\infty}^{\infty} f(x)\,e^{-\,\mathrm{i}\,k\,x}\,dx.$$

(a) What is the formula for the corresponding inverse Fourier transform?
(b) How is $\widehat{f}_1(k)$ related to our Fourier transform $\widehat{f}(k)$?

♢ 7.1.18. Another convention for the Fourier transform is to define $\widehat{f}_2(k) = \displaystyle\int_{-\infty}^{\infty} f(x)\,e^{-2\pi\,\mathrm{i}\,k\,x}\,dx$.
Answer the questions in Exercise 7.1.17 for this version of the Fourier transform.

♡ 7.1.19. The *cosine* and *sine transforms* of a real function $f(x)$ are defined as

$$\widehat{c}\,(k) = \int_{-\infty}^{\infty} f(x)\cos k\,x\,dx, \qquad \widehat{s}\,(k) = \int_{-\infty}^{\infty} f(x)\sin k\,x\,dx. \tag{7.40}$$

(*i*) Prove that $\widehat{f}(k) = \widehat{c}\,(k) - \mathrm{i}\,\widehat{s}\,(k)$. (*ii*) Find the cosine and sine transforms of the functions in Exercise 7.1.1. (*iii*) Show that $\widehat{c}\,(k)$ is an even function, while $\widehat{s}\,(k)$ is an odd function. (*iv*) Show that if f is an even function, then $\widehat{s}\,(k) \equiv 0$, while if f is an odd function, then $\widehat{c}\,(k) \equiv 0$.

♢ 7.1.20. The *two-dimensional Fourier transform* of a function $f(x,y)$ defined for $(x,y) \in \mathbb{R}^2$ is

$$\widehat{f}(k,l) = \frac{1}{2\pi}\int_{-\infty}^{\infty}\int_{-\infty}^{\infty} f(x,y)\,e^{-\,\mathrm{i}\,(k\,x + l\,y)}\,dx\,dy. \tag{7.41}$$

(a) Compute the Fourier transform of the following functions:
(*i*) $e^{-\,|\,x\,|-|\,y\,|}$; (*ii*) $e^{-x^2-y^2}$; (*iii*) the delta function $\delta(x-\xi)\,\delta(y-\eta)$,
(*iv*) $\begin{cases} 1, & |\,x\,|, |\,y\,| \le 1, \\ 0, & \text{otherwise,} \end{cases}$ (*v*) $\begin{cases} 1, & |\,x\,| + |\,y\,| \le 1, \\ 0, & \text{otherwise,} \end{cases}$ (*vi*) $\cos(x-y)$.
(b) Show that if $f(x,y) = g(x)\,h(y)$, then $\widehat{f}(k,l) = \widehat{g}(k)\,\widehat{h}(l)$.
(c) What is the formula for the inverse two-dimensional Fourier transform, i.e., how can you reconstruct $f(x,y)$ from $\widehat{f}(k,l)$?

7.2 Derivatives and Integrals

One of the most significant features of the Fourier transform is that it converts calculus into algebra! More specifically, the two basic operations in calculus — differentiation and integration of functions — are realized as algebraic operations on their Fourier transforms. (The downside is that algebraic operations become more complicated in the frequency domain.)

Differentiation

Let us begin with derivatives. If we differentiate[†] the basic inverse Fourier transform formula

$$f(x) \sim \frac{1}{\sqrt{2\pi}} \int_{-\infty}^{\infty} \widehat{f}(k)\, e^{\,\mathrm{i}\,k\,x}\, dk$$

with respect to x, we obtain

$$f'(x) \sim \frac{1}{\sqrt{2\pi}} \int_{-\infty}^{\infty} \mathrm{i}\,k\,\widehat{f}(k)\, e^{\,\mathrm{i}\,k\,x}\, dk. \tag{7.42}$$

The resulting integral is itself in the form of an inverse Fourier transform, namely of $\mathrm{i}\,k\,\widehat{f}(k)$, which immediately implies the following key result.

Proposition 7.7. *The Fourier transform of the derivative $f'(x)$ of a function is obtained by multiplication of its Fourier transform by* $\mathrm{i}\,k$:

$$\mathcal{F}[\,f'(x)\,] = \mathrm{i}\,k\,\widehat{f}(k). \tag{7.43}$$

Similarly, the Fourier transform of the product function $x\,f(x)$ is obtained by differentiating the Fourier transform of $f(x)$:

$$\mathcal{F}[\,x\,f(x)\,] = \mathrm{i}\,\frac{d\widehat{f}}{dk}\,. \tag{7.44}$$

The second statement follows easily from the first via the Symmetry Principle of Theorem 7.3. While the result is stated for ordinary functions, as noted earlier, the Fourier transform — just like Fourier series — is entirely compatible with the calculus of generalized functions.

Example 7.8. The derivative of the even exponential pulse $f_e(x) = e^{-a\,|\,x\,|}$ is a multiple of the odd exponential pulse $f_o(x) = (\operatorname{sign} x)\, e^{-a\,|\,x\,|}$:

$$f_e'(x) = -\,a\,(\operatorname{sign} x)\, e^{-a\,|\,x\,|} = -\,a f_o(x).$$

Proposition 7.7 says that their Fourier transforms are related by

$$\mathrm{i}\,k\,\widehat{f}_e(k) = \mathrm{i}\sqrt{\frac{2}{\pi}}\,\frac{k\,a}{k^2+a^2} = -\,a\,\widehat{f}_o(k),$$

[†] We are assuming that the integrand is sufficiently nice in order to bring the derivative under the integral sign; see [**37**, **117**] for a fully rigorous justification.

as previously noted in (7.21, 24). On the other hand, the odd exponential pulse has a jump discontinuity of magnitude 2 at $x = 0$, and so its derivative contains a delta function:

$$f_o'(x) = -\,a\,e^{-a\,|\,x\,|} + 2\,\delta(x) = -\,a\,f_e(x) + 2\,\delta(x).$$

This is reflected in the relation between their Fourier transforms. If we multiply (7.24) by $\mathrm{i}\,k$, we obtain

$$\mathrm{i}\,k\,\widehat{f}_o(k) = \sqrt{\frac{2}{\pi}}\,\frac{k^2}{k^2+a^2} = \sqrt{\frac{2}{\pi}} - \sqrt{\frac{2}{\pi}}\,\frac{a^2}{k^2+a^2} = 2\,\widehat{\delta}(k) - a\,\widehat{f}_e(k).$$

Higher-order derivatives are handled by iterating the first-order formula (7.43).

Corollary 7.9. *The Fourier transform of* $f^{(n)}(x)$ *is* $(\,\mathrm{i}\,k)^n\,\widehat{f}(k)$.

This result has an important consequence: the smoothness of the function $f(x)$ is manifested in the rate of decay of its Fourier transform $\widehat{f}(k)$. We already noted that the Fourier transform of a (nice) function must decay to zero at large frequencies: $\widehat{f}(k) \to 0$ as $|\,k\,| \to \infty$. (This result can be viewed as the Fourier transform version of the Riemann–Lebesgue Lemma 3.46.) If the nth derivative $f^{(n)}(x)$ is also a reasonable function, then its Fourier transform $\widehat{f^{(n)}}(k) = (\,\mathrm{i}\,k)^n\,\widehat{f}(k)$ must go to zero as $|\,k\,| \to \infty$. This requires that $\widehat{f}(k)$ go to zero more rapidly than $|\,k\,|^{-n}$. Thus, the smoother $f(x)$, the more rapid the decay of its Fourier transform. As a general rule of thumb, local features of $f(x)$, such as smoothness, are manifested by global features of $\widehat{f}(k)$, such as the rate of decay for large $|\,k\,|$. The Symmetry Principle implies that the reverse is also true: global features of $f(x)$ correspond to local features of $\widehat{f}(k)$. For instance, the degree of smoothness of $\widehat{f}(k)$ governs the rate of decay of $f(x)$ as $x \to \pm\infty$. This local-global duality is one of the major themes of Fourier theory.

Integration

Integration is the inverse operation to differentiation, and so should correspond to division by $\mathrm{i}\,k$ in frequency space. As with Fourier series, this is not completely correct; there is an extra constant involved, which contributes an additional delta function.

Proposition 7.10. *If* $f(x)$ *has Fourier transform* $\widehat{f}(k)$, *then the Fourier transform of its integral* $g(x) = \displaystyle\int_{-\infty}^{x} f(y)\,dy$ *is*

$$\widehat{g}(k) = -\,\frac{\mathrm{i}}{k}\,\widehat{f}(k) + \pi\,\widehat{f}(0)\,\delta(k). \tag{7.45}$$

Proof: First notice that

$$\lim_{x\to-\infty} g(x) = 0, \qquad\qquad \lim_{x\to+\infty} g(x) = \int_{-\infty}^{\infty} f(x)\,dx = \sqrt{2\pi}\,\widehat{f}(0).$$

Therefore, if we subtract a suitable multiple of the step function from the integral, the resulting function

$$h(x) = g(x) - \sqrt{2\pi}\,\widehat{f}(0)\,\sigma(x)$$

decays to 0 at both $\pm\infty$. Consulting our table of Fourier transforms, we find

$$\widehat{h}(k) = \widehat{g}(k) - \pi\,\widehat{f}(0)\,\delta(k) + \frac{\mathrm{i}}{k}\,\widehat{f}(0)\,. \tag{7.46}$$

On the other hand,

$$h'(x) = f(x) - \sqrt{2\pi}\,\widehat{f}(0)\,\delta(x).$$

Since $h(x) \to 0$ as $|\,x\,| \to \infty$, we can apply our differentiation rule (7.43), and conclude that

$$\mathrm{i}\,k\,\widehat{h}(k) = \widehat{f}(k) - \widehat{f}(0). \tag{7.47}$$

Combining (7.46) and (7.47) establishes the desired formula (7.45). *Q.E.D.*

Example 7.11. The Fourier transform of the inverse tangent function

$$f(x) = \tan^{-1} x = \int_0^x \frac{dy}{1+y^2} = \int_{-\infty}^x \frac{dy}{1+y^2} - \frac{\pi}{2}$$

can be computed by combining Proposition 7.10 with (7.28, 34):

$$\widehat{f}(k) = \left(-\frac{\mathrm{i}}{k}\sqrt{\frac{\pi}{2}}\,\frac{e^{-|\,k\,|}}{k} + \frac{\pi^{3/2}}{\sqrt{2}}\,\delta(k) \right) - \frac{\pi^{3/2}}{\sqrt{2}}\,\delta(k) = -\,\mathrm{i}\sqrt{\frac{\pi}{2}}\,\frac{e^{-|\,k\,|}}{k}\,.$$

The singularity at $k = 0$ reflects the lack of decay of the inverse tangent as $|\,x\,| \to \infty$.

Exercises

7.2.1. Determine the Fourier transform of the following functions:
(a) $e^{-x^2/2}$, (b) $x\,e^{-x^2/2}$, (c) $x^2\,e^{-x^2/2}$, (d) x, (e) $x\,e^{-2\,|\,x\,|}$, (f) $x\tan^{-1} x$.

7.2.2. Find the Fourier transform of (a) the error function $\operatorname{erf} x = \dfrac{2}{\sqrt{\pi}}\displaystyle\int_0^x e^{-z^2}\,dz$;
(b) the complementary error function $\operatorname{erfc} x = \dfrac{2}{\sqrt{\pi}}\displaystyle\int_x^\infty e^{-z^2}\,dz$.

7.2.3. Find the inverse Fourier transform of the following functions:
(a) k, (b) $k\,e^{-k^2}$, (c) $\dfrac{k}{(1+k^2)^2}$, (d) $\dfrac{k^2}{k-\mathrm{i}}$, (e) $\dfrac{1}{k^2-k}$.

7.2.4. Is the usual formula $\sigma'(x) = \delta(x)$ relating the step and delta functions compatible with their Fourier transforms? Justify your answer.

7.2.5. Find the Fourier transform of the derivative $\delta'(x)$ of the delta function in three ways: (a) First, directly from the definition of $\delta'(x)$; (b) second, using the formula for the Fourier transform of the derivative of a function; (c) third, as a limit of the Fourier transforms of the derivatives of the functions in Exercise 7.1.8. (d) Are your answers all the same? If not, can you explain any discrepancies?

7.2.6. Show that one can obtain the Fourier transform of the Gaussian function $f(x) = e^{-x^2/2}$ by the following trick. First, prove that $\widehat{f}'(k) = -\,k\,\widehat{f}(k)$. Use this to deduce that $\widehat{f}(k) = c\,e^{-k^2/2}$ for some constant c. Finally, use the Symmetry Principle to determine c.

7.2.7. If $f(x)$ has Fourier transform $\widehat{f}(k)$, which function has Fourier transform $\dfrac{\widehat{f}(k)}{k}$?

◇ 7.2.8. If $f(x)$ has Fourier transform $\widehat{f}(k)$, what is the Fourier transform of $\dfrac{f(x)}{x}$?

7.2.9. Use Exercise 7.2.8 to find the Fourier transform of

(a) $1/x$, (b) $x^{-1}e^{-|x|}$, (c) $x^{-1}e^{-x^2}$, (d) $(x^3+4x)^{-1}$.

7.2.10. Directly justify formula (7.43) by integrating the relevant Fourier transform integral by parts. What do you need to assume about the behavior of $f(x)$ for large $|x|$?

7.2.11. Given the Fourier transform $\widehat{f}(k)$ of $f(x)$, find the Fourier transform of its integral $g(x) = \int_a^x f(y)\,dy$ starting at the point $a \in \mathbb{R}$.

◇ 7.2.12. (a) Explain why the Fourier transform of a 2π–periodic function $f(x)$ is a linear combination of delta functions, $\widehat{f}(k) = \sum_{n=-\infty}^{\infty} c_n\,\delta(k-n)$, where c_n are the (complex) Fourier series coefficients (3.65) of $f(x)$ on $[-\pi,\pi]$.
(b) Find the Fourier transform of the following periodic functions:
(i) $\sin 2x$, (ii) $\cos^3 x$, (iii) the 2π–periodic extension of $f(x) = x$,
(iv) the sawtooth function $h(x) = x \bmod 1$, i.e., the fractional part of x.

7.2.13. Determine the Fourier transforms of (a) $\cos x - 1$, (b) $\dfrac{\cos x - 1}{x}$, (c) $\dfrac{\cos x - 1}{x^2}$.
Hint: Use Exercises 7.2.8 and 7.2.12.

◇ 7.2.14. Write down the formulas for differentiation and integration for the alternative Fourier transforms of Exercises 7.1.17 and 7.1.18.

7.2.15. (a) What is the two-dimensional Fourier transform, (7.41), of the gradient $\nabla f(x,y)$ of a function of two variables?
(b) Use your formula to find the Fourier transform of the gradient of $f(x,y) = e^{-x^2-y^2}$.

7.3 Green's Functions and Convolution

The fact that the Fourier transform converts differentiation in the physical domain into multiplication in the frequency domain is one of its most compelling features. A particularly important consequence is that it effectively transforms differential equations into algebraic equations, and thereby facilitates their solution by elementary algebra. One begins by applying the Fourier transform to both sides of the differential equation under consideration. Solving the resulting algebraic equation will produce a formula for the Fourier transform of the desired solution, which can then be immediately reconstructed via the inverse Fourier transform. In the following chapter, we will use these techniques to solve partial differential equations.

Solution of Boundary Value Problems

The Fourier transform is particularly well adapted to boundary value problems on the entire real line. In place of the boundary conditions used on finite intervals, we look for solutions that decay to zero sufficiently rapidly as $|x| \to \infty$ — in order that their Fourier transform be well defined (in the context of ordinary functions). In quantum mechanics, [**66, 72**], these solutions are known as the *bound states*, and they correspond to subatomic particles that are trapped or localized in a region of space. For example, the electrons in an atom are bound states localized by the electrostatic attraction of the nucleus.

As a specific example, consider the boundary value problem

$$-\frac{d^2u}{dx^2}+\omega^2 u = h(x), \qquad -\infty < x < \infty, \tag{7.48}$$

where $\omega > 0$ is a positive constant. The boundary conditions require that the solution decay: $u(x) \to 0$, as $|x| \to \infty$. We will solve this problem by applying the Fourier transform to both sides of the differential equation. Taking Corollary 7.9 into account, the result is the linear algebraic equation

$$k^2\,\widehat{u}(k) + \omega^2\,\widehat{u}(k) = \widehat{h}(k)$$

relating the Fourier transforms of u and h. Unlike the differential equation, the transformed equation can be immediately solved for

$$\widehat{u}(k) = \frac{\widehat{h}(k)}{k^2+\omega^2}\,. \tag{7.49}$$

Therefore, we can reconstruct the solution by applying the inverse Fourier transform formula (7.9):

$$u(x) = \frac{1}{\sqrt{2\pi}}\int_{-\infty}^{\infty}\frac{\widehat{h}(k)\,e^{\,i\,k\,x}}{k^2+\omega^2}\,dk. \tag{7.50}$$

For example, if the forcing function is an even exponential pulse,

$$h(x) = e^{-\,|\,x\,|} \qquad \text{with} \qquad \widehat{h}(k) = \sqrt{\frac{2}{\pi}}\,\frac{1}{k^2+1}\,,$$

then (7.50) writes the solution as a Fourier integral:

$$u(x) = \frac{1}{\pi}\int_{-\infty}^{\infty}\frac{e^{\,i\,k\,x}}{(k^2+\omega^2)(k^2+1)}\,dk = \frac{1}{\pi}\int_{-\infty}^{\infty}\frac{\cos k\,x}{(k^2+\omega^2)(k^2+1)}\,dk\,,$$

where we note that the imaginary part of the complex integral vanishes because the integrand is an odd function. (Indeed, if the forcing function is real, the solution must also be real.) The Fourier integral can be explicitly evaluated using partial fractions to rewrite

$$\widehat{u}(k) = \sqrt{\frac{2}{\pi}}\,\frac{1}{(k^2+\omega^2)(k^2+1)} = \sqrt{\frac{2}{\pi}}\,\frac{1}{\omega^2-1}\left(\frac{1}{k^2+1}-\frac{1}{k^2+\omega^2}\right), \qquad \omega^2 \neq 1.$$

Thus, according to our table of Fourier transforms, the solution to this boundary value problem is

$$u(x) = \frac{e^{-\,|\,x\,|}-\dfrac{1}{\omega}\,e^{-\,\omega\,|\,x\,|}}{\omega^2-1} \qquad \text{when} \qquad \omega^2 \neq 1. \tag{7.51}$$

The reader may wish to verify that this function is indeed a solution, meaning that it is twice continuously differentiable (which is not so immediately apparent from the formula), decays to 0 as $|\,x\,| \to \infty$, and satisfies the differential equation everywhere. The "resonant" case $\omega^2 = 1$ is left to Exercise 7.3.6.

Remark: The method of partial fractions that you may have learned in first-year calculus is often an effective tool for evaluating (inverse) Fourier transforms of such rational functions.

A particularly important case is that in which the forcing function

$$h(x) = \delta_\xi(x) = \delta(x - \xi)$$

represents a unit impulse concentrated at $x = \xi$. The resulting solution is the Green's function $G(x;\xi)$ for the boundary value problem. According to (7.49), its Fourier transform with respect to x is

$$\widehat{G}(k;\xi) = \frac{1}{\sqrt{2\pi}}\,\frac{e^{-\,i\,k\,\xi}}{k^2 + \omega^2},$$

which is the product of an exponential factor $e^{-\,i\,k\,\xi}$, representing the Fourier transform of $\delta_\xi(x)$, times a multiple of the Fourier transform of the even exponential pulse $e^{-\omega\,|\,x\,|}$. We apply the Shift Theorem 7.4, and conclude that the Green's function for this boundary value problem is an exponential pulse centered at ξ, namely

$$G(x;\xi) = \frac{1}{2\,\omega}\,e^{-\omega\,|\,x - \xi\,|} = g(x - \xi), \qquad \text{where} \qquad g(x) = G(x;0) = \frac{1}{2\,\omega}\,e^{-\omega\,|\,x\,|}. \tag{7.52}$$

Observe that, as with other self-adjoint boundary value problems, the Green's function is symmetric under interchange of x and ξ, so $G(x;\xi) = G(\xi;x)$. As a function of x, it satisfies the homogeneous differential equation $-u'' + \omega^2 u = 0$, except at the point $x = \xi$, where its derivative has a jump discontinuity of unit magnitude. It also decays as $|\,x\,| \to \infty$, as required by the boundary conditions. The fact that $G(x;\xi) = g(x - \xi)$ depends only on the difference $x - \xi$ is a consequence of the translation invariance of the boundary value problem. The superposition principle based on the Green's function tells us that the solution to the inhomogeneous boundary value problem (7.48) under a general forcing can be represented in the integral form

$$u(x) = \int_{-\infty}^{\infty} G(x;\xi)\,h(\xi)\,d\xi = \int_{-\infty}^{\infty} g(x - \xi)\,h(\xi)\,d\xi = \frac{1}{2\omega}\int_{-\infty}^{\infty} e^{-\omega|\,x-\xi\,|}\,h(\xi)\,d\xi. \tag{7.53}$$

The reader may enjoy recovering the particular exponential solution (7.51) from this integral formula.

Exercises

7.3.1. Use partial fractions to compute the inverse Fourier transform of the following rational functions. *Hint*: First solve Exercise 7.1.3.

(a) $\dfrac{1}{k^2 - 5k - 6}$, (b) $\dfrac{e^{\,i\,k}}{k^2 - 1}$, (c) $\dfrac{1}{k^4 - 1}$, (d) $\dfrac{\sin 2k}{k^2 + 2k - 3}$.

7.3.2. Find the inverse Fourier transform of the function $\dfrac{1}{k^2 + 2k + 5}$:
(a) using partial fractions; (b) by completing the square. Are your answers the same?

7.3.3. Use partial fractions to compute the Fourier transform of the following functions:

(a) $\dfrac{1}{x^2 - x - 2}$, (b) $\dfrac{1}{x^3 + x}$, (c) $\dfrac{\cos x}{x^2 - 9}$.

7.3.4. Find a solution to the differential equation $-\dfrac{d^2u}{dx^2} + 4u = \delta(x)$ by using the Fourier transform.

7.3.5. Use the Fourier transform to solve the boundary value problem

$$-u'' + u = \delta'(x-1) \quad \text{for} \quad -\infty < x < \infty, \quad \text{with} \quad u(x) \to 0 \quad \text{as} \quad x \to \pm\infty.$$

♢ 7.3.6. (a) Use the Fourier transform to solve (7.48) with $h(x) = e^{-\,|\,x\,|}$ when $\omega = 1$.
(b) Verify that your solution can be obtained as a limit of (7.51) as $\omega \to 1$.

7.3.7. Use the Fourier transform to find a bounded solution to the differential equation

$$u'''' + u = e^{-2\,|\,x\,|}.$$

7.3.8. Use the Fourier transform to find an integral formula for a bounded solution to the *Airy differential equation* $-\dfrac{d^2u}{dx^2} = x\,u.$

♢ 7.3.9. Prove that (7.51) is a twice continuously differentiable function of x and satisfies the differential equation (7.48).

Convolution

In our solution to the boundary value problem (7.48), we ended up deriving a formula for its Fourier transform (7.49) as the product of two known Fourier transforms. The final Green's function formula (7.53), obtained by applying the inverse Fourier transform, is indicative of a general property, in that it is given by a *convolution product*.

Definition 7.12. The *convolution* of scalar functions $f(x)$ and $g(x)$ is the scalar function $h = f * g$ defined by the formula

$$h(x) = f * g(x) = \int_{-\infty}^{\infty} f(x-\xi)\, g(\xi)\, d\xi. \tag{7.54}$$

We list the basic properties of the convolution product, leaving their verification as exercises for the reader. All of these assume that the implied convolution integrals converge.

(a) *Symmetry*: $f * g = g * f,$

(b) *Bilinearity*: $\begin{cases} f * (a\,g + b\,h) = a\,(f * g) + b\,(f * h), \\ (a\,f + b\,g) * h = a\,(f * h) + b\,(g * h), \end{cases} \qquad a, b \in \mathbb{C},$

(c) *Associativity*: $f * (g * h) = (f * g) * h,$

(d) *Zero function*: $f * 0 = 0,$

(e) *Delta function*: $f * \delta = f.$

One tricky feature is that the constant function 1 is *not* a unit for the convolution product; indeed,

$$f * 1 = 1 * f = \int_{-\infty}^{\infty} f(\xi)\, d\xi$$

is a constant function, namely the total integral of f, and not the original function $f(x)$. In fact, according to the final property, the delta function plays the role of the "convolution unit":

$$f * \delta(x) = \int_{-\infty}^{\infty} f(x-\xi)\, \delta(\xi)\, d\xi = f(x).$$

In particular, our solution (7.52) has the form of a convolution product between an even exponential pulse $g(x) = (2\omega)^{-1} e^{-\omega|x|}$ and the forcing function:

$$u(x) = g * h(x).$$

On the other hand, its Fourier transform (7.49) is, up to a factor, the ordinary multiplicative product

$$\widehat{u}(k) = \sqrt{2\pi}\, \widehat{g}(k)\, \widehat{h}(k)$$

of the Fourier transforms of g and h. In fact, this is a general property of the Fourier transform: convolution in the physical domain corresponds to multiplication in the frequency domain, and conversely.

Theorem 7.13. *The Fourier transform of the convolution* $h(x) = f * g(x)$ *of two functions is a multiple of the product of their Fourier transforms*:

$$\widehat{h}(k) = \sqrt{2\pi}\, \widehat{f}(k)\, \widehat{g}(k). \tag{7.55}$$

Conversely, the Fourier transform of their product $h(x) = f(x)\, g(x)$ *is, up to a multiple, the convolution of their Fourier transforms*:

$$\widehat{h}(k) = \frac{1}{\sqrt{2\pi}}\, \widehat{f} * \widehat{g}(k) = \frac{1}{\sqrt{2\pi}} \int_{-\infty}^{\infty} \widehat{f}(k-\kappa)\, \widehat{g}(\kappa)\, d\kappa. \tag{7.56}$$

Proof: Combining the definition of the Fourier transform with the convolution formula (7.54), we obtain

$$\widehat{h}(k) = \frac{1}{\sqrt{2\pi}} \int_{-\infty}^{\infty} h(x)\, e^{-\,i\,k\,x}\, dx = \frac{1}{\sqrt{2\pi}} \int_{-\infty}^{\infty}\int_{-\infty}^{\infty} f(x-\xi)\, g(\xi)\, e^{-\,i\,k\,x}\, dx\, d\xi.$$

Applying the change of variables $\eta = x - \xi$ in the inner integral produces

$$\begin{aligned}\widehat{h}(k) &= \frac{1}{\sqrt{2\pi}} \int_{-\infty}^{\infty}\int_{-\infty}^{\infty} f(\eta)\, g(\xi)\, e^{-\,i\,k(\xi+\eta)}\, d\xi\, d\eta \\ &= \sqrt{2\pi}\left(\frac{1}{\sqrt{2\pi}} \int_{-\infty}^{\infty} f(\eta)\, e^{-\,i\,k\,\eta}\, d\eta\right)\left(\frac{1}{\sqrt{2\pi}} \int_{-\infty}^{\infty} g(\xi)\, e^{-\,i\,k\,\xi}\, d\xi\right) = \sqrt{2\pi}\, \widehat{f}(k)\, \widehat{g}(k),\end{aligned}$$

proving (7.55). The second formula can be proved in a similar fashion, or by simply noting that it follows directly from the Symmetry Principle of Theorem 7.3. *Q.E.D.*

Example 7.14. We already know, (7.29), that the Fourier transform of

$$f(x) = \frac{\sin x}{x}$$

is the box function

$$\widehat{f}(k) = \sqrt{\frac{\pi}{2}}\,\bigl[\,\sigma(k+1) - \sigma(k-1)\,\bigr] = \begin{cases} \sqrt{\dfrac{\pi}{2}}, & -1 < k < 1, \\ 0, & |\,k\,| > 1. \end{cases}$$

We also know that the Fourier transform of

$$g(x) = \frac{1}{x} \qquad \text{is} \qquad \widehat{g}(k) = -\,i\,\sqrt{\frac{\pi}{2}}\ \operatorname{sign} k.$$

Therefore, the Fourier transform of their product

$$h(x) = f(x)\, g(x) = \frac{\sin x}{x^2}$$

can be obtained by convolution:

$$\widehat{h}(k) = \frac{1}{\sqrt{2\pi}}\, \widehat{f} * \widehat{g}(k) = \frac{1}{\sqrt{2\pi}} \int_{-\infty}^{\infty} \widehat{f}(\kappa)\, \widehat{g}(k-\kappa)\, d\kappa$$

$$= -\,\mathrm{i}\,\sqrt{\frac{\pi}{8}} \int_{-1}^{1} \operatorname{sign}(k-\kappa)\, d\kappa = \begin{cases} \mathrm{i}\,\sqrt{\dfrac{\pi}{2}} & k < -1, \\ -\,\mathrm{i}\,\sqrt{\dfrac{\pi}{2}}\, k, & -1 < k < 1, \\ -\,\mathrm{i}\,\sqrt{\dfrac{\pi}{2}} & k > 1. \end{cases}$$

Exercises

7.3.10. (a) Find the Fourier transform of the convolution $h(x) = f_e * g(x)$ of an even exponential pulse $f_e(x) = e^{-|x|}$ and a Gaussian $g(x) = e^{-x^2}$. (b) What is $h(x)$?

7.3.11. What is the convolution of a Gaussian kernel e^{-x^2} with itself? *Hint*: Use the Fourier transform.

7.3.12. Find the function whose Fourier transform is $\widehat{f}(k) = (k^2+1)^{-2}$.

♡ 7.3.13. (a) Write down the Fourier transform of the box function $f(x) = \begin{cases} 1, & |x| < \frac{1}{2}, \\ 0, & |x| > \frac{1}{2}. \end{cases}$
(b) Graph the hat function $h(x) = f * f(x)$ and find its Fourier transform.
(c) Determine the *cubic B spline* $s(x) = h * h(x)$ and its Fourier transform.

7.3.14. Let $f(x) = \begin{cases} \sin x, & 0 < x < \pi, \\ 0, & \text{otherwise,} \end{cases}$ $\quad g(x) = \begin{cases} \cos x, & 0 < x < \pi, \\ 0, & \text{otherwise.} \end{cases}$

(a) Find the Fourier transforms of $f(x)$ and $g(x)$; (b) compute the convolution $h(x) = f * g(x)$; (c) find its Fourier transform $\widehat{h}(k)$.

7.3.15. Use convolution to find an integral formula for the function whose Fourier transform is

$$\text{(a)}\ \frac{e^{-k^2}}{k^2+1}, \qquad \text{(b)}\ \frac{\sin k}{k(k^2+1)}, \qquad \text{(c)}\ \frac{\sin^2 k}{k^2}, \qquad \text{(d)}\ \frac{\operatorname{sign} k}{1+\mathrm{i}\,k}.$$

If possible, evaluate the resulting convolution integral.

7.3.16. Let $f(x)$ be a smooth function. (a) Find its convolution $\delta' * f$ with the derivative of the delta fiunction. (b) More generally, find $\delta^{(n)} * f$.

7.3.17. According to Proposition 7.7, the Fourier transform of the derivative $f'(x)$ is obtained by multiplying $\widehat{f}(k)$ by $\mathrm{i}\,k$. Can you reconcile this result with the Convolution Theorem 7.13?

♢ 7.3.18. The *Hilbert transform* of a function $f(x)$ is defined as the integral

$$h(x) = \frac{1}{\pi} ⨍_{-\infty}^{\infty} \frac{f(\xi)\, d\xi}{\xi - x}. \tag{7.57}$$

Find a formula for its Fourier transform $\widehat{h}(k)$ in terms of $\widehat{f}(k)$. *Remark*: The bar on the integral indicates the *principal value integral*, [**2**], which is $\displaystyle\lim_{\delta \to 0^+} \left(\int_{-\infty}^{x-\delta} + \int_{x+\delta}^{\infty} \right) \frac{f(\xi)\, d\xi}{\xi - x}$, and is employed to avoid the integral diverging at the singular point $x = \xi$.

7.3.19. Use the Fourier transform to solve the integral equation $\int_{-\infty}^{\infty} e^{-|x-\xi|} u(\xi)\, d\xi = f(x)$. Then verify your solution when $f(x) = e^{-2|x|}$.

7.3.20. Suppose that $f(x)$ and $g(x)$ are identically 0 for all $x < 0$. Prove that their convolution product $h = f * g$ reduces to a finite integral: $h(x) = \begin{cases} \int_0^x f(x-\xi)\, g(\xi)\, d\xi, & x > 0, \\ 0, & x \leq 0. \end{cases}$

7.3.21. Given that the support of $f(x)$ is contained in the interval $[a,b]$ and the support of $g(x)$ is contained in $[c,d]$, what can you say about the support of their convolution $h(x) = f * g(x)$?

♢ 7.3.22. Prove the convolution properties (a–e).

♢ 7.3.23. In this exercise, we explain how convolution can be used to smooth out rough data. Let $g_\varepsilon(x) = \dfrac{\varepsilon}{\pi(\varepsilon^2 + x^2)}$. (a) If $f(x)$ is any (reasonable) function, show that $f_\varepsilon(x) = g_\varepsilon * f(x)$ for $\varepsilon \neq 0$ is a C^∞ function. (b) Show that $\lim_{\varepsilon \to 0} f_\varepsilon(x) = f(x)$.

7.3.24. Explain why the Shift Theorem 7.4 is a special case of the Convolution Theorem 7.13.

♢ 7.3.25. Suppose $f(x)$ and $g(x)$ are 2π–periodic and have respective complex Fourier coefficients c_k and d_k. Prove that the complex Fourier coefficients e_k of the product function $f(x)\,g(x)$ are given by the *convolution summation* $e_k = \sum_{j=-\infty}^{\infty} c_j\, d_{k-j}$. *Hint*: Substitute the formulas for the complex Fourier coefficients into the summation, making sure to use two different integration variables, and then use (6.37).

7.4 The Fourier Transform on Hilbert Space

While we do not possess all the analytic tools to embark on a fully rigorous treatment of the mathematical theory underlying the Fourier transform, it is worth outlining a few of the more important features. We have already noted that the Fourier transform, when defined, is a linear operator, taking functions $f(x)$ on physical space to functions $\widehat{f}(k)$ on frequency space. A critical question is the following: to precisely which function space should the theory be applied? Not every function admits a Fourier transform in the classical sense[†] — the Fourier integral (7.6) is required to converge, and this places restrictions on the function and its asymptotics at large distances.

It turns out the proper setting for the rigorous theory is the *Hilbert space* of complex-valued square-integrable functions — the same infinite-dimensional vector space that lies at the heart of modern quantum mechanics. In Section 3.5, we already introduced the Hilbert space $\mathrm{L}^2[a,b]$ on a finite interval; here we adapt Definition 3.34 to the entire real line. Thus, the Hilbert space $\mathrm{L}^2 = \mathrm{L}^2(\mathbb{R})$ is the infinite-dimensional vector space consisting of all complex-valued functions $f(x)$ that are defined for all $x \in \mathbb{R}$ and have finite L^2 norm:

$$\| f \|^2 = \int_{-\infty}^{\infty} |\, f(x)\,|^2 \, dx < \infty. \qquad (7.58)$$

[†] We leave aside the more advanced issues involving generalized functions.

For example, any piecewise continuous function that satisfies the decay criterion

$$| f(x) | \le \frac{M}{| x |^{1/2+\delta}}, \qquad \text{for all sufficiently large} \qquad | x | \gg 0, \tag{7.59}$$

for some $M > 0$ and $\delta > 0$, belongs to L^2. However, as in Section 3.5, Hilbert space contains many more functions, and the precise definitions and identification of its elements is quite subtle. On the other hand, most nondecaying functions do not belong to L^2, including the constant function $f(x) \equiv 1$ as well as all oscillatory complex exponentials, $e^{\,i k x}$ for $k \in \mathbb{R}$.

The Hermitian inner product on the complex Hilbert space L^2 is prescribed in the usual manner,

$$\langle f , g \rangle = \int_{-\infty}^{\infty} f(x)\, \overline{g(x)}\, dx, \tag{7.60}$$

so that $\| f \|^2 = \langle f , f \rangle$. The Cauchy–Schwarz inequality

$$| \langle f , g \rangle | \le \| f \| \, \| g \| \tag{7.61}$$

ensures that the inner product integral is finite whenever $f, g \in L^2$. Observe that the Fourier transform (7.6) can be regarded as a multiple of the inner product of the function $f(x)$ with the complex exponential functions:

$$\widehat{f}(k) = \frac{1}{\sqrt{2\pi}} \int_{-\infty}^{\infty} f(x)\, e^{-\,i k x}\, dx = \frac{1}{\sqrt{2\pi}} \langle f(x) , e^{\,i k x} \rangle. \tag{7.62}$$

However, when interpreting this formula, one must bear in mind that the exponentials are *not* themselves elements of L^2.

Let us state the fundamental result governing the effect of the Fourier transform on functions in Hilbert space. It can be regarded as a direct analogue of the Pointwise Convergence Theorem 3.8 for Fourier series.

Theorem 7.15. *If $f(x) \in L^2$ is square-integrable, then its Fourier transform $\widehat{f}(k) \in L^2$ is a well-defined, square-integrable function of the frequency variable k. If $f(x)$ is continuously differentiable at a point x, then the inverse Fourier transform integral* (7.9) *equals its value $f(x)$. More generally, if the left- and right-hand limits $f(x^-)$, $f(x^+)$, $f'(x^-)$, $f'(x^+)$ exist, then the inverse Fourier transform integral converges to the average value $\frac{1}{2}\left[f(x^-) + f(x^+) \right]$.*

Thus, the Fourier transform $\widehat{f} = \mathcal{F}[f]$ defines a linear transformation from L^2 functions of x to L^2 functions of k. In fact, the Fourier transform preserves inner products. This important result is known as *Parseval's formula*, whose Fourier series counterpart appeared in (3.122).

Theorem 7.16. *If $\widehat{f}(k) = \mathcal{F}[f(x)]$ and $\widehat{g}(k) = \mathcal{F}[g(x)]$, then $\langle f , g \rangle = \langle \widehat{f} , \widehat{g} \rangle$, i.e.,*

$$\int_{-\infty}^{\infty} f(x)\, \overline{g(x)}\, dx = \int_{-\infty}^{\infty} \widehat{f}(k)\, \overline{\widehat{g}(k)}\, dk. \tag{7.63}$$

Proof: Let us sketch a formal proof that serves to motivate why this result is valid. We use the definition (7.6) of the Fourier transform to evaluate

$$\begin{aligned}
\int_{-\infty}^{\infty} \widehat{f}(k)\, \overline{\widehat{g}(k)}\, dk &= \int_{-\infty}^{\infty} \left(\frac{1}{\sqrt{2\pi}} \int_{-\infty}^{\infty} f(x)\, e^{-\,i k x}\, dx \right) \left(\frac{1}{\sqrt{2\pi}} \int_{-\infty}^{\infty} \overline{g(y)}\, e^{+\,i k y}\, dy \right) dk \\
&= \int_{-\infty}^{\infty} \int_{-\infty}^{\infty} f(x)\, \overline{g(y)} \left(\frac{1}{2\pi} \int_{-\infty}^{\infty} e^{-\,i k (x-y)}\, dk \right) dx\, dy.
\end{aligned}$$

Now according to (7.38), the inner k integral can be replaced by the delta function $\delta(x-y)$, and hence

$$\int_{-\infty}^{\infty} \widehat{f}(k)\,\overline{\widehat{g}(k)}\,dk = \int_{-\infty}^{\infty}\int_{-\infty}^{\infty} f(x)\,\overline{g(y)}\,\delta(x-y)\,dx\,dy = \int_{-\infty}^{\infty} f(x)\,\overline{g(x)}\,dx.$$

This completes our "proof"; see [**37, 68, 117**] for a rigorous version. *Q.E.D.*

In particular, orthogonal functions, satisfying $\langle f\,, g\rangle = 0$, will have orthogonal Fourier transforms, $\langle \widehat{f}\,, \widehat{g}\rangle = 0$. Choosing $f = g$ in Parseval's formula (7.63) produces *Plancherel's formula*

$$\|f\|^2 = \|\widehat{f}\|^2, \qquad \text{or, explicitly,} \qquad \int_{-\infty}^{\infty} |\,f(x)\,|^2\,dx = \int_{-\infty}^{\infty} |\,\widehat{f}(k)\,|^2\,dk. \tag{7.64}$$

Thus, the Fourier transform $\mathcal{F}\colon \mathrm{L}^2 \to \mathrm{L}^2$ defines a norm-preserving, or *unitary*, linear transformation on Hilbert space, mapping L^2 functions of the physical variable x to L^2 functions of the frequency variable k.

Quantum Mechanics and the Uncertainty Principle

In its popularized form, the Heisenberg Uncertainty Principle is a by now familiar philosophical concept. First formulated in the 1920s by the German physicist Werner Heisenberg, one of the founders of modern quantum mechanics, it states that, in a physical system, certain quantities cannot be simultaneously measured with complete accuracy. For instance, the more precisely one measures the position of a particle, the less accuracy there will be in the measurement of its momentum; conversely, the greater the accuracy in the momentum, the less certainty in its position. A similar uncertainty couples energy and time. Experimental verification of the uncertainty principle can be found even in fairly simple situations. Consider a light beam passing through a small hole. The position of the photons is constrained by the hole; the effect of their momenta is observed in the pattern of light diffused on a screen placed beyond the hole. The smaller the hole, the more constrained the photon's position as it passes through, hence, according to the Uncertainty Principle, the less certainty there is in the observed momentum, and, consequently, the wider and more diffuse the resulting image on the screen.

This is not the place to discuss the philosophical and experimental consequences of Heisenberg's Principle. What we will show is that the Uncertainty Principle is, in fact, a mathematical property of the Fourier transform! In quantum theory, each of the paired quantities, e.g., position and momentum, are interrelated by the Fourier transform. Indeed, Proposition 7.7 says that the Fourier transform of the differentiation operator representing momentum is a multiplication operator representing position and vice versa. This Fourier-transform-based duality between position and momentum, that is, between multiplication and differentiation, lies at the heart of the Uncertainty Principle.

In quantum mechanics, the wave functions of a quantum system are characterized as the elements of unit norm, $\|\varphi\| = 1$, belonging to the underlying state space, which, in a one-dimensional model of a single particle, is the Hilbert space $\mathrm{L}^2 = \mathrm{L}^2(\mathbb{R})$ consisting of square-integrable complex-valued functions of x. As we already noted in Section 3.5, the squared modulus of the wave function, $|\,\varphi(x)\,|^2$, represents the probability density of the particle being found at position x. Consequently, the *mean* or *expected value* of any function $f(x)$ of the position variable is given by its integral against the system's probability density and denoted by

$$\langle f(x) \rangle = \int_{-\infty}^{\infty} f(x) \, | \, \varphi(x) \, |^2 \, dx. \tag{7.65}$$

In particular,

$$\langle x \rangle = \int_{-\infty}^{\infty} x \, | \, \varphi(x) \, |^2 \, dx \tag{7.66}$$

is the expected measured position of the particle, while Δx, defined by

$$(\Delta x)^2 = \langle \, (x - \langle x \rangle)^2 \, \rangle = \langle x^2 \rangle - \langle x \rangle^2 , \tag{7.67}$$

which is the probability density's *variance*, is the statistical deviation of the particle's measured position from the mean. We note that the next-to-last term equals

$$\langle x^2 \rangle = \int_{-\infty}^{\infty} x^2 \, | \, \varphi(x) \, |^2 \, dx = \| \, x \, \varphi(x) \, \|^2. \tag{7.68}$$

On the other hand, the momentum variable p is related to the Fourier transform frequency via the *de Broglie relation* $p = \hbar \, k$, where

$$\hbar = \frac{h}{2\pi} \approx 1.055 \times 10^{-34} \ \text{ joule seconds} \tag{7.69}$$

is *Planck's constant*, whose value governs the quantization of physical quantities. Therefore, the mean, or expected value, of any function of momentum $g(p)$ is given by its integral against the squared modulus of the Fourier transformed wave function:

$$\langle g(p) \rangle = \int_{-\infty}^{\infty} g(\hbar \, k) \, | \, \widehat{\varphi}(k) \, |^2 \, dk. \tag{7.70}$$

In particular, the mean of the momentum measurements of the particle is

$$\langle p \rangle = \hbar \int_{-\infty}^{\infty} k \, | \, \widehat{\varphi}(k) \, |^2 \, dk = - \, \mathrm{i} \, \hbar \int_{-\infty}^{\infty} \varphi'(x) \, \overline{\varphi(x)} \, dx = - \, \mathrm{i} \, \hbar \, \langle \, \varphi' \, , \varphi \, \rangle, \tag{7.71}$$

where we used Parseval's formula (7.63) to convert to an integral over position, and (7.43) to infer that $k \, \widehat{\varphi}(k)$ is the Fourier transform of $- \, \mathrm{i} \, \varphi'(x)$. Similarly,

$$(\Delta p)^2 = \langle \, (p - \langle p \rangle)^2 \, \rangle = \langle p^2 \rangle - \langle p \rangle^2 \tag{7.72}$$

is the squared *variance* of the momentum, where, by Plancherel's formula (7.64) and (7.43),

$$\begin{aligned} \langle p^2 \rangle = \hbar^2 \int_{-\infty}^{\infty} k^2 \, | \, \widehat{\varphi}(k) \, |^2 \, dk &= \hbar^2 \int_{-\infty}^{\infty} | \, \mathrm{i} \, k \, \widehat{\varphi}(k) \, |^2 \, dk \\ &= \hbar^2 \int_{-\infty}^{\infty} | \, \varphi'(x) \, |^2 \, dx = \hbar^2 \, \| \, \varphi'(x) \, \|^2. \end{aligned} \tag{7.73}$$

With this interpretation, the *Uncertainty Principle* for position and momentum measurements can be stated.

Theorem 7.17. *If $\varphi(x)$ is a wave function, so $\| \varphi \| = 1$, then the observed variances in position and momentum satisfy the inequality*

$$\Delta x \, \Delta p \geq \tfrac{1}{2} \hbar. \tag{7.74}$$

Now, the smaller the variance of a quantity such as position or momentum, the more accurate will be its measurement. Thus, the Heisenberg inequality (7.74) effectively quantifies the statement that the more accurately we are able to measure the momentum p, the less accurate will be any measurement of its position x, and vice versa. For more details, along with physical and experimental consequences, you should consult an introductory text on mathematical quantum mechanics, e.g., [**66**, **72**].

Proof: For any value of the real parameter t,

$$\begin{aligned} 0 &\leq \| t \, x \, \varphi(x) + \varphi'(x) \|^2 \\ &= t^2 \| x \, \varphi(x) \|^2 + t \left[\langle \varphi'(x) \, , x \, \varphi(x) \rangle + \langle x \, \varphi(x) \, , \varphi'(x) \rangle \right] + \| \varphi'(x) \|^2. \end{aligned} \tag{7.75}$$

The middle term in the final expression can be evaluated as follows:

$$\begin{aligned} \langle \varphi'(x) \, , x \, \varphi(x) \rangle + \langle x \, \varphi(x) \, , \varphi'(x) \rangle &= \int_{-\infty}^{\infty} \left[x \, \varphi'(x) \, \overline{\varphi(x)} + x \, \varphi(x) \, \overline{\varphi'(x)} \right] dx \\ &= \int_{-\infty}^{\infty} x \, \frac{d}{dx} | \varphi(x) |^2 \, dx = - \int_{-\infty}^{\infty} | \varphi(x) |^2 \, dx = -1, \end{aligned}$$

via an integration by parts, noting that the boundary terms vanish, provided $\varphi(x)$ satisfies the L^2 decay criterion (7.59). Thus, in view of (7.68) and (7.73), the inequality in (7.75) reads

$$\langle x^2 \rangle \, t^2 - t + \frac{\langle p^2 \rangle}{\hbar^2} \geq 0 \qquad \text{for all} \qquad t \in \mathbb{R}.$$

The minimum value of the left-hand side occurs at $t_\star = 1/(2 \langle x^2 \rangle)$, where its value is

$$\frac{\langle p^2 \rangle}{\hbar^2} - \frac{1}{4 \langle x^2 \rangle} \geq 0, \qquad \text{which implies} \qquad \langle x^2 \rangle \langle p^2 \rangle \geq \tfrac{1}{4} \hbar^2.$$

To obtain the uncertainty relation (7.74), one performs the selfsame calculation, but with $x - \langle x \rangle$ replacing x and $p - \langle p \rangle$ replacing p. The result is

$$\left\langle (x - \langle x \rangle)^2 \right\rangle t^2 - t + \frac{\left\langle (p - \langle p \rangle)^2 \right\rangle}{\hbar^2} = (\Delta x)^2 \, t^2 - t + \frac{(\Delta p)^2}{\hbar^2} \geq 0. \tag{7.76}$$

Substituting $t = 1/(2 \, (\Delta x)^2)$ produces the Heisenberg inequality (7.74). *Q.E.D.*

Exercises

7.4.1. (a) Write out the Plancherel formula for the square wave pulse $f(x) = \begin{cases} 1, & |x| < 1, \\ 0, & |x| > 1. \end{cases}$

(b) What is $\displaystyle\int_0^\infty \frac{\sin^2 x}{x^2} \, dx$?

7.4.2. Apply the Plancherel formula to the even decaying pulse (7.20) to evaluate $\displaystyle\int_{-\infty}^{\infty} \frac{dx}{(a^2+x^2)^2}$. How would you compute this integral using elementary calculus?

♡ 7.4.3. (a) Find the Fourier transform of the function $f_n(x) = \begin{cases} -n^2 \operatorname{sign} x, & |x| < \frac{1}{n}, \\ 0, & \text{otherwise}, \end{cases}$ where n is a positive integer. (b) Write out the Plancherel formula for $f_n(x)$. (c) Determine the limit, as $n \to \infty$, of the Fourier transform of $f_n(x)$. (d) Explain why the limit should be the Fourier transform of the derivative of the delta function $\delta'(x)$.

7.4.4. Prove that Parseval's formula is a consequence of Plancherel's formula. *Hint*: Use the identity in Exercise 3.5.34(b).

♢ 7.4.5. Prove that the Hilbert space $L^2(\mathbb{R})$ is a complex vector space.

♢ 7.4.6. We did not quite tell the truth when we said that L^2 functions must decay at large distances: Prove that the following function is in L^2 but does not go to zero as $|x| \to \infty$:

$$f(x) = \begin{cases} 1, & n - n^{-2} < x < n + n^{-2} \quad \text{for} \quad n = \pm 1, \pm 2, \pm 3, \ldots, \\ 0, & \text{otherwise.} \end{cases}$$

7.4.7. Modify the function in Exercise 7.4.6 to produce a function $f \in L^2$ that nevertheless satisfies $\lim_{n \to \pm\infty} f(n) = \infty$ for $n \in \mathbb{Z}$.

♢ 7.4.8. Suppose $f \in L^2$ is continuously differentiable, $f \in C^1$, and has bounded derivative: $|f'(x)| \le M$ for all $x \in \mathbb{R}$. Prove that $f(x) \to 0$ as $x \to \pm\infty$.

7.4.9. (a) Find the constant $a > 0$ such that $\varphi(x) = a\, e^{-|x|}$ is a wave function.
(b) Verify the Heisenberg inequality (7.74) for this particular wave function.

7.4.10. Answer Exercise 7.4.9 when (a) $\varphi(x) = a\, e^{-x^2}$; (b) $\varphi(x) = \dfrac{a}{1+x^2}$.

♢ 7.4.11. Write out a detailed derivation of the final inequality (7.76).

Chapter 8
Linear and Nonlinear Evolution Equations

The term *evolution equation* refers to a dynamical partial differential equation that involves both time t and space $\mathbf{x} = (x_1, \ldots, x_n)$ as independent variables and takes the form

$$\frac{\partial u}{\partial t} = K[u], \tag{8.1}$$

whose left-hand side is just the first-order time derivative of the dependent variable u, while the right-hand side, which can be linear or nonlinear, involves only u and its space derivatives and, possibly, t and $\mathbf{x}$. Examples already encountered include the linear and nonlinear transport equations in Chapter 2 and the heat equation. (But not the wave equation or Laplace equation.) In this chapter, we will analyze several important evolution equations, both linear and nonlinear, involving a single spatial variable.

Our first stop is to revisit the heat equation. We introduce the fundamental solution, which, for dynamical partial differential equations, assumes the role of the Green's function, in that its initial condition is a concentrated delta impulse. The fundamental solution leads to an integral superposition formula for the solutions produced by more general initial conditions or by external forcing. For the heat equation on the entire real line, the Fourier transform enables us to construct an explicit formula that identifies its fundamental solution as a Gaussian filter. We next present the Maximum Principle that rigorously justifies the entropic decay of temperature in a heated body and underlies much of the advanced mathematical analysis of parabolic partial differential equations. Finally, we discuss the Black–Scholes equation, the paradigmatic model for investment portfolios, first proposed in the early 1970s and now lying at the heart of the modern financial industry. We will find that the Black–Scholes equation can be transformed into the linear heat equation, whose fundamental solution is applied to establish the celebrated Black–Scholes formula for option pricing.

The following section provides a brief introduction to symmetry-based solution techniques for linear and nonlinear partial differential equations. Knowing a symmetry of a partial differential equation allows one to readily construct additional solutions from any known solution. Solutions that remain invariant under a one-parameter family of symmetries can be found by solving a reduced ordinary differential equation. The most important are the traveling wave solutions, which are invariant under translation symmetries, and similarity solutions, which are invariant under scaling symmetries.

The next evolution equation to appear is a paradigmatic model of nonlinear diffusion known as Burgers' equation. It can be regarded as a very simplified model of fluid dynamics, combining both nonlinear and viscous effects. We discover a remarkable nonlinear change of variables that maps Burgers' equation to the linear heat equation, and thereby facilitates its analysis, allowing us to construct explicit solutions, and investigate how they converge to shock wave solutions of the nonlinear transport equation in the inviscid limit.

Next, we turn our attention to the simplest third-order linear evolution equation, which arises as a model for wave mechanics. Unlike first- and second-order wave equations, its solutions are not simple traveling waves, but instead exhibit dispersion, in which oscillatory waves of different frequencies move at different speeds. As a result, initially localized disturbances will spread out or disperse, even while they conserve the underlying energy. Dispersion implies that the individual wave velocities differ from the group velocity, which measures the speed of propagation of energy in the system. An everyday manifestation of this phenomenon can be observed in the ripples caused by throwing a rock into a pond: the individual waves move faster than the overall disturbance. Finally, we present the remarkable Talbot effect, only recently discovered, in which solutions having discontinuous initial data and subject to periodic boundary conditions exhibit radically different profiles at rational and irrational times.

Our final example is the celebrated Korteweg–de Vries equation, which originally arose in the work of the nineteenth-century French applied mathematician Joseph Boussinesq as a model for surface waves on shallow water. It combines the effects of linear dispersion and nonlinear transport. Unlike the linearly dispersive model, the Korteweg–de Vries equation admits explicit, localized traveling wave solutions, now known as "solitons". Remarkably, despite the potentially complicated nonlinear nature of their interaction, two solitons emerge from a collision with their individual profiles preserved, the only residual effect being a relative phase shift. The Korteweg–de Vries equation is the prototype of a completely integrable partial differential equation, whose many remarkable properties were first discovered in the mid 1960s. A surprising number of such completely integrable nonlinear systems appear in a variety of applications, including dynamical models in fluids, plasmas, optics, and solid mechanics. Their analysis remains an extremely active area of contemporary research, [**2**, **36**].

8.1 The Fundamental Solution to the Heat Equation

One disadvantage of the Fourier series solution to the heat equation is that it is not nearly as explicit as one might desire for practical applications, numerical computations, or even further theoretical investigations and developments. An alternative approach is based on the idea of the *fundamental solution*, which plays the role of the Green's function in solving initial value problems. The fundamental solution measures the effect of a concentrated, instantaneous impulse, either in the initial conditions or as an external force on the system.

We restrict our attention to homogeneous boundary conditions — keeping in mind that these can always be included by use of linear superposition. The basic idea is to analyze the case in which the initial data $u(0,x) = \delta_\xi(x) = \delta(x-\xi)$ is a delta function, which we can interpret as a highly concentrated unit heat source, e.g., a soldering iron or laser beam, that is instantaneously applied at a position ξ along a metal bar. The heat will diffuse away from its initial concentration, and the resulting *fundamental solution* is denoted by

$$u(t,x) = F(t,x;\xi), \qquad \text{with} \qquad F(0,x;\xi) = \delta(x-\xi). \tag{8.2}$$

For each fixed ξ, the fundamental solution, considered as a function of $t > 0$ and x, must satisfy the underlying partial differential equation, and so, for the heat equation,

$$\frac{\partial F}{\partial t} = \gamma \frac{\partial^2 F}{\partial x^2}, \tag{8.3}$$

along with the specified homogeneous boundary conditions.

As with the Green's function, once we have determined the fundamental solution, we can then use linear superposition to reconstruct the general solution to the initial-boundary value problem. Namely, we first write the initial data

$$u(0,x) = f(x) = \int_a^b \delta(x-\xi)\, f(\xi)\, d\xi \tag{8.4}$$

as a superposition of delta functions, as in (6.16). Linearity implies that the solution can be expressed as the corresponding superposition of the responses to those individual concentrated delta profiles:

$$u(t,x) = \int_a^b F(t,x;\xi)\, f(\xi)\, d\xi. \tag{8.5}$$

Assuming that we can differentiate under the integral sign, the fact that $F(t,x;\xi)$ satisfies the differential equation and the homogeneous boundary conditions for each fixed ξ immediately implies that the integral (8.5) is also a solution with the correct initial and (homogeneous) boundary conditions.

Unfortunately, most boundary value problems do not have fundamental solutions that can be written down in closed form. An important exception is the case of an infinitely long homogeneous bar, which requires solving the heat equation on the entire real line:

$$\frac{\partial u}{\partial t} = \frac{\partial^2 u}{\partial x^2}, \qquad \text{for} \qquad -\infty < x < \infty, \qquad t > 0. \tag{8.6}$$

For simplicity, we have chosen units in which the thermal diffusivity is $\gamma = 1$. The solution $u(t,x)$ is defined for all $x \in \mathbb{R}$, and has initial conditions

$$u(0,x) = f(x) \qquad \text{for} \qquad -\infty < x < \infty. \tag{8.7}$$

In order to specify the solution uniquely, we shall require that the temperature be square-integrable, i.e., in L^2, at all times, so that

$$\int_{-\infty}^{\infty} |\, u(t,x)\,|^2\, dx < \infty \qquad \text{for all} \qquad t \geq 0. \tag{8.8}$$

Roughly speaking, square-integrability requires that the temperature be vanishingly small at large distances, and hence plays the role of boundary conditions in this context.

To solve the initial value problem (8.6–7), we apply the Fourier transform, in the x variable, to both sides of the differential equation. In view of the effect of the Fourier transform on derivatives, cf. (7.43), the result is

$$\frac{\partial \widehat{u}}{\partial t} = -\, k^2\, \widehat{u}, \tag{8.9}$$

where

$$\widehat{u}(t,k) = \frac{1}{\sqrt{2\pi}} \int_{-\infty}^{\infty} u(t,x)\, e^{-\,\mathrm{i}\,k\,x}\, dx \tag{8.10}$$

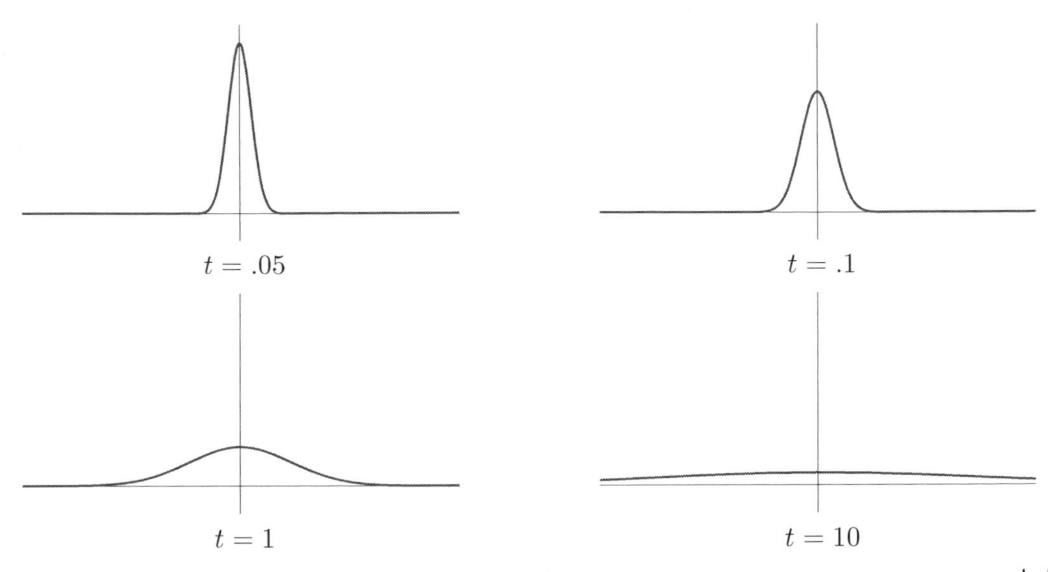

Figure 8.1. The fundamental solution to the one-dimensional heat equation.

is the Fourier transformed solution. For each fixed k, (8.9) can be viewed as a first-order linear ordinary differential equation for $\widehat{u}(t,k)$, with initial conditions

$$\widehat{u}(0,k) = \widehat{f}(k) = \frac{1}{\sqrt{2\pi}} \int_{-\infty}^{\infty} f(x)\, e^{-\,\mathrm{i}\,k\,x}\, dx \tag{8.11}$$

given by Fourier transforming the initial data (8.7). The solution to the initial value problem (8.9, 11) is immediate:

$$\widehat{u}(t,k) = e^{-\,k^2\,t}\, \widehat{f}(k). \tag{8.12}$$

We can thus recover the solution to the initial value problem (8.6–7) by applying the inverse Fourier transform to (8.12), leading to the explicit integral formula

$$u(t,x) = \frac{1}{\sqrt{2\pi}} \int_{-\infty}^{\infty} e^{\,\mathrm{i}\,k\,x}\, \widehat{u}(t,k)\, dk = \frac{1}{\sqrt{2\pi}} \int_{-\infty}^{\infty} e^{\,\mathrm{i}\,k\,x - k^2\,t}\, \widehat{f}(k)\, dk. \tag{8.13}$$

In particular, to construct the fundamental solution, we take the initial temperature profile to be a delta function $\delta_\xi(x) = \delta(x-\xi)$ concentrated at $x=\xi$. According to (7.37), its Fourier transform is

$$\widehat{\delta_\xi}(k) = \frac{e^{-\,\mathrm{i}\,k\,\xi}}{\sqrt{2\pi}}\,.$$

Plugging this into (8.13), and then referring to our table of Fourier transforms, we are led to the following explicit formula for the fundamental solution:

$$F(t,x;\xi) = \frac{1}{2\pi} \int_{-\infty}^{\infty} e^{\,\mathrm{i}\,k(x-\xi) - k^2\,t}\, dk = \frac{1}{2\sqrt{\pi\, t}}\, e^{-\,(x-\xi)^2/(4\,t)} \qquad \text{for} \qquad t>0. \tag{8.14}$$

As you can verify, for each fixed ξ, the function $F(t,x;\xi)$ is indeed a solution to the heat equation for all $t>0$. In addition,

$$\lim_{t\to 0^+} F(t,x;\xi) = \begin{cases} 0, & x \neq \xi, \\ \infty, & x = \xi. \end{cases}$$

Furthermore, its integral

$$\int_{-\infty}^{\infty} F(t,x;\xi)\,dx = 1 \tag{8.15}$$

is constant — in accordance with the law of conservation of thermal energy; see Exercise 8.1.20. Therefore, as $t \to 0^+$, the fundamental solution satisfies the original limiting definition (6.8–9) of the delta function, and so $F(0,x;\xi) = \delta_\xi(x)$ has the desired initial temperature profile.

In Figure 8.1, we graph $F(t,x;0)$ at the indicated times. It starts life as a delta spike concentrated at the origin, and then immediately smooths out into a tall and narrow bell-shaped curve, centered at $x = 0$. As time increases, the solution shrinks and widens, eventually decaying everywhere to zero. Its amplitude is proportional to $t^{-1/2}$, while its overall width is proportional to $t^{1/2}$. The thermal energy (8.15), which is the area under the graph, remains fixed while gradually spreading out over the entire real line.

Remark: In probability, these exponentially bell-shaped curves are known as *normal* or *Gaussian distributions*, [**39**]. The width of the bell curve measures its *standard deviation*. For this reason, the fundamental solution to the heat equation is sometimes referred to as a *Gaussian filter*.

Remark: The fact that the fundamental solution depends only on the difference $x - \xi$, and hence has the same profile at all $\xi \in \mathbb{R}$, is a consequence of the translation invariance of the heat equation, reflecting the fact that it models the thermodynamics of a uniform medium. See Section 8.2 for additional symmetry properties of the heat equation and its solutions.

Remark: One of the striking properties of the heat equation is that thermal energy propagates with *infinite* speed. Indeed, because, at any $t > 0$, the fundamental solution is nonzero for all x, the effect of an initial concentration of heat will immediately be felt along the entire length of an infinite bar. (The graphs in Figure 8.1 are a little misleading because they fail to show the extremely small, but still positive, exponentially decreasing tails.) This effect, while more or less negligible at large distances, is nevertheless in clear violation of physical intuition — not to mention relativity, which postulates that signals cannot propagate faster than the speed of light. Despite this non-physical artifact, the heat equation remains an accurate model for heat propagation and similar diffusive phenomena, and so continues to be successfully used in applications.

With the fundamental solution in hand, we can adapt the linear superposition formula (8.5) to reconstruct the general solution

$$u(t,x) = \frac{1}{2\sqrt{\pi t}} \int_{-\infty}^{\infty} e^{-(x-\xi)^2/(4t)} f(\xi)\,d\xi \tag{8.16}$$

to our initial value problem (8.6). This solution formula is merely a restatement of (8.13) combined with the Fourier transform formula (8.11). Comparing with (7.54), we see that the solutions are obtained by convolution of the initial data with a one-parameter family of progressively wider and shorter Gaussian filters:

$$u(t,x) = F_0(t,x) * f(x), \qquad \text{where} \qquad F_0(t,x) = F(t,x;0) = \frac{e^{-x^2/(4t)}}{2\sqrt{\pi t}}.$$

Since $u(t,x)$ solves the heat equation, we conclude that Gaussian filter convolution has the same smoothing effect on the initial signal $f(x)$. Indeed, the convolution integral (8.16)

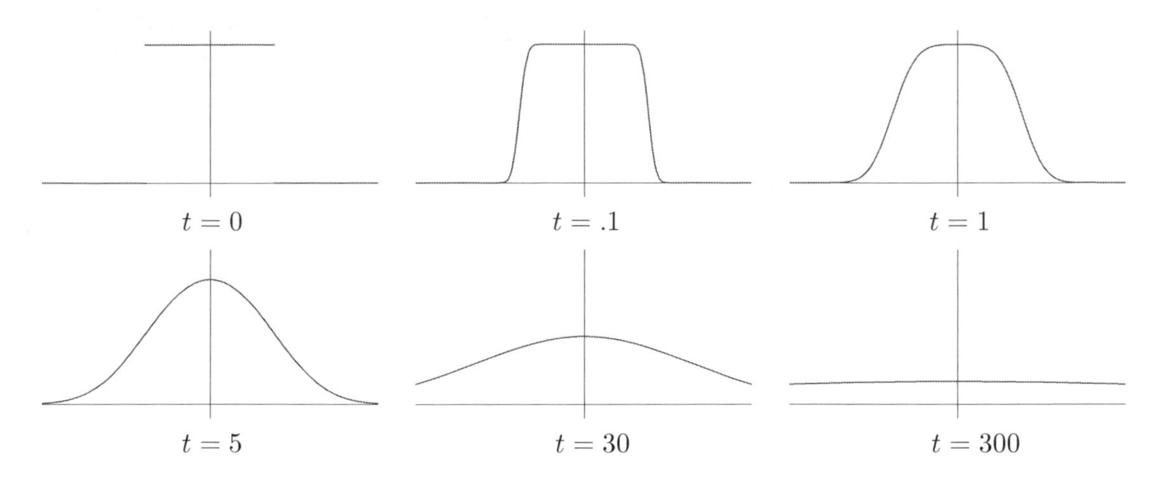

Figure 8.2. Error function solution to the heat equation.

serves to replace each initial value $f(x)$ by a weighted average of nearby values, the weight being determined by the Gaussian distribution. This has the effect of smoothing out high-frequency variations in the signal, and, consequently, the Gaussian convolution formula (8.16) provides an effective method for denoising rough signals and data.

Example 8.1. An infinite bar is initially heated to unit temperature along a finite interval. The initial temperature profile is thus a box function

$$u(0,x) = f(x) = \sigma(x-a) - \sigma(x-b) = \begin{cases} 1, & a < x < b, \\ 0, & \text{otherwise.} \end{cases}$$

The ensuing temperature is provided by the solution to the heat equation obtained by the integral formula (8.16):

$$u(t,x) = \frac{1}{2\sqrt{\pi\, t}} \int_a^b e^{-(x-\xi)^2/(4t)}\, d\xi = \frac{1}{2}\left[\operatorname{erf}\left(\frac{x-a}{2\sqrt{t}} \right) - \operatorname{erf}\left(\frac{x-b}{2\sqrt{t}} \right) \right], \tag{8.17}$$

where erf denotes the error function, as defined in (2.87). Graphs of the solution (8.17) for $a = -5$, $b = 5$, at the indicated times, are displayed in Figure 8.2. Observe the instantaneous smoothing of the sharp interface and instantaneous propagation of the disturbance, followed by a gradual decay to thermal equilibrium, with $u(t,x) \to 0$ as $t \to \infty$.

The Forced Heat Equation and Duhamel's Principle

The fundamental solution approach can be also applied to solve the inhomogeneous heat equation

$$u_t = u_{xx} + h(t,x), \tag{8.18}$$

modeling a bar subject to an external heat source $h(t,x)$, which might depend on both position and time. We begin by solving the particular case

$$u_t = u_{xx} + \delta(t-\tau)\,\delta(x-\xi), \tag{8.19}$$

whose inhomogeneity represents a heat source of unit magnitude that is concentrated at a position $x = \xi$ and applied at a single time $t = \tau > 0$. Physically, this models the effect of instantaneously applying a soldering iron to a single spot on the bar. Let us also impose homogeneous initial conditions

$$u(0, x) = 0 \tag{8.20}$$

as well as homogeneous boundary conditions of one of our standard types. The resulting solution

$$u(t, x) = G(t, x; \tau, \xi) \tag{8.21}$$

will be referred to as the *general fundamental solution* to the heat equation. Since a heat source that is applied at time τ will affect the solution only at later times $t \geq \tau$, we expect that

$$G(t, x; \tau, \xi) = 0 \qquad \text{for all} \qquad t < \tau. \tag{8.22}$$

Indeed, since $u(t, x)$ solves the unforced heat equation at all times $t < \tau$ subject to homogeneous boundary conditions and has zero initial temperature, this follows immediately from the uniqueness of the solution to the initial-boundary value problem.

Once we know the general fundamental solution (8.21), we are able to solve the problem for a general external heat source (8.18). We first write the forcing as a superposition

$$h(t, x) = \int_0^\infty \int_a^b \delta(t - \tau)\, \delta(x - \xi)\, h(\tau, \xi)\, d\xi\, d\tau \tag{8.23}$$

of concentrated instantaneous heat sources. Linearity allows us to conclude that the solution is given by the self-same superposition formula

$$u(t, x) = \int_0^t \int_a^b G(t, x; \tau, \xi)\, h(\tau, \xi)\, d\xi\, d\tau. \tag{8.24}$$

The fact that we only need to integrate over times $0 \leq \tau \leq t$ is a consequence of (8.22).

Remark: If we have a nonzero initial condition, $u(0, x) = f(x)$, then, by linear superposition, the solution

$$u(t, x) = \int_a^b F(t, x; \xi)\, f(\xi)\, d\xi + \int_0^t \int_a^b G(t, x; \tau, \xi)\, h(\tau, \xi)\, d\xi\, d\tau \tag{8.25}$$

is a combination of (*a*) the solution with no external heat source, but nonzero initial conditions, plus (*b*) the solution with homogeneous initial conditions but nonzero heat source.

Let us explicitly solve the forced heat equation on an infinite interval $-\infty < x < \infty$. We begin by computing the general fundamental solution. As before, we take the Fourier transform of both sides of the partial differential equation (8.18) with respect to x. In view of (7.37, 43), we find

$$\frac{\partial \widehat{u}}{\partial t} + k^2\, \widehat{u} = \frac{1}{\sqrt{2\pi}}\, e^{-\,i\,k\,\xi}\, \delta(t - \tau), \tag{8.26}$$

which is an inhomogeneous first-order ordinary differential equation for the Fourier transform $\widehat{u}(t, k)$ of $u(t, x)$, while (8.20) implies the initial condition

$$\widehat{u}(0, k) = 0. \tag{8.27}$$

We solve the initial value problem (8.26–27) by the usual method, [**18**, **23**]. Multiplying the differential equation by the integrating factor $e^{k^2 t}$ yields

$$\frac{\partial}{\partial t}\,(\,e^{k^2 t}\,\widehat{u}\,) = \frac{1}{\sqrt{2\pi}}\,e^{k^2 t - \mathrm{i}\,k\,\xi}\,\delta(t-\tau).$$

Integrating both sides from 0 to t and using the initial condition, we obtain

$$\widehat{u}(t,k) = \frac{1}{\sqrt{2\pi}}\;e^{-\,k^2(t-\tau) - \mathrm{i}\,k\,\xi}\;\sigma(t-\tau),$$

where $\sigma(s)$ is the usual step function (6.23). Finally, we apply the inverse Fourier transform formula (7.9), and then (8.14), to deduce that

$$\begin{aligned} u(t,x) = G(t,x;\tau,\xi) &= \frac{\sigma(t-\tau)}{2\pi}\int_{-\infty}^{\infty} e^{-\,k^2(t-\tau) + \mathrm{i}\,k\,(x-\xi)}\,dk \\ &= \frac{\sigma(t-\tau)}{2\sqrt{\pi(t-\tau)}}\,\exp\left[-\,\frac{(x-\xi)^2}{4(t-\tau)}\right] = \sigma(t-\tau)F(t-\tau,x;\xi)\,. \end{aligned} \tag{8.28}$$

Thus, the general fundamental solution is obtained by translating the fundamental solution $F(t,x;\xi)$ for the initial value problem to a starting time of $t=\tau$ instead of $t=0$. Finally, the superposition principle (8.24) produces the solution,

$$u(t,x) = \int_0^t\int_{-\infty}^{\infty}\frac{h(\tau,\xi)}{2\,\sqrt{\pi(t-\tau)}}\;\exp\left[-\,\frac{(x-\xi)^2}{4\,(t-\tau)}\right]d\xi\,d\tau, \tag{8.29}$$

to the heat equation with source term and zero initial condition on an infinite bar. A nonzero initial condition $u(0,x) = f(x)$ leads, as in the superposition formula (8.25), to an additional term of the form (8.16) in the solution formula.

Remark: The fact that an initial condition has the same aftereffect on the temperature as an instantaneous applied heat source of the same magnitude, thus implying the identification (8.28) of the two types of fundamental solution, is known as *Duhamel's Principle*, named after the nineteenth-century French mathematician Jean–Marie Duhamel. Duhamel's Principle remains valid over a broad range of linear evolution equations.

Example 8.2. An infinitely long bar with unit thermal diffusivity starts out uniformly at zero degrees. Beginning at time $t=0$, a concentrated heat source of unit magnitude is continually applied at the origin. The resulting temperature is the solution $u(t,x)$ to the initial value problem

$$u_t = u_{xx} + \delta(x), \qquad u(0,x) = 0, \qquad t>0, \qquad -\infty<x<\infty.$$

According to (8.29), the solution is given by

$$\begin{aligned} u(t,x) &= \int_0^t\int_{-\infty}^{\infty}\frac{\delta(\xi)}{2\,\sqrt{\pi(t-\tau)}}\;\exp\left[-\,\frac{(x-\xi)^2}{4\,(t-\tau)}\right]d\xi\,d\tau \\ &= \int_0^t\frac{1}{2\,\sqrt{\pi(t-\tau)}}\;\exp\left[-\,\frac{x^2}{4\,(t-\tau)}\right]d\tau = \sqrt{\frac{t}{\pi}}\;\exp\left[-\,\frac{x^2}{4\,t}\right] + \frac{x\,\mathrm{erf}\left(\dfrac{x}{2\,\sqrt{t}}\right) - |\,x\,|}{2}\,. \end{aligned}$$

Three snapshots can be seen in Figure 8.3. Observe that the solution is even in x and monotonically decreasing as $|\,x\,|\to\infty$. Moreover, it has a corner at the origin with limiting

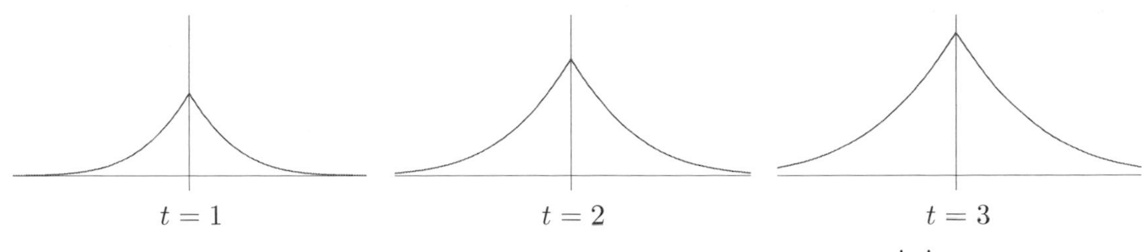

Figure 8.3. Effect of a concentrated heat source.

tangent lines of slopes $\pm\frac{1}{2}$, which implies that its second x derivative produces the delta-function forcing term. At each time t, the solution can be viewed as the linear superposition of a continuous family of fundamental solutions, corresponding to the cumulative effect of individual heat sources applied at each previous time $0 \le \tau \le t$. Moreover, it is not difficult to see that, at each fixed x, the temperature is monotonically increasing in t, with $u(t,x) \to \infty$ as $t \to \infty$, and hence the continuous heat source eventually produces an unbounded temperature in the entire infinite bar.

The Black–Scholes Equation and Mathematical Finance

The most important and influential partial differential equation in financial modeling and investment is the celebrated *Black–Scholes equation*

$$\frac{\partial u}{\partial t} + \frac{\sigma^2}{2}\,x^2\,\frac{\partial^2 u}{\partial x^2} + r\,x\,\frac{\partial u}{\partial x} - r\,u = 0, \tag{8.30}$$

first proposed in 1973 by the American economists Fischer Black and Myron Scholes, [**19**], and Robert Merton, [**71**]. The dependent variable $u(t,x)$ represents the monetary value of a single financial *option*, meaning a contract to either buy or sell an asset at a specified *exercise price* p at a certain future time $t_\star$. The value $u(t,x)$ of the option will depend on the current time $t \le t_\star$ and the current price $x \ge 0$ of the underlying asset. As with many financial models, one assumes the absence of arbitrage, meaning that there is no way to make a riskless profit. The constant $\sigma > 0$ represents the asset's *volatility*, while r denotes the (assumed fixed) *interest rate* for bank deposits, where investors could place their money with a guaranteed rate of return instead of buying the option. (Investors borrowing money to buy the asset would use a negative value of r.) The derivation of the Black–Scholes equation from basic financial modeling relies on the theory of stochastic differential equations, [**83**], which would take us too far afield to explain here; instead, we refer the interested reader to [**123**]. The Black–Scholes equation and its generalizations form the basis of much of the modern financial world, and, increasingly, the insurance industry.

Observe first that the Black–Scholes equation is a *backwards* diffusion process, since, upon solving for

$$\frac{\partial u}{\partial t} = -\,\frac{\sigma^2}{2}\,x^2\,\frac{\partial^2 u}{\partial x^2} - r\,x\,\frac{\partial u}{\partial x} + r\,u, \tag{8.31}$$

the coefficient of the diffusion term u_{xx} is *negative*. This implies that the initial value problem is well-posed only when time runs *backwards*. In other words, given a prescribed value of the option at some specified time in the future, we can use the Black–Scholes equation to determine its current value. However, ill-posedness implies that we cannot predict future values from the current worth of the portfolio.

The "final value problem" for the Black–Scholes equation is to determine the option's value $u(t,x)$ at the current time t and asset value $x \geq 0$, given the *final condition*

$$u(t_\star, x) = f(x) \tag{8.32}$$

at the exercise time $t_\star > t$. For a so-called *European call option*, whereby the asset is to be bought at the exercise price $p > 0$ at the specified time, the final condition is

$$u(t_\star, x) = \max\{\, x - p,\ 0 \,\}, \tag{8.33}$$

representing the investor's profit when $x > p$, or, when $x \leq p$, the option not being exercised so as to avoid a loss. Analogously, for a *put option*, where the asset is to be sold, the final condition is

$$u(t_\star, x) = \max\{\, p - x,\ 0 \,\}. \tag{8.34}$$

The solution $u(t,x)$ will be defined for all $t < t_\star$ and all $x > 0$, subject to the boundary conditions

$$u(t,0) = 0, \qquad u(t,x) \sim x \quad \text{as} \quad x \to \infty,$$

where the asymptotic boundary condition means that the ratio $u(t,x)/x$ tends to a constant as $x \to \infty$.

Fortunately, the Black–Scholes equation can be solved explicitly by transforming it into the heat equation. The first step is to convert it to a forward diffusion process, by setting

$$\tau = \tfrac{1}{2}\sigma^2\,(t_\star - t), \qquad v(\tau, x) = u(t_\star - 2\tau/\sigma^2, x),$$

so that τ effectively runs forward from 0 as the actual time t runs backwards from $t_\star$. This substitution has the effect of converting the final condition (8.32) into an initial condition $v(0,x) = f(x)$. Moreover, a straightforward chain rule computation shows that v satisfies

$$\frac{\partial v}{\partial \tau} = x^2\,\frac{\partial^2 v}{\partial x^2} + \kappa\, x\,\frac{\partial v}{\partial x} - \kappa\, v, \qquad \text{where} \qquad \kappa = \frac{2\,r}{\sigma^2}\,.$$

The next step is to remove the explicit dependence on the independent variable x. The hint is that the right-hand side has the form of an Euler ordinary differential equation, [**23**, **89**]. According to Exercise 4.3.23, these terms can be placed into constant-coefficient form by the change of independent variables $x = e^y$. Indeed, writing

$$w(\tau, y) = v(\tau, e^y) = v(\tau, x) \qquad \text{when} \qquad x = e^y,$$

we apply the chain rule to compute the derivatives

$$\frac{\partial w}{\partial \tau} = \frac{\partial v}{\partial \tau}\,, \qquad \frac{\partial w}{\partial y} = e^y\,\frac{\partial v}{\partial x} = x\,\frac{\partial v}{\partial x}\,, \qquad \frac{\partial^2 w}{\partial y^2} = e^{2\,y}\,\frac{\partial^2 v}{\partial x^2} + e^y\,\frac{\partial v}{\partial x} = x^2\,\frac{\partial^2 v}{\partial x^2} + x\,\frac{\partial v}{\partial x}\,.$$

As a result, we find that w solves the partial differential equation

$$\frac{\partial w}{\partial \tau} = \frac{\partial^2 w}{\partial y^2} + (\kappa - 1)\,\frac{\partial w}{\partial y} - \kappa\, w. \tag{8.35}$$

This is getting closer to the heat equation, and, in fact, can be changed into it by setting

$$w(\tau, y) = e^{\alpha\,\tau + \beta\, y}\, z(\tau, y)$$

for suitable constants α, β. Indeed, differentiating and substituting into (8.35) yields

$$\frac{\partial z}{\partial \tau} + \alpha\, z = \frac{\partial^2 z}{\partial y^2} + 2\,\beta\, \frac{\partial z}{\partial y} + \beta^2 z + (\kappa - 1)\left(\frac{\partial z}{\partial y} + \beta\, z\right) - \kappa\, z.$$

The terms involving $\partial z/\partial y$ and z are eliminated by setting

$$\alpha = -\tfrac{1}{4}\,(\kappa + 1)^2, \qquad \beta = -\tfrac{1}{2}\,(\kappa - 1). \tag{8.36}$$

We conclude that the function

$$z(\tau, y) = e^{(\kappa+1)^2\tau/4 + (\kappa-1)y/2}\, w(\tau, y) \tag{8.37}$$

satisfies the heat equation

$$\frac{\partial z}{\partial \tau} = \frac{\partial^2 z}{\partial y^2}\,. \tag{8.38}$$

Unwinding the preceding argument, we have managed to prove the following:

Proposition 8.3. *If $z(\tau, y)$ is the solution to the initial value problem*

$$\frac{\partial z}{\partial \tau} = \frac{\partial^2 z}{\partial y^2}\,, \qquad z(0, y) = h(y) = e^{(\kappa-1)y/2} f(e^y), \tag{8.39}$$

for $\tau > 0,\ -\infty < y < \infty$, then

$$u(t, x) = x^{-(\kappa-1)/2} e^{-(\kappa+1)^2\sigma^2(t_\star - t)/8}\, z\big(\tfrac{1}{2}\sigma^2(t_\star - t), \log x\big) \tag{8.40}$$

solves the final value problem (8.30, 32) for the Black–Scholes equation for $t < t_\star$ and $0 < x < \infty$.

Now, according to (8.16), the solution to the initial value problem (8.39) can be written as a convolution integral of the initial data with the heat equation's fundamental solution:

$$z(\tau, y) = \frac{1}{2\sqrt{\pi\,\tau}} \int_{-\infty}^{\infty} e^{-(y-\eta)^2/(4\tau)}\, h(\eta)\, d\eta = \frac{1}{2\sqrt{\pi\,\tau}} \int_{-\infty}^{\infty} e^{-(y-\eta)^2/(4\tau) + (\kappa-1)\eta/2} f(e^\eta)\, d\eta. \tag{8.41}$$

Combining this formula with (8.40) produces an explicit solution formula for the general final value problem for the Black–Scholes equation. In particular, for the European call option (8.33), the initial condition is

$$z(0, y) = h(y) = e^{(\kappa-1)y/2} \max\{\, e^y - p,\, 0\,\},$$

and so

$$z(\tau, y) = \frac{1}{2\sqrt{\pi\,\tau}} \int_{\log p}^{\infty} e^{-(y-\eta)^2/(4\tau) + (\kappa-1)\eta/2} (e^\eta - p)\, d\eta.$$

The integral can evaluated by completing the square inside the exponential, producing

$$\begin{aligned} z(\tau, y) = \frac{1}{2}\Bigg[& e^{(\kappa+1)^2\tau/4 + (\kappa+1)y/2} \operatorname{erfc}\left(\frac{\log p - (\kappa+1)\tau - y}{2\sqrt{\tau}}\right) \\ & - p\, e^{(\kappa-1)^2\tau/4 + (\kappa-1)y/2} \operatorname{erfc}\left(\frac{\log p - (\kappa-1)\tau - y}{2\sqrt{\tau}}\right)\Bigg], \end{aligned} \tag{8.42}$$

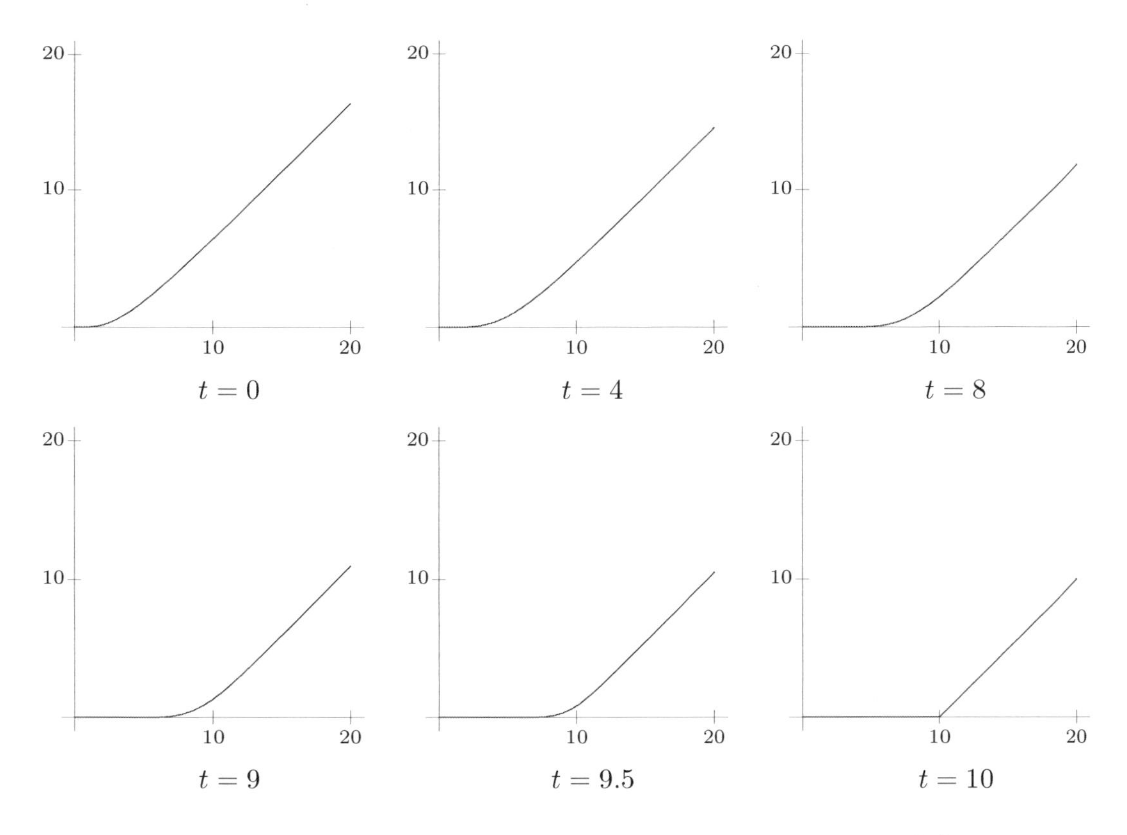

Figure 8.4. Solution to the Black–Scholes equation.

where

$$\operatorname{erfc} x = \frac{2}{\sqrt{\pi}} \int_x^\infty e^{-z^2}\, dz = 1 - \operatorname{erf} x \tag{8.43}$$

is the *complementary error function*, cf. (2.87). Substituting (8.42) into (8.40) results in the celebrated *Black–Scholes formula* for a European call option:

$$\begin{aligned} u(t,x) = \frac{1}{2}\Bigg[\, x \operatorname{erfc}\left(-\, \frac{\left(r + \frac{1}{2}\sigma^2\right)(t_\star - t) + \log(x/p)}{\sqrt{2\sigma^2 (t_\star - t)}} \right) \\ - p\, e^{-r(t_\star - t)} \operatorname{erfc}\left(-\, \frac{\left(r - \frac{1}{2}\sigma^2\right)(t_\star - t) + \log(x/p)}{\sqrt{2\sigma^2 (t_\star - t)}} \right) \Bigg]. \end{aligned} \tag{8.44}$$

A graph of the solution for the specific values $t_\star = 10$, $r = .1$, $\sigma = .2$, $p = 10$ appears in Figure 8.4. Observe that the option's value slowly decreases as the time gets closer and closer to the exercise time $t_\star$, thereby lessening any chances of further profit stemming from the option's underlying price volatility.

Exercises

8.1.1. Find the solution to the heat equation $u_t = u_{xx}$ on the real line having the following initial condition at time $t = 0$. Then sketch graphs of the resulting temperature distribution at times $t = 0, 1$, and 5.
(a) e^{-x^2}, (b) the step function $\sigma(x)$, (c) $e^{-|x|}$, (d) $\begin{cases} 1-|x|, & |x|<1, \\ 0, & \text{otherwise.} \end{cases}$

8.1.2. On an infinite bar with unit thermal diffusivity, a concentrated unit heat source is instantaneously applied at the origin at time $t = 0$. A heat sensor measures the resulting temperature in the bar at position $x = 1$. Determine the maximum temperature measured by the sensor. At what time is the maximum achieved?

8.1.3. (a) Find the solution to the heat equation (8.6) whose initial data corresponds to a pair of unit heat sources placed at positions $x = \pm 1$. (b) Graph the solution at times $t = .1, .25, .5, 1$. (c) At what time(s) does the origin experience its maximum overall temperature? What is the maximum temperature at the origin?

8.1.4. (a) Use the Fourier transform to solve the initial value problem

$$\frac{\partial u}{\partial t} = \frac{\partial^2 u}{\partial x^2}, \qquad u(0,x) = \delta'(x-\xi), \qquad -\infty < x < \infty, \quad t > 0,$$

whose initial data is the derivative of the delta function at a fixed position ξ.
(b) Show that your solution can be written as the derivative $\partial F/\partial x$ of the fundamental solution $F(t,x;\xi)$. Explain why this observation should be valid.

8.1.5. Suppose that the initial data $u(0,x) = f(x)$ is real. Explain why the Fourier transform solution formula (8.13) defines a real function $u(t,x)$ for all $t > 0$.

8.1.6. (a) What is the maximum value of the fundamental solution at time t?
(b) Can you justify the claim that its width is proportional to $\sqrt{t}$?

8.1.7. Prove directly that (8.5) is indeed a solution to the heat equation, and, moreover, has the correct initial and boundary conditions.

8.1.8. Show, by a direct computation, that the final formula in (8.14) is a solution to the heat equation for all $t > 0$.

♢ 8.1.9. Justify formula (8.15).

8.1.10. According to Exercises 4.1.11–12, both the t and x partial derivatives of the fundamental solution solve the heat equation. (a) Write down the initial value problem satisfied by these two solutions. (b) Set $\xi = 0$ and then sketch graphs of each solution at several selected times. (c) Reconstruct each solution as a Fourier integral.

8.1.11. Let $u(t,x) = \dfrac{\partial F}{\partial x}(t,x;0)$ denote the x derivative of the fundamental solution (8.14).
(a) Prove that $u(t,x)$ is a solution to the heat equation $u_t = u_{xx}$ on the domain $\{-\infty < x < \infty,\ t > 0\}$. (b) For fixed x, prove that $\lim_{t\to 0^+} u(t,x) = 0$. (c) Explain why, despite the results in parts (a) and (b), $u(t,x)$ is *not* a classical solution to the initial value problem $u_t = u_{xx}$, $u(0,x) = 0$. What is the classical solution? (d) What initial value problem does $u(t,x)$ satisfy?

8.1.12. Justify all the statements in Example 8.2.

♡ 8.1.13. (a) Solve the heat equation on an infinite bar when the initial temperature is equal to 1 for $|x| < 1$ and 0 elsewhere, while a unit heat source is applied to the same part of the bar $|x| < 1$ for a unit time period $0 < t < 1$. (b) At what time and what location is the bar the hottest? (c) What is the final equilibrium temperature of the bar?

8.1.14. An insulated bar 1 meter long, with constant diffusivity $\gamma = 1$, is taken from a freezer that is kept at -10° C, and then has its ends kept at room temperature of 20° C. A soldering iron with temperature 350° C is continually held at the midpoint of the bar.
(a) Set up an initial value problem modeling the temperature distribution in the bar.
(b) Find the corresponding equilibrium temperature distribution.

♡ 8.1.15. Consider the heat equation with unit thermal diffusivity on the interval $0 < x < 1$ subject to homogeneous Dirichlet boundary conditions.
(a) Find a Fourier series representation for the fundamental solution $\widehat{F}(t,x;\xi)$ that solves the initial-boundary value problem
$$u_t = u_{xx}, \quad t > 0, \quad 0 < x < 1, \qquad u(0,x) = \delta(x-\xi), \quad u(t,0) = 0 = u(t,1).$$
Your solution should depend on t, x and the point ξ where the initial delta impulse is applied.
(b) For the value $\xi = .3$, use a computer program to sum the first few terms in the series and graph the result at times $t = .0001, .001, .01$, and $.1$. Make sure you have included enough terms to obtain a reasonably accurate graph.
(c) Compare your graphs with those of the fundamental solution $F(t,x;.3)$ on an infinite interval at the same times. What is the maximum deviation between the two solutions on the entire interval $0 \le x \le 1$?
(d) Use your fundamental solution $\widehat{F}(t,x;\xi)$ to construct a series solution to the general initial value problem $u(0,x) = f(x)$. Is your series the same as the usual Fourier series solution? If not, explain any discrepancy.

8.1.16. *True or false*: Periodic forcing of the heat equation at a particular frequency can produce resonance. Justify your answer.

8.1.17. Find the fundamental solution for the *cable equation* $v_t = \gamma v_{xx} - \alpha v$ on the real line. *Hint*: See Exercise 4.1.16.

8.1.18. The partial differential equation $u_t + c\,u_x = \gamma\,u_{xx}$ models transport of a diffusing pollutant in a fluid flow. Assuming that the speed c is constant, write down a solution to the initial value problem $u(0,x) = f(x)$ for $-\infty < x < \infty$. *Hint*: Look at Exercise 4.1.17.

◇ 8.1.19. Use the Fourier transform to solve the initial value problem $\mathrm{i}\,u_t = u_{xx}$, $u(0,x) = f(x)$, for the one-dimensional Schrödinger equation on the real line $-\infty < x < \infty$.

◇ 8.1.20. Let $u(t,x)$ be a solution to the heat equation having finite thermal energy, $E(t) = \int_{-\infty}^{\infty} u(t,x)\,dx < \infty$, and satisfying $u_x(t,x) \to 0$ as $x \to \pm\infty$, for all $t \ge 0$. Prove the law of *conservation of thermal energy*: $E(t) =$ constant.

8.1.21. Explain in your own words how a function $u(t,x)$ can satisfy $u(t,x) \to 0$ uniformly as $t \to \infty$ while maintaining the constancy of $\int_{-\infty}^{\infty} u(t,x)\,dx = 1$ for all t. Discuss what this signifies regarding the interchange of limits and integrals.

8.1.22. (a) Prove that if $\widehat{f}(k) \in \mathrm{L}^2$ is square-integrable, then so is $e^{-a k^2}\,\widehat{f}(k)$ for any $a > 0$.
(b) Prove that when the initial data $f(x) \in \mathrm{L}^2$ is square integrable, so is the Fourier integral solution (8.13) for all $t \ge 0$.

8.1.23. Find the solution to the Black–Scholes equation for a put option (8.34).

8.1.24. (a) If we increase the interest rate r, does the value of a call option (*i*) increase; (*ii*) decrease; (*iii*) stay the same; (*iv*) could do any of the above? Justify your answer.
(b) Answer the same question when rate stays fixed, but the volatility σ is increased.

◇ 8.1.25. Justify formula (8.42).

8.2 Symmetry and Similarity

The geometric approach to partial differential equations enables one to exploit their symmetry properties to construct explicit solutions of both mathematical and physical interest. Unlike separation of variables, which is restricted to special types of linear partial differential equations,[†] symmetry methods can also be successfully applied to a broad range of nonlinear partial differential equations. While we do not have the mathematical tools to develop the full range of symmetry techniques, we will learn how to exploit some of the most basic symmetry properties: translations, leading to traveling wave solutions; scalings, leading to similarity solutions; and, in subsequent chapters, rotational symmetries.

In general, by a *symmetry* of an equation, we mean a transformation that takes solutions to solutions. Thus, knowing a symmetry transformation, if we are in possession of one solution, then we can construct a second solution by applying the symmetry. And, possibly, a third solution by applying the symmetry yet again. And so on. If we know lots of symmetries, then we can produce lots of solutions by this simple device.

Remark: General symmetry techniques are founded on the theory of *Lie groups*, named after the influential nineteenth-century Norwegian mathematician Sophus Lie (pronounced "Lee"). Lie's theory is a profound synthesis of group theory and differential geometry, and provides an algorithm for completely determining all the (continuous) symmetries of a given differential equation. Although the theory lies beyond the scope of this introductory text, direct inspection and/or physical intuition will often produce the most important symmetries of the system, which can then be directly exploited. Modern applications of Lie's symmetry methods to partial differential equations arising in physics and engineering can be traced back to an influential book on hydrodynamics by the author's thesis advisor, Garrett Birkhoff, [**17**]. A complete and comprehensive treatment of Lie symmetry methods can be found in the author's first book [**87**], and, at a more introductory level, in the recent books [**27**, **58**], the first having a particular emphasis on applications in fluid mechanics.

The heat equation serves as an excellent testing ground for the general methodology, since it admits a rich variety of symmetry transformations that take solutions to solutions. The simplest are the translations. Moving the space and time coordinates by a fixed amount,

$$t \longmapsto t+a, \qquad\qquad x \longmapsto x+b, \tag{8.45}$$

where a, b are constants, changes the function $u(t,x)$ into the translated function[‡]

$$U(t,x) = u(t-a, x-b). \tag{8.46}$$

A simple application of the chain rule proves that the partial derivatives of U with respect to t and x agree with the corresponding partial derivatives of u, so

$$\frac{\partial U}{\partial t} = \frac{\partial u}{\partial t}, \qquad \frac{\partial U}{\partial x} = \frac{\partial u}{\partial x}, \qquad \frac{\partial^2 U}{\partial x^2} = \frac{\partial^2 u}{\partial x^2},$$

[†] This is not entirely fair: separation of variables can also be applied to certain nonlinear partial differential equations such as Hamilton–Jacobi equations, [**73**].

[‡] The minus signs arise because when we set $\widehat{t} = t+a$, $\widehat{x} = x+b$, then the translated function is $U(\widehat{t},\widehat{x}) = u(t,x) = u(\widehat{t}-a, \widehat{x}-b)$. Dropping the hats produces the stated formula.

and so on. In particular, the function $U(t,x)$ is a solution to the heat equation $U_t = \gamma U_{xx}$ whenever $u(t,x)$ also solves $u_t = \gamma u_{xx}$. Physically, translation symmetry formalizes the property that the heat equation models a homogeneous medium, and hence the solution does not depend on the choice of reference point or origin of our coordinate system.

As a consequence, each solution to the heat equation will produce an infinite family of translated solutions. For example, starting with the separable solution

$$u(t,x) = e^{-\gamma t} \sin x,$$

we immediately produce the additional translated solutions

$$U(t,x) = e^{-\gamma (t-a)} \sin(x-b),$$

valid for any choice of constants a, b.

Warning: Typically, the symmetries of a differential equation do not respect initial or boundary conditions. For instance, if $u(t,x)$ is defined for $t \geq 0$ and in the domain $0 \leq x \leq \ell$, then its translated version (8.46) is defined for $t \geq a$ and in the translated domain $b \leq x \leq \ell + b$, and so will solve a translated initial-boundary value problem.

A second important class of symmetries consists of the scaling invariances. We already know that if $u(t,x)$ is a solution, then so is the scalar multiple $c\,u(t,x)$ for any constant c; this is a simple consequence of linearity of the heat equation. We can also add an arbitrary constant to the temperature, noting that

$$U(t,x) = c\,u(t,x) + k \tag{8.47}$$

is a solution for any choice of constants c, k. Physically, the transformation (8.47) amounts to a change in the scale used to measure temperature. For instance, if u is measured in degrees Celsius, and we set $c = \frac{9}{5}$ and $k = 32$, then $U = \frac{9}{5}u + 32$ will be measured in degrees Fahrenheit. Thus, reassuringly, the physical processes described by the heat equation do not depend on our choice of thermometer.

More interestingly, suppose we rescale the space and time variables:

$$t \longmapsto \alpha\, t, \qquad x \longmapsto \beta\, x, \tag{8.48}$$

where $\alpha, \beta \neq 0$ are nonzero constants. The effect of such a scaling transformation is to convert $u(t,x)$ into a rescaled function†

$$U(t,x) = u(\alpha^{-1} t, \beta^{-1} x). \tag{8.49}$$

The derivatives of U are related to those of u according to the formulas

$$\frac{\partial U}{\partial t} = \frac{1}{\alpha}\frac{\partial u}{\partial t}, \qquad \frac{\partial U}{\partial x} = \frac{1}{\beta}\frac{\partial u}{\partial x}, \qquad \frac{\partial^2 U}{\partial x^2} = \frac{1}{\beta^2}\frac{\partial^2 u}{\partial x^2}.$$

Therefore, if u satisfies the heat equation $u_t = \gamma u_{xx}$, then U satisfies the rescaled heat equation

$$U_t = \frac{1}{\alpha} u_t = \frac{\gamma}{\alpha} u_{xx} = \frac{\beta^2 \gamma}{\alpha} U_{xx},$$

† As before, setting $\hat{t} = \alpha t$, $\hat{x} = \beta x$, produces the rescaled function $U(\hat{t},\hat{x}) = u(t,x) = u(\alpha^{-1}\hat{t}, \beta^{-1}\hat{x})$, and we then drop the hats.

which we rewrite as

$$U_t = \Gamma\, U_{xx}, \qquad \text{where} \qquad \Gamma = \frac{\beta^2\,\gamma}{\alpha}\,. \tag{8.50}$$

Thus, the net effect of scaling space and time is merely to rescale the diffusion coefficient. Physically, the scaling symmetry (8.48) corresponds to a change in the physical units used to measure time and distance. For instance, to change from minutes to seconds, set $\alpha = 60$, and from yards to meters, set $\beta = .9144$. The net effect (8.50) on the diffusion coefficient γ is a reflection of its physical units, namely distance^2/time.

In particular, if we choose

$$\alpha = \gamma, \qquad \beta = 1,$$

then the rescaled diffusion coefficient becomes $\Gamma = 1$. This observation has the following important consequence. If $U(t,x)$ solves the heat equation for a unit diffusivity, $\Gamma = 1$, then

$$u(t,x) = U(\gamma\, t, x) \tag{8.51}$$

solves the heat equation for the diffusivity $\gamma > 0$. Thus, the only effect of the diffusion coefficient is to speed up or slow down time. A body with diffusivity $\gamma = 2$ will cool down twice as fast as a body (of the same shape subject to similar boundary conditions and initial conditions) with diffusivity $\gamma = 1$. Note that this particular rescaling has not altered the space coordinates, and so $U(t,x)$ is defined on the same spatial domain as $u(t,x)$.

On the other hand, if we set $\alpha = \beta^2$, then the rescaled diffusion coefficient is exactly the same as the original: $\Gamma = \gamma$. Thus, the transformation

$$t \longmapsto \beta^2\, t, \qquad x \longmapsto \beta\, x, \tag{8.52}$$

does not alter the equation, and hence defines a *scaling symmetry* — also known as a *similarity transformation* — for the heat equation. Combining (8.52) with the linear rescaling $u \mapsto c\,u$, we make the elementary, but important, observation that if $u(t,x)$ is any solution to the heat equation, then so is the function

$$U(t,x) = c\, u(\beta^{-2}\, t, \beta^{-1}\, x), \tag{8.53}$$

for the *same* diffusion coefficient γ. For example, rescaling the solution

$$u(t,x) = e^{-\gamma\, t}\cos x \qquad \text{leads to the solution} \qquad U(t,x) = c\, e^{-\gamma\, t/\beta^2}\cos\frac{x}{\beta}\,.$$

Warning: As in the case of translations, rescaling space by a factor $\beta \neq 1$ will alter the domain of definition of the solution. If $u(t,x)$ is defined for $a \le x \le b$, then $U(t,x)$, as given in (8.53), is defined for $\beta\, a \le x \le \beta\, b$ (or, when $\beta < 0$, for $\beta\, b \le x \le \beta\, a$).

For example, suppose that we have solved the heat equation for the temperature $u(t,x)$ on a bar of length 1, subject to certain initial and boundary conditions. We are then given a bar composed of the same material of length 2. Since the diffusivity coefficient has not changed, we can directly construct the new solution $U(t,x)$ by rescaling. Setting $\beta = 2$ will serve to double the length. If we also rescale time by a factor $\alpha = \beta^2 = 4$, then the rescaled function $U(t,x) = u\big(\frac{1}{4}\, t, \frac{1}{2}\, x\big)$ will be a solution of the heat equation on the longer bar with the same diffusivity constant. The net effect is that the rescaled solution will be evolving four times as slowly as the original, and hence it effectively takes a bar that is twice the length four times as long to cool down.

Similarity Solutions

A *similarity solution* of a partial differential equation is one that remains unchanged (invariant) under a one-parameter family[†] of scaling symmetries. For a partial differential equation in two variables — say t and x — the similarity solutions can be found by solving an *ordinary differential equation*.

Suppose our partial differential equation admits the scaling symmetries

$$t \longmapsto \beta^a t, \qquad x \longmapsto \beta^b x, \qquad u \longmapsto \beta^c u, \qquad \beta \neq 0, \tag{8.54}$$

where a, b, c are fixed constants with a, b not both zero. As above, this means that if $u(t,x)$ is a solution to the differential equation, so is the rescaled function

$$U(t,x) = \beta^c u(\beta^{-a} t, \beta^{-b} x) \tag{8.55}$$

for all values of $\beta \neq 0$. Checking that this indeed defines a symmetry is a simple matter of applying the chain rule, which implies that the derivatives scale according to

$$u_t \longmapsto \beta^{c-a} u_t, \quad u_x \longmapsto \beta^{c-b} u_x, \quad u_{tt} \longmapsto \beta^{c-2a} u_{tt}, \quad u_{xt} \longmapsto \beta^{c-a-b} u_{xt}, \tag{8.56}$$

and so on. Products of derivatives scale multiplicatively, e.g., $x^4 u\, u_{xt} \mapsto \beta^{2c-a+3b} x^4 u\, u_{xt}$. In order that a (polynomial) differential equation admit such a scaling symmetry, each of its terms must scale by the *same* overall power of β.

By definition, $u(t,x)$ is called a *similarity solution* if it remains unchanged (invariant) under the scaling symmetries (8.54), so that

$$u(t,x) = \beta^c u(\beta^{-a} t, \beta^{-b} x) \tag{8.57}$$

for all $\beta > 0$. Let us, for specificity, assume that $a \neq 0$, leaving the case $a = 0$, $b \neq 0$, for the reader to complete in Exercise 8.2.13. Since the left-hand side of (8.57) does not depend on β, we can fix its value to be[‡] $\beta = t^{1/a}$, and conclude that the similarity solution must have the form

$$u(t,x) = t^{c/a} v(\xi), \qquad \text{where} \qquad \xi = x\, t^{-b/a} \qquad \text{and} \qquad v(\xi) = u(1,\xi), \tag{8.58}$$

are referred to as the *similarity variables*, since they remain invariant when subjected to the scaling transformations (8.54). We then use the chain rule to find the formulas for the partial derivatives of u in terms of the ordinary derivatives of v with respect to ξ. Substituting these expressions into the scale-invariant partial differential equation for $u(t,x)$, and then canceling a common factor of t, will effectively reduce it to an *ordinary differential equation* for the function $v(\xi)$. Each solution to the resulting ordinary differential equation then gives rise to a scale-invariant solution to the original partial differential equation through the similarity ansatz (8.58).

Example 8.4. As a first example, let us return to the nonlinear transport equation

$$u_t + u\, u_x = 0, \tag{8.59}$$

[†] Or, more accurately, a one-parameter group, [**87**].

[‡] This assumes $t > 0$; for $t < 0$, just replace t by $-t$.

which we studied in Section 2.3. Under (8.54, 56), the equation rescales to

$$\beta^{c-a} u_t + \beta^{2c-b} u\, u_x = 0,$$

which is unchanged, provided $c - a = 2c - b$, and hence $c = b - a$. Setting $a = 1$, $c = b - 1$, we conclude that if $u(t,x)$ is any solution, then so is the rescaled function

$$U(t,x) = \beta^{b-1}\, u(\beta^{-1} t, \beta^{-b} x)$$

for any b and any $\beta \neq 0$.

To find the associated similarity solutions, we use (8.58) to introduce the ansatz

$$u(t,x) = t^{b-1}\, v(\xi), \qquad \text{where} \qquad \xi = x\, t^{-b}. \tag{8.60}$$

Differentiating, we obtain

$$u_t = -b\, x\, t^{-2}\, v'(\xi) + (b-1)\, t^{b-2}\, v(\xi) = t^{b-2}\big[-b\,\xi\, v'(\xi) + (b-1)\, v(\xi)\big], \qquad u_x = t^{-1}\, v'(\xi).$$

Substituting these expressions into the transport equation (8.59) yields

$$0 = u_t + u\, u_x = t^{b-2}\big[(v - b\,\xi)\, v' + (b-1)\, v\big],$$

and so

$$(v - b\,\xi)\, \frac{dv}{d\xi} + (b-1)\, v = 0. \tag{8.61}$$

Any solution to this nonlinear first-order ordinary differential equation will, when substituted into (8.60), produce a similarity solution to the nonlinear transport equation.

If $b = 1$, then either $v = b\,\xi$, producing the particular similarity solution $u(t,x) = x/t$ that we earlier used to construct the rarefaction wave (2.54), or v is constant, and so is u. Otherwise, we can, in fact, linearize (8.61) by treating ξ as a function of v, whence

$$(b-1)\, v\, \frac{d\xi}{dv} - b\,\xi = -\, v.$$

The general solution to such a linear first-order ordinary differential equation is found by the standard method, [**18**, **23**], resulting in

$$\xi = v + k\, v^{b/(b-1)},$$

where k is the constant of integration. Recalling (8.60), we find that the similarity solutions $u(t,x)$ are defined by an implicit equation

$$x = k\, u^{b/(b-1)} + t\, u.$$

For example, if $b = 2$, the (multi-valued) solution is a sideways-moving parabola:

$$x = k\, u^2 + t\, u, \qquad \text{so that} \qquad u = \frac{-\,t \pm \sqrt{t^2 + 4\,k\,x}}{2\,k}.$$

Example 8.5. Consider the linear heat equation

$$u_t = u_{xx}. \tag{8.62}$$

Under the rescaling (8.54), the equation becomes $\beta^{c-a} u_t = \beta^{c-2b} u_{xx}$, and thus (8.54) represents a symmetry if and only if $a = 2b$. Therefore, if $u(t,x)$ is any solution, so is the rescaled function

$$U(t,x) = \beta^{c}\, u(\beta^{-2} t, \beta^{-1} x).$$

Of course, the initial scaling factor stems from the linearity of the equation.

The scale-invariant solutions are constructed through the similarity ansatz

$$u(t,x) = t^{c/2}\, v(\xi), \qquad \text{where} \qquad \xi = x/\sqrt{t}\,.$$

Differentiation yields

$$\begin{aligned} u_t &= -\tfrac{1}{2}\, x\, t^{c/2-3/2}\, v'(\xi) + \tfrac{1}{2}\, c\, t^{c/2-1}\, v(\xi) = t^{c/2-1}\big[-\tfrac{1}{2}\,\xi\, v'(\xi) + \tfrac{1}{2}\, c\, v(\xi)\big], \\ u_{xx} &= t^{c/2-1}\, v''(\xi). \end{aligned}$$

Substituting these expressions into the heat equation and canceling a common power of t, we find that v must satisfy the linear ordinary differential equation

$$v'' + \tfrac{1}{2}\,\xi\, v' - \tfrac{1}{2}\, c\, v = 0. \tag{8.63}$$

If $c = 0$, then (8.63) is effectively a linear first-order ordinary differential equation for $v'(\xi)$, which can be readily solved by the usual method, thereby producing the solution

$$v(\xi) = c_1 + c_2 \operatorname{erf}\big(\tfrac{1}{2}\,\xi\big),$$

where c_1, c_2 are arbitrary constants and erf is the error function (2.87). The corresponding similarity solution to the heat equation is

$$u(t,x) = c_1 + c_2 \operatorname{erf}\left(\frac{x}{\sqrt{t}}\right).$$

The error function solutions that we encountered in (8.17) can be built up as a linear combination of translations of this similarity solution.

If $c \neq 0$, most solutions to the ordinary differential equation (8.63) are not elementary functions.[†] One is in need of more sophisticated techniques, e.g., the method of power series to be developed in Section 11.3, to understand its solutions, and hence the resulting similarity solutions to the heat equation.

Exercises

8.2.1. If it takes a 2 cm long insulated bar 23 minutes to cool down to room temperature, how long does it take a 4 cm bar?

8.2.2. If it takes a 5 centimeter long insulated iron bar 10 minutes to cool down so as not to burn your hand, how long does it take a 20 centimeter bar made out of the same material to cool down to the same temperature?

◇ 8.2.3. (a) Given $\gamma > 0$, use a scaling transformation to write down the formula for the fundamental solution for the general heat equation $u_t = \gamma\, u_{xx}$ for $x \in \mathbb{R}$. (b) Write down the corresponding integral formula for the solution to the initial value problem.

[†] According to [**87**; Example 3.3], the general solution can be written in terms of parabolic cylinder functions, [**86**].

8.2.4. Use scaling to construct the series solution for a heated circular ring of radius r and thermal diffusivity γ. Does scaling also give the correct formulas for the Fourier coefficients in terms of the initial temperature distribution?

8.2.5. A solution $u(t,x)$ to the heat equation is measured in degrees Fahrenheit. What is the corresponding temperature in degrees Kelvin? Which symmetry transformation takes the first solution to the second solution, and how does it affect the diffusion coefficient?

8.2.6. Is time reversal, $t \mapsto -t$, a symmetry of the heat equation? Write down a physical explanation, and then a mathematical justification.

8.2.7. According to Exercise 4.1.17, the partial differential equation $u_t + c\,u_x = \gamma\,u_{xx}$ models diffusion in a convective flow. Show how to use scaling to place the differential equation in the form $u_t + u_x = P^{-1}\,u_{xx}$, where P is called the *Péclet number*, and controls the rate of mixing. Is there a scaling that will reduce the problem to the case $P = 1$?

8.2.8. Suppose you know a solution $u^\star(t,x)$ to the heat equation that satisfies $u^\star(1,x) = f(x)$. Explain how to solve the initial value problem with $u(0,x) = f(x)$.

8.2.9. Solve the following initial value problems for the heat equation $u_t = u_{xx}$ for $x \in \mathbb{R}$:
(a) $u(0,x) = e^{-x^2/4}$. *Hint*: Use Exercise 8.2.8. (b) $u(0,x) = e^{-4x^2}$.
(c) $u(0,x) = x^2\,e^{-x^2/4}$. *Hint*: Use Exercise 4.1.12.

8.2.10. Define the functions $H_n(x)$ for $n = 0, 1, 2, \ \ldots\,$, by the formula

$$\frac{d^n}{dx^n}\,e^{-x^2} = (-1)^n H_n(x)\,e^{-x^2}. \tag{8.64}$$

(a) Prove that $H_n(x)$ is a polynomial of degree n, known as the nth *Hermite polynomial*.
(b) Calculate the first four Hermite polynomials.
(c) Assuming $\gamma = 1$, find the solution to the heat equation for $-\infty < x < \infty$ and $t > 0$, given the initial data $u(0,x) = H_n(x)\,e^{-x^2}$. *Hint*: Combine Exercises 4.1.11, 8.2.8.

8.2.11. Find the scaling symmetries and corresponding similarity solutions of the following partial differential equations: (a) $u_t = x^2\,u_x$, (b) $u_t + u^2 u_x = 0$, (c) $u_{tt} = u_{xx}$.

8.2.12. Show that the wave equation $u_{tt} = c^2 u_{xx}$ has the following invariance properties: if $u(t,x)$ is a solution, so is (a) any time translate: $u(t-a,x)$, where a is fixed; (b) any space translate: $u(t,x-b)$, where b is fixed; (c) the dilated function $u(\beta t, \beta x)$ for $\beta \neq 0$; (d) any derivative: say $\partial u/\partial x$ or $\partial^2 u/\partial t^2$, provided u is sufficiently smooth.

♢ 8.2.13. Suppose $a = 0$, $b \neq 0$ in the scaling transformation (8.57).
(a) Discuss how to reduce the partial differential equation to an ordinary differential equation for the corresponding similarity solutions.
(b) Illustrate your method with the partial differential equation $t\,u_t = u\,u_{xx}$.

8.2.14. *True or false*: (a) A homogeneous polynomial solution to a partial differential equation is always a similarity solution. (b) An inhomogeneous polynomial solution to a partial differential equation can never be a similarity solution.

8.2.15. (a) Find all scaling symmetries of the two-dimensional Laplace equation $u_{xx} + u_{yy} = 0$. (b) Write down the ordinary differential equation for the similarity solutions. (c) Can you find an explicit formula for the similarity solutions? *Hint*: Look at Exercise 8.2.14(a).

♡ 8.2.16. Besides the translations and scalings, Lie symmetry methods, [**87**], produce two other classes of symmetry transformations for the heat equation $u_t = u_{xx}$. Given that $u(t,x)$ is a solution to the heat equation:
(a) Prove that $U(t,x) = e^{c^2 t - c x}\,u(t, x - 2ct)$ is also a solution to the heat equation for any $c \in \mathbb{R}$. What solution do you obtain if $u(t,x) = a$ is a constant solution? *Remark*: This transformation can be interpreted as the effect of a *Galilean boost* to a coordinate frame that is moving with speed c.

(*b*) Prove that $U(t,x) = \dfrac{e^{-cx^2/(4(1+ct))}}{\sqrt{1+ct}}\; u\left(\dfrac{t}{1+ct}, \dfrac{x}{1+ct}\right)$ is a solution to the heat equation for any $c \in \mathbb{R}$. What solution do you obtain if $u(t,x) = a$ is a constant?

8.3 The Maximum Principle

We have already noted the temporal decay of temperature, as governed by the heat equation, to thermal equilibrium. While the temperature at any individual point in a physical medium can fluctuate — depending on what is happening elsewhere, thermodynamics tells us that the overall heat content of an isolated body must continually decrease. The *Maximum Principle* is the mathematical formulation of this physical law, and states that the temperature of a body cannot, in the absence of external heat sources, ever become larger than its initial or boundary values. This can be viewed as a dynamical counterpart to the Maximum Principle for the Laplace equation, as formulated in Theorem 4.9, stating that the maximum temperature of a body in equilibrium is achieved only on its boundary.

The proof of the Maximum Principle will be facilitated if we analyze the more general situation in which heat energy is being continually extracted throughout the body.

Theorem 8.6. *Let $\gamma > 0$. Suppose $u(t,x)$ is a solution to the forced heat equation*

$$\frac{\partial u}{\partial t} = \gamma \frac{\partial^2 u}{\partial x^2} + F(t,x) \tag{8.65}$$

on the rectangular domain

$$R = \{a < x < b,\ 0 < t < c\}.$$

Assume that the forcing term is nowhere positive: $F(t,x) \leq 0$ for all $(t,x) \in R$. Then the maximum of $u(t,x)$ on the closed rectangle $\overline{R}$ is attained at $t = 0$ or $x = a$ or $x = b$.

In other words, if no new heat is being introduced, the maximum overall temperature occurs either at the initial time or on the body's boundary. In particular, in the fully insulated case $F(t,x) \equiv 0$, (8.65) reduces to the heat equation, and Theorem 8.6 applies as stated.

Proof: First let us first prove the result under the stronger assumption $F(t,x) < 0$, which implies that

$$\frac{\partial u}{\partial t} < \gamma \frac{\partial^2 u}{\partial x^2} \tag{8.66}$$

everywhere in the rectangle R. Suppose first that $u(t,x)$ has a (local) maximum at a point $(t^\star, x^\star)$ in the interior of R. Then, by multivariable calculus, [**8**, **108**], its gradient must vanish there, $\nabla u(t^\star, x^\star) = \mathbf{0}$, and hence

$$u_t(t^\star, x^\star) = u_x(t^\star, x^\star) = 0. \tag{8.67}$$

Our assumption implies that the scalar function $h(x) = u(t^\star, x)$ has a maximum at $x = x^\star$. Thus, by the second derivative test for functions of a single variable,

$$h''(x^\star) = u_{xx}(t^\star, x^\star) \leq 0. \tag{8.68}$$

But the requirements (8.67–68) are clearly incompatible with the initial inequality (8.66). We conclude that the solution $u(t,x)$ cannot have a local maximum at any point in the interior of R.

We still need to exclude the possibility of a maximum occurring at a non-corner point $(t^\star, x^\star) = (c, x^\star)$, $a < x^\star < b$, on the right-hand edge of the rectangle. If such were to occur, then the function $g(t) = u(t, x^\star)$ would be nondecreasing at $t = c$, and hence $g'(t) = u_t(c, x^\star) \geq 0$ there. The preceding argument also implies that $u_{xx}(c, x^\star) \leq 0$, and again these two requirements are incompatible with (8.66). We conclude that any (local) maximum must occur on one of the other three sides of the rectangle, in accordance with the statement of the theorem.

To generalize the argument to the case $F(t,x) \leq 0$ — which includes the heat equation — requires a little trick. Starting with the solution $u(t,x)$ to (8.65), we set

$$v(t,x) = u(t,x) + \varepsilon\, x^2, \qquad \text{where} \qquad \varepsilon > 0.$$

Then,

$$\frac{\partial v}{\partial t} = \frac{\partial u}{\partial t} = \gamma\, \frac{\partial^2 u}{\partial x^2} + F(t,x) = \gamma\, \frac{\partial^2 v}{\partial x^2} - 2\gamma\,\varepsilon + F(t,x) = \gamma\, \frac{\partial^2 v}{\partial x^2} + \widetilde{F}(t,x),$$

where, by our original assumption on $F(t,x)$,

$$\widetilde{F}(t,x) = F(t,x) - 2\gamma\,\varepsilon < 0$$

everywhere in R. Thus, by the previous argument, a local maximum of $v(t,x)$ can occur only when $t = 0$ or $x = a$ or $x = b$. Now we let $\varepsilon \to 0$ and conclude the same for u. More rigorously, let M denote the maximum value of $u(t,x)$ on the indicated three sides of the rectangle. Then

$$v(t,x) \leq M + \varepsilon \max\{\, a^2, b^2 \,\}$$

there, and hence, by the preceding argument,

$$u(t,x) \leq v(t,x) \leq M + \varepsilon \max\{\, a^2, b^2 \,\} \qquad \text{for all} \qquad (t,x) \in R.$$

Now, letting $\varepsilon \to 0^+$ proves that $u(t,x) \leq M$ everywhere in R. *Q.E.D.*

For the unforced heat equation, we can bound the solution from both above and below by its boundary and initial temperatures:

Corollary 8.7. *Suppose $u(t,x)$ solves the heat equation $u_t = \gamma\, u_{xx}$, with $\gamma > 0$, for $a < x < b$, $0 < t < c$. Set*

$$B = \{\, (0,x) \mid a \leq x \leq b \,\} \cup \{\, (t,a) \mid 0 \leq t \leq c \,\} \cup \{\, (t,b) \mid 0 \leq t \leq c \,\},$$

and let

$$M = \max\{\, u(t,x) \mid (t,x) \in B \,\}, \qquad m = \min\{\, u(t,x) \mid (t,x) \in B \,\}, \tag{8.69}$$

be, respectively, the maximum and minimum values for the initial and boundary temperatures. Then $m \leq u(t,x) \leq M$ for all $a \leq x \leq b$, $0 \leq t \leq c$.

Proof: The upper bound $u(t,x) \leq M$ follows from the Maximum Principle of Theorem 8.6. To establish the lower bound, we note that $\widetilde{u}(t,x) = -\,u(t,x)$ also solves the heat equation, satisfying $\widetilde{u}(t,x) \leq -\,m$ on B, and hence, by the Maximum Principle, everywhere in the rectangle. But this implies $u(t,x) = -\,\widetilde{u}(t,x) \geq m$. *Q.E.D.*

Remark: Theorem 8.6 is sometimes referred to as the *Weak Maximum Principle* for the heat equation. The *Strong Maximum Principle* states that, provided the solution $u(t,x)$ is not constant, its value at any non-initial, non-boundary point $(t,x) \in \widehat{R} = \{a < x < b,\ 0 < t \leq c\}$ is *strictly* less than its maximum initial and boundary values; in other words, $u(t,x) < M$ for $(t,x) \in \widehat{R}$, where M is given in (8.69). Similarly, the Strong Maximum Principle implies that, for nonconstant solutions to the heat equation, the inequalities in Corollary 8.7 are strict: $m < u(t,x) < M$ for all $(t,x) \in \widehat{R}$. Proofs of the Strong Maximum Principle are more delicate, and can be found in [**38, 61**].

One immediate application of the Maximum Principle is to prove uniqueness of solutions to the heat equation.

Theorem 8.8. *There is at most one solution to the Dirichlet initial-boundary value problem for the forced heat equation.*

Proof: Suppose u and $\widetilde{u}$ are any two solutions with the same initial and boundary values. Then their difference $v = u - \widetilde{u}$ solves the homogeneous initial-boundary value problem for the unforced heat equation, with minimum and maximum boundary values $m = 0 \leq v(t,x) \leq 0 = M$ for $t = 0$, $a \leq x \leq b$, and also $x = a$ or b, $0 \leq t \leq c$. But then Corollary 8.7 implies that $0 \leq v(t,x) \leq 0$ everywhere, which implies that $u \equiv \widetilde{u}$, thereby establishing uniqueness. *Q.E.D.*

Remark: Existence of the solution follows from the convergence of our Fourier series — assuming that the initial and boundary data and the forcing function are sufficiently nice.

Exercises

8.3.1. *True or false*: Assuming no external heat source, if the initial and boundary temperatures of a one-dimensional body are always positive, the temperature within the body is necessarily positive.

8.3.2. Suppose $u(t,x)$ and $v(t,x)$ are two solutions to the heat equation such that $u \leq v$ when $t = 0$ and when $x = a$ or $x = b$. Prove that $u(t,x) \leq v(t,x)$ for all $a \leq x \leq b$ and all $t \geq 0$. Provide a physical interpretation of this result.

8.3.3. For $t > 0$, let $u(t,x)$ be a solution to the unforced heat equation on an interval $a < x < b$, subject to homogeneous Dirichlet boundary conditions. Prove that $M(t) = \max\{\, u(t,x) \mid a \leq x \leq b \,\}$ is a nonincreasing function of t.

8.3.4. (a) State and prove a Maximum Principle for the *convection-diffusion equation* $u_t = u_{xx} + u_x$. (b) Does the equation $u_t = u_{xx} - u_x$ also admit a Maximum Principle?

8.3.5. Consider the parabolic equation $\dfrac{\partial u}{\partial t} = x\,\dfrac{\partial^2 u}{\partial x^2} + \dfrac{\partial u}{\partial x}$ on the interval $1 < x < 2$, with initial and boundary conditions $u(0,x) = f(x)$, $u(t,1) = \alpha(t)$, $u(t,2) = \beta(t)$.
(a) State and prove a version of the Maximum Principle for this problem.
(b) Establish uniqueness of the solution to this initial-boundary value problem.

8.3.6. (a) Show that $u(t,x) = -x^2 - 2xt$ is a solution to the diffusion equation $u_t = x\,u_{xx}$.
(b) Explain why this differential equation does not admit a Maximum Principle.

8.3.7. Suppose that $u(t,x)$ is a nonconstant solution to the heat equation on the interval $0 < x < \ell$ when subject to either homogeneous (a) Dirichlet, (b) Neumann, or (c) mixed boundary conditions. Prove that the function $E(t) = \int_0^\ell u(t,x)^2\, dx$ is everywhere decreasing: $E(t_1) > E(t_2)$ whenever $t_1 < t_2$.

8.3.8. *True or false*: The wave equation $u_{tt} = c^2 u_{xx}$ satisfies a Maximum Principle. If true, clearly state the principle; if false, explain why not.

8.4 Nonlinear Diffusion

First-order partial differential equations serve to model conservative wave motion, beginning with the basic one-dimensional scalar transport equations that we studied in Chapter 2, and progressing on to higher-dimensional systems, the equations of gas dynamics, the full-blown Euler equations of fluid mechanics, and yet more complicated systems of partial differential equations modeling plasmas, magneto-hydrodynamics, etc. However, such systems fail to account for frictional and viscous effects, which are typically modeled by parabolic diffusion equations such as the heat equation and its generalizations, both linear and nonlinear. In this section, we investigate the consequences of combining nonlinear wave motion with linear diffusion by analyzing the simplest such model. As we will see, the dissipative term has the effect of smoothing out abrupt shock discontinuities, and the result is a well-determined, smooth dynamical process with classical solutions. Moreover, in the inviscid limit, the smooth solutions converge (nonuniformly) to a discontinuous shock wave, leading to the method of viscosity solutions that has been successfully employed to analyze such nonlinear dynamical processes.

Burgers' Equation

The simplest nonlinear diffusion equation is known as[†] *Burgers' equation*

$$u_t + u\,u_x = \gamma\, u_{xx}, \tag{8.70}$$

which is obtained by appending a simple linear diffusion term to the nonlinear transport equation (2.31). As with the heat equation, the diffusion coefficient $\gamma \geq 0$ must be nonnegative in order that the initial value problem be well-posed in forwards time. In fluid and gas dynamics, one interprets the right-hand side as modeling the effect of viscosity, and so Burgers' equation represents a very simplified version of the equations of viscous fluid flows, including the celebrated and widely applied Navier–Stokes equations (1.4), [**122**]. When the viscosity coefficient vanishes, $\gamma = 0$, Burgers' equation reduces to the nonlinear transport equation (2.31), which, as a consequence, is often referred to as the *inviscid Burgers' equation*.

[†] The equation is named after the Dutch physicist Johannes Martinus Burgers, [**26**], and so the apostrophe goes after the "s". Burgers' equation was apparently first studied as a physical model by the British (later American) applied mathematician Harry Bateman, [**13**], in the early twentieth century.

Since Burgers' equation is of first order in t, we expect that its solutions will be uniquely prescribed by their initial values

$$u(0,x) = f(x), \qquad -\infty < x < \infty. \tag{8.71}$$

(For simplicity, we will ignore boundary effects here.) Small, slowly varying solutions — more specifically, those for which both $|\,u(t,x)\,|$ and $|\,u_x(t,x)\,|$ are small — tend to act like solutions to the heat equation, smoothing out and decaying to 0 as time progresses. On the other hand, when the solution is large or rapidly varying, the nonlinear term tends to play the dominant role, and we might expect the solution to behave like nonlinear transport waves, perhaps steepening into some sort of shock. But, as we will learn, the smoothing effect of the diffusion term, no matter how small, ultimately prevents the appearance of a discontinuous shock wave. Indeed, it can be proved that, under rather mild assumptions on the initial data, the solution to the initial value problem (8.70–71) remains smooth and well defined for all subsequent times, [**122**].

The simplest explicit solutions are the *traveling waves*, for which

$$u(t,x) = v(\xi) = v(x - c\,t), \qquad \text{where} \qquad \xi = x - c\,t, \tag{8.72}$$

indicates a fixed profile, moving to the right with constant speed c. By the chain rule,

$$\frac{\partial u}{\partial t} = -\,c\,v'(\xi), \qquad \frac{\partial u}{\partial x} = v'(\xi), \qquad \frac{\partial^2 u}{\partial x^2} = v''(\xi).$$

Substituting these expressions into Burgers' equation (8.70), we conclude that $v(\xi)$ must satisfy the nonlinear second-order ordinary differential equation

$$-\,c\,v' + v\,v' = \gamma\,v''.$$

This equation can be solved by first integrating both sides with respect to ξ, and so

$$\gamma\,v' = k - c\,v + \tfrac{1}{2}\,v^2,$$

where k is a constant of integration. Following the analysis after Proposition 2.3, as $\xi \to \pm\infty$, the bounded solutions to such an autonomous first-order ordinary differential equation tend to one of the fixed points provided by the roots of the quadratic polynomial on the right-hand side. Therefore, for there to be a *bounded* traveling-wave solution $v(\xi)$, the quadratic polynomial must have two real roots, which requires $k < \frac{1}{2}c^2$. Assuming this holds, we rewrite the equation in the form

$$2\,\gamma\,\frac{dv}{d\xi} = (v-a)(v-b), \qquad \text{where} \qquad c = \tfrac{1}{2}\,(a+b), \qquad k = \tfrac{1}{2}\,a\,b. \tag{8.73}$$

To obtain bounded solutions, we must require $a < v < b$. Integrating (8.73) by the usual method, cf. (2.19), we find

$$\int \frac{2\,\gamma\,dv}{(v-a)(v-b)} = \frac{2\,\gamma}{b-a}\,\log\left(\frac{b-v}{v-a}\right) = \xi - \delta,$$

where δ is another constant of integration. Solving for

$$v(\xi) = \frac{a\,e^{(b-a)(\xi-\delta)/(2\gamma)} + b}{e^{(b-a)(\xi-\delta)/(2\gamma)} + 1}\,,$$

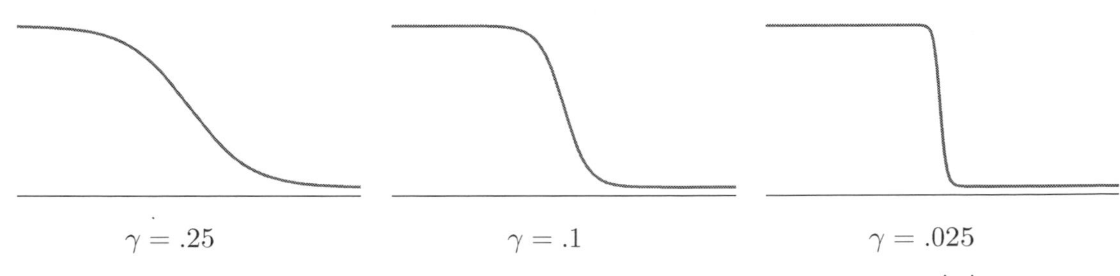

Figure 8.5. Traveling-wave solutions to Burgers' equation.

and recalling (8.73), we conclude that the bounded traveling-wave solutions to Burgers' equation all have the explicit form

$$u(t,x) = \frac{a\,e^{(b-a)(x-c\,t-\delta)/(2\,\gamma)} + b}{e^{(b-a)(x-c\,t-\delta)/(2\,\gamma)} + 1}, \tag{8.74}$$

where $a < b$ and δ are arbitrary constants. Observe that our solution is a monotonically decreasing function of x, with asymptotic values

$$\lim_{x\to-\infty} u(t,x) = b, \qquad\qquad \lim_{x\to\infty} u(t,x) = a,$$

at large distances. The wave travels to the right, unchanged in form, with speed $c = \frac{1}{2}(a+b)$ equal to the average of its asymptotic values. In particular, if $a = -b$, the result is a stationary-wave solution. In Figure 8.5 we graph sample profiles, corresponding to $a = .1$, $b = 1$, for three different values of the diffusion coefficient. Note that the smaller γ is, the sharper the transition layer between the two asymptotic values of the solution.

In the *inviscid limit* as the diffusion becomes vanishingly small, $\gamma \to 0$, the traveling-wave solutions (8.74) converge to the step shock-wave solutions (2.51) of the nonlinear transport equation. Indeed, this can be proved to hold in general: as $\gamma \to 0$, solutions to Burgers' equation (8.70) converge to the corresponding solutions to the nonlinear transport equation (2.31) that are subject to the Rankine–Hugoniot and entropy conditions (2.53, 55). Thus, the method of vanishing viscosity allows one to monitor solutions to the nonlinear transport equation as they evolve into regimes where multiple shocks interact and merge. This approach also reconfirms our physical intuition, in that most physical systems retain a very small dissipative component that serves to mollify abrupt discontinuities that might appear in a theoretical model that fails to take friction or viscous effects into account. In the modern theory of partial differential equations, the resulting *viscosity solution method* has been successfully used to characterize the discontinuous solutions to a broad range of inviscid nonlinear wave equations as limits of classical solutions to a viscously regularized system. We refer the interested reader to [**64**, **107**, **122**] for further details.

The Hopf–Cole Transformation

By a remarkable stroke of good fortune, the nonlinear Burgers' equation can be converted into the linear heat equation and thereby explicitly solved. The transformation that *linearizes* the nonlinear Burgers' equation first appeared in an obscure exercise in a nineteenth-century differential equations textbook, [**41**; vol. 6, p. 102]. Its rediscovery by the applied mathematicians Eberhard Hopf, [**56**], and Julian Cole, [**32**], was a milestone in the modern era of nonlinear partial differential equations, and it is now named the Hopf–Cole transformation in their honor.

In general, *linearization* — that is, converting a given nonlinear differential equation into a linear equation — is extremely challenging, and, in most instances, impossible. On the other hand, the reverse process — "nonlinearizing" a linear equation — is trivial: any nonlinear change of dependent variables will do the trick! However, the resulting nonlinear equation, while evidently linearizable by inverting the change of variables, is rarely of independent interest. But sometimes there is a lucky accident, and the resulting linearization of a physically relevant nonlinear differential equation can have a profound impact on our understanding of more complicated nonlinear systems.

In the present context, our starting point is the linear heat equation

$$v_t = \gamma v_{xx}. \tag{8.75}$$

Among all possible nonlinear changes of dependent variable, one of the simplest that might spring to mind is an exponential function. Let us, therefore, investigate the effect of an exponential change of variables

$$v(t,x) = e^{\alpha\varphi(t,x)}, \qquad \text{so} \qquad \varphi(t,x) = \frac{1}{\alpha}\log v(t,x), \tag{8.76}$$

where α is a nonzero constant. The function $\varphi(t,x)$ is real, provided $v(t,x)$ is a *positive* solution to the heat equation. Fortunately, this is not hard to arrange: if the initial data $v(0,x) > 0$ is strictly positive, then, as a consequence of the Maximum Principle in Corollary 8.7, the resulting solution $v(t,x) > 0$ is positive for all $t > 0$.

To determine the differential equation satisfied by the function φ, we invoke the chain and product rules to differentiate (8.76):

$$v_t = \alpha\varphi_t\, e^{\alpha\varphi}, \qquad v_x = \alpha\varphi_x\, e^{\alpha\varphi}, \qquad v_{xx} = (\alpha\varphi_{xx} + \alpha^2\varphi_x^2)\, e^{\alpha\varphi}.$$

Substituting the first and last formulas into the heat equation (8.75) and canceling a common exponential factor, we conclude that $\varphi(t,x)$ satisfies the nonlinear partial differential equation

$$\varphi_t = \gamma\varphi_{xx} + \gamma\alpha\varphi_x^2, \tag{8.77}$$

known as the *potential Burgers' equation*, for reasons that will soon become apparent.

The second step in the process is to differentiate the potential Burgers' equation with respect to x; the result is

$$\varphi_{tx} = \gamma\varphi_{xxx} + 2\gamma\alpha\varphi_x\varphi_{xx}. \tag{8.78}$$

If we now set

$$\frac{\partial\varphi}{\partial x} = u, \tag{8.79}$$

so that φ acquires the status of a *potential function*, then the resulting partial differential equation

$$u_t = \gamma u_{xx} + 2\gamma\alpha u u_x$$

coincides with Burgers' equation (8.70) when $\alpha = -1/(2\gamma)$. In this manner, we have arrived at the famous *Hopf–Cole transformation*.

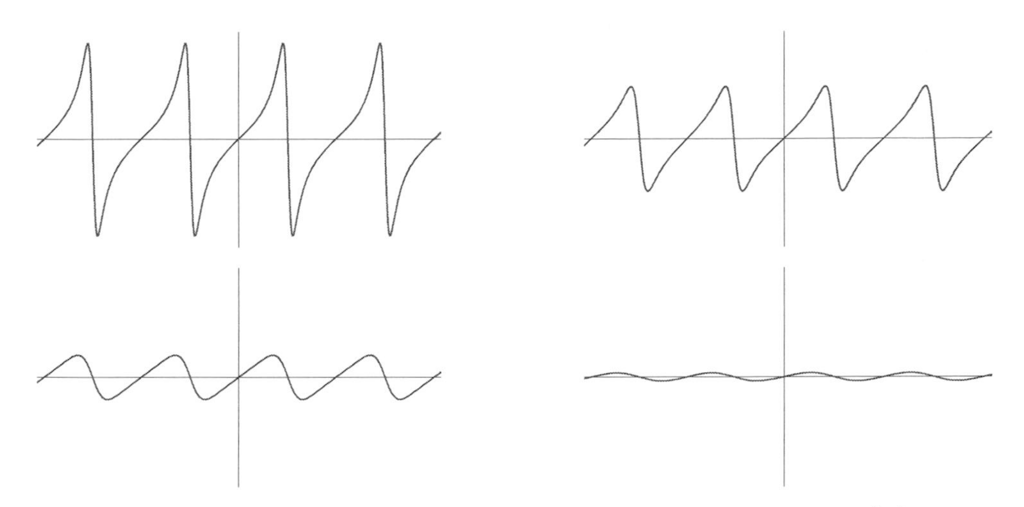

Figure 8.6. Trignometric solution to Burgers' equation.

Theorem 8.9. *If $v(t,x) > 0$ is any positive solution to the linear heat equation $v_t = \gamma v_{xx}$, then*

$$u(t,x) = \frac{\partial}{\partial x}\left[-2\gamma \log v(t,x) \right] = -2\gamma \frac{v_x}{v} \tag{8.80}$$

solves Burgers' equation $u_t + u u_x = \gamma u_{xx}$.

Do all solutions to Burgers' equation arise in this way? In order to answer this question, we run the argument in reverse. First, choose a potential function $\widetilde{\varphi}(t,x)$ that satisfies (8.79); for example,

$$\widetilde{\varphi}(t,x) = \int_0^x u(t,y)\, dy.$$

If $u(t,x)$ is any solution to Burgers' equation, then $\widetilde{\varphi}(t,x)$ satisfies (8.78). Integrating both sides of the latter equation with respect to x, we conclude that

$$\widetilde{\varphi}_t = \gamma \widetilde{\varphi}_{xx} + \gamma \alpha \widetilde{\varphi}_x^2 + g(t),$$

for some integration "constant" $g(t)$. Thus, unless $g(t) \equiv 0$, our potential function $\widetilde{\varphi}$ doesn't satisfy the potential Burgers' equation (8.77), but that is because we chose the "wrong" potential. Indeed, if we define

$$\varphi(t,x) = \widetilde{\varphi}(t,x) - G(t), \qquad \text{where} \qquad G'(t) = g(t),$$

then

$$\varphi_t = \widetilde{\varphi}_t - g(t) = \gamma \widetilde{\varphi}_{xx} + \gamma \alpha \widetilde{\varphi}_x^2 = \gamma \varphi_{xx} + \gamma \alpha \varphi_x^2,$$

and hence the modified potential $\varphi(t,x)$ *is* a solution to the potential Burgers' equation (8.77). From this it easily follows that

$$v(t,x) = e^{-\varphi(t,x)/(2\gamma)} \tag{8.81}$$

is a positive solution to the heat equation, from which the Burgers' solution $u(t,x)$ can be recovered through (8.80). We conclude that *every* solution to Burgers' equation comes from a positive solution to the heat equation via the Hopf–Cole transformation.

Example 8.10. As a simple example, the separable solution

$$v(t,x) = a + b\,e^{-\gamma\omega^2 t}\cos\omega x$$

to the heat equation leads to the following solution to Burgers' equation:

$$u(t,x) = \frac{2\gamma\,b\,\omega\,\sin\omega x}{a\,e^{\gamma\omega^2 t} + b\cos\omega x}\,. \tag{8.82}$$

A representative example is plotted in Figure 8.6. We should require that $a > |\,b\,|$ in order that $v(t,x) > 0$ be a positive solution to the heat equation for $t \geq 0$; otherwise the resulting solution to Burgers' equation will have singularities at the roots of u — as in the first graph in Figure 8.6. This family of solutions is primarily affected by the viscosity term, and rapidly decays to zero.

To solve the initial value problem (8.70–71) for Burgers' equation, we note that, under the Hopf–Cole transformation (8.80),

$$v(0,x) = \exp\left(-\frac{\varphi(0,x)}{2\gamma}\right) = \exp\left(-\frac{1}{2\gamma}\int_0^x f(y)\,dy\right) \equiv h(x). \tag{8.83}$$

Remark: The lower limit of the integral can be changed from 0 to any other convenient value. The only effect is to multiply $v(t,x)$ by an overall constant, which does not change the final form of $u(t,x)$ in (8.80).

According to formula (8.16) (adapted to general diffusivity, as in Exercise 8.2.3), the solution to the initial value problem (8.75, 83) for the heat equation can be expressed as a convolution integral with the fundamental solution

$$v(t,x) = \frac{1}{2\sqrt{\pi\gamma t}}\int_{-\infty}^{\infty} e^{-(x-\xi)^2/(4\gamma t)}\,h(\xi)\,d\xi.$$

Therefore, setting $\widehat{v}(t,x) = 2\sqrt{\pi\gamma t}\,v(t,x)$, the solution to the Burgers' initial value problem (8.70–71), valid for $t > 0$, is given by

$$u(t,x) = -\frac{2\gamma}{\widehat{v}(t,x)}\,\frac{\partial\widehat{v}}{\partial x}\,, \qquad \text{where} \qquad \begin{cases} \widehat{v}(t,x) = \displaystyle\int_{-\infty}^{\infty} e^{-H(t,x;\xi)}\,d\xi, \\[2ex] H(t,x;\xi) = \dfrac{(x-\xi)^2}{4\gamma t} + \dfrac{1}{2\gamma}\displaystyle\int_0^{\xi} f(\eta)\,d\eta. \end{cases} \tag{8.84}$$

Example 8.11. To demonstrate the smoothing effect of the diffusion terms, let us see what happens to the initial data

$$u(0,x) = \begin{cases} a, & x < 0, \\ b, & x > 0, \end{cases} \tag{8.85}$$

in the form of a step function. We assume that $a > b$, which corresponds to a shock wave in the inviscid limit $\gamma = 0$. (In Exercise 8.4.4, the reader is asked to analyze the case $a < b$, which corresponds to a rarefaction wave.) In this case,

$$H(t,x;\xi) = \frac{(x-\xi)^2}{4\gamma t} + \begin{cases} \dfrac{a\,\xi}{2\gamma}, & \xi < 0, \\[2ex] \dfrac{b\,\xi}{2\gamma}, & \xi > 0. \end{cases} \tag{8.86}$$

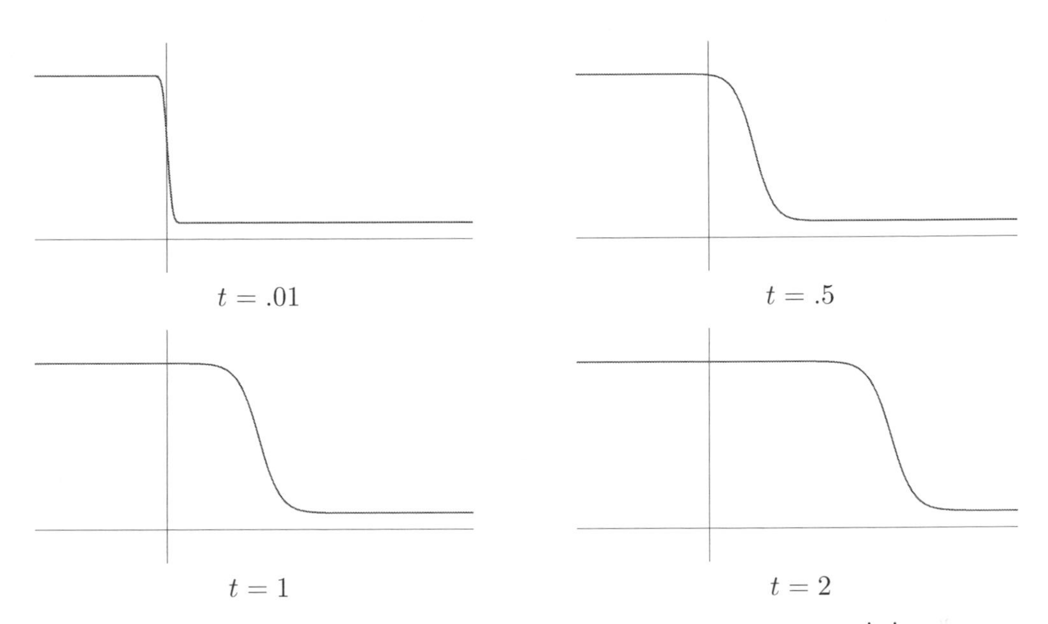

Figure 8.7. Shock-wave solution to Burgers' equation.

After some algebraic manipulations, the solution (8.84) is found to have the explicit form

$$u(t,x) = a + \frac{b-a}{1+\exp\left(\dfrac{b-a}{2\gamma}(x-ct)\right)\operatorname{erfc}\left(\dfrac{x-at}{2\sqrt{\gamma t}}\right)\Big/\operatorname{erfc}\left(\dfrac{bt-x}{2\sqrt{\gamma t}}\right)}, \tag{8.87}$$

with $c = \frac{1}{2}(a+b)$, where $\operatorname{erfc} z = 1 - \operatorname{erf} z$ denotes the complementary error function (8.43). The solution, for $a = 1$, $b = .1$, and $\gamma = .03$, is plotted at various times in Figure 8.7. Observe that, as with the heat equation, the jump discontinuity is immediately smoothed out, and the solution soon assumes the form of a smoothly varying transition between its two original heights. The larger the diffusion coefficient in relation to the jump magnitude, the more pronounced the smoothing effect. Moreover, as $\gamma \to 0$, the solution $u(t,x)$ converges to the shock-wave solution (2.51) to the transport equation, in which the speed of the shock is c, the average of the step heights — in accordance with the Rankine–Hugoniot shock rule. Indeed, in view of (2.88),

$$\lim_{z\to\infty} \operatorname{erfc} z = 0, \qquad \lim_{z\to-\infty} \operatorname{erfc} z = 2. \tag{8.88}$$

Thus, for $t > 0$, as $\gamma \to 0$, the ratio of the two complementary error functions in (8.87) tends to ∞ when $x < bt$, to 1 when $bt < x < at$, and to 0 when $x > at$. On the other hand, since $a > b$, the exponential term tends to ∞ when $x < ct$, and to 0 when $x > ct$. Put together, these imply that the solution $u(t,x) \to a$ when $x < ct$, while $u(t,x) \to b$, when $x > ct$, thus proving convergence to the shock-wave solution.

Example 8.12. Consider the case in which the initial data $u(0,x) = \delta(x)$ is a concentrated delta function impulse at the origin. In the solution formula (8.84), starting the integral for $H(t,x;\xi)$ at 0 is problematic, but as noted earlier, we are free to select any

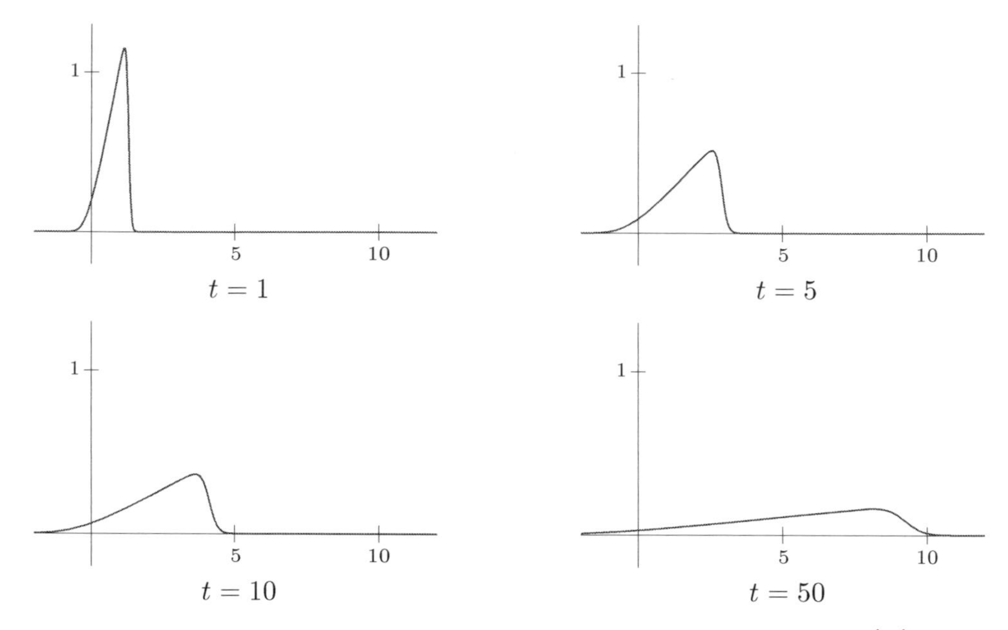

Figure 8.8. Triangular-wave solution to Burgers' equation.

other starting point, e.g., $-\infty$. Thus, we take

$$H(t,x;\xi) = \frac{(x-\xi)^2}{4\gamma t} + \frac{1}{2\gamma}\int_{-\infty}^{\xi} \delta(\eta)\,d\eta = \begin{cases} \dfrac{(x-\xi)^2}{4\gamma t}, & \xi < 0, \\ \dfrac{1}{2\gamma} + \dfrac{(x-\xi)^2}{4\gamma t}, & \xi > 0. \end{cases}$$

We then evaluate

$$\widehat{v}(t,x) = \int_{-\infty}^{\infty} e^{-H(t,x;\xi)}\,d\xi = \sqrt{\pi\gamma t}\left[1 - \operatorname{erf}\left(\frac{x}{2\sqrt{\gamma t}}\right) + e^{-1/(2\gamma)}\left\{1 + \operatorname{erf}\left(\frac{x}{2\sqrt{\gamma t}}\right)\right\}\right].$$

Therefore, the solution to the initial value problem is

$$u(t,x) = -\frac{2\gamma}{\widehat{v}(t,x)}\frac{\partial \widehat{v}}{\partial x} = 2\sqrt{\frac{\gamma}{\pi t}}\;\frac{e^{-x^2/(4\gamma t)}}{\coth\left(\dfrac{1}{4\gamma}\right) - \operatorname{erf}\left(\dfrac{x}{2\sqrt{\gamma t}}\right)}, \tag{8.89}$$

where

$$\coth z = \frac{\cosh z}{\sinh z} = \frac{e^z + e^{-z}}{e^z - e^{-z}} = \frac{e^{2z}+1}{e^{2z}-1}$$

is the hyperbolic cotangent function. A graph of this solution when $\gamma = .02$ and $a = 1$ appears in Figure 8.8. As you can see, the initial concentration diffuses out, but, in contrast to the heat equation, does not remain symmetric, since the nonlinear advection term causes the wave to steepen in front. Eventually, as the effect of the diffusion accumulates, the propagating triangular wave becomes vanishingly small.

Exercises

8.4.1. Find the solution to Burgers' equation that has the following initial data:

$$u(0,x) = \quad (a)\ \sigma(x), \qquad (b)\ \sigma(-x), \qquad (c)\ \begin{cases} 1, & 0 < x < 1, \\ 0, & \text{otherwise.} \end{cases}$$

8.4.2. Starting with the heat equation solution $v(t,x) = 1 + t^{-1/2}\, e^{-x^2/(4\gamma t)}$, find the corresponding solution to Burgers' equation and discuss its behavior.

8.4.3. Justify the solution formula (8.87).

♢ 8.4.4. (a) Prove that $\lim_{z \to \infty} z\, e^{z^2} \operatorname{erfc} z = 1/\sqrt{\pi}$. (b) Show that when $a < b$, the Burgers' solution (8.87) converges to the rarefaction wave (2.54) in the inviscid limit $\gamma \to 0^+$.

8.4.5. *True or false*: If $u(t,x)$ solves Burgers' equation for the step function initial condition $u(0,x) = \sigma(x)$, then $v(t,x) = u_x(t,x)$ solves the initial value problem with $v(0,x) = \delta(x)$.

8.4.6. *True or false*: If $\widehat{v}(t,x)$ is as given in (8.84), then

$$\frac{\partial \widehat{v}}{\partial x} = \int_{-\infty}^{\infty} \frac{\xi - x}{2\gamma t}\, e^{-H(t,x;\xi)}\, d\xi,$$

and hence the solution to the Burgers' initial value problem (8.70–71) can be written as

$$u(t,x) = \frac{\displaystyle\int_{-\infty}^{\infty} \frac{x-\xi}{t}\, e^{-H(t,x;\xi)}\, d\xi}{\displaystyle\int_{-\infty}^{\infty} e^{-H(t,x;\xi)}\, d\xi}, \qquad \text{where} \qquad H(t,x;\xi) = \frac{(x-\xi)^2}{4\gamma t} + \frac{1}{2\gamma}\int_0^{\xi} f(\eta)\, d\eta.$$

8.4.7. Show that if $u(t,x)$ solves Burgers' equation, then $U(t,x) = u(t, x - ct) + c$ is also a solution. What is the physical interpretation of this symmetry?

8.4.8. (a) What is the effect of a scaling transformation $(t,x,u) \longmapsto (\alpha t, \beta x, \lambda u)$ on Burgers' equation? (b) Use your result to solve the initial value problem for the rescaled Burgers' equation $U_t + \rho\, U\, U_x = \sigma\, U_{xx}$, $U(0,x) = F(x)$.

♡ 8.4.9. (a) Find all scaling symmetries of Burgers' equation. (b) Determine the ordinary differential equation satisfied by the similarity solutions. (c) *True or false*: The Hopf–Cole transformation maps similarity solutions of the heat equation to similarity solutions of Burgers' equation.

8.4.10. What happens if you nonlinearize the heat equation (8.75) using the change of variables

$$(a)\ v = \varphi^2; \qquad (b)\ v = \sqrt{\varphi}; \qquad (c)\ v = \log \varphi\,?$$

8.4.11. What partial differential equation results from applying the exponential change of variables (8.76) to:

(a) the wave equation $v_{tt} = c^2 v_{xx}$? (b) the Laplace equation $v_{xx} + v_{yy} = 0$?

8.5 Dispersion and Solitons

In this section, we finally venture beyond the by now familiar terrain of second-order partial differential equations. While considerably less common than those of first and second order, higher-order equations arise in certain applications, particularly third-order dispersive models for wave motion, [**2**, **122**], and fourth-order systems modeling elastic plates and shells, [**7**]. We will focus our attention on two basic third-order evolution equations. The first is a simple linear equation with a third derivative term. It arises as a simplified model for unidirectional wave motion, and thus has more in common with first-order transport equations than with the second-order dissipative heat equation. The third-order derivative induces a process of *dispersion*, in which waves of different frequencies propagate at different speeds. Thus, unlike the first- and second-order wave equations, in which waves maintain their initial profile as they move, dispersive waves will spread out and decay even while conserving energy. Waves on the surface of a liquid are familiar examples of dispersive waves — an initially concentrated disturbance, caused by, say, throwing a rock in a pond, spreads out over the surface as its different vibrational components move off at different speeds.

Our second example is a remarkable nonlinear third-order evolution equation known as the Korteweg–de Vries equation, which combines dispersive effects with nonlinear transport. As with Burgers' equation (but for very different mathematical reasons), the dispersive term thwarts the tendency for solutions to break into shock waves, and, in fact, classical solutions exist for all time. Moreover, a general localized initial disturbance will break up into a finite number of solitary waves; the taller the wave, the faster it moves. Even more remarkable are the interactive properties of these solitary waves. One ordinarily expects nonlinearity to induce very complicated and not easily predictable behavior. However, when two solitary-wave solutions to the Korteweg–de Vries equation collide, they eventually emerge from the interaction unchanged, save for a phase shift. This unexpected and remarkable phenomenon was first detected through numerical simulations in the 1960s and distinguished with the neologism *soliton*. It was then found that solitons appear in a surprising number of basic nonlinear physical models. The investigation of their mathematical properties has had deep ramifications, not just within partial differential equations and fluid mechanics, but throughout applied mathematics and theoretical physics; it has even contributed to the solution of long-outstanding problems in complex function theory. Further development of the modern theory and amazing properties of integrable soliton equations can be found in [**2**, **36**].

Linear Dispersion

The simplest nontrivial third-order partial differential equation is the linear equation

$$u_t + u_{xxx} = 0, \tag{8.90}$$

which models the unidirectional[†] propagation of linear dispersive waves. To avoid complications engendered by boundary conditions, we shall initially look only at solutions on the entire line, so $-\infty < x < \infty$. Since the equation involves only a first-order time derivative, one expects its solutions to be uniquely specified by a single initial condition

$$u(0,x) = f(x), \qquad -\infty < x < \infty. \tag{8.91}$$

[†] Bidirectional propagation, as we saw in the wave equation, requires a second-order time derivative. As in the d'Alembert solution to the second-order wave equation, the reduction to a unidirectional model is based on an (approximate) factorization of the bidirectional operator.

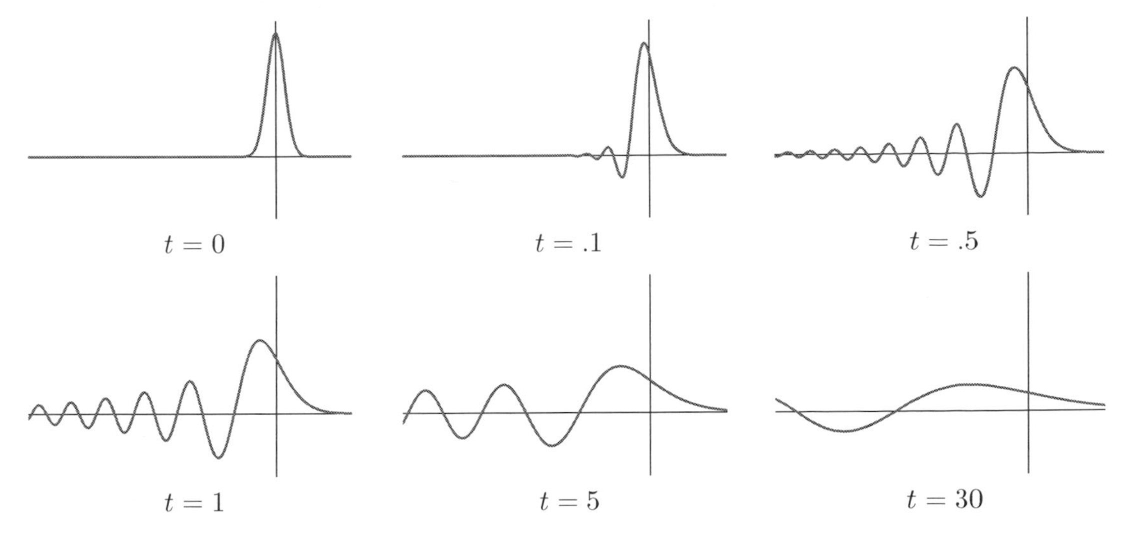

Figure 8.9. Gaussian solution to the dispersive wave equation.

In wave mechanics, $u(t,x)$ represents the height of the fluid at time t and position x, and the initial condition (8.91) specifies the initial disturbance.

As with the heat equation (and, indeed, any linear constant-coefficient evolution equation), the Fourier transform is an effective tool for solving the initial value problem on the real line. Assuming that the solution $u(t,\cdot) \in \mathrm{L}^2(\mathbb{R})$ remains square integrable at all times t (a fact that can be justified a priori — see Exercise 8.5.18(*b*)), let

$$\widehat{u}(t,k) = \frac{1}{\sqrt{2\pi}} \int_{-\infty}^{\infty} u(t,x)\, e^{-\,\mathrm{i}\,k x}\, dx$$

be its spatial Fourier transform. Owing to its effect on derivatives, the Fourier transform converts the partial differential equation (8.90) into a first-order linear ordinary differential equation:

$$\frac{\partial \widehat{u}}{\partial t} + (\mathrm{i}\,k)^3\, \widehat{u} = \frac{\partial \widehat{u}}{\partial t} - \mathrm{i}\,k^3\, \widehat{u} = 0, \tag{8.92}$$

in which the frequency variable k appears as a parameter. The corresponding initial conditions

$$\widehat{u}(0,k) = \widehat{f}(k) = \frac{1}{\sqrt{2\pi}} \int_{-\infty}^{\infty} f(x)\, e^{-\,\mathrm{i}\,k x}\, dx \tag{8.93}$$

are provided by the Fourier transform of (8.91). The solution to the initial value problem (8.92–93) is

$$\widehat{u}(t,k) = \widehat{f}(k)\, e^{\,\mathrm{i}\,k^3 t}.$$

Inverting the Fourier transform yields the explicit formula for the solution

$$u(t,x) = \frac{1}{\sqrt{2\pi}} \int_{-\infty}^{\infty} \widehat{f}(k)\, e^{\,\mathrm{i}\,(k x + k^3 t)}\, dk \tag{8.94}$$

to the initial value problem (8.90–91) for the dispersive wave equation.

Example 8.13. Suppose that the initial profile

$$u(0,x) = f(x) = e^{-x^2}$$

is a Gaussian. According to our table of Fourier transforms (see page 272),

$$\widehat{f}(k) = \frac{e^{-k^2/4}}{\sqrt{2}},$$

and hence the corresponding solution to the dispersive wave equation (8.90) is

$$u(t,x) = \frac{1}{2\sqrt{\pi}} \int_{-\infty}^{\infty} e^{\,\mathrm{i}\,(k\,x+k^3\,t)-k^2/4}\, dk = \frac{1}{2\sqrt{\pi}} \int_{-\infty}^{\infty} e^{-k^2/4} \cos(k\,x + k^3\,t)\, dk;$$

the imaginary part vanishes thanks to the oddness of the integrand. (Indeed, the solution must be real, since the initial data is real.) A plot of the solution at various times appears in Figure 8.9. Note the propagation of initially rapid oscillations to the rear (negative x) of the initial disturbance. The dispersion causes the oscillations to gradually spread out and decrease in amplitude, with the effect that $u(t,x) \to 0$ uniformly as $t \to \infty$, even though, according to Exercise 8.5.7, both the mass $M = \displaystyle\int_{-\infty}^{\infty} u(t,x)\, dx$ and the energy $E = \displaystyle\int_{-\infty}^{\infty} u(t,x)^2\, dx$ of the wave are conserved, i.e., are both constant in time.

Example 8.14. The *fundamental solution* to the dispersive wave equation is generated by a concentrated initial disturbance:

$$u(0,x) = \delta(x).$$

The Fourier transform of the delta function is just $\widehat{\delta}(k) = 1/\sqrt{2\pi}$. Therefore, the corresponding solution (8.94) is

$$u(t,x) = \frac{1}{2\pi} \int_{-\infty}^{\infty} e^{\,\mathrm{i}\,(k\,x+k^3\,t)}\, dk = \frac{1}{\pi} \int_{0}^{\infty} \cos(k\,x + k^3\,t)\, dk, \tag{8.95}$$

since the solution is real (or, equivalently, the imaginary part of the integrand is odd), while the real part of the integrand is even.

A priori, it appears that the integral (8.95) does not converge, because the integrand does not go to zero as $|\,k\,| \to \infty$. However, the increasingly rapid oscillations induced by the cubic term tend to cancel each other out and allow convergence. To prove this, given $l > 0$, we perform a (non-obvious) integration by parts:

$$\begin{aligned} \int_0^l \cos(k\,x + k^3\,t)\, dk &= \int_0^l \frac{1}{x + 3\,k^2\,t}\, \frac{d}{dk} \sin(k\,x + k^3\,t)\, dk \qquad (8.96) \\ &= \frac{\sin(k\,x + k^3\,t)}{x + 3\,k^2\,t} \,\bigg|_{k=0}^{l} - \int_0^l \frac{d}{dk}\left(\frac{1}{x + 3\,k^2\,t}\right) \sin(k\,x + k^3\,t)\, dk \\ &= \frac{\sin(l\,x + l^3\,t)}{x + 3\,l^2\,t} + \int_0^l \frac{6\,k\,t \sin(k\,x + k^3\,t)}{(x + 3\,k^2\,t)^2}\, dk. \end{aligned}$$

Provided $t \neq 0$, as $l \to \infty$, the first term on the right goes to zero, while the final integral converges absolutely due to the rapid decay of the integrand.

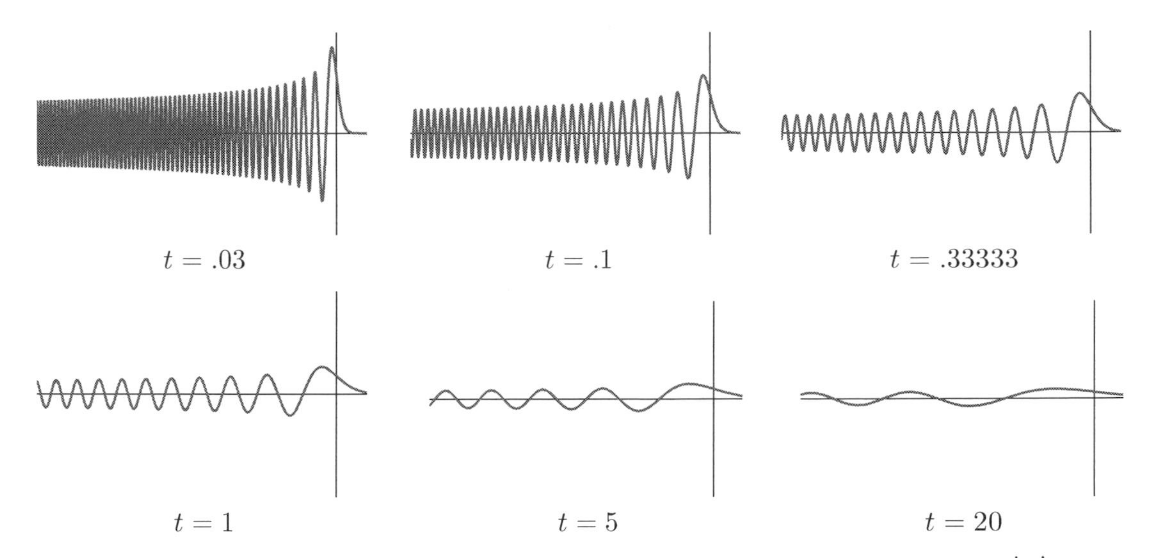

Figure 8.10. Fundamental solution to the dispersive wave equation.

While the integral in the solution formula (8.95) cannot be evaluated in terms of elementary functions, it is related to the integral defining the *Airy function*

$$\operatorname{Ai}(z) = \frac{1}{\pi} \int_0^\infty \cos\left(s\,z + \tfrac{1}{3}\,s^3\right) ds, \tag{8.97}$$

an important special function, [**86**], that was first employed by the nineteenth-century British applied mathematician George Airy in his studies of optical caustics (the focusing of light waves through a lens, e.g., a magnifying glass) and rainbows, [**4**]. Indeed, applying the change of variables

$$s = k\,\sqrt[3]{3\,t}\,, \qquad z = \frac{x}{\sqrt[3]{3\,t}}\,,$$

to the Airy function integral (8.97), we deduce that the fundamental solution to the dispersive wave equation (8.90) can be written as

$$u(t,x) = \frac{1}{\sqrt[3]{3\,t}}\ \operatorname{Ai}\left(\frac{x}{\sqrt[3]{3\,t}}\right). \tag{8.98}$$

See Figure 8.10 for a graph of the solution at several times; in particular, at $t = 1/3$ the solution is exactly the Airy function. We see that the immediate effect of the initial delta impluse is to spawn a highly oscillatory wave trailing off to $-\infty$. (As with the heat equation, signals propagate with infinite speed.) As time progresses, the dispersive effects cause the oscillations to spread out, with their overall amplitude decaying in proportion to $t^{-1/3}$. On the other hand, as $t \to 0^+$, the solution becomes more and more oscillatory for negative x, and so converges *weakly* to the initial delta function. We also note that (8.98) has the form of a similarity solution, since it is invariant under the scaling symmetry

$$(t, x, u) \longmapsto (\lambda^{-3} t, \lambda^{-1}\,x, \lambda\,u).$$

Equation (8.98) gives the response to an initial delta function concentrated at the

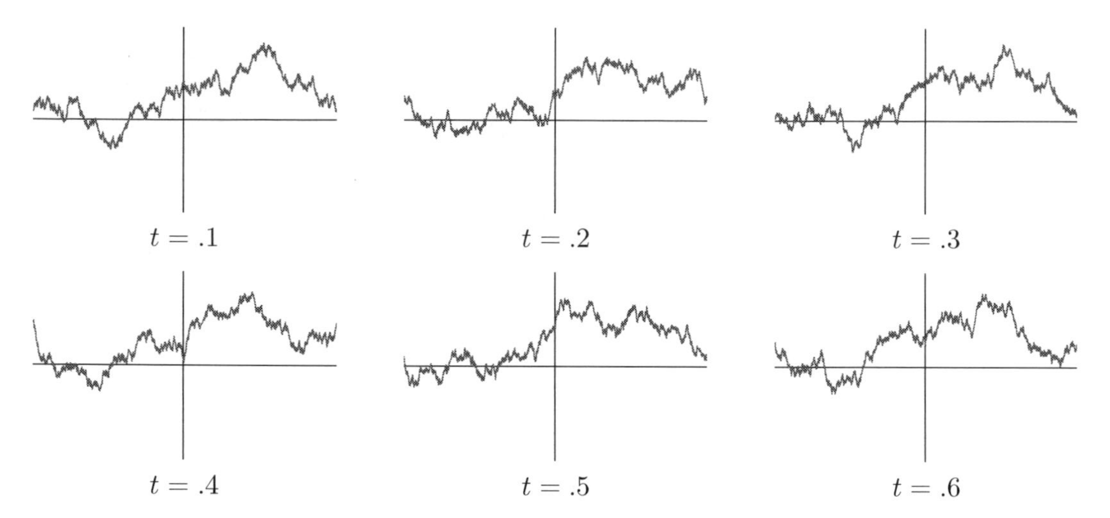

Figure 8.11. Periodic dispersion at irrational (with respect to π) times.

origin. By translation invariance, we immediately deduce that

$$F(t,x;\xi) = \frac{1}{\sqrt[3]{3t}}\,\mathrm{Ai}\left(\frac{x-\xi}{\sqrt[3]{3t}}\right)$$

is the *fundamental solution* corresponding to an initial delta impulse at $x = \xi$. Therefore, we can use linear superposition to find an explicit formula for the solution to the initial value problem that bypasses the Fourier transform. Namely, writing the general initial data as a superposition of delta functions,

$$u(0,x) = f(x) = \int_{-\infty}^{\infty} f(\xi)\,\delta(x-\xi)\,d\xi,$$

we conclude that the resulting solution is the selfsame combination of fundamental solutions:

$$u(t,x) = \frac{1}{\sqrt[3]{3t}}\int_{-\infty}^{\infty} f(\xi)\,\mathrm{Ai}\left(\frac{x-\xi}{\sqrt[3]{3t}}\right)d\xi. \tag{8.99}$$

Example 8.15. *Dispersive Quantization.* Let us investigate the periodic initial-boundary value problem for our basic linear dispersive equation on the interval $-\pi \le x \le \pi$:

$$u_t + u_{xxx} = 0, \quad u(t,-\pi) = u(t,\pi), \quad u_x(t,-\pi) = u_x(t,\pi), \quad u_{xx}(t,-\pi) = u_{xx}(t,\pi), \tag{8.100}$$

with initial data $u(0,x) = f(x)$. The Fourier series formula for the resulting solution is straightforwardly constructed:

$$u(t,x) = \sum_{k=-\infty}^{\infty} c_k\, e^{\,\mathrm{i}\,(kx+k^3t)}, \tag{8.101}$$

where c_k are the usual (complex) Fourier coefficients (3.65) of the initial data $f(x)$.

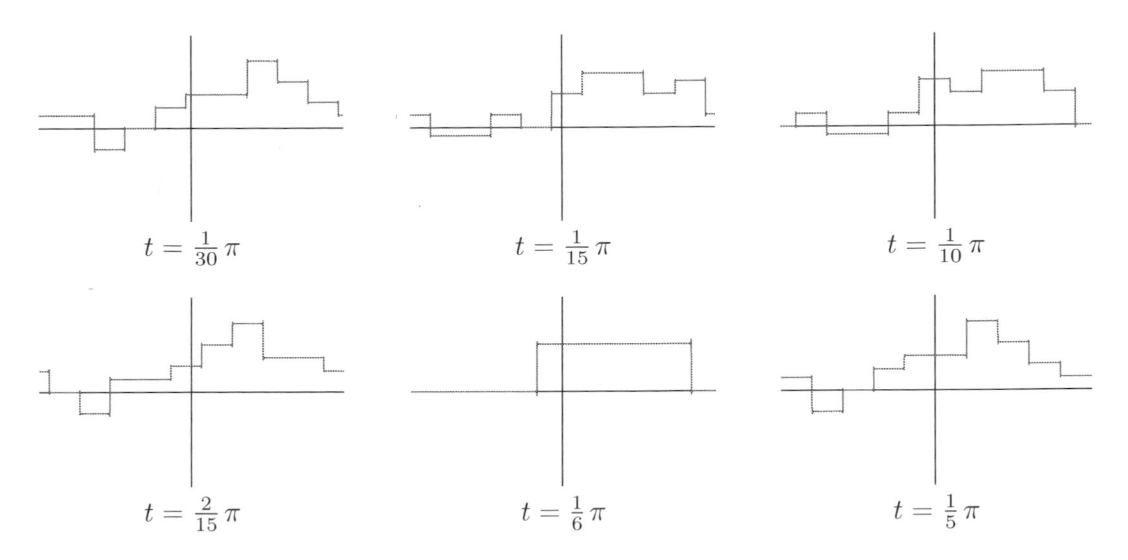

Figure 8.12. Periodic dispersion at rational (with respect to π) times.

Let us take the initial data to be the unit step function: $u(0,x) = \sigma(x)$. In view of its Fourier series (3.67), the resulting solution formula (8.101) becomes

$$\begin{aligned} u(t,x) &= \frac{1}{2} - \frac{\mathrm{i}}{\pi} \sum_{l=-\infty}^{\infty} \frac{e^{\,\mathrm{i}\,[\,(2l+1)x+(2l+1)^3 t\,]}}{2l+1} \\ &= \frac{1}{2} + \frac{2}{\pi} \sum_{l=0}^{\infty} \frac{\sin\big[\,(2l+1)\,x + (2l+1)^3 t\,\big]}{2l+1}. \end{aligned} \tag{8.102}$$

Let us graph this solution. At times uniformly spaced by $\Delta t = .1$, the resulting solution profiles are plotted in Figure 8.11. The solution appears to have a continuous but fractal-like structure, reminiscent of Weierstrass' continuous but nowhere differentiable function, [**55**; pp. 401–421]. The temporal evolution continues in this fashion until the initial data are formed again at $t = 2\pi$, after which the process periodically repeats.

However, when the times are spaced by $\Delta t = \frac{1}{30}\pi \approx .10472$, the resulting solution profiles, as plotted in Figure 8.12, are strikingly different! Indeed, as you are asked to prove in Exercise 8.5.8, at each rational time $t = 2\pi p/q$, where p, q are integers, the solution (8.102) to the initial-boundary value problem is discontinuous but constant on subintervals of length $2\pi/q$. This remarkable behavior, in which the solution profiles of linearly dispersive periodic boundary value problems have markedly different behaviors at rational and irrational times (with respect to π), was first observed, in the 1990's, in optics and quantum mechanics by the British physicist Michael Berry, [**16, 115**], and named the *Talbot effect*, after an optical experiment conducted by the inventor of the photographic negative, William Henry Fox Talbot. While writing this book, I rediscovered the effect, which I like to call *dispersive quantization*, [**88**], and found that it arises in a wide range of linearly dispersive periodic initial-boundary value problems, [**30**].

The Dispersion Relation

As noted earlier, a key feature of the third-order wave equation (8.90) is that waves disperse, in the sense that those of different frequencies move at different speeds. Our goal now is to better understand the dispersion process. To this end, consider a solution whose initial profile

$$u(0,x) = e^{\,\mathrm{i}\,k\,x}$$

is a complex oscillatory function. Since the initial data does not decay as $|\,x\,| \to \infty$, we cannot use the Fourier integral solution formula (8.94) directly. Instead, anticipating the induced wave to exhibit temporal oscillations, let us try an exponential solution ansatz

$$u(t,x) = e^{\,\mathrm{i}\,(k\,x - \omega\,t)} \tag{8.103}$$

representing a complex oscillatory wave of temporal *frequency* ω and *wave number* (spatial frequency) k. Since

$$\frac{\partial u}{\partial t} = -\,\mathrm{i}\,\omega\,e^{\,\mathrm{i}\,(k\,x-\omega\,t)}, \qquad\qquad \frac{\partial^3 u}{\partial x^3} = -\,\mathrm{i}\,k^3\,e^{\,\mathrm{i}\,(k\,x-\omega\,t)},$$

(8.103) satisfies the partial differential equation (8.90) if and only if its frequency and wave number satisfy the *dispersion relation*

$$\omega = -\,k^3. \tag{8.104}$$

Therefore, the exponential solution (8.103) of wave number k takes the form

$$u(t,x) = e^{\,\mathrm{i}\,(k\,x + k^3\,t)}. \tag{8.105}$$

Our Fourier transform formula (8.94) for the solution can thus be viewed as a (continuous) linear superposition of these elementary exponential solutions. In general, to find the dispersion relation for a linear constant-coefficient partial differential equation, one substitutes the exponential ansatz (8.103). On cancellation of the common exponential factors, the result is an equation expressing the frequency ω as a function of the wave number k.

Any exponential solution (8.103) is automatically in the form of a traveling wave, since we can write

$$u(t,x) = e^{\,\mathrm{i}\,(k\,x-\omega\,t)} = e^{\,\mathrm{i}\,k\,(x-c_p\,t)}, \qquad \text{where} \qquad c_p = \frac{\omega}{k} \tag{8.106}$$

is the *wave speed* or, as it is more usually called, the *phase velocity*. If the dispersion relation is linear in the wave number, $\omega = c\,k$, as occurs in the linear transport equation $u_t + c\,u_x = 0$, then all waves move at an identical speed $c_p = c$, and hence localized disturbances stay localized as they propagate through the medium. In the dispersive case, ω is no longer a linear function of k, and so waves of different spatial frequencies move at different speeds. In the particular case (8.90), those with wave number k move at speed $c_p = \omega/k = -\,k^2$, and so the higher the wave number, the faster the wave propagates to the left. As the individual exponential constituents separate, the overall effect is the dispersive decay of an initially localized wave, with slowly diminishing amplitude and increasingly rapid oscillation as $x \to -\infty$.

The general solution to the linear partial differential equation under consideration is then built up by linear superposition of the exponential solutions,

$$u(t,x) = \int_{-\infty}^{\infty} e^{\,\mathrm{i}\,(k\,x-\omega\,t)}\,g(k)\,dk, \tag{8.107}$$

where $\omega = \omega(k)$ is determined by the relevant dispersion relation. While the evolution of the individual waves is an immediate consequence of the dispersion relation, the evolution of the localized wave packet represented by (8.107) is less evident. To determine its speed of propagation, let us switch to a moving coordinate frame of speed c by setting $x = c\,t + \xi$. The solution formula (8.107) then becomes

$$u(t, c\,t + \xi) = \int_{-\infty}^{\infty} e^{\,\mathrm{i}\,(ck-\omega)t} e^{\,\mathrm{i}\,k\,\xi}\, g(k)\, dk. \tag{8.108}$$

For a fixed value of ξ, the integral is of the general oscillatory form

$$H(t) = \int_{-\infty}^{\infty} e^{\,\mathrm{i}\,\varphi(k)\,t}\, h(k)\, dk, \tag{8.109}$$

where, in our case, $\varphi(k) = c\,k - \omega(k)$ and $h(k) = e^{\,\mathrm{i}\,k\,\xi}\, g(k)$. We are interested in understanding the behavior of such an oscillatory integral as $t \to \infty$. Now, if $\varphi(k) = k$, then (8.109) is just a Fourier integral, (7.9), and, as we learned in Chapter 7, $H(t) \to 0$ as $t \to \infty$, for any reasonable function $h(k)$. Intuitively, the increasingly rapid oscillations of the exponential factor tend to cancel each other out in the high-frequency limit. A similar result holds wherever $\varphi(k)$ has no stationary points, i.e., $\varphi'(k) \neq 0$, since one can then perform a local change of variables $\widetilde{k} = \varphi(k)$ to convert that part of the oscillatory integral to Fourier form, and again the increasingly rapid oscillations cause the limit to vanish. In this fashion, we arrive at the key insight of Stokes and Kelvin that produced the powerful *Method of Stationary Phase*. Namely, for large $t \gg 0$, the primary contribution to the highly oscillatory integral (8.109) occurs at the *stationary points* of the phase function, that is, where $\varphi'(k) = 0$. A rigorous justification of the method, along with precise error bounds, can be found in [**85**].

In the present context, the Method of Stationary Phase implies that the most significant contribution to the integral (8.108) occurs when

$$0 = \frac{d}{dk}\,(\omega - c\,k) = \frac{d\omega}{dk} - c. \tag{8.110}$$

Thus, surprisingly, the principal contribution of the components at wave number k is felt when moving at the *group velocity*

$$c_g = \frac{d\omega}{dk}\,. \tag{8.111}$$

Interestingly, unless the dispersion relation is linear in the wave number, the group velocity (8.111), which determines the speed of propagation of the energy, is *not the same* as the phase velocity (8.106), which governs the speed of propagation of an individual oscillatory wave. For example, in the case of the dispersive wave equation (8.90), $\omega = -k^3$, and so $c_g = -3\,k^2$, which is three times as fast as the phase velocity, $c_p = \omega/k = -k^2$. Thus, the energy propagates faster than the individual waves. This can be observed in Figure 8.9: while the bulk of the disturbance is spreading out rather rapidly to the left, the individual wave crests are moving slower.

On the other hand, the dispersion relation associated with deep water waves is (ignoring physical constants) $\omega = \sqrt{k}$, [**122**]. Now, the phase velocity is $c_p = \omega/k = 1/\sqrt{k}$, whereas the group velocity is $c_g = d\omega/dk = 1/(2\sqrt{k}\,) = \frac{1}{2}\,c_p$, and so the individual waves move twice as fast as the speed of propagation of the underlying wave energy. For an experimental verification, just throw a stone in a still pond. An individual wave crest emerges in back and then steadily grows as it moves through the disturbance, eventually subsiding and disappearing into the still water ahead of the expanding wave packet triggered by the stone. The distinction between group velocity and phase velocity is also well understood by surfers, who know that the largest waves seen out to sea are not the largest when they break upon the shore.

Exercises

8.5.1. Sketch a picture of the solution for the initial value problem in Example 8.13 at times $t = -.1, -.5$, and -1.

♠ 8.5.2. (a) Write down an integral formula for the solution to the dispersive wave equation (8.90) with initial data $u(0,x) = \begin{cases} 1, & 0 < x < 1, \\ 0, & \text{otherwise.} \end{cases}$ (b) Use a computer package to plot your solution at several times and discuss what you observe.

8.5.3. (a) Write down an integral formula for the solution to the initial value problem

$$u_t + u_x + u_{xxx} = 0, \qquad u(0,x) = f(x).$$

(b) Based on the results in Example 8.13, discuss the behavior of the solution to the initial value problem $u(0,x) = e^{-x^2}$ as t increases.

8.5.4. Find the (*i*) dispersion relation, (*ii*) phase velocity, and (*iii*) group velocity for the following partial differential equations. Which are dispersive? (a) $u_t + u_x + u_{xxx} = 0$, (b) $u_t = u_{xxxxx}$, (c) $u_t + u_x - u_{xxt} = 0$, (d) $u_{tt} = c^2 u_{xx}$, (e) $u_{tt} = u_{xx} - u_{xxxx}$.

8.5.5. Find all linear evolution equations for which the group velocity equals the phase velocity. Justify your answer.

8.5.6. Show that the phase velocity is greater than the group velocity if and only if the phase velocity is a decreasing function of k for $k > 0$ and an increasing function of k for $k < 0$. How would you observe this in a physical system?

◇ 8.5.7. (a) *Conservation of Mass*: Prove that $T = u$ is a density associated with a conservation law of the dispersive wave equation (8.90). What is the corresponding flux? Under what conditions is total mass conserved? (b) *Conservation of Energy*: Establish the same result for the energy density $T = u^2$. (c) Is u^3 the density of a conservation law?

◇ 8.5.8. Prove that when $t = \pi p/q$, where p, q are integers, the solution (8.102) is constant on each interval $\pi j/q < x < \pi(j+1)/q$ for integers $j \in \mathbb{Z}$. *Hint*: Use Exercise 6.1.29(d). *Remark*: The proof that the solution is continuous and fractal at irrational times is considerably more difficult, [**90**].

◇ 8.5.9. (a) Find the complex Fourier series representing the fundamental solution $F(t,x;\xi)$ to the periodic initial-boundary value problem (8.100). (b) Prove that at time $t = 2\pi p/q$, where p, q are relatively prime integers, $F(t,x;\xi)$ is a linear combination of delta functions based at the points $\xi + 2\pi j/q$. *Hint*: Use Exercise 6.1.29(c). (c) Let $u(t,x)$ be any solution to (8.100). Prove that $u(2\pi p/q, x)$ is a linear combination of a finite number of translates, $f(x - x_j)$, of the initial data.

The Korteweg–de Vries Equation

The simplest wave model that combines dispersion with nonlinearity is the celebrated *Korteweg–de Vries equation*

$$u_t + u_{xxx} + u\,u_x = 0. \tag{8.112}$$

It was first derived, in 1872, by the French applied mathematician Joseph Boussinesq, [**21**; eq. (30)], [**22**; eqs. (283, 291)], as a model for surface waves on shallow water. Two decades later, it was rediscovered by the Dutch applied mathematician Diederik Korteweg and his student Gustav de Vries, [**65**], and, despite Boussinesq's priority, it is nowadays named after them. In the early 1960s, the American mathematical physicists Martin Kruskal and Norman Zabusky, [**125**], used the Korteweg–de Vries equation as a continuum model for a one-dimensional chain of masses interconnected by nonlinear springs: the Fermi–Pasta–Ulam problem, [**40**]. Numerical experimentation revealed its many remarkable properties, which were soon rigorously established. Their work sparked the rapid development of one of the most remarkable and far-reaching discoveries of the modern era: integrable nonlinear partial differential equations, [**2, 36**].

The most important special solutions to the Korteweg–de Vries equation are the *traveling waves*. We seek solutions

$$u = v(\xi) = v(x - ct), \qquad \text{where} \qquad \xi = x - ct,$$

that have a fixed profile while moving with speed c. By the chain rule,

$$\frac{\partial u}{\partial t} = -\,c\,v'(\xi), \qquad \frac{\partial u}{\partial x} = v'(\xi), \qquad \frac{\partial^3 u}{\partial x^3} = v'''(\xi).$$

Substituting these expressions into the Korteweg–de Vries equation (8.112), we conclude that $v(\xi)$ must satisfy the nonlinear third-order ordinary differential equation

$$v''' + v\,v' - c\,v' = 0. \tag{8.113}$$

Let us further assume that the traveling wave is *localized*, meaning that the solution and its derivatives are vanishingly small at large distances:

$$\lim_{x \to \pm\infty} u(t,x) = \lim_{x \to \pm\infty} \frac{\partial u}{\partial x}(t,x) = \lim_{x \to \pm\infty} \frac{\partial^2 u}{\partial x^2}(t,x) = 0. \tag{8.114}$$

This implies that we should impose the boundary conditions

$$\lim_{\xi \to \pm\infty} v(\xi) = \lim_{\xi \to \pm\infty} v'(\xi) = \lim_{\xi \to \pm\infty} v''(\xi) = 0. \tag{8.115}$$

The ordinary differential equation (8.113) can, in fact, be solved in closed form. First, note that it has the form

$$\frac{d}{d\xi}\left(v'' + \tfrac{1}{2}v^2 - c\,v\right) = 0, \qquad \text{and hence} \qquad v'' + \tfrac{1}{2}v^2 - c\,v = a,$$

where a indicates the constant of integration. The localizing boundary conditions (8.115) imply that $a = 0$. Multiplying the resulting equation by v' allows us to integrate a second time:

$$0 = v'\left(v'' + \tfrac{1}{2}v^2 - c\,v\right) = \frac{d}{d\xi}\left[\,\tfrac{1}{2}\,(v')^2 + \tfrac{1}{6}v^3 - \tfrac{1}{2}c\,v^2\,\right] = 0.$$

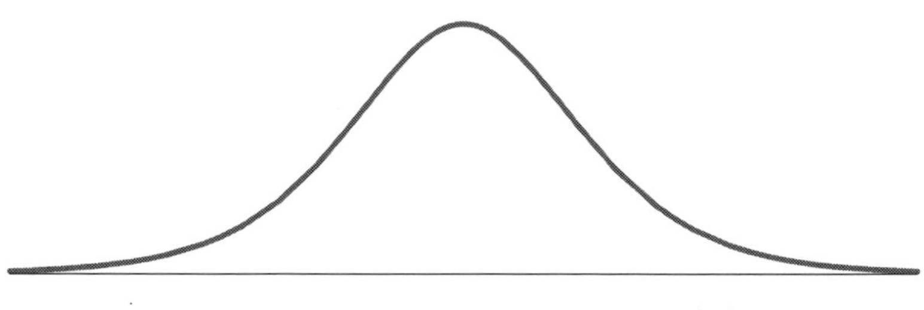

Figure 8.13. Solitary wave/soliton.

Thus,

$$\tfrac{1}{2}\,(v')^2 + \tfrac{1}{6}\,v^3 - \tfrac{1}{2}\,c\,v^2 = b,$$

where b is a second constant of integration, which, again by the boundary conditions (8.115), is also zero. Setting $b = 0$, and solving for v', we conclude that $v(\xi)$ satisfies the autonomous first-order ordinary differential equation

$$\frac{dv}{d\xi} = v\,\sqrt{c - \tfrac{1}{3}\,v}\,,$$

which is integrated by the standard method:

$$\int \frac{dv}{v\,\sqrt{c - \tfrac{1}{3}\,v}} = \xi + \delta,$$

where δ is constant. Consulting a table of integrals, e.g., [**48**], and then solving for v, we conclude that the solution has the form

$$v(\xi) = 3\,c\,\operatorname{sech}^2\!\left(\tfrac{1}{2}\sqrt{c}\,\xi + \delta\right), \tag{8.116}$$

where

$$\operatorname{sech} y = \frac{1}{\cosh y} = \frac{2}{e^y + e^{-y}}$$

is the *hyperbolic secant function*. The solution has the form graphed in Figure 8.13. It is a symmetric, monotone, exponentially decreasing function on either side of its maximum height of $3c$. (Despite its suggestive profile, it is *not* a Gaussian.) The resulting localized traveling-wave solutions to the Korteweg–de Vries equation are thus

$$u(t,x) = 3\,c\,\operatorname{sech}^2\left[\,\tfrac{1}{2}\sqrt{c}\,(x - c\,t) + \delta\,\right], \tag{8.117}$$

where $c > 0$ represents the wave speed — which is necessarily positive, and so all such solutions move to the right — while δ represents an overall phase shift. The amplitude of the wave is three times its speed, while its width is proportional to $1/\sqrt{c}$. Thus, the taller (and narrower) the wave, the faster it moves.

Localized traveling waves are commonly known as *solitary waves*. They were first observed in nature by the British engineer J. Scott Russell, [**104**], who recounts how one was triggered by the sudden motion of a barge along an Edinburgh canal. Scott Russell ended up chasing the propagating wave on horseback for several miles — a physical indication of its stability. Russell's observations were dismissed by his contemporary Airy, who, relying on his linearly dispersive model for surface waves (8.90), claimed that such localized

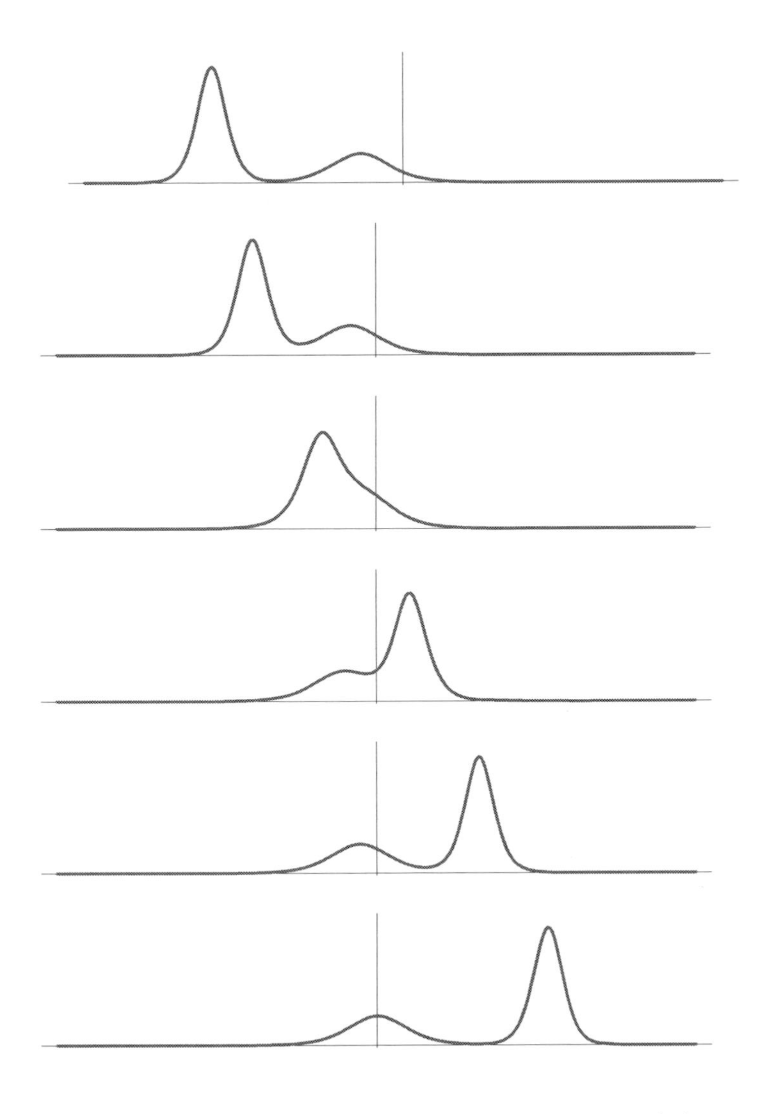

Figure 8.14. Interaction of two solitons.

disturbances could not exist. Much later, Boussinesq derived the proper nonlinear surface wave model (8.112), valid for long waves in shallow water, along with its solitary wave solutions (8.117), thereby fully exonerating Russell's physical observations and insight.

It took almost a century before all the remarkable properties of these solutions came to light. The most striking is how two such solitary waves interact. While linear equations always admit a superposition principle, one cannot naïvely combine two solutions to a nonlinear equation. However, in the case of the Korteweg–de Vries equation, suppose the initial data represent a taller solitary wave to the left of a shorter one. As time evolves, the taller wave will move faster, and eventually catch up to the shorter one. They then experience a complicated nonlinear interaction, as expected. But, remarkably, after a while, they emerge from the interaction unscathed! The smaller wave is now in back and the larger one in front, and both unchanged in speed, amplitude, and profile. They then proceed independently, with the smaller solitary wave lagging farther and farther behind the faster, taller wave. The only effect of their encounter is an overall phase shift, so that the taller wave is a bit behind where it would be if it had not encountered the shorter wave, while the shorter wave is a little ahead of its unhindered position. Figure 8.14 plots a typical such interaction.

Owing to this "particle-like" behavior under interaction, these solutions were given a special name: *soliton*. An explicit formula for a *two-soliton solution* to the Korteweg–de Vries equation can be written in the following form:

$$u(t,x) = 12\,\frac{\partial^2}{\partial x^2}\log\Delta(t,x), \tag{8.118}$$

where

$$\Delta(t,x) = \det\begin{pmatrix} 1+\varepsilon_1(t,x) & \dfrac{2\,b_1}{b_1+b_2}\,\varepsilon_2(t,x) \\ \dfrac{2\,b_2}{b_1+b_2}\,\varepsilon_1(t,x) & 1+\varepsilon_2(t,x) \end{pmatrix}, \tag{8.119}$$

where $0 < b_1 < b_2$, and

$$\varepsilon_j(t,x) = \exp\big[\,b_j(x-b_j^2\,t)+d_j\,\big], \qquad j=1,2. \tag{8.120}$$

The constants $c_j = b_j^2$ represent the wave speeds, while the d_j correspond to phase shifts of the individual solitons. Proving that (8.118) is indeed a solution to the Korteweg–de Vries equation is a straightforward, albeit tedious, exercise in differentiation. In Exercise 8.5.14, the reader is asked to investigate its asymptotic behavior, as $t \to \pm\infty$, and prove that the solution does, indeed, break up into two solitons, having the same profiles, speeds, and amplitudes in both the distant past and future.

A similar dynamic occurs when there are multiple collisions among solitons. Faster solitons catch up to slower ones moving to their right. After the various solitons finish colliding and interacting, they emerge in order, from smallest to largest, each moving at its characteristic speed and becoming more and more separated from its peers. An explicit formula for the n–soliton solution is provided by the same logarithmic derivative (8.118) in which $\Delta(t,x)$ now represents the determinant of an $n\times n$ matrix whose i^{th} diagonal entry is $1+\varepsilon_i(t,x)$, while the off-diagonal (i,j) entry, $i\neq j$, is $\dfrac{2\,b_i}{b_i+b_j}\,\varepsilon_j(t,x)$, using the same formula (8.120) for the ε_j's, and where $0 < b_1 < \;\cdots\; < b_n$ correspond to the n different soliton wave speeds $c_j = b_j^2$. Furthermore, it can be shown that, starting with an *arbitrary* localized initial disturbance $u(0,x) = f(x)$ that decays sufficiently rapidly as $|\,x\,| \to \infty$, the resulting solution eventually emits a finite number of solitons of different heights, moving off at their respective speeds to the right, and so arranged in order from smallest to largest, followed by a small, asymptotically self-similar dispersive tail that gradually disappears.

The source of these highly non-obvious facts and formulas lies beyond the scope of this introductory text. Soon after the initial numerical studies, Gardner, Green, Kruskal, and Miura, [**45**], discovered a profound connection between the solutions to the Korteweg–de Vries equation and the eigenvalues λ of the Sturm–Liouville boundary value problem

$$-\frac{d^2\psi}{dx^2}+6\,u(t,x)\,\psi=\lambda\,\psi, \quad -\infty<x<\infty, \qquad \text{with} \qquad \psi(t,x)\longrightarrow 0 \quad \text{as} \quad |\,x\,|\longrightarrow\infty. \tag{8.121}$$

Their remarkable result is that whenever $u(t,x)$ is a localized solution to the Korteweg–de Vries equation (8.112), the eigenvalues of (8.121) are constant, meaning that they do not vary with the time t, while the continuous spectrum has a very simple temporal evolution. In physical applications of the stationary Schrödinger equation (8.121), in which $u(t,x)$ represents a quantum-mechanical potential, the eigenvalues correspond to bound states, while the continuous spectrum governs its scattering behavior. The solution to the so-called *inverse scattering problem* reconstructs the potential $u(t,x)$ from its spectrum, and can be viewed as a nonlinear version of the Fourier transform, in that it effectively linearizes the Korteweg–de Vries equation and thereby reveals its many remarkable properties. In particular, the eigenvalues are responsible for the preceding determinantal formulae for the multi-soliton solutions, while, when present, the continuous spectrum governs the dispersive tail. See [**2**, **36**] for additional details.

Exercises

8.5.10. Justify the statement that the width of a soliton is proportional to the inverse of the square root of its speed.

8.5.11. Prove that the function (8.116) is a symmetric, monotone, exponentially decreasing function on either side of its maximum height of $3c$.

8.5.12. Let $u(t,x)$ solve the Korteweg–de Vries equation.
(*a*) Show that $U(t,x) = u(t, x - ct) + c$ is also a solution.
(*b*) Give a physical interpretation of this symmetry.

8.5.13. (*a*) Find all scaling symmetries of the Korteweg–de Vries equation.
(*b*) Write down an ansatz for the similarity solutions, and then find the corresponding reduced ordinary differential equation. (Unfortunately, the similarity solutions cannot be written in terms of elementary functions, [**2**].)

♡ 8.5.14. (*a*) Let $u(t,x)$ be the two-soliton solution defined in (8.118). Let $\widetilde{u}(t,\xi) = u(t, \xi + ct)$ represent the solution as viewed in a coordinate frame moving with speed c. Prove that

$$\lim_{t\to\infty} \widetilde{u}(t,\xi) = \begin{cases} 3c_1 \operatorname{sech}^2\left[\frac{1}{2}\sqrt{c_1}\,\xi + \delta_1\right], & c = c_1, \\ 3c_2 \operatorname{sech}^2\left[\frac{1}{2}\sqrt{c_2}\,\xi + \delta_2\right], & c = c_2, \\ 0, & \text{otherwise}, \end{cases}$$

for suitable constants δ_1, δ_2. Explain why this justifies the statement that the solution indeed breaks up into two individual solitons as $t \to \infty$. (*b*) Explain why $\widetilde{u}(t,\xi)$ has a similar limiting behavior as $t \to -\infty$, but with possibly different constants $\widehat{\delta}_1, \widehat{\delta}_2$. (*c*) Use your formulas to discuss how the solitons are affected by the collision.

8.5.15. Let $\alpha, \beta \neq 0$. Find the soliton solutions to the rescaled Korteweg–de Vries equation $u_t + \alpha\, u_{xxx} + \beta\, u\, u_x = 0$. How are their speed, amplitude, and width interrelated?

8.5.16. (*a*) Find the solitary wave solutions to the *modified Korteweg–de Vries equation* $u_t + u_{xxx} + u^2 u_x = 0$. (*b*) Discuss how the amplitude and width of a solitary wave is related to its speed. *Note*: The modified Korteweg–de Vries equation is also integrable, and its solitary wave solutions are solitons, cf. [**36**].

8.5.17. Answer Exercise 8.5.16 for the *Benjamin–Bona–Mahony equation* $u_t - u_{xxt} + u\, u_x = 0$, [**14**]. *Note*: The BBM equation is *not* integrable, and collisions between its solitary waves produce a small, but measurable, inelastic effect, [**1**].

♢ 8.5.18. (*a*) Show that $T_1 = u$ is the density for a conservation law for the Korteweg–de Vries equation. (*b*) Show that $T_2 = u^2$ is also a conserved density. (*c*) Find a conserved density of the form $T_3 = u_x^2 + \mu u^3$ for a suitable constant μ. *Remark*: The Korteweg–de Vries equation in fact has *infinitely many* conservation laws, whose densities depend on higher and higher-order derivatives of the solution, [**76**, **87**]. It was this discovery that unlocked the door to all its remarkable integrability properties, [**2**, **36**].

8.5.19. Find two conservation laws of

(a) the modified Korteweg–de Vries equation $u_t + u_{xxx} + u^2 u_x = 0$;

(b) the Benjamin–Bona–Mahony equation $u_t - u_{xxt} + u\, u_x = 0$.

Chapter 9
A General Framework for Linear Partial Differential Equations

Before pressing on to the higher-dimensional manifestations of the heat, wave, and Laplace/ Poisson equations, it is worth pausing to develop a general, abstract, linear-algebraic framework that underlies many of the linear partial differential equations arising throughout the subject and its applications. The power of mathematical abstraction is that concentrating on the essential features and not being distracted by the at times messy particular details enables one to establish, relatively painlessly, very general results that can be applied throughout the subject and beyond. Each abstract concept has, as its source, an elementary finite-dimensional version valid for linear algebraic systems and matrices, which is then generalized and extended to include linear boundary value problems and then initial-boundary value problems governed by differential equations. All of the abstract definitions and results contained here will be immediately applicable to the boundary and initial value problems of physical interest, and serve to deepen our understanding of the underlying commonalities among systems and solution techniques. Nevertheless, a more applications-oriented reader may prefer to skip ahead to the more concrete developments contained in the following chapters, referring to the background material presented here as necessary.

Most equilibrium systems are modeled as boundary value problems involving a linear differential operator that satisfies the two key conditions of being "self-adjoint" and either "positive definite" or, slightly more generally, "positive semi-definite". So, our first task is to introduce the adjoint of a linear function in general, and, for our specific purposes, a linear differential operator. The adjoint is a far-reaching generalization of the elementary matrix transpose. Its formulation relies on the specification of inner products on both the domain and target spaces of the operator, and, when one is dealing with linear differential operators, the imposition of suitable homogeneous boundary conditions on the spaces of allowable functions. In applications, the relevant inner products are typically dictated by the underlying physics. One immediate application of the adjoint is the Fredholm Alternative, which delineates the constraints required for the existence of solutions to linear systems, including linear boundary value problems.

A linear operator that equals its own adjoint is called self-adjoint. The simplest example is the linear function defined by a symmetric matrix. The most important subclasses are the positive definite and positive semi-definite operators, which are the natural analogues of positive (semi-)definite matrices. We will learn how to construct self-adjoint positive (semi-)definite operators in a canonical manner. Almost all of the linear differential operators studied in this text, including the Laplacian, are, when subject to suitable boundary conditions, self-adjoint and either positive definite or positive semi-definite. The key distinction is that positive definite linear systems and boundary value problems admit unique solutions, whereas in the positive semi-definite case, the solution either does not exist, since the Fredholm constraints are not satisfied, or, when it exists, is not unique. In their dynamical manifestations, positive definite operators induce stable vibrational systems, whereas the positive semi-definite cases contain unstable modes that can lead to disastrous physical consequences.

A critically important fact is that the solution to a positive definite linear system can be characterized by a minimization principle, provided by a certain quadratic function or, in the infinite-dimensional function-space version, quadratic functional. In physical contexts, the function(al) often represents the potential energy of the system, and the solution minimizes said energy among all possible configurations satisfying the prescribed boundary conditions, thereby quantifying the maxim that Nature is inherently conservative and seeks to minimize energy. In mathematics, minimization principles underlie advanced functional-analytic methods used to establish existence theorems, as well as the finite element numerical schemes to be presented in Chapter 10.

For linear dynamical systems like the heat and wave equations, separation of variables leads to an eigenvalue problem for the linear differential operator governing the corresponding equilibrium system. In the simple one-dimensional cases discussed in Chapter 4, the eigenfunctions are trigonometric, producing the classical Fourier expansions for the solutions. The effectuality of the Fourier method relies on the eigenfunctions' orthogonality, and we already hinted that this is no accident. Rather, it is a consequence of their status as the eigenfunctions of a self-adjoint linear operator. Not only are such eigenfunctions automatically mutually orthogonal with respect to the underlying inner product, the eigenvalues are necessarily real and, when the operator is positive definite, also positive.

Orthogonality underlies the Fourier-like expansion of quite general functions as series in the eigenfunctions, whose convergence, in general, requires that the eigenfunctions form a complete system. For positive definite boundary value problems on bounded domains, we will establish completeness by combining the eigenfunction expansion for the associated Green's function with a basic minimization principle for the eigenvalues based on the Rayleigh quotient. On the other hand, problems on unbounded domains do not typically admit complete systems of eigenfunctions and require the more advanced analytical concepts of continuous spectrum and generalized Fourier transforms that lie beyond the scope of this text.

The chapter concludes by describing a general framework for dynamics that produces time-dependent series solutions, in terms of the eigenfunctions of the underlying equilibrium operator, for diffusion equations, vibration equations, and quantum-mechanical systems. The final two chapters will then specialize these general theories and constructions to analyze initial-boundary value problems for the two- and three-dimensional heat, wave, and Schrödinger equations in simple geometries. More advanced developments and further applications can be found in higher-level texts, including [**35, 38, 44, 61, 99**].

9.1 Adjoints

Our starting point is a linear operator

$$L : U \longrightarrow V \tag{9.1}$$

that maps a vector space U to another vector space V. For most of the development, we deal with real vector spaces, although the final discussion of the Schrödinger equation requires us to venture into the complex realm. For our purposes, L represents a linear differential operator, and the elements of the domain space U and the target space V are suitable scalar- or vector-valued functions. In elastomechanics, the elements of U are displacements of a deformable body, while the elements of V are the associated strains. In electromagnetism and gravitation, elements of U represent potentials, and elements of V are electric or magnetic or gravitational fields. In thermodynamics, U contains temperature distributions, while V contains temperature gradients. In fluid mechanics, U is the space of potential functions, while V is the space of fluid velocities. And so on.

The abstract definition of the adjoint of a linear operator relies on an inner product structure on both its domain and target spaces. We distinguish the inner products on U and V (which may be different even when U and V happen to be the same vector space) by using a single angle bracket

$$\langle\, u\,,\widetilde{u}\,\rangle \quad \text{to denote the inner product between } u,\widetilde{u}\in U,$$

and a double angle bracket

$$\langle\!\langle\, v\,,\widetilde{v}\,\rangle\!\rangle \quad \text{to denote the inner product between } v,\widetilde{v}\in V.$$

In applications, the appropriate inner products are often based on the underlying physics.

Definition 9.1. Let U, V be inner product spaces, and let $L\colon U \to V$ be a linear operator. The *adjoint* of L is the unique linear operator $L^*\colon V \to U$ that satisfies

$$\langle\!\langle\, L[u]\,,v\,\rangle\!\rangle = \langle\, u\,, L^*[v]\,\rangle \qquad \text{for all} \qquad u\in U, \quad v\in V. \tag{9.2}$$

Observe that the adjoint goes in the *reverse* direction, that is, from V back to U. To master the definition, let us first look at the finite-dimensional case.

Example 9.2. According to Theorem B.33, every linear function $L\colon \mathbb{R}^n \to \mathbb{R}^m$ is given by matrix multiplication, so that $L[\mathbf{u}] = A\mathbf{u}$ for $\mathbf{u}\in\mathbb{R}^n$, where A is an $m\times n$ matrix. The adjoint function $L^*\colon\mathbb{R}^m\to\mathbb{R}^n$ is also linear, so it is also represented by matrix multiplication, $L^*[\mathbf{v}] = A^*\mathbf{v}$ for $\mathbf{v}\in\mathbb{R}^m$, by an $n\times m$ matrix A^*.

Suppose first that we impose the ordinary Euclidean dot products

$$\langle\,\mathbf{u}\,,\widetilde{\mathbf{u}}\,\rangle = \mathbf{u}\cdot\widetilde{\mathbf{u}} = \mathbf{u}^T\widetilde{\mathbf{u}}, \qquad \mathbf{u},\widetilde{\mathbf{u}}\in\mathbb{R}^n, \qquad\qquad \langle\!\langle\,\mathbf{v}\,,\widetilde{\mathbf{v}}\,\rangle\!\rangle = \mathbf{v}\cdot\widetilde{\mathbf{v}} = \mathbf{v}^T\widetilde{\mathbf{v}}, \qquad \mathbf{v},\widetilde{\mathbf{v}}\in\mathbb{R}^m,$$

as our inner products on both $\mathbb{R}^n$ and $\mathbb{R}^m$. Evaluation of both sides of the adjoint identity (9.2) yields

$$\begin{aligned}\langle\!\langle\, L[\mathbf{u}]\,,\mathbf{v}\,\rangle\!\rangle &= \langle\!\langle\, A\mathbf{u}\,,\mathbf{v}\,\rangle\!\rangle = (A\mathbf{u})^T\mathbf{v} = \mathbf{u}^T A^T\,\mathbf{v},\\ \langle\,\mathbf{u}\,, L^*[\mathbf{v}]\,\rangle &= \langle\,\mathbf{u}\,, A^*\,\mathbf{v}\,\rangle = \mathbf{u}^T A^*\,\mathbf{v}.\end{aligned} \tag{9.3}$$

Since these expressions must agree for all $\mathbf{u},\mathbf{v}$, we conclude (see Exercise 9.1.6) that the matrix A^* representing L^* is equal to the transposed matrix A^T. Therefore, *the adjoint of a matrix with respect to the Euclidean dot product is its transpose*: $A^* = A^T$. So one can regard the adjoint as a vast generalization of the elementary operation of transposing a matrix.

More generally, suppose we take weighted inner products on the domain and target spaces:

$$\langle\,\mathbf{u}\,,\widetilde{\mathbf{u}}\,\rangle = \mathbf{u}^T M\,\widetilde{\mathbf{u}}, \qquad \mathbf{u},\widetilde{\mathbf{u}}\in\mathbb{R}^n, \qquad \langle\!\langle\,\mathbf{v}\,,\widetilde{\mathbf{v}}\,\rangle\!\rangle = \mathbf{v}^T C\,\widetilde{\mathbf{v}}, \qquad \mathbf{v},\widetilde{\mathbf{v}}\in\mathbb{R}^m, \tag{9.4}$$

where M and C are symmetric, positive definite matrices of respective sizes $n \times n$ and $m \times m$, cf. Proposition B.13. Then, repeating the previous calculation (9.3), we find

$$\begin{aligned} \langle\!\langle\, L[\,\mathbf{u}\,]\,,\mathbf{v}\,\rangle\!\rangle &= \langle\!\langle\, A\,\mathbf{u}\,,\mathbf{v}\,\rangle\!\rangle = (A\,\mathbf{u})^T C\,\mathbf{v} = \mathbf{u}^T A^T C\,\mathbf{v}, \\ \langle\, \mathbf{u}\,, L^*[\,\mathbf{v}\,]\,\rangle &= \langle\, \mathbf{u}\,, A^*\,\mathbf{v}\,\rangle = \mathbf{u}^T M A^*\,\mathbf{v}. \end{aligned} \tag{9.5}$$

Comparing these expressions, we conclude that the *weighted adjoint matrix* is

$$A^* = M^{-1} A^T C. \tag{9.6}$$

Therefore, the adjoint does indeed depend on which inner products are being used on both the domain and target spaces.

Differential Operators

For applications to linear differential equations, our attention is focused on adjoints of differential operators defined on infinite-dimensional function spaces. Let us begin with the simplest example.

Example 9.3. Consider the derivative $v = D[u] = du/dx$, which defines a linear operator $D : U \to V$ mapping a vector space U of differentiable functions $u(x)$ to a vector space containing their derivatives $v(x) = u'(x)$. We assume that the functions in question are defined on a fixed bounded interval $a \leq x \leq b$.

In order to compute its adjoint, we need to impose inner products on both the domain space U and the target space V. The simplest context is to adopt the standard L^2 inner product on both:

$$\langle\, u\,, \widetilde{u}\,\rangle = \int_a^b u(x)\,\widetilde{u}(x)\,dx, \qquad\qquad \langle\!\langle\, v\,, \widetilde{v}\,\rangle\!\rangle = \int_a^b v(x)\,\widetilde{v}(x)\,dx. \tag{9.7}$$

According to the defining equation (9.2), the adjoint operator $D^* : V \to U$ must satisfy the inner product identity

$$\langle\!\langle\, D[\,u\,]\,, v\,\rangle\!\rangle = \langle\, u\,, D^*[\,v\,]\,\rangle \qquad \text{for all} \qquad u \in U, \quad v \in V. \tag{9.8}$$

First, we compute the left-hand side:

$$\langle\!\langle\, D[\,u\,]\,, v\,\rangle\!\rangle = \left\langle\!\!\left\langle\, \frac{du}{dx}\,, v\,\right\rangle\!\!\right\rangle = \int_a^b \frac{du}{dx}\, v\; dx. \tag{9.9}$$

On the other hand, the right-hand side should equal

$$\langle\, u\,, D^*[\,v\,]\,\rangle = \int_a^b u\, D^*[\,v\,]\,dx. \tag{9.10}$$

Now, in the latter integral, we see u multiplying the result of applying the linear operator D^* to v. To identify this integrand with that in (9.9), we need to somehow remove the derivative from u. The secret is *integration by parts*, which allows us to rewrite the first integral in the form

$$\int_a^b \frac{du}{dx}\, v\,dx = \big[\, u(b)\, v(b) - u(a)\, v(a)\,\big] - \int_a^b u\, \frac{dv}{dx}\, dx. \tag{9.11}$$

Ignoring the two boundary terms for a moment, we observe that the remaining integral has the form of an inner product

$$- \int_a^b u \, \frac{dv}{dx} \, dx = \int_a^b u \left[- \frac{dv}{dx} \right] dx = \left\langle u \, , - \frac{dv}{dx} \right\rangle = \langle \, u \, , - D[\, v \,] \, \rangle. \tag{9.12}$$

Equating (9.9) and (9.12), we deduce that

$$\langle\!\langle \, D[\, u \,] \, , v \, \rangle\!\rangle = \left\langle\!\!\left\langle \, \frac{du}{dx} \, , v \, \right\rangle\!\!\right\rangle = \left\langle u \, , - \frac{dv}{dx} \right\rangle = \langle \, u \, , - D[\, v \,] \, \rangle.$$

Thus, to satisfy the adjoint equation (9.8), we must have

$$\langle \, u \, , D^*[\, v \,] \, \rangle = \langle \, u \, , - D[\, v \,] \, \rangle \qquad \text{for all} \qquad u \in U, \quad v \in V,$$

and so the adjoint of the derivative operator is its negative:

$$D^* = - D. \tag{9.13}$$

However, the preceding argument is valid *only* if the boundary terms in the integration by parts formula (9.11) vanish:

$$u(b) \, v(b) - u(a) \, v(a) = 0, \tag{9.14}$$

which necessitates imposing suitable boundary conditions on the functions u and v. For example, imposing Dirichlet boundary conditions

$$u(a) = 0, \qquad \qquad u(b) = 0, \tag{9.15}$$

will ensure that (9.14) holds, and therefore validates (9.13). In this case, the domain space of $D \colon U \to V$ is the vector space

$$U = \{ \, u(x) \mid u(a) = u(b) = 0 \, \},$$

while no boundary conditions need be imposed on the functions $v(x)$ in the target space V. An evident alternative is to require that $v(a) = v(b) = 0$. In this case, the target space

$$V = \{ \, v(x) \mid v(a) = v(b) = 0 \, \}$$

consists of all functions that vanish at the endpoints. Since the derivative $D \colon U \to V$ is required to map a function $u(x) \in U$ to an *allowable* function $v(x) \in V$, the domain space now consists of functions satisfying the Neumann boundary conditions:

$$U = \{ \, u(x) \mid u'(a) = u'(b) = 0 \, \}.$$

These are evidently not the only two possibilities. Let us list the most important combinations of boundary conditions that imply the vanishing of the boundary terms (9.14), and so ensure the validity of the adjoint equation (9.13):

(*a*) Dirichlet boundary conditions: $u(a) = u(b) = 0$.

(*b*) Mixed boundary conditions: $u(a) = u'(b) = 0$, or $u'(a) = u(b) = 0$.

(*c*) Neumann boundary conditions: $u'(a) = u'(b) = 0$.

(*d*) Periodic boundary conditions: $u(a) = u(b)$ and $u'(a) = u'(b)$.

In all cases, the boundary conditions impose restrictions on the domain space U and, in cases (*b*–*d*) when we are identifying $v(x) = u'(x)$, the target space V also.

Remark: In the preceding discussion, we were purposely vague about the required differentiability of the functions. In finite dimensions, every linear function $L: \mathbb{R}^n \to \mathbb{R}^m$ is given by matrix multiplication $L[u] = A\mathbf{u}$, and hence is defined on all of the underlying vector space $\mathbb{R}^n$. Linear operators on infinite-dimensional function spaces are typically not defined on all possible functions. For example, the derivative operator $L = D: U \to V$ requires the function $u \in U$ to be differentiable. However, the target function $v = D[u] = u'$ is not necessarily as smooth, and so may belong to a different function space; for instance if $u \in \mathrm{C}^1[a,b]$, then $v = u' \in \mathrm{C}^0[a,b]$. On the other hand, the adjoint $D^* = -D$ is defined only on differentiable functions v, so if $v \in \mathrm{C}^1[a,b]$, then $u = -v' \in \mathrm{C}^0[a,b]$. Keeping a detailed account of the various smoothness requirements quickly becomes distracting.

To circumvent this technical annoyance, we will always deal with a fixed class of functions, e.g., continuous functions or, more generally, L^2 functions, that are constrained only by the imposed boundary conditions. When we write $L: U \to V$, we allow the possibility that the linear operator L may be defined only on a "dense" subspace of the domain space U. For instance, we will write $D: U \to V$ with $U = V = \mathrm{C}^0[a,b]$, even though $D[u] = u' \in V$ only if u belongs to the dense subspace $\mathrm{C}^1[a,b] \subset U = \mathrm{C}^0[a,b]$. Similarly, $D^*: V \to U$ is also defined only on the dense subspace $\mathrm{C}^1[a,b] \subset V = \mathrm{C}^0[a,b]$. The term *dense* refers to the fact that any continuous function in the full space $U = \mathrm{C}^0[a,b]$ can be arbitrarily closely approximated in norm by a continuously differentiable function in the subspace $\mathrm{C}^1[a,b]$. Or, to put it another way, given a continuous function $u \in \mathrm{C}^0[a,b]$, there exists a sequence of continuously differentiable functions $u_1, u_2, u_3, \ldots \in \mathrm{C}^1[a,b]$ such that $\| u_k - u \| \to 0$ as $k \to \infty$. A similar density result can be proved for $U = \mathrm{L}^2[a,b]$; see [**37**, **96**, **98**] for details.

Warning: In more advanced treatments, our notion of adjoint is usually called the *formal adjoint*. A true adjoint requires more subtle technical hypotheses on the operator and its domain, cf. [**95**].

Example 9.4. Let us recompute the adjoint of the derivative operator $D: U \to V$, this time with respect to the weighted L^2 inner products

$$\langle u, \widetilde{u} \rangle = \int_a^b u(x)\, \widetilde{u}(x)\, \rho(x)\, dx, \qquad \langle\!\langle v, \widetilde{v} \rangle\!\rangle = \int_a^b v(x)\, \widetilde{v}(x)\, \kappa(x)\, dx, \tag{9.16}$$

where $\rho(x) > 0$ and $\kappa(x) > 0$ are strictly positive functions that, physically, might represent the density and stiffness of a nonuniform bar. Now we need to compare

$$\langle\!\langle D[u], v \rangle\!\rangle = \int_a^b \frac{du}{dx}\, v(x)\, \kappa(x)\, dx, \qquad \text{with} \qquad \langle u, D^*[v] \rangle = \int_a^b u(x)\, D^*[v]\, \rho(x)\, dx.$$

Integrating the first expression by parts, we obtain

$$\begin{aligned} \int_a^b \frac{du}{dx}\, v\, \kappa\, dx &= \big[\, u(b)\, v(b)\, \kappa(b) - u(a)\, v(a)\, \kappa(a) \,\big] - \int_a^b u\, \frac{d(\kappa\, v)}{dx}\, dx \\ &= \int_a^b u \left(-\frac{1}{\rho}\, \frac{d(\kappa\, v)}{dx} \right) \rho\, dx, \end{aligned} \tag{9.17}$$

provided that we select our boundary conditions so that

$$u(b)\, v(b)\, \kappa(b) - u(a)\, v(a)\, \kappa(a) = 0. \tag{9.18}$$

As you can check, this follows from any of the listed boundary conditions: Dirichlet, Neumann, or mixed, as well as periodic, provided $\kappa(a) = \kappa(b)$. We conclude that, in such situations, the weighted adjoint of the derivative operator D is the differential operator

$$D^*[\,v(x)\,] = -\,\frac{1}{\rho(x)}\,\frac{d}{dx}\,\big[\,\kappa(x)\,v(x)\,\big] = -\,\frac{\kappa(x)}{\rho(x)}\,\frac{dv}{dx} - \frac{\kappa'(x)}{\rho(x)}\,v(x). \tag{9.19}$$

As with matrices, the adjoint of a differential operator depends crucially on the specification of inner products.

The following basic results are left as exercises for the reader. The first generalizes the fact that transposing a transposed matrix reverts to the original.

Proposition 9.5. *The adjoint of the adjoint is the original operator*: $L = (L^*)^*$.

The second generalizes the fact that the transpose of the product of two matrices is the product of the transposes, but in the reverse order.

Proposition 9.6. *If* $L\colon U \to V$ *and* $M\colon V \to W$ *are linear operators on inner product spaces, with* $L^*\colon V \to U$ *and* $M^*\colon W \to V$ *their respective adjoints, then the composite linear operator* $M \circ L\colon U \to W$ *has adjoint* $(M \circ L)^* = L^* \circ M^*\colon W \to U$.

Example 9.7. Let us compute the adjoint of the second derivative operator $D^2 = D \circ D$ with respect to the standard L^2 inner products on both the domain and target spaces. According to Proposition 9.6 and equation (9.13), at least on a formal level,

$$(D^2)^* = D^* \circ D^* = (-\,D) \circ (-\,D) = D^2, \tag{9.20}$$

and hence D^2 equals its own adjoint. However, the validity of (9.13) required that the functions in the domain and target spaces of both D's satisfy appropriate boundary conditions. For example, the domain of the first $D\colon U \to V$ could be $U = \{\,u(x)\,|\,u(a) = u(b) = 0\,\}$, while its target space V is unconstrained; the second D could then map V to $W = \{\,w(x)\,|\,w(a) = w(b) = 0\,\}$, which will thus also require that $u''(a) = u''(b) = 0$ in order that $D^2 = D \circ D$ map U to W. Another option would be to impose Neumann conditions on the first D, with $U = \{\,u'(a) = u'(b) = 0\}$ and thus $V = \{\,v(a) = v(b) = 0\,\}$, while W remains unconstrained. Under either these or other suitably compatible constraints, both adjoint identifications $D^* = -\,D$ are valid, thus justifying (9.20). Keep in mind that, according to our earlier remark, the differentiation operators are, in fact, defined only on the dense subspaces containing sufficiently smooth functions.

Higher–Dimensional Operators

The most natural multi-dimensional analogue of the derivative is the *gradient* operator, which, on a two-dimensional space, is given by

$$\nabla u = \operatorname{grad} u = \begin{pmatrix} \partial u/\partial x \\ \partial u/\partial y \end{pmatrix}.$$

The gradient ∇ defines a linear operator that takes a scalar-valued function $u(x,y)$ to the vector-valued function consisting of its two first-order partial derivatives. Thus, the domain space U consists of scalar-valued functions $u(x,y)$, or *scalar fields*, defined for $(x,y) \in \Omega$, where the domain $\Omega \subset \mathbb{R}^2$ is assumed to be both bounded and connected, and with a nice boundary $\partial\Omega$. (Similar considerations apply to three- and even higher-dimensional problems.) The target space V consists of vector-valued functions, or *vector fields*, $\mathbf{v}(x,y) = (\, v_1(x,y), v_2(x,y)\,)^T$ defined on Ω. As in the preceding subsection, the gradient operator $\nabla\colon U \to V$ is well defined only on the dense subspace $\mathrm{C}^1(\Omega) \subset U$ consisting of continuously differentiable scalar fields.

In accordance with the general Definition 9.1, the adjoint of the gradient must go in the reverse direction,

$$\nabla^*\colon V \longrightarrow U,$$

mapping a vector field $\mathbf{v}(x,y)$ to a scalar field $w(x,y) = \nabla^*\mathbf{v}$. The defining equation (9.2) for the adjoint, namely

$$\langle\!\langle\, \nabla u\,, \mathbf{v}\,\rangle\!\rangle = \langle\, u\,, \nabla^*\mathbf{v}\,\rangle, \tag{9.21}$$

relies on the choice of inner products on the two vector spaces. Let us start with the L^2 *inner product* between scalar fields:

$$\langle\, u\,, \widetilde{u}\,\rangle = \iint_\Omega u(x,y)\, \widetilde{u}(x,y)\, dx\, dy. \tag{9.22}$$

Similarly, the L^2 inner product between vector fields defined on Ω is obtained by integrating their usual dot product:

$$\langle\!\langle\, \mathbf{v}\,, \widetilde{\mathbf{v}}\,\rangle\!\rangle = \iint_\Omega \mathbf{v}(x,y)\cdot \widetilde{\mathbf{v}}(x,y)\, dx\, dy = \iint_\Omega \big[\, v_1(x,y)\,\widetilde{v}_1(x,y) + v_2(x,y)\,\widetilde{v}_2(x,y)\,\big]\, dx\, dy. \tag{9.23}$$

The adjoint identity (9.21) is supposed to hold for all appropriate scalar fields u and vector fields $\mathbf{v}$. For the L^2 inner products (9.22, 23), the two sides of the identity read

$$\begin{aligned}
\langle\!\langle\, \nabla u\,, \mathbf{v}\,\rangle\!\rangle &= \iint_\Omega \nabla u\cdot \mathbf{v}\, dx\, dy = \iint_\Omega \left(\frac{\partial u}{\partial x}\, v_1 + \frac{\partial u}{\partial y}\, v_2 \right) dx\, dy,\\
\langle\, u\,, \nabla^*\mathbf{v}\,\rangle &= \iint_\Omega u\, \nabla^*\mathbf{v}\, dx\, dy.
\end{aligned}$$

Thus, to compare these two double integrals, we must somehow remove the derivatives from the scalar field u. As in the one-dimensional computation (9.8), the mechanism is an *integration by parts* formula for double integrals:

$$\iint_\Omega \nabla u\cdot \mathbf{v}\, dx\, dy = \oint_{\partial\Omega} u\, (\mathbf{v}\cdot \mathbf{n})\, ds \;-\; \iint_\Omega u\, (\nabla\cdot \mathbf{v})\, dx\, dy, \tag{9.24}$$

which was already noted in (6.83). The left-hand side is just $\langle\!\langle\, \nabla u\,, \mathbf{v}\,\rangle\!\rangle$. If the boundary line integral vanishes,

$$\oint_{\partial\Omega} u\, (\mathbf{v}\cdot \mathbf{n})\, ds = 0, \tag{9.25}$$

then the right-hand side of formula (9.24) reduces to

$$-\iint_\Omega u\, (\nabla\cdot \mathbf{v})\, dx\, dy = -\,\langle\, u\,, \nabla\cdot \mathbf{v}\,\rangle = \langle\, u\,, -\,\nabla\cdot \mathbf{v}\,\rangle.$$

Therefore, subject to the boundary constraint (9.25), we deduce the L^2 inner product identity

$$\langle\!\langle\, \nabla u\,, \mathbf{v}\,\rangle\!\rangle = \langle\, u\,, -\,\nabla\cdot \mathbf{v}\,\rangle, \tag{9.26}$$

which implies that the L^2 adjoint of the gradient operator is minus the *divergence operator*:

$$\nabla^* \mathbf{v} = -\nabla \cdot \mathbf{v}. \tag{9.27}$$

The vanishing of the boundary integral (9.25) will be ensured by the imposition of suitable homogeneous boundary conditions on the scalar field u and/or the vector field $\mathbf{v}$. Clearly the line integral will vanish if either $u = 0$ or $\mathbf{v} \cdot \mathbf{n} = 0$ at each point on the boundary. These possibilities lead immediately to the three principal types of (homogeneous) boundary conditions. The first are the *Dirichlet boundary conditions*, which require

$$u = 0 \qquad \text{on} \qquad \partial\Omega. \tag{9.28}$$

Alternatively, we can set

$$\mathbf{v} \cdot \mathbf{n} = 0 \qquad \text{on} \qquad \partial\Omega, \tag{9.29}$$

which requires that $\mathbf{v}$ be everywhere tangent to the boundary. Since ∇ must map the scalar field $u \in U$ to an admissible vector field $\mathbf{v} = \nabla u \in V$, the boundary condition (9.29) requires that u satisfy the homogeneous *Neumann boundary conditions*

$$\frac{\partial u}{\partial \mathbf{n}} = \nabla u \cdot \mathbf{n} = 0 \qquad \text{on} \qquad \partial\Omega. \tag{9.30}$$

One can evidently also mix the boundary conditions, imposing Dirichlet conditions on part of the boundary and Neumann conditions on the complementary part:

$$u = 0 \quad \text{on} \quad D \subset \partial\Omega, \qquad \mathbf{v} \cdot \mathbf{n} = \frac{\partial u}{\partial \mathbf{n}} = 0 \quad \text{on} \quad N = \partial\Omega \setminus D, \tag{9.31}$$

with neither D nor N empty.

More generally, when modeling deflections of nonuniform membranes, heat flow through heterogeneous media, and similar physical equilibria, we replace the L^2 inner product between scalar and vector fields (9.23) by suitably weighted versions[†]

$$\begin{aligned} \langle u, \widetilde{u} \rangle &= \iint_\Omega u(x,y)\, \widetilde{u}(x,y)\, \rho(x,y)\, dx\, dy, \\ \langle\!\langle \mathbf{v}, \widetilde{\mathbf{v}} \rangle\!\rangle &= \iint_\Omega \big[v_1(x,y)\, \widetilde{v}_1(x,y)\, \kappa_1(x,y) + v_2(x,y)\, \widetilde{v}_2(x,y)\, \kappa_2(x,y) \big]\, dx\, dy, \end{aligned} \tag{9.32}$$

in which $\rho(x,y), \kappa_1(x,y), \kappa_2(x,y) > 0$ are strictly positive functions for $(x,y) \in \Omega$. In applications, ρ represents a density, while κ_1, κ_2 represent stiffnesses or thermal conductivities. To compute the weighted adjoint of the gradient operator, we apply a similar integration by parts argument based on (6.83):

$$\begin{aligned} \langle\!\langle \nabla u, \mathbf{v} \rangle\!\rangle &= \iint_\Omega \left(\kappa_1 v_1 \frac{\partial u}{\partial x} + \kappa_2 v_2 \frac{\partial u}{\partial y} \right) dx\, dy \\ &= \oint_{\partial\Omega} u \big(-\kappa_2 v_2\, dx + \kappa_1 v_1\, dy \big) - \iint_\Omega u \left(\frac{\partial(\kappa_1 v_1)}{\partial x} + \frac{\partial(\kappa_2 v_2)}{\partial y} \right) dx\, dy \\ &= \iint_\Omega u \left[-\frac{1}{\rho} \left(\frac{\partial(\kappa_1 v_1)}{\partial x} + \frac{\partial(\kappa_2 v_2)}{\partial y} \right) \right] \rho\, dx\, dy, \end{aligned} \tag{9.33}$$

[†] Exercise 9.2.14 treats an even more general pair of inner products.

provided the boundary integral vanishes. Equating the left-hand side to

$$\langle u , \nabla^* \mathbf{v} \rangle = \iint_\Omega u \, (\nabla^* \mathbf{v}) \, \rho \, dx \, dy,$$

we deduce that the adjoint of the gradient operator with respect to the weighted inner products (9.32) is minus the "weighted divergence operator":

$$\nabla^* \mathbf{v} = -\frac{1}{\rho} \left(\frac{\partial(\kappa_1 v_1)}{\partial x} + \frac{\partial(\kappa_2 v_2)}{\partial y} \right) = -\frac{\kappa_1}{\rho} \frac{\partial v_1}{\partial x} - \frac{\kappa_2}{\rho} \frac{\partial v_2}{\partial y} - \frac{1}{\rho} \frac{\partial \kappa_1}{\partial x} \, v_1 - \frac{1}{\rho} \frac{\partial \kappa_2}{\partial y} \, v_2. \tag{9.34}$$

The vanishing of the boundary integral,

$$0 = \oint_{\partial\Omega} u \big(-\kappa_2 v_2 \, dx + \kappa_1 v_1 \, dy \big) = \oint_{\partial\Omega} u \, \widetilde{\mathbf{v}} \cdot \mathbf{n} \, ds, \qquad \text{where} \qquad \widetilde{\mathbf{v}} = \begin{pmatrix} \kappa_1 v_1 \\ \kappa_2 v_2 \end{pmatrix},$$

is ensured if either $u = 0$ or $\widetilde{\mathbf{v}} \cdot \mathbf{n} = 0$ on $\partial\Omega$. The former is the usual homogeneous Dirichlet condition, but the latter is a "weighted" version of the homogeneous Neumann boundary condition, requiring that $\widetilde{\nabla} u \cdot \mathbf{n} = 0$ on the boundary, where $\widetilde{\nabla} u = \big(\kappa_1 u_x, \kappa_2 u_y \big)^T$ represents a "weighted normal flux vector".

Example 9.8. Let us compute the adjoint of the second-order Laplacian operator $\Delta = \partial^2/\partial x^2 + \partial^2/\partial y^2$ with respect to the L^2 inner products on both its domain and target spaces. The computation is a simple consequence of the double integral identity (6.88), which we rewrite as

$$\langle \Delta u , v \rangle = \iint_\Omega v \, \Delta u \, dx \, dy = \oint_{\partial\Omega} \left(u \, \frac{\partial v}{\partial \mathbf{n}} - v \, \frac{\partial u}{\partial \mathbf{n}} \right) ds \; + \iint_\Omega u \, \Delta v \, dx \, dy = \langle u , \Delta v \rangle.$$

Thus, provided the boundary integral vanishes, we can conclude that the Laplacian equals its own adjoint: $\Delta^* = \Delta$. This is assured when $u \, \partial v / \partial \mathbf{n} = v \, \partial u / \partial \mathbf{n}$ at each point in $\partial\Omega$. For example, the adjoint computation is valid if either $u = v = 0$ or $\partial u / \partial \mathbf{n} = \partial v / \partial \mathbf{n} = 0$ at every point of the boundary of the domain. Keep in mind that if we require $v = 0$ on some or all of $\partial\Omega$, then this imposes the condition $\Delta u = 0$ there in order that Δ map u to an admissible v; similar considerations apply when $\partial v / \partial \mathbf{n} = 0$.

Exercises

9.1.1. Choose one from the following list of inner products on $\mathbb{R}^2$. Then find the adjoint of $A = \begin{pmatrix} 1 & 2 \\ -1 & 3 \end{pmatrix}$ when your inner product is used on both its domain and target space.
(a) The Euclidean dot product; (b) the weighted inner product $\langle \mathbf{v} , \mathbf{w} \rangle = 2 v_1 w_1 + 3 v_2 w_2$;
(c) the inner product $\langle \mathbf{v} , \mathbf{w} \rangle = \mathbf{v}^T C \mathbf{w}$ defined by the symmetric positive definite matrix $C = \begin{pmatrix} 2 & -1 \\ -1 & 4 \end{pmatrix}$.

9.1.2. From the list in Exercise 9.1.1, choose a different inner product on the domain and the target space, and then determine the adjoint of the matrix A.

9.1.3. Choose one from the following list of inner products on $\mathbb{R}^3$ for both the domain and target space, and find the adjoint of $A = \begin{pmatrix} 1 & 1 & 0 \\ -1 & 0 & 1 \\ 0 & -1 & 2 \end{pmatrix}$. (*a*) The Euclidean dot product on $\mathbb{R}^3$; (*b*) the weighted inner product $\langle \mathbf{v}, \mathbf{w} \rangle = v_1 w_1 + 2 v_2 w_2 + 3 v_3 w_3$; (*c*) the inner product $\langle \mathbf{v}, \mathbf{w} \rangle = \mathbf{v}^T C \mathbf{w}$ defined by the symmetric positive definite matrix $C = \begin{pmatrix} 2 & 1 & 0 \\ 1 & 2 & 1 \\ 0 & 1 & 2 \end{pmatrix}$.

9.1.4. From the list in Exercise 9.1.3, choose different inner products on the domain and target space, and then compute the adjoint of the matrix A.

9.1.5. Choose an inner product on $\mathbb{R}^2$ from the list in Exercise 9.1.1 and an inner product on $\mathbb{R}^3$ from the list in Exercise 9.1.3, and then compute the adjoint of $A = \begin{pmatrix} 1 & 3 \\ 0 & 2 \\ -1 & 1 \end{pmatrix}$.

♢ 9.1.6. (*a*) Let C be an $m \times n$ matrix. Suppose $\mathbf{u}^T C \mathbf{v} = 0$ for all $\mathbf{u} \in \mathbb{R}^m$ and $\mathbf{v} \in \mathbb{R}^n$. Prove that $C = \mathrm{O}$ must be the zero matrix. (*b*) Let A, B be $m \times n$ matrices such that $\mathbf{u}^T A \mathbf{v} = \mathbf{u}^T B \mathbf{v}$ for all $\mathbf{u} \in \mathbb{R}^m$ and $\mathbf{v} \in \mathbb{R}^n$. Prove that $A = B$. (*c*) Find an $n \times n$ matrix $C \neq \mathrm{O}$ such that $\mathbf{u}^T C \mathbf{u} = 0$ for all $\mathbf{u} \in \mathbb{R}^n$.

9.1.7. Let $U = \mathrm{C}^0[0,1]$. Find the adjoint I^* of the identity operator $\mathrm{I}: U \to U$ under the weighted inner products (9.16).

9.1.8. Compute the adjoint of the derivative operator $v = D[u] = u'$ under the weighted inner products $\langle u, \widetilde{u} \rangle = \int_0^1 e^x \, u(x) \, \widetilde{u}(x) \, dx$, $\langle\langle v, \widetilde{v} \rangle\rangle = \int_0^1 (1+x) \, v(x) \, \widetilde{v}(x) \, dx$. Clearly state any boundary conditions that you are imposing.

9.1.9. Let $L[u] = x u'(x) + u(x)$ and $0 < a < x < b$. When subject to homogeneous Dirichlet boundary conditions $u(a) = u(b) = 0$, determine the adjoint $L^*[v]$ with respect to (*a*) the L^2 inner products (9.7); (*b*) the weighted inner products (9.16).

9.1.10. Consider the linear operator $L[u] = \begin{pmatrix} u' \\ u \end{pmatrix}$ that maps $u(x) \in \mathrm{C}^1$ to the vector-valued function whose components consist of the function and its first derivative. Imposing the boundary conditions $u(0) = u(1)$, compute the adjoint L^* with respect to the L^2 inner products on both the domain and target spaces.

9.1.11. *True or false*: The adjoint of the divergence operator $\nabla \cdot \mathbf{v}$ with respect to the L^2 inner products (9.22, 23) is minus the gradient operator: $(\nabla \cdot)^* u = -\nabla u$. If true, what boundary conditions do you need to assume? If false, what is the adjoint?

9.1.12. Find the adjoint of the two-dimensional curl operator $\nabla \times \mathbf{v}$, as defined in (6.73), with respect to the L^2 inner products (9.22, 23). Carefully state any required boundary conditions.

♢ 9.1.13. Prove that (*a*) the adjoint of a linear operator is also a linear operator; (*b*) the adjoint is unique.

♢ 9.1.14. Let $L, M: U \to V$ be linear operators on the same inner product spaces. Prove that (*a*) $(L+M)^* = L^* + M^*$, (*b*) $(cL)^* = c L^*$ for $c \in \mathbb{R}$.

♢ 9.1.15. Prove Proposition 9.5.

♢ 9.1.16. Prove Proposition 9.6.

9.1.17. *True or false*: If $L: U \to U$ is invertible, then $(L^{-1})^* = (L^*)^{-1}$.

The Fredholm Alternative

Given a linear operator $L: U \to V$ between inner product spaces U, V, a fundamental problem is to solve the associated inhomogeneous linear system

$$L[u] = f \tag{9.35}$$

for various forcing functions $f \in V$. In finite dimensions, this reduces to a linear algebraic system, $A\mathbf{u} = \mathbf{f}$, defined by a coefficient matrix A. For the linear ordinary and partial differential operators of interest to us, (9.35) represents a linear boundary value problem. In general, an inhomogeneous linear system will not be solvable unless its right-hand side satisfies certain constraints, ensuring that f belongs to the range of L. These conditions can be readily characterized using the adjoint operator via the so-called *Fredholm Alternative*, named after the early-twentieth-century Swedish mathematician Ivar Fredholm. Fredholm's primary interest was in solving linear integral equations, but his solvability criterion was then recognized to be a completely general property of linear systems, including linear algebraic systems, linear differential equations, linear boundary value problems, and so on.

Recall that the *kernel* of a linear operator L is the set of solutions to the homogeneous linear system $L[u] = 0$.

Definition 9.9. The *cokernel* of a linear operator $L: U \to V$ between inner product spaces is defined as the kernel of its adjoint:

$$\operatorname{coker} L = \ker L^* = \left\{ v \in V \;\middle|\; L^*[v] = 0 \right\}. \tag{9.36}$$

We can now state and prove the *Fredholm Alternative*.

Theorem 9.10. *If the linear system $L[u] = f$ has a solution, then the right-hand side must be orthogonal to the cokernel of L, i.e.,*

$$\langle\!\langle\, v\,, f \,\rangle\!\rangle = 0 \qquad \textit{for all} \qquad v \in \operatorname{coker} L. \tag{9.37}$$

Proof: If $L[u] = f$, then, given $v \in \operatorname{coker} L$, the adjoint equation (9.2) implies

$$\langle\!\langle\, v\,, f \,\rangle\!\rangle = \langle\!\langle\, v\,, L[u] \,\rangle\!\rangle = \langle\, L^*[v]\,, u \,\rangle = 0,$$

since $L^*[v] = 0$ by the definition of the cokernel. *Q.E.D.*

Remark: In practice, one needs to check the orthogonality constraints (9.37) only when v runs through a basis of the cokernel. In particular, if the only solution to the homogeneous adjoint system $L^*[v] = 0$ is the trivial solution $v = 0$, then there are no constraints, and we expect that the inhomogeneous linear system (9.35) can be solved for any "reasonable" forcing function f. In finite dimensions, this is certainly the case, [**89**]. For boundary value problems defined by linear differential operators, one needs to determine what "reasonable" means, and then prove an appropriate existence theorem. Although valid for all of the boundary value problems presented here, when subject to continuous or even piecewise continuous forcing functions f, rigorous proofs of the existence of solutions for partial differential equations involve the advanced mathematical machinery of functional analysis — see, e.g., [**38, 44, 61, 99**] — and lie beyond the scope of this introductory text.

Example 9.11. Consider the linear algebraic system

$$u_1 - u_3 = f_1, \qquad u_2 - 2u_3 = f_2, \qquad u_1 - 2u_2 + 3u_3 = f_3. \tag{9.38}$$

Using Gaussian Elimination (or by inspection), one easily sees that (9.38) admits a solution if and only if the compatibility condition

$$-f_1 + 2f_2 + f_3 = 0 \tag{9.39}$$

holds. Moreover, when this occurs, a solution exists but is not unique. To connect this to the Fredholm Alternative, we write the system in matrix form $L[\mathbf{u}] = \mathbf{f}$, where $L[\mathbf{u}] = A\mathbf{u}$ represents multiplication by the coefficient matrix

$$A = \begin{pmatrix} 1 & 0 & -1 \\ 0 & 1 & -2 \\ 1 & -2 & 3 \end{pmatrix}.$$

Using the dot product on $\mathbb{R}^3$, the adjoint linear function $L^*[\mathbf{v}] = A^T\mathbf{v}$ is represented by the transposed matrix

$$A^T = \begin{pmatrix} 1 & 0 & 1 \\ 0 & 1 & -2 \\ -1 & -2 & 3 \end{pmatrix}.$$

Therefore, the cokernel is found by solving the homogeneous adjoint linear system $A^T\mathbf{y} = 0$, i.e.,

$$v_1 + v_3 = 0, \qquad v_2 - 2v_3 = 0, \qquad -v_1 - 2v_2 + 3v_3 = 0,$$

whose solutions consist of all scalar multiples of $\mathbf{v} = (-1, 2, 1)^T$. We recognize the compatibility condition (9.39) as requiring that the right-hand side be orthogonal (under the dot product) to the cokernel basis vector,

$$\mathbf{v} \cdot \mathbf{f} = -f_1 + 2f_2 + f_3 = 0,$$

in accordance with the Fredholm Alternative constraint (9.37).

Example 9.12. Let us solve the boundary value problem

$$u'' = f(x), \qquad u'(0) = 0, \qquad u'(\ell) = 0, \tag{9.40}$$

modeling the displacement, under an external force, of a uniform elastic bar of length ℓ both of whose ends are free. Solving the differential equation by direct integration, we find that

$$u(x) = ax + b + \int_0^x \left(\int_0^y f(z)\, dz \right) dy,$$

where the constants a, b are to be determined by the boundary conditions. Since

$$u'(x) = a + \int_0^x f(z)\, dz,$$

the boundary condition $u'(0) = 0$ implies that $a = 0$. The second boundary condition requires

$$u'(\ell) = \int_0^\ell f(x)\, dx = 0. \tag{9.41}$$

If this fails, then the boundary value problem has no solution. On the other hand, if the forcing function $f(x)$ satisfies the constraint (9.41), then the resulting solution of the boundary value problem has the form

$$u(x) = b + \int_0^x \left(\int_0^y f(z)\,dz \right) dy, \tag{9.42}$$

where the constant b is arbitrary. Thus, when it exists, the solution to the boundary value problem is not unique. The constant b solves the corresponding homogeneous problem, and represents a rigid translation of the entire bar by a distance b.

The solvability constraint (9.41) follows from the Fredholm Alternative. Indeed, according to Example 9.7, under the L^2 inner products and the given boundary conditions, $(D^2)^* = D^2$, and hence the adjoint system is the unforced homogeneous boundary value problem

$$v'' = 0, \qquad v'(0) = 0, \qquad v'(\ell) = 0,$$

with solution $v(x) = c$ for any constant c. Thus, the cokernel consists of all scalar multiples of the constant function $v_\star(x) \equiv 1$. The Fredholm Alternative requires that the forcing function in the original boundary value problem be orthogonal to the cokernel functions, and so

$$\langle 1, f \rangle = \int_0^\ell f(x)\,dx = 0,$$

which is precisely the condition (9.41) required for existence of a (nonunique) equilibrium solution.

Example 9.13. Consider the homogeneous Neumann boundary value problem for the Poisson equation on a bounded domain $\Omega \subset \mathbb{R}^2$, namely,

$$-\Delta u = f \quad \text{in} \quad \Omega, \qquad \frac{\partial u}{\partial \mathbf{n}} = 0 \quad \text{on} \quad \partial\Omega. \tag{9.43}$$

According to Example 9.8, the Laplacian is self-adjoint under the L^2 inner product and the prescribed boundary conditions: $\Delta^* = \Delta$. Thus, the homogeneous adjoint system is merely

$$-\Delta v = 0 \quad \text{in} \quad \Omega, \qquad \frac{\partial v}{\partial \mathbf{n}} = 0 \quad \text{on} \quad \partial\Omega.$$

Theorem 6.15 tells us that the only solutions to the adjoint problem are the constant functions, $v(x,y) \equiv c$. Thus, a basis for the cokernel consists of the function $v(x,y) \equiv 1$, and so the Fredholm Alternative requires that the forcing function in (9.43) satisfy

$$\langle 1, f \rangle = \iint_\Omega f(x,y)\,dx\,dy = 0, \tag{9.44}$$

reproducing our earlier constraint (6.90) for the homogeneous Neumann case.

Exercises

9.1.18. Use the Fredholm Alternative to determine whether the following linear systems are compatible. When compatible, write down the general solution.

(a) $\begin{aligned} 2x - 4y &= -2, \\ -x + 2y &= 3, \end{aligned}$ (b) $\begin{aligned} 6x - 3y + 9z &= 6, \\ 2x - y + 3z &= 2, \end{aligned}$ (c) $\begin{aligned} 2x + 3y &= -1, \\ 3x + 7y &= 1, \\ x + 4y &= 2, \\ -x + y &= 3, \end{aligned}$

(d) $\begin{aligned} 2x_1 - 3x_2 - x_3 &= -1, \\ 3x_1 - x_2 &= 1, \\ 4x_1 + x_2 + x_3 &= 2, \end{aligned}$ (e) $\begin{aligned} 2x_1 + 3x_2 - x_4 &= -1, \\ 3x_1 + 2x_3 - x_4 &= 0, \\ x_1 - x_2 + x_3 &= 1. \end{aligned}$

9.1.19. Use the Fredholm Alternative to find the compatibility conditions for the following systems of linear equations.
(a) $2x + y = a,\ \ x + 4y = b,\ \ -3x + 2y = c;$
(b) $x + 2y + 3z = a,\ \ -x + 5y - 2z = b,\ \ 2x - 3y + 5z = c;$
(c) $x_1 + 2x_2 + 3x_3 = b_1,\ \ x_2 + 2x_3 = b_2,\ \ 3x_1 + 5x_2 + 7x_3 = b_3,\ \ -2x_1 + x_2 + 4x_3 = b_4;$
(d) $x - 3y + 2z + w = a,\ \ 4x - 2y + 2z + 3w = b,\ \ 5x - 5y + 4z + 4w = c,\ \ 2x + 4y - 2z + w = d.$

9.1.20. Suppose A is a symmetric matrix. Show that the linear system $A\mathbf{x} = \mathbf{b}$ has a solution if and only if $\mathbf{b}$ is orthogonal to $\ker A$.

9.1.21. Use the Fredholm Alternative to determine whether or not there exists a solution to the following boundary value problem: $x u'' + u' = 1 - \frac{2}{3}x,\ \ u'(1) = u'(2) = 0$. If so, write down all solutions.

9.1.22. Analyze the periodic boundary value problem

$$-u'' = f(x), \qquad u(0) = u(2\pi), \qquad u'(0) = u'(2\pi),$$

along the same lines as in Example 9.12. Characterize the forcing functions for which the problem has a solution. Explain why the constraints, if any, are in accordance with the Fredholm Alternative. Write down a forcing function $f(x)$ that satisfies all your constraints, and then find all corresponding solutions.

9.1.23. Answer Exercise 9.1.22 for the boundary value problems:
(a) $u'''' = f(x), \quad u''(0) = u'''(0) = 0, \quad u''(1) = u'''(1) = 0;$
(b) $u'''' = f(x), \quad u''(0) = u'''(0) = 0, \quad u(1) = u''(1) = 0.$

♡ 9.1.24. Let λ be a real parameter. (a) For which values of λ does the boundary value problem $u'' + \lambda u = h(x),\ u(0) = 0,\ u(1) = 0$, have a unique solution? (b) Construct the Green's function for all such λ. (c) In the nonunique cases, use the Fredholm Alternative to find conditions on the forcing function $h(x)$ that are required for the existence of a solution.

9.1.25. Let $\Omega \subset \mathbb{R}^2$ be a bounded, connected domain. Using the L^2 inner products (9.22, 23) on scalar and vector fields, write out the Fredholm Alternative constraints for the solvability of the boundary value problem $\nabla \cdot \mathbf{v} = f$ in Ω, subject to the homogeneous boundary conditions $\mathbf{v} \cdot \mathbf{n} = 0$ on $\partial\Omega$.

9.1.26. Let $\Omega \subset \mathbb{R}^2$ be a bounded simply connected domain. Using the L^2 inner products (9.22, 23) on scalar and vector fields on a domain $\Omega \subset \mathbb{R}^2$, write out the Fredholm Alternative constraints for the solvability of the boundary value problem $\nabla u = \mathbf{f}$ in Ω, subject to the homogeneous boundary conditions $u = 0$ on $\partial\Omega$.

9.2 Self–Adjoint and Positive Definite Linear Functions

In finite-dimensional linear algebra, there are two particularly important classes of matrices: symmetric, equal to their own transpose, and positive definite, as prescribed by Definition B.12. The goal of this section is to adapt both concepts to more general linear operators, paying particular attention to the case of linear differential operators. The resulting classes of self-adjoint and positive (semi-)definite differential operators are ubiquitous in applications of ordinary and partial differential equations.

Self–Adjointness

Throughout this section, U will be a fixed inner product space. We have already seen that the transpose of a matrix is a very special case of the adjoint operation. Thus, the natural analogue of a symmetric matrix is a linear operator that equals its own adjoint.

Definition 9.14. A linear operator $S: U \to U$ is called *self-adjoint* if $S^* = S$.

Thus, according to (9.2), S is self-adjoint if and only if

$$\langle\, S[u]\,, \widetilde{u}\,\rangle = \langle\, u\,, S[\widetilde{u}\,]\,\rangle \qquad \text{for all} \qquad u, \widetilde{u} \in U. \tag{9.45}$$

Example 9.15. In the finite-dimensional case, a linear function $S: \mathbb{R}^n \to \mathbb{R}^n$ is realized by matrix multiplication: $S[\mathbf{u}] = K\mathbf{u}$, where K is a square matrix of size $n \times n$. If we use the ordinary dot product on $\mathbb{R}^n$, then, according to Example 9.2, the adjoint function $S^*: \mathbb{R}^n \to \mathbb{R}^n$ is given by multiplication by the transposed matrix: $S^*[\mathbf{u}] = K^T\mathbf{u}$. Thus, a linear function is self-adjoint with respect to the dot product if and only if it is represented by a symmetric matrix: $K^T = K$.

On the other hand, if we adopt the weighted inner product $\langle\, \mathbf{u}\,, \widetilde{\mathbf{u}}\,\rangle = \mathbf{u}^T C\, \widetilde{\mathbf{u}}$ provided by the symmetric positive definite matrix $C > 0$, then, according to (9.6), the adjoint function S^* has matrix representative $C^{-1}K^T C$, and hence S is self-adjoint under the weighted inner product if and only if the matrix K satisfies $K = C^{-1}K^T C$.

Example 9.16. In Example 9.7, we argued that the second-order derivative operator $S = D^2$ is self-adjoint with respect to the L^2 inner product, when subject to suitable homogeneous boundary conditions. A direct verification of this result is instructive. According to the general adjoint equation (9.2), we need to equate

$$\int_a^b S[u]\,\widetilde{u}\,dx = \langle\, S[u]\,, \widetilde{u}\,\rangle = \langle\, u\,, S^*[\widetilde{u}]\,\rangle = \int_a^b u\, S^*[\widetilde{u}]\,dx. \tag{9.46}$$

As before, the computation relies on (in this case two) integration by parts:

$$\begin{aligned}\langle\, S[u]\,, \widetilde{u}\,\rangle = \int_a^b \frac{d^2u}{dx^2}\,\widetilde{u}\,dx &= \frac{du}{dx}\,\widetilde{u}\,\bigg|_{x=a}^{b} - \int_a^b \frac{du}{dx}\,\frac{d\widetilde{u}}{dx}\,dx \\ &= \left[\frac{du}{dx}\,\widetilde{u} - u\,\frac{d\widetilde{u}}{dx}\right]\bigg|_{x=a}^{b} + \int_a^b u\,\frac{d^2\widetilde{u}}{dx^2}\,dx.\end{aligned}$$

Comparing with (9.46), we conclude that $S^* = D^2 = S$, *provided* the boundary terms vanish:

$$\left[\frac{du}{dx}\,\widetilde{u} - u\,\frac{d\widetilde{u}}{dx}\right]\bigg|_{x=a}^{b} = \big[\,u'(b)\,\widetilde{u}(b) - u(b)\,\widetilde{u}'(b)\,\big] - \big[\,u'(a)\,\widetilde{u}(a) - u(a)\,\widetilde{u}'(a)\,\big] = 0. \tag{9.47}$$

This requires that we impose suitable boundary conditions at the endpoints, which will serve to characterize the underlying vector space U on which $S = D^2$ acts. One possibility is to set $U = \{\, u(a) = u(b) = 0\,\}$, thereby imposing homogeneous Dirichlet boundary conditions. Since $\widetilde{u} \in U$ also, $\widetilde{u}(a) = \widetilde{u}(b) = 0$, and hence (9.47) holds, proving self-adjointness. Alternatively, one can impose homogeneous Neumann, mixed, or periodic boundary conditions to specify the space U and similarly establish self-adjointness of $S = D^2$.

Positive Definiteness

Let us turn to the characterization of positive definite, and, slightly less stringently, positive semi-definite linear operators. These serve to extend the notions of positive definite and semi-definite matrices to linear differential operators defining boundary value problems.

Definition 9.17. A linear operator $S: U \to U$ on an inner product space is called *positive definite*, written $S > 0$, if

$$\langle u \,, S[u] \rangle > 0 \qquad \text{for all} \qquad u \neq 0. \tag{9.48}$$

The operator S is *positive semi-definite*, written $S \geq 0$, if

$$\langle u \,, S[u] \rangle \geq 0 \qquad \text{for all} \qquad u. \tag{9.49}$$

Observe that, on the finite-dimensional space $U = \mathbb{R}^n$ equipped with the dot product, the linear function $S[\mathbf{u}] = K\mathbf{u}$ is positive (semi-)definite if and only if K is a positive (semi-)definite matrix, as per Definition B.12. (However, changing the inner product on $\mathbb{R}^n$ will result in an alternative notion of positive definiteness for the matrix K; see Exercise 9.2.5.) In the infinite-dimensional situations involving differential operators, the domain of the operator may be only a dense subspace of the full inner product space U, and one imposes the positivity condition (9.48) or (9.49) only on those functions u lying in the domain of S. Fortunately, this technicality has no serious effect on the subsequent development.

Example 9.18. Consider the operator $S = -D^2$ acting on the space U consisting of all C^2 functions defined on a bounded interval $[a, b]$ and subject to homogeneous Dirichlet boundary conditions $u(a) = u(b) = 0$. To establish positive definiteness, we evaluate

$$\langle S[u] \,, u \rangle = \int_a^b \left(-\,\frac{d^2u}{dx^2}\,u \right) dx = -\,\frac{du}{dx}\,u \,\bigg|_{x=a}^{b} + \int_a^b \left(\frac{du}{dx} \right)^2 dx = \int_a^b \left(\frac{du}{dx} \right)^2 dx,$$

where we integrated by parts and then used the boundary conditions to eliminate the boundary terms. The final expression is clearly ≥ 0, and hence S is at least positive semi-definite. Moreover, since $u'(x)$ is continuous, the only way the final integral could vanish is if $u'(x) \equiv 0$, which means $u(x) \equiv c$ is constant. However, the only constant function satisfying the homogeneous Dirichlet boundary conditions is $u(x) \equiv 0$. Thus, $\langle S[u] \,, u \rangle > 0$ for all $0 \neq u \in U$, which implies $S > 0$. A similar argument implies positive definiteness when the functions are subject to the mixed boundary conditions $u(a) = u'(b) = 0$. On the other hand, any constant function satisfies the Neumann boundary conditions $u'(a) = u'(b) = 0$, and hence in this case $S \geq 0$ is only positive semi-definite.

Proposition 9.19. *If* $S > 0$, *then* $\ker S = \{0\}$. *As a consequence, a positive definite linear system* $S[u] = f$ *with* f *in the range of* S, *so* $f \in \operatorname{rng} S$, *must have a unique solution.*

Proof: If $S[u] = 0$, then $\langle u \,, S[u] \rangle = 0$, which, according to (9.48), is possible only if $u = 0$. The second statement follows directly from Theorem 1.6. *Q.E.D.*

Thus, in the finite-dimensional case, positive definiteness implies that the coefficient matrix of $S[\mathbf{u}] = K\mathbf{u}$ is nonsingular, and hence existence of a solution is automatic. In the infinite-dimensional cases of boundary value problems, existence of solutions usually requires some further analysis, [**63**].

The most common means of producing self-adjoint, positive (semi-)definite linear operators is provided by the following general construction. From here on, in order to distinguish the possibly different norms resulting from the inner products on the domain and target spaces of a linear operator $L: U \to V$, we employ, respectively, the following double and triple bar notation:

$$\| u \| = \langle u , u \rangle, \qquad u \in U, \qquad ||| v ||| = \langle\!\langle v , v \rangle\!\rangle, \qquad v \in V. \tag{9.50}$$

Theorem 9.20. *Let $L: U \to V$ be a linear map between inner product spaces with adjoint $L^*: V \to U$. Then the composite map*

$$S = L^* \circ L : U \longrightarrow U$$

is always self-adjoint, $S = S^*$, and positive semi-definite, $S \geq 0$, with $\ker S = \ker L$. Moreover, $S > 0$ is positive definite if and only if $\ker L = \{0\}$.

Proof: First, by Propositions 9.5 and 9.6,

$$S^* = (L^* \circ L)^* = L^* \circ (L^*)^* = L^* \circ L = S,$$

proving self-adjointness. Furthermore,

$$\langle u , S[u] \rangle = \langle u , L^*[L[u]] \rangle = \langle\!\langle L[u] , L[u] \rangle\!\rangle = ||| L[u] |||^2 \geq 0 \tag{9.51}$$

for all u, proving positive semi-definiteness. Moreover, the result is > 0 as long as $L[u] \neq 0$. Thus, if $\ker L = \{ u \mid L[u] = 0 \} = \{0\}$, then $\langle u , S[u] \rangle > 0$ for all $u \neq 0$, and hence S is positive definite. Finally, the same computation proves that $\ker S = \ker L$. Indeed, if $L[u] = 0$, then $S[u] = L^*[L[u]] = L^*[0] = 0$. On the other hand, if $S[u] = 0$, then $0 = \langle u , S[u] \rangle = ||| L[u] |||^2$, and hence $L[u] = 0$. *Q.E.D.*

We are particularly interested in linear systems that are based on the construction of Theorem 9.20, namely

$$S[u] = L^*[L[u]] = f. \tag{9.52}$$

We will refer to the system (9.52) as *positive definite* or *positive semi-definite* according to the status of its defining operator S. Thus, the system is positive definite if and only if $\ker S = \ker L = \{0\}$, i.e., the only solution to the homogeneous system $S[z] = 0$ is the trivial solution $z = 0$. In this case, the solution to (9.52) (provided it exists) is unique. On the other hand, if there are nonzero solutions to $S[z] = 0$, then (9.52) is only positive semi-definite, and does not admit a unique solution. Moreover, unless the Fredholm Alternative constraints (9.37) hold, then there are no solutions. By Theorem 9.20, we can identify

$$\operatorname{coker} S = \ker S^* = \ker S = \ker L, \tag{9.53}$$

which thus implies the following:

Theorem 9.21. *Let $S = L^* \circ L$. If the linear system $S[u] = f$ has a solution, then $\langle z , f \rangle = 0$ for all $z \in \ker L$. Moreover, if $S[u] = f$ and $S[\widetilde{u}] = f$ are two solutions to the same linear system, then $\widetilde{u} = u + z$, where $z \in \ker L$ is any solution to $L[z] = 0$.*

Example 9.22. In the finite-dimensional case, any linear function $L: \mathbb{R}^n \to \mathbb{R}^m$ is represented by matrix multiplication: $L[\mathbf{u}] = A\mathbf{u}$. For the dot product on both the domain and target spaces, $L^*[\mathbf{v}] = A^T\mathbf{v}$, and so the self-adjoint combination $S = L^* \circ L: \mathbb{R}^n \to \mathbb{R}^n$ is represented by the $n \times n$ symmetric matrix $K = A^T A$. According to Theorem 9.20, the matrix K is always positive semi-definite, and is positive definite if and only if the only solution to the homogeneous linear system $A\mathbf{z} = \mathbf{0}$ is the trivial solution $\mathbf{z} = \mathbf{0}$. In the positive semi-definite case, the Fredholm Alternative of Theorem 9.21 states that the linear system $K\mathbf{u} = \mathbf{f}$ has a solution if and only if $\mathbf{z} \cdot \mathbf{f} = 0$ for all $\mathbf{z} \in \ker A$. (As noted before, existence of solutions in the finite-dimensional case is not an issue.) Moreover, if $\mathbf{u}$ is any solution, so is $\widetilde{\mathbf{u}} = \mathbf{u} + \mathbf{z}$ for any $\mathbf{z} \in \ker A$.

More generally, if we adopt the weighted inner products (9.4) on the domain and target spaces represented by the respective positive definite matrices $M > 0$ and $C > 0$, then the adjoint map L^* has matrix representative $M^{-1}A^T C$, and hence $S = L^* \circ L$ is given by multiplication by the (not necessarily symmetric) $n \times n$ matrix $K = M^{-1}A^T C\, A$. Again, $K \geq 0$ in all cases, and $K > 0$ if and only if $\ker A = \{\mathbf{0}\}$. Now, the Fredholm Alternative states that the linear system $K\mathbf{u} = M^{-1}A^T C\, A\mathbf{u} = \mathbf{f}$ has a solution if and only if $\langle\, \mathbf{z} \,,\mathbf{f}\,\rangle = \mathbf{z}^T M \mathbf{f} = 0$ for all $\mathbf{z} \in \ker A$. See [**89, 112**] for applications of this construction in mechanics, electrical networks, and the stability of structures.

Example 9.23. Consider next the differentiation operator $D[\,u\,] = u'$. According to Example 9.3, if we impose suitable homogeneous boundary conditions on the space of allowable functions — Dirichlet, Neumann, mixed, or periodic — and use the L^2 inner products on both domain and target space, then $D^*[\,v\,] = -\,v'$. Therefore, the self-adjoint operator of Theorem 9.20 is given by $S = D^* \circ D = -\,D^2$.

According to Theorem 9.20, the resulting boundary value problem

$$S[\,u\,] = -\,u'' = f$$

is always positive semi-definite, and is positive definite if and only if $\ker D = \{0\}$, i.e., the only function that satisfies $D[\,u\,] = u' = 0$ along with the boundary conditions is the zero function. Consider first the Dirichlet boundary conditions $u(a) = u(b) = 0$. On a connected interval, $u' = 0$ if and only if $u = c$ is a constant function. However, the boundary conditions require that $c = 0$, and hence only the zero function appears in the kernel. We conclude that the Dirichlet boundary value problem is positive definite, and its solution unique. A similar argument applies to the mixed boundary conditions, e.g., $u(a) = u'(b) = 0$, since the condition at $x = a$ is enough to ensure that the constant function must be zero. On the other hand, *any* constant function satisfies the Neumann boundary conditions $u'(a) = u'(b) = 0$, and hence in this case, $\ker D$ consists of all constant functions. Therefore, the Neumann boundary value problem is only positive semi-definite. And, as we saw, the solution, when it exists, is not unique, since we can add any constant function to a solution and obtain another solution. A similar argument proves that the periodic boundary value problem, with $u(a) = u(b)$, $u'(a) = u'(b)$, is also positive semi-definite, with the same kinds of existence and uniqueness properties.

More generally, if we use weighted inner products (9.16) on the domain and target spaces, then, again subject to suitable boundary conditions, the adjoint is given by (9.19), and so the self-adjoint boundary value problem $S[\,u\,] = D^* \circ D[\,u\,] = f$ is based on the more general differential equation

$$S[\,u\,] = -\,\frac{1}{\rho(x)}\left(\frac{d}{dx}\,\kappa(x)\,\frac{du}{dx}\right) = f(x). \tag{9.54}$$

Such boundary value problems model the deformations of a nonuniform elastic bar with density $\rho(x)$ and stiffness $\kappa(x)$, when subject to the external forcing function $f(x)$. Again, the positive definiteness of the problem depends on whether $\ker D = \{0\}$, and so the exact same classification holds as in the unweighted case: the Dirichlet and mixed boundary value problems are positive definite and have a unique solution, whereas the Neumann and periodic boundary value problems are only positive semi-definite, and the existence of a solution requires the Fredholm conditions to be satisfied.

Self-adjointness underlies the symmetry of the associated Green's function. As a function of x, the Green's function $G_\xi(x) = G(x;\xi)$ satisfies the boundary value problem with delta function forcing concentrated at position $x = \xi$:

$$S[G_\xi] = \delta_\xi, \qquad \text{or, explicitly,} \qquad -\frac{1}{\rho(x)}\frac{\partial}{\partial x}\left(\kappa(x)\frac{\partial G}{\partial x}\right) = \delta(x-\xi), \tag{9.55}$$

along with the required homogeneous boundary conditions. Suppose first that we are using the L^2 inner product on the interval $[a,b]$, so that $\rho(x) \equiv 1$. Using the definition of the delta function $\delta_\xi(x) = \delta(x-\xi)$ and the self-adjointness of S, we have, for any $a < x, \xi < b$,

$$\begin{aligned} G(x;\xi) = G_\xi(x) &= \int_a^b G_\xi(y)\,\delta_x(y)\,dy = \langle\, G_\xi\,,\delta_x\,\rangle = \langle\, G_\xi\,, S[G_x]\,\rangle \\ &= \langle\, S[G_\xi]\,, G_x\,\rangle = \langle\, \delta_\xi\,, G_x\,\rangle = \int_a^b \delta_\xi(y)\,G_x(y)\,dy = G_x(\xi) = G(\xi;x). \end{aligned} \tag{9.56}$$

This establishes† the symmetry equation

$$G(x;\xi) = G(\xi;x) \tag{9.57}$$

for the Green's function of a self-adjoint boundary value problem under the L^2 inner product. This can be regarded as the differential operator version of the fact that the inverse of a symmetric matrix is also symmetric.

On the other hand, if we adopt a weighted inner product

$$\langle\, u\,, \widetilde{u}\,\rangle = \int_a^b u(y)\,\widetilde{u}(y)\,\rho(y)\,dy,$$

then the preceding argument must be slightly modified:

$$\begin{aligned} \rho(x)\,G(x;\xi) = \rho(x)\,G_\xi(x) &= \int_a^b \rho(y)\,G_\xi(y)\,\delta_x(y)\,dy = \langle\, G_\xi\,,\delta_x\,\rangle = \langle\, G_\xi\,, S[G_x]\,\rangle \\ &= \langle\, S[G_\xi]\,, G_x\,\rangle = \langle\, \delta_\xi\,, G_x\,\rangle = \int_a^b \delta_\xi(y)\,G_x(y)\,\rho(y)\,dy = \rho(\xi)\,G_x(\xi) = \rho(\xi)\,G(\xi;x), \end{aligned}$$

and so the Green's function associated with a weighted self-adjoint boundary value problem satisfies a "weighted symmetry condition"

$$\rho(x)\,G(x;\xi) = \rho(\xi)\,G(\xi;x). \tag{9.58}$$

Remark: Equation (9.58) implies that the *modified Green's function*

$$\widehat{G}(x;\xi) = \frac{G(x;\xi)}{\rho(\xi)} \qquad \text{is genuinely symmetric:} \qquad \widehat{G}(x;\xi) = \widehat{G}(\xi;x). \tag{9.59}$$

† Symmetry at the endpoints is a consequence of continuity.

The modified Green's function also has the advantage of recasting the superposition formula for the solution to the boundary value problem $S[u] = f$ as the appropriate weighted inner product:

$$u(x) = \int_a^b G(x;\xi)\, f(\xi)\, d\xi = \int_a^b \widehat{G}(x;\xi)\, f(\xi)\, \rho(\xi)\, d\xi = \langle\, \widehat{G}_x\,, f\,\rangle, \quad \text{where} \quad \widehat{G}_x(\xi) = \widehat{G}(x;\xi).$$

Two–Dimensional Boundary Value Problems

Let us next apply the self-adjoint formalism to study boundary value problems on a bounded, connected, two-dimensional domain $\Omega \subset \mathbb{R}^2$. We take $L = \nabla$ to be the gradient operator, mapping a scalar field u to a vector field $\mathbf{v} = \nabla u$. We impose a suitable set of homogeneous boundary conditions, i.e., Dirichlet, Neumann, or mixed. According to the calculation in Section 9.1, if we adopt the basic L^2 inner products (9.22, 23) between scalar and vector fields, then the adjoint of the gradient is the negative of the divergence: $\nabla^* \mathbf{v} = -\nabla \cdot \mathbf{v}$. Therefore, the self-adjoint combination of Theorem 9.20 yields

$$\nabla^* \circ \nabla[\,u\,] = -\nabla \cdot (\nabla u) = -\Delta u,$$

where Δ is the Laplacian operator. In this manner, we are able to write the two-dimensional Poisson equation in self-adjoint form

$$-\Delta u = -\nabla \cdot (\nabla u) = \nabla^* \circ \nabla u = f, \tag{9.60}$$

as always subject to the selected boundary conditions.

According to Theorem 9.20, $-\Delta = \nabla^* \circ \nabla$ is positive definite if and only if the kernel of the gradient operator — restricted to the appropriate space of scalar fields — is trivial: $\ker \nabla = \{0\}$. Since we are assuming that the domain Ω is connected, Lemma 6.16 tells us that the only functions that could show up in $\ker \nabla$, and thus prevent positive definiteness, are the constants. The boundary conditions will tell us whether this occurs. The only constant function that satisfies either homogeneous Dirichlet or homogeneous mixed boundary conditions is the zero function, and thus, just as in the one-dimensional case, the boundary value problem for the Poisson equation subject to Dirichlet or mixed boundary conditions is positive definite. In particular, this means that its solution is uniquely defined. On the other hand, any constant function satisfies the homogeneous Neumann boundary condition $\partial u/\partial \mathbf{n} = 0$, and hence such boundary value problems are only positive semi-definite. Existence of a solution relies on the Fredholm Alternative, as we discussed in Example 9.13; moreover, when it exists, the solution is no longer unique, because one can add in any constant without affecting either the equation or the boundary conditions.

More generally, if we impose weighted inner products (9.32) on our spaces of scalar and vector fields, then, recalling (9.34), the corresponding self-adjoint boundary value problem takes the more general form

$$\nabla^* \circ \nabla u = -\frac{1}{\rho(x,y)}\frac{\partial}{\partial x}\left(\kappa_1(x,y)\,\frac{\partial u}{\partial x}\right) - \frac{1}{\rho(x,y)}\frac{\partial}{\partial y}\left(\kappa_2(x,y)\,\frac{\partial u}{\partial y}\right) = f(x,y), \tag{9.61}$$

along with the chosen boundary conditions on $\partial\Omega$. Again, the Dirichlet and mixed boundary value problems are positive definite, with unique solutions, while the (suitably weighted) Neumann problem is only positive semi-definite.

The partial differential equation (9.61) arises in various physical contexts. For example, consider a steady-state fluid flow moving in a domain $\Omega \subset \mathbb{R}^2$ described by a vector field $\mathbf{v}$. The flow is called *irrotational* if it has zero curl, $\nabla \times \mathbf{v} = \mathbf{0}$, and hence, assuming that Ω is simply connected, is a gradient $\mathbf{v} = \nabla u$, where $u(x,y)$ is known as the fluid *velocity potential*. The constitutive assumptions connect the fluid velocity with its rate of flow $\mathbf{w} = \kappa\,\mathbf{v}$, where $\kappa(x,y) > 0$ is the scalar density of the fluid. Conservation of mass provides the final equation, namely $\nabla \cdot \mathbf{w} + f = 0$, where $f(x,y)$ represents fluid sources ($f > 0$) or sinks ($f < 0$). Therefore, the basic equilibrium equations take the form

$$-\nabla\cdot(\kappa\,\nabla u) = f, \qquad \text{or} \qquad -\frac{\partial}{\partial x}\left(\kappa(x,y)\,\frac{\partial u}{\partial x}\right) - \frac{\partial}{\partial y}\left(\kappa(x,y)\,\frac{\partial u}{\partial y}\right) = f(x,y), \tag{9.62}$$

which is (9.61) with $\rho \to 1$ and $\kappa_1, \kappa_2 \to \kappa$. The case of a homogeneous (constant density) fluid thus reduces to the Poisson equation (4.84), with f replaced by f/κ.

Symmetry of the Green's function for the Poisson equation and the more general boundary value problems (9.61, 62) follows by an evident adaptation of the one-dimensional argument presented above. Details are left as Exercise 9.2.17.

Exercises

9.2.1. Which of the following matrices define self-adjoint linear functions $S\colon \mathbb{R}^2 \to \mathbb{R}^2$ relative to the dot product? (a) $\begin{pmatrix} 1 & 0 \\ 0 & 1 \end{pmatrix}$, (b) $\begin{pmatrix} 0 & 3 \\ 2 & 2 \end{pmatrix}$, (c) $\begin{pmatrix} 1 & 0 \\ 2 & -5 \end{pmatrix}$, (d) $\begin{pmatrix} 3 & 2 \\ 2 & 1 \end{pmatrix}$.

9.2.2. Answer Exercise 9.2.1 for the inner products

$$(i)\ \langle \mathbf{u}, \tilde{\mathbf{u}} \rangle = 2u_1\tilde{u}_1 + 3u_2\tilde{u}_2; \qquad (ii)\ \langle \mathbf{u}, \tilde{\mathbf{u}} \rangle = \mathbf{u}^T C\,\tilde{\mathbf{u}}, \text{ where } C = \begin{pmatrix} 2 & -1 \\ -1 & 3 \end{pmatrix}.$$

9.2.3. *True or false*: Given an inner product $\langle \mathbf{u}, \mathbf{v} \rangle$ on $\mathbb{R}^n$:
(a) The inverse of a nonsingular self-adjoint $n \times n$ matrix is self-adjoint.
(b) The inverse of a nonsingular positive definite $n \times n$ matrix is positive definite.

9.2.4. Prove that $K > 0$ is a positive definite $n \times n$ matrix if and only if $J = K^T + K$ is a symmetric positive definite matrix.

♢ 9.2.5. (a) Prove that the $n \times n$ matrix K defines a self-adjoint linear function on $\mathbb{R}^n$ with respect to the inner product $\langle \mathbf{u}, \tilde{\mathbf{u}} \rangle = \mathbf{u}^T C\,\tilde{\mathbf{u}}$ for C a symmetric positive definite matrix if and only if the matrix $J = CK$ is symmetric, and hence defines a self-adjoint linear function with respect to the dot product. (b) Prove that $K > 0$ under the given inner product if and only if $J > 0$ under the dot product.

9.2.6. Let $D[u] = u'$ be the derivative operator acting on the vector space of C^2 scalar functions $u(x)$ defined for $0 \le x \le 1$ and satisfying the boundary conditions $u(0) = 0$, $u(1) = 0$.
(a) Given the weighted inner product $\langle u, \tilde{u} \rangle = \int_0^1 u(x)\,\tilde{u}(x)\,e^x\,dx$ on both its domain and target spaces, determine the corresponding adjoint operator D^*.
(b) Let $S = D^* \circ D$. Write down and solve the boundary value problem $S[u] = 2e^x$.

9.2.7. Let $c(x) \in \mathrm{C}^0[a,b]$ be a continuous function. Prove that the linear multiplication operator $S[u] = c(x)\,u(x)$ is self-adjoint with respect to the L^2 inner product. What sort of boundary conditions need to be imposed?

9.2.8. *True or false*: The Neumann boundary value problem $-u'' + u = x$, $u'(0) = u'(\pi) = 0$, admits a unique solution.

9.2.9. Prove that the complex differential operator $L[u] = \mathrm{i}\,\dfrac{du}{dx}$ is self-adjoint with respect to the L^2 Hermitian inner product $\langle u, v \rangle = \displaystyle\int_{-\pi}^{\pi} u(x)\,\overline{v(x)}\,dx$ on the space of continuously differentiable complex-valued 2π–periodic functions: $u(x + 2\pi) = u(x)$.

9.2.10. Let $L = D^2$. Using the L^2 inner products on both the domain and target spaces, write down a set of homogeneous boundary conditions that makes $L^* = D^2$. Then set $S = L^* \circ L = D^4$. Do your boundary conditions lead to a boundary value problem $S[u] = f$ that is (*i*) positive definite; (*ii*) positive semi-definite; or (*iii*) neither?

9.2.11. Let β be a real constant. *True or false*: The second derivative operator $S[u] = u''$ is self-adjoint with respect to the L^2 inner product on the space of functions

$$U = \left\{ u(x) \in \mathrm{C}^2[0,1] \;\middle|\; u(0) = 0,\ u'(1) + \beta\, u(1) = 0 \right\}$$

subject to Dirichlet boundary conditions at the left-hand endpoint and Robin boundary conditions at the right-hand endpoint.

♡ 9.2.12. Let β be a real constant. Consider the differential operator $S[u] = -\,u''$ acting on the space of functions

$$U = \left\{ u(x) \in \mathrm{C}^2[0,1] \;\middle|\; u(0) = 0,\ u'(1) + \beta\, u(1) = 0 \right\}$$

subject to Dirichlet boundary conditions at the left-hand endpoint and Robin boundary conditions at the right-hand endpoint. Prove that $S > 0$ is positive definite with respect to the L^2 inner product if and only if $\beta > -1$. *Hint*: Use the analysis following (4.48).

♡ 9.2.13. The equilibrium equations for a toroidal membrane (an inner tube) lead to the Poisson equation $-\,u_{xx} - u_{yy} = f(x,y)$ on a rectangle $0 < x < a$, $0 < y < b$, subject to periodic boundary conditions

$$u(x,0) = u(x,b), \qquad u_y(x,0) = u_y(x,b), \qquad u(0,y) = u(a,y), \qquad u_x(0,y) = u_x(a,y).$$

(*a*) Prove that the toroidal boundary value problem is self-adjoint. (*b*) Is it positive definite, positive semi-definite, or neither? (*c*) Are there any conditions that must be imposed on the forcing function $f(x,y)$ in order that a solution exist?

◇ 9.2.14. Find the adjoint of the gradient operator ∇ with respect to the L^2 inner product (9.22) between scalar fields, and the following weighted inner product between (column) vector fields $\mathbf{v} = (\,v_1(x,y), v_2(x,y)\,)^T$, $\widetilde{\mathbf{v}} = (\,\widetilde{v}_1(x,y), \widetilde{v}_2(x,y)\,)^T$:

$$\langle\!\langle\, \mathbf{v}, \widetilde{\mathbf{v}} \,\rangle\!\rangle = \iint_\Omega \mathbf{v}(x,y)^T C(x,y)\, \widetilde{\mathbf{v}}(x,y)\,dx\,dy,$$

where the 2×2 matrix $C(x,y) = \begin{pmatrix} \alpha(x,y) & \beta(x,y) \\ \beta(x,y) & \gamma(x,y) \end{pmatrix} > 0$ is symmetric, positive definite at all points $(x,y) \in \Omega$. What sort of boundary conditions do you need to impose? Write out the corresponding boundary value problem for the equilibrium equation $\nabla^* \circ \nabla u = f$.

9.2.15. Let $\Omega \subset \mathbb{R}^2$ be a bounded domain. Construct a set of homogeneous boundary conditions on $\partial\Omega$ that make the *biharmonic equation* $\Delta^2 u = f$: (*a*) self-adjoint, (*b*) positive definite, (*c*) positive semi-definite, but not positive definite.

9.2.16. Write down the boundary value problem $\widehat{S}_\xi[\widehat{G}_\xi] = \delta_\xi$ satisfied by the modified Green's function $\widehat{G}_\xi(x) = \widehat{G}(x;\xi)$ given in (9.59). Is the underlying linear operator $\widehat{S}_\xi$, which may depend on ξ, self-adjoint with respect to a suitable inner product?

◇ 9.2.17. Prove symmetry of the Green's function, $G(\boldsymbol{\xi};\mathbf{x}) = G(\mathbf{x};\boldsymbol{\xi})$, for the Poisson equation on a bounded domain $\Omega \subset \mathbb{R}^2$ subject to homogeneous Dirichlet boundary conditions. *Hint*: Look at how we established (9.56).

9.2.18. Generalize Exercise 9.2.17 to the partial differential equation (9.61).

9.3 Minimization Principles

One of the most important features of positive definite linear problems is that their solution can be characterized by a quadratic minimization priniciple. In many physical contexts, equilibrium configuration(s) serve to minimize the potential energy of the system. Think of a small ball rolling around in a bowl. After frictional effects have stopped its motion, the ball will be left sitting in equilibrium at the bottom of the bowl — the position that minimizes the gravitational potential energy. Minimization priniciples are employed in functional analytic proofs of existence of solutions, as well as providing a foundation for the powerful finite element numerical method to be studied in Chapter 10.

The basic theorem on quadratic minimization principles is as follows.

Theorem 9.24. *Let $S: U \to U$ be a self-adjoint and positive definite linear operator on an inner product space U. Suppose that the linear system*

$$S[u] = f \tag{9.63}$$

admits a (necessarily unique) solution $u_\star$. Then $u_\star$ minimizes the value of the associated quadratic function(al)

$$Q[u] = \tfrac{1}{2}\langle u, S[u]\rangle - \langle f, u\rangle, \tag{9.64}$$

meaning that $Q[u_\star] < Q[u]$ for all admissible $u \neq u_\star$ in U.

Proof: We are given that $S[u_\star] = f$, and so, for any $u \in U$,

$$Q[u] = \tfrac{1}{2}\langle u, S[u]\rangle - \langle u, S[u_\star]\rangle = \tfrac{1}{2}\langle u - u_\star, S[u - u_\star]\rangle - \tfrac{1}{2}\langle u_\star, S[u_\star]\rangle, \tag{9.65}$$

where we used linearity, along with our assumption that S is self-adjoint, to identify the terms $\langle u, S[u_\star]\rangle = \langle u_\star, S[u]\rangle$. Since $S > 0$, the first term on the right-hand side of (9.65) is always ≥ 0; moreover it equals 0 if and only if $u = u_\star$. On the other hand, the second term does not depend on u at all. Thus, to minimize $Q[u]$, we must make the first term as small as possible, which is accomplished by setting $u = u_\star$. *Q.E.D.*

Example 9.25. Consider the the problem of minimizing a *quadratic function*

$$Q(u_1, \ldots, u_n) = \frac{1}{2} \sum_{i,j=1}^{n} k_{ij}\, u_i\, u_j - \sum_{i=1}^{n} f_i\, u_i + c, \tag{9.66}$$

depending on n variables $\mathbf{u} = (u_1, u_2, \ldots, u_n)^T \in \mathbb{R}^n$, with fixed real coefficients k_{ij}, f_i, and c. Since $u_i u_j = u_j u_i$, we can assume, without loss of generality, that the coefficients of the quadratic terms are symmetric: $k_{ij} = k_{ji}$. We rewrite (9.66) in matrix notation as

$$Q(\mathbf{u}) = \tfrac{1}{2}\,\mathbf{u} \cdot K\mathbf{u} - \mathbf{f} \cdot \mathbf{u} + c, \tag{9.67}$$

which, apart from the inessential constant term, agrees with (9.64) once we set $S[\mathbf{u}] = K\mathbf{u}$ and use the dot product $\langle \mathbf{u}, \widetilde{\mathbf{u}}\rangle = \mathbf{u} \cdot \widetilde{\mathbf{u}}$ as the inner product on $\mathbb{R}^n$. Thus, according to Theorem 9.24, if K is a symmetric positive definite matrix, then the quadratic function (9.67) has a unique minimizer $\mathbf{u}^\star = (u_1^\star, \ldots, u_n^\star)^T$, which is the solution to the linear system $K\mathbf{u}^\star = \mathbf{f}$.

If the positive definite linear operator in Theorem 9.24 comes from the self-adjoint construction of Theorem 9.20, so $S = L^* \circ L$, then, by (9.51), the quadratic term can be re-expressed as $\langle u, S[u]\rangle = |||\, L[u]\, |||^2$, using our notational convention (9.50) for the norm on the target space V of L. We can thus rephrase the minimization principle as follows.

Theorem 9.26. *Suppose $L: U \to V$ is a linear operator between inner product spaces with adjoint $L^*: V \to U$. Assume that* $\ker L = \{\mathbf{0}\}$, *and let $S = L^* \circ L: U \to U$ be the associated positive definite linear operator. If $f \in \operatorname{rng} S$, then the quadratic function*

$$Q[u] = \tfrac{1}{2} \,|||\, L[u] \,|||^2 - \langle f , u \rangle \tag{9.68}$$

has a unique minimizer $u_\star$, which is the solution to the linear system $S[u] = f$.

Warning: In (9.68), the first term $|||\, L[u] \,|||^2$ is computed using the norm based on the inner product on V, while the second term $\langle f , u \rangle$ employs the inner product on U.

One of the most important applications of minimization is the method of least squares, which is extensively applied in data analysis and approximation theory. We refer the interested reader to [**89**] for developments in this direction. Here we will concentrate on applications to differential equations.

Example 9.27. Consider the boundary value problem

$$-u'' = f(x), \qquad u(a) = 0, \qquad u(b) = 0. \tag{9.69}$$

The underlying differential operator $S = D^* \circ D = -D^2$, when acting on the space of functions satisfying the homogeneous Dirichlet boundary conditions, is self-adjoint and, in fact, positive definite, since $\ker D = \{0\}$. Explicitly, positive definiteness requires

$$\langle S[u] , u \rangle = \int_a^b \left[-u''(x)\, u(x) \right] dx = \int_a^b u'(x)^2 \, dx > 0 \tag{9.70}$$

for all nonzero $u(x) \not\equiv 0$ with $u(a) = u(b) = 0$. Notice how we used an integration by parts, invoking the boundary conditions to eliminate the boundary contributions, to expose the positivity of the integral. The associated quadratic functional is, using (9.68),

$$Q[u] = \tfrac{1}{2} |||\, u' \,|||^2 - \langle f , u \rangle = \int_a^b \left[\tfrac{1}{2} u'(x)^2 - f(x)\, u(x) \right] dx.$$

Its minimum value, taken over all C^2 functions that satisfy the homogeneous Dirichlet boundary conditions, occurs precisely when $u = u_\star$ is the solution to the boundary value problem.

Sturm–Liouville Boundary Value Problems

The most important class of boundary value problems governed by second-order ordinary differential equations was first systematically investigated by the nineteenth-century French mathematicians Jacques Sturm and Joseph Liouville. A *Sturm–Liouville boundary value problem* is based on a second-order ordinary differential equation of the form

$$S[u] = -\frac{d}{dx}\left(p(x)\, \frac{du}{dx} \right) + q(x)\, u = -p(x)\, \frac{d^2u}{dx^2} - p'(x)\, \frac{du}{dx} + q(x)\, u = f(x), \tag{9.71}$$

on a bounded interval $a \le x \le b$, supplemented by Dirichlet, Neumann, mixed, or periodic boundary conditions. To avoid singular points of the differential equation (although we will later discover that most cases of interest have one or more singular points), we assume here that $p(x) > 0$ and, to ensure positive definiteness, $q(x) > 0$ for all $a \le x \le b$.

Sturm–Liouville equations and boundary value problems appear in a remarkably broad range of applications, and particularly in the analysis of partial differential equations by the method of separation of variables. Moreover, most of the important special functions, including Airy functions, Bessel functions, Legendre functions, hypergeometric functions, and so on, naturally appear as solutions to particular Sturm–Liouville equations, [**85, 86**]. In the final two chapters, the analysis of basic linear partial differential equations in curvilinear coordinates, in both two and three dimensions, will require us to solve several particular examples, including the Bessel, Legendre, and Laguerre equations. For now, though, we concentrate on understanding how Sturm–Liouville boundary value problems fit into our self-adjoint and positive definite framework.

Our starting point is the linear operator

$$L[u] = \begin{pmatrix} u' \\ u \end{pmatrix} \tag{9.72}$$

that maps a scalar function $u(x) \in U$ to a vector-valued function $\mathbf{v}(x) = (v_1(x), v_2(x))^T \in V$, whose components are $v_1 = u'$, $v_2 = u$. To compute the adjoint of $L: U \to V$, we use the standard L^2 inner product (9.7) on U, but adopt the following weighted inner product on V:

$$\langle\!\langle \mathbf{v}, \widetilde{\mathbf{v}} \rangle\!\rangle = \int_a^b \left[p(x)\, v_1(x)\, \widetilde{v}_1(x) + q(x)\, v_2(x)\, \widetilde{v}_2(x) \right] dx, \quad \mathbf{v} = \begin{pmatrix} v_1 \\ v_2 \end{pmatrix}, \quad \widetilde{\mathbf{v}} = \begin{pmatrix} \widetilde{v}_1 \\ \widetilde{v}_2 \end{pmatrix}. \tag{9.73}$$

The positivity assumptions on the weight functions p, q ensure that the latter is a bona fide inner product. As usual, the adjoint computation relies on integration by parts. Here, we only need to manipulate the first summand:

$$\begin{aligned} \langle\!\langle L[u], \mathbf{v} \rangle\!\rangle &= \int_a^b (p\, u' v_1 + q\, u\, v_2)\, dx \\ &= p(b)\, u(b)\, v_1(b) - p(a)\, u(a)\, v_1(a) + \int_a^b u \left[-(p v_1)' + q v_2 \right] dx. \end{aligned}$$

The boundary terms will disappear, provided that, at each endpoint, either u or v_1 vanishes. Since for the linear operator $\mathbf{v} = L[u]$ given by (9.72), we can identify $v_1 = u'$, we conclude that any of our usual boundary conditions — Dirichlet, mixed, or Neumann — remain valid here. Under any of these conditions,

$$\langle\!\langle L[u], \mathbf{v} \rangle\!\rangle = \int_a^b u \left[-(p v_1)' + q v_2 \right] dx = \langle u, L^*[\mathbf{v}] \rangle,$$

and so the adjoint operator is given by

$$L^*[\mathbf{v}] = -\frac{d(p v_1)}{dx} + q v_2 = -p v_1' - p' v_1 + q v_2.$$

The canonical self-adjoint combination

$$S[u] = L^* \circ L[u] = L^* \begin{pmatrix} u' \\ u \end{pmatrix} = -\frac{d}{dx}\left(p\, \frac{du}{dx} \right) + q u \tag{9.74}$$

then reproduces the Sturm–Liouville differential operator (9.71). Moreover, since $\ker L = \{0\}$ is trivial (why?), the boundary value problem is positive definite for *all boundary conditions*, not only Dirichlet and mixed, but also Neumann!

A proof of the following general existence theorem can be found in [**63**].

Theorem 9.28. *Let $p(x) > 0$ and $q(x) > 0$ for $a \le x \le b$. Then, for any choice of boundary conditions (including Neumann), the Sturm–Liouville boundary value problem (9.71) admits a unique solution.*

Theorem 9.26 tells us that the solution to the Sturm–Liouville boundary value problem (9.71) can be characterized as the unique minimizer of the quadratic functional

$$Q[u] = \tfrac{1}{2} ||| \, L[u] \, |||^2 - \langle f, u \rangle = \int_a^b \left[\tfrac{1}{2} p(x) u'(x)^2 + \tfrac{1}{2} q(x) u(x)^2 - f(x) u(x) \right] dx \qquad (9.75)$$

among all C^2 functions satisfying the prescribed homogeneous boundary conditions.

Example 9.29. Let $\omega > 0$. Consider the constant-coefficient Sturm–Liouville problem

$$-u'' + \omega^2 u = f(x), \qquad u(0) = u(1) = 0,$$

which we studied earlier in Example 6.10. Theorem 9.28 guarantees the existence of a unique solution. The solution achieves the minimum possible value for the quadratic functional

$$Q[u] = \int_0^1 \left[\tfrac{1}{2} u'^2 + \tfrac{1}{2} \omega^2 u^2 - f u \right] dx$$

among all C^2 functions satisfying the given boundary conditions.

More generally, suppose we adopt a weighted inner product

$$\langle u, \widetilde{u} \rangle = \int_a^b u(x) \, \widetilde{u}(x) \, \rho(x) \, dx \qquad (9.76)$$

on the domain space U, where $\rho(x) > 0$ on $[a, b]$. The same integration by parts computation proves that, when subject to the homogeneous boundary conditions,

$$L^*[\mathbf{v}] = \frac{1}{\rho} \left(-\frac{d(p v_1)}{dx} + q v_2 \right) = -\frac{p}{\rho} v_1' - \frac{p'}{\rho} v_1 + \frac{q}{\rho} v_2,$$

and so the weighted Sturm–Liouville differential operator is

$$S[u] = L^* \circ L[u] = \frac{1}{\rho} \left[-\frac{d}{dx} \left(p \frac{du}{dx} \right) + q u \right]. \qquad (9.77)$$

The corresponding weighted Sturm–Liouville equation $S[u] = f$ has the form

$$S[u] = \frac{1}{\rho(x)} \left[-\frac{d}{dx} \left(p(x) \frac{du}{dx} \right) + q(x) u \right] = -\frac{p(x)}{\rho(x)} \frac{d^2 u}{dx^2} - \frac{p'(x)}{\rho(x)} \frac{du}{dx} + \frac{q(x)}{\rho(x)} u = f(x), \qquad (9.78)$$

which is, in fact, identical to the ordinary Sturm–Liouville equation (9.71) after we replace f by ρf. Be that as it may, the weighted generalization will become important when we study the associated eigenvalue problems.

Example 9.30. Let $m > 0$ be a fixed positive number. Consider the differential equation

$$B[u] = -u'' - \frac{1}{x} u' + \frac{m^2}{x^2} u = f(x), \qquad (9.79)$$

where B is known as the *Bessel differential operator* of *order* m. To place it in weighted Sturm–Liouville form (9.78), we must find $p(x), q(x)$, and $\rho(x)$ that satisfy

$$\frac{p(x)}{\rho(x)} = 1, \qquad \frac{p'(x)}{\rho(x)} = \frac{1}{x}, \qquad \frac{q(x)}{\rho(x)} = \frac{m^2}{x^2}.$$

Dividing the second equation by the first, we see that $p'(x)/p(x) = 1/x$, and hence we can set

$$p(x) = x, \qquad q(x) = \frac{m^2}{x}, \qquad \rho(x) = x.$$

Thus, when subject to homogeneous Dirichlet, mixed, or even Neumann boundary conditions on an interval $0 < a \le x \le b$, the Bessel operator B is positive definite and self-adjoint with respect to the weighted inner product

$$\langle u, \widetilde{u} \rangle = \int_a^b u(x)\, \widetilde{u}(x)\, x\, dx. \tag{9.80}$$

Exercises

9.3.1. Consider the boundary value problem $-u'' = x,\ u(0) = u(1) = 0$. (*i*) Find the solution. (*ii*) Write down a minimization principle that characterizes the solution. (*iii*) What is the value of the minimized quadratic functional on the solution? (*iv*) Write down at least two other functions that satisfy the boundary conditions and check that they produce larger values for the energy.

9.3.2. Answer Exercise 9.3.1 for the boundary value problems
(a) $\dfrac{d}{dx}\left(\dfrac{1}{1+x^2}\dfrac{du}{dx}\right) = x^2,\ u(-1) = u(1) = 0$; (b) $-(e^x u')' = e^{-x},\ u(0) = u'(1) = 0$;
(c) $x^2 u'' + 2x u' = 3x^2,\ u'(1) = u(2) = 0$; (d) $x u'' + 3u' = 1,\ u(-2) = u(-1) = 0$.

9.3.3. Let $Q[u] = \int_0^1 \left[\frac{1}{2}(u')^2 - 5u\right] dx$. (a) Find the function $u_\star(x)$ that minimizes $Q[u]$ among all C^2 functions that satisfy $u(0) = u(1) = 0$.
(b) Test your answer by computing $Q[u_\star]$ and then comparing with the value of $Q[u]$ when $u(x) =$ (*i*) $x - x^2$, (*ii*) $\frac{3}{2}x - \frac{3}{2}x^3$, (*iii*) $\frac{2}{3}\sin \pi x$, (*iv*) $x^2 - x^4$.

9.3.4. For each of the following functionals and associated boundary conditions: (*i*) write down a boundary value problem satisfied by the minimizing function, and (*ii*) find the minimizing function $u_\star(x)$:

(a) $\displaystyle\int_0^1 \left[\tfrac{1}{2}(u')^2 - 3u\right] dx, \quad u(0) = u(1) = 0,$

(b) $\displaystyle\int_0^1 \left[\tfrac{1}{2}(x+1)(u')^2 - 5u\right] dx, \quad u(0) = u(1) = 0,$

(c) $\displaystyle\int_1^3 \left[x(u')^2 + 2u\right] dx, \quad u(1) = u(3) = 0,$

(d) $\displaystyle\int_0^1 \left[\tfrac{1}{2}e^x(u')^2 - (1+e^x)u\right] dx, \quad u(0) = u(1) = 0,$

(e) $\displaystyle\int_{-1}^1 \frac{(x^2+1)(u')^2 + xu}{(x^2+1)^2}\, dx, \quad u(-1) = u(1) = 0.$

9.3.5. Which of the following quadratic functionals possess a unique minimizer among all C^2 functions satisfying the indicated boundary conditions? Find the minimizer if it exists.

(a) $\int_1^2 \left[\,\frac{1}{2}x\,(u')^2 + 2(x-1)\,u\,\right] dx, \quad u(1) = u(2) = 0;$

(b) $\int_{-\pi}^{\pi} \left[\,\frac{1}{2}x\,(u')^2 - u\cos x\,\right] dx, \quad u(-\pi) = u(\pi) = 0;$

(c) $\int_{-1}^{1} \left[\,(u')^2\cos x - u\sin x\,\right] dx, \quad u(-1) = u'(1) = 0;$

(d) $\int_{-2}^{2} \left[\,(1-x^2)\,(u')^2 - u\,\right] dx, \quad u(-2) = u(2) = 0;$

(e) $\int_0^1 \left[\,(x+1)\,(u')^2 - u\,\right] dx, \quad u'(0) = u'(1) = 0.$

9.3.6. Let $D[u] = u'$ be the derivative operator acting on the vector space of C^2 scalar functions $u(x)$ defined for $0 \le x \le 1$ and satisfying the boundary conditions $u(0) = 0$, $u'(1) = 0$.
(a) Given the weighted inner product $\langle\, u\,, \widetilde{u}\,\rangle = \int_0^1 u(x)\,\widetilde{u}(x)\,e^x\,dx$ on both its domain and target spaces, determine the corresponding adjoint operator D^*.
(b) Let $S = D^* \circ D$. Write down and solve the boundary value problem $S[u] = 3e^x$.
(c) Write down a minimization principle that characterizes the solution you found in part (b), or explain why none exists.

9.3.7. Solve the Sturm–Liouville boundary value problem $-4u'' + 9u = 1$, $u(0) = 0$, $u(2) = 0$. Is your solution unique?

9.3.8. Answer Exercise 9.3.7 for the Neumann boundary conditions $u'(0) = 0$, $u'(2) = 0$.

9.3.9. (*i*) Write the following differential equations in Sturm–Liouville form. (*ii*) If possible, write down a minimization principle that characterizes the solutions to the Dirichlet boundary value problem on the interval $[1,2]$. (a) $-e^x u'' - e^x u' = e^{2x}$, (b) $-x u'' - u' + 2u = 1$, (c) $-u'' - 2u' + u = e^x$, (d) $-x^2 u'' + 2x u' + 3u = 1$, (e) $x u'' + (1-x)\,u' + u = 0$.

9.3.10. *True or false*: The Sturm–Liouville operator (9.71) is self-adjoint and positive definite when subject to periodic boundary conditions $u(a) = u(b)$, $u'(a) = u'(b)$.

9.3.11. Does the quadratic functional $Q[u] = \int_0^1 \left[\,\frac{1}{2}(u')^2 - \left(x - \frac{1}{2}\right)u\,\right] dx$ have a minimum value when $u(x)$ is subject to the homogeneous Neumann boundary value conditions $u'(0) = u'(1) = 0$? If so, determine the minimum value and find all minimizing functions.

♡ 9.3.12. (*a*) Determine the adjoint of the differential operator $L[u] = u' + 2x\,u$ with respect to the L^2 inner products on $[0,1]$ when subject to the fixed boundary conditions $u(0) = u(1) = 0$. (*b*) Is the self-adjoint operator $S = L^* \circ L$ positive definite? Explain your answer. (*c*) Write out the boundary value problem represented by $S[u] = f$. (*d*) Find the solution to the boundary value problem when $f(x) = e^{x^2}$. *Hint*: To integrate the differential equation, work with the factored form of the differential operator. (*e*) Discuss what happens if you instead impose the Neumann boundary conditions $u'(0) = u'(1) = 0$.

9.3.13. Discuss the self-adjointness and positive definiteness of boundary value problems associated with the Bessel operator (9.79) of order $m = 0$.

9.3.14. Let $u_\star(x)$ be the solution to the self-adjoint positive definite boundary value problem $S[u_\star] = f$. Prove that if $f(x) \not\equiv 0$, then the minimum of the associated quadratic functional is strictly negative: $Q[u_\star] < 0$.

9.3.15. Find a function $u(x)$ such that $\int_0^1 u''(x)\,u(x)\,dx > 0$. How do you reconcile this with the claimed positivity in (9.70)?

9.3.16. Does the inequality (9.70) hold when $u(x) \not\equiv 0$ is subject to the Neumann boundary conditions $u'(a) = u'(b) = 0$?

9.3.17. *True or false*: When subject to homogeneous Dirichlet boundary conditions on an interval $[a, b]$, every nonsingular second-order linear ordinary differential equation $a(x)\,u'' + b(x)\,u' + c(x)\,u = f(x)$ is (a) self-adjoint, (b) positive definite, (c) positive semi-definite, with respect to some weighted inner product (9.76).

The Dirichlet Principle

Let us now apply these ideas to boundary value problems governed by the Poisson equation

$$-\Delta u = \nabla^* \circ \nabla u = f. \tag{9.81}$$

In the positive definite cases in which the partial differential equation is supplemented by either homogeneous Dirichlet or homogeneous mixed boundary conditions, our general Minimization Theorem 9.24 implies that the solution can be characterized by the justly famous *Dirichlet Principle*.

Theorem 9.31. *The function $u(x, y)$ that minimizes the* Dirichlet integral

$$Q[u] = \tfrac{1}{2}\,|||\,\nabla u\,|||^2 - \langle\, f\,, u\,\rangle = \iint_\Omega \left(\tfrac{1}{2}\,u_x^2 + \tfrac{1}{2}\,u_y^2 - f\,u\right) dx\,dy \tag{9.82}$$

among all C^2 functions that satisfy the prescribed homogeneous Dirichlet or mixed boundary conditions is the solution to the corresponding boundary value problem for the Poisson equation $-\Delta u = f$.

The fact that a minimizer to the Dirichlet integral (9.82) satisfies the Poisson equation is an immediate consequence of our general Minimization Theorem 9.26. On the other hand, proving the *existence* of a C^2 minimizing function is a nontrivial issue. Indeed, the need for a rigorous existence proof was not immediately recognized: arguing from the finite-dimensional situation, Dirichlet deemed existence to be self-evident, but it was not until 50 years later that Hilbert supplied the first rigorous proof — which was one of his primary motivations for introducing the mathematical machinery of Hilbert space.

The Dirichlet principle (9.82) was derived under the assumption that the boundary conditions are homogeneous — either pure Dirichlet or mixed. As it turns out, the minimization principle, as stated, also applies to the inhomogeneous Dirichlet boundary value problem. However, the minimizing functional that characterizes the solution to a mixed boundary value problem with inhomogeneous Neumann conditions on part of the boundary acquires an additional boundary term.

Theorem 9.32. *The solution $u(x, y)$ to the boundary value problem*

$$-\Delta u = f \quad \text{in} \quad \Omega, \qquad u = h \quad \text{on} \quad D \subset \partial\Omega, \qquad \frac{\partial u}{\partial \mathbf{n}} = k \quad \text{on} \quad N = \partial\Omega \setminus D, \tag{9.83}$$

with $D \neq \varnothing$, is characterized as the unique function that minimizes the modified Dirichlet integral

$$\widehat{Q}[u] = \iint_\Omega \left(\tfrac{1}{2}\,u_x^2 + \tfrac{1}{2}\,u_y^2 - f\,u\right) dx\,dy \; - \; \int_N k\,u\,ds \tag{9.84}$$

among all C^2 functions that satisfy the prescribed boundary conditions.

In particular, the inhomogeneous Dirichlet problem has $N = \varnothing$, in which case the extra boundary integral does not appear.

Proof: Write $u(x,y) = \widetilde{u}(x,y) + v(x,y)$, where v is any function that satisfies the given boundary conditions: $v = h$ on D, while $\partial v/\partial \mathbf{n} = k$ on N. (We specifically do not require that v satisfy the Poisson equation.) Their difference $\widetilde{u} = u - v$ satisfies the corresponding homogeneous boundary conditions, along with the modified Poisson equation

$$-\Delta \widetilde{u} = \widetilde{f} \equiv f + \Delta v \quad \text{in} \quad \Omega, \qquad \widetilde{u} = 0 \quad \text{on} \quad D, \qquad \frac{\partial \widetilde{u}}{\partial \mathbf{n}} = 0 \quad \text{on} \quad N.$$

Theorem 9.31 implies that $\widetilde{u}$ minimizes the Dirichlet functional

$$\widetilde{Q}[\,\widetilde{u}\,] = \tfrac{1}{2}\,|||\,\nabla \widetilde{u}\,|||^2 - \langle\!\langle\, \widetilde{f}\,, \widetilde{u}\,\rangle\!\rangle = \iint_\Omega \left(\tfrac{1}{2}\widetilde{u}_x^2 + \tfrac{1}{2}\widetilde{u}_y^2 - \widetilde{f}\,\widetilde{u}\right) dx\,dy$$

among all functions satisfying the homogeneous boundary conditions. We compute

$$\begin{aligned}
\widetilde{Q}[\,\widetilde{u}\,] &= \widetilde{Q}[\,u - v\,] = \tfrac{1}{2}\,|||\,\nabla(u-v)\,|||^2 - \langle\, f + \Delta v\,, u - v\,\rangle \\
&= \tfrac{1}{2}\,|||\,\nabla u\,|||^2 - \langle\, \nabla u\,, \nabla v\,\rangle + \tfrac{1}{2}\,|||\,\nabla v\,|||^2 - \langle\, f\,, u\,\rangle - \langle\, \Delta v\,, u\,\rangle + \langle\, f + \Delta v\,, v\,\rangle \\
&= Q[\,u\,] - \iint_\Omega (\nabla u \cdot \nabla v + u\,\Delta v)\,dx\,dy + C_0,
\end{aligned}$$

where

$$C_0 = \tfrac{1}{2}\,|||\,\nabla v\,|||^2 + \langle\, f + \Delta v\,, v\,\rangle$$

does not depend on u. We then apply formula (6.83) to evaluate the middle terms:

$$\iint_\Omega (\nabla u \cdot \nabla v + u\,\Delta v)\,dx\,dy = \oint_{\partial\Omega} u\,\frac{\partial v}{\partial \mathbf{n}}\,ds = \int_D h\,\frac{\partial v}{\partial \mathbf{n}}\,ds \;+\; \int_N u\,k\,ds.$$

Thus,

$$\widetilde{Q}[\,\widetilde{u}\,] = Q[\,u\,] - \int_N k\,u\,ds + C_1 = \widehat{Q}[\,u\,] + C_1,$$

where the final term $C_1 = C_0 + \displaystyle\int_D h\,\frac{\partial v}{\partial \mathbf{n}}\,ds$ is fixed by the boundary conditions and the choice of v, and so its value does not change when the function u is varied. We conclude that $\widetilde{u}$ minimizes $\widetilde{Q}[\,\widetilde{u}\,]$ if and only if $u = \widetilde{u} + v$ minimizes $\widehat{Q}[\,u\,]$. *Q.E.D.*

Exercises

♡ 9.3.18. (a) Show that the function $u(x,y) = \frac{1}{2}(-x\,y + x\,y^2 + x^2\,y - x^2\,y^2)$ solves the homogeneous Dirichlet boundary value problem for the Poisson equation $-\Delta u = x^2 + y^2 - x - y$ on the unit square $S = \{0 \le x \le 1,\ 0 \le y \le 1\}$. (*b*) Write down the Dirichlet integral (9.82) for this boundary value problem. What is its value for your solution? (*c*) Write down three other functions that satisfy the homogeneous Dirichlet boundary conditions on S, and check that all three have larger Dirichlet integrals.

9.3.19. (a) Suppose $u(x,y)$ solves the boundary value problem $-\Delta u = f$ in Ω and $u = 0$ on $\partial\Omega$, with $f(x,y) \not\equiv 0$. Prove that its Dirichlet integral (9.82) is strictly negative: $Q[\,u\,] < 0$. (*b*) Does this result hold for the inhomogeneous boundary value problem $u = h$ on $\partial\Omega$?

♡ 9.3.20. Consider the boundary value problem $-\Delta u = 1$, $x^2+y^2<1$, $u=0$, $x^2+y^2=1$. (*a*) Find all solutions. (*b*) Formulate the Dirichlet minimization principle for this problem. Carefully indicate the function space over which you are minimizing. Make sure your solution belongs to the function space. (*c*) Which of the following functions belong to your function space? (*i*) $1-x^2-y^2$, (*ii*) $1-\frac{1}{2}x^2-\frac{1}{2}y^2$, (*iii*) $x-x^3-xy^2$, (*iv*) $x^4-x^2y^2+y^4$, (*v*) $\frac{1}{2}e^{-x^2-y^2}-\frac{1}{2}e^{-1}$. (*d*) For each function in part (*c*) that does belong to your function space, verify that its Dirichlet integral is larger than your solution's value.

9.3.21. Suppose $\lambda>0$. Under what conditions does the inhomogeneous Neumann problem $-\Delta u+\lambda u=f$ in Ω, $\partial u/\partial \mathbf{n}=k$ on $\partial\Omega$, for the Helmholtz equation have a solution? Is the solution unique? *Hint*: Is the boundary value problem positive definite?

◇ 9.3.22. Suppose $\kappa(x)>0$ for all $a\le x\le b$.

(*a*) Prove that the solution $u_\star(x)$ to the inhomogeneous Dirichlet boundary value problem

$$-\frac{d}{dx}\left(\kappa(x)\,\frac{du}{dx}\right)=f(x),\qquad u(a)=\alpha,\qquad u(b)=\beta,$$

minimizes the functional $Q[u]=\int_a^b\left[\,\frac{1}{2}\,\kappa(x)\,u'(x)^2-f(x)\,u(x)\,\right]dx.$

Hint: Mimic the proof of Theorem 9.32.

(*b*) Construct a minimization principle for the mixed boundary value problem

$$-\frac{d}{dx}\left(\kappa(x)\,\frac{du}{dx}\right)=f(x),\qquad u(a)=\alpha,\qquad u'(b)=\beta.$$

9.3.23. Use the result of Exercise 9.3.22 to find the C^2 function $u_\star(x)$ that minimizes the integral $Q[u]=\displaystyle\int_1^2\left[\frac{x}{2}\left(\frac{du}{dx}\right)^2+x^2u\right]dx$ when subject to the boundary conditions $u(1)=0$, $u(2)=1$.

9.3.24. Find the function $u(x)$ that minimizes the integral $Q[u]=\displaystyle\int_1^2\,[\,x(u')^2+x^2u\,]\,dx$ subject to the boundary conditions $u(1)=1$, $u'(2)=0$. *Hint*: Use Exercise 9.3.22(*b*).

9.3.25. Prove that the functional $Q[u]=\displaystyle\int_0^1(u')^2\,dx$, when subject to the mixed boundary conditions $u(0)=0$, $u'(1)=1$, has no minimizer.

♡ 9.3.26. Let $p_1(x,y), p_2(x,y), q(x,y)>0$ be strictly positive functions on a closed, bounded, connected domain $\overline{\Omega}\subset\mathbb{R}^2$. Consider the boundary value problem for the second-order partial differential equation

$$-\frac{\partial}{\partial x}\left(p_1(x,y)\,\frac{\partial u}{\partial x}\right)-\frac{\partial}{\partial y}\left(p_2(x,y)\,\frac{\partial u}{\partial y}\right)+q(x,y)\,u=f(x,y),\qquad (x,y)\in\Omega,\qquad (9.85)$$

subject to homogeneous Dirichlet boundary conditions $u=0$ on $\partial\Omega$.
(*a*) *True or false*: Equation (9.85) is an elliptic partial differential equation. (*b*) Write the boundary value problem in self-adjoint form $L^*\circ L[u]=f$. *Hint*: Regard (9.85) as a "two-dimensional Sturm–Liouville equation". (*c*) Prove that this boundary value problem is positive definite, and then find a minimization principle that characterizes the solution. (*d*) Find suitable homogeneous Neumann-type boundary conditions involving the values of the derivatives of u on $\partial\Omega$ that make the resulting boundary value problem for (9.85) self-adjoint. Is your boundary value problem positive definite? Why or why not?

9.4 Eigenvalues and Eigenfunctions

We have already come to appreciate the value of eigenfunctions for constructing separable solutions to dynamical partial differential equations such as the one-dimensional heat and wave equations. In both cases, the eigenfunctions are trigonometric, and are used to write the solution to the initial value problem in the form of a Fourier series. The most important feature is that the Fourier eigenfunctions are orthogonal with respect to the underlying L^2 inner product. As we remarked earlier, orthogonality is not an accident. Rather, it is a direct consequence of the self-adjointness of the linear differential operator prescribing the eigenvalue equation. The goal of this section is, in preparation for extending the eigenfunction method to higher-dimensional and more general dynamical problems, to establish the orthogonality property of eigenfunctions in general, discuss how positive (semi-)definiteness affects the eigenvalues, and present the basic theory of eigenfunction series expansions, thereby significantly generalizing basic Fourier series. As an application, we deduce a general formula for the Green's function of a positive definite boundary value problem as an infinite series in the eigenfunctions, and use this to formulate a condition that guarantees completeness of the eigenfunctions. Along the way, we also need to introduce an important minimization principle, the Rayleigh quotient, that characterizes the eigenvalues of a positive definite linear system.

We begin with the *eigenvalue problem*

$$S[v] = \lambda v \tag{9.86}$$

for a linear operator $S: U \to U$ on† a real or complex vector space U. Clearly, $v = 0$ solves the eigenvalue equation no matter what the scalar λ is. If the homogeneous linear system (9.86) admits a *nonzero* solution $0 \neq v \in U$, then $\lambda \in \mathbb{C}$ is called an *eigenvalue* of the operator S and v a corresponding *eigenvector* or *eigenfunction*, depending on the context. If λ is an eigenvalue, then the corresponding *eigenspace* is the subspace

$$V_\lambda = \ker(S - \lambda\, \mathrm{I}) = \{\, v \mid S[v] = \lambda v \,\} \subset U, \tag{9.87}$$

consisting of all the eigenvectors/eigenfunctions along with the zero element. To avoid technical difficulties, we will work under the assumption that all the eigenspaces are finite-dimensional, and we call $1 \leq \dim V_\lambda < \infty$ the *geometric multiplicity* of the eigenvalue λ. Finite-dimensionality is almost always valid, and indeed, will be later established for regular boundary value problems on bounded domains.

Self–Adjoint Operators

In the applications considered here, the vector space U comes equipped with an inner product, and S is a self-adjoint linear operator. In such instances, one can readily establish the basic orthogonality property of the eigenvectors/eigenfunctions.

Theorem 9.33. *If $S = S^*$ is a self-adjoint linear operator on an inner product space U, then all its eigenvalues are real. Moreover, the eigenvectors/eigenfunctions associated with different eigenvalues are automatically orthogonal.*

† As discussed earlier, in the infinite-dimensional case, the differential operator S might be only defined on a dense subspace of U consisting of sufficiently smooth functions.

Proof: To prove the first part of the theorem, suppose λ is a complex eigenvalue, so that $S[v] = \lambda v$ for some complex eigenvector/eigenfunction $v \neq 0$. Then, using the sesquilinearity (B.19) of the underlying Hermitian inner product[‡] and self-adjointness (9.45) of S, we find

$$\lambda \| v \|^2 = \langle \lambda v , v \rangle = \langle S[v] , v \rangle = \langle v , S[v] \rangle = \langle v , \lambda v \rangle = \overline{\lambda} \| v \|^2.$$

Since $v \neq 0$, this immediately implies that $\lambda = \overline{\lambda}$, its complex conjugate, and hence λ must necessarily be real.

To prove orthogonality, suppose $S[u] = \lambda u$ and $S[v] = \mu v$. Again by self-adjointness,

$$\lambda \langle u , v \rangle = \langle \lambda u , v \rangle = \langle S[u] , v \rangle = \langle u , S[v] \rangle = \langle u , \mu v \rangle = \mu \langle u , v \rangle,$$

where the final equality relies on the fact that the eigenvalue μ is real. Therefore, the assumption that $\lambda \neq \mu$ immediately implies orthogonality: $\langle u , v \rangle = 0$. *Q.E.D.*

Thus, the eigenvalues of self-adjoint linear operators are necessarily real. If, in addition, the operator is positive definite, then its eigenvalues must, in fact, be positive.

Theorem 9.34. *If $S > 0$ is a self-adjoint positive definite linear operator, then all its eigenvalues are strictly positive*: $\lambda > 0$. *If $S \geq 0$ is self-adjoint and positive semi-definite, then its eigenvalues are nonnegative*: $\lambda \geq 0$.

Proof: Self-adjointness assures us that all of the eigenvalues are real. Suppose $S[u] = \lambda u$ with $u \neq 0$ a real eigenfunction. Then

$$\lambda \| u \|^2 = \lambda \langle u , u \rangle = \langle \lambda u , u \rangle = \langle S[u] , u \rangle > 0,$$

by positive definiteness. Since $\| u \|^2 > 0$, this immediately implies that $\lambda > 0$. The same argument implies that $\lambda \geq 0$ in the positive semi-definite case. *Q.E.D.*

All the linear operators to be considered in this text are real, and, at the very least, self-adjoint, and often either positive definite or semi-definite. Thus, we will restrict our attention from here on (at least until we reach the Schrödinger equation in the final subsection) to real operators defined on real vector spaces, knowing a priori that we are not overlooking any eigenvalues or eigenfunctions by this restriction.

Example 9.35. In finite dimensions, if we equip $U = \mathbb{R}^n$ with the dot product, then any self-adjoint linear function is given by multiplication by an $n \times n$ symmetric matrix: $S[u] = K\mathbf{u}$, where $K^T = K$. Theorem 9.33 implies the well-known result that a symmetric matrix has only real eigenvalues. Moreover, the eigenvectors associated with different eigenvalues are mutually orthogonal.

In fact, it can be proved that, in general, the eigenvectors of a symmetric matrix are complete, [**89**]. In other words, there exists an orthogonal basis $\mathbf{v}_1, \ldots, \mathbf{v}_n$ of $\mathbb{R}^n$ consisting of eigenvectors of K, so $K\mathbf{v}_j = \lambda_j \mathbf{v}_j$ for $j = 1, \ldots, n$. If the eigenvalues $\lambda_1, \ldots, \lambda_n$ are all simple, so $\lambda_i \neq \lambda_j$ for $i \neq j$, then the basis eigenvectors are automatically orthogonal. When K has repeated eigenvalues, this requires selecting an orthogonal basis of each of the associated eigenspaces $V_\lambda = \ker(K - \lambda \mathrm{I})$, e.g., using the Gram–Schmidt process.

[‡] We are temporarily working in the vector space of complex-valued functions. Once we establish reality of the eigenvalues and eigenfunctions, we can shift our focus back to the real function space.

Completeness implies that the number of linearly independent eigenvectors associated with an eigenvalue, i.e., its geometric multiplicity, is the same as the eigenvalue's algebraic multiplicity. If, furthermore, the matrix $K > 0$ is symmetric and positive definite, then Theorem 9.34 implies that all its eigenvalues are positive: $\lambda_j > 0$. In this case, thanks to completeness, the converse is also valid: a symmetric matrix is positive definite if and only if it has all positive eigenvalues. These results can all be immediately generalized to self-adjoint matrices under general inner products on $\mathbb{R}^n$.

Example 9.36. Consider the Dirichlet eigenvalue problem

$$-\frac{d^2 v}{dx^2} = \lambda\, v, \qquad v(0) = 0, \qquad v(\ell) = 0,$$

for the differential operator $S = -D^2$ on an interval of length $\ell > 0$. As we know — see, for instance, Section 4.1 — the eigenvalues and eigenfunctions are

$$\lambda_n = \left(\frac{n\,\pi}{\ell}\right)^2, \qquad v_n(x) = \sin\frac{n\pi x}{\ell}, \qquad n = 1, 2, 3, \ldots.$$

We now understand this example in our general framework. The fact that the eigenvalues are real and positive follows from the fact that the boundary value problem is defined by the self-adjoint positive definite operator

$$S[u] = D^* \circ D[u] = -D^2[u] = -u'',$$

acting on the vector space $U = \{u(0) = u(\ell) = 0\}$, equipped with the L^2 inner product:

$$\langle u, v \rangle = \int_0^\ell u(x)\, v(x)\, dx.$$

The orthogonality of the Fourier sine functions,

$$\langle v_m, v_n \rangle = \int_0^\ell \sin\frac{m\pi x}{\ell}\, \sin\frac{n\pi x}{\ell}\, dx = 0 \qquad \text{for} \qquad m \neq n,$$

is also an automatic consequence of their status as eigenfunctions of this self-adjoint boundary value problem.

Example 9.37. Similarly, the periodic boundary value problem

$$-v'' = \lambda v, \qquad v(-\pi) = v(\pi), \qquad v'(-\pi) = v'(\pi), \tag{9.88}$$

has eigenvalues $\lambda_0 = 0$, with eigenfunction $v_0(x) \equiv 1$, and $\lambda_n = n^2$, for $n = 1, 2, 3, \ldots$, each possessing two independent eigenfunctions: $v_n(x) = \cos n x$ and $\widetilde{v}_n(x) = \sin n x$. In this case, a zero eigenvalue appears because $S = D^* \circ D = -D^2$ is only positive semi-definite on the space of periodic functions. Theorem 9.33 implies the all-important orthogonality of the Fourier eigenfunctions corresponding to *different* eigenvalues: $\langle v_m, v_n \rangle = \langle v_m, \widetilde{v}_n \rangle = \langle \widetilde{v}_m, \widetilde{v}_n \rangle = 0$ for $m \neq n$, under the L^2 inner product on $[-\pi, \pi]$. However, since they have the same eigenvalue, the orthogonality of $v_n(x) = \cos n x$ and $\widetilde{v}_n(x) = \sin n x$, while true, is not ensured and must be checked by hand.

Example 9.38. On the other hand, the self-adjoint boundary value problem

$$-\frac{d^2 u}{dx^2} = \lambda\, u, \qquad \lim_{x \to \infty} u(x) = 0, \qquad \lim_{x \to -\infty} u(x) = 0, \tag{9.89}$$

on the real line has no eigenvalues: no matter what the value of λ, the only solution decaying to 0 at both $\pm\infty$ is the zero solution. Indeed, exponential solutions that decay at one end become infinitely large at the other. The trigonometric functions $u(x) = \cos\omega x$ and $\sin\omega x$ satisfy the differential equation when $\lambda = \omega^2 > 0$, but do not go to zero as $|x| \to \infty$, and so do not qualify as bona fide eigenfunctions. Rather, because they are bounded on the entire line, they represent the "continuous spectrum" of the underlying self-adjoint differential operator, [**95**]. In this particular context, the continuous spectrum leads directly to the Fourier transform.

Example 9.39. The eigenvalue problem for the Bessel differential operator of order m, given in (9.79), is governed by the following differential equation:

$$S[u] = -u'' - \frac{1}{x}u' + \frac{m^2}{x^2}u = \lambda u, \tag{9.90}$$

or, equivalently,

$$x^2\frac{d^2u}{dx^2} + x\frac{du}{dx} + (\lambda x^2 - m^2)u = 0,$$

supplemented by appropriate homogeneous boundary conditions at the endpoints of the interval $0 \le a < b$. Its eigenfunctions are not elementary, but, as we will learn in Chapter 11, can be expressed in terms of Bessel functions. Nevertheless, no matter what their eventual formula, Theorem 9.33 guarantees the orthogonality of any two eigenfunctions $v, \widetilde{v}$ associated with distinct eigenvalues $\lambda \neq \widetilde{\lambda}$ under the weighted inner product (9.80):

$$\langle v, \widetilde{v} \rangle = \int_a^b v(x)\,\widetilde{v}(x)\,x\,dx = 0.$$

Example 9.40. According to equation (9.60), on a bounded domain $\Omega \subset \mathbb{R}^2$, the (negative) Laplacian $-\Delta$ forms a self-adjoint positive (semi-)definite operator under the L^2 inner product (9.22) when subject to one of the usual sets of homogeneous boundary conditions. Let us, for specificity, concentrate on the Dirichlet case. The *eigenfunctions* of the Laplacian are the nonzero solutions to the following boundary value problem:

$$-\Delta v = \lambda v \quad \text{in} \quad \Omega, \qquad u = 0 \quad \text{on} \quad \partial\Omega. \tag{9.91}$$

The underlying partial differential equation, namely

$$\frac{\partial^2 v}{\partial x^2} + \frac{\partial^2 v}{\partial y^2} + \lambda v = 0,$$

is known as the *Helmholtz equation*, named after the influential and wide-ranging German applied mathematician Hermann von Helmholtz. As we will see, the Helmholtz equation plays a central role in the solution of the two-dimensional heat, wave, and Schrödinger equations.

Only in a few special cases, e.g., rectangles and circular disks, can the eigenfunctions and eigenvalues be determined exactly; see Chapter 11 for details. Nevertheless, Theorem 9.34 guarantees that, for all domains, the eigenvalues are always nonnegative, $\lambda \ge 0$, with $\lambda_0 = 0$ being an eigenvalue only in positive semi-definite cases, e.g., Neumann boundary conditions. Moreover, Theorem 9.33 ensures the orthogonality of any two eigenfunctions,

$$\langle v, \widetilde{v} \rangle = \iint_\Omega v(x,y)\,\widetilde{v}(x,y)\,dx\,dy = 0,$$

that are associated with distinct eigenvalues $\lambda \neq \widetilde{\lambda}$.

The Rayleigh Quotient

We have already learned how to characterize the solutions of positive definite boundary value problems by a minimization principle. One can also characterize their eigenvalues by a minimization principle, named after the prolific nineteenth-century English applied mathematician Lord Rayleigh (John Strutt).

Definition 9.41. Let $S: U \to U$ be a self-adjoint linear operator on an inner product space. The *Rayleigh quotient* of S is defined as

$$R[u] = \frac{\langle u, S[u] \rangle}{\| u \|^2} \qquad \text{for} \qquad 0 \neq u \in U. \tag{9.92}$$

We are, in fact, primarily interested in the Rayleigh quotient of positive definite operators, for which $R[u] > 0$ for all $u \neq 0$. If $S = L^* \circ L$, then, using (9.51), we can rewrite the Rayleigh quotient in the alternative form

$$R[u] = \frac{||| L[u] |||^2}{\| u \|^2}, \tag{9.93}$$

keeping in mind our notational convention (9.50) for the respective norms on U and V.

Theorem 9.42. *Let S be a self-adjoint linear operator. Then the minimum value of its Rayleigh quotient,*

$$\lambda_\star = \min \{ R[u] \mid u \neq 0 \}, \tag{9.94}$$

is the smallest eigenvalue of the operator S. Moreover, any $0 \neq v_\star \in U$ that achieves this minimum value, $R[v_\star] = \lambda_\star$, is an associated eigenvector/eigenfunction: $S[v_\star] = \lambda_\star v_\star$.

Proof: Suppose that $v_\star \in U$ is a minimizing element, and

$$\lambda_\star = R[v_\star] = \frac{\langle v_\star, S[v_\star] \rangle}{\| v_\star \|^2} \tag{9.95}$$

the minimum value. Given any $u \in U$, define the scalar function[†]

$$\begin{aligned} g(t) = R[v_\star + t\,u] &= \frac{\langle v_\star + t\,u, S[v_\star + t\,u] \rangle}{\| v_\star + t\,u \|^2} \\ &= \frac{\langle v_\star, S[v_\star] \rangle + 2t \langle u, S[v_\star] \rangle + t^2 \langle u, S[u] \rangle}{\| v_\star \|^2 + 2t \langle u, v_\star \rangle + t^2 \| u \|^2}, \end{aligned}$$

where we used the self-adjointness of S and the fact that we are working in a real inner product space to identify the terms

$$\langle u, S[v_\star] \rangle = \langle S[u], v_\star \rangle = \langle v_\star, S[u] \rangle.$$

Since

$$g(0) = R[v_\star] \le R[v_\star + t\,u] = g(t),$$

the function $g(t)$ will attain its minimum value at $t = 0$. Elementary calculus tells us that

$$0 = g'(0) = 2\, \frac{\langle u, S[v_\star] \rangle \| v_\star \|^2 - \langle v_\star, S[v_\star] \rangle \langle u, v_\star \rangle}{\| v_\star \|^4}.$$

[†] $g(t)$ is not defined if $v_\star + t\,u = 0$, but this does not affect the argument.

Therefore, using (9.95) to replace $\langle v_\star , S[v_\star] \rangle$ by $\lambda_\star \| v_\star \|^2$, we must have

$$\langle u , S[v_\star] \rangle - \lambda_\star \langle u , v_\star \rangle = \langle u , S[v_\star] - \lambda_\star v_\star \rangle = 0. \tag{9.96}$$

The only way the inner product in (9.96) can vanish for all possible $u \in U$ is if

$$S[v_\star] = \lambda_\star v_\star, \tag{9.97}$$

which means that $0 \neq v_\star$ is an eigenfunction and $\lambda_\star$ its associated eigenvalue.

On the other hand, if v is any eigenfunction, so $S[v] = \lambda v$, where, by self-adjointness, the eigenvalue λ is necessarily real, then the value of its Rayleigh quotient is

$$R[v] = \frac{\langle v , S[v] \rangle}{\| v \|^2} = \frac{\langle v , \lambda v \rangle}{\| v \|^2} = \lambda. \tag{9.98}$$

Since $\lambda_\star$ was, by definition, the smallest possible value of the Rayleigh quotient, it thus must necessarily be the smallest eigenvalue. *Q.E.D.*

Remark: The existence of a minimizing function is not addressed in this result, and, indeed, there may be no minimum eigenvalue; the infimum of the set of eigenvalues could be $-\infty$ or, even if finite, not an eigenvalue. However, for the positive definite boundary value problems considered here, the eigenvalues are all strictly positive, and one can, with some additional analysis, [**44**], prove the existence of a minimizing eigenfunction, and hence a smallest positive eigenvalue.

We label the eigenvalues in increasing order, so that, assuming positive definiteness, $0 < \lambda_1 \leq \lambda_2 \leq \lambda_3 \leq \cdots$, with λ_1 the minimum eigenvalue and hence the minimum value of the Rayleigh quotient. To characterize the other eigenvalues, we need to restrict the class of functions over which one minimizes. Indeed, since the nth eigenfunction v_n must be orthogonal to all its predecessors $v_1, \ldots, v_{n-1}$, it makes sense to try minimizing the Rayleigh quotient over such elements.

Theorem 9.43. *Let $v_1, \ldots, v_{n-1}$ be eigenfunctions corresponding to the first $n-1$ eigenvalues $0 < \lambda_1 \leq \cdots \leq \lambda_{n-1}$ of the positive definite self-adjoint linear operator S. Let*

$$U_{n-1} = \{ \, u \mid \langle u , v_1 \rangle = \cdots = \langle u , v_{n-1} \rangle = 0 \, \} \subset U \tag{9.99}$$

be the set of functions that are orthogonal to the indicated eigenfunctions. Then the minimum value of the Rayleigh quotient function restricted to the subspace U_{n-1} is the nth eigenvalue of S, that is,

$$\lambda_n = \min \{ \, R[u] \mid 0 \neq u \in U_{n-1} \, \}, \tag{9.100}$$

and any minimizer is an associated eigenfunction v_n.

Proof: We follow the preceding proof, but now restrict $v_\star$ and u to belong to the subspace U_{n-1}. Observe that $S[u] \in U_{n-1}$ whenever $u \in U_{n-1}$, because, by self-adjointness,

$$\langle S[u] , v_j \rangle = \langle u , S[v_j] \rangle = \lambda_j \langle u , v_j \rangle = 0 \qquad \text{for} \qquad j = 1, \ldots, n-1.$$

Thus, if $0 \neq v_\star \in U_{n-1}$ minimizes the Rayleigh quotient, then (9.96) holds for arbitrary $u \in U_{n-1}$. In particular, choosing $u = S[v_\star] - \lambda_\star v_\star$, we conclude that $v_\star$ satisfies the eigenvalue equation (9.97), and hence must be an eigenfunction that is orthogonal to the first $n-1$ eigenfunctions. This means that $\lambda_\star = \lambda_n$ must be the next-lowest eigenvalue and $v_\star = v_n$ one of its associated eigenfunctions. *Q.E.D.*

Example 9.44. Return to the Dirichlet eigenvalue problem on the interval $[0,\ell]$ for the self-adjoint (under the L^2 inner product) differential operator $-D^2 = D^* \circ D$ discussed in Example 9.36. Its Rayleigh quotient can be written as

$$R[u] = \frac{\langle u\,, -u''\rangle}{\|u\|^2} = -\frac{\displaystyle\int_0^\ell u(x)\,u''(x)\,dx}{\displaystyle\int_0^\ell u(x)^2\,dx} = \frac{\displaystyle\int_0^\ell u'(x)^2\,dx}{\displaystyle\int_0^\ell u(x)^2\,dx} = \frac{|||\,u'\,|||^2}{\|u\|^2},$$

where the second expression, based on the alternative form (9.93), can be readily deduced from the first via an integration by parts. (Here, both domain and target space of $L = D$ use the same L^2 norm.) According to Theorem 9.42, the minimum value of $R[u]$ over all nonzero functions $u(x) \not\equiv 0$ satisfying the boundary conditions $u(0) = u(\ell) = 0$ is the lowest eigenvalue, namely

$$\lambda_1 = \frac{\pi^2}{\ell^2} = \min\left\{ R[u] \;\middle|\; u(0) = u(\ell) = 0,\ u(x) \not\equiv 0 \right\},$$

which is achieved if and only if $u(x)$ is a nonzero constant multiple of $\sin(\pi x/\ell)$, the corresponding eigenfunction. The reader is invited to numerically test this result by fixing a value of ℓ, and then evaluating $R[u]$ on various functions $u(x)$ satisfying the boundary conditions to check that the numerical value is always larger than π^2/ℓ^2, the smallest eigenvalue. The second eigenvalue can be found by minimizing over all nonzero functions that are orthogonal to the first eigenfunction:

$$\lambda_2 = \frac{4\,\pi^2}{\ell^2} = \min\left\{ R[u] \;\middle|\; u(0) = u(\ell) = 0,\ \int_0^\ell u(x)\sin\frac{\pi}{\ell}\,x\,dx = 0,\ u(x) \not\equiv 0 \right\},$$

and similarly for the higher eigenvalues.

Example 9.45. Consider the Helmholtz eigenvalue problem (9.91) on a bounded domain $\Omega \subset \mathbb{R}^2$, subject to Dirichlet boundary conditions. The associated Rayleigh quotient (9.93) can be written in the form

$$R[u] = \frac{|||\,\nabla u\,|||^2}{\|u\|^2} = \frac{\displaystyle\iint_\Omega \left[\left(\frac{\partial u}{\partial x}\right)^2 + \left(\frac{\partial u}{\partial y}\right)^2\right] dx\,dy}{\displaystyle\iint_\Omega u(x,y)^2\,dx\,dy}. \tag{9.101}$$

Its minimum value among all nonzero functions $u(x,y) \not\equiv 0$ subject to the boundary conditions $u = 0$ on $\partial\Omega$ is the smallest eigenvalue λ_1, and the minimizing function is any nonzero constant multiple of the associated eigenfunction $v_1(x,y)$. To obtain a higher eigenvalue λ_n, one minimizes $R[u]$, where $u(x,y) \not\equiv 0$ again satisfies the boundary conditions and, in addition, is orthogonal to the preceding $n-1$ eigenfunctions:

$$0 = \langle u\,, v_k\rangle = \iint_\Omega u(x,y)\,v_k(x,y)\,dx\,dy, \qquad \text{for} \qquad k = 1, \ldots, n-1.$$

It can be proved, [**34, 44**], that, as long as the domain is bounded with, as always, a reasonably nice boundary, there is a solution to each of these minimization problems, and hence the Helmholtz equation admits an infinite sequence of positive eigenvalues $0 < \lambda_1 \leq \lambda_2 \leq \lambda_3 \leq \cdots$, with $\lambda_n \to \infty$ becoming arbitrarily large as $n \to \infty$; see also Theorem 9.47 below.

Eigenfunction Series

For our applications to dynamical partial differential equations, we will be particularly interested in expanding more general functions in terms of the orthogonal eigenfunctions, the simplest case being the classical Fourier series. To fix notation, we will proceed as if we were treating a one-dimensional boundary value problem, although the formulas are equally valid for higher-dimensional problems, e.g., those governed by the Helmholtz equation. Thus, we consider an eigenvalue problem of the form $S[v] = \lambda v$, where S is a positive definite or semi-definite operator that is self-adjoint relative to a weighted L^2 inner product

$$\langle v , \widetilde{v} \rangle = \int_a^b v(x)\, \widetilde{v}(x)\, \rho(x)\, dx, \tag{9.102}$$

with $\rho(x) > 0$ on the bounded interval $a \leq x \leq b$.

Let $0 \leq \lambda_1 \leq \lambda_2 \leq \lambda_3 \leq \cdots$ be the eigenvalues, and $v_1, v_2, v_3, \ldots$, the corresponding eigenfunctions. Theorem 9.33 assures us that those corresponding to different eigenvalues are mutually orthogonal:

$$\langle v_j , v_k \rangle = 0, \qquad j \neq k. \tag{9.103}$$

Orthogonality is not automatic if v_j and v_k belong to the same eigenvalue, but it can be ensured by selecting an orthogonal basis of each eigenspace V_λ, if necessary by applying the Gram–Schmidt orthogonalization process, [**89**].

Let $f \in U$ be an arbitrary function in our inner product space. The *eigenfunction series* of f is, by definition, its generalized Fourier series:

$$f \sim \sum_k c_k\, v_k, \qquad \text{where the coefficient} \qquad c_k = \frac{\langle f , v_k \rangle}{\| v_k \|^2} \tag{9.104}$$

is found by formally taking the inner product of both sides of (9.104) with the eigenfunction v_k and invoking their mutual orthogonality. (Note that our earlier eigenfunction series formula (3.108) assumed orthonormality; here, it will be convenient to not necessarily impose the condition $\| v_k \| = 1$.) For example, in the case covered by Example 9.36, (9.104) becomes the usual Fourier sine series for the function f, whereas for Example 9.37, it represents its full periodic Fourier series. In a similar fashion, Example 9.40 leads to series in the eigenfunctions of the Laplacian operator on a bounded domain subject to appropriate homogeneous boundary conditions; explicit examples of the latter can be found in Chapters 11 and 12.

As we learned in Section 3.5, convergence (in norm) of the series (9.104) requires completeness of the eigenfunctions. (Pointwise and uniform convergence are then implied by more restrictive hypotheses on the function and the domain, e.g., $f \in \mathrm{C}^1$.) In the finite-dimensional context, when $S\colon \mathbb{R}^n \to \mathbb{R}^n$ is given by matrix multiplication, $S[\mathbf{u}] = K\mathbf{u}$, there are only finitely many eigenvectors, and so the summation (9.104) has only finitely many terms. There are, hence, no convergence considerations, and completeness is automatic. For boundary value problems in infinite-dimensional function space, the completeness of the resulting eigensolutions is a more delicate issue. In Example 9.36, the eigenvalue problem for $S = -D^2$ subject to homogeneous Dirichlet boundary conditions on a bounded interval leads to the Fourier sine eigenfunctions, which we know to be complete. On the other hand, the corresponding eigenvalue problem on the real line, treated in Example 9.38, has *no* eigenfunctions, and so completeness is out of the question. As we will see, the eigenfunctions associated with regular boundary value problems on bounded domains are automatically complete, whereas singular problems and problems on unbounded domains require additional analysis.

Whether or not the eigenfunctions are complete, we always have *Bessel's inequality*[†] (3.117):

$$\sum_k c_k^2 \, \| \, v_k \, \|^2 \;\leq\; \| \, f \, \|^2. \tag{9.105}$$

Theorem 3.43 says that the eigenfunctions are complete if and only if Bessel's inequality is an equality, which is then the Plancherel formula for the eigenfunction expansion.

Green's Functions and Completeness

We now combine two of our principal themes. Remarkably, the key to the completeness of eigenfunctions for boundary value problems lies in the eigenfunction expansion of the Green's function! Assume that S is both self-adjoint and positive definite. Thus, by Theorem 9.34, all its eigenvalues are positive. We index them in increasing order:

$$0 < \lambda_1 \leq \lambda_2 \leq \lambda_3 \leq \; \cdots \; , \tag{9.106}$$

where each eigenvalue is repeated according to its multiplicity.

By positive definiteness, the boundary value problem $S[u] = f$ has a unique solution.[‡] Therefore, it admits a Green's function $G_\xi(x) = G(x;\xi)$, which satisfies the boundary value problem

$$S[G_\xi] = \delta_\xi, \tag{9.107}$$

with a delta function impulse on the right-hand side. For each fixed ξ, let us write down the eigenfunction series (9.104) for the Green's function:

$$G(x;\xi) = \sum_{k=1}^{\infty} c_k(\xi)\, v_k(x), \qquad \text{where the coefficient} \qquad c_k(\xi) = \frac{\langle \, G_\xi \, , v_k \, \rangle}{\| \, v_k \, \|^2} \tag{9.108}$$

depends on the impulse point ξ. Since $S[v_k] = \lambda_k v_k$, the coefficients can be explicitly evaluated by means of the following calculation:

$$\begin{aligned} \lambda_k \, c_k(\xi) \, \| \, v_k \, \|^2 &= \langle \, G_\xi \, , \lambda_k v_k \, \rangle = \langle \, G_\xi \, , S[v_k] \, \rangle \\ &= \langle \, S[G_\xi] \, , v_k \, \rangle = \langle \, \delta_\xi \, , v_k \, \rangle = \int_a^b \delta(x-\xi)\, v_k(x)\, \rho(x)\, dx = v_k(\xi)\, \rho(\xi), \end{aligned}$$

where $\rho(x)$ is the weight function of our inner product (9.102), and we invoked the self-adjointness of S. Solving for

$$c_k(\xi) = \frac{v_k(\xi)\, \rho(\xi)}{\lambda_k \, \| \, v_k \, \|^2} \tag{9.109}$$

† Formula (3.117) assumed orthonormality of the functions; here we are stating the analogous result for orthogonal elements. Moreover, here, the eigenfunctions and hence the coefficients c_k are all real, so we don't need absolute value signs.

‡ As usual, we are assuming existence of the solution; Proposition 9.19 guarantees uniqueness.

and then substituting back into (9.108), we deduce the explicit eigenfunction series

$$G(x;\xi) \sim \sum_{k=1}^{\infty} \frac{v_k(x)\, v_k(\xi)\, \rho(\xi)}{\lambda_k \,\|\, v_k \,\|^2} \tag{9.110}$$

for the Green's function. Observe that this expression is compatible with the weighted symmetry equation (9.58).

Example 9.46. According to Example 6.9, the Green's function for the L^2 self-adjoint boundary value problem

$$-u'' = f(x), \qquad u(0) = 0 = u(1),$$

is

$$G(x;\xi) = \begin{cases} x(1-\xi), & x \le \xi, \\ \xi(1-x), & x \ge \xi. \end{cases} \tag{9.111}$$

On the other hand, the eigenfunctions for

$$-v'' = \lambda\, v, \qquad v(0) = 0 = v(1),$$

are $v_k(x) = \sin k\pi x$, with corresponding eigenvalues $\lambda_k = k^2\pi^2$, for $k = 1, 2, 3, \ldots$. Since

$$\|\, v_k \,\|^2 = \int_0^1 \sin^2 k\pi x\, dx = \tfrac{1}{2},$$

formula (9.110) implies the eigenfunction expansion

$$G(x;\xi) = \sum_{k=1}^{\infty} \frac{2 \sin k\pi x\, \sin k\pi\xi}{k^2\pi^2}\,. \tag{9.112}$$

This result can be checked by a direct computation of the Fourier sine series of (9.111).

Let us now apply Bessel's inequality (9.105) to the eigenfunction series (9.108) for the Green's function; using (9.109), the result is

$$\sum_{k=1}^{n} c_k(\xi)^2 \,\|\, v_k \,\|^2 = \sum_{k=1}^{n} \frac{v_k(\xi)^2 \rho(\xi)^2}{\lambda_k^2 \,\|\, v_k \,\|^2} \le \|\, G_\xi \,\|^2 = \int_a^b G(x;\xi)^2 \rho(x)\, dx. \tag{9.113}$$

We divide by $\rho(\xi) > 0$, and then integrate both sides of the resulting inequality from a to b. On the left-hand side, the integrated summands are

$$\int_a^b \frac{v_k(\xi)^2 \rho(\xi)}{\lambda_k^2 \,\|\, v_k \,\|^2}\, d\xi = \frac{1}{\lambda_k^2 \,\|\, v_k \,\|^2} \int_a^b v_k(\xi)^2 \rho(\xi)\, d\xi = \frac{1}{\lambda_k^2}\,.$$

Substituting back into (9.113) establishes the interesting inequality

$$\sum_{k=1}^{n} \frac{1}{\lambda_k^2} \le \int_a^b \!\! \int_a^b G(x;\xi)^2\, \frac{\rho(x)}{\rho(\xi)}\, dx\, d\xi. \tag{9.114}$$

To make the right-hand side look less strange, we can replace $G(x;\xi)$ by the symmetric modified Green's function $\widehat{G}(x;\xi) = G(x;\xi)/\rho(\xi) = \widehat{G}(\xi;x)$, cf. (9.59), whence

$$\int_a^b \int_a^b G(x;\xi)^2 \,\frac{\rho(x)}{\rho(\xi)}\, dx\, d\xi = \int_a^b \int_a^b \widehat{G}(x;\xi)^2 \rho(x)\, \rho(\xi)\, dx\, d\xi \equiv \|\, \widehat{G} \,\|^2, \tag{9.115}$$

which we can interpret as a "double weighted L^2 norm" of the modified Green's function $\widehat{G}(x;\xi)$. Since the summands in (9.114) are all positive, we can let $n \to \infty$, and conclude that

$$\sum_{k=1}^{\infty} \frac{1}{\lambda_k^2} \leq \|\, \widehat{G} \,\|^2. \tag{9.116}$$

Thus, assuming that the right-hand side of this inequality is finite, the summation on the left converges. This implies that its summands must go to zero: $\lambda_k^{-2} \to 0$ as $k \to \infty$. We have thus proved the first statement of the following important result.

Theorem 9.47. *If* $\|\, \widehat{G} \,\|^2 < \infty$, *then the eigenvalues of the positive definite self-adjoint operator* S *are unbounded:* $0 < \lambda_k \to \infty$ *as* $k \to \infty$. *Moreover, the associated orthogonal eigenfunctions* $v_1, v_2, v_3, \ldots$, *are complete.*

Proof: Our remaining task is to prove completeness — that is, that the eigenfunction series (9.104) of any function $f \in U$ converges in norm. For $n = 2, 3, 4, \ldots$, consider the function

$$g_{n-1} = f - \sum_{k=1}^{n-1} c_k v_k,$$

i.e., the difference between the function f and the $(n-1)^{\text{st}}$ partial sum of its eigenfunction series. Completeness requires that

$$\|\, g_{n-1} \,\| \longrightarrow 0 \qquad \text{as} \qquad n \to \infty. \tag{9.117}$$

We can assume that $g_{n-1} \neq 0$, since otherwise, the eigenfunction series terminates, with $0 = g_{n-1} = g_n = g_{n+1} = \cdots$ (why?), and so (9.117) holds trivially.

First, note that, for any $j = 1, \ldots, n-1$,

$$\langle\, g_{n-1}\,, v_j \,\rangle = \langle\, f\,, v_j \,\rangle - \sum_{k=1}^{n-1} c_k \langle\, v_k\,, v_j \,\rangle = \langle\, f\,, v_j \,\rangle - c_j \|\, v_j \,\|^2 = 0,$$

by the orthogonality of the eigenfunctions combined with the formula (9.104) for the coefficient c_j. Thus, $g_{n-1} \in V_{n-1}$, the subspace (9.99) of functions orthogonal to the first $n-1$ eigenfunctions used in the Rayleigh Minimization Theorem 9.43. Since, according to (9.100), λ_n is the *minimum* value of the Rayleigh quotient among all nonzero elements of V_{n-1}, we must have

$$\lambda_n \leq R[\, g_{n-1}\,] = \frac{\langle\, g_{n-1}\,, S[\, g_{n-1}\,] \,\rangle}{\|\, g_{n-1} \,\|^2},$$

and hence

$$\begin{aligned}
\lambda_n \| g_{n-1} \|^2 &\leq \langle g_{n-1} , S[g_{n-1}] \rangle \\
&= \left\langle f - \sum_{k=1}^{n-1} c_k v_k , S\left[f - \sum_{k=1}^{n-1} c_k v_k \right] \right\rangle \\
&= \left\langle f - \sum_{k=1}^{n-1} c_k v_k , S[f] - \sum_{k=1}^{n-1} c_k S[v_k] \right\rangle \\
&= \left\langle f - \sum_{k=1}^{n-1} c_k v_k , S[f] - \sum_{k=1}^{n-1} c_k \lambda_k v_k \right\rangle \\
&= \langle f , S[f] \rangle - \sum_{k=1}^{n-1} \lambda_k c_k \langle f , v_k \rangle - \sum_{k=1}^{n-1} c_k \langle v_k , S[f] \rangle + \sum_{k=1}^{n-1} \lambda_k c_k^2 \| v_k \|^2 \\
&= \langle f , S[f] \rangle - \sum_{k=1}^{n-1} \lambda_k \frac{\langle f , v_k \rangle^2}{\| v_k \|^2} .
\end{aligned}$$

In the final equality, we used the self-adjointness of S to identify

$$\langle v_k , S[f] \rangle = \langle S[v_k] , f \rangle - \lambda_k \langle v_k , f \rangle - \lambda_k \langle f , v_k \rangle,$$

coupled with the formula in (9.104) for the coefficients c_k. Since the summands in the final expression are all positive, we conclude that

$$\| g_{n-1} \|^2 \leq \frac{\langle f , S[f] \rangle}{\lambda_n} .$$

Since we already know that $\lambda_n \to \infty$, the right-hand side of the final inequality goes to 0 as $n \to \infty$. This implies (9.117) and hence establishes completeness. *Q.E.D.*

One important corollary of this theorem is that, since each eigenvalue is repeated according to its geometric multiplicity, the multiplicity cannot be infinite (why?), and hence each eigenspace of such an S is necessarily finite-dimensional.

Example 9.48. For the eigenvalue problem considered in Example 9.46, since $\rho(x) \equiv 1$, the double norm of the (modified) Green's function $G(x;\xi) = \widehat{G}(x;\xi)$ is

$$\| G \|^2 = \int_0^1 \int_0^1 G(x;\xi)^2 \, dx \, d\xi = 2 \int_0^1 \int_0^\xi x^2 (1-\xi)^2 \, dx \, d\xi = \frac{1}{90} < \infty.$$

Thus, Theorem 9.47 re-establishes the completeness of the sine eigenfunctions, meaning that the eigenfunction series, which is just the ordinary Fourier sine series on $[0,1]$, converges in norm.

Indeed, for any regular Sturm–Liouville boundary value problem on a bounded interval, the (modified) Green's function is automatically continuous, and hence its double weighted norm is finite. Thus, Theorem 9.47 implies the completeness of the Sturm–Liouville eigenfunctions. In Chapters 11 and 12, we will extend this result to some important singular boundary value problems.

Example 9.49. The completeness result of Theorem 9.47 doesn't directly apply to the periodic boundary value problem of Example 9.37, because it is not positive definite, and hence there is no Green's function. However, we can convert it into a positive definite problem by a simple trick. As you are asked to prove in Exercise 9.4.4, if $S \geq 0$ is any positive semi-definite operator and $\mu > 0$ any positive constant, then $\widehat{S} = S + \mu\,\mathrm{I}$ is positive definite, where $\mathrm{I}[u] = u$ is the identity operator. Thus, we replace the original periodic boundary value problem (9.88) by the following modification:

$$-v'' + \mu v = \lambda v, \qquad v(-\pi) = v(\pi), \qquad v'(-\pi) = v'(\pi). \tag{9.118}$$

This does not alter the eigenfunctions, while adding μ to each of the eigenvalues, and hence the modified problem has eigenvalues $\lambda_0 = \mu$, with eigenfunction $v_0(x) \equiv 1$, and $\lambda_n = n^2 + \mu$, with two independent eigenfunctions: $v_n(x) = \cos nx$ and $\widetilde{v}_n(x) = \sin nx$.

The Green's function for the periodic boundary value problem

$$-v'' + \mu v = \delta(x - \xi), \qquad v(-\pi) = v(\pi), \qquad v'(-\pi) = v'(\pi),$$

where $\mu > 0$ is a fixed constant, is derived along the same lines as in Example 6.10. Setting $\mu = \omega^2$, the result is

$$G(x;\xi) = \frac{\cosh \omega \left(\pi - |\,x - \xi\,|\right)}{2\,\omega \sinh \pi\omega}. \tag{9.119}$$

Its double L^2 norm is clearly finite, and, although unnecessary, can even be computed:

$$\|\,G\,\|^2 = \int_{-\pi}^{\pi}\int_{-\pi}^{\pi} G(x;\xi)^2\,dx\,d\xi = \frac{\pi\left(2\pi\omega + \sinh 2\pi\omega\right)}{4\omega^3 \sinh^2 \pi\omega} < \infty.$$

As a result, Theorem 9.47 reconfirms the completeness of the trigonometric eigenfunctions.

Example 9.50. According to (6.120), the Green's function $G(\mathbf{x};\boldsymbol{\xi})$ for the Dirichlet boundary value problem for the Poisson equation on a domain $\Omega \subset \mathbb{R}^2$ is the sum of a logarithmic potential (6.106) and a harmonic function. Thus $G(\mathbf{x};\boldsymbol{\xi})^2$ is a sum of three terms: the first two, involving $(\log r)^2$ and $\log r$ with $r = \|\,\mathbf{x} - \boldsymbol{\xi}\,\|$, have mild singularities when $\mathbf{x} = \boldsymbol{\xi}$, while the last term is smooth (indeed analytic) everywhere. Using this information, it is not hard to prove that its double L^2 norm

$$\|\,G\,\|^2 = \iint_{\Omega}\left[\iint_{\Omega} G(x,y;\xi,\eta)^2\,dx\,dy\right] d\xi\,d\eta < \infty$$

is finite. Indeed, the only problematic point is the logarithmic singularity at $\mathbf{x} = \boldsymbol{\xi}$, but a polar coordinate computation, similar to that used in the proof of Theorem 6.17, shows that such logarithmic singularities still have finite integrals. Therefore, Theorem 9.47 implies that the Helmholtz eigenvalues $\lambda_n \to \infty$, and the corresponding Helmholtz eigenfunctions $v_n(x,y)$ form a complete orthogonal system.

Remark: In problems involving unbounded domains, such as the Schrödinger equation for the hydrogen atom to be discussed in Section 12.7, the eigenfunctions are typically not complete, and one needs to introduce additional solutions corresponding to what is known as the *continuous spectrum* of the operator. Functions are now represented by combinations of discrete Fourier-like sums over the eigenfunctions (the bound states in the quantum-mechanical system) plus a Fourier integral-like term involving the continuous spectrum (the scattering states), [**66, 72**]. A full discussion of completeness and convergence in such cases must be deferred to an advanced course in analysis, [**95**].

Exercises

9.4.1. Find the eigenvalues and an orthonormal eigenvector basis for the following symmetric matrices:

(a) $\begin{pmatrix} 2 & 6 \\ 6 & -7 \end{pmatrix}$, (b) $\begin{pmatrix} 5 & -2 \\ -2 & 5 \end{pmatrix}$, (c) $\begin{pmatrix} 2 & -1 \\ -1 & 5 \end{pmatrix}$, (d) $\begin{pmatrix} 1 & 0 & 4 \\ 0 & 1 & 3 \\ 4 & 3 & 1 \end{pmatrix}$, (e) $\begin{pmatrix} 6 & -4 & 1 \\ -4 & 6 & -1 \\ 1 & -1 & 11 \end{pmatrix}$.

9.4.2. Determine whether the following symmetric matrices are positive definite by computing their eigenvalues.

(a) $\begin{pmatrix} 2 & -2 \\ -2 & 3 \end{pmatrix}$ (b) $\begin{pmatrix} -2 & 3 \\ 3 & 6 \end{pmatrix}$, (c) $\begin{pmatrix} 1 & -1 & 0 \\ -1 & 2 & -1 \\ 0 & -1 & 1 \end{pmatrix}$, (d) $\begin{pmatrix} 4 & -1 & -2 \\ -1 & 4 & -1 \\ -2 & -1 & 4 \end{pmatrix}$.

9.4.3. Suppose $S[\mathbf{u}] = K\mathbf{u}$, where $K = \begin{pmatrix} 0 & 1 \\ -1 & 0 \end{pmatrix}$. (a) Show that $S: \mathbb{R}^2 \to \mathbb{R}^2$ is positive semi-definite under the dot product. (b) Find the eigenvalues of S. (c) Explain why your result in part (b) does not contradict Theorem 9.34.

◇ 9.4.4. Suppose that $S: U \to U$ is a positive semi-definite linear operator. Let $\mathrm{I}: U \to U$ be the identity operator, so $\mathrm{I}[u] = u$. (a) Prove that, for any positive scalar $\mu > 0$, the operator $S_\mu = S + \mu\,\mathrm{I}$ is positive definite. (b) Show that S and S_μ have the same eigenfunctions. Do they have the same eigenvalues? If not, how are their eigenvalues related?

9.4.5. Find the minimum value of $R[v] = \dfrac{\displaystyle\int_0^1 v'^2\,dx}{\displaystyle\int_0^1 v^2\,dx}$ on the space of C^2 functions $v(x)$ defined on $0 \le x \le 1$ that are subject to one of the following pairs of boundary conditions:

$$(a)\ v(0) = v(1) = 0, \qquad (b)\ v(0) = v'(1) = 0, \qquad (c)\ v'(0) = v'(1) = 0.$$

9.4.6. Find the minimum value of $R[v] = \dfrac{\displaystyle\int_1^e x^2 v'^2\,dx}{\displaystyle\int_1^e v^2\,dx}$ on the space of C^2 functions defined on $[1, e]$ subject to the boundary conditions $v(1) = v(e) = 0$.

9.4.7. Show that the Rayleigh quotient $R[v]$ has the same value for all nonzero scalar multiples of an element $0 \ne v \in U$, i.e., $R[cv] = R[v]$ for all $c \ne 0$.

9.4.8. Prove that the minimum value of the Rayleigh quotient of a positive semi-definite, but not positive definite, operator is 0.

♡ 9.4.9. (a) Find the eigenfunctions and eigenvalues for the boundary value problem

$$-x^2u'' - x\,u' = \lambda\,u, \qquad u(1) = u(e) = 0.$$

(b) Under which inner product are the eigenfunctions orthogonal? Justify your answer by direct computation.
(c) Write down the eigenfunction expansion of a function $f(x)$ defined for $1 \le x \le e$.
(d) Find the Green's function for

$$-x^2u'' - x\,u' = f(x), \qquad u(1) = u(e) = 0,$$

both in closed form and as a series in the eigenfunctions you found in part (a).
(e) Is your Green's function symmetric? Discuss.
(f) Prove the completeness of the eigenfunctions.

9.4.10. Discuss completeness of the eigenfunctions of the boundary value problem

$$-x^2u'' - 2\,x\,u' = \lambda\,u, \qquad |\,u(0)\,| < \infty, \qquad u(1) = 0.$$

9.4.11. Consider the eigenvalue problem $-u'' = \lambda u$, $u(0) = 0$, $u'(1) = 0$. (a) Is the problem self-adjoint? positive definite? Which inner product are you referring to? (b) Find all eigenvalues and eigenfunctions. (c) Write down the explicit formula for the eigenfunction expansion of a function $f(x)$ defined on $[0,1]$. (d) Find the Green's function and use it to prove completeness of the eigenfunctions.

♡ 9.4.12. (a) Find the eigenfunctions and eigenvalues for the *Chebyshev boundary value problem*

$$(x^2-1)u'' + x u' = \lambda u, \qquad u(-1) = u(1) = 0.$$

Hint: Let $x = \cos\theta$. (b) Under what inner product are the eigenfunctions orthogonal? Justify your answer by direct computation. (c) Find the Green's function for

$$(x^2-1)u'' + x u' = f(x), \qquad u(-1) = u(1) = 0,$$

both in closed form and as a series in the eigenfunctions you found in part (a). (d) Discuss completeness of the eigenfunctions.

9.4.13. Consider the differential operator $S[u] = -u'' + u$ on the space of C^2 functions $u(x)$ defined for all x and subject to the boundary conditions $\lim_{x\to\infty} u(x) = \lim_{x\to-\infty} u(x) = 0$. (a) Find the Green's function $G(x;\xi)$. (b) Compute its double L^2 norm: $\| G \|^2$. What does this indicate about the completeness of the eigenfunctions of S? (c) Justify your conclusion in part (b) by determining the eigenfunctions.

9.4.14. Find all (real and complex) eigenvalues of the first-derivative operator $D = d/dx$ on the interval $[0,1]$ subject to the single periodic boundary condition $v(0) = v(1)$. Are the corresponding eigenfunctions orthogonal? For which inner product?

♡ 9.4.15. Consider the Dirichlet boundary value problem

$$-\Delta u = h(x,y), \quad u(x,0)=0, \quad u(x,1)=0, \quad u(0,y)=0, \quad u(1,y)=0, \quad 0<x,y<1,$$

for the Poisson equation on the unit square. (a) Find the eigenfunction series expansion for the Green's function of this problem. (b) Does your series coincide with that derived in Exercise 6.3.22? Explain any discrepancies. (c) For the impulse points $(\xi,\eta) = (.5,.5)$ and $(.7,.8)$, graph the result of summing the first 9, 25, and 100 terms in your series, and discuss what you observe in light of what you expect the Green's function to look like.

9.4.16. Find the eigenfunction series expansion for the Green's function of the following mixed boundary value problems:
(a) $-\Delta u = h(x,y), \quad u(x,0)=0, \quad u(x,1)=0, \quad u_x(0,y)=0, \quad u_x(1,y)=0, \quad 0<x,y<1;$
(b) $-\Delta u = h(x,y), \quad u(x,0)=0, \quad u_y(x,1)=0, \quad u(0,y)=0, \quad u_x(1,y)=0, \quad 0<x,y<1.$

9.4.17. Find the eigenfunction series expansion for the Green's function of the following Helmholtz boundary value problem:

$$-\Delta u + u = h(x,y), \qquad u(x,0) = u(x,\pi) = u(0,y) = u(\pi,y) = 0, \qquad 0<x,y<\pi.$$

♢ 9.4.18. If the eigenvalues of a self-adjoint linear operator satisfy $\lambda_n \to \infty$ as $n \to \infty$, explain why each eigenspace is necessarily finite-dimensional.

9.4.19. *True or false*: If $S\colon \mathbb{R}^n \to \mathbb{R}^n$ is any linear function, then one can find an inner product on $\mathbb{R}^n$ that makes S self-adjoint.

9.5 A General Framework for Dynamics

In this final section, we show how to use general eigenfunction expansions to analyze three important classes of linear dynamical systems: parabolic diffusion equations such as the heat equation, hyperbolic vibration equations such as the wave equation, and the Schrödinger equation, a complex evolution equation that governs the dynamical processes of quantum mechanics. In all three cases we can, assuming completeness, write the general solution to the initial-boundary value problem as a convergent eigenfunction series with time-dependent coefficients, and thereby establish several general properties governing their dynamics.

Evolution Equations

In all cases, our starting point is the basic *equilibrium equation*, which is a linear system of the form

$$S[u] = f, \tag{9.120}$$

where f represents an external forcing. The linear operator S is assumed to be of the usual self-adjoint form

$$S = L^* \circ L, \tag{9.121}$$

which is either positive definite, when $\ker L = \{0\}$, or positive semi-definite, the latter case being characterized by the existence of null eigenfunctions $0 \neq v \in \ker L = \ker S$. In finite dimensions, (9.120) represents a linear algebraic system consisting of n equations in n unknowns with positive (semi-)definite coefficient matrix. In infinite-dimensional function space, it represents a self-adjoint positive (semi-)definite boundary value problem for the unknown function u.

With the equilibrium operator in hand, there are two principal classical dynamical systems of importance as physical models. The first are the (unforced) *diffusion processes* modeled by an evolution equation of the form

$$\frac{\partial u}{\partial t} = -\, S[u] = -\, L^* \circ L[u]. \tag{9.122}$$

In the discrete case, this represents a first-order system of ordinary differential equations, known as a linear *gradient flow*. In the continuous case, S is a linear differential operator equipped with homogeneous boundary conditions, and (9.122) represents a linear partial differential equation for the time-varying function $u = u(t,x)$, the heat equation being the prototypical example. (As in the preceding section, the notation employed below indicates that we are working in a single space dimension, but the methods and results apply equally well to higher-dimensional problems.) The addition of external forcing to the diffusion process is treated in Exercise 9.5.6.

The basic separation of variables solution technique was already outlined in Section 3.1. To recap, the separable solutions are of exponential form

$$u(t,x) = e^{-\lambda t}\, v(x), \tag{9.123}$$

where $v \in U$ is a fixed function. Since the operator S is linear and does not involve t differentiation, we find

$$\frac{\partial u}{\partial t} = -\,\lambda\, e^{-\lambda t}\, v, \qquad \text{while} \qquad S[u] = e^{-\lambda t}\, S[v].$$

Substituting back into (9.122) and canceling the common exponential factors, we are led to the eigenvalue problem

$$S[v] = \lambda\, v. \tag{9.124}$$

Thus, (9.123) defines a solution if and only if v is an eigenfunction for the linear operator S, with λ the corresponding eigenvalue.

We let $v_k(x)$, $k = 1, 2, \ldots$, be the orthogonal eigenfunctions and $0 \le \lambda_1 \le \lambda_2 \le \lambda_3 \le \cdots \to \infty$ the corresponding eigenvalues. Assuming completeness, the solution to the initial value problem

$$u(0, x) = f(x) \tag{9.125}$$

can be expanded in terms of the eigensolutions:

$$u(t, x) = \sum_{k=1}^{\infty} e^{-\lambda_k t} c_k v_k(x), \qquad \text{where} \qquad c_k = \frac{\langle f, v_k \rangle}{\| v_k \|^2} \tag{9.126}$$

are the eigenfunction coefficients of the initial data. In particular, the *fundamental solution* of the diffusion equation is defined as the solution $u = F(t, x; \xi)$ to the initial value problem

$$u(0, x) = \delta_\xi(x) \tag{9.127}$$

induced by an initial delta impulse at the point ξ. Its eigenfunction coefficients are

$$c_k = \frac{\langle \delta_\xi, v_k \rangle}{\| v_k \|^2} = \frac{1}{\| v_k \|^2} \int_a^b \delta(x - \xi)\, v_k(x)\, \rho(x)\, dx = \frac{v_k(\xi)\, \rho(\xi)}{\| v_k \|^2}.$$

Thus,

$$F(t, x; \xi) = \sum_{k=1}^{\infty} e^{-\lambda_k t} \frac{v_k(x)\, v_k(\xi)\, \rho(\xi)}{\| v_k \|^2}, \tag{9.128}$$

where the denominator denotes the appropriately weighted L^2 norm of the eigenfunction:

$$\| v_k \|^2 = \int_a^b v_k(x)^2\, \rho(x)\, dx.$$

As with the one-dimensional heat equation, if the equilibrium operator is positive definite, $S > 0$, then all the eigenvalues are strictly positive, and hence, generically, solutions decay to 0 at the exponential rate prescribed by the smallest eigenvalue, which can be characterized as the minimum value of the Rayleigh quotient. On the other hand, if S is only positive semi-definite, then the solution will tend to a null eigenmode, that is, an element of $\ker S = \ker L$, as its asymptotic equilibrium state. If $\dim \ker S = p$, the first p eigenvalues are all $0 = \lambda_1 = \cdots = \lambda_p < \lambda_{p+1}$, and the solution

$$u(t, x) \longrightarrow \sum_{k=1}^{p} c_k v_k(x) \qquad \text{as} \qquad t \longrightarrow \infty$$

will tend to its eventual equilibrium configuration at an exponential rate determined by the smallest positive eigenvalue $\lambda_{p+1} > 0$. In almost all applications, $p = 1$ and there is a single, constant null eigenfunction. The Neumann and periodic boundary value problems for the heat equation are prototypical examples.

Exercises

9.5.1. Find the eigenfunction series of the fundamental solution for the heat equation $u_t = \gamma\, u_{xx}$ on the interval $0 \le x \le 1$ subject to homogeneous Dirichlet boundary conditions.

9.5.2. Solve Exercise 9.5.1 for (a) the mixed boundary conditions $u(t,0) = u_x(t,1) = 0$; (b) homogeneous Neumann boundary conditions.

9.5.3. Let $D[u] = u'$ be the derivative operator acting on the vector space of C^1 scalar functions $u(x)$ defined for $0 \le x \le 1$ and satisfying the boundary conditions $u(0) = u'(1) = 0$.
(a) Given the L^2 inner product on its domain space and the weighted inner product $\langle v, \widetilde{v} \rangle = \int_0^1 v(x)\, \widetilde{v}(x)\, x\, dx$ on its target space, determine the adjoint operator D^*.
(b) Let $S = D^* \circ D$. Write out the diffusion equation $u_t = -\, S[u]$ explicitly, as a partial differential equation plus boundary conditions.
(c) Given the initial condition $u(0,x) = x - x^2$, what is the asymptotic equilibrium $u_\star(x) = \lim_{t \to \infty} u(t,x)$ of the resulting solution to the diffusion equation?

9.5.4. Write down an eigenfunction series for the solution $u(t,x)$ to the initial value problem $u(0,x) = f(x)$ for the fourth-order evolution equation $u_t = -\, u_{xxxx}$ subject to the boundary conditions $u(t,0) = u_{xx}(t,0) = u(t,1) = u_{xx}(t,1) = 0$. Does your solution tend to an equilibrium state? If so, at what rate?

9.5.5. Answer Exercise 9.5.4 for the boundary conditions
$$u_x(t,0) = u_{xxx}(t,0) = u_x(t,1) = u_{xxx}(t,1) = 0.$$

♢ 9.5.6. Explain how to solve the forced diffusion equation $u_t = -\, S[u] + f$, subject to homogeneous boundary conditions, when $f(x)$ does not depend on time t. Does the solution tend to equilibrium as $t \to \infty$? If so, what is the rate of decay, and what is the equilibrium?

9.5.7. Show that if $u(t,x)$ solves the diffusion equation (9.122), then $\| \, u(t,\cdot\,) \, \| \ge \| \, u(s,\cdot\,) \, \|$ whenever $t \le s$.

♢ 9.5.8. Let $S > 0$ be a positive definite operator. Suppose $F(t,x;\xi)$ is the fundamental solution for the diffusion equation (9.122). Prove that $G(x;\xi) = \int_0^\infty F(t,x;\xi)\, dt$ is the Green's function for the corresponding equilibrium equation $S[u] = f$.

Vibration Equations

The second important class of dynamical systems comprises the second-order (in time) *vibration equations*
$$\frac{\partial^2 u}{\partial t^2} = -\, S[u], \tag{9.129}$$
which we initially analyze in the absence of external forcing. Vibrational systems arise as a consequence of Newton's equations of motion in the absence of frictional forces. Their continuum versions model the propagation of waves in solids and fluids, electromagnetic waves, plasma waves, and many other related physical systems.

For a general vibration equation, the separable solutions are of trigonometric form
$$u(t,x) = \cos(\omega\, t)\, v(x) \qquad \text{or} \qquad \sin(\omega\, t)\, v(x). \tag{9.130}$$

Substituting either ansatz back into (9.129) results in the same eigenvalue problem (9.124) for $v(x)$ with eigenvalue $\lambda = \omega^2$ equal to the square of the vibrational frequency. We conclude that the *normal modes* or *eigensolutions* take the form

$$u_k(t,x) = \cos(\omega_k t)\, v_k(x), \qquad \widetilde{u}_k(t,x) = \sin(\omega_k t)\, v_k(x),$$

provided $\lambda_k = \omega_k^2 > 0$ is a nonzero eigenvalue and v_k an associated eigenfunction. Thus, the natural vibrational frequencies of the system are the square roots of the nonzero eigenvalues, a fact that we already observed in the context of the one-dimensional wave equation.

In the positive definite case, the eigenvalues are all strictly positive, and so the general solution is built up as a linear combination of vibrational eigenmodes:

$$\begin{aligned} u(t,x) &= \sum_{k=1}^{\infty} \big[\, c_k u_k(t,x) + d_k \widetilde{u}_k(t,x) \,\big] \\ &= \sum_{k=1}^{\infty} \big[\, c_k \cos(\omega_k t) + d_k \sin(\omega_k t) \,\big]\, v_k(x) = \sum_{k=1}^{\infty} r_k \cos(\omega_k t + \delta_k)\, v_k, \end{aligned} \tag{9.131}$$

where

$$r_k = \sqrt{c_k^2 + d_k^2}\,, \qquad \delta_k = \tan^{-1}\frac{d_k}{c_k}\,. \tag{9.132}$$

The initial conditions

$$g(x) = u(0,x) = \sum_{k=1}^{\infty} c_k v_k(x), \qquad h(x) = u_t(0,x) = \sum_{k=1}^{\infty} d_k\, \omega_k v_k(x), \tag{9.133}$$

are used to specify the coefficients:

$$c_k = \frac{\langle\, g\,, v_k \,\rangle}{\|\, v_k \,\|^2}\,, \qquad d_k = \frac{\langle\, h\,, v_k \,\rangle}{\omega_k \,\|\, v_k \,\|^2}\,. \tag{9.134}$$

In the unstable, positive semi-definite cases, any null eigenfunction $v_0 \in \ker S = \ker L$ contributes two aperiodic eigensolutions:

$$u_0(t,x) = v_0(x), \qquad \widetilde{u}_0(t,x) = t\, v_0(x),$$

as can be readily checked. The first is constant in time, while the second is an unstable, linearly growing mode, which is excited if and only if the initial velocity is *not* orthogonal to the null eigenfunction: $\langle\, h\,, v_0 \,\rangle \neq 0$.

If, as occurred in the one-dimensional wave equation, the natural frequencies happen to be integer multiples of a common frequency, $\omega_k = n_k \omega_\star$ for $n_k \in \mathbb{N}$, then the solution (9.131) is a periodic function of t with period $p_\star = 2\pi/\omega_\star$. On the other hand, in most cases the frequencies are not rationally related, and the solution is only *quasiperiodic*. Although it is the sum of individually periodic modes, it is not periodic, and never exactly reproduces its initial behavior; see the illustrative Example 2.20 for additional details.

Forcing and Resonance

Periodically forcing an undamped mechanical structure, modeled by a vibrational system of ordinary differential equations, at a frequency that is distinct from its natural vibrational frequencies, leads, in general, to a quasiperiodic response. The solution is a sum of the unforced vibrational modes superimposed with an additional component that vibrates at the forcing frequency. However, if forced at one of its natural frequencies, the system may experience a catastrophic resonance. See [**89**; §9.6] for details.

The same type of quasiperiodic/resonant response is also observed in the partial differential equations governing the vibrations of continuous media. Consider the forced vibrational equation

$$\frac{\partial^2 u}{\partial t^2} = -\,S[\,u\,] + F(t,x), \tag{9.135}$$

subject to specified homogeneous boundary conditions. The external forcing function $F(t,x)$ may depend on both time t and position x. We will be particularly interested in a periodically varying external force of the form

$$F(t,x) = \cos(\omega\,t)\;h(x), \tag{9.136}$$

where ω is the forcing frequency, while the forcing profile $h(x)$ is unvarying.

As always, the solution to an inhomogeneous linear equation can be written as a combination,

$$u(t,x) = u_\star(t,x) + z(t,x), \tag{9.137}$$

of a particular solution $u_\star(t,x)$ to the inhomogeneous forced equation combined with the general solution $z(t,x)$ to the homogeneous equation, namely

$$\frac{\partial^2 z}{\partial t^2} = -\,S[\,z\,]. \tag{9.138}$$

The boundary and initial conditions will serve to uniquely prescribe the solution $u(t,x)$, but there is some flexibility in its two constituents (9.137). For instance, we may ask that the particular solution $u_\star$ satisfy the homogeneous boundary conditions along with zero (homogeneous) initial conditions and thus represent the pure response of the system to the forcing. The homogeneous solution $z(t,x)$ will then reflect the effect of the initial and boundary conditions unadulterated by the external forcing. The final solution is the combined sum of the two individual responses.

In the case of periodic forcing (9.136), we look for a particular solution

$$u_\star(t,x) = \cos(\omega\,t)\;v_\star(x) \tag{9.139}$$

that vibrates at the forcing frequency. Substituting the ansatz (9.139) into the equation (9.135) and canceling the common cosine factors, we discover that $v_\star(x)$ must satisfy the boundary value problem prescribed by a forced differential equation

$$S[\,v_\star\,] - \omega^2\,v_\star = h(x), \tag{9.140}$$

supplemented by the relevant homogeneous boundary conditions: Dirichlet, Neumann, mixed, or periodic.

At this juncture, there are two possibilities. If the unforced homogeneous boundary value problem

$$S[\,v\,] - \omega^2\,v = 0 \tag{9.141}$$

has only the trivial solution $v \equiv 0$, then, according to the Fredholm Alternative Theorem 9.10, a solution to the forced boundary value problem will exist† for any form of the

† Existence is immediate in finite-dimensional systems. For boundary value problems, this relies on an analytic existence theorem, e.g., Theorem 9.28.

forcing function $h(x)$. In other words, if ω^2 is *not* an eigenvalue, then the particular solution (9.139) will vibrate with the forcing frequency, and the general solution will be a periodic or quasiperiodic combination (9.137) of the natural vibrational modes along with the vibrational response to the periodic forcing.

On the other hand, if $\omega^2 = \lambda_k$ is an eigenvalue, and so $\omega = \omega_k$ coincides with one of the natural vibrational frequencies of the homogeneous problem, then (9.141) admits nontrivial solutions, namely the eigenfunction† $v_k(x)$. In such cases, the Fredholm Alternative tells us that the boundary value problem (9.140) admits a solution if and only if the forcing function is orthogonal to the eigenfunction:

$$\langle h , v_k \rangle = 0. \tag{9.142}$$

If this holds, then the resulting particular solution (9.139) still vibrates with the forcing frequency, and resonance doesn't occur.

If we force in a resonant manner — meaning that the Fredholm condition (9.142) does *not* hold — then the solution will be a resonantly growing vibration of the form

$$u_\star(t,x) = a\, t \sin(\omega_k t)\, v_k(x) + \cos(\omega_k t)\, v_\star(x), \tag{9.143}$$

in which a is a constant to be specified as follows. By direct calculation,

$$\begin{aligned}\frac{\partial^2 u_\star}{\partial t^2} + S[u_\star] &= a\, t \sin(\omega_k t) \left(S[v_k] - \omega_k^2 v_k(x) \right) \\ &\qquad + \cos(\omega_k t) \left(S[v_\star] - \omega_k^2 v_\star(x) + 2 a \omega_k v_k(x) \right).\end{aligned}$$

The first term vanishes, since $v_k(x)$ is an eigenfunction with eigenvalue $\lambda_k = \omega_k^2$. Therefore, (9.143) satisfies the forced boundary value problem if and only if $v_\star(x)$ satisfies the forced boundary value problem

$$S[v_\star] - \omega_k^2 v_\star(x) = h(x) - 2 a \omega_k v_k(x). \tag{9.144}$$

Again, the Fredholm Alternative implies that (9.144) admits a solution $v_\star(x)$ if and only if

$$0 = \langle h - 2a\omega_k v_k , v_k \rangle = \langle h , v_k \rangle - 2a \, \| v_k \|^2, \qquad \text{and hence} \qquad a = \frac{\langle h , v_k \rangle}{2 \, \| v_k \|^2}, \tag{9.145}$$

which serves to fix the value of the constant in the resonant solution ansatz (9.143). In a real-world situation, such large resonant (or even near resonant) vibrations will, if unchecked, eventually either leads to a catastrophic breakdown of the system or to a transition into the nonlinear regime.

Example 9.51. As a specific example, consider the initial-boundary value problem modeling the forced vibrations of a uniform string of unit length that is fixed at both ends:

$$\begin{gathered} u_{tt} = c^2 u_{xx} + \cos(\omega t)\, h(x), \\ u(t,0) = 0 = u(t,1), \qquad u(0,x) = f(x), \qquad u_t(0,x) = g(x). \end{gathered} \tag{9.146}$$

† For simplicity, we assume that the eigenvalue λ_k is simple, and so there is a unique, up to constant multiple, eigenfunction v_k. Modifications for multiple eigenvalues proceed analogously.

The particular solution $u_\star(t,x)$ will have the nonresonant form (9.139), provided there exists a solution $v_\star(x)$ to the boundary value problem

$$S[v_\star] - \omega^2 v_\star = -c^2 v_\star'' - \omega^2 v_\star = h(x), \qquad v_\star(0) = 0 = v_\star(1). \tag{9.147}$$

The natural frequencies and associated eigenfunctions of the unforced Dirichlet boundary value problem are

$$\omega_k = k c \pi, \qquad v_k(x) = \sin k\pi x, \qquad k = 1,2,3,\ldots .$$

Thus, the boundary value problem (9.147) will admit a solution, and hence the forcing is not resonant, if either $\omega \neq \omega_k$ is not a natural frequency or $\omega = \omega_k$ for some k but the forcing profile is orthogonal to the associated eigenfunction:

$$0 = \langle h, v_k \rangle = \int_0^1 h(x) \sin k\pi x \, dx. \tag{9.148}$$

Otherwise, the system will undergo a resonant response.

For example, under periodic forcing of frequency ω with trigonometric sine profile $h(x) \equiv \sin k\pi x$, for k a positive integer, the particular solution to (9.147) is

$$v_\star(x) = \frac{\sin k\pi x}{\omega^2 - k^2\pi^2 c^2}, \qquad \text{so that} \qquad u_\star(t,x) = \frac{\cos \omega t \, \sin k\pi x}{\omega^2 - k^2\pi^2 c^2}, \tag{9.149}$$

which is valid provided $\omega \neq \omega_k = k\pi c$. Observe that we may allow the forcing frequency to coincide with any of the other natural frequencies, $\omega = \omega_n$ for $n \neq k$, because the sine profiles are mutually orthogonal, and so the nonresonance condition (9.148) holds. On the other hand, if $\omega = \omega_k = k\pi c$, then the particular solution

$$u_\star(t,x) = \frac{t \, \sin k\pi c t \, \sin k\pi x}{2k\pi c} \tag{9.150}$$

is resonant and grows linearly in time.

To obtain the full solution to the initial-boundary value problem, we write $u = u_\star + z$, where $z(t,x)$ must satisfy

$$z_{tt} - c^2 z_{xx} = 0, \qquad z(t,0) = 0 = z(t,1),$$

along with the modified initial conditions

$$z(0,x) = f(x) - \frac{\sin k\pi x}{\omega^2 - k^2\pi^2 c^2}, \qquad \frac{\partial z}{\partial t}(0,x) = g(x),$$

stemming from the fact that the particular solution (9.149) has a nontrivial initial displacement. (In the resonant case (9.150), there is no extra term in the initial data.) Note that the closer ω is to the resonant frequency, the larger the modification of the initial data, and hence the larger the response of the system to the periodic forcing. As before, the solution $z(t,x)$ to the homogeneous equation can be written as a Fourier sine series (4.68). The final formulas are left for the reader to write out in detail; see Exercise 9.5.14.

Exercises

9.5.9. Which of the following forcing functions $F(t,x)$ excites resonance in the wave equation $u_{tt} = u_{xx} + F(t,x)$ when subject to homogeneous Dirichlet boundary conditions on the interval $0 \le x \le 1$? (a) $\sin 3t$, (b) $\sin 3\pi t$, (c) $\sin \frac{3}{2}\pi t$, (d) $\sin \pi t \ \sin \pi x$, (e) $\sin \pi t \ \sin 2\pi x$, (f) $\sin 2\pi t \ \cos \pi x$, (g) $x(1-x)\sin 2\pi t$.

9.5.10. Answer Exercise 9.5.9 when the solution is subject to the mixed boundary conditions $u(t,0) = u_x(t,1) = 0$.

♡ 9.5.11. Let $\omega > 0$. Find the solution to the initial-boundary value problem

$$u_{tt} = u_{xx} + \cos \omega t, \qquad u(t,0) = 0 = u(t,1), \qquad u(0,x) = 0 = u_t(0,x).$$

9.5.12. Answer Exercise 9.5.11 for homogeneous Neumann boundary conditions.

9.5.13. A piano wire of length 1 m and wave speed $c = 2$ m/sec can support a maximal deflection of 5 cm before breaking. Suppose the wire starts at rest, with both ends fixed, and then is subject to a uniform periodic force $F(t,x) = \frac{1}{10}\cos \omega t \ \sin \pi x$. What range of frequencies will cause the wire to break?

♢ 9.5.14. Write out the eigenfunction series solution to the initial-boundary value problem in Example 9.51 with $h(x) \equiv \sin k\pi x$.

9.5.15. How should the solution formulas (9.131, 134) be modified when there are unstable modes? Write down explicit conditions on the initial data that prevent an instability from being excited.

♢ 9.5.16. Explain how to convert the homogeneous wave equation with inhomogeneous Dirichlet boundary conditions $u(t,0) = \alpha(t)$, $u(t,\ell) = \beta(t)$, into a homogeneous boundary value problem for the forced wave equation. *Hint*: Mimic (4.46).

♡ 9.5.17. Two children hold a jump rope taut, while one of them periodically shakes their end of the rope. Use the inhomogeneous boundary value problem

$$\frac{\partial^2 u}{\partial t^2} = \frac{\partial^2 u}{\partial x^2}, \qquad u(t,0) = 0, \qquad u(t,1) = \sin \omega t,$$

to model the motion of the rope, adopting units in which the wave speed $c = 1$.
(a) What are the resonant frequencies of this system?
(b) Apply the method of Exercise 9.5.16 to find a particular solution to the boundary value problem when ω is a nonresonant frequency.
(c) Suppose the rope starts at rest. Find a series solution to the corresponding initial-boundary value problem when ω is a nonresonant frequency.
(d) Answer parts (b,c) when ω is a resonant frequency. *Hint*: Use the ansatz (9.143).

9.5.18. Explain how to solve the periodically forced *telegrapher's equation*

$$u_{tt} + a\,u_t = c^2\, u_{xx} + h(x)\cos \omega t$$

on the interval $0 \le x \le 1$ when subject to homogeneous Dirichlet boundary conditions. At which frequencies does the forcing function excite a resonant response?
Hint: First solve Exercise 4.2.9.

9.5.19. The fourth-order evolution equation $u_{tt} = -c^2 u_{xxxx}$, subject to the boundary conditions $u(t,0) = u_{xx}(t,0) = u(t,1) = u_{xx}(t,1) = 0$, models the transverse vibrations of a simply supported uniform thin elastic beam, in which $c > 0$ represents the wave speed. Write down an eigenfunction series for the solution to the initial value problem $u(0,x) = f(x)$, $u_t(0,x) = 0$. Is the solution
(*i*) periodic, (*ii*) quasiperiodic, (*iii*) chaotic, (*iv*) none of the above?

The Schrödinger Equation

The fundamental dynamical system that governs all quantum-mechanical systems is known as the *Schrödinger equation*, first written down by the great twentieth-century German physicist Erwin Schrödinger, one of the preeminent founders of modern quantum physics. His original series of papers in which, by fits and starts, he arrives at his fundamental equation, makes for fascinating reading, [**101**].

Unlike classical mechanics, quantum mechanics is a completely linear theory, governed by linear systems of partial differential equations. The abstract form of the linear *Schrödinger equation* is

$$\mathrm{i}\,\hbar\,\frac{\partial\psi}{\partial t} = S[\psi], \tag{9.151}$$

where S is a linear operator of the usual self-adjoint form (9.121). In this equation, $\mathrm{i} = \sqrt{-1}$, while $\hbar$ is Planck's constant (7.69). The operator S is known as the *Hamiltonian* for the quantum-mechanical system, and, typically, represents the quantum energy operator. For physical systems such as atoms and nuclei, the relevant Hamiltonian operator is constructed from the classical energy through the rather mysterious process of "quantization".

At each time t, the solution $\psi(t,x)$ to the Schrödinger equation represents the wave function of the quantum system, and so should be a complex-valued square-integrable function having unit L^2 norm: $\|\,\psi\,\| = 1$. (The reader may wish to revisit Sections 3.5 and 7.1 for a discussion of the basics of quantum mechanics and Hilbert space.) We interpret the wave function as a probability density on the possible quantum states, and so the Schrödinger equation governs the dynamical evolution of quantum probabilities. The interested reader should consult a basic text on quantum mechanics, e.g., [**66, 72, 115**], for full details on both the physics and underlying mathematics.

Proposition 9.52. *If $\psi(t,x)$ is a solution to the Schrödinger equation, its Hermitian L^2 norm $\|\,\psi(t,\cdot)\,\|$ is fixed for all time.*

Proof: Since the solution is complex-valued, we use the sesquilinearity of the underlying Hermitian inner product, as in (B.19), to compute

$$\begin{aligned}\frac{d}{dt}\,\|\,\psi(t,\cdot)\,\|^2 &= \left\langle \frac{\partial\psi}{\partial t}\,,\,\psi\right\rangle + \left\langle \psi\,,\,\frac{\partial\psi}{\partial t}\right\rangle = \left\langle -\frac{\mathrm{i}}{\hbar}\,S[\psi]\,,\,\psi\right\rangle + \left\langle \psi\,,\,-\frac{\mathrm{i}}{\hbar}\,S[\psi]\right\rangle \\ &= -\frac{\mathrm{i}}{\hbar}\,\langle\, S[\psi]\,,\psi\,\rangle + \frac{\mathrm{i}}{\hbar}\,\langle\,\psi\,,S[\psi]\,\rangle = 0,\end{aligned}$$

which vanishes because S is self-adjoint. This implies that $\|\,\psi(t,\cdot)\,\|^2$ is constant. *Q.E.D.*

As a result, if the initial data $\psi(t_0,x) = \psi_0(x)$ is a quantum-mechanical wave function, meaning that $\|\,\psi_0\,\| = 1$, then, at each time t, the solution $\psi(t,x)$ to the Schrödinger equation also has norm 1, and hence remains a wave function throughout the evolutionary process.

Apart from the extra factor of $\mathrm{i}\,\hbar$, the Schrödinger equation looks like a diffusion equation (9.122). This inspires us to seek separable solutions with an exponential ansatz:

$$\psi(t,x) = e^{\alpha\,t}\,v(x).$$

Substituting this expression into the Schrödinger equation (9.151) and canceling the common exponential factors reduces us to the usual eigenvalue problem

$$S[v] = \lambda v, \qquad \text{with eigenvalue} \qquad \lambda = \mathrm{i}\,\hbar\,\alpha.$$

By self-adjointness, the eigenvalues are necessarily real. Let v_k denote the normalized eigenfunction, so $\| v_k \| = 1$, associated with the kth eigenvalue λ_k. The corresponding eigensolution of the Schrödinger equation is the complex-valued function

$$\psi_k(t,x) = e^{-\mathrm{i}\,\lambda_k t/\hbar}\, v_k(x).$$

Observe that, in contrast to the exponentially decaying solutions to the diffusion equation, the eigensolutions to the Schrödinger equation are periodic, with vibrational frequencies $\omega_k = -\lambda_k/\hbar$ proportional to the eigenvalues. (Along with constant solutions corresponding to the null eigenmodes, if any.) The general solution is a (quasi)periodic series in the fundamental eigensolutions,

$$\psi(t,x) = \sum_k c_k \psi_k(t,x) = \sum_k c_k\, e^{-\mathrm{i}\,\lambda_k t/\hbar}\, v_k(x), \tag{9.152}$$

whose coefficients are prescribed by the initial conditions. The periodicity of the summands has the additional implication that, again unlike the diffusion equation, the Schrödinger equation can be run backwards in time, i.e., it remains well-posed in the past. Consequently, we can determine both the past and future behavior of a quantum system from its present configuration.

The eigenvalues represent the energy levels of the system described by the Schrödinger equation and can be experimentally detected by exciting the system. For instance, when an excited electron orbiting a nucleus jumps back to a lower energy level, it emits a photon whose observed electromagnetic spectral line corresponds to the difference between the energies of the two quantum levels. This motivates the use of the term *spectrum* to describe the eigenvalues of a linear Hamiltonian operator.

Example 9.53. The simplest version of the Schrödinger equation is based on the derivative operator $L = D$, leading to the self-adjoint combination $S = L^* \circ L = -D^2$ when subject to appropriate boundary conditions. In this case, the Schrödinger equation (9.151) reduces to the second-order partial differential equation

$$\mathrm{i}\,\hbar\,\frac{\partial \psi}{\partial t} = -\frac{\partial^2 \psi}{\partial x^2}\,. \tag{9.153}$$

If we impose the Dirichlet boundary conditions $\psi(t,0) = \psi(t,\ell) = 0$, then the Schrödinger equation (9.153) governs the dynamics of a quantum particle that is confined to the interval $0 < x < \ell$; the boundary conditions imply that there is zero probability of the particle escaping from the interval.

According to Section 4.1, the eigenfunctions of the Dirichlet eigenvalue problem

$$v'' + \lambda v = 0, \qquad v(0) = v(\ell) = 0,$$

are

$$v_k(x) = \sqrt{\frac{2}{\ell}}\,\sin\frac{k\pi}{\ell}\,x, \quad \text{with eigenvalue} \quad \lambda_k = \frac{k^2\pi^2}{\ell^2}, \quad \text{for} \quad k = 1, 2, \dots\,,$$

where the initial factor ensures that v_k has unit L^2 norm, and hence is a bona fide wave function. The corresponding oscillatory eigenmodes are

$$\psi_k(t,x) = \sqrt{\frac{2}{\ell}}\, \exp\left(-\,\mathrm{i}\,\frac{k^2\pi^2}{\hbar\,\ell^2}\, t \right)\, \sin \frac{k\,\pi}{\ell}\, x. \tag{9.154}$$

Since the temporal frequencies $\omega_k = -\,k^2\pi^2/(\hbar\,\ell^2)$ depend nonlinearly on the wave number $k\,\pi/\ell$, the Schrödinger equation, is, in fact, *dispersive*, sharing many similarities with the third-order linear equation (8.90); see, for instance, Exercises 9.5.25, 27.

Exercises

9.5.20. (a) Solve the following initial boundary value problem:

$$\mathrm{i}\,\hbar\,\psi_t = -\,\psi_{xx}, \qquad \psi(t,0) = \psi(t,1) = 0, \qquad \psi(0,x) = 1.$$

(b) Using your solution formula, verify that $\|\,\psi(t,\cdot\,)\,\| = 1$ for all t.

9.5.21. Answer Exercise 9.5.20 for the initial condition $\psi(0,x) = \sqrt{30}\,x(1-x)$.

9.5.22. Answer Exercise 9.5.20 when the solution is subject to Neumann boundary conditions $\psi_x(t,0) = \psi_x(t,1) = 0$.

9.5.23. Write down the eigenseries solution for the Schrödinger equation on a bounded interval $[0,\ell]$ when subject to homogeneous Neumann boundary conditions.

9.5.24. Given the solution formula (9.152), and assuming completeness of the eigenfunctions, prove that $\|\,\psi(t,\cdot\,)\,\|^2 = \sum_k |\,c_k\,|^2$ for all t.

♢ 9.5.25. Write down the dispersion relation, phase velocity, and group velocity for the one-dimensional Schrödinger equation (9.153).

9.5.26. Show that the real and imaginary parts of the solution $\psi(t,x) = u(t,x) + \mathrm{i}\,v(t,x)$ to the one-dimensional Schrödinger equation (9.153) are solutions to the beam equation of Exercise 9.5.19. What is the wave speed?

♢ 9.5.27. *The Talbot effect for the linear Schrödinger equation*: Let $u(t,x)$ solve the periodic initial-boundary value problem

$$\mathrm{i}\,u_t = u_{xx}, \quad u(t,-\pi) = u(t,\pi), \quad u_x(t,-\pi) = u_x(t,\pi),$$

with initial data $u(0,x) = \sigma(x)$ given by the unit step function. Prove that when $t = \pi\,p/q$, where p,q are integers, the solution $u(t,x)$ is constant on each interval $\hbar\pi\,j/q < x < \hbar\pi(j+1)/q$ for integers $j \in \mathbb{Z}$. *Hint*: Use Exercise 6.1.29(*d*).

9.5.28. The wave function $\psi(t,x)$ of a one-dimensional free quantum particle of mass m satisfies the Schrödinger equation $\mathrm{i}\,\psi_t = -\,\hbar\,\psi_{xx}/(2m)$ on the real line $-\infty < x < \infty$. Assuming that ψ and its x derivatives decay reasonably rapidly to zero as $|\,x\,| \to \infty$, prove that the particle's expected position $\langle\, x\,\rangle = \int_{-\infty}^{\infty} x\,|\,\psi(t,x)\,|^2\, dx$ moves on a straight line.

Hint: Prove that $\dfrac{d^2\langle\, x\,\rangle}{dt^2} = 0$.

♡ 9.5.29. Consider the periodically forced Schrödinger equation $\mathrm{i}\,\hbar\,\psi_t = -\psi_{xx} + e^{\,\mathrm{i}\,\omega\,t}$ on the interval $0 \le x \le 1$, subject to homogeneous Dirichlet boundary conditions. (a) At which frequencies ω does the forcing function excite a resonant response? (*b*) Find the solution to the general initial value problem for a nonresonant forcing frequency. (*c*) Find the solution to the general initial value problem for a resonant forcing frequency. What are the conditions on $\hbar$ that ensure that the resulting solution remains a wave function?

♡ 9.5.30. The Schrödinger equation for the *harmonic oscillator* is $\mathrm{i}\,\hbar\,\psi_t = \psi_{xx} - x^2\psi$. Write this equation in the self-adjoint form (9.151) under a suitable choice of boundary conditions. Write down the self-adjoint boundary value problem for the eigenfunctions.
Remark: The eigenfunctions are not elementary functions. After studying Section 11.3, you may wish to return here to investigate its solutions.

Chapter 10
Finite Elements and Weak Solutions

In Chapter 5, we studied the oldest, and in many ways the simplest, class of numerical algorithms for approximating the solutions to partial differential equations: those based on finite difference approximations. In the present chapter, we introduce the second of the two major numerical paradigms: the finite element method. Finite elements are of more recent vintage, having first appeared soon after the Second World War; historical details can be found in [**113**]. As a consequence of their ability to adapt to complicated geometries, finite elements have, in many situations, become the method of choice for solving equilibrium boundary value problems governed by elliptic partial differential equations. Finite elements can also be adapted to dynamical problems, but lack of space prevents us from pursuing such extensions in this text.

Finite elements rely on a more sophisticated understanding of the partial differential equation, in that, unlike finite differences, they are not obtained by simply replacing derivatives by their numerical approximations. Rather, they are initially founded on an associated minimization principle that, as we learned in Chapter 9, characterizes the unique solution to a positive definite boundary value problem. The basic idea is to restrict the minimizing functional to an appropriately chosen finite-dimensional subspace of functions. Such a restriction produces a finite-dimensional minimization problem, which can then be solved by numerical linear algebra. When properly formulated, the restricted finite-dimensional minimization problem will have a solution that well approximates the true minimizer, and hence the solution to the original boundary value problem. To gain familiarity with the underlying principles, we will first illustrate the basic constructions in the context of boundary value problems for ordinary differential equations. The following section extends finite element analysis to boundary value problems associated with the two-dimensional Laplace and Poisson equations, thereby revealing the key features used in applications to the numerical solution of multidimensional equilibrium boundary value problems.

An alternative approach to the finite element method, one that can be applied even in situations in which no minimum principle is available, is founded on the concept of a weak solution to the differential equation, a construction of independent analytical importance. The term "weak" refers to the fact that one is able to relax the differentiability requirements imposed on classical solutions. Indeed, as we will show, discontinuous shock wave solutions as well as the nonsmooth, and hence nonclassical, solutions to the wave equation that we encountered in Chapters 2 and 4 can all be rigorously characterized through the weak solution formulation. For the finite element approximation, rather than impose the weak solution criterion on the entire infinite-dimensional function space, one again restricts to a suitably chosen finite-dimensional subspace. For positive definite boundary value problems, which necessarily admit a minimization principle, the weak solution approach leads to the same finite element equations.

A rigorous justification and proof of convergence of the finite element approximations requires further analysis, and we refer the interested reader to more specialized texts, such as [**6, 113, 126**]. In this chapter, we shall focus our effort on understanding how to formulate and implement the finite element method in practical contexts.

10.1 Minimization and Finite Elements

To explain the principal ideas underpinning the finite element method, we return to the abstract framework for boundary value problems that was developed in Chapter 9. Recall Theorem 9.26, which characterizes the unique solution to a positive definite linear system as the minimizer, $u_\star \in U$, of an associated quadratic functional $Q: U \to \mathbb{R}$. For boundary value problems governed by differential equations, U is an infinite-dimensional function space containing all sufficiently smooth functions that satisfy the prescribed homogeneous boundary conditions. (Modifications to deal with inhomogeneous boundary conditions will be discussed in due course.)

This framework sets the stage for the first key idea of the finite element method. Instead of trying to minimize the functional $Q[u]$ over the entire infinite-dimensional function space, we will seek to minimize it over a *finite-dimensional subspace* $W \subset U$. The effect is to reduce a problem in analysis — a boundary value problem for a differential equation — to a problem in linear algebra, and hence one that a computer is capable of solving. On the surface, the idea seems crazy: how could one expect to come close to finding the minimizer in a gigantic infinite-dimensional function space by restricting the search to a mere finite-dimensional subspace? But this is where the magic of infinite dimensions comes into play. One can, in fact, approximate all (reasonable) functions arbitrarily closely by functions belonging to finite-dimensional subspaces. Indeed, you are already familiar with two examples: Fourier series, where one approximates rather general periodic functions by trigonometric polynomials, and interpolation theory, in which one approximates functions by ordinary polynomials, or, more sophisticatedly, by splines, [**89, 102**]. Thus, the finite element idea perhaps is not as outlandish as it might initially seem.

To be a bit more explicit, let us begin with a linear operator $L: U \to V$ between real inner product spaces, where, as in Section 9.1, $\langle u , \widetilde{u} \rangle$ is used to denote the inner product in U, and $\langle\!\langle v , \widetilde{v} \rangle\!\rangle$ the inner product in V. To ensure uniqueness of solutions, we always assume that L has trivial kernel: $\ker L = \{0\}$. According to Theorem 9.26, the element $u_\star \in U$ that minimizes the quadratic function(al)

$$Q[u] = \tfrac{1}{2} \,|\!|\!|\, L[u] \,|\!|\!|^2 - \langle f , u \rangle, \tag{10.1}$$

where $|\!|\!| \cdot |\!|\!|$ denotes the norm in V, is the solution to the linear system

$$S[u] = f, \qquad \text{where} \qquad S = L^* \circ L, \tag{10.2}$$

with $L^*: V \to U$ denoting the adjoint operator. The hypothesis that L has trivial kernel implies that S is a self-adjoint positive definite linear operator, which implies that the solution to (10.2), and hence the minimizer of $Q[u]$, is unique. In our applications, L is a linear differential operator between function spaces, e.g., the gradient, while $Q[u]$ represents a quadratic functional, e.g., the Dirichlet principle, and the associated linear system (10.2) forms a positive definite boundary value problem, e.g., the Poisson equation along with suitable boundary conditions.

To form a finite element approximation to the solution $u_\star \in U$, rather than try to minimize $Q[u]$ on the entire function space U, we now seek to minimize it on a suitably chosen finite-dimensional subspace $W \subset U$. We will specify W by selecting a set of linearly independent functions $\varphi_1, \ldots, \varphi_n \in U$, and letting W be their span. Thus, $\varphi_1, \ldots, \varphi_n$ form a basis of W, whereby $\dim W = n$, and the general element of W is a (uniquely determined) linear combination

$$w(x) = c_1 \varphi_1(x) + \cdots + c_n \varphi_n(x) \tag{10.3}$$

of the basis functions. Our goal is to minimize $Q[w]$ over all possible $w \in W$; in other words, we need to determine the coefficients $c_1, \ldots, c_n \in \mathbb{R}$ such that

$$Q[w] = Q[c_1 \varphi_1 + \cdots + c_n \varphi_n] \tag{10.4}$$

is as small as possible. Substituting (10.3) back into (10.1) and then expanding, using the linearity of L and then the bilinearity of the inner product, we find that the resulting expression is the quadratic function

$$P(\mathbf{c}) = \frac{1}{2} \sum_{i,j=1}^{n} k_{ij} c_i c_j - \sum_{i=1}^{n} b_i c_i = \tfrac{1}{2} \mathbf{c}^T K \mathbf{c} - \mathbf{c}^T \mathbf{b}, \tag{10.5}$$

in which

- $\mathbf{c} = (c_1, c_2, \ldots, c_n)^T \in \mathbb{R}^n$ is the vector of unknown coefficients in (10.3);
- $K = (k_{ij})$ is the symmetric $n \times n$ matrix with entries

$$k_{ij} = \langle\!\langle L[\varphi_i], L[\varphi_j] \rangle\!\rangle, \qquad i, j = 1, \ldots, n; \tag{10.6}$$

- $\mathbf{b} = (b_1, b_2, \ldots, b_n)^T$ is the vector with entries

$$b_i = \langle f, \varphi_i \rangle, \qquad i = 1, \ldots, n. \tag{10.7}$$

Note that formula (10.6) uses the inner product on the target space V, whereas (10.7) relies on the inner product on the domain space U.

Thus, once we specify the basis functions φ_i, the coefficients k_{ij} and b_i are all known quantities. We have effectively reduced our original problem to the finite-dimensional problem of minimizing the quadratic function (10.5) over all possible vectors $\mathbf{c} \in \mathbb{R}^n$. The symmetric matrix K is, in fact, positive definite, since, by the preceding computation,

$$\mathbf{c}^T K \mathbf{c} = \sum_{i,j=1}^{n} k_{ij} c_i c_j = |||\, L[c_1 \varphi_1(x) + \cdots + c_n \varphi_n] \,|||^2 = |||\, L[w] \,|||^2 > 0, \tag{10.8}$$

as long as $L[w] \neq 0$. Moreover, our initial assumption tells us that $L[w] = 0$ if and only if $w = 0$, which, by linear independence, occurs only when $\mathbf{c} = \mathbf{0}$. Thus, (10.8) is indeed positive for all $\mathbf{c} \neq \mathbf{0}$. We can now invoke the finite-dimensional minimization result contained in Example 9.25 to conclude that the unique minimizer to (10.5) is obtained by solving the associated linear system

$$K \mathbf{c} = \mathbf{b}, \qquad \text{whereby} \qquad \mathbf{c} = K^{-1} \mathbf{b}. \tag{10.9}$$

Remark: When of moderate size, the linear system (10.9) can be solved by basic Gaussian Elimination. When the size (i.e., the dimension, n, of the subspace W) becomes too large, as is often the case in dealing with partial differential equations, it is better to rely on an iterative linear system solver, e.g., Gauss–Seidel or Successive Over–Relaxation (SOR); see [**89, 118**] for details.

This summarizes the basic abstract setting for the finite element method. The key issue, then, is how to effectively choose the finite-dimensional subspace W. Two candidates that might spring to mind are the space of polynomials of degree $\leq n$ and the space of trigonometric polynomials (truncated Fourier series) of degree $\leq n$. However, for a variety of reasons, neither is well suited to the finite element method. One constraint is that the functions in W must satisfy the relevant boundary conditions — otherwise, W would not be a subspace of U. More importantly, in order to obtain sufficient accuracy of the approximate solution, the linear algebraic system (10.9) will typically — especially when dealing with partial differential equations — be quite large, and hence it is desirable that the coefficient matrix K be as sparse as possible, i.e., have lots of zero entries. Otherwise, computing the solution may well be too time-consuming to be of much practical value.

With this in mind, the second innovative contribution of the finite element method is to first (paradoxically) *enlarge* the space U of allowable functions upon which to minimize the quadratic functional $Q[u]$. The governing differential equation requires its (classical) solutions to have a certain degree of smoothness, whereas the associated minimization principle typically requires that they possess only half as many derivatives. Thus, for second-order boundary value problems, the differential equation requires continuous second-order derivatives, while the quadratic functional $Q[u]$ involves only first-order derivatives. It fact, it can be rigorously shown that, under rather mild hypotheses, the functional retains the *same* minimizing solution, even when one allows functions that fail to qualify as classical solutions to the differential equation. We will proceed to develop the method in the context of particular, fairly elementary examples.

Exercises

10.1.1. Let $U = \{ u(x) \in \mathrm{C}^2[0,\pi] \mid u(0) = u(\pi) = 0 \}$ and $V = \{ v(x) \in \mathrm{C}^1[0,\pi] \}$ both be equipped with the L^2 inner product. Let $L: U \to V$ be given by $L[u] = D[u] = u'$, and $f(x) = x - 1$. (a) Write out the quadratic functional $Q[u]$ given by (10.1). (b) Write out the associated boundary value problem (10.2). (c) Find the function $u_\star(x) \in U$ that minimizes $Q[u]$. What is the value of $Q[u_\star]$? (d) Let $W \subset U$ be the subspace spanned by $\sin x$ and $\sin 2x$. Write out the corresponding finite-dimensional minimization problem (10.8). (e) Find the function $w_\star(x) \in W$ that minimizes $Q[w]$. Is $Q[w_\star] \geq Q[u_\star]$? If not, why not? How close is your finite element minimizer $w_\star(x)$ to the actual minimizer $u_\star(x)$?

10.1.2. Let $U = \{ u(x) \in \mathrm{C}^2[0,1] \mid u(0) = u(1) = 0 \}$ and $V = \{ v(x) \in \mathrm{C}^1[0,1] \}$ both have the L^2 inner product. Let $L: U \to V$ be given by $L[u] = u'(x) - u(x)$, and $f(x) = 1$ for all x. (a) Write out the quadratic functional $Q[u]$ given by (10.1). (b) Write out the associated boundary value problem (10.2). (c) Find the function $u_\star(x) \in U$ that minimizes $Q[u]$. What is the value of $Q[u_\star]$? (d) Let $W \subset U$ be the subspace containing all cubic polynomials $p(x)$ that satisfy the boundary conditions: $p(0) = p(1) = 0$. Find a basis of W and then write out the corresponding finite-dimensional minimization problem (10.8). (e) Find the polynomial $p_\star(x) \in W$ that minimizes $Q[p]$ for $p \in W$. Is $Q[p_\star] \geq Q[u_\star]$? If not, why not? How close is your finite element minimizer $p_\star(x)$ to the minimizer $u_\star(x)$?

10.1.3. Let $U = \{ u(x) \in \mathrm{C}^2[1,2] \mid u(1) = u(2) = 0 \}$, $V = \{ (v_1(x), v_2(x))^T \mid v_1, v_2 \in \mathrm{C}^1[1,2] \}$, both be endowed with the L^2 inner product. Let $L: U \to V$ be given by $L[u] = \begin{pmatrix} x\,u'(x) \\ \sqrt{2}\,u(x) \end{pmatrix}$, and let $f(x) = 2$ for all $1 \le x \le 2$. (a) Write out the quadratic functional $Q[u]$ given by (10.1). (b) Write out the associated boundary value problem (10.2). (c) Find the function $u_\star(x) \in U$ that minimizes $Q[u]$. What is the value of $Q[u_\star]$? (d) Let $W \subset U$ be the subspace containing all cubic polynomials $p(x)$ that satisfy the boundary conditions $p(1) = p(2) = 0$. Find a basis of W and then write out the corresponding finite-dimensional minimization problem (10.8). (e) Find the polynomial $p_\star(x) \in W$ that minimizes $Q[p]$ for $p \in W$. Is $Q[p_\star] \ge Q[u_\star]$? If not, why not? How close is your finite element minimizer $p_\star(x)$ to the actual minimizer $u_\star(x)$?

♡ 10.1.4. (a) Find the solution to the boundary value problem $-u'' = x^2 - x$, $u(-1) = u(1) = 0$. (b) Write down a quadratic functional $Q[u]$ that is minimized by your solution. (c) Let W be the subspace spanned by the two functions $(1-x^2)$, $x(1-x^2)$. Find the function $w_\star(x) \in W$ that minimizes the restriction of your quadratic functional to W. Compare $w_\star$ with your solution from part (a). (d) Answer part (c) for the subspace W spanned by $\sin \pi x, \sin 2\pi x$. Which of the two approximations is the better?

♡ 10.1.5. (a) Find the function $u_\star(x)$ that minimizes $Q[u] = \int_0^1 \left[\frac{1}{2}(x+1)\,u'(x)^2 - u(x) \right] dx$ over the vector space U consisting of C^2 functions satisfying $u(0) = u(1) = 0$. (b) Let $W_3 \subset U$ be the subspace consisting of all cubic polynomials $w(x)$ that satisfy the same boundary conditions. Find the function $w_\star(x)$ that minimizes the restriction $Q[w]$ for $w \in W_3$. Compare $w_\star(x)$ and $u_\star(x)$: how close are they in the L^2 norm? What is the maximal discrepancy $|\,w_\star(x) - u_\star(x)\,|$ for $0 \le x \le 1$? (c) Suppose you enlarge your finite-dimensional subspace $W_4 \subset U$ to contain all quartic polynomials that satisfy the boundary conditions. Is your new finite element approximation better? Discuss.

♡ 10.1.6. (a) Find the function $u_\star(x)$ that minimizes $Q[u] = \int_0^1 \left[\frac{1}{2} e^x\, u'(x)^2 - 3u(x) \right] dx$ over the space U consisting of C^2 functions satisfying the boundary conditions $u(0) = u'(1) = 0$. (b) Let $W \subset U$ be the subspace containing all cubic polynomials $w(x)$ that satisfy the boundary conditions. Find the polynomial $w_\star(x)$ that minimizes the restriction $Q[w]$ for $w \in W$. Compare $w_\star(x)$ and $u_\star(x)$: how close are they in the L^2 norm? What is the maximal discrepancy $|\,w_\star(x) - u_\star(x)\,|$ for $0 \le x \le 1$?

10.1.7. Consider the Dirichlet boundary value problem

$$-\Delta u = x(1-x) + y(1-y), \qquad u(x,0) = u(x,1) = u(0,y) = u(1,y) = 0,$$

on the unit square $\{0 < x, y < 1\}$.

(a) Find the exact solution $u_\star(x,y)$. *Hint*: It is a polynomial.

(b) Write down a minimization principle $Q[u]$ that characterizes the solution. Be careful to specify the function space U over which the minimization takes place.

(c) Let $W \subset U$ be the subspace spanned by the four functions $\sin \pi x \sin \pi y$, $\sin 2\pi x \sin \pi y$, $\sin \pi x \sin 2\pi y$, and $\sin 2\pi x \sin 2\pi y$. Find the function $w_\star \in W$ that minimizes the restriction of $Q[w]$ to $w \in W$. How close is $w_\star$ to the solution you found in part (a)?

♢ 10.1.8. Justify the identification of (10.4) with the quadratic function (10.5).

10.2 Finite Elements for Ordinary Differential Equations

To understand the preceding abstract formulation in concrete terms, let us focus our attention on boundary value problems governed by a second-order ordinary differential equation.

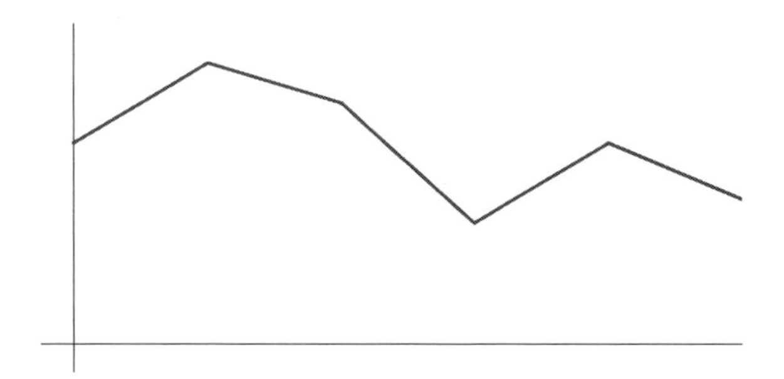

Figure 10.1. A continuous piecewise affine function.

For example, we might be interested in solving a Sturm–Liouville problem (9.71) subject to, say, homogeneous Dirichlet boundary conditions. Once we understand how the finite element constructions work in this relatively simple context, we will be in a good position to extend the techniques to much more general linear boundary value problems governed by elliptic partial differential equations.

For such one-dimensional boundary value problems, a popular and effective choice of the finite-dimensional subspace W is to employ continuous, piecewise affine functions. Recall that a function is *affine* if its graph is a straight line: $f(x) = a x + b$. (The function is *linear*, in accordance with Definition B.32, if and only if $b = 0$.) A function is called *piecewise affine* if its graph consists of a finite number of straight line segments; a typical example is plotted in Figure 10.1. Continuity requires that the individual segments be connected together end to end.

Given a boundary value problem on a bounded interval $[a, b]$, let us fix a finite collection of *nodes*

$$a = x_0 < x_1 < x_2 < \cdots < x_{n-1} < x_n = b.$$

The formulas simplify if one uses equally spaced nodes, but this is not necessary for the construction to be carried out. Let W denote the vector space consisting of all continuous functions $w(x)$ that are defined on the interval $a \le x \le b$, satisfy the homogeneous boundary conditions, and are affine when restricted to each subinterval $[x_j, x_{j+1}]$. On each subinterval, we write

$$w(x) = c_j + b_j(x - x_j), \qquad \text{for} \qquad x_j \le x \le x_{j+1}, \qquad j = 0, \ldots, n-1,$$

for certain constants c_j, b_j. Continuity of $w(x)$ requires

$$c_j = w(x_j^+) = w(x_j^-) = c_{j-1} + b_{j-1} h_{j-1}, \qquad j = 1, \ldots, n-1, \tag{10.10}$$

where $h_{j-1} = x_j - x_{j-1}$ denotes the length of the jth subinterval. The homogeneous Dirichlet boundary conditions at the endpoints require

$$w(a) = c_0 = 0, \qquad w(b) = c_{n-1} + b_{n-1} h_{n-1} = 0. \tag{10.11}$$

Observe that the function $w(x)$ involves a total of $2n$ unspecified coefficients $c_0, \ldots, c_{n-1}$, $b_0, \ldots, b_{n-1}$. The continuity conditions (10.10) and the second boundary condition (10.11) uniquely determine the b_j. The first boundary condition specifies c_0, while the remaining $n-1$ coefficients $c_1 = w(x_1), \ldots, c_{n-1} = w(x_{n-1})$ are arbitrary, specifying the values of $w(x)$ at the interior nodes. We conclude that the finite element subspace W has dimension $n-1$, the number of interior nodes.

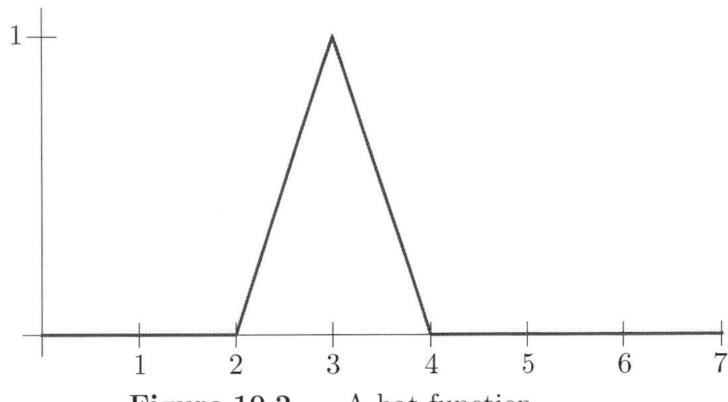

Figure 10.2. A hat function.

Remark: Every function $w(x)$ in our subspace has piecewise constant first derivative $w'(x)$. However, the jump discontinuities in $w'(x)$ imply that its second derivative $w''(x)$ may well include delta function impulses at the nodes, and hence $w(x)$ is far from being a solution to the differential equation. Nevertheless, in practice, the finite element minimizer $w_\star(x) \in W$ will (under suitable assumptions) provide a reasonable approximation to the actual solution $u_\star(x)$.

The most convenient basis for W consists of the *hat functions*, which are continuous, piecewise affine functions satisfying

$$\varphi_j(x_k) = \begin{cases} 1, & j = k, \\ 0, & j \neq k, \end{cases} \qquad \text{for} \qquad j = 1, \ldots, n-1, \qquad k = 0, \ldots, n. \tag{10.12}$$

The graph of a typical hat function appears in Figure 10.2. The explicit formula is easily established:

$$\varphi_j(x) = \begin{cases} \dfrac{x - x_{j-1}}{x_j - x_{j-1}}, & x_{j-1} \leq x \leq x_j, \\ \dfrac{x_{j+1} - x}{x_{j+1} - x_j}, & x_j \leq x \leq x_{j+1}, \\ 0, & x \leq x_{j-1} \quad \text{or} \quad x \geq x_{j+1}, \end{cases} \qquad j = 1, \ldots, n-1. \tag{10.13}$$

One advantage of using these basis functions is that, thanks to (10.12), the coefficients in the linear combination

$$w(x) = c_1 \varphi_1(x) + \cdots + c_n \varphi_n(x)$$

coincide with its values at the nodes:

$$c_j = w(x_j), \qquad j = 1, \ldots, n. \tag{10.14}$$

Example 10.1. Let $\kappa(x) > 0$ for $0 \leq x \leq \ell$. Consider the equilibrium equations

$$S[u] = -\frac{d}{dx}\left(\kappa(x)\,\frac{du}{dx}\right) = f(x), \qquad 0 < x < \ell, \qquad u(0) = u(\ell) = 0,$$

for a nonuniform bar with fixed ends and variable stiffness $\kappa(x)$, that is subject to an external forcing $f(x)$. In order to find a finite element approximation to the resulting displacement $u(x)$, we begin with the minimization principle based on the quadratic functional

$$Q[u] = \int_0^\ell \left[\tfrac{1}{2} \kappa(x)\, u'(x)^2 - f(x)\, u(x) \right] dx,$$

which is a special case of (9.75). We divide the interval $[0,\ell]$ into n equal subintervals, each of length $h = \ell/n$. The resulting uniform mesh has nodes

$$x_j = j\,h = \frac{j\,\ell}{n}, \qquad j = 0, \ldots, n.$$

The corresponding finite element basis hat functions are explicitly given by

$$\varphi_j(x) = \begin{cases} (x - x_{j-1})/h, & x_{j-1} \le x \le x_j, \\ (x_{j+1} - x)/h, & x_j \le x \le x_{j+1}, \\ 0, & \text{otherwise,} \end{cases} \qquad j = 1, \ldots, n-1. \tag{10.15}$$

The associated linear system (10.9) has coefficient matrix entries

$$k_{ij} = \langle\!\langle\, \varphi_i' \,, \varphi_j' \,\rangle\!\rangle = \int_0^\ell \varphi_i'(x)\, \varphi_j'(x)\, \kappa(x)\, dx, \qquad i,j = 1, \ldots, n-1.$$

Since the function $\varphi_i(x)$ vanishes except on the interval $x_{i-1} < x < x_{i+1}$, while $\varphi_j(x)$ vanishes outside $x_{j-1} < x < x_{j+1}$, the integral will vanish unless $i = j$ or $i = j \pm 1$. Moreover,

$$\varphi_i'(x) = \begin{cases} 1/h, & x_{j-1} \le x \le x_j, \\ -1/h, & x_j \le x \le x_{j+1}, \\ 0, & \text{otherwise,} \end{cases} \qquad j = 1, \ldots, n-1.$$

Therefore, the finite element coefficient matrix assumes the tridiagonal form

$$K = \frac{1}{h^2} \begin{pmatrix} s_0 + s_1 & -\,s_1 & & & & \\ -\,s_1 & s_1 + s_2 & -\,s_2 & & & \\ & -\,s_2 & s_2 + s_3 & -\,s_3 & & \\ & & \ddots & \ddots & \ddots & \\ & & & -\,s_{n-3} & s_{n-3} + s_{n-2} & -\,s_{n-2} \\ & & & & -\,s_{n-2} & s_{n-2} + s_{n-1} \end{pmatrix}, \tag{10.16}$$

where

$$s_j = \int_{x_j}^{x_{j+1}} \kappa(x)\, dx \tag{10.17}$$

is the total stiffness of the jth subinterval. The corresponding right-hand side has entries

$$\begin{aligned} b_j = \langle\, f \,, \varphi_j \,\rangle &= \int_0^\ell f(x)\, \varphi_j(x)\, dx \\ &= \frac{1}{h} \left[\int_{x_{j-1}}^{x_j} (x - x_{j-1}) f(x)\, dx + \int_{x_j}^{x_{j+1}} (x_{j+1} - x) f(x)\, dx \right]. \end{aligned} \tag{10.18}$$

In practice, we do not have to explicitly evaluate the integrals (10.17, 18), but may replace them by suitably close numerical approximations. When the step size $h \ll 1$ is small, then the integrals are taken over small intervals, and so the elementary trapezoid rule, [**24**, **108**], produces sufficiently accurate approximations:

$$s_j \approx \frac{h}{2}\left[\kappa(x_j)+\kappa(x_{j+1})\right], \qquad b_j \approx h\,f(x_j). \tag{10.19}$$

The resulting finite element system $K\mathbf{c}=\mathbf{b}$ is then solved for $\mathbf{c}$, whose entries, according to (10.14), coincide with the values of the finite element approximation to the solution at the nodes: $c_j = w(x_j) \approx u(x_j)$. Indeed, the tridiagonal Gaussian Elimination algorithm, [**89**], will rapidly produce the desired solution. Since the accuracy of the finite element solution increases with the number of nodes, this numerical scheme allows us to easily compute very accurate approximations to the solution to the boundary value problem.

In particular, in the homogeneous case $\kappa(x) \equiv 1$, the coefficient matrix (10.16) reduces to the special form

$$K = \frac{1}{h}\begin{pmatrix} 2 & -1 & & & & \\ -1 & 2 & -1 & & & \\ & -1 & 2 & -1 & & \\ & & \ddots & \ddots & \ddots & \\ & & & -1 & 2 & -1 \\ & & & & -1 & 2 \end{pmatrix}. \tag{10.20}$$

In this case, the jth equation in the finite element linear system is, upon dividing by h,

$$-\frac{c_{j+1}-2c_j+c_{j-1}}{h^2} = f(x_j). \tag{10.21}$$

Since $c_j \approx u(x_j)$, the left-hand side coincides with the standard finite difference approximation to minus the second derivative $-u''(x_j)$ at the node x_j, cf. (5.5). As a result, in this particular case the finite element and finite difference numerical solution schemes happen to coincide.

The sparse tridiagonal nature of the finite element matrix is a consequence of the fact that the basis functions are zero on much of the interval, or, in more mathematical language, that they have *small support*, in the following sense.

Definition 10.2. The *support* of a function $f(x)$, written $\operatorname{supp} f$, is the closure of the set where $f(x) \neq 0$.

Thus, a point x will belong to the support, provided f is not zero there, or at least is not zero at nearby points. For example, the support of the hat function (10.13) is the (small) interval $[x_{j-1}, x_{j+1}]$. The key property, ensuring sparseness, is that the integral of the product of two functions will be zero if their supports have empty intersection, or, slightly more generally, have only a finite number of points in common.

Example 10.3. Consider the boundary value problem

$$-\frac{d}{dx}\,(x+1)\,\frac{du}{dx} = 1, \qquad u(0)=0, \qquad u(1)=0. \tag{10.22}$$

The explicit solution is easily found by direct integration:

$$u(x) = -\,x + \frac{\log(x+1)}{\log 2}\,. \tag{10.23}$$

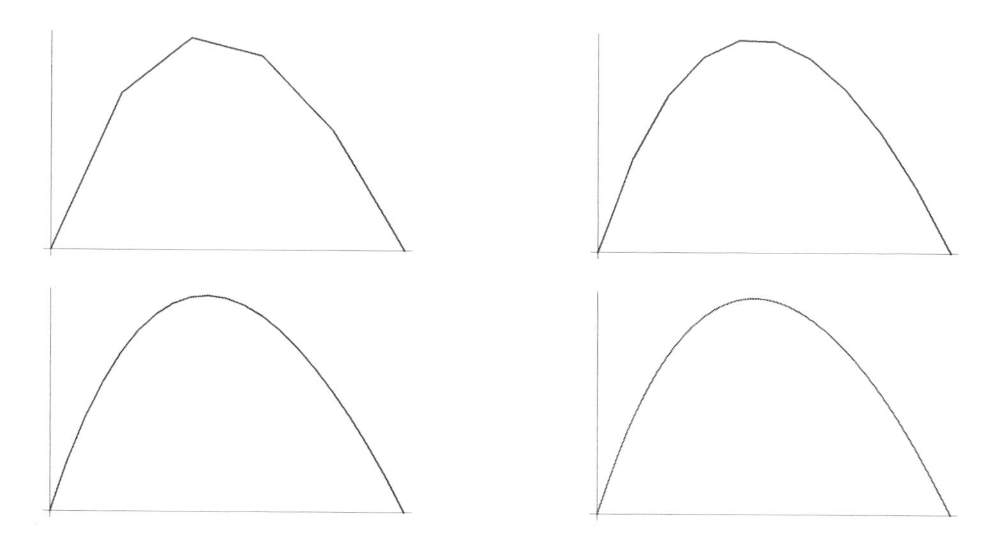

Figure 10.3. Finite element solution to (10.22).

It minimizes the associated quadratic functional

$$Q[u] = \int_0^\ell \left[\, \tfrac{1}{2}\,(x+1)\,u'(x)^2 - u(x) \,\right] dx \tag{10.24}$$

over the space of all C^2 functions $u(x)$ that satisfy the given boundary conditions. The finite element system (10.9) has coefficient matrix given by (10.16) and right-hand side (10.18), where

$$s_j = \int_{x_j}^{x_{j+1}} (1+x)\,dx = h\,(1+x_j) + \tfrac{1}{2}\,h^2 = h + h^2\big(j+\tfrac{1}{2}\big), \qquad b_j = \int_{x_j}^{x_{j+1}} 1\,dx = h.$$

The resulting piecewise affine approximation to the solution is plotted in Figure 10.3. The first three graphs contain, respectively, 5, 10, 20 nodes, so that $h = .2, .1, .05$, while the last plots the exact solution (10.23). The maximal errors at the nodes are, respectively, .000298, .000075, .000019, while the maximal overall errors between the exact solution and its piecewise affine finite element approximations are .00611, .00166, .00043. (One can more closely fit the solution curve by employing a cubic spline to interpolate the computed nodal values, [**89, 102**], which has the effect of reducing the preceding maximal overall errors by a factor of, approximately, 20.) Thus, even when computed on rather coarse meshes, the finite element approximation gives quite respectable results.

Remark: One can obtain a smoother, and hence more realistic, approximation to the solution by smoothly interpolating the finite element approximations $c_j \approx u(x_j)$ at the nodes, e.g., by use of cubic splines, [**89, 102**]. Alternatively, one can require that the finite element functions themselves be smoother, e.g., by making the finite element subspace consist of piecewise cubic splines that satisfy the boundary conditions.

Exercises

♣ 10.2.1. Use the finite element method to approximate the solution to the boundary value problem $-\dfrac{d}{dx}\left(e^{-x}\dfrac{du}{dx}\right) = 1$, $u(0) = u(2) = 0$. Carefully explain how you are setting up the calculation. Plot the resulting solutions and compare your answer with the exact solution. You should use an equally spaced mesh, but try at least three different mesh spacings and compare your results. By inspecting the errors in your various approximations, can you predict how many nodes would be required for six-digit accuracy of the numerical approximation?

♠ 10.2.2. For each of the following boundary value problems: (*i*) Solve the problem exactly. (*ii*) Approximate the solution using the finite element method based on ten equally spaced nodes. (*iii*) Compare the graphs of the exact solution and its piecewise affine finite element approximation. What is the maximal error in your approximation at the nodes? on the entire interval?

(a) $-u'' = \begin{cases} 1 & x > 1, \\ 0, & x < 1, \end{cases}$ $u(0) = u(2) = 0$; (b) $-\dfrac{d}{dx}\left((1+x)\dfrac{du}{dx}\right) = 1$, $u(0) = u(1) = 0$;

(c) $-\dfrac{d}{dx}\left(x^2\dfrac{du}{dx}\right) = -x$, $u(1) = u(3) = 0$; (d) $-\dfrac{d}{dx}\left(e^x\dfrac{du}{dx}\right) = e^x$, $u(-1) = u(1) = 0$.

♣ 10.2.3. (a) Find the exact solution to the boundary value problem $-u'' = 3x$, $u(0) = u(1) = 0$. (*b*) Use the finite element method based on five equally spaced nodes to approximate the solution. (*c*) Compare the graphs of the exact solution and its piecewise affine finite element approximation. (*d*) What is the maximal error (*i*) at the nodes? (*ii*) on the entire interval?

♣ 10.2.4. Use finite elements to approximate the solution to the Sturm–Liouville boundary value problem $-u'' + (x+1)u = x\,e^x$, $u(0) = 0$, $u(1) = 0$, using 5, 10, and 20 equally spaced nodes.

♣ 10.2.5. (a) Devise a finite element scheme for numerically approximating the solution to the mixed boundary value problem

$$-\frac{d}{dx}\left(\kappa(x)\frac{du}{dx}\right) = f(x), \qquad a < x < b, \qquad u(a) = 0, \qquad u'(b) = 0.$$

(*b*) Test your method on the particular boundary value problem

$$-\frac{d}{dx}\left((1+x)\frac{du}{dx}\right) = 1, \qquad 0 < x < 1, \qquad u(0) = 0, \qquad u'(1) = 0,$$

using 10 equally spaced nodes. Compare your approximation with the exact solution.

♠ 10.2.6. Consider the periodic boundary value problem

$$-u'' + u = x, \qquad u(0) = u(2\pi), \qquad u'(0) = u'(2\pi).$$

(*a*) Write down the analytic solution. (*b*) Write down a minimization principle. (*c*) Divide the interval $[0, 2\pi]$ into $n = 5$ equal subintervals, and let W_n denote the subspace consisting of all piecewise affine functions that satisfy the boundary conditions. What is the dimension of W_n? Write down a basis. (*d*) Construct the finite element approximation to the solution to the boundary value problem by minimizing the functional from part (*b*) on the subspace W_n. Graph the result and compare with the exact solution. What is the maximal error on the interval? (*e*) Repeat part (*d*) for $n = 10, 20$, and 40 subintervals, and discuss the convergence of your solutions.

♠ 10.2.7. Answer Exercise 10.2.6 when the finite element subspace W_n consists of all periodic piecewise affine functions of period 1, so $w(x+1) = w(x)$. Which approximation is better?

♣ 10.2.8. Use the method of Exercise 10.2.7 to approximate the solution to the following periodic boundary value problem for the *Mathieu equation*:

$$-u'' + (1 + \cos x)\,u = 1, \qquad u(0) = u(2\pi), \qquad u'(0) = u'(2\pi).$$

♠ 10.2.9. Consider the boundary value problem solved in Example 10.3. Let W_n be the subspace consisting of all polynomials $u(x)$ of degree $\leq n$ satisfying the boundary conditions $u(0) = u(1) = 0$. In this project, we will try to approximate the exact solution to the boundary value problem by minimizing the functional (10.24) on the polynomial subspace W_n. For $n = 5, 10$, and 20: (*a*) First, determine a basis for W_n. (*b*) Set up the minimization problem as a system of linear equations for the coefficients of the polynomial minimizer relative to your basis. (*c*) Solve the polynomial minimization problem and compare your "polynomial finite element" solution with the exact solution and the piecewise affine finite element solution graphed in Figure 10.3.

♠ 10.2.10. Consider the boundary value problem $-u'' + \lambda u = x$, for $0 < x < \pi$, with $u(0) = 0$, $u(1) = 0$. (*a*) For what values of λ does the system have a unique solution? (*b*) For which values of λ can you find a minimization principle that characterizes the solution? Is the minimizer unique for all such values of λ? (*c*) Using n equally spaced nodes, write down the finite element equations for approximating the solution to the boundary value problem. *Note*: Although the finite element construction is supposed to work only when there is a minimization principle, we will consider the resulting linear algebraic system for any value of λ. (*d*) Select a value of λ for which the solution can be characterized by a minimization principle and verify that the finite element approximation with $n = 10$ approximates the exact solution. (*e*) Experiment with other values of λ. Does your finite element solution give a good approximation to the exact solution when it exists? What happens at values of λ for which the solution does not exist or is not unique?

10.3 Finite Elements in Two Dimensions

The same basic framework underlies the adaptation of finite element techniques for numerically approximating the solution to boundary value problems governed by elliptic partial differential equations. In this section, we concentrate on the simplest case: the two-dimensional Poisson equation. Having mastered this, the reader will be well equipped to carry over the method to more general equations and higher dimensions. As before, we concentrate on the practical design of the finite element procedure, and refer the reader to more advanced texts, e.g., [**6**, **113**, **126**], for the analytical details and proofs of convergence. Most of the multi-dimensional complications lie not in the underlying theory, but rather in the realm of data management and organization.

For specificity, consider the homogeneous Dirichlet boundary value problem

$$-\Delta u = f \quad \text{in} \quad \Omega, \qquad\qquad u = 0 \quad \text{on} \quad \partial\Omega, \tag{10.25}$$

on a bounded domain $\Omega \subset \mathbb{R}^2$. According to Theorem 9.31, the solution $u_\star(x, y)$ is characterized as the unique minimizer of the Dirichlet functional

$$Q[u] = \tfrac{1}{2} \, ||| \, \nabla u \, |||^2 - \langle \, u \, , f \, \rangle = \iint_\Omega \left(\tfrac{1}{2} u_x^2 + \tfrac{1}{2} u_y^2 - f \, u \right) dx \, dy \tag{10.26}$$

among all C^2 functions $u(x, y)$ that satisfy the prescribed boundary conditions.

To construct a finite element approximation, we restrict the Dirichlet functional to a suitably chosen finite-dimensional subspace. As in the one-dimensional version, the most effective subspaces contain functions that may lack the requisite degree of smoothness that qualifies them as candidate solutions to the partial differential equation. Nevertheless, they will provide good approximations to the actual classical solution. Another important practical consideration, ensuring sparseness of the finite element matrix, is to employ functions

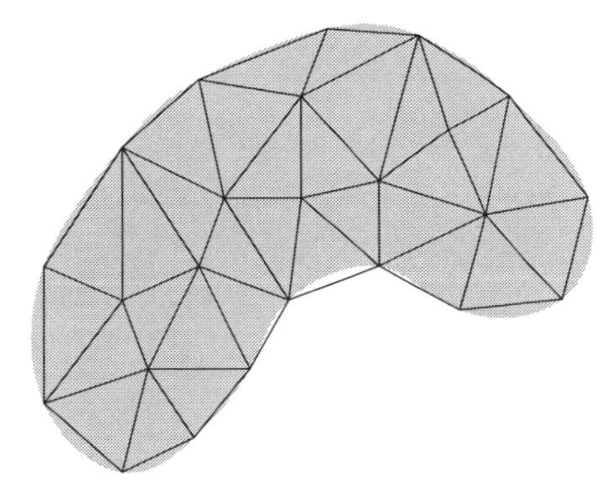

Figure 10.4. Triangulation of a planar domain.

that have small support, meaning that they vanish on most of the domain. Sparseness has the benefit that the solution to the linear finite element system can be relatively rapidly calculated, usually by application of an iterative numerical scheme such as the Gauss–Seidel or SOR methods discussed in [**89, 118**].

Triangulation

The first step is to introduce a *mesh* consisting of a finite number of *nodes* $\mathbf{x}_l = (x_l, y_l)$, $l = 1, \ldots, m$, usually lying inside the domain $\Omega \subset \mathbb{R}^2$. Unlike finite difference schemes, finite element methods are not tied to a rectangular mesh, thus endowing them with considerably more flexibility in the allowable discretizations of the domain. We regard the nodes as the vertices of a *triangulation* of the domain, consisting of a collection of non-overlapping small triangles, which we denote by $T_1, \ldots, T_N$, whose union $T_\star = \bigcup_\nu T_\nu$ approximates Ω; see Figure 10.4 for a typical example. The nodes are split into two categories — *interior nodes* and *boundary nodes*, the latter lying on or close to $\partial\Omega$. A curved boundary will thus be approximated by the polygonal boundary $\partial T_\star$ of the triangulation, whose vertexvertices are the boundary nodes. Thus, in any practical implementation of a finite element scheme, the first requirement is a routine that will automatically triangulate a specified domain in some "reasonable" manner, as explained below.

As in our one-dimensional construction, the functions $w(x, y)$ in the finite-dimensional subspace W will be continuous and *piecewise affine*, which means that, on each triangle, the graph of w is a flat plane and hence has the formula[†]

$$w(x, y) = \alpha^\nu + \beta^\nu x + \gamma^\nu y \qquad \text{when} \qquad (x, y) \in T_\nu, \tag{10.27}$$

for certain constants $\alpha^\nu, \beta^\nu, \gamma^\nu$. Continuity of w requires that its values on a common edge between two triangles must agree, and this will impose compatibility constraints on the coefficients $\alpha^\mu, \beta^\mu, \gamma^\mu$ and $\alpha^\nu, \beta^\nu, \gamma^\nu$ associated with adjacent pairs of triangles T_μ and T_ν. The full graph of the piecewise affine function $z = w(x, y)$ forms a connected polyhedral surface whose triangular faces lie above the triangles T_ν; see Figure 10.5 for an illustration. In addition, we require that the piecewise affine function $w(x, y)$ vanish at the boundary nodes, which implies that it vanishes on the entire polygonal boundary of the triangulation,

[†] Here and subsequently, the index ν is a superscript, not a power.

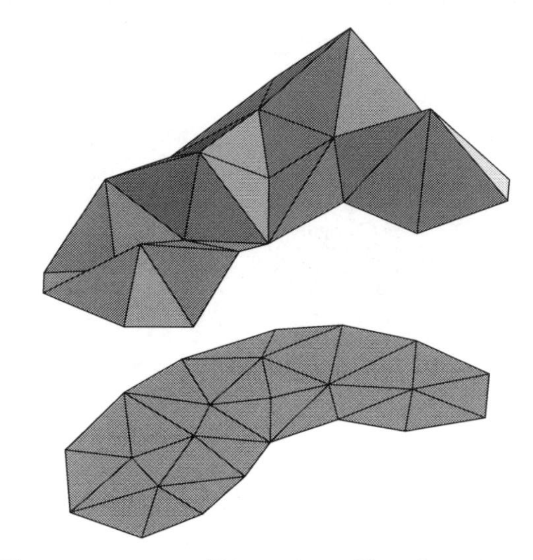

Figure 10.5. Piecewise affine function.

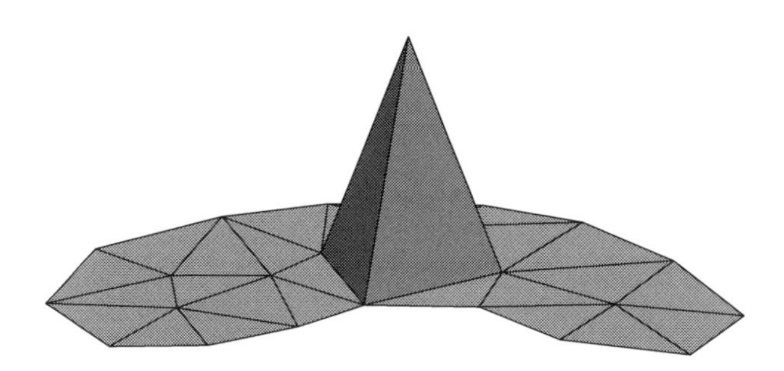

Figure 10.6. Finite element pyramid function.

$\partial T_\star$, and hence (approximately) satisfies the homogeneous Dirichlet boundary conditions on the curved boundary of the original domain, $\partial\Omega$.

The next step is to choose a basis of the subspace of piecewise affine functions associated with the given triangulation and subject to the imposed homogeneous Dirichlet boundary conditions. The analogue of the one-dimensional hat function (10.12) is the *pyramid function* $\varphi_l(x,y)$, which has the value 1 at a single node $\mathbf{x}_l = (x_l, y_l)$, and vanishes at all the other nodes:

$$\varphi_l(x_i, y_i) = \begin{cases} 1, & i = l, \\ 0, & i \neq l. \end{cases} \tag{10.28}$$

Because, on any triangle, the pyramid function $\varphi_l(x,y)$ is uniquely determined by its values at the vertices, it will be nonzero only on those triangles that have the node $\mathbf{x}_l$ as one of their vertices. Hence, as its name implies, the graph of φ_l forms a pyramid of unit height sitting on a flat plane; a typical example appears in Figure 10.6.

The pyramid functions $\varphi_l(x,y)$ associated with the *interior nodes* $\mathbf{x}_l$ automatically satisfy the homogeneous Dirichlet boundary conditions on the boundary of the domain — or, more correctly, on the polygonal boundary of the triangulated domain. Thus, the finite element subspace W is the span of the interior node pyramid functions, and so a general piecewise affine function $w \in W$ is a linear combination thereof:

$$w(x,y) = \sum_{l=1}^{n} c_l \, \varphi_l(x,y), \tag{10.29}$$

where the sum ranges over the n interior nodes of the triangulation. Owing to the original specification (10.28) of the pyramid functions, the coefficients

$$c_l = w(x_l, y_l) \;\approx\; u(x_l, y_l), \qquad\qquad l = 1, \ldots, n, \tag{10.30}$$

are the *same* as the values of the finite element approximation $w(x,y)$ at the interior nodes. This immediately implies linear independence of the pyramid functions, since the only linear combination that vanishes at all nodes is the trivial one $c_1 = \cdots = c_n = 0$.

Determining the explicit formulas for the pyramid functions is not difficult. On one of the triangles T_ν that has $\mathbf{x}_l$ as a vertex, $\varphi_l(x,y)$ will be the unique affine function (10.27) that takes the value 1 at the vertex $\mathbf{x}_l$ and 0 at its other two vertices $\mathbf{x}_i$ and $\mathbf{x}_j$. Thus, we seek a formula for an affine function or *element*

$$\omega_l^\nu(x,y) = \alpha_l^\nu + \beta_l^\nu \, x + \gamma_l^\nu \, y, \qquad\qquad (x,y) \in T_\nu, \tag{10.31}$$

that takes the prescribed values

$$\begin{aligned} \omega_l^\nu(x_i,y_i) &= \alpha_l^\nu + \beta_l^\nu \, x_i + \gamma_l^\nu \, y_i = 0, \\ \omega_l^\nu(x_j,y_j) &= \alpha_l^\nu + \beta_l^\nu \, x_j + \gamma_l^\nu \, y_j = 0, \\ \omega_l^\nu(x_l,y_k) &= \alpha_l^\nu + \beta_l^\nu \, x_l + \gamma_l^\nu \, y_l = 1. \end{aligned} \tag{10.32}$$

Solving this linear system for the coefficients — using either Cramer's Rule or direct Gaussian Elimination — produces the explicit formulas

$$\alpha_l^\nu = \frac{x_i \, y_j - x_j \, y_i}{\Delta_\nu}, \qquad \beta_l^\nu = \frac{y_i - y_j}{\Delta_\nu}, \qquad \gamma_l^\nu = \frac{x_j - x_i}{\Delta_\nu}, \tag{10.33}$$

where the denominator

$$\Delta_\nu = \det \begin{pmatrix} 1 & x_i & y_i \\ 1 & x_j & y_j \\ 1 & x_l & y_l \end{pmatrix} = \pm 2 \,\text{area}\, T_\nu \tag{10.34}$$

is, up to sign, twice the area of the triangle T_ν; see Exercise 10.3.5.

Example 10.4. Consider an isosceles right triangle T with vertices

$$\mathbf{x}_1 = (0,0), \qquad \mathbf{x}_2 = (1,0), \qquad \mathbf{x}_3 = (0,1).$$

Using (10.33–34) (or solving the linear system (10.32) directly), we immediately produce the three corresponding affine elements

$$\omega_1(x,y) = 1 - x - y, \qquad \omega_2(x,y) = x, \qquad \omega_3(x,y) = y. \tag{10.35}$$

As required, each ω_l equals 1 at the vertex $\mathbf{x}_l$ and is zero at the other two vertices.

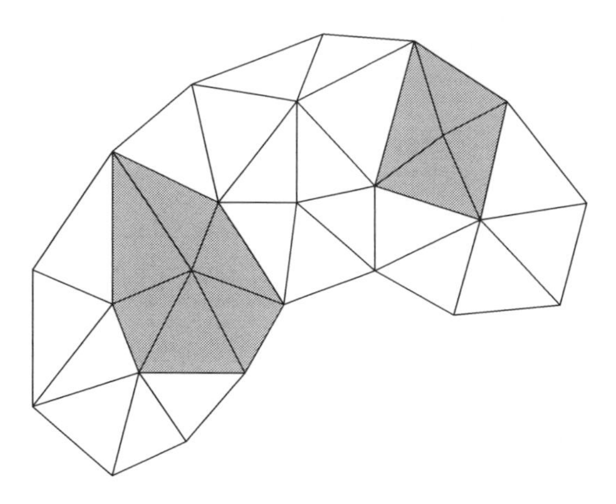

Figure 10.7. Vertex polygons.

A pyramid function is then obtained by piecing together the individual affine elements:

$$\varphi_l(x,y) = \begin{cases} \omega_l^\nu(x,y), & \text{if } (x,y) \in T_\nu \text{ and } \mathbf{x}_l \text{ is a vertex of } T_\nu, \\ 0, & \text{otherwise.} \end{cases} \tag{10.36}$$

Continuity of $\varphi_l(x,y)$ is assured, since the constituent affine elements have the same values at common vertices, and hence also along common edges. The support of the pyramid function (10.36) is the *vertex polygon*

$$\operatorname{supp} \varphi_l = P_l = \bigcup_\nu T_\nu \tag{10.37}$$

consisting of all the triangles T_ν that have the node $\mathbf{x}_l$ as a vertex. In other words, $\varphi_l(x,y) = 0$ whenever $(x,y) \notin P_l$. The node $\mathbf{x}_l$ lies on the interior of its vertex polygon P_l, while the vertices of P_l are all the nodes connected to $\mathbf{x}_l$ by a single edge of the triangulation. In Figure 10.7, the shaded regions indicate two of the vertex polygons for the triangulation in Figure 10.4.

Example 10.5. The simplest, and most common, triangulations are based on regular meshes. For example, suppose that the nodes lie on a square grid, and so are of the form $\mathbf{x}_{i,j} = (i\,h + a, j\,h + b)$, where (i,j) run over a collection of integer pairs, $h > 0$ is the inter-node spacing, and (a,b) represents an overall offset. If we choose the triangles to all have the same orientation, as in the first picture in Figure 10.8, then the vertex polygons all have the same shape, consisting of six triangles of total area $3\,h^2$ — the shaded region. On the other hand, if we choose an alternating triangulation, as in the second picture, then there are two types of vertex polygons. The first, consisting of four triangles, has area $2\,h^2$, while the second, containing eight triangles, has twice the area, $4\,h^2$. In practice, there are good reasons to prefer the former triangulation.

In general, to ensure convergence of the finite element solution to the true minimizer, one should choose triangulations that satisfy the following properties:

- The three side lengths of any individual triangle should be of comparable size, and so long, skinny triangles and obtuse triangles should be avoided.
- The areas of nearby triangles T_ν should not vary too much.
- The areas of nearby vertex polygons P_l should also not vary too much.

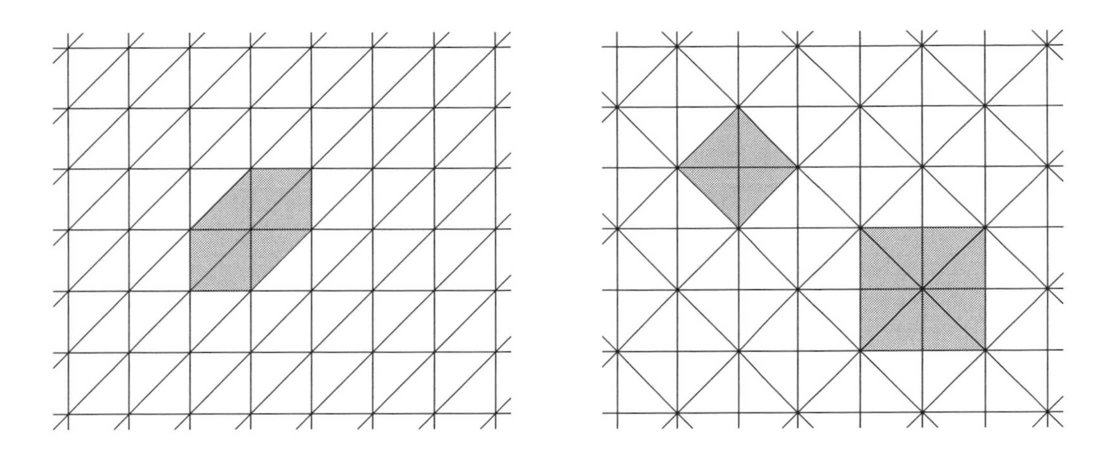

Figure 10.8. Square mesh triangulations.

While the nearby triangles should be of comparable size, one might very well allow wide variations over the entire domain, with small triangles in regions where the solution is changing rapidly, and large triangles in less active regions.

Exercises

10.3.1. Sketch a triangulation of the following domains so that all triangles have side length at most .5: (*a*) a unit square; (*b*) an isosceles triangle with vertices $(-.5, 0), (.5, 0)$ and $(0, 1)$; (*c*) the square $\{\, |\, x\, |, |\, y\, | \leq 2 \,\}$ with the hole $\{\, |\, x\, |, |\, y\, | < 1 \,\}$ removed; (*d*) the unit disk; (*e*) the annulus $1 \leq \|\, \mathbf{x}\, \| \leq 2$.

10.3.2. Describe the vertex polygons for a triangulation that uses regular equilateral triangles.

10.3.3. Are there any restrictions on the number of sides a vertex polygon can have?

10.3.4. Find the three finite element functions $\omega_1(x, y)$, $\omega_2(x, y)$, $\omega_3(x, y)$, associated with
(*a*) the triangle having vertices $(1, 0), (0, 1)$, and $(1, 1)$;
(*b*) the triangle having vertices $(0, 1), (1, -1)$, and $(-1, -1)$;
(*c*) an equilateral triangle centered at the origin having one vertex at $(1, 0)$.

◇ 10.3.5. (a) Prove that the area of a planar triangle T with vertices $(a, b), (c, d), (e, f)$ is equal to $\frac{1}{2}\,|\,\Delta\,|$, where $\Delta = \det \begin{pmatrix} 1 & a & b \\ 1 & c & d \\ 1 & e & f \end{pmatrix}$. (*b*) Prove that $\Delta > 0$ if and only if the vertices of the triangle are listed in counterclockwise order.

◇ 10.3.6. Give a detailed justification of the continuity of the pyramid function (10.36).

♡ 10.3.7. An alternative to triangular elements is to employ piecewise *bi-affine functions*, meaning $\omega(x, y) = \alpha + \beta\, x + \gamma\, y + \delta\, x\, y$, on rectangles. (a) Suppose R is a rectangle with vertices $(x_1, y_1), (x_2, y_2), (x_3, y_3), (x_4, y_4)$, whose sides are parallel to the coordinate axes. Prove that, for each $l = 1, \ldots, 4$, there is a unique bi-affine function $\omega_l(x, y)$ defined on R that has the value $\omega_l(x_l, y_l) = 1$ at one vertex while $\omega_l(x_i, y_i) = 0,\ i \neq l$, at the other three vertices.

(*b*) Write out the four bi-affine functions $\omega_1(x,y),\ldots,\omega_4(x,y)$, when (*i*) $R=\{0\le x,y\le 1\}$, (*ii*) $R=\{-1\le x,y\le 1\}$. (*c*) Does the result in part (a) hold for rectangles whose sides are not aligned with the axes? For general quadrilaterals?

The Finite Element Equations

We now seek to approximate the solution to the homogeneous Dirichlet boundary value problem by restricting the Dirichlet functional (10.26) to the selected finite element subspace W. Using the general framework of Section 10.1, we substitute the formula (10.29) for a general element of W into the quadratic Dirichlet functional (9.82). Expanding, we obtain

$$\begin{aligned} Q[w] = Q\left[\sum_{i=1}^{n} c_i\,\varphi_i\right] &= \iint_\Omega \left[\left(\sum_{i=1}^{n} c_i\,\nabla\varphi_i\right)^2 - f(x,y)\left(\sum_{i=1}^{n} c_i\,\varphi_i\right)\right] dx\,dy \\ &= \frac{1}{2}\sum_{i,j=1}^{n} k_{ij}\,c_i\,c_j - \sum_{i=1}^{n} b_i\,c_i = \tfrac12\,\mathbf{c}^T K\,\mathbf{c} - \mathbf{b}^T\mathbf{c}. \end{aligned} \tag{10.38}$$

Here $K=(k_{ij})$ is a symmetric $n\times n$ matrix, while $\mathbf{b}=(\,b_1,b_2,\ldots,b_n\,)^T$ is a vector in $\mathbb{R}^n$, with respective entries

$$\begin{aligned} k_{ij} &= \langle\!\langle\,\nabla\varphi_i\,,\nabla\varphi_j\,\rangle\!\rangle = \iint_\Omega \nabla\varphi_i\cdot\nabla\varphi_j\,dx\,dy, \\ b_i &= \langle\,f\,,\varphi_i\,\rangle = \iint_\Omega f\,\varphi_i\,dx\,dy, \end{aligned} \tag{10.39}$$

which also follow directly from the general formulas (10.6–7). Thus, the finite element approximation (10.29) will minimize the quadratic function

$$P(\mathbf{c}) = \tfrac12\,\mathbf{c}^T K\,\mathbf{c} - \mathbf{b}^T\mathbf{c} \tag{10.40}$$

over all possible choices of coefficients $\mathbf{c}=(\,c_1,c_2,\ldots,c_n\,)^T\in\mathbb{R}^n$, i.e., over all possible function values at the interior nodes. As above, the minimizer's coefficients are obtained by solving the associated linear system

$$K\,\mathbf{c} = \mathbf{b}, \tag{10.41}$$

using either Gaussian Elimination or a suitable iterative linear systems solver.

To find explicit formulas for the matrix coefficients k_{ij} in (10.39), we begin by noting that the gradient of the affine element (10.31) is equal to

$$\mathbf{g}_l^\nu = \nabla\omega_l^\nu(x,y) = \begin{pmatrix} \partial\omega_l^\nu/\partial x \\ \partial\omega_l^\nu/\partial y \end{pmatrix} = \begin{pmatrix} \beta_l^\nu \\ \gamma_l^\nu \end{pmatrix} = \frac{1}{\Delta_\nu}\begin{pmatrix} y_i - y_j \\ x_j - x_i \end{pmatrix}, \qquad (x,y)\in T_\nu, \tag{10.42}$$

which is a constant vector inside the triangle T_ν, while $\nabla\omega_l^\nu = \mathbf{0}$ outside T_ν. Therefore,

$$\nabla\varphi_l(x,y) = \begin{cases} \mathbf{g}_l^\nu, & \text{if } (x,y)\in T_\nu \text{ that has } \mathbf{x}_l \text{ as a vertex}, \\ \mathbf{0}, & \text{otherwise}. \end{cases} \tag{10.43}$$

Actually, (10.43) is not quite correct, since the gradient is not well defined on the boundary of a triangle T_ν, but this will not cause us any difficulty in evaluating the ensuing integrals.

We will approximate integrals over the domain Ω by summing the corresponding integrals over the individual triangles — which relies on our assumption that the polygonal boundary of the triangulation $\partial T_\star$ is a reasonably close approximation to the true boundary $\partial\Omega$. In particular,

$$k_{ij} \approx \sum_\nu \iint_{T_\nu} \nabla\varphi_i \cdot \nabla\varphi_j \, dx\, dy \;\equiv\; \sum_\nu k_{ij}^\nu. \tag{10.44}$$

Now, according to (10.43), one or the other gradient in the integrand will vanish on the entire triangle T_ν *unless* both $\mathbf{x}_i$ and $\mathbf{x}_j$ are vertices. Therefore, the only terms contributing to the sum are those triangles T_ν that have both $\mathbf{x}_i$ and $\mathbf{x}_j$ as vertices. If $i \neq j$, there are only two such triangles, having a common edge, while if $i = j$, every triangle in the ith vertex polygon P_i contributes. The individual summands are easily evaluated, since the gradients are constant on the triangles, and so, by (10.43),

$$k_{ij}^\nu = \iint_{T_\nu} \mathbf{g}_i^\nu \cdot \mathbf{g}_j^\nu \, dx\, dy = \mathbf{g}_i^\nu \cdot \mathbf{g}_j^\nu \text{ area } T_\nu = \tfrac{1}{2}\, \mathbf{g}_i^\nu \cdot \mathbf{g}_j^\nu \,|\, \Delta_\nu \,|\,.$$

Let T_ν have vertices $\mathbf{x}_i, \mathbf{x}_j, \mathbf{x}_l$. Then, by (10.34, 42, 43),

$$\begin{aligned} k_{ij}^\nu &= \frac{1}{2}\,\frac{(y_j - y_l)(y_l - y_i) + (x_l - x_j)(x_i - x_l)}{(\Delta_\nu)^2}\,|\,\Delta_\nu\,| = -\,\frac{(\mathbf{x}_i - \mathbf{x}_l)\cdot(\mathbf{x}_j - \mathbf{x}_l)}{2\,|\,\Delta_\nu\,|}, \qquad i \neq j, \\ k_{ii}^\nu &= \frac{1}{2}\,\frac{(y_j - y_l)^2 + (x_l - x_j)^2}{(\Delta_\nu)^2}\,|\,\Delta_\nu\,| = \frac{\|\,\mathbf{x}_j - \mathbf{x}_l\,\|^2}{2\,|\,\Delta_\nu\,|} \\ &= -\,\frac{(\mathbf{x}_i - \mathbf{x}_l)\cdot(\mathbf{x}_i - \mathbf{x}_j) + (\mathbf{x}_i - \mathbf{x}_l)\cdot(\mathbf{x}_j - \mathbf{x}_l)}{2\,\Delta_\nu} = -\,k_{ij}^\nu - k_{il}^\nu\,. \end{aligned} \tag{10.45}$$

In this manner, each triangle T_ν specifies a collection of six different coefficients, $k_{ij}^\nu = k_{ji}^\nu$, indexed by its vertices, and known as the *elemental stiffnesses* of T_ν. Interestingly, the elemental stiffnesses depend only on the three vertex *angles* in the triangle and not on its size. Thus, similar triangles have the *same* elemental stiffnesses. Indeed, according to Exercise 10.3.13,

$$k_{ii}^\nu = \tfrac{1}{2}(\cot\theta_j^\nu + \cot\theta_l^\nu), \qquad \text{while} \qquad k_{ij}^\nu = k_{ji}^\nu = -\tfrac{1}{2}\cot\theta_l^\nu, \quad i \neq j, \tag{10.46}$$

where $0 < \theta_l^\nu < \pi$ denotes the angle in T_ν at the vertex $\mathbf{x}_l$.

Example 10.6. The right triangle with vertices $\mathbf{x}_1 = (0,0)$, $\mathbf{x}_2 = (1,0)$, $\mathbf{x}_3 = (0,1)$ has elemental stiffnesses

$$k_{11} = 1, \qquad k_{22} = k_{33} = \tfrac{1}{2}, \qquad k_{12} = k_{21} = k_{13} = k_{31} = -\tfrac{1}{2}, \qquad k_{23} = k_{32} = 0. \tag{10.47}$$

The same holds for any other isosceles right triangle, provided its vertices are labeled in the same manner. Similarly, an equilateral triangle has all 60° angles, and so its elemental stiffnesses are

$$\begin{aligned} k_{11} &= k_{22} = k_{33} = \tfrac{1}{\sqrt{3}} \approx .5774, \\ k_{12} &= k_{21} = k_{13} = k_{31} = k_{23} = k_{32} = -\tfrac{1}{2\sqrt{3}} \approx -.2887. \end{aligned} \tag{10.48}$$

Exercises

10.3.8. Write down the elemental stiffnesses for: (*a*) the triangle with vertices $(0,1)$, $(-1,2)$, $(0,-1)$; (*b*) the triangle with vertices $(1,1)$, $(-1,1)$, $(0,-2)$; (*c*) a $30-60-90$ degree right triangle; (*d*) a right triangle with side lengths $3,4,5$; (*e*) an isosceles triangle of height 3 and base 2; (*f*) a "golden" isosceles triangle with angles $36°, 72°, 72°$.

♢ 10.3.9. A *rectangular mesh* has nodes $\mathbf{x}_{i,j} = (i\,\Delta x + a, j\,\Delta y + b)$, where $\Delta x, \Delta y > 0$ are, respectively, the horizontal and vertical step sizes. Find the elemental stiffnesses for the triangles associated with such a rectangular mesh.

10.3.10. *True or false*: Let T be a triangle, and $\widetilde{T}$ a triangle obtained by rotating T by $60°$. Then T and $\widetilde{T}$ have the same elemental stiffnesses.

10.3.11. Prove that the gradient (10.42) of the affine element is equal to $\nabla \omega_l^\nu = \|\,\mathbf{a}_l^\nu\,\|^{-2}\,\mathbf{a}_l^\nu$, where $\mathbf{a}_l^\nu$ is the *altitude vector* that goes to the vertex $\mathbf{x}_l$ from its opposite side, as indicated in the figure.

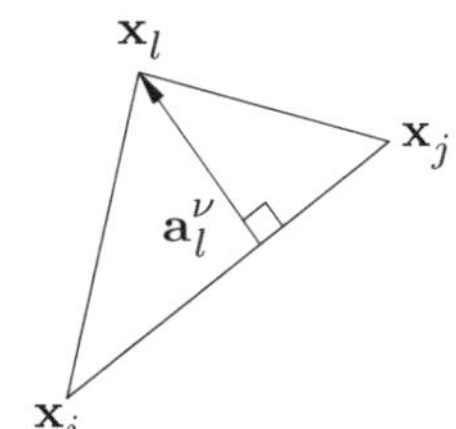

10.3.12. Explain why the pyramid functions are linearly independent.

♢ 10.3.13. Prove formulas (10.46).

Assembling the Elements

The elemental stiffnesses of each triangle will contribute, through the summation (10.44), to the finite element coefficient matrix K. We begin by constructing a larger matrix $\widehat{K}$, which we call the *full finite element matrix*, of size $m \times m$, where m is the total number of nodes in our triangulation, including both interior and boundary nodes. The rows and columns of $\widehat{K}$ are labeled by the nodes $\mathbf{x}_1, \ldots, \mathbf{x}_m$. Let $K_\nu = (k_{ij}^\nu)$ be the corresponding $m \times m$ matrix containing the elemental stiffnesses k_{ij}^ν of T_ν in the rows and columns indexed by its vertices, and all other entries equal to 0. Thus, K_ν will have (at most) nine nonzero entries. The resulting $m \times m$ matrices are summed together over all the triangles $T_1, \ldots, T_N$, whereby

$$\widehat{K} = \sum_{\nu=1}^{N} K_\nu, \tag{10.49}$$

in accordance with (10.44).

The full finite element matrix $\widehat{K}$ is too large, since its rows and columns include all the nodes, whereas the finite element matrix K appearing in (10.41) refers only to the n interior nodes. The *reduced* $n \times n$ *finite element matrix* K is simply obtained from $\widehat{K}$ by deleting all rows and columns indexed by boundary nodes, retaining only the elements k_{ij} for which both $\mathbf{x}_i$ and $\mathbf{x}_j$ are interior nodes. For the homogeneous boundary value problem, this is all we require. As we will subsequently see, inhomogeneous boundary conditions are most easily handled by retaining (another part of) the full matrix $\widehat{K}$.

The easiest way to absorb the construction is by working through a particular example.

Example 10.7. A metal plate has the shape of an oval running track, consisting of a rectangle, with side lengths 1 m by 2 m, and two semi-circular disks glued onto its

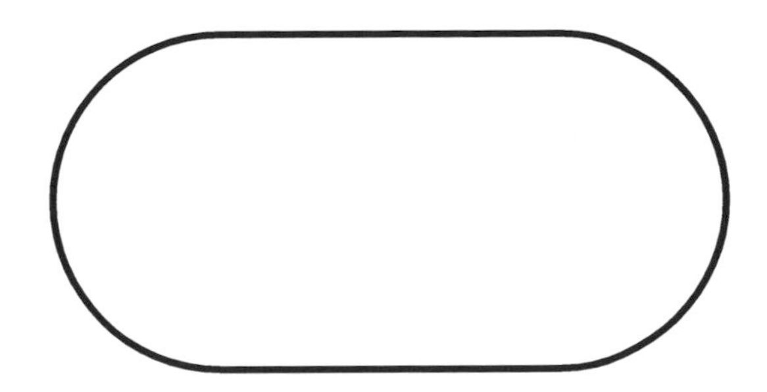

Figure 10.9. The oval plate.

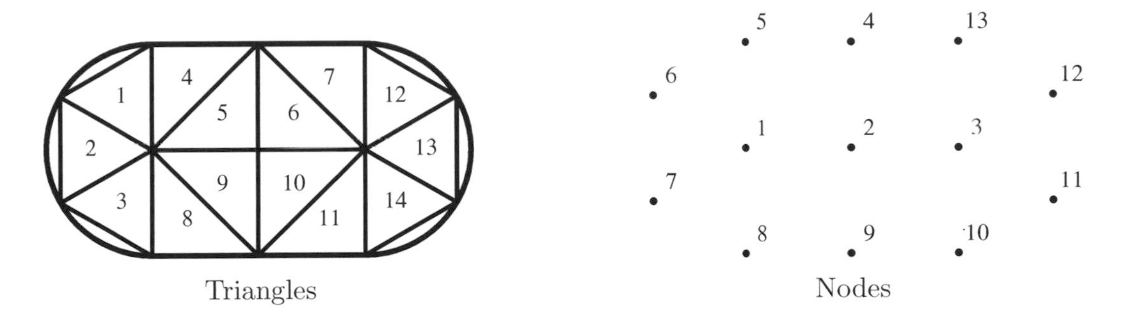

Figure 10.10. A coarse triangulation of the oval plate.

shorter ends, as sketched in Figure 10.9. The plate is subject to a heat source, while its edges are held at a fixed temperature. The problem is to find the equilibrium temperature distribution within the plate. Mathematically, we must solve the planar Poisson equation, subject to Dirichlet boundary conditions, for the equilibrium temperature $u(x, y)$.

Let us describe how to set up the finite element approximation. We begin with a very coarse triangulation of the plate, which will not give particularly accurate results, but serves to illustrate how to go about assembling the finite element matrix. We divide the rectangular part of the plate into eight right triangles, while each semicircular end will be approximated by three equilateral triangles. The triangles are numbered from 1 to 14 as indicated in Figure 10.10. There are 13 nodes in all, numbered as in the second figure. Only nodes $1, 2, 3$ are interior, while the boundary nodes are labeled 4 through 13 in counterclockwise order starting at the top. The full finite element matrix $\widehat{K}$ will have size 13×13, its rows and columns labeled by all the nodes, while the reduced matrix K appearing in the finite element equations (10.41) consists of the upper left 3×3 submatrix of $\widehat{K}$ corresponding to the three interior nodes.

For each $\nu = 1, \ldots, 14$, the triangle T_ν will contribute its elemental stiffnesses, as indexed by its vertices, to the matrix $\widehat{K}$ through a summand K_ν. For example, the first triangle T_1 is equilateral, and so has elemental stiffnesses (10.48). Its vertices are labeled 1, 5, and 6, and therefore we place the stiffnesses in the rows and columns numbered $1, 5, 6$

to form the summand

$$K_1 = \begin{pmatrix} .5774 & 0 & 0 & 0 & -.2887 & -.2887 & 0 & 0 & \ldots \\ 0 & 0 & 0 & 0 & 0 & 0 & 0 & 0 & \ldots \\ 0 & 0 & 0 & 0 & 0 & 0 & 0 & 0 & \ldots \\ 0 & 0 & 0 & 0 & 0 & 0 & 0 & 0 & \ldots \\ -.2887 & 0 & 0 & 0 & .5774 & -.2887 & 0 & 0 & \ldots \\ -.2887 & 0 & 0 & 0 & -.2887 & .5774 & 0 & 0 & \ldots \\ 0 & 0 & 0 & 0 & 0 & 0 & 0 & 0 & \ldots \\ 0 & 0 & 0 & 0 & 0 & 0 & 0 & 0 & \ldots \\ \vdots & \vdots & \vdots & \vdots & \vdots & \vdots & \vdots & \vdots & \ddots \end{pmatrix},$$

where all the undisplayed entries in the full 13×13 matrix are 0. The next triangle T_2 has the same equilateral elemental stiffness matrix (10.48), but now its vertices are $1, 6, 7$, and so it will contribute

$$K_2 = \begin{pmatrix} .5774 & 0 & 0 & 0 & 0 & -.2887 & -.2887 & 0 & \ldots \\ 0 & 0 & 0 & 0 & 0 & 0 & 0 & 0 & \ldots \\ 0 & 0 & 0 & 0 & 0 & 0 & 0 & 0 & \ldots \\ 0 & 0 & 0 & 0 & 0 & 0 & 0 & 0 & \ldots \\ 0 & 0 & 0 & 0 & 0 & 0 & 0 & 0 & \ldots \\ -.2887 & 0 & 0 & 0 & 0 & .5774 & -.2887 & 0 & \ldots \\ -.2887 & 0 & 0 & 0 & 0 & -.2887 & .5774 & 0 & \ldots \\ 0 & 0 & 0 & 0 & 0 & 0 & 0 & 0 & \ldots \\ \vdots & \vdots & \vdots & \vdots & \vdots & \vdots & \vdots & \vdots & \ddots \end{pmatrix}.$$

Similarly for K_3, with vertices $1, 7, 8$. On the other hand, T_4 is an isosceles right triangle, and so has elemental stiffnesses (10.47). Its vertices are labeled 1, 4, and 5, with vertex 5 at the right angle. Therefore, its contribution is

$$K_4 = \begin{pmatrix} .5 & 0 & 0 & 0 & -.5 & 0 & 0 & 0 & \ldots \\ 0 & 0 & 0 & 0 & 0 & 0 & 0 & 0 & \ldots \\ 0 & 0 & 0 & 0 & 0 & 0 & 0 & 0 & \ldots \\ 0 & 0 & 0 & .5 & -.5 & 0 & 0 & 0 & \ldots \\ -.5 & 0 & 0 & -.5 & 1.0 & 0 & 0 & 0 & \ldots \\ 0 & 0 & 0 & 0 & 0 & 0 & 0 & 0 & \ldots \\ 0 & 0 & 0 & 0 & 0 & 0 & 0 & 0 & \ldots \\ 0 & 0 & 0 & 0 & 0 & 0 & 0 & 0 & \ldots \\ \vdots & \vdots & \vdots & \vdots & \vdots & \vdots & \vdots & \vdots & \ddots \end{pmatrix}.$$

Continuing in this manner, we assemble 14 contributions $K_1, \ldots, K_{14}$, each with at most 9 nonzero entries. The full finite element matrix is their sum

$$
\begin{aligned}
\widehat{K} &= K_1 + K_2 + \cdots + K_{14} \\
&= \begin{pmatrix}
3.732 & -1 & 0 & 0 & -.7887 & -.5774 & -.5774 & -.7887 & 0 & 0 & 0 & 0 & 0 \\
-1 & 4 & -1 & -1 & 0 & 0 & 0 & 0 & -1 & 0 & 0 & 0 & 0 \\
0 & -1 & 3.732 & 0 & 0 & 0 & 0 & 0 & 0 & -.7887 & -.5774 & -.5774 & -.7887 \\
0 & -1 & 0 & 2 & -.5 & 0 & 0 & 0 & 0 & 0 & 0 & 0 & -.5 \\
-.7887 & 0 & 0 & -.5 & 1.577 & -.2887 & 0 & 0 & 0 & 0 & 0 & 0 & 0 \\
-.5774 & 0 & 0 & 0 & -.2887 & 1.155 & -.2887 & 0 & 0 & 0 & 0 & 0 & 0 \\
-.5774 & 0 & 0 & 0 & 0 & -.2887 & 1.155 & -.2887 & 0 & 0 & 0 & 0 & 0 \\
-.7887 & 0 & 0 & 0 & 0 & 0 & -.2887 & 1.577 & -.5 & 0 & 0 & 0 & 0 \\
0 & -1 & 0 & 0 & 0 & 0 & 0 & -.5 & 2 & -.5 & 0 & 0 & 0 \\
0 & 0 & -.7887 & 0 & 0 & 0 & 0 & 0 & -.5 & 1.577 & -.2887 & 0 & 0 \\
0 & 0 & -.5774 & 0 & 0 & 0 & 0 & 0 & 0 & -.2887 & 1.155 & -.2887 & 0 \\
0 & 0 & -.5774 & 0 & 0 & 0 & 0 & 0 & 0 & 0 & -.2887 & 1.155 & -.2887 \\
0 & 0 & -.7887 & -.5 & 0 & 0 & 0 & 0 & 0 & 0 & 0 & -.2887 & 1.577
\end{pmatrix}.
\end{aligned}
\tag{10.50}
$$

Since nodes $1, 2, 3$ are interior, the reduced finite element matrix

$$
K = \begin{pmatrix} 3.732 & -1 & 0 \\ -1 & 4 & -1 \\ 0 & -1 & 3.732 \end{pmatrix}
\tag{10.51}
$$

uses only the upper left 3×3 block of $\widehat{K}$. Clearly, it would not be difficult to directly construct K, bypassing $\widehat{K}$ entirely.

For a finer triangulation, the construction is similar, but the matrices become much larger. The procedure can, of course, be automated. Fortunately, if we choose a very regular triangulation, then we do not need to be nearly as meticulous in assembling the stiffness matrices, since many of the entries are the same. The simplest case employs a uniform square mesh, and so triangulates the domain into isosceles right triangles. This is accomplished by laying out a relatively dense square grid over the domain $\Omega \subset \mathbb{R}^2$. The interior nodes are the grid points that fall inside the oval domain, while the boundary nodes are all those grid points lying adjacent to one or more of the interior nodes, and are near but not necessarily precisely on the boundary $\partial\Omega$. Figure 10.11 shows the nodes in a square grid with intermesh spacing $h = .2$. While a bit crude in its approximation of the boundary of the domain, this procedure does have the advantage of making the construction of the associated finite element matrix relatively painless.

For such a mesh, all the triangles are isosceles right triangles, with elemental stiffnesses (10.47). Summing the corresponding matrices K_ν over all the triangles, as in (10.49), we find that the rows and columns of $\widehat{K}$ corresponding to the interior nodes all have the same

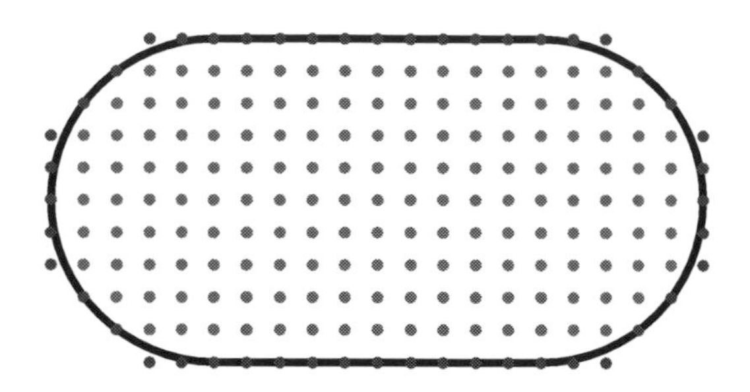

Figure 10.11. A square mesh for the oval plate.

form. Namely, if i labels an interior node, then the corresponding diagonal entry is $k_{ii} = 4$, while the off-diagonal entries $k_{ij} = k_{ji}$, $i \neq j$, are equal to -1 when node i is adjacent to node j on the grid, and are equal to 0 in all other cases. Node j is allowed to be a boundary node. (Interestingly, the result does not depend on how one orients the pair of triangles making up each square of the grid, which plays a role only in the computation of the right-hand side of the finite element equation.) Observe that the same computation applies even to our coarse triangulation. The interior node 2 belongs to all right isosceles triangles, and the corresponding nonzero entries in (10.50) are $k_{22} = 4$ and $k_{21} = k_{23} = k_{24} = k_{29} = -1$, indicating the four adjacent nodes.

Remark: The coefficient matrix constructed from the finite element method on a square (or even rectangular) grid is the *same* as the coefficient matrix arising from a finite difference solution to the Laplace or Poisson equation, as described in Example 5.7. The finite element approach has the advantage of readily adapting to much more general discretizations of the domain, and is not restricted to rectangular grids.

The Coefficient Vector and the Boundary Conditions

So far, we have been concentrating on assembling the finite element coefficient matrix K. We also need to compute the forcing vector $\mathbf{b} = (b_1, b_2, \ldots, b_n)^T$ appearing on the right-hand side of the fundamental linear equation (10.41). According to (10.39), the entries b_i are found by integrating the product of the forcing function and the finite element basis function. As before, we will approximate the integral over the domain Ω by an integral over the triangles, and so

$$b_i = \iint_\Omega f(x,y)\,\varphi_i(x,y)\,dx\,dy \approx \sum_\nu \iint_{T_\nu} f(x,y)\,\omega_i^\nu(x,y)\,dx\,dy \equiv \sum_\nu b_i^\nu. \qquad (10.52)$$

Typically, an exact computation of the various triangular double integrals is not so convenient, and so we resort to a numerical approximation. Since we are assuming that the individual triangles are small, we can get away with a very crude numerical integration scheme. If the function $f(x,y)$ does not vary much over the triangle T_ν — which will certainly be the case if T_ν is sufficiently small — we may approximate $f(x,y) \approx c_i^\nu$ for

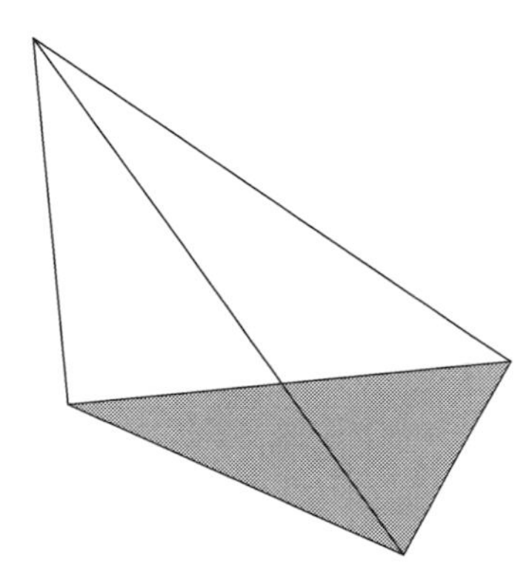

Figure 10.12. Finite element tetrahedron.

$(x, y) \in T_\nu$ by a constant. The integral (10.52) is then approximated by

$$b_i^\nu = \iint_{T_\nu} f(x,y)\,\omega_i^\nu(x,y)\,dx\,dy \approx c_i^\nu \iint_{T_\nu} \omega_i^\nu(x,y)\,dx\,dy = \tfrac{1}{3}\,c_i^\nu \text{ area } T_\nu = \tfrac{1}{6}\,c_i^\nu\,|\,\Delta_\nu\,|. \tag{10.53}$$

The formula for the integral of the affine element $\omega_i^\nu(x, y)$ follows from solid geometry: it equals the volume under its graph, a tetrahedron of height 1 and base T_ν, as illustrated in Figure 10.12.

How to choose the constant c_i^ν? In practice, the simplest choice is to let $c_i^\nu = f(x_i, y_i)$ be the value of the function at the ith vertex. With this choice, the sum in (10.52) becomes

$$b_i \approx \sum_\nu \tfrac{1}{3}\,f(x_i, y_i) \text{ area } T_\nu = \tfrac{1}{3}\,f(x_i, y_i) \text{ area } P_i, \tag{10.54}$$

where P_i is the vertex polygon (10.37) corresponding to the node $\mathbf{x}_i$. In particular, for the square mesh with the uniform choice of triangles, as in the first plot in Figure 10.8,

$$\text{area } P_i = 3\,h^2 \qquad \text{for all } i, \text{ and so} \qquad b_i \approx f(x_i, y_i)\,h^2 \tag{10.55}$$

is well approximated by just h^2 times the value of the forcing function at the node. This is the underlying reason to choose the uniform triangulation for the square mesh; the alternating version would give unequal values for the b_i over adjacent nodes, and this could give rise to unnecessary errors in the final approximation.

Example 10.8. For the coarsely triangulated oval plate, the reduced stiffness matrix is (10.51). The Poisson equation

$$-\Delta u = 4$$

models a constant external heat source of magnitude 4° over the entire plate. If we keep the edges of the plate fixed at 0°, then we need to solve the finite element equation $K\mathbf{c} = \mathbf{b}$, where K is the coefficient matrix (10.51). The entries of $\mathbf{b}$ are, by (10.54), equal to 4 (the right-hand side of the differential equation) times one-third the area of the corresponding vertex polygon, which for node 2 is the square consisting of four right triangles, each of area $\frac{1}{2}$, whereas for nodes 1 and 3 it consists of four right triangles of area $\frac{1}{2}$ plus three

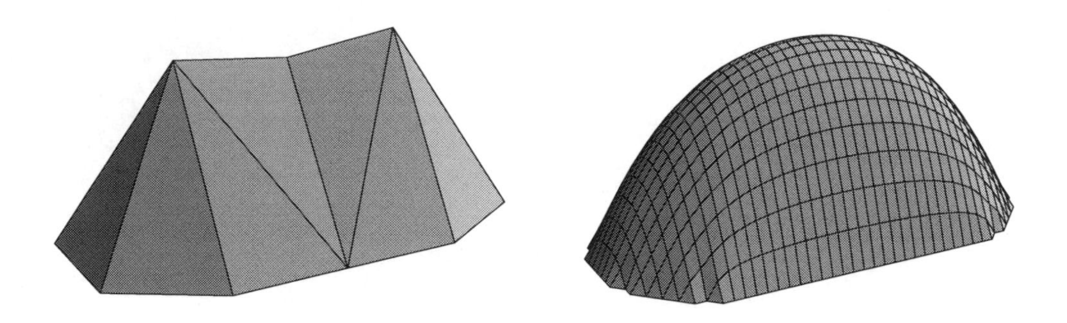

Figure 10.13. Finite element solutions to Poisson's equation for an oval plate.

equilateral triangles, each of area $\frac{\sqrt{3}}{4}$; see Figure 10.10. Thus,

$$\mathbf{b} = \tfrac{4}{3}\left(2 + \tfrac{3\sqrt{3}}{4},\ 2,\ 2 + \tfrac{3\sqrt{3}}{4}\right)^T = (\,4.3987, 2.6667, 4.3987\,)^T.$$

The solution to the final linear system $K\,\mathbf{c} = \mathbf{b}$ is easily found:

$$\mathbf{c} = (\,1.5672, 1.4503, 1.5672\,)^T.$$

Its entries are the values of the finite element approximation at the three interior nodes. The piecewise affine finite element solution is plotted in the first illustration in Figure 10.13. A more accurate approximation, based on a square grid triangulation of size $h = .1$, appears in the second figure. Here, the largest errors are concentrated near the poorly approximated corners of the oval, and could be improved by a more sophisticated triangulation.

Inhomogeneous Boundary Conditions

So far, we have restricted our attention to problems with homogeneous Dirichlet boundary conditions. According to Theorem 9.32, the solution to the inhomogeneous Dirichlet problem

$$-\Delta u = f \quad \text{in} \quad \Omega, \qquad\qquad u = h \quad \text{on} \quad \partial\Omega,$$

is also obtained by minimizing the Dirichlet functional (9.82). However, now the minimization takes place over the set of functions that satisfy the inhomogeneous boundary conditions. It is not difficult to fit this problem into the finite element scheme.

The elements corresponding to the interior nodes of our triangulation remain as before, but now we need to include additional elements to ensure that our approximation satisfies the boundary conditions. Note that if $\mathbf{x}_l$ is a boundary node, then the corresponding *boundary element* $\varphi_l(x,y)$ satisfies (10.28), and so has the same piecewise affine form (10.36). The corresponding finite element approximation

$$w(x,y) = \sum_{l=1}^{m} c_l\,\varphi_l(x,y) \tag{10.56}$$

has the same form as before, (10.29), but now the sum is over *all* nodes, both interior and boundary. As before, the coefficients $c_l = w(x_l,y_l) \approx u(x_l,y_l)$ are the values of the finite element approximation at the nodes. Therefore, in order to satisfy the boundary conditions, we require

$$c_j = h_j = h(x_j, y_j) \qquad \text{whenever} \quad \mathbf{x}_j = (x_j, y_j) \quad \text{is a boundary node.} \tag{10.57}$$

If the boundary node $\mathbf{x}_j$ does not lie precisely on the boundary $\partial\Omega$, then $h(x_j, y_j)$ is not defined, and so we need to approximate the value h_j appropriately, e.g., using the value of $h(x, y)$ at a nearby boundary point $(x, y) \in \partial\Omega$.

The derivation of the finite element equations proceeds as before, but now there are additional terms arising from the nonzero boundary values. Leaving the intervening details to Exercise 10.3.23, the final outcome can be written as follows. Let $\widehat{K}$ denote the full $m \times m$ finite element matrix constructed as above. The reduced coefficient matrix K is obtained by retaining the rows and columns corresponding to only interior nodes, and so will have size $n \times n$, where n is the number of interior nodes. The *boundary coefficient matrix* $\widetilde{K}$ is the $n \times (m - n)$ matrix consisting of those entries of the interior rows that do not appear in K, i.e., those lying in the columns indexed by the boundary nodes. For instance, in the coarse triangulation of the oval plate, the full finite element matrix is given in (10.50), and the upper 3×3 subblock is the reduced matrix (10.51). The remaining entries of the first three rows form the boundary coefficient matrix

$$\widetilde{K} = \begin{pmatrix} 0 & -.7887 & -.5774 & -.5774 & -.7887 & 0 & 0 & 0 & 0 & 0 \\ -1 & 0 & 0 & 0 & 0 & -1 & 0 & 0 & 0 & 0 \\ 0 & 0 & 0 & 0 & 0 & 0 & -.7887 & -.5774 & -.5774 & -.7887 \end{pmatrix}. \tag{10.58}$$

We similarly split the coefficients c_i of the finite element function (10.56) into two groups. We let $\mathbf{c} = (c_1, c_2, \ldots, c_n)^T \in \mathbb{R}^n$ denote the as yet unknown coefficients corresponding to the values of the approximation at the interior nodes $\mathbf{x}_i$, while $\mathbf{h} = (h_1, h_2, \ldots, h_{m-n})^T \in \mathbb{R}^{m-n}$ will be the vector containing the boundary values (10.57). The solution to the finite element approximation (10.56) is then obtained by solving the associated linear system

$$K\mathbf{c} + \widetilde{K}\,\mathbf{h} = \mathbf{b}, \qquad \text{or, equivalently,} \qquad K\mathbf{c} = \mathbf{f} = \mathbf{b} - \widetilde{K}\,\mathbf{h}. \tag{10.59}$$

Example 10.9. For the oval plate discussed in Example 10.7, suppose the right-hand semicircular edge is held at 10°, the left-hand semicircular edge at -10°, while the two straight edges have a linearly varying temperature distribution ranging from -10° at the left to 10° at the right, as illustrated in Figure 10.14. Our task is to compute its equilibrium temperature, assuming no internal heat source. Thus, for the coarse triangulation we have the boundary node values

$$\mathbf{h} = (h_4, \ldots, h_{13})^T = (0, -10, -10, -10, -10, 0, 10, 10, 10, 10)^T.$$

Using the previously computed formulas (10.51, 58) for the interior and boundary coefficient matrices $K, \widetilde{K}$, we approximate the solution to the Laplace equation by solving (10.59). We are assuming that there is no external forcing function, $f(x, y) \equiv 0$, and hence $\mathbf{b} = \mathbf{0}$, and so we must solve $K\mathbf{c} = \mathbf{f} = -\widetilde{K}\,\mathbf{h} = (2.1856, 3.6, 7.6497)^T$. The finite element function corresponding to the solution $\mathbf{c} = (1.0679, 1.8, 2.5320)^T$ is plotted in the first illustration in Figure 10.14. Even on such a coarse mesh, the approximation is not too bad, as evidenced by the second illustration, which plots the finite element solution for the finer square mesh of Figure 10.11.

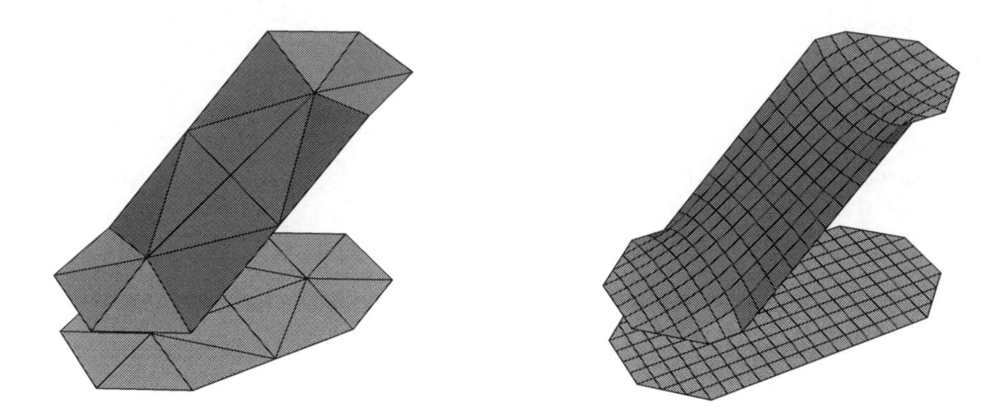

Figure 10.14. Solution to the Dirichlet problem for the oval plate.

Exercises

♠ 10.3.14. Consider the Dirichlet boundary value problem $\Delta u = 0$, $u(x,0) = \sin x$, $u(x,\pi) = 0$, $u(0,y) = 0$, $u(\pi,y) = 0$, on the square $S = \{0 < x, y < \pi\}$. (a) Find the exact solution. (b) Set up and solve the finite element equations based on a square mesh with $n = 2$ squares on each side of S. Write out the reduced finite element matrix, the boundary coefficient matrix, and the value of your approximation at the middle of the unit square. How close is this value to the exact solution there? (c) Repeat part (b) for $n = 4$ squares per side. Is the value of your approximation at the center of the unit square closer to the true solution? (d) Use a computer to find a finite element approximation to $u\left(\frac{1}{2}\pi, \frac{1}{2}\pi\right)$ using $n = 8$ squares per side. Is your approximation converging to the exact solution as the mesh becomes finer and finer?

♣ 10.3.15. Approximate the solution to the Dirichlet problem $\Delta u = 0$, $u(x,0) = x$, $u(x,1) = 1 - x$, $u(0,y) = y$, $u(1,y) = 1 - y$, by use of finite elements with mesh sizes $\Delta x = \Delta y = .25$ and $.1$. Compare your approximations with the solution you obtained in Exercise 4.3.12(d). What is the maximal error at the nodes in each case?

♠ 10.3.16. A metal plate has the shape of an equilateral triangle with unit sides. One side is heated to $100°$, while the other two are kept at $0°$. In order to approximate the equilibrium temperature distribution, the plate is divided into smaller equilateral triangles, with n triangles on each side, and the corresponding finite element approximation is then computed. (a) How many triangles are in the triangulation? How many interior nodes? How many edge nodes? (b) For $n = 2$, set up and solve the finite element linear system to find an approximation to the temperature at the center of the triangle. (c) Answer part (b) when $n = 3$. (d) Use a computer to find the finite element approximation to the temperature at the center when $n = 5, 10$, and 15. Are your values converging to the actual temperature? (e) Plot the finite element approximations you constructed in the previous parts.

10.3.17. Find the equilibrium temperature distribution in a unit equilateral triangle when one side is heated to $100°$, while the other two are insulated.

♠ 10.3.18. A metal plate has the shape of a 3 cm square with a 1 cm square hole cut out of the middle. The plate is heated by fixing the inner edge at temperature $100°$ while keeping the outer edge at $0°$. (a) Find the (approximate) equilibrium temperature using finite elements with a mesh width of $\Delta x = \Delta y = .5$ cm. Plot your approximate solution using a three-dimensional graphics program. (*b*) Let C denote the square contour lying midway between the inner and outer square boundaries of the plate. Using your finite element approximation, at what point(s) on C is the temperature a (*i*) minimum? (*ii*) maximum? (*iii*) equal to $50°$, the average of the two boundary temperatures? (*c*) Repeat part (*a*) using a smaller mesh width of $h = .2$. How much does this affect your answers in part (*b*)?

♣ 10.3.19. Answer Exercise 10.3.18 when the plate is additionally subjected to a constant heat source $f(x,y) = 600\,x + 800\,y - 2400$.

♠ 10.3.20. (*a*) Construct a finite element approximation to the solution, using a maximal mesh size of .1, to the following boundary value problem on the unit disk:

$$\Delta u = 0, \qquad x^2 + y^2 < 1, \qquad u = \begin{cases} 1, & x^2+y^2 = 1, \;\; y > 0, \\ 0, & x^2+y^2 = 1, \;\; y < 0. \end{cases}$$

(*b*) Compare your solution with the exact solution given in Example 4.7.

♣ 10.3.21. (*a*) Use finite elements to approximate the solution to the boundary value problem

$$-\Delta u + u = 0, \qquad 0 < x, y < 1, \qquad u(x,0) = u(x,1) = u(0,y) = 0, \qquad u(1,y) = 1.$$

(*b*) Compare your result with the first 5 and 10 summands in the series solution obtained via separation of variables.

♢ 10.3.22. (*a*) Justify the construction of the finite element matrix for a square mesh described in the text. (*b*) How would you modify the matrix for a rectangular mesh, as in Exercise 10.3.9?

♢ 10.3.23. Justify the inhomogeneous finite element construction in the text.

♡ 10.3.24. (*a*) Explain how to adapt the finite element method to a mixed boundary value problem with inhomogeneous Neumann conditions. (*b*) Apply your method to the problem

$$\Delta u = 0, \qquad \frac{\partial u}{\partial y}(x,0) = x, \qquad u(x,1) = 0, \qquad u(0,y) = 0, \qquad u(1,y) = 0.$$

(*c*) Solve the boundary value problem via separation of variables. Compare the values of your solutions at the center of the square.

10.4 Weak Solutions

An alternative route to the finite element method, which avoids the requirement of a minimization principle, rests upon the notion of a weak solution to a differential equation — a concept of considerable independent interest, since it includes many of the nonclassical solutions that we encountered earlier in this book. In particular, the discontinuous shock waves of Section 2.3 are, in fact, weak solutions to the nonlinear transport equation, as are the continuous but only piecewise smooth solutions to the wave equation that resulted from applying d'Alembert's formula to nonsmooth initial data. Weak solutions have become an incredibly powerful idea in the modern theory of partial differential equations, and we have space to present only the very basics here. They are particularly appropriate in the study of discontinuous and nonsmooth physical phenomena, including shock waves, cracks and dislocations in elastic media, singularities in liquid crystals, and so on. In the mathematical analysis of partial differential equations, it is often easier to prove the existence of a weak solution, for which one can then try to establish sufficient smoothness in order that it qualify as a classical solution. Further developments along with a range of applications can be found in more advanced texts, including [**38, 44, 61, 99, 107, 122**].

Weak Formulations of Linear Systems

The key idea behind the concept of a weak solution begins with a rather trivial observation: the only element in an inner product space that is orthogonal to every other element is the zero element.

Lemma 10.10. *Let V be an inner product space with inner product*[†] $\langle\!\langle \cdot\,,\cdot \rangle\!\rangle$. *An element $v_\star \in V$ satisfies $\langle\!\langle v_\star\,, v \rangle\!\rangle = 0$ for all $v \in V$ if and only if $v_\star = 0$.*

Proof: In particular, $v_\star$ must be orthogonal to itself, so $0 = \langle\!\langle v_\star\,, v_\star \rangle\!\rangle = |\!|\!| v_\star |\!|\!|^2$, which immediately implies $v_\star = 0$. *Q.E.D.*

Thus, one method of solving a linear — or even nonlinear — equation $F[u] = 0$ is to write it in the form

$$\langle\!\langle F[u]\,, v \rangle\!\rangle = 0 \qquad \text{for all} \qquad v \in V, \tag{10.60}$$

where V is the target space of $F: U \to V$. In particular, for an inhomogeneous linear system, $L[u] = f$, with $L: U \to V$ a linear operator between inner product spaces, the condition (10.60) takes the form

$$0 = \langle\!\langle L[u] - f\,, v \rangle\!\rangle = \langle\!\langle L[u]\,, v \rangle\!\rangle - \langle\!\langle f\,, v \rangle\!\rangle \qquad \text{for all} \qquad v \in V,$$

or, equivalently,

$$\langle\, u\,, L^*[v]\,\rangle - \langle\!\langle f\,, v \rangle\!\rangle = 0 \qquad \text{for all} \qquad v \in V, \tag{10.61}$$

where $L^*: V \to U$ denotes the adjoint of the operator L, as defined in (9.2). We will call (10.61) the *weak formulation* of the original linear system.

So far we have not really done anything of substance, and, indeed, for linear systems of algebraic equations, this more complicated characterization of solutions is of scant help. However, this is no longer the case for differential equations, because, thanks to the integration by parts argument used to determine the adjoint operator, the solution u to the weak form (10.61) is not restricted by the degree of smoothness required of a classical solution. A simple example will illustrate the basic construction.

Example 10.11. On a bounded interval $a \le x \le b$, consider the elementary boundary value problem

$$-\frac{d^2u}{dx^2} = f(x), \qquad u(a) = u(b) = 0.$$

The underlying vector space is $U = \{\, u(x) \in \mathrm{C}^2[a,b] \mid u(a) = u(b) = 0 \,\}$. To obtain a weak formulation, we multiply the differential equation by a test function $v(x) \in U$ and integrate:

$$\int_a^b \big[-u''(x) - f(x) \,\big] v(x)\, dx = 0. \tag{10.62}$$

The left-hand integral can be identified with the L^2 inner product between the left-hand side of the equation $L[u] - f = -u'' - f = 0$ and the test function v. According to Lemma 10.10, condition (10.62) holds for all $v(x) \in U$ if and only if $u(x) \in U$ satisfies the

[†] Shortly, as in the general framework developed in Chapter 9, V will be identified as the target space of a linear operator $L: U \to V$, and hence the choice of notation for its inner product.

boundary value problem. However, suppose that we integrate the first term by parts once. The boundary conditions on v imply that the boundary terms vanish, and the result is

$$\int_a^b \left[\, u'(x)\, v'(x) - f(x)\, v(x) \,\right] dx = 0. \tag{10.63}$$

A function $u(x)$ that satisfies the latter integral condition for all smooth test functions $v(x)$ will be called a *weak solution* to the original boundary value problem. The key observation is that the original differential equation, as well as the integral reformulation (10.62), requires that $u(x)$ be twice differentiable, whereas the weak version (10.63) requires only that its first derivative be defined.

Of course, one need not stop at (10.63). Performing another integration by parts on its first term and invoking the boundary conditions on u produces

$$\int_a^b \left[-\, u(x)\, v''(x) - f(x)\, v(x) \,\right] dx = 0. \tag{10.64}$$

Now $u(x)$ need only be (piecewise) continuous in order that the integral be defined — keeping in mind that the test function $v(x)$ is still required to be smooth. Equation (10.64) is sometimes referred to as the *fully weak formulation* of the boundary value problem, while the intermediate integral (10.63), in which the derivatives are evenly distributed among u and v, is then known as the *semi-weak formulation*.

Remark: Recall also the Definition 6.5 of weak convergence, which similarly involves integrating the standard convergence criterion against a suitable test function. Both are part and parcel of a general weak analytical framework that plays an essential role in all of modern advanced analysis, including partial differential equations.

The preceding example is a particular case of a general construction based on the abstract formulation of self-adjoint linear systems in Chapter 9. Let $L\colon U \to V$ be a linear map between inner product spaces, and let $S = L^* \circ L : U \to U$ be the associated self-adjoint operator. We further assume that $\ker L = \{0\}$, which implies that $S > 0$ is positive definite and, provided $f \in \operatorname{rng} S$, the associated linear system

$$S[\,u\,] = L^* \circ L[\,u\,] = f \tag{10.65}$$

has a unique solution.

In order to construct a weak formulation of the linear system (10.65), we begin by taking its inner product with a test function $v \in U$, whereby

$$0 = \langle\, S[\,u\,] - f\,, v \,\rangle = \langle\, S[\,u\,]\,, v \,\rangle - \langle\, f\,, v \,\rangle = \langle\, L^* \circ L[\,u\,]\,, v \,\rangle - \langle\, f\,, v \,\rangle.$$

Integration by parts, as in the preceding example, amounts to moving the adjoint operator so that it acts on the test function v, and in this manner we obtain the *weak formulation*

$$\langle\!\langle\, L[\,u\,]\,, L[\,v\,] \,\rangle\!\rangle = \langle\, f\,, v \,\rangle \qquad \text{for all} \qquad v \in U, \tag{10.66}$$

where we use our usual notation conventions regarding the inner products on U and V.

Warning: Unlike the minimization principle (10.1), the weak formulation (10.66) does not have a factor of $\frac{1}{2}$ on the left-hand side. Since, in the applications treated here, L is a differential operator of order, say, k, the weak formulation requires only that $u \in \mathrm{C}^k$ be k times differentiable, whereas, since S has order $2k$, the classical formulation (10.65) requires $u \in \mathrm{C}^{2k}$ to have twice as many derivatives.

Similarly, the fully weak formulation involves an additional integration by parts, realized in the abstract framework by moving the linear operator L acting on u so as to act on the test element v, and so

$$\langle u , L^* \circ L[v] \rangle = \langle u , S[v] \rangle = \langle f , v \rangle \qquad \text{for all} \qquad v \in U. \tag{10.67}$$

In practice, it is often advantageous to restrict the class of test functions in order to avoid technicalities involving smoothness and boundary behavior. This requires replacing the simple argument used to establish Lemma 10.10 by a more sophisticated result, named after the nineteenth-century German analyst Paul du Bois–Reymond.

Lemma 10.12. *Let $f(x)$ be a continuous function for $a \leq x \leq b$. Then*

$$\int_a^b f(x)\, v(x)\, dx = 0$$

for every C^1 *function $v(x)$ with compact support in the open interval (a, b) if and only if $f(x) \equiv 0$.*

Proof: Suppose $f(x_0) > 0$ for some $a < x_0 < b$. Then, by continuity, $f(x) > 0$ for all x in some interval $a < x_0 - \varepsilon < x < x_0 + \varepsilon < b$ around x_0. Choose $v(x)$ to be a C^1 function that is strictly positive in this interval and vanishes outside. An example is

$$v(x) = \begin{cases} \left((x - x_0)^2 - \varepsilon^2 \right)^2, & |\, x - x_0 \,| \leq \varepsilon, \\ 0, & \text{otherwise}. \end{cases} \tag{10.68}$$

Then $f(x)\, v(x) > 0$ when $|\, x - x_0 \,| < \varepsilon$ and $= 0$ everywhere else. This implies

$$\int_a^b f(x)\, v(x)\, dx = \int_{x_0 - \varepsilon}^{x_0 + \varepsilon} f(x)\, v(x)\, dx > 0,$$

which contradicts the original assumption. An analogous argument rules out $f(x_0) < 0$ for some $a < x_0 < b$. *Q.E.D.*

Finite Elements Based on Weak Solutions

To characterize weak solutions, one imposes the appropriate integral criterion on the entire infinite-dimensional space of smooth test functions. Thus, an evident approximation strategy is to restrict the criterion to a suitable finite-dimensional subspace, thereby seeking an approximate weak solution that belongs to the subspace.

More precisely, concentrating on the self-adjoint framework discussed at the end of the preceding subsection, we restrict the weak formulation (10.66) of the linear system (10.65) to a finite-dimensional subspace $W \subset U$, and thus seek $w \in W$ such that

$$\langle\!\langle\, L[w] , L[v] \,\rangle\!\rangle = \langle f , v \rangle \qquad \text{for all} \qquad v \in W. \tag{10.69}$$

In this fashion, we characerize the *finite element approximation* to the weak solution u as the element $w \in W$ such that (10.69) holds for all $v \in W$.

To analyze this condition, as in (10.3), we now specify a basis $\varphi_1, \ldots, \varphi_n$ of W, and thus can write both w and v as linear combinations thereof:

$$w = c_1 \varphi_1 + \cdots + c_n \varphi_n, \qquad v = d_1 \varphi_1 + \cdots + d_n \varphi_n.$$

Substituting these expressions into (10.69) produces the bilinear function

$$B(\mathbf{c},\mathbf{d}) = \sum_{i,j=1}^{n} k_{ij} c_i d_j - \sum_{i=1}^{n} b_i d_i = \mathbf{c}^T K\,\mathbf{d} - \mathbf{b}^T \mathbf{d} = (K\,\mathbf{c} - \mathbf{b})^T \mathbf{d} = 0, \tag{10.70}$$

where

$$k_{ij} = \langle\!\langle\, L[\varphi_i]\,, L[\varphi_j]\,\rangle\!\rangle, \qquad b_i = \langle\, f\,, \varphi_i\,\rangle, \qquad i,j = 1,\ldots,n, \tag{10.71}$$

are the *same* as our earlier specifications (10.6, 7), and we used the fact that $K^T = K$ is a symmetric matrix to arrive at the final expression in (10.70). The condition that (10.69) hold for all $v \in W$ is equivalent to the requirement that (10.70) hold for all $\mathbf{d} = (\,d_1, d_2, \ldots, d_n\,)^T \in \mathbb{R}^n$, which, in turn, implies that $\mathbf{c} = (\,c_1, c_2, \ldots, c_n\,)^T$ must satisfy the linear system

$$K\,\mathbf{c} = \mathbf{b}.$$

But we immediately recognize that this is exactly the same as the finite element linear system (10.9)! We therefore conclude that *for a positive definite linear system constructed as above, the weak finite element approximation to the solution is the same as the minimizing finite element approximation.* In other words, it does not matter whether we characterize the solutions through the minimization principle or the weak reformulation; the resulting finite element approximations are exactly the same. There is thus no need to present any additional examples illustrating this construction.

In general, while the weak formulation is of much wider applicability, outside of boundary value problems with well-defined minimization principles, the rigorous underpinning that guarantees that the numerical solution is close to the actual solution is harder to establish and, in fact, not always valid. Indeed, one can find boundary value problems without analytic solutions that have spurious finite element numerical solutions, and, conversely, boundary value problems with solutions for which some finite element approximations do not exist because the resulting coefficient matrix is singular, [**113, 126**].

Shock Waves as Weak Solutions

Finally, let us return to our earlier analysis, in Section 2.3, of shock waves, but now in the context of weak solutions. We begin by writing the nonlinear transport equation in the conservative form

$$\frac{\partial u}{\partial t} + \frac{\partial}{\partial x}\left(\tfrac{1}{2}u^2\right) = 0. \tag{10.72}$$

Since shock waves are discontinuous functions, they do not qualify as classical solutions. However, they can be rigorously characterized as weak solutions, a formulation that will, reassuringly, lead to the Rankine–Hugoniot Equal Area Rule for shock dynamics.

To construct a weak formulation of the nonlinear transport equation, we follow the general framework, and hence begin by multiplying the equation (10.72) by a smooth test function $v(t,x)$ and integrating over a domain $\Omega \subset \mathbb{R}^2$:

$$\iint_\Omega \left[\frac{\partial u}{\partial t} + \frac{\partial}{\partial x}\left(\tfrac{1}{2}u^2\right)\right] v(t,x)\,dt\,dx = 0. \tag{10.73}$$

As a direct consequence of the two-dimensional version of the du Bois–Reymond Lemma, cf. Exercise 10.4.7, if $u(t,x) \in \mathrm{C}^1$ and condition (10.73) holds for all C^1 functions $v(t,x)$

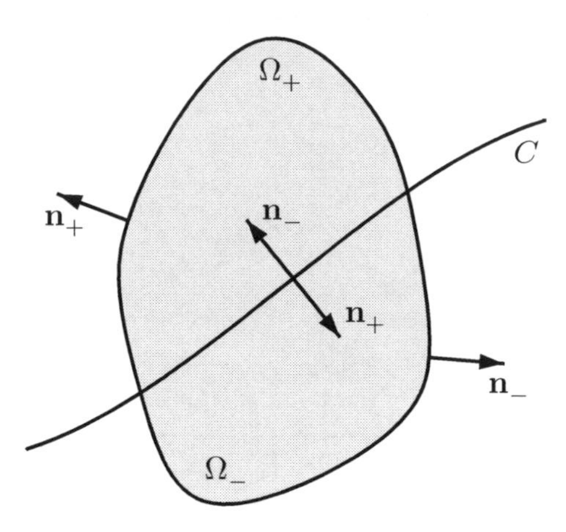

Figure 10.15. Integration domain for weak shock-wave solution.

with compact support contained in Ω, then $u(t,x)$ is necessarily a classical solution to the partial differential equation (10.72). The next step is to integrate by parts in order to remove the derivatives from u, and this is accomplished by appealing to *Green's formula* (6.82), which we rewrite in the form

$$\iint_\Omega \left(u_1 \frac{\partial v}{\partial t} + u_2 \frac{\partial v}{\partial x} \right) dt\, dx = \oint_{\partial\Omega} (\mathbf{u}\cdot\mathbf{n})\, v\, ds - \iint_\Omega \left(\frac{\partial u_1}{\partial t} + \frac{\partial u_2}{\partial x} \right) v\, dt\, dx, \tag{10.74}$$

where $\mathbf{u} = (u_1, u_2)^T$. In our case, we identify the integral in (10.73) with the left-hand side of (10.74) by setting $u_1 = u$, $u_2 = \frac{1}{2}u^2$. Since v has compact support, the boundary integral vanishes, and thus we arrive at the weak formulation of the equation.

Definition 10.13. A function $u(t,x)$ is said to be a *weak solution* to the nonlinear transport equation (10.72) on $\Omega \subset \mathbb{R}^2$ if

$$\iint_\Omega \left(u \frac{\partial v}{\partial t} + \tfrac{1}{2} u^2 \frac{\partial v}{\partial x} \right) dt\, dx = 0 \tag{10.75}$$

for all C^1 functions $v(t,x)$ with compact support: $\operatorname{supp} v \subset \Omega$.

The key point is that, in the weak formulation (10.75), the derivatives are acting solely on $v(t,x)$, which we assume to be smooth, and not on our prospective solution $u(t,x)$, which now need not even be continuous for the integral to be well defined.

Let us derive the Rankine–Hugoniot shock condition (2.53) as a consequence of the weak formulation. Suppose $u(t,x)$ is a weak solution, defined on a domain $\Omega \subset \mathbb{R}^2$, that has a single jump discontinuity along a curve C parametrized by $x = \sigma(t)$ that separates Ω into two subdomains, say Ω_+ and Ω_-, such that its restriction to either subdomain, denoted by $u_+ = u \mid \Omega_+$ and $u_- = u \mid \Omega_-$, are each classical solutions on their respective domains, while the separating curve $C = \{ x = \sigma(t) \}$ represents a shock-wave discontinuity. For specificity, we assume that Ω_+ lies above and Ω_- lies below C in the (t,x)–plane; see Figure 10.15. Let us investigate what the preceding weak formulation implies in this situation. We split the integral (10.75) into two parts, and then apply the integration by parts formula (10.74) to each individual double integral, keeping in mind that, when restricted to Ω_+ or Ω_-, the integrand is sufficiently smooth to justify application of the formula:

$$\begin{aligned}
0 &= \iint_\Omega \left(u\,\frac{\partial v}{\partial t} + \tfrac{1}{2}u^2\,\frac{\partial v}{\partial x} \right) dt\,dx \\
&= \iint_{\Omega_+} \left(u_+\,\frac{\partial v}{\partial t} + \tfrac{1}{2}u_+^2\,\frac{\partial v}{\partial x} \right) dt\,dx \;+ \iint_{\Omega_-} \left(u_-\,\frac{\partial v}{\partial t} + \tfrac{1}{2}u_-^2\,\frac{\partial v}{\partial x} \right) dt\,dx \\
&= \oint_{\partial\Omega_+} (\widetilde{\mathbf{u}}_+ \cdot \mathbf{n}_+)\,v\,ds \;- \iint_{\Omega_+} \left[\frac{\partial u_+}{\partial t} + \frac{\partial}{\partial x}\left(\tfrac{1}{2}u_+^2\right) \right] v\,dt\,dx \;+ \\
&\qquad + \oint_{\partial\Omega_-} (\widetilde{\mathbf{u}}_- \cdot \mathbf{n}_-)\,v\,ds \;- \iint_{\Omega_-} \left[\frac{\partial u_-}{\partial t} + \frac{\partial}{\partial x}\left(\tfrac{1}{2}u_-^2\right) \right] v\,dt\,dx \\
&= \int_C (\widetilde{\mathbf{u}}_+ \cdot \mathbf{n}_+ + \widetilde{\mathbf{u}}_- \cdot \mathbf{n}_-)\,v\,ds.
\end{aligned}$$

Here

$$\widetilde{\mathbf{u}}_+ = \begin{pmatrix} u_+ \\ \tfrac{1}{2}u_+^2 \end{pmatrix}, \qquad \widetilde{\mathbf{u}}_- = \begin{pmatrix} u_- \\ \tfrac{1}{2}u_-^2 \end{pmatrix},$$

while $\mathbf{n}_+, \mathbf{n}_-$ are the unit outwards normals on, respectively, $\partial\Omega_+$ and $\partial\Omega_-$. The final equality follows from the fact that the support of v is contained strictly inside Ω, and hence vanishes on those parts of the boundaries of Ω_+ and Ω_- that do not lie on the curve C. In particular, since C is the graph of $x = \sigma(t)$, the unit normals along it are, respectively,

$$\mathbf{n}_+ = \frac{1}{\sqrt{1 + \left(\dfrac{d\sigma}{dt}\right)^2}} \begin{pmatrix} \dfrac{d\sigma}{dt} \\ -1 \end{pmatrix}, \qquad \mathbf{n}_- = -\,\mathbf{n}_+ = \frac{1}{\sqrt{1 + \left(\dfrac{d\sigma}{dt}\right)^2}} \begin{pmatrix} -\dfrac{d\sigma}{dt} \\ 1 \end{pmatrix},$$

keeping in mind our convention that Ω_+ lies above and Ω_- lies below C, while

$$ds = \sqrt{1 + \left(\frac{d\sigma}{dt}\right)^2}\; dt.$$

Thus, the final line integral reduces to

$$\int_C \left[(u_- - u_+)\,\frac{d\sigma}{dt} - \tfrac{1}{2}(u_-^2 - u_+^2) \right] v\,dt = 0. \tag{10.76}$$

Since (10.76) vanishes for all C^1 functions $v(t,x)$ with compact support, the du Bois–Reymond Lemma 10.12 implies that

$$(u_- - u_+)\,\frac{d\sigma}{dt} = \tfrac{1}{2}(u_-^2 - u_+^2) \qquad \text{on} \qquad C,$$

thereby re-establishing the Rankine–Hugoniot shock condition (2.53). The upshot is that the shock-wave solutions produced in Section 2.3 are bona fide weak solutions.

Another computation shows that the rarefaction wave (2.54) also qualifies as a weak solution. However, so does the non-physical reverse shock solution discussed in Example 2.11. Thus, although the weak formulation recovers the Rankine–Hugoniot condition, it does not address the problem of causality, which must be additionally imposed to single out a unique, physically meaningful weak solution. Further developments of these ideas can be found in more advanced monographs, e.g., [**107, 122**].

Exercises

10.4.1. Write out semi-weak and fully weak formulations for the following boundary value problems: (a) $-u'' + 2u = x - x^2$, $u(0) = u(1) = 0$; (b) $e^x u'' + u = \cos x$, $u'(0) = u'(2) = 0$; (c) $x u'' + u' + x u = 0$, $u(1) = u(2) = 0$.

10.4.2. (a) Write down a weak formulation for the boundary value problem $-u'' + 3u = x$, $u(0) = u(1) = 0$. (b) Based on your weak formulation, construct a finite element approximation to the solution, using $n = 10$ nodes.

10.4.3. (a) Write down a weak formulation of the transport equation $u_t + 3u_x = 0$ on the real line. (b) Solve the initial value problem $u(0,x) = \begin{cases} 1 - |x|, & |x| \le 1, \\ 0, & \text{otherwise}. \end{cases}$

(c) Explain why the result of part (b) is not a classical solution to the wave equation. Is it a weak solution according to your formulation in part (a)?

10.4.4. (a) Write down a semi-weak formulation of the wave equation $u_{tt} = 4u_{xx}$ on the real line. (b) Solve the initial value problem $u(0,x) = \rho(x)$, $u_t(0,x) = 0$, where the initial displacement is a ramp function (6.25). (c) Explain why the result of part (b) is not a classical solution to the wave equation. Does it satisfy the semi-weak formulation of part (a)? Explain your answer.

♢ 10.4.5. (a) Starting with the nonlinear transport equation written in the alternative conservative form (2.56), find a corresponding weak formulation.
(b) Prove that your weak formulation produces the alternative entropy condition (2.58) for the motion of a shock discontinuity.

♢ 10.4.6. Prove that the du Bois–Reymond Lemma 10.12 remains valid even when $v(x) \in \mathrm{C}^\infty$ is required to be infinitely differentiable.

♢ 10.4.7. *The Two-dimensional du Bois–Reymond Lemma*: Let $\Omega \subset \mathbb{R}^2$ be a domain, and $f(t,x)$ a continuous function defined thereon. Prove that $\iint_\Omega f(t,x)\, v(t,x)\, dt\, dx = 0$ for every C^1 function $v(t,x)$ with compact support in Ω if and only if $f(t,x) \equiv 0$.

♠ 10.4.8. (a) Investigate the ability of finite elements to approximate a solution to the non-positive-definite boundary value problem $\Delta u + \lambda u = 0$, $0 < x < \pi$, $0 < y < \pi$, $u(x,0) = 1$, $u(x,\pi) = u(0,y) = u(\pi,y) = 0$, when (i) $\lambda = 1$, (ii) $\lambda = 2$. Use separation of variables to find a series solution and use it to determine the accuracy of your finite element solution in part (a).

Chapter 11
Dynamics of Planar Media

In previous chapters, we studied the equilibrium configurations of planar media — plates and membranes — governed by the two-dimensional Laplace and Poisson equations. In this chapter, we analyze their dynamics, modeled by the two-dimensional heat and wave equations. The heat equation describes diffusion of, say, heat energy in a thin metal plate, an animal population dispersing over a region, or a pollutant spreading out into a shallow lake. The wave equation models small vibrations of a two-dimensional membrane such as a drum. Since both equations fit into the general framework for dynamics that we established in Section 9.5, their solutions share many of the general qualitative and analytic properties possessed by their respective one-dimensional counterparts.

Although the increase in dimension may tax our analytical prowess, we have, in fact, already mastered the principal solution techniques: separation of variables, eigenfunction series, and fundamental solutions. When applied to partial differential equations in higher dimensions, separation of variables in curvilinear coordinates often leads to new linear, but non-constant-coefficient, ordinary differential equations, whose solutions are no longer elementary functions. Rather, they are expressed in terms of a variety of important *special functions*, which include the error and Airy functions we encountered earlier; the Bessel functions, which play a starring role in the present chapter; and the Legendre and Ferrers functions, spherical harmonics, and spherical Bessel functions arising in three-dimensional problems. Special functions are ubiquitous in more advanced applications in physics, chemistry, mechanics, and mathematics, and, over the last two hundred and fifty years, many prominent mathematicians have devoted significant effort to establishing their fundamental properties, to the extent that they are now, by and large, well understood, [**86**]. To acquire the requisite familiarity with special functions, in preparation for employing them to solve higher-dimensional partial differential equations, we must first learn basic series solution techniques for linear second-order ordinary differential equations.

11.1 Diffusion in Planar Media

As we learned in Chapter 4, the equilibrium temperature $u(x, y)$ of a thin, uniform, isotropic plate is governed by the two-dimensional Laplace equation

$$\Delta u = u_{xx} + u_{yy} = 0.$$

Working by analogy, the dynamical diffusion of the plate's temperature should be modeled by the two-dimensional heat equation

$$u_t = \gamma\,\Delta u = \gamma\,(u_{xx} + u_{yy}). \tag{11.1}$$

The coefficient $\gamma > 0$, assumed constant, measures the relative speed of diffusion of heat energy throughout the plate; its positivity is required on physical grounds, and also serves to avoid ill-posedness inherent in running diffusion processes backwards in time. In this model, we are assuming that the plate is uniform and isotropic, and experiences no loss of heat or external heat sources other than at its edge — which can be arranged by covering its top and bottom with insulation.

The solution $u(t, \mathbf{x}) = u(t, x, y)$ to the heat equation measures the temperature, at time t, at each point $\mathbf{x} = (x, y)$ in the (bounded) domain $\Omega \subset \mathbb{R}^2$ occupied by the plate. To uniquely specify the solution $u(t, x, y)$, we must impose suitable initial and boundary conditions. The initial data is the temperature of the plate

$$u(0, x, y) = f(x, y), \qquad (x, y) \in \Omega, \tag{11.2}$$

at an initial time, which for simplicity, we take to be $t_0 = 0$. The most important boundary conditions are as follows:

- *Dirichlet boundary conditions*: Specifying

$$u = h \qquad \text{on} \qquad \partial\Omega \tag{11.3}$$

fixes the temperature along the edge of the plate.

- *Neumann boundary conditions*: Let $\mathbf{n}$ be the unit outwards normal on the boundary of the domain. Specifying the normal derivative of the temperature,

$$\frac{\partial u}{\partial \mathbf{n}} = k \qquad \text{on} \qquad \partial\Omega, \tag{11.4}$$

effectively prescribes the heat flux along the boundary. Setting $k = 0$ corresponds to an insulated boundary.

- *Mixed boundary conditions*: More generally, we can impose Dirichlet conditions on part of the boundary $D \subsetneq \partial\Omega$ and Neumann conditions on its complement $N = \partial\Omega \setminus D$. For instance, homogeneous mixed boundary conditions

$$u = 0 \qquad \text{on} \qquad D, \qquad\qquad \frac{\partial u}{\partial \mathbf{n}} = 0 \qquad \text{on} \qquad N, \tag{11.5}$$

correspond to freezing a portion of the boundary and insulating the remainder.

- *Robin boundary conditions*:

$$\frac{\partial u}{\partial \mathbf{n}} + \beta\, u = \tau \qquad \text{on} \qquad \partial\Omega, \tag{11.6}$$

where the edge of the plate sits in a heat bath at temperature τ.

Under reasonable assumptions on the domain, the initial data, and the boundary data, a general theorem, [**34, 38, 99**], guarantees the existence of a unique solution $u(t, x, y)$ to any of these initial-boundary value problems for all subsequent times $t > 0$. Our practical goal is to both compute and understand the behavior of the solution in specific situations.

Derivation of the Diffusion and Heat Equations

The physical derivation of the two-dimensional (and three-dimensional) heat equation relies on the same basic thermodynamic laws that were used, in Section 4.1, to establish the one-dimensional version. The first principle is that heat energy flows from hot to cold as rapidly as possible. According to multivariable calculus, [**8**, **108**], the negative temperature gradient $-\nabla u$ points in the direction of the steepest decrease in the temperature function u at a point, and so heat energy will flow in that direction. Therefore, the heat flux vector $\mathbf{w}$, which measures the magnitude and direction of the flow of heat energy, should be proportional to the temperature gradient:

$$\mathbf{w}(t,x,y) = -\,\kappa(x,y)\,\nabla u(t,x,y). \tag{11.7}$$

The scalar quantity $\kappa(x,y) > 0$ measures the *thermal conductivity* of the material, so (11.7) is the multi-dimensional form of *Fourier's Law of Cooling* (4.5). We are assuming that the thermal conductivity depends only on the position $(x,y) \in \Omega$, which means that the material in the plate

(*a*) is not changing in time;

(*b*) is *isotropic*, meaning that its thermal conductivity is the same in all directions;

(*c*) and, moreover, its thermal conductivity is not affected by any change in temperature.

Dropping either assumption (*b*) or (*c*) would result in a considerably more challenging nonlinear diffusion equation.

The second thermodynamic principle is that, in the absence of external heat sources, heat can enter any subregion $R \subset \Omega$ only through its boundary ∂R. (Keep in mind that the plate is insulated from above and below.) Let $\varepsilon(t,x,y)$ denote the heat energy density at each time and point in the domain, so that

$$H_R(t) = \iint_R \varepsilon(t,x,y)\,dx\,dy$$

represents the total heat energy contained within the subregion R at time t. The amount of additional heat energy entering R at a boundary point $\mathbf{x} \in \partial R$ is given by the normal component of the heat flux vector, namely $-\,\mathbf{w}\cdot\mathbf{n}$, where, as always, $\mathbf{n}$ denotes the *outward* unit normal to the boundary ∂R. Thus, the total heat flux entering the region R is obtained by integration along the boundary of R, resulting in the line integral $-\oint_{\partial R} \mathbf{w}\cdot\mathbf{n}\;ds$. Equating the rate of change of heat energy to the heat flux yields

$$\frac{dH_R}{dt} = \iint_R \frac{\partial \varepsilon}{\partial t}(t,x,y)\,dx\,dy = -\oint_{\partial R} \mathbf{w}\cdot\mathbf{n}\;ds = -\iint_R \nabla\cdot\mathbf{w}\;dx\,dy,$$

where we applied the divergence form of Green's Theorem, (6.80), to convert the flux line integral into a double integral. Thus,

$$\iint_R \left(\frac{\partial \varepsilon}{\partial t} + \nabla\cdot\mathbf{w} \right) dx\,dy = 0. \tag{11.8}$$

Keep in mind that this result must hold for *any* subdomain $R \subset \Omega$. Now, according to Exercise 11.1.13, the only way in which an integral of a continuous function can vanish for *all* subdomains is if the integrand is identically zero, and so

$$\frac{\partial \varepsilon}{\partial t} + \nabla\cdot\mathbf{w} = 0. \tag{11.9}$$

In this manner, we arrive at the basic *conservation law* relating the heat energy density ε and the heat flux vector $\mathbf{w}$.

As in our one-dimensional model, cf. (4.3), the heat energy density $\varepsilon(t,x,y)$ is proportional to the temperature, so

$$\varepsilon(t,x,y) = \sigma(x,y)\,u(t,x,y), \qquad \text{where} \qquad \sigma(x,y) = \rho(x,y)\,\chi(x,y) \tag{11.10}$$

is the product of the *density* ρ and the *specific heat capacity* χ of the material at the point $(x,y) \in \Omega$. Combining this with the Fourier Law (11.7) and the energy balance equation (11.10) leads to the general two-dimensional *diffusion equation*

$$\frac{\partial u}{\partial t} = \frac{1}{\sigma}\nabla \cdot \big(\kappa\,\nabla u\big) \tag{11.11}$$

governing the thermodynamics of an isotropic medium in the absence of external heat sources or sinks. In full detail, this second-order partial differential equation is

$$\frac{\partial u}{\partial t} = \frac{1}{\sigma(x,y)}\left[\frac{\partial}{\partial x}\left(\kappa(x,y)\,\frac{\partial u}{\partial x}\right) + \frac{\partial}{\partial y}\left(\kappa(x,y)\,\frac{\partial u}{\partial y}\right)\right]. \tag{11.12}$$

Such diffusion equations are also used to model movements of populations, e.g., bacteria in a petri dish or wolves in the Canadian Rockies, [**81**, **84**]. Here the solution $u(t,x,y)$ represents the population density at position (x,y) at time t, which diffuses over the domain due to random motions of the individuals. Similar diffusion processes model the mixing of solutes in liquids, with the diffusion induced by the random Brownian motion from molecular collisions. More generally, diffusion processes in the presence of chemical reactions and convection due to fluid motion are modeled by the more general class of *reaction-diffusion* and *convection-diffusion equations*, [**107**].

In particular, if the body (or the environment or the solvent) is uniform, then both σ and κ are constant, and so (11.11) reduces to the heat equation (11.1) with *thermal diffusivity*

$$\gamma = \frac{\kappa}{\sigma} = \frac{\kappa}{\rho\,\chi}\,. \tag{11.13}$$

Both the heat and more general diffusion equations are examples of *parabolic* partial differential equations, the terminology being adapted from Definition 4.12 to apply to partial differential equations in more than two variables. As we will see, all the basic qualitative features of solutions to the one-dimensional heat equation carry over to parabolic partial differential equations in higher dimensions.

Indeed, the general diffusion equation (11.12) can be readily fit into the self-adjoint dynamical framework of Section 9.5, taking the form

$$u_t = -\,\nabla^* \circ \nabla u. \tag{11.14}$$

The gradient operator ∇ maps scalar fields u to vector fields $\mathbf{v} = \nabla u$; its adjoint ∇^*, which goes in the reverse direction, is taken with respect to the weighted inner products

$$\langle\, u\,,\widetilde{u}\,\rangle = \iint_\Omega u(x,y)\,\widetilde{u}(x,y)\,\sigma(x,y)\,dx\,dy, \quad \langle\!\langle\, \mathbf{v}\,,\widetilde{\mathbf{v}}\,\rangle\!\rangle = \iint_\Omega \mathbf{v}(x,y)\cdot\widetilde{\mathbf{v}}(x,y)\,\kappa(x,y)\,dx\,dy, \tag{11.15}$$

between, respectively, scalar and vector fields. As in (9.33), a straightforward integration by parts tells us that

$$\nabla^*\mathbf{v} = -\,\frac{1}{\sigma}\,\nabla\cdot(\kappa\,\mathbf{v}) = -\,\frac{1}{\sigma}\left[\frac{\partial(\kappa\,v_1)}{\partial x} + \frac{\partial(\kappa\,v_2)}{\partial y}\right], \qquad \text{when} \qquad \mathbf{v} = \begin{pmatrix} v_1 \\ v_2 \end{pmatrix}. \tag{11.16}$$

Therefore, the right-hand side of (11.14) equals

$$- \nabla^* \circ \nabla u = \frac{1}{\sigma} \nabla \cdot (\kappa \nabla u), \tag{11.17}$$

which thereby recovers the general diffusion equation (11.11). As always, the validity of the adjoint formula (11.16) rests on the imposition of suitable homogeneous boundary conditions: Dirichlet, Neumann, mixed, or Robin.

In particular, to obtain the heat equation, we take σ and κ to be constant, and so the inner products (11.15) reduce, up to a constant factor, to the usual L^2 inner products between scalar and vector fields. In this case, the adjoint of the gradient is, up to a scale factor, minus the divergence: $\nabla^* = -\gamma \nabla \cdot$, where $\gamma = \kappa/\sigma$. In this scenario, (11.14) reduces to the two-dimensional heat equation (11.1).

Separation of Variables

Let us now discuss analytical solution techniques. According to Section 9.5, the separable solutions to any linear evolution equation

$$u_t = - S[u] \tag{11.18}$$

are of exponential form

$$u(t,x,y) = e^{-\lambda t} v(x,y). \tag{11.19}$$

Since the linear operator S involves differentiation with respect to only the spatial variables x, y, we obtain

$$\frac{\partial u}{\partial t} = - \lambda e^{-\lambda t} v(x,y), \qquad \text{while} \qquad S[u] = e^{-\lambda t} S[v].$$

Substituting back into the diffusion equation (11.18) and canceling the exponentials, we conclude that

$$S[v] = \lambda v. \tag{11.20}$$

Thus, $v(x,y)$ must be an eigenfunction for the linear operator S, subject to the relevant homogeneous boundary conditions.

In the case of the heat equation (11.1),

$$S[u] = - \gamma \Delta u,$$

and hence, as in Example 9.40, the eigenvalue equation (11.20) is the two-dimensional *Helmholtz equation*

$$\gamma \Delta v + \lambda v = 0, \qquad \text{or, in detail,} \qquad \gamma \left(\frac{\partial^2 v}{\partial x^2} + \frac{\partial^2 v}{\partial y^2} \right) + \lambda v = 0. \tag{11.21}$$

According to Theorem 9.34, self-adjointness implies that the eigenvalues are all real and nonnegative: $\lambda \geq 0$. In the positive definite cases — Dirichlet and mixed boundary conditions — they are strictly positive, while the Neumann boundary value problem admits a zero eigenvalue $\lambda_0 = 0$ corresponding to the constant eigenfunction $v_0(x,y) \equiv 1$.

Let us index the eigenvalues in increasing order:

$$0 < \lambda_1 \leq \lambda_2 \leq \lambda_3 \leq \cdots , \tag{11.22}$$

repeated according to their multiplicities, where $\lambda_0 = 0$ is an eigenvalue only in the Neumann case, and $\lambda_k \to \infty$ as $k \to \infty$. For each eigenvalue λ_k, let $v_k(x,y)$ be an independent eigenfunction. The corresponding separable solution is

$$u_k(t,x,y) = e^{-\lambda_k t}\, v_k(x,y).$$

Those corresponding to positive eigenvalues are exponentially decaying in time, while a zero eigenvalue produces a constant solution $u_0(t,x,y) \equiv 1$. The general solution to the homogeneous boundary value problem can then be built up as an infinite series in these basic eigensolutions

$$u(t,x,y) = \sum_{k=1}^{\infty} c_k\, u_k(t,x,y) = \sum_{k=1}^{\infty} c_k\, e^{-\lambda_k t}\, v_k(x,y). \tag{11.23}$$

The coefficients c_k are prescribed by the initial conditions, which require

$$\sum_{k=1}^{\infty} c_k\, v_k(x,y) = f(x,y). \tag{11.24}$$

Since S is self-adjoint, Theorem 9.33 guarantees orthogonality[†] of the eigenfunctions under the L^2 inner product on the domain Ω:

$$\langle\, v_j\,, v_k\,\rangle = \iint_{\Omega} v_j(x,y)\, v_k(x,y)\, dx\, dy = 0, \qquad j \neq k. \tag{11.25}$$

As a consequence, the coefficients in (11.24) are given by the standard orthogonality formula (9.104), namely

$$c_k = \frac{\langle\, f\,, v_k\,\rangle}{\|\, v_k\,\|^2} = \frac{\displaystyle\iint_{\Omega} f(x,y)\, v_k(x,y)\, dx\, dy}{\displaystyle\iint_{\Omega} v_k(x,y)^2\, dx\, dy}. \tag{11.26}$$

(For the more general diffusion equation (11.11), one uses the appropriately weighted inner product.) The exponential decay of the eigenfunction coefficients implies that the resulting eigensolution series (11.23) converges and thus produces the solution to the initial-boundary value problem for the diffusion equation. See [**34**; p. 369] for a precise statement and proof of the general theorem.

Qualitative Properties

Before tackling examples in which we are able to construct explicit formulas for the eigenfunctions and eigenvalues, let us see what the eigenfunction series solution (11.23) can tell us about general diffusion processes. Based on our experience with the case of a one-dimensional bar, the final conclusions will not be especially surprising. Indeed, they also apply, word for word, to diffusion processes in three-dimensional solid bodies. A reader who is impatient to see the explicit formulas may wish to skip ahead to the following section, returning here as needed.

† As usual, in the case of a repeated eigenvalue, one chooses an orthogonal basis of the associated eigenspace to ensure orthogonality of all the basis eigenfunctions.

Keep in mind that we are still dealing with the solution to the homogeneous boundary value problem. The first observation is that all terms in the series solution (11.23), with the possible exception of a null eigenfunction term that appears in the semi-definite Neumann case, are tending to zero exponentially fast. Since most eigenvalues are large, all the higher-order terms in the series become almost instantaneously negligible, and hence the solution can be accurately approximated by a finite sum over the first few eigenfunction modes. As time goes on, more and more of the modes can be neglected, and the solution decays to thermal equilibrium at an exponentially fast rate. The rate of convergence to thermal equilibrium is, for most initial data, governed by the smallest positive eigenvalue $\lambda_1 > 0$ for the Helmholtz boundary value problem on the domain.

In the positive definite cases of homogeneous Dirichlet or mixed boundary conditions, thermal equilibrium is $u(t,x,y) \to u_\star(x,y) \equiv 0$. Here, the equilibrium temperature is equal to the zero boundary temperature — even if this temperature is fixed on only a small part of the boundary. The initial heat is eventually dissipated away through the uninsulated part of the boundary. In the semi-definite Neumann case, corresponding to a completely insulated plate, the general solution has the form

$$u(t,x,y) = c_0 + \sum_{k=1}^{\infty} c_k\, e^{-\lambda_k t}\, v_k(x,y), \tag{11.27}$$

where the sum is over the positive eigenmodes, $\lambda_k > 0$. Since all the summands are exponentially decaying, the final equilibrium temperature $u_\star = c_0$ is the same as the constant term in the eigenfunction expansion. We evaluate this term using the orthogonality formula (11.26), and so, as $t \to \infty$,

$$u(t,x,y) \longrightarrow c_0 = \frac{\langle f, 1 \rangle}{\| 1 \|^2} = \frac{\displaystyle\iint_\Omega f(x,y)\, dx\, dy}{\displaystyle\iint_\Omega dx\, dy} = \frac{1}{\text{area}\,\Omega} \iint_\Omega f(x,y)\, dx\, dy. \tag{11.28}$$

We conclude that the equilibrium temperature is equal to the average initial temperature distribution. Thus, when the plate is fully insulated, the heat energy cannot escape, and instead redistributes itself in a uniform manner over the domain.

Diffusion has a smoothing effect on the initial temperature distribution $f(x,y)$. Assume that the eigenfunction coefficients are uniformly bounded, so $|\, c_k \,| \le M$ for some constant M. This will certainly be the case if $f(x,y)$ is piecewise continuous or, more generally, belongs to L^2, since Bessel's inequality, (3.117), which holds for general orthogonal systems, implies that $c_k \to 0$ as $k \to \infty$. Many distributions, including delta functions, also have bounded Fourier coefficients. Then, at any time $t > 0$ after the initial instant, the coefficients $c_k\, e^{-\lambda_k t}$ in the eigenfunction series solution (11.23) are exponentially small as $k \to \infty$, which is enough to ensure smoothness of the solution $u(t,x,y)$ for each $t > 0$. Therefore, the diffusion process serves to immediately smooth out jumps, corners, and other discontinuities in the initial data. As time progresses, the local variations in the solution become less and less pronounced, as it asymptotically reaches a constant equilibrium state.

As a result, diffusion processes can be effectively applied to smooth and denoise planar images. The initial data $u(0,x,y) = f(x,y)$ represents the gray-scale value of the image at position (x,y), so that $0 \le f(x,y) \le 1$, with 0 representing black and 1 representing white. As time progresses, the solution $u(t,x,y)$ represents a more and more smoothed version

   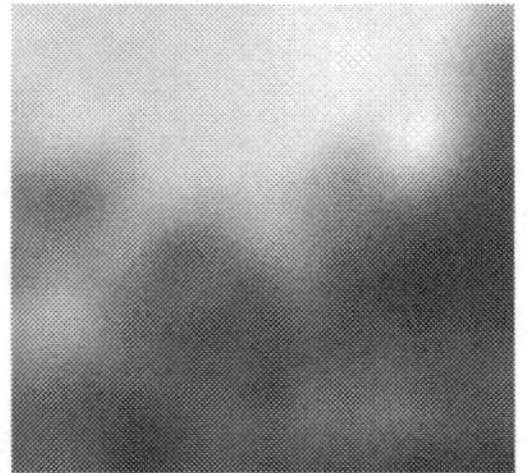

Figure 11.1. Smoothing a gray scale image.

of the image. Although this has the effect of removing unwanted high-frequency noise, there is also a gradual blurring of the actual features. Thus, the "time" or "multiscale" parameter t needs to be chosen to optimally balance between the two effects — the larger t is the more noise is removed, but the more noticeable the blurring. A representative illustration appears in Figure 11.1. The blurring affects small-scale features first, then, gradually, those at larger and larger scales, until eventually the entire image is blurred to a uniform gray. To further suppress undesirable blurring effects, modern image-processing filters are based on anisotropic (and thus *nonlinear*) diffusion equations; see [**100**] for a survey of recent progress in this active field.

Since the forward heat equation effectively blurs the features in an image, we might be tempted to reverse "time" in order to sharpen the image. However, the argument presented in Section 4.1 tells us that the backwards heat equation is ill-posed, and hence cannot be used directly for this purpose. Various "regularization" strategies have been devised to circumvent this mathematical barrier, and thereby design effective image enhancement algorithms, [**46**].

Inhomogeneous Boundary Conditions and Forcing

Let us next briefly discuss how to incorporate inhomogeneous boundary conditions and external heat sources into the general solution framework. Consider, as a specific example, the forced heat equation

$$u_t = \gamma\,\Delta u + F(x,y) \qquad \text{for} \qquad (x,y) \in \Omega, \tag{11.29}$$

where $F(x,y)$ represents an unvarying external heat source or sink, subject to inhomogeneous Dirichlet boundary conditions

$$u(x,y) = h(x,y) \qquad \text{for} \qquad (x,y) \in \partial\Omega, \tag{11.30}$$

that fixes the temperature of the plate on its boundary. When the external forcing does not vary in time, we expect the solution to eventually settle down to an equilibrium configuration: $u(t,x,y) \to u_\star(x,y)$ as $t \to \infty$. This will be justified below.

The time-independent equilibrium temperature $u_\star(x,y)$ satisfies the equation obtained by setting $u_t = 0$ in the evolution equation (11.29), which reduces it to the Poisson equation

$$-\gamma\,\Delta u_\star = F \qquad \text{for} \qquad (x,y) \in \Omega. \tag{11.31}$$

The equilibrium solution is subject to the same inhomogeneous Dirichlet boundary conditions (11.30). Positive definiteness of the Dirichlet boundary value problem implies that there is a unique equilibrium solution, which can be characterized as the sole minimizer of the associated Dirichlet principle; for details see Section 9.3.

With the equilibrium solution in hand, we let

$$v(t,x,y) = u(t,x,y) - u_\star(x,y)$$

measure the deviation of the dynamical solution u from its eventual equilibrium. By linearity $v(t,x,y)$ satisfies the unforced heat equation subject to homogeneous boundary conditions:

$$v_t = \gamma\,\Delta v, \qquad (x,y) \in \Omega, \qquad\qquad v = 0, \qquad (x,y) \in \partial\Omega. \tag{11.32}$$

Therefore, v can be expanded in an eigenfunction series (11.23), and will decay to zero, $v(t,x,y) \to 0$, at an exponentially fast rate prescribed by the smallest eigenvalue λ_1 of the associated homogeneous Helmholtz boundary value problem. (Special initial data can decay at a faster rate, prescribed by a larger eigenvalue.) Consequently, the solution to the forced inhomogeneous problem (11.29–30) will approach thermal equilibrium,

$$u(t,x,y) = v(t,x,y) + u_\star(x,y) \quad \longrightarrow \quad u_\star(x,y),$$

at exactly the same exponential rate as its homogeneous counterpart.

The Maximum Principle

Finally, let us state and prove the (Weak) Maximum Principle for the two-dimensional heat equation. As in the one-dimensional situation described in Section 8.3, it states that the maximum temperature in a body that is either insulated or having heat removed from its interior must occur either at the initial time or on its boundary. Observe that there are no conditions imposed on the boundary temperatures.

Theorem 11.1. *Suppose $u(t,x,y)$ is a solution to the forced heat equation*

$$u_t = \gamma\,\Delta u + F(t,x,y), \qquad \text{for} \qquad (x,y) \in \Omega, \quad 0 < t < c,$$

where Ω is a bounded domain, and $\gamma > 0$. Suppose $F(t,x,y) \le 0$ for all $(x,y) \in \overline{\Omega}$ and $0 \le t \le c$. Then the global maximum of u on the set $\{\,(t,x,y) \mid (x,y) \in \overline{\Omega},\ 0 \le t \le c\}$ occurs either when $t = 0$ or at a boundary point $(x,y) \in \partial\Omega$.

Proof: First, let us prove the result under the assumption that $F(t,x,y) < 0$ everywhere. At a local interior maximum, $u_t = 0$, and, since its Hessian matrix $\nabla^2 u = \begin{pmatrix} u_{xx} & u_{xy} \\ u_{xy} & u_{yy} \end{pmatrix}$ must be negative semi-definite, both diagonal entries $u_{xx}, u_{yy} \le 0$ there. This would imply that $u_t - \gamma\,\Delta u \ge 0$, resulting in a contradiction. If the maximum were to occur when $t = c$, then $u_t \ge 0$ there, and also $u_{xx}, u_{yy} \le 0$, leading again to a contradiction.

To generalize to the case $F(t,x,y) \le 0$, which includes the heat equation when $F(t,x,y) \equiv 0$, set

$$v(t,x,y) = u(t,x,y) + \varepsilon\,(x^2 + y^2), \qquad \text{where} \qquad \varepsilon > 0.$$

Then,

$$\frac{\partial v}{\partial t} = \gamma\,\Delta v - 4\,\gamma\,\varepsilon + F(t,x,y) = \gamma\,\Delta v + \widetilde{F}(t,x,y),$$

where

$$\widetilde{F}(t,x,y) = F(t,x,y) - 4\gamma\varepsilon < 0.$$

Thus, by the previous paragraph, the maximum of v occurs either when $t = 0$ or at a boundary point $(x,y) \in \partial\Omega$. We then let $\varepsilon \to 0$ and conclude the same for u. More precisely, let $u(t,x,y) \leq M$ on $t = 0$ or $(x,y) \in \partial\Omega$. Then

$$v(t,x,y) \leq M + C\,\varepsilon, \qquad \text{where} \qquad C = \max\left\{\, x^2 + y^2 \;\middle|\; (x,y) \in \partial\Omega \,\right\} < \infty,$$

since Ω is a bounded domain. Thus,

$$u(t,x,y) \leq v(t,x,y) \leq M + C\,\varepsilon.$$

Letting $\varepsilon \to 0$ proves that $u(t,x,y) \leq M$ at all $(x,y) \in \overline{\Omega}$, $0 \leq t \leq c$, which completes the proof. *Q.E.D.*

Remark: The preceding proof can be readily adapted to general diffusion equations (11.12) — assuming that the coefficients σ, κ remain strictly positive throughout the domain.

Exercises

11.1.1. A homogeneous, isotropic circular metal disk of radius 1 meter has its entire boundary insulated. The initial temperature at a point is equal to the distance of the point from the center. Formulate an initial-boundary value problem governing the disk's subsequent temperature dynamics. What is the eventual equilibrium temperature of the disk?

11.1.2. A homogeneous, isotropic, circular metal disk of radius 2 cm has half its boundary fixed at $100°$ and the other half insulated. Given a prescribed initial temperature distribution, set up the initial-boundary value problem governing its subsequent temperature profile. What is the eventual equilibrium temperature of the disk? Does your answer depend on the initial temperature?

11.1.3. Given the initial temperature distribution $f(x,y) = x\,y\,(1-x)(1-y)$ on the unit square $\Omega = \{\,0 \leq x, y \leq 1\,\}$, determine the equilibrium temperature when subject to homogeneous (a) Dirichlet boundary conditions; (b) Neumann boundary conditions.

11.1.4. A square plate with side lengths 1 meter has its right and left edges insulated, its top edge held at $100°$, and its bottom edge held at $0°$. Assuming that the plate is made out of a homogeneous, isotropic material, formulate an appropriate initial-boundary value problem describing the temperature dynamics of the plate. Then find its eventual equilibrium temperature.

11.1.5. A square plate with side lengths 1 meter has initial temperature $5°$ throughout, and evolves subject to the Neumann boundary conditions $\partial u/\partial \mathbf{n} = 1$ on its entire boundary. What is the eventual equilibrium temperature?

♡ 11.1.6. Let $u(t,x,y)$ be a solution to the heat equation on a bounded domain Ω subject to homogeneous Neumann conditions on its boundary $\partial\Omega$. (a) Prove that the total heat $H(t) = \iint_\Omega u(t,x,y)\,dx\,dy$ is conserved, i.e., is constant in time. (b) Use part (a) to prove that the eventual equilibrium solution is everywhere equal to the average of the initial temperature $u(0,x,y)$. (c) What can you say about the behavior of the total heat for the homogeneous Dirichlet boundary value problem? (d) What about an inhomogeneous Dirichlet boundary value problem?

11.1.7. Let $u(t,x,y)$ be a nonconstant solution to the heat equation on a connected, bounded domain Ω subject to homogeneous Dirichlet boundary conditions on $\partial\Omega$. (a) Prove that its L^2 norm $N(t) = \sqrt{\iint_\Omega u(t,x,y)^2\, dx\, dy}$ is a strictly decreasing function of t. (b) Is this also true for mixed boundary conditions? (c) For Neumann boundary conditions?

11.1.8. Are the conclusions in Exercises 11.1.6 and 11.1.7 valid for the general diffusion equation (11.12)?

♢ 11.1.9. Write out the eigenvalue equation governing the separable solutions to the general diffusion equation (11.11), subject to appropriate boundary conditions. Given a complete system of eigenfunctions, write down the eigenfunction series solution to the initial value problem $u(0,x,y) = f(x,y)$, including the formulas for the coefficients.

11.1.10. *True or false*: The equilibrium temperature of a fully insulated nonuniform plate whose thermodynamics are governed by the general diffusion equation (11.12) equals the average initial temperature.

11.1.11. Let $\alpha > 0$, and consider the initial-boundary value problem $u_t = \Delta u - \alpha\, u$, $u(0,x,y) = f(x,y)$ on a bounded domain $\Omega \subset \mathbb{R}^2$, with boundary conditions $\partial u/\partial \mathbf{n} = 0$ on $\partial\Omega$.
(a) Write the equation in self-adjoint form (9.122). *Hint*: Look at Exercise 9.3.26.
(b) Prove that the problem has a unique equilibrium solution.

11.1.12. Write each of the following linear evolution equations in the self-adjoint form (9.122) by choosing suitable inner products and a suitable set of homogeneous boundary conditions. Is the operator you construct positive definite?

(a) $u_t = u_{xx} + u_{yy} - u$, (b) $u_t = y\, u_{xx} + x\, u_{yy}$, (c) $u_t = \Delta^2 u$.

♢ 11.1.13. Prove that if $f(x,y)$ is continuous and $\iint_R f(x,y)\, dx\, dy = 0$ for all $R \subset \Omega$, then $f(x,y) \equiv 0$ for $(x,y) \in \Omega$. *Hint*: Adapt the method in Exercise 6.1.23.

11.2 Explicit Solutions of the Heat Equation

Solving the two-dimensional heat equation in series form requires knowing the eigenfunctions for the associated Helmholtz boundary value problem. Unfortunately, as with the vast majority of partial differential equations, explicit solution formulas are few and far between. In this section, we discuss two specific cases in which the required eigenfunctions can be found in closed form. The calculations rely on a further separation of variables, which, as we know, works in only a very limited class of domains. Nevertheless, interesting solution features can be gleaned from these particular geometries.

The first example is a rectangular domain, and the eigensolutions can be expressed in terms of elementary functions — trigonometric functions and exponentials. We then study the heating of a circular disk. In this case, the eigenfunctions are no longer elementary functions, but, rather, are expressed in terms of Bessel functions. Understanding their basic properties will require us to take a detour to develop the fundamentals of power series solutions to ordinary differential equations.

Heating of a Rectangle

A homogeneous rectangular plate

$$R = \{\, 0 < x < a,\ 0 < y < b \,\}$$

is heated to a prescribed initial temperature,

$$u(0, x, y) = f(x, y), \qquad \text{for} \qquad (x, y) \in R. \tag{11.33}$$

Then its top and bottom are insulated, while its sides are held at zero temperature. Our task is to understand the thermodynamic evolution of the plate's temperature.

The temperature $u(t, x, y)$ evolves according to the two-dimensional heat equation

$$u_t = \gamma\,(u_{xx} + u_{yy}), \qquad \text{for} \qquad (x, y) \in R, \qquad t > 0, \tag{11.34}$$

where $\gamma > 0$ is the plate's thermal diffusivity, while subject to homogeneous Dirichlet conditions along the boundary of the rectangle at all subsequent times:

$$u(t, 0, y) = u(t, a, y) = u(t, x, 0) = u(t, x, b) = 0, \qquad 0 < x < a, \qquad 0 < y < b, \qquad t > 0. \tag{11.35}$$

As in (11.19), the eigensolutions to the heat equation are obtained from the usual exponential ansatz $u(t, x, y) = e^{-\lambda t}\, v(x, y)$. Substituting this expression into the heat equation, we conclude that the function $v(x, y)$ solves the Helmholtz eigenvalue problem

$$\gamma\,(v_{xx} + v_{yy}) + \lambda\, v = 0, \qquad (x, y) \in R, \tag{11.36}$$

subject to the same homogeneous Dirichlet boundary conditions:

$$v(0, y) = v(a, y) = v(x, 0) = v(x, b) = 0, \qquad 0 < x < a, \quad 0 < y < b. \tag{11.37}$$

To tackle the rectangular Helmholtz eigenvalue problem (11.36–37), we shall, as in (4.89), introduce a further separation of variables, writing the solution

$$v(x, y) = p(x)\, q(y)$$

as the product of functions depending on the individual Cartesian coordinates. Substituting this expression into the Helmholtz equation (11.36), we find

$$\gamma\, p''(x)\, q(y) + \gamma\, p(x)\, q''(y) + \lambda\, p(x)\, q(y) = 0.$$

To effect the variable separation, we collect all terms involving x on one side and all terms involving y on the other side of the equation, which is accomplished by dividing by $v = p\,q$ and rearranging the terms:

$$\gamma\,\frac{p''(x)}{p(x)} = -\,\gamma\,\frac{q''(y)}{q(y)} - \lambda \equiv -\,\mu.$$

The left-hand side of this equation depends only on x, whereas the middle term depends only on y. As before, this requires that the expressions equal a common *separation constant*, denoted by $-\,\mu$. (The minus sign is for later convenience.) In this manner, we reduce our partial differential equation to a pair of one-dimensional eigenvalue problems

$$\gamma\,\frac{d^2 p}{dx^2} + \mu\, p = 0, \qquad \gamma\,\frac{d^2 q}{dy^2} + (\lambda - \mu)\, q = 0, \tag{11.38}$$

each of which is subject to homogeneous Dirichlet boundary conditions

$$p(0) = p(a) = 0, \qquad q(0) = q(b) = 0, \tag{11.39}$$

stemming from the boundary conditions (11.37). To obtain a nontrivial separable solution to the Helmholtz equation, we seek nonzero solutions to these two supplementary eigenvalue problems.

We have already solved these particular two boundary value problems (11.38–39) many times; see, for instance, (4.21). The eigenfunctions are, respectively,

$$p_m(x) = \sin\frac{m\pi x}{a}, \qquad m = 1,2,3,\ldots, \qquad\qquad q_n(y) = \sin\frac{n\pi y}{b}, \qquad n = 1,2,3,\ldots,$$

with

$$\mu = \frac{m^2\pi^2\gamma}{a^2}, \qquad \lambda - \mu = \frac{n^2\pi^2\gamma}{b^2}, \qquad \text{so that} \qquad \lambda = \frac{m^2\pi^2\gamma}{a^2} + \frac{n^2\pi^2\gamma}{b^2}.$$

Therefore, the separable eigenfunction solutions to the Helmholtz boundary value problem (11.35–36) have the doubly trigonometric form

$$v_{m,n}(x,y) = \sin\frac{m\pi x}{a}\,\sin\frac{n\pi y}{b}, \qquad \text{for} \qquad m,n = 1,2,3,\ \ldots\ , \tag{11.40}$$

with associated eigenvalues

$$\lambda_{m,n} = \frac{m^2\pi^2\gamma}{a^2} + \frac{n^2\pi^2\gamma}{b^2} = \left(\frac{m^2}{a^2} + \frac{n^2}{b^2}\right)\pi^2\gamma. \tag{11.41}$$

Each of these corresponds to an exponentially decaying eigensolution

$$u_{m,n}(t,x,y) = e^{-\lambda_{m,n}t}\,v_{m,n}(x,y) = \exp\left[-\left(\frac{m^2}{a^2} + \frac{n^2}{b^2}\right)\pi^2\gamma t\right]\,\sin\frac{m\pi x}{a}\,\sin\frac{n\pi y}{b} \tag{11.42}$$

to the original rectangular Dirichlet boundary value problem for the heat equation.

Using the fact that the univariate sine functions form a complete system, it is not hard to prove, [**120**], that the separable eigenfunction solutions (11.42) are complete, and so there are no non-separable eigenfunctions.† As a consequence, the general solution to the initial-boundary value problem can be expressed as a linear combination

$$u(t,x,y) = \sum_{m,n=1}^{\infty} c_{m,n}\,u_{m,n}(t,x,y) = \sum_{m,n=1}^{\infty} c_{m,n}\,e^{-\lambda_{m,n}t}\,v_{m,n}(x,y) \tag{11.43}$$

of the eigenmodes. The coefficients $c_{m,n}$ are prescribed by the initial conditions, which take the form of a double Fourier sine series

$$f(x,y) = u(0,x,y) = \sum_{m,n=1}^{\infty} c_{m,n}v_{m,n}(x,y) = \sum_{m,n=1}^{\infty} c_{m,n}\,\sin\frac{m\pi x}{a}\,\sin\frac{n\pi y}{b}.$$

Self-adjointness of the Laplacian operator coupled with the boundary conditions implies that‡ the eigenfunctions $v_{m,n}(x,y)$ are orthogonal with respect to the L^2 inner product

† This appears to be a general fact, true in all known examples, but I know of no general proof. Theorem 9.47 can be used to establish completeness of the eigenfunctions, but does not guarantee that they can all be constructed by separation of variables.

‡ Technically, orthogonality is guaranteed only when the eigenvalues are distinct: $\lambda_{m,n} \neq \lambda_{k,l}$. However, by a direct computation, one finds that orthogonality continues to hold even when the indicated eigenfunctions are associated with equal eigenvalues. See the final subsection of this chapter for a discussion of when such "accidental degeneracies" arise.

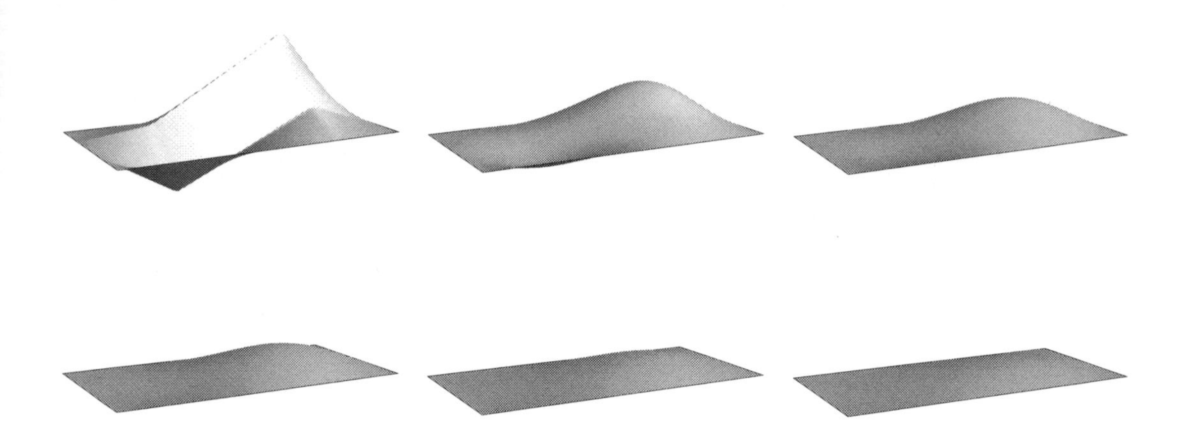

Figure 11.2. Heat diffusion in a rectangle.

on the rectangle:

$$\langle v_{k,l}, v_{m,n} \rangle = \int_0^b \int_0^a v_{k,l}(x,y)\, v_{m,n}(x,y)\, dx\, dy = 0 \qquad \text{unless} \qquad k = m \quad \text{and} \quad l = n.$$

(The skeptical reader can verify the orthogonality relations directly from the eigenfunction formulas (11.40).) Thus, we can appeal to our usual orthogonality formula (11.26) to evaluate the coefficients

$$c_{m,n} = \frac{\langle f, v_{m,n} \rangle}{\| v_{m,n} \|^2} = \frac{4}{a\,b} \int_0^b \int_0^a f(x,y) \, \sin \frac{m \pi x}{a} \, \sin \frac{n \pi y}{b} \, dx\, dy, \tag{11.44}$$

where the formula for the norms of the eigenfunctions

$$\| v_{m,n} \|^2 = \int_0^b \int_0^a v_{m,n}(x,y)^2 \, dx\, dy = \int_0^b \int_0^a \sin^2 \frac{m \pi x}{a} \, \sin^2 \frac{n \pi y}{b} \, dx\, dy = \tfrac{1}{4} a\, b \tag{11.45}$$

follows from a direct evaluation of the double integral. Unfortunately, while orthogonality is (mostly) automatic, computation of the norms must inevitably be done “by hand”.

For generic initial temperature distributions, the rectangle approaches thermal equilibrium at a rate equal to the smallest eigenvalue:

$$\lambda_{1,1} = \left(\frac{1}{a^2} + \frac{1}{b^2} \right) \pi^2 \gamma, \tag{11.46}$$

i.e., the sum of the reciprocals of the squared lengths of its sides multiplied by the diffusion coefficient. The larger the rectangle, or the smaller the diffusion coefficient, the smaller the value of $\lambda_{1,1}$, and hence the slower the return to thermal equilibrium. The exponentially fast decay rate of the Fourier series implies that the solution immediately smooths out any discontinuities in the initial temperature profile. Indeed, the higher modes, with m and n large, decay to zero almost instantaneously, and so the solution quickly behaves like a finite sum over a few low-order modes. Assuming that $c_{1,1} \neq 0$, the slowest-decaying mode in the Fourier series (11.43) is

$$c_{1,1}\, u_{1,1}(t,x,y) = c_{1,1} \exp\left[-\left(\frac{1}{a^2} + \frac{1}{b^2} \right) \pi^2 \gamma\, t \right] \sin \frac{\pi x}{a} \sin \frac{\pi y}{b}. \tag{11.47}$$

Thus, in the long run, the temperature becomes entirely of one sign — either positive or negative depending on the sign of $c_{1,1}$ — throughout the rectangle. This observation is, in fact, indicative of the general phenomenon that an eigenfunction associated with the smallest positive eigenvalue of a self-adjoint elliptic operator is necessarily of one sign throughout the domain, [**34**]. A typical solution is plotted at several times in Figure 11.2. Non-generic initial conditions, with $c_{1,1} = 0$, decay more rapidly, and their asymptotic temperature profiles are not of one sign.

Exercises

11.2.1. A rectangle of size 2 cm by 1 cm has initial temperature $f(x,y) = \sin \pi x \sin \pi y$ for $0 \le x \le 2$, $0 \le y \le 1$. All four sides of the rectangle are held at 0°. Assuming that the thermal diffusivity of the plate is $\gamma = 1$, write down a formula for its subsequent temperature $u(t,x,y)$. What is the rate of decay to thermal equilibrium?

11.2.2. Solve Exercise 11.2.1 when the initial temperature $f(x,y)$ is
(a) $x\,y$, (b) $\begin{cases} 1, & 0 < x < 1, \\ 0, & 1 < x < 2; \end{cases}$ (c) $\left(1 - |\,1 - x\,|\right)\left(\frac{1}{2} - |\,\frac{1}{2} - y\,|\right)$.

11.2.3. Solve the initial-boundary value problem for the heat equation $u_t = 2\,\Delta u$ on the rectangle $-1 < x < 1$, $0 < y < 1$ when the two short sides are kept at 0°, the two long sides are insulated, and the initial temperature distribution is $u(0,x,y) = \begin{cases} -1, & x < 0, \\ +1, & x > 0, \end{cases}$ $0 < y < 1$.

11.2.4. Answer Exercise 11.2.3 when the two long sides are kept at 0° and the two short sides are insulated.

♡ 11.2.5. A rectangular plate of size 1 meter by 3 meters is made out a metal with unit diffusivity. The plate is taken from a 0° freezer, and, from then on, one of its long sides is heated to 100°, the other is held at 0°, while its top, bottom, and both of the short sides are fully insulated. (*a*) Set up the initial-boundary value problem governing the time-dependent temperature of the plate. (*b*) What is the equilibrium temperature? (*c*) Use your answer from part (*b*) to construct an eigenfunction series for the solution. (*d*) How long until the temperature of the plate is everywhere within 1° of its eventual equilibrium?
Hint: Once t is no longer small, you can approximate the series solution by its first term.

11.2.6. Among all rectangular plates of a prescribed area, which one returns to thermal equilibrium the slowest when subject to Dirichlet boundary conditions? The fastest? Use your physical intuition to explain your answer, but justify it mathematically.

11.2.7. Answer Exercise 11.2.6 for a fully insulated rectangular plate, i.e., subject to Neumann boundary conditions.

♡ 11.2.8. A square metal plate is taken from an oven, and then set out to cool, with its top and bottom insulated. Find the rate of cooling, in terms of the side length and the thermal diffusivity, if (*a*) all four sides are held at 0°; (*b*) one side is insulated and the other three sides are held at 0°; (*c*) two adjacent sides are insulated and the other two are held at 0°; (*d*) two opposite sides are insulated and the other two are held at 0°; (*e*) three sides are insulated and the remaining side is held at 0°. Order the cooling rates of the plates from fastest to slowest. Do your results confirm your intuition?

♡ 11.2.9. Two square plates are made out of the same homogeneous material, and both are initially heated to $100°$. All four sides of the first plate are held at $0°$, whereas one of the sides of the second plate is insulated while the other three sides are held at $0°$. Which plate cools down the fastest? How much faster? Assuming the thermal diffusivity $\gamma = 1$, how long do you have to wait until every point on each plate is within $1°$ of its equilibrium temperature? *Hint*: Once t is no longer small, the series solution is well approximated by its first term.

♡ 11.2.10. *Multiple choice*: On a unit square that is subject to Dirichlet boundary conditions, the eigenvalues of the Laplace operator are
(a) all simple, (b) at most double, or (c) can have arbitrarily large multiplicity.

♡ 11.2.11. The thermodynamics of a thin circular cylindrical shell of radius a and height h, e.g., the side of a tin can after its top and bottom are removed, is modeled by the heat equation $\dfrac{\partial u}{\partial t} = \gamma\left(\dfrac{1}{a^2}\dfrac{\partial^2 u}{\partial \theta^2} + \dfrac{\partial^2 u}{\partial z^2}\right)$, in which $u(t,\theta,z)$ measures the temperature of the point on the cylinder at time $t > 0$, angle $-\pi < \theta \leq \pi$, and height $0 < z < h$. Keep in mind that $u(t,\theta,z)$ must be a 2π–periodic function of the angular coordinate θ. Assume that the cylinder is everywhere insulated, while its two circular ends are held at $0°$. Given an initial temperature distribution at time $t = 0$, write down a series formula for the cylinder's temperature at subsequent times. What is the eventual equilibrium temperature? How fast does the cylinder return to equilibrium?

♡ 11.2.12. Consider the initial-boundary value problem
$$u_t = u_{xx} + u_{yy}, \qquad u(0,x,y) = 0, \qquad 0 < x, y < \pi, \qquad t > 0,$$
for the heat equation in a square subject to the Dirichlet conditions
$$u(0,y) = u(\pi,y) = 0 = u(x,0), \qquad u(x,\pi) = f(x), \qquad 0 < x, y < \pi.$$
Write out an eigenfunction series formulas for
(a) the equilibrium solution $u_\star(x,y) = \lim_{t\to\infty} u(t,x,y)$; (b) the solution $u(t,x,y)$.

11.2.13. Solve Exercise 11.2.1 when one long side of the plate is held at $100°$.
Hint: See Exercise 11.2.12.

Heating of a Disk — Preliminaries

Let us perform a similar analysis of the thermodynamics of a circular disk. For simplicity (or by choice of suitable physical units), we will assume that the disk
$$D = \{\, x^2 + y^2 \leq 1 \,\} \subset \mathbb{R}^2$$
has unit radius and unit diffusivity $\gamma = 1$. We shall solve the heat equation on D subject to homogeneous Dirichlet boundary values of zero temperature at the circular edge
$$\partial D = C = \{\, x^2 + y^2 = 1 \,\}.$$
Thus, the full initial-boundary value problem is
$$\begin{aligned} &\frac{\partial u}{\partial t} = \frac{\partial^2 u}{\partial x^2} + \frac{\partial^2 u}{\partial y^2}, && x^2 + y^2 < 1, \\ &u(t,x,y) = 0, && x^2 + y^2 = 1, \qquad t > 0, \\ &u(0,x,y) = f(x,y), && x^2 + y^2 \leq 1. \end{aligned} \tag{11.48}$$

We remark that a simple rescaling of space and time, as outlined in Exercise 11.4.7, can be used to recover the solution for an arbitrary diffusion coefficient and a disk of arbitrary radius from this particular case.

Since we are working in a circular domain, we instinctively pass to polar coordinates (r, θ). In view of the polar coordinate formula (4.105) for the Laplace operator, the heat equation and boundary and initial conditions assume the form

$$\frac{\partial u}{\partial t} = \frac{\partial^2 u}{\partial r^2} + \frac{1}{r}\frac{\partial u}{\partial r} + \frac{1}{r^2}\frac{\partial^2 u}{\partial \theta^2}, \qquad u(t,1,\theta) = 0, \qquad u(0,r,\theta) = f(r,\theta), \tag{11.49}$$

where the solution $u(t, r, \theta)$ is defined for all $0 \le r \le 1$ and $t \ge 0$. To ensure that the solution represents a single-valued function on the entire disk, it is required to be a 2π–periodic function of the angular variable:

$$u(t, r, \theta + 2\pi) = u(t, r, \theta).$$

To obtain the separable solutions

$$u(t, r, \theta) = e^{-\lambda t}\, v(r, \theta), \tag{11.50}$$

we need to solve the polar coordinate form of the Helmholtz equation

$$\frac{\partial^2 v}{\partial r^2} + \frac{1}{r}\frac{\partial v}{\partial r} + \frac{1}{r^2}\frac{\partial^2 v}{\partial \theta^2} + \lambda\, v = 0, \qquad \begin{array}{c} 0 \le r < 1, \\ -\pi < \theta \le \pi, \end{array} \tag{11.51}$$

subject to the boundary conditions

$$v(1, \theta) = 0, \qquad v(r, \theta + 2\pi) = v(r, \theta). \tag{11.52}$$

To solve the polar Helmholtz boundary value problem (11.51–52), we invoke a further separation of variables by writing

$$v(r, \theta) = p(r)\, q(\theta). \tag{11.53}$$

Substituting this ansatz into (11.51), collecting all terms involving r and all terms involving θ, and then equating both to a common separation constant, we are led to the pair of ordinary differential equations

$$r^2 \frac{d^2 p}{dr^2} + r\frac{dp}{dr} + (\lambda r^2 - \mu)\, p = 0, \qquad \frac{d^2 q}{d\theta^2} + \mu\, q = 0, \tag{11.54}$$

where λ is the Helmholtz eigenvalue, and μ the separation constant.

Let us start with the equation for $q(\theta)$. The second boundary condition in (11.52) requires that $q(\theta)$ be 2π–periodic. Therefore, the required solutions are the elementary trigonometric functions

$$q(\theta) = \cos m\theta \qquad \text{or} \qquad \sin m\theta, \qquad \text{where} \qquad \mu = m^2, \tag{11.55}$$

with $m = 0, 1, 2, \ldots$ a nonnegative integer.

Substituting the formula for the separation constant, $\mu = m^2$, the differential equation for $p(r)$ takes the form

$$r^2 \frac{d^2 p}{dr^2} + r\frac{dp}{dr} + (\lambda r^2 - m^2)\, p = 0, \qquad 0 \le r \le 1. \tag{11.56}$$

Ordinarily, one imposes two boundary conditions in order to pin down a solution to such a second-order ordinary differential equation. But our Dirichlet condition, namely $p(1) = 0$, specifies its value at only one of the endpoints. The other endpoint is a *singular point* for the ordinary differential equation, because the coefficient of the highest-order derivative, namely r^2, vanishes at $r = 0$. This situation might remind you of our solution to the Euler differential equation (4.111) in the context of separable solutions to the Laplace equation on the disk. As there, we require that the solution be bounded at $r = 0$, and so seek eigensolutions that satisfy the boundary conditions

$$| p(0) | < \infty, \qquad\qquad p(1) = 0. \tag{11.57}$$

While (11.56) appears in a variety of applications, it is more challenging than any ordinary differential equation we have encountered so far. Indeed, most solutions cannot be written in terms of the elementary functions (rational functions, trigonometric functions, exponentials, logarithms, etc.) you see in first-year calculus. Nevertheless, owing to their ubiquity in physical applications, its solutions have been extensively studied and tabulated, and so are, in a sense, well known, [**85, 86, 119**].

To simplify the subsequent analysis, we make a preliminary rescaling of the independent variable, replacing r by

$$z = \sqrt{\lambda}\, r.$$

(We know the eigenvalue $\lambda > 0$, since we are dealing with a positive definite boundary value problem.) Note that, by the chain rule,

$$\frac{dp}{dr} = \sqrt{\lambda}\,\frac{dp}{dz}, \qquad\qquad \frac{d^2p}{dr^2} = \lambda\,\frac{d^2p}{dz^2},$$

and hence

$$r\,\frac{dp}{dr} = z\,\frac{dp}{dz}, \qquad\qquad r^2\,\frac{d^2p}{dr^2} = z^2\,\frac{d^2p}{dz^2}.$$

The net effect is to eliminate the eigenvalue parameter λ (or, rather, hide it in the change of variables), so that (11.56) assumes the slightly simpler form

$$z^2\,\frac{d^2p}{dz^2} + z\,\frac{dp}{dz} + (z^2 - m^2)\,p = 0. \tag{11.58}$$

The resulting ordinary differential equation (11.58) is known as *Bessel's equation*, named after the early-nineteenth-century German astronomer Wilhelm Bessel, who first encountered its solutions, now known as *Bessel functions*, in his study of planetary orbits. Special cases had already appeared in the investigations of Daniel Bernoulli on vibrations of a hanging chain, and in those of Fourier on the thermodynamics of a cylindrical body. To make further progress, we need to take time out to study their basic properties, and this will require us to develop the method of power series solutions of ordinary differential equations. With this in hand, we can then return to complete our solution to the heat equation on a disk.

11.3 Series Solutions of Ordinary Differential Equations

When confronted with a novel ordinary differential equation, we have several available options for deriving and understanding its solutions. For instance, the "look-up" method relies on published handbooks. One of the most useful references that collects many solved differential equations is the classic German compendium by Kamke, [**62**]. Two more recent English-language handbooks are [**93, 127**]. In addition, many symbolic computer algebra programs, including MATHEMATICA and MAPLE, will produce solutions, when expressible in terms of both elementary and special functions, to a wide range of differential equations.

Of course, use of numerical integration to approximate solutions, [**24, 60, 80**], is always an option. Numerical methods do, however, have their limitations, and are best accompanied by some understanding of the underlying theory, coupled with qualitative or quantitative expectations of how the solutions should behave. Furthermore, numerical methods provide less than adequate insight into the nature of the special functions that regularly appear as solutions of the particular differential equations arising in separation of variables. A numerical approximation cannot, in itself, establish rigorous mathematical properties of the solutions of the differential equation.

A more classical means of constructing and approximating the solutions of differential equations is based on their power series expansions, a.k.a. Taylor series. The Taylor expansion of a solution at a point x_0 is found by substituting a general power series into the differential equation and equating coefficients of the various powers of $x - x_0$. The initial conditions at x_0 serve to uniquely determine the coefficients and hence all the derivatives of the solution at the initial point. The Taylor expansion of a special function is an effective tool for deducing some of its key properties, as well as providing a means of computing reasonable numerical approximations to its values within the radius of convergence of the series. (However, serious numerical computations more often rely on nonconvergent asymptotic expansions, [**85**].)

In this section, we provide a brief introduction to the basic series solution techniques for ordinary differential equations, concentrating on second-order linear differential equations, since these form by far the most important class of examples arising in applications. At a regular point, the method will produce a standard Taylor expansion for the solution, while so-called regular singular points require a slightly more general type of series expansion. Generalizations to irregular singular points, higher-order equations, nonlinear equations, and even linear and nonlinear systems are deferred to more advanced texts, including [**54, 59**].

The Gamma Function

Before delving into the machinery of series solutions and special functions, we need to introduce the gamma function, which effectively generalizes the factorial operation to non-integers. Recall that the *factorial* of a nonnegative integer $n \geq 0$ is defined inductively by the iterative formula

$$n! = n \cdot (n-1)!, \qquad \text{starting with} \qquad 0! = 1. \tag{11.59}$$

When n is a positive integer, the iteration terminates, yielding the familiar expression

$$n! = n(n-1)(n-2) \cdots 3 \cdot 2 \cdot 1. \tag{11.60}$$

However, for more general values of n, the iteration never stops, and it cannot be used to compute its factorial. Our goal is to circumvent this difficulty, and introduce a function $f(x)$ that is defined for *all* values of x and will play the role of such a factorial. First, mimicking (11.59), the function should satisfy the functional equation

$$f(x) = x\, f(x-1) \tag{11.61}$$

where defined. If, in addition, $f(0) = 1$, then we know that $f(n) = n!$ whenever n is a nonnegative integer, and hence such a function will extend the definition of the factorial to more general real and complex numbers.

A moment's thought should convince the reader that there are many possible ways to construct such a function; see Exercise 11.3.6 for a nonstandard example. The most important version is due to Euler. The modern definition of Euler's gamma function relies on an integral formula discovered by the eighteenth-century French mathematician Adrien–Marie Legendre, who will play a starring role in Chapter 12.

Definition 11.2. The *gamma function* is defined by

$$\Gamma(x) = \int_0^\infty e^{-t}\, t^{x-1}\, dt. \tag{11.62}$$

The first fact is that, for real x, the gamma function integral converges only when $x > 0$; otherwise the singularity of t^{x-1} at $t = 0$ is too severe. The key property that turns the gamma function into a substitute for the factorial function relies on an elementary integration by parts:

$$\Gamma(x+1) = \int_0^\infty e^{-t}\, t^{x}\, dt = -\,e^{-t}\, t^{x}\,\Big|_{t=0}^{\infty} + x \int_0^\infty e^{-t}\, t^{x-1}\, dt.$$

The boundary terms vanish whenever $x > 0$, while the final integral is merely $\Gamma(x)$. Therefore, the gamma function satisfies the *recurrence relation*

$$\Gamma(x+1) = x\,\Gamma(x). \tag{11.63}$$

If we set $f(x) = \Gamma(x+1)$, then (11.63) becomes (11.61). Moreover, by direct integration,

$$\Gamma(1) = \int_0^\infty e^{-t}\, dt = 1.$$

Combining this with the recurrence relation (11.63), we deduce that

$$\Gamma(n+1) = n! \tag{11.64}$$

whenever $n \geq 0$ is a nonnegative integer. Therefore, we can identify $x!$ with the value $\Gamma(x+1)$ whenever $x > -1$ is *any* real number.

Remark: The reader may legitimately ask why not replace t^{x-1} by t^x in the definition of $\Gamma(z)$, which would avoid the $n+1$ in (11.64). There is no good answer; we are merely following a well-established precedent set by Legendre and enshrined in all subsequent works.

Thus, at integer values of x, the gamma function agrees with the elementary factorial. A few other values can be computed exactly. One important case is at $x = \frac{1}{2}$. Using the substitution $t = s^2$, with $dt = 2\,s\,ds$, we obtain

$$\Gamma\left(\tfrac{1}{2}\right) = \int_0^\infty e^{-t}\, t^{-1/2}\, dt = \int_0^\infty 2\, e^{-s^2}\, ds = \sqrt{\pi}, \tag{11.65}$$

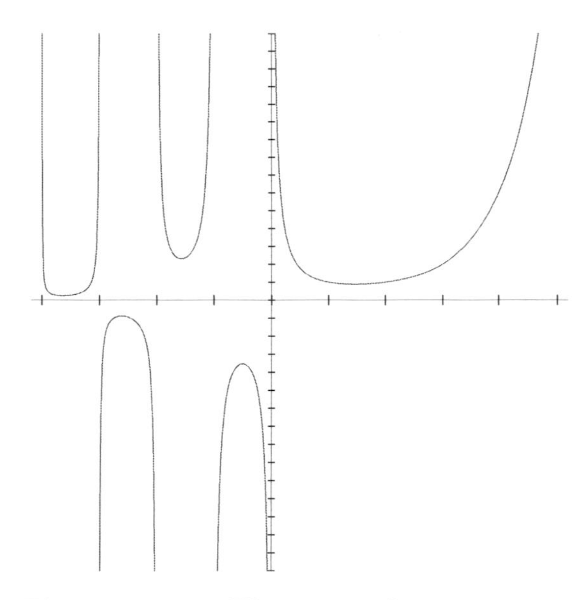

Figure 11.3. The gamma function.

where the final integral was evaluated in (2.100). Thus, using the identification with the factorial function, we identify this value with $\left(-\frac{1}{2}\right)! = \sqrt{\pi}$. The recurrence relation (11.63) will then produce the value of the gamma function at all half-integers $\frac{1}{2}, \frac{3}{2}, \frac{5}{2}, \ldots$. For example,

$$\Gamma\left(\tfrac{3}{2}\right) = \tfrac{1}{2}\,\Gamma\left(\tfrac{1}{2}\right) = \tfrac{1}{2}\sqrt{\pi}, \qquad (11.66)$$

and hence $\frac{1}{2}! = \frac{1}{2}\sqrt{\pi}$. The recurrence relation can also be employed to extend the definition of $\Gamma(x)$ to (most) negative values of x. For example, setting $x = -\frac{1}{2}$ in (11.63), we have

$$\Gamma\left(\tfrac{1}{2}\right) = -\tfrac{1}{2}\,\Gamma\left(-\tfrac{1}{2}\right), \qquad \text{so} \qquad \Gamma\left(-\tfrac{1}{2}\right) = -2\,\Gamma\left(\tfrac{1}{2}\right) = -2\sqrt{\pi}\,.$$

The only points at which this device fails are the negative integers, and indeed, $\Gamma(x)$ has a singularity when $x = -1, -2, -3, \ldots$. A graph[†] of the gamma function is displayed in Figure 11.3.

Remark: Most special functions of importance for applications arise as solutions to fairly simple ordinary differential equations. The gamma function is a significant exception. Indeed, it can be proved, [**11**], that the gamma function *does not* satisfy *any* algebraic differential equation!

Regular Points

We are now ready to develop the method of series solutions to ordinary differential equations. Before we proceed to develop the general computational machinery, a naïve calculation in an elementary example will be enlightening.

[†] The axes are at different scales; the tick marks are at integer values.

Example 11.3. Consider the initial value problem

$$\frac{d^2u}{dx^2} + u = 0, \qquad u(0) = 1, \qquad u'(0) = 0. \tag{11.67}$$

Let us investigate whether we can construct an analytic solution in the form of a convergent power series

$$u(x) = u_0 + u_1 x + u_2 x^2 + u_3 x^3 + \cdots = \sum_{n=0}^{\infty} u_n x^n \tag{11.68}$$

that is based at the initial point $x_0 = 0$. Term-by-term differentiation yields the following series expansions[†] for its derivatives:

$$\begin{aligned} \frac{du}{dx} &= u_1 + 2u_2 x + 3u_3 x^2 + 4u_4 x^3 + \cdots &&= \sum_{n=0}^{\infty} (n+1)\, u_{n+1} x^n, \\ \frac{d^2u}{dx^2} &= 2u_2 + 6u_3 x + 12u_4 x^2 + 20u_5 x^3 + \cdots &&= \sum_{n=0}^{\infty} (n+1)(n+2)\, u_{n+2} x^n. \end{aligned} \tag{11.69}$$

The next step is to substitute the series (11.68–69) into the differential equation and collect common powers of x:

$$\frac{d^2u}{dx^2} + u = (2u_2 + u_0) + (6u_3 + u_1)\, x + (12u_4 + u_2)\, x^2 + (20u_5 + u_3)\, x^3 + \cdots = 0.$$

At this point, one focuses attention on the individual coefficients, appealing to the following basic observation:

Two convergent power series are equal if and only if all their coefficients are equal.

In particular, a power series represents the zero function[‡] if and only if all its coefficients are 0. In this manner we obtain the following infinite sequence of algebraic *recurrence relations* among the coefficients:

$$\begin{array}{lr} 1 & 2u_2 + u_0 = 0, \\ x & 6u_3 + u_1 = 0, \\ x^2 & 12u_4 + u_2 = 0, \\ x^3 & 20u_5 + u_3 = 0, \\ x^4 & 30u_6 + u_4 = 0, \\ \vdots & \vdots \\ x^n & (n+1)(n+2)\, u_{n+2} + u_n = 0. \end{array} \tag{11.70}$$

Now, the initial conditions serve to prescribe the first two coefficients:

$$u(0) = u_0 = 1, \qquad u'(0) = u_1 = 0.$$

[†] When working with the series in summation form, it helps to re-index in order to display the term of degree n.

[‡] Here it is essential that we work with analytic functions, since this result is *not* true for C^∞ functions! For example, the function e^{-1/x^2} has identically zero power series at $x_0 = 0$; see Exercise 11.3.21.

We then solve the recurrence relations in order: The first determines $u_2 = -\frac{1}{2}u_0 = -\frac{1}{2}$; the second, $u_3 = -\frac{1}{6}u_1 = 0$; next, $u_4 = -\frac{1}{12}u_2 = \frac{1}{24}$; then $u_5 = -\frac{1}{20}u_3 = 0$; then $u_6 = -\frac{1}{30}u_4 = -\frac{1}{720}$; and so on. In general, it is not hard to see that

$$u_{2k} = \frac{(-1)^k}{(2k)!}, \qquad u_{2k+1} = 0, \qquad k = 0, 1, 2, \ldots .$$

Hence, the required series solution is

$$u(x) = 1 - \tfrac{1}{2}x^2 + \tfrac{1}{24}x^3 - \tfrac{1}{720}x^6 + \cdots = \sum_{k=0}^{\infty} \frac{(-1)^k}{k!}\,x^{2k},$$

which, by the ratio test, converges for all x. We have thus recovered the well-known Taylor series for $\cos x$, which is indeed the solution to the initial value problem. Changing the initial conditions to $u(0) = u_0 = 0$, $u'(0) = u_1 = 1$, will similarly produce the usual Taylor expansion of $\sin x$. Note that the generation of the Taylor series does not rely on any a priori knowledge of trigonometric functions or the direct solution method for linear constant-coefficient ordinary differential equations.

Building on this experience, let us describe the general method. We shall concentrate on solving a second-order homogeneous linear differential equation

$$p(x)\,\frac{d^2u}{dx^2} + q(x)\,\frac{du}{dx} + r(x)\,u = 0. \tag{11.71}$$

The coefficients $p(x), q(x), r(x)$ are assumed to be analytic functions on some common domain. This means that, at a point x_0 within the domain, they admit convergent power series expansions

$$\begin{aligned} p(x) &= p_0 + p_1\,(x - x_0) + p_2\,(x - x_0)^2 + \cdots, \\ q(x) &= q_0 + q_1\,(x - x_0) + q_2\,(x - x_0)^2 + \cdots, \\ r(x) &= r_0 + r_1\,(x - x_0) + r_2\,(x - x_0)^2 + \cdots. \end{aligned} \tag{11.72}$$

We expect that solutions to the differential equation are also analytic. This expectation is justified, provided that the equation is *regular* at the point x_0, in the following sense.

Definition 11.4. A point $x = x_0$ is a *regular point* of a second-order linear ordinary differential equation (11.71) if the leading coefficient does not vanish there:

$$p_0 = p(x_0) \neq 0.$$

A point where $p(x_0) = 0$ is known as a *singular point*.

In short, at a regular point, the second-order derivative term does not disappear, and so the equation is "genuinely" of second order.

Remark: The definition of a singular point assumes that the other two coefficients do not also vanish there, so that either $q(x_0) \neq 0$ or $r(x_0) \neq 0$. If all three functions happen to vanish at x_0, we can cancel any common factor $(x - x_0)^k$, and hence, without loss of generality, assume that at least one of the coefficient functions is nonzero at x_0.

Proofs of the basic existence theorem for differential equations at regular points can be found in [**18, 54, 59**].

Theorem 11.5. *Let x_0 be a regular point for the second-order homogeneous linear ordinary differential equation* (11.71)*. Then there exists a unique, analytic solution $u(x)$ to the initial value problem*

$$u(x_0) = a, \qquad u'(x_0) = b. \tag{11.73}$$

The radius of convergence of the power series for $u(x)$ is at least as large as the distance from the regular point x_0 to the nearest singular point of the differential equation in the complex plane.

Thus, every solution to an analytic differential equation at a regular point x_0 can be expanded in a convergent power series

$$u(x) = u_0 + u_1(x - x_0) + u_2(x - x_0)^2 + \cdots = \sum_{n=0}^{\infty} u_n(x - x_0)^n. \tag{11.74}$$

Since the power series necessarily coincides with the Taylor series for $u(x)$, its coefficients[†]

$$u_n = \frac{u^{(n)}(x_0)}{n!}$$

are multiples of the derivatives of the function at the point x_0. In particular, the first two coefficients,

$$u_0 = u(x_0) = a, \qquad u_1 = u'(x_0) = b, \tag{11.75}$$

are fixed by the initial conditions. The remaining coefficients will then be uniquely prescribed thanks to the uniqueness of solutions to initial value problems.

Near a regular point, the second-order differential equation (11.71) admits two linearly independent analytic solutions, which we denote by $\widehat{u}(x)$ and $\widetilde{u}(x)$. The general solution can be written as a linear combination of the two basis solutions:

$$u(x) = a\,\widehat{u}(x) + b\,\widetilde{u}(x). \tag{11.76}$$

A convenient choice is to have the first satisfy the initial conditions

$$\widehat{u}(x_0) = 1, \qquad \widehat{u}'(x_0) = 0, \tag{11.77}$$

and the second satisfy

$$\widetilde{u}(x_0) = 0, \qquad \widetilde{u}'(x_0) = 1, \tag{11.78}$$

although other conventions may be used depending on the circumstances. Given (11.77–78), the linear combination (11.76) automatically satisfies the initial conditions (11.73).

The basic computational strategy to construct the power series solution to the initial value problem is a straightforward adaptation of the method used in Example 11.3. One substitutes the known power series (11.72) for the coefficient functions and the unknown power series (11.74) for the solution into the differential equation (11.71). Multiplying out the formulas and collecting the common powers of $x - x_0$ will result in a (complicated) power series whose individual coefficients must be equated to zero. The lowest-order terms are multiples of $(x - x_0)^0 = 1$, i.e., the constant terms. They produce a linear relation

$$u_2 = R_2(u_0, u_1) = R_2(a, b)$$

[†] Some authors prefer to include the $n!$'s in the original power series; this is purely a matter of personal taste.

that prescribes the coefficient u_2 in terms of the initial data (11.75). The coefficient of $(x - x_0)$ leads to a relation

$$u_3 = R_3(u_0, u_1, u_2) = R_3(a, b, R_2(a, b))$$

that prescribes u_3 in terms of the initial data and the previously computed coefficient u_2. And so on. At the nth stage of the procedure, the coefficient of $(x - x_0)^n$ produces the linear *recurrence relation*

$$u_{n+2} = R_n(u_0, u_1, \ldots, u_{n+1}), \qquad n = 0, 1, 2, \ldots\,, \tag{11.79}$$

that will prescribe the $(n + 2)^{\text{nd}}$ order coefficient in terms of the previously computed coefficients. In this fashion, we will have constructed a formal power series solution to the differential equation at a regular point. The one remaining issue is whether the resulting power series actually converges. The full analysis can be found in [**54**, **59**], and will serve to complete the proof of the general Existence Theorem 11.5.

Rather than continue on in general, the best way to learn the method is to work through another, less trivial, example.

The Airy Equation

We will illustrate the procedure by constructing power series solutions to the *Airy equation*

$$\frac{d^2u}{dx^2} = x\,u. \tag{11.80}$$

This second-order linear ordinary differential equation, which arises in applications to optics, rainbows, and dispersive waves, has solutions that cannot be expressed in terms of elementary functions.

For the Airy equation (11.80), the leading coefficient is constant, and so every point is a regular point. For simplicity, we will look only for power series based at the origin $x_0 = 0$, and therefore of the form (11.68). Equating the two series

$$\begin{aligned} u''(x) &= 2\,u_2 + 6\,u_3\,x + 12\,u_4\,x^2 + 20\,u_5 x^3 + \;\cdots &&= \sum_{n=0}^{\infty} (n+1)(n+2)\,u_{n+2}\,x^n, \\ x\,u(x) &= u_0\,x + u_1\,x^2 + u_2\,x^3 + \;\cdots &&= \sum_{n=1}^{\infty} u_{n-1}\,x^n, \end{aligned}$$

leads to the following recurrence relations relating the coefficients:

$$\begin{array}{cr} 1 & 2\,u_2 = 0, \\ x & 6\,u_3 = u_0, \\ x^2 & 12\,u_4 = u_1, \\ x^3 & 20\,u_5 = u_2, \\ x^4 & 30\,u_6 = u_3, \\ \vdots & \vdots \\ x^n & (n+1)(n+2)\,u_{n+2} = u_{n-1}. \end{array}$$

As before, we solve them in order: The first equation determines u_2. The second prescribes $u_3 = \frac{1}{6} u_0$ in terms of u_0. Next, we find $u_4 = \frac{1}{12} u_1$ in terms of u_1, followed by $u_5 = \frac{1}{20} u_2 = 0$; then $u_6 = \frac{1}{30} u_3 = \frac{1}{180} u_0$ is first given in terms of u_3, but we already know the latter in terms of u_0. And so on.

Let us now construct two basis solutions. The first has the initial conditions

$$u_0 = \widehat{u}(0) = 1, \qquad u_1 = \widehat{u}'(0) = 0.$$

The recurrence relations imply that the only nonzero coefficients c_n occur when $n = 3k$ is a multiple of 3. Moreover,

$$u_{3k} = \frac{u_{3k-3}}{3k(3k-1)}.$$

A straightforward induction proves that

$$u_{3k} = \frac{1}{3k(3k-1)(3k-3)(3k-4)\cdots 6\cdot 5\cdot 3\cdot 2}.$$

The resulting solution is

$$\widehat{u}(x) = 1 + \tfrac{1}{6}x^3 + \tfrac{1}{180}x^6 + \cdots = 1 + \sum_{k=1}^{\infty} \frac{x^{3k}}{3k(3k-1)(3k-3)(3k-4)\cdots 6\cdot 5\cdot 3\cdot 2}. \tag{11.81}$$

Note that the denominator is similar to a factorial, except every third term is omitted. A straightforward application of the ratio test confirms that the series converges for all (complex) x, in conformity with the general Theorem 11.5, which guarantees an infinite radius of convergence because the Airy equation has no singular points.

Similarly, starting with the initial conditions

$$u_0 = \widetilde{u}(0) = 0, \qquad u_1 = \widetilde{u}'(0) = 1,$$

we find that the only nonzero coefficients u_n occur when $n = 3k+1$. The recurrence relation

$$u_{3k+1} = \frac{u_{3k-2}}{(3k+1)(3k)} \qquad \text{yields} \qquad u_{3k+1} = \frac{1}{(3k+1)(3k)(3k-2)(3k-3)\cdots 7\cdot 6\cdot 4\cdot 3}.$$

The resulting solution is

$$\widetilde{u}(x) = x + \tfrac{1}{12}x^4 + \tfrac{1}{504}x^7 + \cdots = x + \sum_{k=1}^{\infty} \frac{x^{3k+1}}{(3k+1)(3k)(3k-2)(3k-3)\cdots 7\cdot 6\cdot 4\cdot 3}. \tag{11.82}$$

Again, the denominator skips every third term in the product. Every solution to the Airy equation can be written as a linear combination of these two basis power series solutions:

$$u(x) = a\,\widehat{u}(x) + b\,\widetilde{u}(x), \qquad \text{where} \qquad a = u(0), \qquad b = u'(0).$$

Both power series (11.81, 82), converge quite rapidly, and so the first few terms will provide a reasonable approximation to the solutions for moderate values of x.

We have, in fact, already encountered another solution to the Airy equation. According to formula (8.97), the integral

$$\mathrm{Ai}(x) = \frac{1}{\pi}\int_0^{\infty} \cos\left(s\,x + \tfrac{1}{3}s^3\right) ds \tag{11.83}$$

defines the *Airy function of the first kind*. Let us prove that it satisfies the Airy differential equation (11.80):

$$\frac{d^2}{dx^2}\,\mathrm{Ai}(x) = x\,\mathrm{Ai}(x).$$

Before differentiating, we recall the integration by parts argument in (8.96) to re-express the Airy integral in absolutely convergent form:

$$\mathrm{Ai}(x) = \frac{2}{\pi}\int_0^\infty \frac{s\,\sin\!\left(s\,x + \frac{1}{3}\,s^3\right)}{(x+s^2)^2}\,ds.$$

We are now permitted to differentiate under the integral sign, producing (after some algebra)

$$\frac{d^2}{dx^2}\,\mathrm{Ai}(x) - x\,\mathrm{Ai}(x) = \frac{2}{\pi}\int_0^\infty \frac{d}{ds}\left[\frac{s\,(x+s^2)\cos\!\left(s\,x+\frac{1}{3}\,s^3\right) - \sin\!\left(s\,x+\frac{1}{3}\,s^3\right)}{(x+s^2)^3}\right] ds = 0.$$

Thus, the Airy function must be a certain linear combination of the two basic series solutions:

$$\mathrm{Ai}(x) = \mathrm{Ai}(0)\,\widehat{u}(x) + \mathrm{Ai}'(0)\,\widetilde{u}(x).$$

Its values at $x = 0$ are, in fact, given by

$$\begin{aligned}
\mathrm{Ai}(0) &= \frac{1}{\pi}\int_0^\infty \cos\!\left(\tfrac{1}{3}\,s^3\right) ds = \frac{\Gamma\!\left(\frac{1}{3}\right)}{2\,\pi\,3^{1/6}} = \frac{1}{3^{2/3}\,\Gamma\!\left(\frac{2}{3}\right)} \approx .355028\,, \\
\mathrm{Ai}'(0) &= -\,\frac{1}{\pi}\int_0^\infty s\,\sin\!\left(\tfrac{1}{3}\,s^3\right) ds = -\,\frac{3^{1/6}\,\Gamma\!\left(\frac{2}{3}\right)}{2\,\pi} = -\,\frac{1}{3^{1/3}\,\Gamma\!\left(\frac{1}{3}\right)} \approx -\,.258819.
\end{aligned} \tag{11.84}$$

The second and third expressions involve the gamma function (11.62); a proof, based on complex integration, can be found in [**85**; p. 54].

Exercises

11.3.1. Find (a) $\Gamma\left(\frac{5}{2}\right)$, (b) $\Gamma\left(\frac{7}{2}\right)$, (c) $\Gamma\left(-\frac{3}{2}\right)$, (d) $\Gamma\left(-\frac{5}{2}\right)$.

11.3.2. Prove that $\Gamma\left(n+\frac{1}{2}\right) = \dfrac{\sqrt{\pi}\,(2n)!}{2^{2n}\,n!}$ for every positive integer n.

11.3.3. Let $x \in \mathbb{C}$ be complex. (a) Prove that the gamma function integral (11.62) converges, provided $\mathrm{Re}\,x > 0$. (b) Is formula (11.63) valid when x is complex?

♢ 11.3.4. Prove that $\Gamma(x) = \displaystyle\int_0^1 (-\log s)^{x-1}\,ds$, and hence, for $0 \le n \in \mathbb{Z}$, we have $n! = \displaystyle\int_0^1 (-\log s)^n\,ds$. *Remark*: Euler first established the latter identity directly, and used it to define the gamma function.

11.3.5. Evaluate $\displaystyle\int_0^\infty \sqrt{x}\,e^{-x^3}\,dx$.

♢ 11.3.6. Can you construct a function $f(x)$ that satisfies the factorial functional equation (11.61) and has the values $f(x) = 1$ for $0 \le x \le 1$? If so, is $f(x) = \Gamma(x+1)$?

11.3.7. Explain how to construct the power series for $\sin x$ by solving the differential equation (11.67).

11.3.8. Construct two independent power series solutions to the Euler equation $x^2 u'' - 2u = 0$ based at the point $x_0 = 1$.

11.3.9. Construct two independent power series solutions to the equation $u'' + x^2 u = 0$ based at the point $x_0 = 0$.

11.3.10. Consider the ordinary differential equation $u'' + 2x u' + 2u = 0$. (a) Find two linearly independent power series solutions in powers of x. (*b*) What is the radius of convergence of your power series? (*c*) By inspection of your series, find one solution to the equation expressible in terms of elementary functions. (*d*) Find an explicit (non-series) formula for the second independent power series solution.

11.3.11. Answer Exercise 11.3.10 for the equation $u'' + \frac{1}{2} x u' - \frac{1}{2} u = 0$, which is a special case of equation (8.63).

11.3.12. Consider the ordinary differential equation $u'' + x u' + 2u = 0$. (a) Find two linearly independent power series solutions based at $x_0 = 0$. (b) Write down the power series for the solution to the initial value problem $u(0) = 1,\ u'(0) = -1$. (c) What is the radius of convergence of your power series solution in part (a)? Can you justify this by direct inspection of your power series?

♢ 11.3.13. The *Hermite equation* of order n is

$$\frac{d^2 u}{dx^2} - 2x \frac{du}{dx} + 2n u = 0. \tag{11.85}$$

Assuming $n \in \mathbb{N}$ is a nonnegative integer: (a) Find two linearly independent power series solutions based at $x_0 = 0$, and then show that one of your solutions is a polynomial of degree n. (*b*) Prove that the Hermite polynomial $H_n(x)$ defined in (8.64) solves the Hermite equation (11.85) and hence is a multiple of the polynomial solution you found in part (a). What is the multiple? (*c*) Prove that the Hermite polynomials are orthogonal with respect to the inner product $\langle u , v \rangle = \int_{-\infty}^{\infty} u(x)\, v(x)\, e^{-x^2}\, dx$.

11.3.14. Use the ratio test to directly determine the radius of convergence of the series solutions (11.81, 82) to the Airy equation.

11.3.15. Write down the general solution to the following ordinary differential equations:
(a) $u'' + (x - c)\, u = 0$, where c is a fixed constant;
(*b*) $u'' = \lambda x u$, where $\lambda \neq 0$ is a fixed nonzero constant.

♢ 11.3.16. The *Airy function of the second kind* is defined by

$$\operatorname{Bi}(x) = \frac{1}{\pi} \int_0^\infty \left[\exp\left(s x - \tfrac{1}{3} s^3 \right) + \sin\left(s x + \tfrac{1}{3} s^3 \right) \right] ds. \tag{11.86}$$

(a) Prove that $\operatorname{Bi}(x)$ is well defined and a solution to the Airy equation. (*b*) Given that[†]

$$\operatorname{Bi}(0) = \frac{1}{3^{1/6}\, \Gamma\left(\frac{2}{3}\right)}, \qquad \operatorname{Bi}'(0) = \frac{3^{1/6}}{\Gamma\left(\frac{1}{3}\right)}, \tag{11.87}$$

explain why every solution to the Airy equation can be written as a linear combination of $\operatorname{Ai}(x)$ and $\operatorname{Bi}(x)$. (*c*) Write the two series solutions (11.81, 82) in terms of $\operatorname{Ai}(x)$ and $\operatorname{Bi}(x)$.

11.3.17. Use the Fourier transform to construct an L^2 solution to the Airy equation. Can you identify your solution?

♢ 11.3.18. Apply separation of variables to the Tricomi equation (4.137), and write down all separable solutions. *Hint*: See Exercise 11.3.15(*b*) and Exercise 11.3.16.

† See [**85**; p. 54] for a proof.

♡ 11.3.19. (a) Show that $u(x) = \sum_{n=1}^{\infty} (n-1)!\, x^n$ is a power series solution to the first-order linear ordinary differential equation $x^2 u' - u + x = 0$. (b) For which x does the series converge? (c) Find an analytic formula for the general solution to the equation. (d) Find a second-order homogeneous linear ordinary differential equation that has this power series as a (formal) solution. *Remark*: The lesson of this exercise is that not all power series solutions to ordinary differential equations converge. Theorem 11.5 guarantees convergence at a regular point, but in this example the power series is based at the singular point $x_0 = 0$.

11.3.20. *True or false*: The only function $f(x)$ that has identically zero Taylor series is the zero function.

♢ 11.3.21. Define $f(x) = \begin{cases} e^{-1/x^2}, & x \neq 0, \\ 0, & x = 0. \end{cases}$ (a) Prove that f is a C^∞ function for all $x \in \mathbb{R}$. (b) Prove that $f(x)$ is not analytic by showing that its Taylor series at $x_0 = 0$ does not converge to $f(x)$ when $x \neq 0$.

Regular Singular Points

As we have just seen, constructing power series solutions at regular points is a reasonably straightforward computational exercise: one writes down a power series with arbitrary coefficients, substitutes into the differential equation along with a pair of initial conditions, and recursively solves for the coefficients. Finding a general formula for the coefficients might be challenging, but producing their successive numerical values, degree by degree, is a mechanical exercise.

However, at a singular point, the solutions cannot be typically written as an ordinary power series, and one needs to be cleverer. Of course, you may object — why not just solve the equation away from the singular point and be done with it. But there are multiple reasons not to do this. First, one may be unable to discover a general formula for the power series coefficients at regular points. Second, the most informative and interesting behavior of solutions is typically found at the singular points, and so series solutions based at singular points are particularly enlightening. And finally, one of the boundary conditions required for us to complete our construction of separable solutions to partial differential equations often occurs at a singular point.

Singular points appear in two guises. The easier to handle, and, fortunately, the ones that arise in almost all applications, are known as "regular singular points". Irregular singular points are nastier, and we will not make any attempt to understand them in this text; the curious reader is referred to [**54, 59**].

Definition 11.6. A second-order linear homogeneous ordinary differential equation that can be written the form

$$(x - x_0)^2\, a(x)\, \frac{d^2u}{dx^2} + (x - x_0)\, b(x)\, \frac{du}{dx} + c(x)\, u = 0, \tag{11.88}$$

where $a(x), b(x)$, and $c(x)$ are analytic at $x = x_0$ and, moreover, $a(x_0) \neq 0$, is said to have a *regular singular point* at x_0.

The simplest example of a second-order equation with a regular singular point at $x_0 = 0$ is the *Euler equation*

$$a\,x^2 u'' + b\,x\,u' + c\,u = 0, \tag{11.89}$$

with a, b, c all constant and $a \neq 0$. Note that all other points are regular points. Euler equations can be readily solved by substituting the power ansatz $u(x) = x^r$. We find

$$a x^2 u'' + b x u' + c u = a r(r-1) x^r + b r x^r + c x^r = 0,$$

provided the exponent r satisfies the *indicial equation*

$$a r(r-1) + b r + c = 0.$$

If this quadratic equation has two distinct roots $r_1 \neq r_2$, we obtain two linearly independent (possibly complex) solutions $\widehat{u}(x) = x^{r_1}$ and $\widetilde{u}(x) = x^{r_2}$. The general solution $u(x) = c_1 x^{r_1} + c_2 x^{r_2}$ is a linear combination thereof. Note that unless r_1 or r_2 is a nonnegative integer, all nonzero solutions have a singularity at the singular point $x = 0$. A repeated root, $r_1 = r_2$, has only one power solution, $\widehat{u}(x) = x^{r_1}$, and requires an additional logarithmic term, $\widetilde{u}(x) = x^{r_1} \log x$, for the second independent solution. In this case, the general solution has the form $u(x) = c_1 x^{r_1} + c_2 x^{r_1} \log x$.

The series solution method at more general regular singular points is modeled on the simple example of the Euler equation. One now seeks a solution that has a series expansion of the form

$$u(x) = (x - x_0)^r \sum_{n=0}^{\infty} u_n (x - x_0)^n = u_0 (x - x_0)^r + u_1 (x - x_0)^{r+1} + u_2 (x - x_0)^{r+2} + \cdots. \tag{11.90}$$

The exponent r is known as the *index*. If $r = 0$, or, more generally, if r is a positive integer, then (11.90) is an ordinary power series, but we allow the possibility of a non-integral, or even complex, index r. We can assume, without any loss of generality, that the leading coefficient $u_0 \neq 0$. Indeed, if $u_k \neq 0$ is the first nonzero coefficient, then the series begins with the term $u_k (x - x_0)^{r+k}$, and we merely replace r by $r + k$ to write it in the form (11.90). Since any scalar multiple of a solution is a solution, we can further assume that $u_0 = 1$, in which case we call (11.90) a *normalized Frobenius series* in honor of the German mathematician Georg Frobenius, who systematically established the calculus of series solutions at regular singular points in the late 1800s. The index r, and the higher-order coefficients $u_1, u_2, \ldots$, are then found by substituting the normalized Frobenius series into the differential equation (11.88) and equating the coefficients of the powers of $x - x_0$ to zero.

Warning: Unlike those in ordinary power series expansions, the coefficients $u_0 = 1$ and u_1 are *not* prescribed by the initial conditions at the point x_0.

Since

$$\begin{aligned} u(x) &= (x - x_0)^r + u_1 (x - x_0)^{r+1} + \cdots, \\ (x - x_0)\, u'(x) &= r\,(x - x_0)^r + (r+1) u_1 (x - x_0)^{r+1} + \cdots, \\ (x - x_0)^2\, u''(x) &= r\,(r-1)\,(x - x_0)^r + (r+1)\, r\, u_1 (x - x_0)^{r+1} + \cdots, \end{aligned}$$

the terms of lowest order in the equation are multiples of $(x - x_0)^r$. Equating their coefficients to zero produces a quadratic equation of the form

$$s_0\, r\,(r-1) + t_0\, r + r_0 = 0, \tag{11.91}$$

where

$$s_0 = s(x_0) = \tfrac{1}{2} p''(x_0), \qquad t_0 = t(x_0) = q'(x_0), \qquad r_0 = r(x_0),$$

are the leading coefficients in the power series expansions of the individual coefficient functions. The quadratic equation (11.91) is known as the *indicial equation*, since it determines the possible indices r in the Frobenius expansion (11.90) of a solution.

As with the Euler equation, the quadratic indicial equation usually has two roots, say r_1 and r_2, which provide two allowable indices, and one thus expects to find two independent Frobenius expansions. Usually, this expectation is realized, but there is an important exception. The general result is summarized in the following list:

(*i*) If $r_2 - r_1$ is not an integer, then there are two linearly independent solutions $\widehat{u}(x)$ and $\widetilde{u}(x)$, each having convergent normalized Frobenius expansions of the form (11.90).

(*ii*) If $r_1 = r_2$, then there is only one solution $\widehat{u}(x)$ with a normalized Frobenius expansion (11.90). One can construct a second independent solution of the form

$$\widetilde{u}(x) = \log(x - x_0)\, \widehat{u}(x) + v(x), \qquad \text{where} \qquad v(x) = \sum_{n=1}^{\infty} v_n (x - x_0)^{n + r_1} \tag{11.92}$$

is a convergent Frobenius series.

(*iii*) Finally, if $r_2 = r_1 + k$, where $k > 0$ is a positive integer, then there is a nonzero solution $\widehat{u}(x)$ with a convergent Frobenius expansion corresponding to the smaller index r_1. One can construct a second independent solution of the form

$$\widetilde{u}(x) = c \log(x - x_0)\, \widehat{u}(x) + v(x), \qquad \text{where} \qquad v(x) = x^{r_2} + \sum_{n=1}^{\infty} v_n (x - x_0)^{n + r_2} \tag{11.93}$$

is a convergent Frobenius series, and c is a constant, which may be 0, in which case the second solution $\widetilde{u}(x)$ is also of Frobenius form.

Thus, in every case, the differential equation has at least one nonzero solution with a convergent Frobenius expansion. If the second independent solution does not have a Frobenius expansion, then it requires an additional logarithmic term of a well-prescribed form. Rather than try to develop the general theory in any more detail here, we will content ourselves to work through a couple of particular examples.

Example 11.7. Consider the second-order ordinary differential equation

$$\frac{d^2 u}{dx^2} + \left(\frac{1}{x} + \frac{x}{2} \right) \frac{du}{dx} + u = 0. \tag{11.94}$$

We look for series solutions based at $x = 0$. Note that, upon multiplying by x^2, the equation takes the form

$$x^2 u'' + x\big(1 + \tfrac{1}{2} x^2\big) u' + x^2 u = 0,$$

and hence $x_0 = 0$ is a regular singular point, with $a(x) = 1$, $b(x) = 1 + \frac{1}{2} x^2$, $c(x) = x^2$. We thus look for a solution that can be represented by a Frobenius expansion:

$$\begin{aligned} u(x) &= x^r + u_1 x^{r+1} + \cdots + u_n x^{n+r} + \cdots, \\ x\, u'(x) &= r x^r + (r+1) u_1 x^{r+1} + \cdots + (n+r) u_n x^{n+r} + \cdots, \\ \tfrac{1}{2} x^3 u'(x) &= \tfrac{1}{2} r x^{r+2} + \tfrac{1}{2}(r+1) u_1 x^{r+3} + \cdots + \tfrac{1}{2}(n + r - 2) u_{n-2} x^{n+r} + \cdots, \\ x^2 u''(x) &= r(r-1) x^r + (r+1) r u_1 x^{r+1} + \cdots + (n+r)(n+r-1) u_n x^{n+r} + \cdots. \end{aligned} \tag{11.95}$$

Substituting into the differential equation, we find that the coefficient of x^r leads to the indicial equation

$$r^2 = 0.$$

There is only one root, $r = 0$, and hence, even though we are at a singular point, the Frobenius expansion reduces to an ordinary power series. The coefficient of $x^{r+1} = x$ tells us that $u_1 = 0$. The general recurrence relation, for $n \geq 2$, is

$$n^2 u_n + \tfrac{1}{2} n u_{n-2} = 0,$$

and hence

$$u_n = -\frac{u_{n-2}}{2n}.$$

Therefore, the odd coefficients $u_{2k+1} = 0$ are all zero, while the even ones are

$$u_{2k} = -\frac{u_{2k-2}}{4k} = \frac{u_{2k-4}}{4k(4k-4)} = -\frac{u_{2k-6}}{4k(4k-4)(4k-8)} = \cdots = \frac{(-1)^k}{4^k\, k!}, \qquad \text{since} \quad u_0 = 1.$$

The resulting power series assumes a recognizable form:

$$\widehat{u}(x) = \sum_{k=1}^{\infty} u_{2k}\, x^{2k} = \sum_{k=1}^{\infty} \frac{1}{k!} \left(-\frac{x^2}{4} \right)^k = e^{-x^2/4},$$

which is an explicit elementary solution to the ordinary differential equation (11.94).

Since there is only one root to the indicial equation, the second solution $\widetilde{u}(x)$ will require a logarithmic term. It can be constructed by a second application of the Frobenius method using the more complicated form (11.92). Alternatively, since the first solution is known, we can use a well-known reduction trick, [**23**]. Given one solution $\widehat{u}(x)$ to a second-order linear ordinary differential equation, the general solution can be found by substituting the ansatz

$$u(x) = v(x)\, \widehat{u}(x) = v(x)\, e^{-x^2/4} \tag{11.96}$$

into the equation. In this case,

$$u'' + \left(\frac{1}{x} + \frac{x}{2} \right) u' + u = v \left[\widehat{u}'' + \left(\frac{1}{x} + \frac{x}{2} \right) \widehat{u}' + \widehat{u} \right] + v' \left[2\,\widehat{u}' + \left(\frac{1}{x} + \frac{x}{2} \right) \widehat{u} \right] + v''\, \widehat{u}$$
$$= e^{-x^2/4} \left(v'' + \frac{v'}{x} \right).$$

If u is to be a solution, v' must satisfy a linear first-order ordinary differential equation:

$$v'' + \frac{v'}{x} = 0, \qquad \text{and hence} \qquad v' = \frac{c}{x}, \qquad v = c \log x + d,$$

where c, d are arbitrary constants. We conclude that the general solution to the original differential equation is

$$\widetilde{u}(x) = v(x)\, \widehat{u}(x) = (c \log x + d)\, e^{-x^2/4}. \tag{11.97}$$

Bessel's Equation

Perhaps the most important "non-elementary" ordinary differential equation is

$$x^2 u'' + x u' + (x^2 - m^2)\, u = 0, \tag{11.98}$$

known as *Bessel's equation* of *order* m. We assume here that the order $m \geq 0$ is a nonnegative real number. (Exercise 11.3.30 investigates Bessel equations of imaginary order.) The Bessel equation arises from separation of variables in a variety of partial differential equations, including the Laplace, heat, and wave equations on a disk, a cylinder, and a spherical ball.

The Bessel equation cannot (except in a few particular instances) be solved in terms of elementary functions, and so the use of power series is essential. The leading coefficient, $p(x) = x^2$, is nonzero *except* when $x = 0$, and so all points except the origin are regular. Therefore, at any $x_0 \neq 0$, the standard power series construction can be used to produce the solutions of the Bessel equation. However, the recurrence relations for the coefficients are not particularly easy to solve in closed form. Moreover, applications tend to demand understanding the behavior of solutions at the singular point $x_0 = 0$.

Comparison with (11.88) immediately shows that $x_0 = 0$ is a regular singular point, and so we seek solutions in Frobenius form. We substitute the first, second, and fourth expressions in (11.95) into the Bessel equation and then equate the coefficients of the various powers of x to zero. The lowest power, x^r, provides the indicial equation

$$r(r-1) + r - m^2 = r^2 - m^2 = 0.$$

It has two solutions, $r = \pm m$, except when $m = 0$, for which $r = 0$ is the only index.

The higher powers of x lead to recurrence relations for the coefficients u_n in the Frobenius series. Replacing m^2 by r^2 produces

$$\begin{aligned}
x^{r+1}: &\quad \left[\,(r+1)^2 - r^2\,\right]u_1 = (2r+1)u_1 = 0, && u_1 = 0,\\
x^{r+2}: &\quad \left[\,(r+2)^2 - r^2\,\right]u_2 + 1 = (4r+4)u_2 + 1 = 0, && u_2 = -\frac{1}{4r+4},\\
x^{r+3}: &\quad \left[\,(r+3)^2 - r^2\,\right]u_3 + u_1 = (6r+9)u_3 + u_1 = 0, && u_3 = -\frac{u_1}{6r+9} = 0,
\end{aligned}$$

and, in general,

$$x^{r+n}: \quad \left[\,(r+n)^2 - r^2\,\right]u_n + u_{n-2} = n(2r+n)u_n + u_{n-2} = 0.$$

Thus, the general recurrence relation is

$$u_n = -\frac{1}{n(2r+n)}\,u_{n-2}, \qquad n = 2, 3, 4, \ldots . \tag{11.99}$$

Starting with $u_0 = 1$, $u_1 = 0$, it is easy to deduce that all $u_n = 0$ for all odd $n = 2k+1$, while for even $n = 2k$,

$$\begin{aligned}
u_{2k} &= -\frac{u_{2k-2}}{4k(k+r)} = \frac{u_{2k-4}}{16k(k-1)(r+k)(r+k-1)} = \cdots\\
&= \frac{(-1)^k}{2^{2k}\,k(k-1)\cdots 3\cdot 2\,(r+k)(r+k-1)\cdots(r+2)(r+1)}.
\end{aligned}$$

We have thus found the series solution

$$\widehat{u}(x) = \sum_{k=0}^{\infty} u_{2k}\,x^{r+2k} = \sum_{k=0}^{\infty} \frac{(-1)^k\,x^{r+2k}}{2^{2k}\,k!\,(r+k)(r+k-1)\cdots(r+2)(r+1)}. \tag{11.100}$$

So far, we have not paid attention to the precise values of the indices $r = \pm m$. In order to continue the recurrence, we need to ensure that the denominators in (11.99) are never 0. Since $n > 0$, a vanishing denominator will appear whenever $2r + n = 0$, and so $r = -\frac{1}{2}n$ is either a negative integer $-1, -2, -3, \ldots$ or half-integer $-\frac{1}{2}, -\frac{3}{2}, -\frac{5}{2}, \ldots$. This will occur when the order $m = -r = \frac{1}{2}n$ is either an integer or a half-integer. Indeed, these are precisely the situations in which the two indices, namely $r_1 = -m$ and $r_2 = m$, differ by an integer, $r_2 - r_1 = n$, and so we are in the tricky case (*iii*) of the Frobenius method.

There is, in fact, a major difference between the integral and the half-integral cases. Recall that the odd coefficients $u_{2k+1} = 0$ in the Frobenius series automatically vanish, and so we only have to worry about the recurrence relation (11.99) for *even* values of n. When $n = 2k$, the factor $2r + n = 2(r + k) = 0$ vanishes only when $r = -k$ is a negative integer; the half-integral values do not, in fact cause problems. Therefore, if the order $m \geq 0$ is *not* an integer, then the Bessel equation of order m admits two linearly independent Frobenius solutions, given by the expansions (11.100) with exponents $r = +m$ and $r = -m$. On the other hand, if m is an integer, there is only one Frobenius solution, namely the expansion (11.100) for the positive index $r = +m$. The Frobenius recurrence with index $r = -m$ breaks down, and the second independent solution must include a logarithmic term; details appear below.

By convention, the standard *Bessel function* of order m is obtained by multiplying the Frobenius solution (11.100) with $r = m$ by

$$\frac{1}{2^m\, m!}, \qquad \text{or, more generally,} \qquad \frac{1}{2^m\, \Gamma(m+1)}, \tag{11.101}$$

where the first factorial form can be used if m is a nonnegative integer, while the more general gamma function expression must be employed for non-integral values of m. The result is

$$\begin{aligned} J_m(x) &= \sum_{k=0}^{\infty} \frac{(-1)^k\, x^{m+2k}}{2^{2k+m}\, k!\, (m+k)!} \\ &= \frac{1}{2^m\, m!}\left[x^m - \frac{x^{m+2}}{4\,(m+1)} + \frac{x^{m+4}}{32\,(m+1)(m+2)} - \frac{x^{m+6}}{384\,(m+1)(m+2)(m+3)} + \cdots \right]. \end{aligned} \tag{11.102}$$

When m is non-integral, the $(m+k)!$ should be replaced by $\Gamma(m+k+1)$, and $m!$ by $\Gamma(m+1)$. With this convention, the series is well defined for all real m except when $m = -1, -2, -3, \ldots$ is a negative integer. Actually, if m is a negative integer, the first m terms in the series vanish, because, at negative integer values, $\Gamma(-n) = \infty$. With this convention, one can prove that

$$J_{-m}(x) = (-1)^m J_m(x), \qquad m = 1, 2, 3, \ldots. \tag{11.103}$$

A simple application of the ratio test tells us that the power series converges for all (complex) values of x, and hence $J_m(x)$ is everywhere analytic. Indeed, the convergence is quite rapid when x is of moderate size, and so summing the series is a reasonably effective method for computing the Bessel function $J_m(x)$ — although in serious applications one adopts more sophisticated numerical techniques based on asymptotic expansions and integral formulas, [**85**, **86**]. In particular, we note that

$$J_0(0) = 1, \qquad J_m(0) = 0, \qquad m > 0. \tag{11.104}$$

Figure 11.4 displays graphs of the first four Bessel functions for $0 \leq x \leq 20$; the vertical axes range from $-.5$ to 1.0. Most software packages, both symbolic and numeric, include

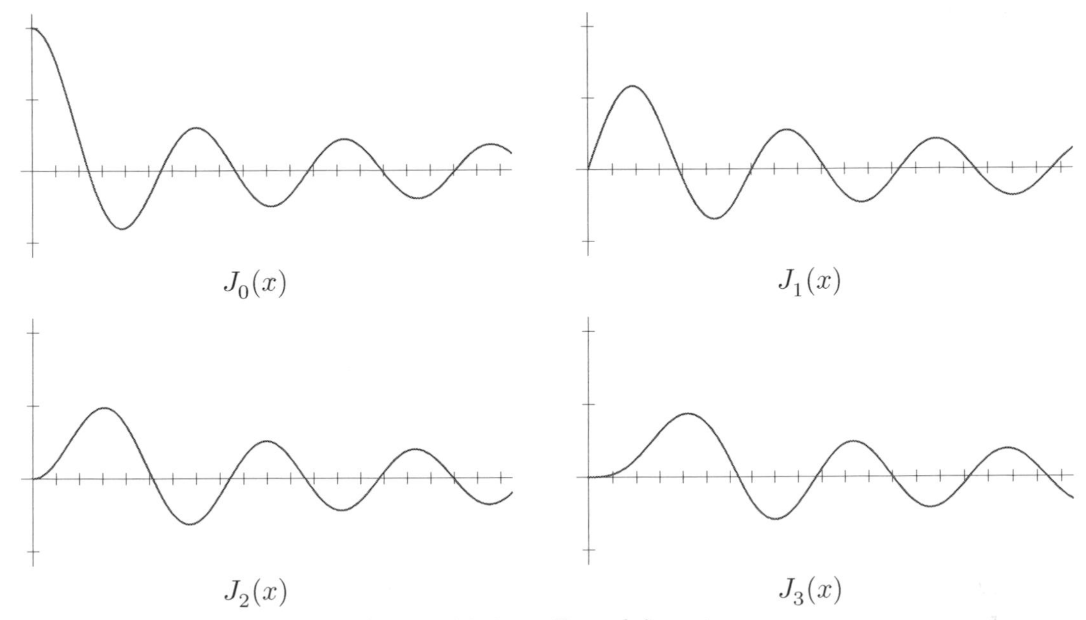

Figure 11.4. Bessel functions.

routines for accurately evaluating and graphing Bessel functions, and their properties can be regarded as well known.

Example 11.8. Consider the Bessel equation of order $m = \frac{1}{2}$. There are two indices, $r = \pm\frac{1}{2}$, and the Frobenius method yields two independent solutions: $J_{1/2}(x)$ and $J_{-1/2}(x)$. For the first, with $r = \frac{1}{2}$, the recurrence relation (11.99) takes the form

$$u_n = -\frac{u_{n-2}}{(n+1)\,n}.$$

Starting with $u_0 = 1$ and $u_1 = 0$, the general formula is easily found to be

$$u_n = \begin{cases} \dfrac{(-1)^k}{(n+1)!}, & n = 2k \text{ even}, \\ 0 & n = 2k+1 \text{ odd}. \end{cases}$$

Therefore, the resulting solution is

$$\widehat{u}(x) = \sqrt{x}\,\sum_{k=0}^{\infty} \frac{(-1)^k}{(2k+1)!}\,x^{2k} = \frac{1}{\sqrt{x}}\,\sum_{k=0}^{\infty} \frac{(-1)^k}{(2k+1)!}\,x^{2k+1} = \frac{\sin x}{\sqrt{x}}.$$

According to (11.101), the Bessel function of order $\frac{1}{2}$ is obtained by dividing this function by

$$\sqrt{2}\,\Gamma\left(\tfrac{3}{2}\right) = \sqrt{\frac{\pi}{2}},$$

where we used (11.66) to evaluate the gamma function at $\frac{3}{2}$. Therefore,

$$J_{1/2}(x) = \sqrt{\frac{2}{\pi x}}\,\sin x. \tag{11.105}$$

Similarly, for the other index $r = -\frac{1}{2}$, the recurrence relation

$$u_n = -\frac{u_{n-2}}{n(n-1)}$$

leads to the formula

$$u_n = \begin{cases} \dfrac{(-1)^k}{n!}, & n = 2k \text{ even}, \\ 0 & n = 2k+1 \text{ odd}, \end{cases}$$

for its coefficients, corresponding to the solution

$$\widetilde{u}(x) = x^{-1/2} \sum_{k=0}^{\infty} \frac{(-1)^k}{(2k)!} x^{2k} = \frac{\cos x}{\sqrt{x}}.$$

Therefore, in view of (11.101) and (11.65), the Bessel function of order $-\frac{1}{2}$ is

$$J_{-1/2}(x) = \frac{\sqrt{2}}{\Gamma\left(\frac{1}{2}\right)} \frac{\cos x}{\sqrt{x}} = \sqrt{\frac{2}{\pi x}} \cos x. \tag{11.106}$$

As we noted above, if m is not an integer, the two independent solutions to the Bessel equation of order m are $J_m(x)$ and $J_{-m}(x)$. However, when m is an integer, (11.103) implies that these two solutions are constant multiples of each other, and so one must look elsewhere for a second independent solution. One method is to use a generalized Frobenius expansion involving a logarithmic term, i.e., (11.92) when $m = 0$ (see Exercise 11.3.33) or (11.93) when $m > 0$. A second approach is to employ the reduction procedure used in Example 11.7. Yet another option relies on the following limiting procedure; see [**85**, **119**] for full details.

Theorem 11.9. *If $m > 0$ is not an integer, then the Bessel functions $J_m(x)$ and $J_{-m}(x)$ provide two linearly independent solutions to the Bessel equation of order m. On the other hand, if $m = 0, 1, 2, 3, \ldots$ is an integer, then a second independent solution, traditionally denoted by $Y_m(x)$ and called the* Bessel function of the second kind *of order m, can be found as a limiting case*

$$Y_m(x) = \lim_{\nu \to m} \frac{J_\nu(x) \cos \nu\pi - J_{-\nu}(x)}{\sin \nu\pi} \tag{11.107}$$

of a certain linear combination of Bessel functions of non-integral order ν.

With some further analysis, it can be shown that the Bessel function of the second kind of order m has the logarithmic Frobenius expansion

$$Y_m(x) = \frac{2}{\pi}\left(\gamma + \log \frac{x}{2}\right) J_m(x) + \sum_{k=0}^{\infty} b_k x^{2k-m}, \qquad m = 0, 1, 2, \ldots, \tag{11.108}$$

with coefficients

$$b_k = \begin{cases} -\dfrac{(m-k-1)!}{\pi\, 2^{2k-m}\, k!}, & 0 \le k \le m-1, \\ \dfrac{(-1)^{k-m-1}(h_{k-m} + h_k)}{\pi\, 2^{2k-m}\, k!\,(k-m)!}, & k \ge m, \end{cases}$$

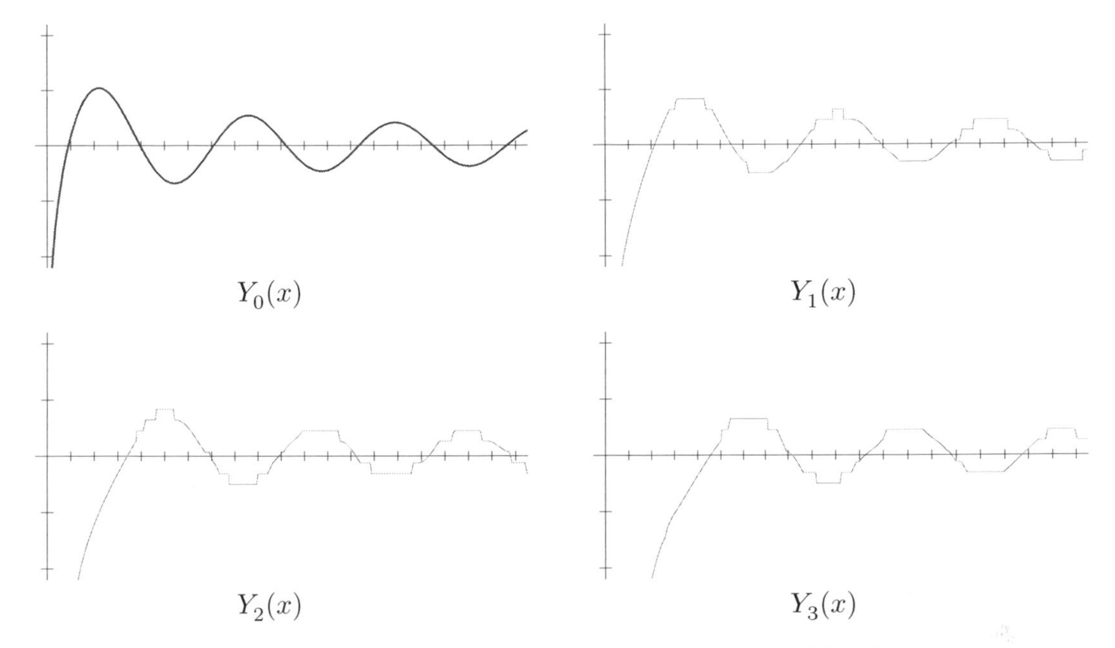

Figure 11.5. Bessel functions of the second kind.

where

$$h_0 = 0, \qquad h_k = 1 + \frac{1}{2} + \frac{1}{3} + \cdots + \frac{1}{k}, \qquad k > 0,$$

while

$$\gamma = \lim_{k \to \infty} \big(h_k - \log k \big) \approx .5772156649\ldots \tag{11.109}$$

is known as the *Euler* or *Euler–Mascheroni constant*. All Bessel functions of the second kind have a singularity at the origin $x = 0$; indeed, by inspection of (11.108), we find that the leading asymptotics as $x \to 0$ are

$$Y_0(x) \sim \frac{2}{\pi} \log x, \qquad Y_m(x) \sim -\frac{2^m (m-1)!}{\pi x^m}, \qquad m > 0. \tag{11.110}$$

Figure 11.5 contains graphs of the first four Bessel function of the second kind on the interval $0 < x \leq 20$; the vertical axis ranges from -1 to 1.

Finally, we show how Bessel functions of different orders are interconnected by two important recurrence relations.

Proposition 11.10. *The Bessel functions are related by the following formulae:*

$$\frac{dJ_m}{dx} + \frac{m}{x} J_m(x) = J_{m-1}(x), \qquad -\frac{dJ_m}{dx} + \frac{m}{x} J_m(x) = J_{m+1}(x). \tag{11.111}$$

Proof: Differentiating the power series

$$x^m J_m(x) = \sum_{k=0}^{\infty} \frac{(-1)^k x^{2m+2k}}{2^{2k+m}\, k!\, (m+k)!}$$

produces

$$\frac{d}{dx}\left[x^m J_m(x)\right] = \sum_{k=0}^{\infty} \frac{(-1)^k\, 2\,(m+k) x^{2m+2k-1}}{2^{2k+m}\, k!\,(m+k)!}$$
$$= x^m \sum_{k=0}^{\infty} \frac{(-1)^k x^{m-1+2k}}{2^{2k+m-1}\, k!\,(m-1+k)!} = x^m J_{m-1}(x). \tag{11.112}$$

Expansion of the left-hand side of this formula leads to

$$x^m \frac{dJ_m}{dx} + m\, x^{m-1} J_m(x) = \frac{d}{dx}\left[x^m J_m(x)\right] = x^m J_{m-1}(x),$$

which establishes the first recurrence formula (11.111). The second is proved by a similar manipulation involving differentiation of $x^{-m} J_m(x)$. *Q.E.D.*

For example, using the second recurrence formula (11.111) along with (11.105), we can write the Bessel function of order $\frac{3}{2}$ in elementary terms:

$$J_{3/2}(x) = -\frac{dJ_{1/2}(x)}{dx} + \frac{1}{2x} J_{1/2}(x)$$
$$= -\sqrt{\frac{2}{\pi}}\left(\frac{\cos x}{x^{1/2}} - \frac{\sin x}{2\,x^{3/2}}\right) + \sqrt{\frac{2}{\pi}}\,\frac{\sin x}{2\,x^{3/2}} = \sqrt{\frac{2}{\pi}}\;\frac{\sin x - x\cos x}{x^{3/2}}. \tag{11.113}$$

Iterating, one concludes that Bessel functions of half-integral order, $m = \pm\frac{1}{2}, \pm\frac{3}{2}, \pm\frac{5}{2}, \ldots,$ are all elementary functions, in that they can be written in terms of trigonometric functions and powers of $\sqrt{x}$. We will make use of these functions in our treatment of the three-dimensional heat and wave equations in spherical geometry. On the other hand, all of the other Bessel functions are non-elementary special functions.

With this, we conclude our brief introduction to the method of Frobenius and the basics of Bessel functions. The reader interested in delving further into either the general method or the host of additional properties of Bessel functions is encouraged to consult a more specialized text, e.g., [**59, 85, 119**].

Exercises

11.3.22. Consider the ordinary differential equation $2x\,u'' + u' + x\,u = 0$. (a) Prove that $x = 0$ is a regular singular point. (*b*) Find two independent series solutions in powers of x.

♡ 11.3.23. Consider the differential equation $\dfrac{u''}{2-x} = \dfrac{u}{x^2}$. (a) Classify all $x_0 \in \mathbb{R}$ as either a (*i*) regular point; (*ii*) regular singular point; and/or (*iii*) irregular singular point. Explain your answers. (*b*) Find a series solution to the equation based at the point $x_0 = 0$, or explain why none exists. What is the radius of convergence of your series?

11.3.24. Consider the differential equation $u'' + \left(1 - \dfrac{1}{x}\right) u' + u = 0$.

(a) Classify all $x_0 \in \mathbb{R}$ as either (*i*) a regular point; (*ii*) a regular singular point; (*iii*) an irregular singular point; (*iv*) none of the above. Explain your answers.

(*b*) Write out the first five nonzero terms in a series solution.

11.3.25. Consider the differential equation $4x u'' + 2u' + u = 0$. (a) Classify the values of x for which the equation has regular points, regular singular points, and irregular singular points. (b) Find two independent series solutions, in powers of x. For what values of x do your series converge? (c) By inspection of your series, write the general solution to the equation in terms of elementary functions.

♡ 11.3.26. The *Chebyshev differential equation* is $(1-x^2)u'' - x u' + m^2 u = 0$. (a) Find all (*i*) regular points; (*ii*) regular singular points; (*iii*) irregular singular points. (b) Show that if m is an integer, the equation has a polynomial solution of degree m, known as a *Chebyshev polynomial*. Write down the Chebyshev polynomials of degrees 1, 2, and 3. (c) For $m = 1$, find two linearly independent series solutions based at the point $x_0 = 1$.

11.3.27. Write the following Bessel functions in terms of elementary functions:

$$\text{(a)}\ J_{5/2}(x), \qquad \text{(b)}\ J_{7/2}(x), \qquad \text{(c)}\ J_{-3/2}(x).$$

◇ 11.3.28. Prove the identity (11.103).

11.3.29. Suppose that $u(x)$ solves Bessel's equation. (a) Find a second order ordinary differential equation satisfied by the function $w(x) = \sqrt{x}\, u(x)$. (b) Use this result to rederive the formulas for $J_{1/2}(x)$ and $J_{-1/2}(x)$.

◇ 11.3.30. Let $m \geq 0$ be real, and consider the *modified Bessel equation* of order m:

$$x^2 u'' + x u' - (x^2 + m^2)\, u = 0. \tag{11.114}$$

(a) Explain why $x_0 = 0$ is a regular singular point.
(b) Use the method of Frobenius to construct a series solution based at $x_0 = 0$. Can you relate your solutions to the Bessel function $J_m(x)$?

◇ 11.3.31. (a) Let a, b, c be constants with $b, c \neq 0$. Show that the function $u(x) = x^a J_0(b x^c)$ solves the ordinary differential equation

$$x^2 \frac{d^2u}{dx^2} + (1-2a)\, x \frac{du}{dx} + (b^2 c^2 x^{2c} + a^2)\, u = 0.$$

What is the general solution to this equation?
(b) Find the general solution to the ordinary differential equation

$$x^2 \frac{d^2u}{dx^2} + \alpha x \frac{du}{dx} + (\beta x^{2c} + \gamma)\, u = 0,$$

for constants α, β, γ, c with $\beta, c \neq 0$.

♡ 11.3.32. Let $k > 0$ be a constant. The ordinary differential equation $\dfrac{d^2u}{dt^2} + e^{-2t} u = 0$ describes the vibrations of a weakening spring whose stiffness $k(t) = e^{-2t}$ is exponentially decaying in time. (a) Show that this equation can be solved in terms of Bessel functions of order 0. *Hint*: Perform a change of variables. (b) Does the solution tend to 0 as $t \to \infty$?

♡ 11.3.33. We know that $\widehat{u}(x) = J_0(x)$ is a solution to the Bessel equation of order 0, namely

$$x u'' + u' + x u = 0. \tag{11.115}$$

In accordance with the general Frobenius method, construct a second solution of the form

$$\widetilde{u}(x) = J_0(x) \log x + \sum_{n=1}^{\infty} v_n x^n.$$

11.3.34. Is it possible to have all solutions to an ordinary differential equation bounded at a regular singular point? If not, explain why not. If true, give an example where this happens.

11.4 The Heat Equation in a Disk, Continued

Now that we have acquired some familiarity with the solutions to Bessel's ordinary differential equation, we are ready to analyze the separable solutions to the heat equation in a polar geometry. At the end of Section 11.2, we were left with the task of solving the Bessel equation (11.58) of integer order m. As we now know, there are two independent solutions, namely the Bessel function of the first kind J_m, (11.102), and the more complicated Bessel function of the second kind Y_m, (11.107), and hence the general solution has the form

$$p(z) = c_1 J_m(z) + c_2 Y_m(z),$$

for constants c_1, c_2. Reverting to our original radial coordinate $r = z/\sqrt{\lambda}$, we conclude that every solution to the radial equation (11.56) has the form

$$p(r) = c_1 J_m\big(\sqrt{\lambda}\, r\big) + c_2 Y_m\big(\sqrt{\lambda}\, r\big).$$

Now, the singular point $r = 0$ represents the center of the disk, and the solutions must remain bounded there. While this is true for $J_m(z)$, the second Bessel function $Y_m(z)$ has, according to (11.110), a singularity at $z = 0$ and so is unsuitable for the present purposes. (On the other hand, it plays a role in other situations, e.g., the heat equation on an annular ring.) Thus, every separable solution that is bounded at $r = 0$ comes from the rescaled Bessel function of the first kind of order m:

$$p(r) = J_m\big(\sqrt{\lambda}\, r\big). \tag{11.116}$$

The Dirichlet boundary condition at the disk's rim $r = 1$ requires

$$p(1) = J_m\big(\sqrt{\lambda}\,\big) = 0.$$

Therefore, in order that λ be a bona fide eigenvalue, $\sqrt{\lambda}$ must be a *root* of the mth order Bessel function J_m.

Remark: We already know, thanks to the positive definiteness of the Dirichlet boundary value problem, that the Helmholtz eigenvalues must all be positive, $\lambda > 0$, and so there will be no difficulty in taking its square root.

The graphs of $J_m(z)$ strongly indicate, and, indeed, it can be rigorously proved, [**85, 119**], that as z increases above 0, each Bessel function oscillates, with slowly decreasing amplitude, between positive and negative values. In fact, asymptotically,

$$J_m(z) \sim \sqrt{\frac{2}{\pi z}}\, \cos\big[\, z - \big(\tfrac{1}{2} m + \tfrac{1}{4}\big)\pi\,\big] \qquad \text{as} \qquad z \longrightarrow \infty, \tag{11.117}$$

and so the oscillations become essentially the same as a (phase-shifted) cosine whose amplitude decreases like $z^{-1/2}$. As a consequence, there exists an infinite sequence of *Bessel roots*, which we number in increasing order:

$$\begin{aligned} &J_m(\zeta_{m,n}) = 0, \qquad \text{where} \\ &\qquad 0 < \zeta_{m,1} < \zeta_{m,2} < \zeta_{m,3} < \ \cdots \qquad \text{with} \quad \zeta_{m,n} \longrightarrow \infty \quad \text{as} \quad n \longrightarrow \infty. \end{aligned} \tag{11.118}$$

It is worth emphasizing that the Bessel functions are *not* periodic, and so their roots are not evenly spaced. However, as a consequence of (11.117), the large Bessel roots are asymptotically close to the evenly spaced roots of the shifted cosine:

$$\zeta_{m,n} \sim \big(n + \tfrac{1}{2} m - \tfrac{1}{4}\big)\pi \qquad \text{as} \qquad n \longrightarrow \infty. \tag{11.119}$$

Owing to their physical importance in a wide range of problems, the Bessel roots have been extensively tabulated. The accompanying table displays all Bessel roots that are < 12 in magnitude. The columns of the table are indexed by m, the order of the Bessel function, and the rows by n, the root number.

Table of Bessel Roots $\zeta_{m,n}$

$n \backslash m$	0	1	2	3	4	5	6	7	...
1	2.4048	3.8317	5.1356	6.3802	7.5883	8.7715	9.9361	11.0864	...
2	5.5201	7.0156	8.4172	9.7610	11.0647	⋮	⋮	⋮	
3	8.6537	10.1735	11.6198	⋮	⋮				
4	11.7915	⋮	⋮						
⋮	⋮								

Remark: According to (11.102),

$$J_m(0) = 0 \qquad \text{for} \qquad m > 0, \qquad \text{while} \qquad J_0(0) = 1.$$

However, we do not count 0 as a bona fide Bessel root, since it does not lead to a valid eigenfunction for the Helmholtz boundary value problem.

Summarizing our progress so far, the eigenvalues

$$\lambda_{m,n} = \zeta_{m,n}^2, \qquad n = 1, 2, 3, \ldots, \qquad m = 0, 1, 2, \ldots, \tag{11.120}$$

of the Bessel boundary value problem (11.56–57) are the squares of the roots of the Bessel function of order m. The corresponding eigenfunctions are

$$w_{m,n}(r) = J_m(\zeta_{m,n} r), \qquad n = 1, 2, 3, \ldots, \qquad m = 0, 1, 2, \ldots, \tag{11.121}$$

defined for $0 \le r \le 1$. Combining (11.121) with the formula (11.55) for the angular components, we conclude that the separable solutions (11.53) to the polar Helmholtz boundary value problem (11.51) are

$$\begin{aligned} v_{0,n}(r) &= J_0(\zeta_{0,n} r), \\ v_{m,n}(r,\theta) &= J_m(\zeta_{m,n} r) \cos m\theta, \qquad \text{where} \qquad m, n = 1, 2, 3, \ldots . \\ \widehat{v}_{m,n}(r,\theta) &= J_m(\zeta_{m,n} r) \sin m\theta, \end{aligned} \tag{11.122}$$

These solutions define the *normal modes* for the unit disk; Figure 11.6 plots the first few of them. The eigenvalues $\lambda_{0,n}$ are simple, and contribute radially symmetric eigenfunctions, whereas the eigenvalues $\lambda_{m,n}$ for $m > 0$ are double, and produce two linearly independent separable eigenfunctions, with trigonometric dependence on the angular variable.

Recalling the original ansatz (11.50), we have at last produced the basic separable eigensolutions

$$\begin{aligned} u_{0,n}(t,r) &= e^{-\zeta_{0,n}^2 t}\, v_{0,n}(r) = e^{-\zeta_{0,n}^2 t} J_0(\zeta_{0,n} r), \\ u_{m,n}(t,r,\theta) &= e^{-\zeta_{m,n}^2 t}\, v_{m,n}(r,\theta) = e^{-\zeta_{m,n}^2 t} J_m(\zeta_{m,n} r) \cos m\theta, \\ \widehat{u}_{m,n}(t,r,\theta) &= e^{-\zeta_{m,n}^2 t}\, \widehat{v}_{m,n}(r,\theta) = e^{-\zeta_{m,n}^2 t} J_m(\zeta_{m,n} r) \sin m\theta, \qquad m, n = 1, 2, 3, \ldots , \end{aligned} \tag{11.123}$$

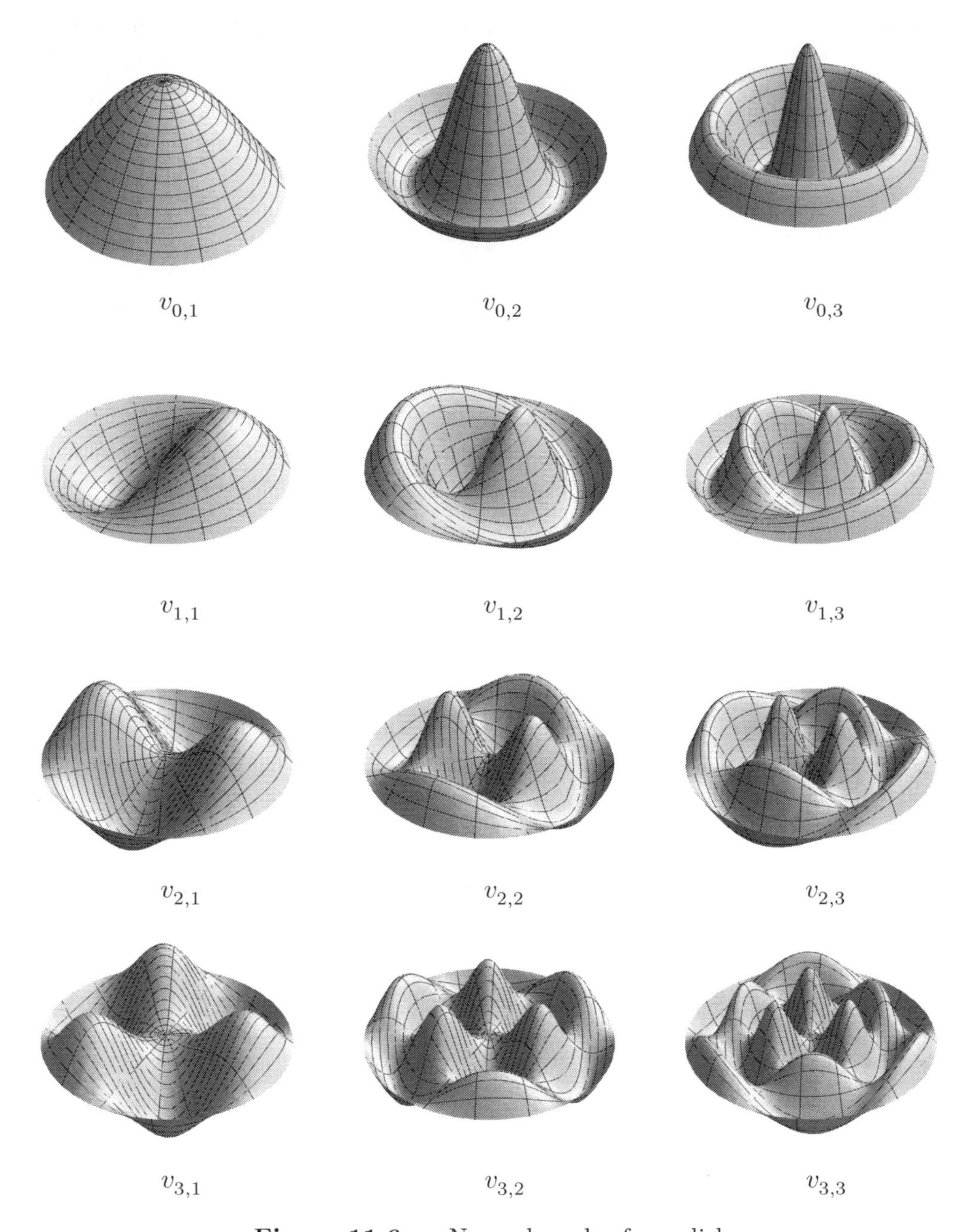

Figure 11.6. Normal modes for a disk.

to the homogeneous Dirichlet boundary value problem for the heat equation on the unit disk. The general solution is obtained by linear superposition, in the form of an infinite series

$$u(t,r,\theta) = \frac{1}{2}\sum_{n=1}^{\infty} a_{0,n}\, u_{0,n}(t,r) + \sum_{m,n=1}^{\infty} \left[\, a_{m,n}\, u_{m,n}(t,r,\theta) + b_{m,n}\, \widehat{u}_{m,n}(t,r,\theta)\,\right], \quad (11.124)$$

where the initial factor of $\frac{1}{2}$ is included, as with ordinary Fourier series, for later convenience. As usual, the coefficients $a_{m,n}, b_{m,n}$ are determined by the initial condition

$$u(0,r,\theta) = \frac{1}{2}\sum_{n=1}^{\infty} a_{0,n}\, v_{0,n}(r) + \sum_{m,n=1}^{\infty} \big[\, a_{m,n}\, v_{m,n}(r,\theta) + b_{m,n}\, \widehat{v}_{m,n}(r,\theta) \,\big] = f(r,\theta). \tag{11.125}$$

This requires that we expand the initial data into a *Fourier–Bessel series* in the eigenfunctions. As before, it is possible to prove, [**34**], that the separable eigenfunctions are *complete* — there are no other eigenfunctions — and hence every (reasonable) function defined on the unit disk can be written as a convergent series in the Bessel eigenfunctions.

Theorem 9.33 gurantees that the eigenfunctions are orthogonal† with respect to the standard L^2 inner product

$$\langle\, u\,, v\,\rangle = \iint_D u(x,y)\, v(x,y)\, dx\, dy = \int_0^1 \int_{-\pi}^{\pi} u(r,\theta)\, v(r,\theta)\; r\, d\theta\, dr$$

on the unit disk. (Note the extra factor of r coming from the polar coordinate form of the area element $dx\,dy = r\,dr\,d\theta$.) The L^2 norms of the Fourier–Bessel eigenfunctions are given by the interesting formulae

$$\|\, v_{0,n} \,\| = \sqrt{\pi}\; \big|\, J_1(\zeta_{0,n}) \,\big|\,, \qquad \|\, v_{m,n} \,\| = \|\, \widehat{v}_{m,n} \,\| = \sqrt{\frac{\pi}{2}}\; \big|\, J_{m+1}(\zeta_{m,n}) \,\big|\,, \tag{11.126}$$

which involve the value of the Bessel function of the next-higher order at the appropriate Bessel root. A proof of (11.126) can be found in Exercise 11.4.22, while numerical values are provided in the accompanying table.

Norms of the Fourier–Bessel Eigenfunctions $\|\, v_{m,n} \,\| = \|\, \widehat{v}_{m,n} \,\|$

$n \backslash m$	0	1	2	3	4	5	6	7
1	.9202	.5048	.4257	.3738	.3363	.3076	.2847	.2658
2	.6031	.3761	.3401	.3126	.2906	.2725	.2572	.2441
3	.4811	.3130	.2913	.2736	.2586	.2458	.2347	.2249
4	.4120	.2737	.2589	.2462	.2352	.2255	.2169	.2092
5	.3661	.2462	.2353	.2257	.2171	.2095	.2025	.1962

Orthogonality of the eigenfunctions implies that the coefficients in the Fourier–Bessel

† For the two independent eigenfunctions corresponding to one of the double eigenvalues, orthogonality must be verified by hand, but, in this case, it follows easily from the orthogonality of their trigonometric components.

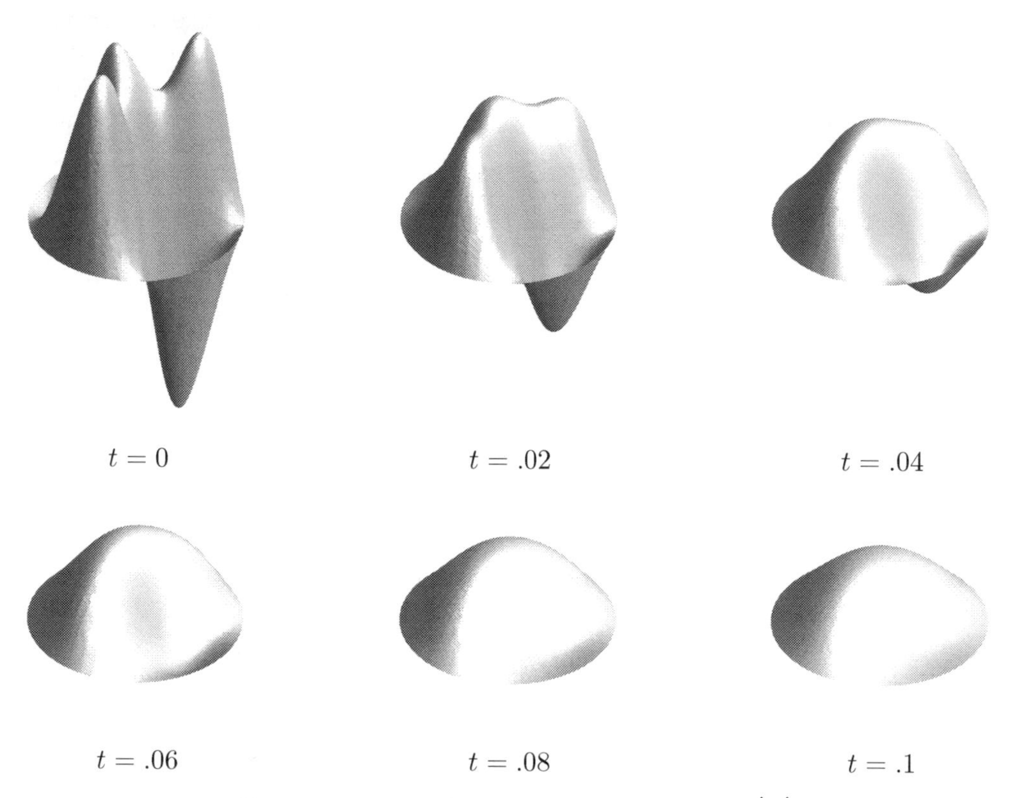

Figure 11.7. Heat diffusion in a disk.

series (11.125) are given by the inner product formulae

$$\begin{aligned} a_{0,n} &= 2\,\frac{\langle f, v_{0,n}\rangle}{\| v_{0,n}\|^2} = \frac{2}{\pi\, J_1(\zeta_{0,n})^2} \int_0^1 \int_{-\pi}^{\pi} f(r,\theta)\, J_0(\zeta_{0,n}\, r)\, r\, d\theta\, dr, \\ a_{m,n} &= \frac{\langle f, v_{m,n}\rangle}{\| v_{m,n}\|^2} = \frac{2}{\pi\, J_{m+1}(\zeta_{m,n})^2} \int_0^1 \int_{-\pi}^{\pi} f(r,\theta)\, J_m(\zeta_{m,n}\, r)\, r \cos m\,\theta\, d\theta\, dr, \\ b_{m,n} &= \frac{\langle f, \widehat{v}_{m,n}\rangle}{\| \widehat{v}_{m,n}\|^2} = \frac{2}{\pi\, J_{m+1}(\zeta_{m,n})^2} \int_0^1 \int_{-\pi}^{\pi} f(r,\theta)\, J_m(\zeta_{m,n}\, r)\, r \sin m\,\theta\, d\theta\, dr. \end{aligned} \tag{11.127}$$

In accordance with the general theory, each individual separable solution (11.123) to the heat equation decays exponentially fast, at a rate $\lambda_{m,n} = \zeta_{m,n}^2$ prescribed by the square of the corresponding Bessel root. In particular, the dominant mode, meaning the one that persists the longest, is

$$u_{0,1}(t,r,\theta) = e^{-\zeta_{0,1}^2\, t}\, J_0(\zeta_{0,1}\, r). \tag{11.128}$$

Its decay rate is prescribed by the smallest positive eigenvalue:

$$\zeta_{0,1}^2 \approx 5.783, \tag{11.129}$$

which is the square of the smallest root of the Bessel function $J_0(z)$. Since $J_0(z) > 0$ for $0 \le z < \zeta_{0,1}$, the dominant eigenfunction $v_{0,1}(r,\theta) = J_0(\zeta_{0,1}\, r) > 0$ is radially symmetric and strictly positive within the entire disk. Consequently, for most initial conditions (specifically those for which $a_{0,1} \neq 0$), the disk's temperature distribution eventually becomes entirely of one sign and radially symmetric, while decaying exponentially fast to zero at the rate given by (11.129). See Figure 11.7 for a plot of a typical solution. Note how, in accordance with the theory, the solution soon acquires a radial symmetry as it decays to thermal equilibrium.

Exercises

11.4.1. At the initial time $t_0 = 0$, a concentrated unit heat source is instantaneously applied at position $x = \frac{1}{2}$, $y = 0$, to a circular metal disk of unit radius and unit thermal diffusivity whose outside edge is held at 0°. Write down an eigenfunction series for the resulting temperature distribution at time $t > 0$. *Hint*: Be careful working with the delta function in polar coordinates; see Exercise 6.3.6.

11.4.2. Solve Exercise 11.4.1 when the concentrated unit heat source is instantaneously applied at the center of the disk.

♡ 11.4.3. (a) Write down the Fourier–Bessel series for the solution to the heat equation on a unit disk with $\gamma = 1$, whose circular edge is held at 0° and subject to the initial conditions $u(0, x, y) \equiv 1$ for $x^2 + y^2 \leq 1$. *Hint*: Use (11.112) to evaluate the integrals for the coefficients. (b) Approximate the time $t_\star \geq 0$ after which the temperature of the disk is everywhere $\leq .5^\circ$.

♣ 11.4.4. (a) Write down the first three nonzero terms in the Fourier–Bessel series for the solution to the heat equation on a unit disk with $\gamma = 1$ whose circular edge is held at 0° subject to the initial conditions $u(0, r, \theta) = 1 - r$ for $r \leq 1$. Use numerical integration to evaluate the coefficients. (b) Use your approximation to determine at which times $t \geq 0$ the temperature of the disk is everywhere $\leq .5^\circ$.

11.4.5. Prove that every separable eigenfunction of the Dirichlet boundary value problem for the Helmholtz equation in the unit disk can be written in the form

$$c\, J_m(\zeta_{m,n} r)\, \cos(m\theta - \alpha) \quad \text{for fixed } c \neq 0 \text{ and } -\pi < \alpha \leq \pi.$$

11.4.6. Suppose the initial data $f(r, \theta)$ in (11.49) satisfies $\displaystyle\int_0^1 \int_{-\pi}^{\pi} f(r, \theta)\, J_0(\zeta_{0,1} r)\, r\, d\theta\, dr = 0$.
(a) What is the decay rate to equilibrium of the resulting heat equation solution $u(t, r, \theta)$? (b) Prove that, generically, the asymptotic temperature distribution has half the disk above the equilibrium temperature and the other half below. Can you predict the diameter that separates the two halves? (c) If you know that $a_{0,1} = 0$, and also that the long-time temperature distribution is radially symmetric, what is the (generic) decay rate? What is the asymptotic temperature distribution?

◇ 11.4.7. Show how to use a scaling symmetry to solve the heat equation in a disk of radius R knowing the solution in a disk of radius 1.

11.4.8. Use rescaling, as in Exercise 11.4.7, to produce the solution to the Dirichlet initial-boundary value problem for a disk of radius 2 with diffusion coefficient $\gamma = 5$.

11.4.9. If it takes a disk of unit radius 3 minutes to reach (approximate) thermal equilibrium, how long will it take a disk of radius 2 made out of the same material and subject to the same homogeneous boundary conditions to reach equilibrium?

11.4.10. Assuming Dirichlet boundary conditions, does a square or a circular disk of the same area reach thermal equilibrium faster? Use your intuition first, and then check using the explicit formulas.

11.4.11. Answer Exercise 11.4.10 when the square and circle have the same perimeter.

11.4.12. Which reaches thermal equilibrium faster: a disk whose edge is held at 0° or a disk of the same radius that is fully insulated?

11.4.13. A circular metal disk is removed from an oven and then fully insulated. *True or false*: (*a*) The eventual equilibrium temperature is constant. (*b*) For large $t \gg 0$, the temperature $u(t,x,y)$ becomes more and more radially symmetric. If false, what can you say about the temperature profile at large times?

♡ 11.4.14. (*a*) Write down an eigenfunction series formula for the temperature dynamics of a disk of radius 1 that has an insulated boundary. (*b*) What is the eventual equilibrium temperature? (*c*) Is the rate of decay to thermal equilibrium (*i*) faster, (*ii*) slower, or (*iii*) the same as a disk with Dirichlet boundary conditions?

♡ 11.4.15. Write out a series solution for the temperature in a half-disk of radius 1, subject to (*a*) homogeneous Dirichlet boundary conditions on its entire boundary; (*b*) homogeneous Dirichlet conditions on the circular part of its boundary and homogeneous Neumann conditions on the straight part. (*c*) Which of the two boundary conditions results in a faster return to equilibrium temperature? How much faster?

11.4.16. A large sheet of metal is heated to 100°. A circular disk and a semi-circular half-disk of the same radius are cut out of it. Their edges are then held at 0°, while being fully insulated from above and below.
(*a*) *True or false*: The half-disk goes to thermal equilibrium twice as fast as the disk.
(*b*) If you need to wait 20 minutes for the circular disk to cool down enough to be picked up in your bare hands, how long do you need to wait to pick up the semi-circular disk?

♣ 11.4.17. Two identical plates have the shape of an annular ring $\{1 < r < 2\}$ with inner radius 1 and outer radius 2. The first has an insulated inner boundary and outer boundary held at 0°, while the second has an insulated outer boundary and inner boundary held at 0°. If both start out at the same temperature, which reaches thermal equilibrium faster? Quantify the rates of decay.

♡ 11.4.18. Let $m \geq 0$ be a nonnegative integer. In this exercise, we investigate the completeness of the eigenfunctions of the Bessel boundary value problem (11.56–57). To this end, define the Sturm–Liouville linear differential operator

$$S[u] = -\frac{1}{x}\frac{d}{dx}\left(x\frac{du}{dx}\right) + \frac{m^2}{x^2}\,u,$$

subject to the boundary conditions $|\,u'(0)\,| < \infty$, $u(1) = 0$, and either $|\,u(0)\,| < \infty$ when $m = 0$, or $u(0) = 0$ when $m > 0$.
(*a*) Show that S is self-adjoint relative to the inner product $\langle\, f\,, g\,\rangle = \int_0^1 f(x)\,g(x)\,x\,dx$.
(*b*) Prove that the eigenfunctions of S are the rescaled Bessel functions $J_m(\zeta_{m,n}\,x)$ for $n = 1, 2, 3, \ldots$. What are the orthogonality relations?
(*c*) Find the Green's function $G(x;\xi)$ and modified Green's function $\widehat{G}(x;\xi)$, cf. (9.59), associated with the boundary value problem $S[u] = 0$.
(*d*) Use the criterion of Theorem 9.47 to prove that the eigenfunctions are complete.

11.4.19. Determine the Bessel roots $\zeta_{1/2,n}$. Do they satisfy the asymptotic formula (11.119)?

♣ 11.4.20. Use a numerical root finder to compute the first 10 Bessel roots $\zeta_{3/2,n}$, $n = 1, \ldots, 10$. Compare your values with the asymptotic formula (11.119).

◇ 11.4.21. Prove that $J_{m-1}(\zeta_{m,n}) = -\,J_{m+1}(\zeta_{m,n})$.

◇ 11.4.22. In this exercise, we prove formula (11.126).
(*a*) First, use the recurrence formulae (11.111) to prove

$$\frac{d}{dx}\left[\,x^2\left(\,J_m(x)^2 - J_{m-1}(x)\,J_{m+1}(x)\,\right)\,\right] = 2x\,J_m(x)^2.$$

(*b*) Integrate both sides of the previous formula from 0 to the Bessel zero $\zeta_{m,n}$ and then use Exercise 11.4.21 to show that

$$\int_0^{\zeta_{m,n}} x\, J_m(x)^2\, dx = -\frac{\zeta_{m,n}^2}{2}\, J_{m-1}(\zeta_{m,n})\, J_{m+1}(\zeta_{m,n}) = \frac{\zeta_{m,n}^2}{2}\, J_{m+1}(\zeta_{m,n})^2.$$

(*c*) Next, use a change of variables to establish the identity

$$\int_0^1 z\, J_m(\zeta_{m,n}\, z)^2\, dz = \tfrac{1}{2}\, J_{m+1}(\zeta_{m,n})^2.$$

(*d*) Finally, use the formulae for $v_{m,n}$ and $\widehat{v}_{m,n}$ to complete the proof of (11.126).

♢ 11.4.23. Prove directly that the eigenfunctions $v_{m,n}(r,\theta)$ and $\widehat{v}_{m,n}(r,\theta)$ in (11.122) are orthogonal with respect to the L^2 inner product on the unit disk.

11.4.24. Establish the following alternative formulae for the eigenfunction norms:

$$\|\, v_{0,n} \,\| = \sqrt{\pi}\, \left|\, J_0'(\zeta_{0,n}) \,\right|, \qquad \|\, v_{m,n} \,\| = \|\, \widehat{v}_{m,n} \,\| = \sqrt{\frac{\pi}{2}}\, \left|\, J_m'(\zeta_{m,n}) \,\right|.$$

11.5 The Fundamental Solution to the Planar Heat Equation

As we learned in Section 8.1, the fundamental solution to the heat equation measures the temperature distribution resulting from a concentrated initial heat source, e.g., a hot soldering iron applied instantaneously at a single point on a metal plate. The physical problem is modeled mathematically by taking a delta function as the initial data along with the relevant homogeneous boundary conditions. Once the fundamental solution is known, one is able to use linear superposition to recover the solution generated by any other initial data.

As in our one-dimensional analysis, we shall concentrate on the most tractable case, in which the domain is the entire plane: $\Omega = \mathbb{R}^2$. Thus, our first goal is to solve the initial value problem

$$u_t = \gamma\, \Delta u, \qquad u(0,x,y) = \delta(x-\xi)\,\delta(y-\eta), \tag{11.130}$$

for $t > 0$ and $(x,y) \in \mathbb{R}^2$. The solution $u = F(t,\mathbf{x};\boldsymbol{\xi}) = F(t,x,y;\xi,\eta)$ to this initial value problem is known as the *fundamental solution* for the heat equation on $\mathbb{R}^2$.

The quickest route to the desired formula relies on the following means of combining solutions of the one-dimensional heat equation to produce solutions of the two-dimensional version.

Lemma 11.11. *Let $v(t,x)$ and $w(t,x)$ be any two solutions to the one-dimensional heat equation $u_t = \gamma\, u_{xx}$. Then their product*

$$u(t,x,y) = v(t,x)\, w(t,y) \tag{11.131}$$

is a solution to the two-dimensional heat equation $u_t = \gamma\,(u_{xx} + u_{yy})$.

Proof: Our assumptions imply that $v_t = \gamma\, v_{xx}$, while $w_t = \gamma\, w_{yy}$ when we write $w(t,y)$ as a function of t and y. Therefore, differentiating (11.131), we find

$$\frac{\partial u}{\partial t} = \frac{\partial v}{\partial t}\, w + v\, \frac{\partial w}{\partial t} = \gamma\, \frac{\partial^2 v}{\partial x^2}\, w + \gamma\, v\, \frac{\partial^2 w}{\partial y^2} = \gamma \left(\frac{\partial^2 u}{\partial x^2} + \frac{\partial^2 u}{\partial y^2} \right),$$

and hence $u(t,x,y)$ solves the two-dimensional heat equation. *Q.E.D.*

For example, if

$$v(t,x) = e^{-\gamma \alpha^2 t} \sin \alpha x, \qquad w(t,y) = e^{-\gamma \beta^2 t} \sin \beta y,$$

are separable solutions of the one-dimensional heat equation, then

$$u(t,x,y) = e^{-\gamma(\alpha^2+\beta^2)t} \sin \alpha x \, \sin \beta y$$

are the separable solutions we used to solve the heat equation on a rectangle. A more interesting case is to choose

$$v(t,x) = \frac{1}{2\sqrt{\pi\gamma t}}\, e^{-(x-\xi)^2/(4\gamma t)}, \qquad w(t,y) = \frac{1}{2\sqrt{\pi\gamma t}}\, e^{-(y-\eta)^2/(4\gamma t)}, \tag{11.132}$$

to be the fundamental solutions (8.14) to the one-dimensional heat equation at respective locations $x = \xi$ and $y = \eta$. Multiplying these two solutions together produces the fundamental solution for the two-dimensional problem.

Theorem 11.12. *The* fundamental solution *to the heat equation $u_t = \gamma \Delta u$ corresponding to a unit delta function placed at position $(\xi,\eta) \in \mathbb{R}^2$ at the initial time $t_0 = 0$ is*

$$F(t,x,y;\xi,\eta) = \frac{1}{4\pi\gamma t}\, e^{-[(x-\xi)^2+(y-\eta)^2]/(4\gamma t)}. \tag{11.133}$$

Proof: Since we already know that both functions (11.132) are solutions to the one-dimensional heat equation, Lemma 11.11 guarantees that their product, which equals (11.133), solves the two-dimensional heat equation for $t > 0$. Moreover, at the initial time,

$$u(0,x,y) = v(0,x)\, w(0,y) = \delta(x-\xi)\, \delta(y-\eta)$$

is a product of delta functions, and hence the result follows. Indeed, the total heat

$$\iint u(t,x,y)\, dx\, dy = \int_{-\infty}^{\infty} v(t,x)\, dx \int_{-\infty}^{\infty} w(t,y)\, dy = 1, \qquad t \geq 0,$$

remains constant, while

$$\lim_{t\to 0^+} u(t,x,y) = \begin{cases} \infty, & (x,y) = (\xi,\eta), \\ 0, & \text{otherwise}, \end{cases}$$

has the standard delta function limit at the initial time instant. *Q.E.D.*

Figure 11.8 depicts the evolution of the fundamental solution when $\gamma = 1$ at the indicated times. Observe that the initially concentrated temperature spreads out in a radially symmetric manner, while the total amount of heat remains constant. At any individual point $(x,y) \neq (0,0)$, the initially zero temperature rises slightly at first, but then decays monotonically back to zero at a rate proportional to $1/t$. As in the one-dimensional case, since the fundamental solution is > 0 for all $t > 0$, the heat energy has an infinite speed of propagation.

Both the one- and two-dimensional fundamental solutions have bell-shaped profiles known as *Gaussian filters*. The most important difference is the initial factor. In a one-dimensional medium, the fundamental solution decays in proportion to $1/\sqrt{t}$, whereas in the plane the decay is more rapid, being proportional to $1/t$. The physical explanation is that the heat energy is able to spread out in two independent directions, and hence diffuses

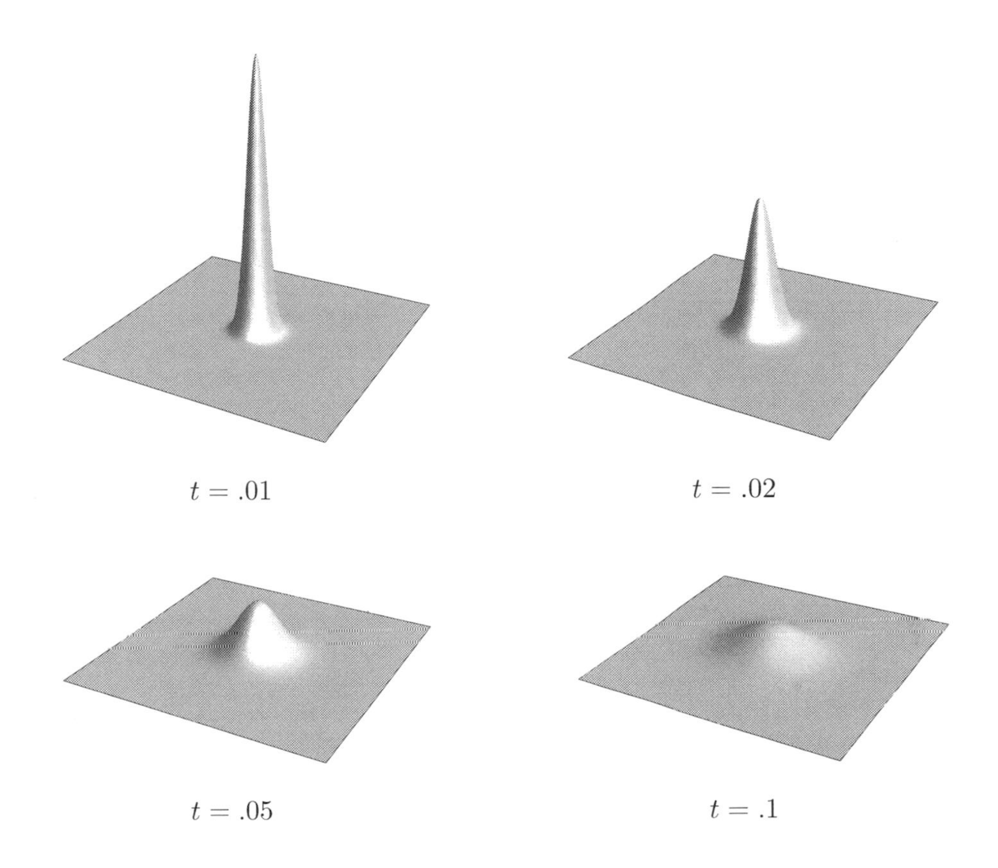

Figure 11.8. The fundamental solution to the planar heat equation.

away from its initial source more rapidly. As we shall see, the decay in three-dimensional space is more rapid still, being proportional to $t^{-3/2}$ for similar reasons; see (12.120).

The principal use of the fundamental solution is for solving the general initial value problem. We express the initial temperature distribution as a superposition of delta function impulses,

$$u(0,x,y) = f(x,y) = \iint f(\xi,\eta)\,\delta(x-\xi,y-\eta)\,d\xi\,d\eta,$$

where, at the point $(\xi,\eta) \in \mathbb{R}^2$, the impulse has magnitude $f(\xi,\eta)$. Linearity implies that the solution is then given by the same superposition of fundamental solutions.

Theorem 11.13. *The solution to the initial value problem*

$$u_t = \gamma\,\Delta u, \qquad u(0,x,y) = f(x,y), \qquad (x,y) \in \mathbb{R}^2,$$

for the planar heat equation is given by the linear superposition formula

$$u(t,x,y) = \frac{1}{4\pi\gamma t} \iint f(\xi,\eta)\,e^{-[(x-\xi)^2+(y-\eta)^2]/(4\gamma t)}\,d\xi\,d\eta. \tag{11.134}$$

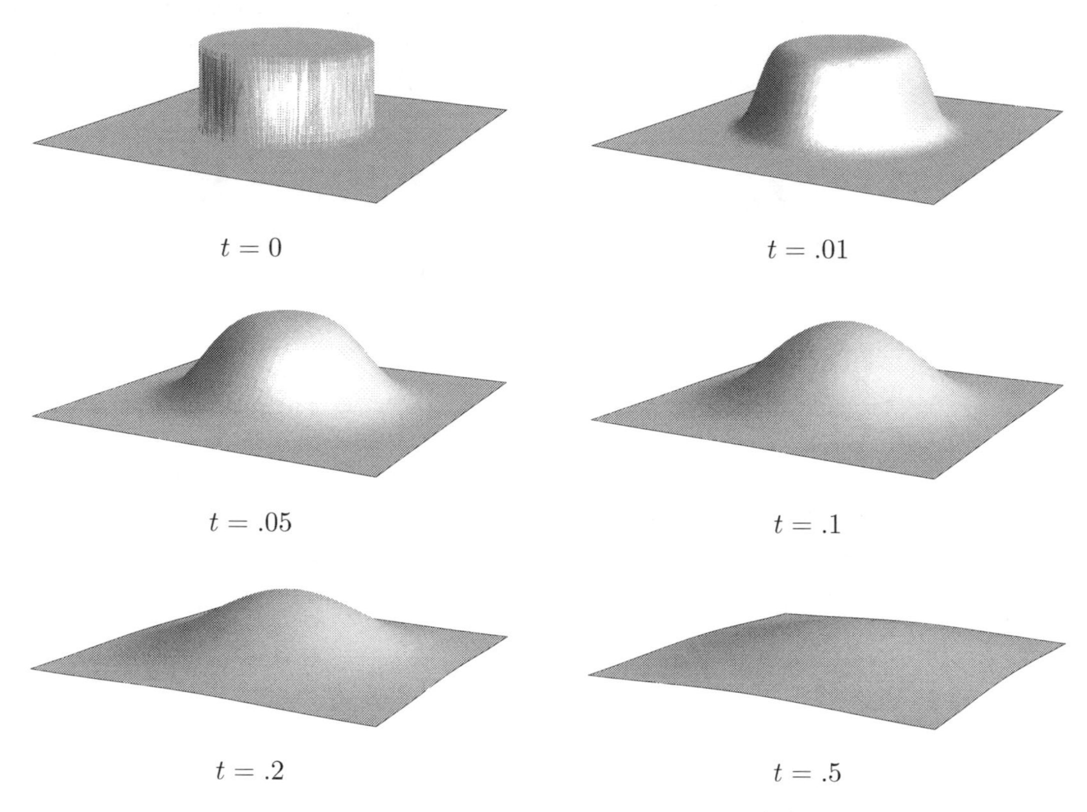

Figure 11.9. Diffusion of a disk.

We can interpret the solution formula (11.134) as a two-dimensional *convolution*

$$u(t,x,y) = F(t,x,y) * f(x,y) \tag{11.135}$$

of the initial data with a one-parameter family of progressively wider and shorter Gaussian filters

$$F(t,x,y) = F(t,x,y;0,0) = \frac{1}{4\pi\gamma t}\, e^{-(x^2+y^2)/(4\gamma t)}. \tag{11.136}$$

As in (7.54), such a convolution can be interpreted as a Gaussian weighted averaging of the function $f(x,y)$, which has the effect of smoothing out the initial data.

Example 11.14. If our initial temperature distribution is constant on a circular region, say

$$u(0,x,y) = \begin{cases} 1 & x^2+y^2<1, \\ 0, & \text{otherwise}, \end{cases}$$

then the solution can be evaluated using (11.134), as follows:

$$u(t,x,y) = \frac{1}{4\pi t}\iint_D e^{-[(x-\xi)^2+(y-\eta)^2]/(4t)}\, d\xi\, d\eta,$$

where the integral is over the unit disk $D = \{\xi^2+\eta^2 \le 1\}$. Unfortunately, the integral cannot be expressed in terms of elementary functions. On the other hand, numerical evaluation of the integral is straightforward. A plot of the resulting radially symmetric solution appears in Figure 11.9. One could also interpret this solution as the diffusion of an animal population in a uniform isotropic environment or bacteria in a similarly uniform large petri dish that are initially confined to a small circular region.

Exercises

11.5.1. Solve the following initial value problem: $u_t = 5\,(u_{xx} + u_{yy}), \quad u(0,x,y) = e^{-(x^2+y^2)}$.

11.5.2. Write down an integral formula for the solution to the following initial value problem:
$$u_t = 3(u_{xx} + u_{yy}), \qquad u(0,x,y) = (1 + x^2 + y^2)^{-2}.$$

11.5.3. At the initial time $t = 0$, a unit heat source is instantaneously applied at the origin of the (x,y)–plane. For $t > 0$, what is the maximum temperature experienced at a point $(x,y) \neq \mathbf{0}$? At what time is the maximum temperature achieved? Does the temperature approach an equilibrium value as $t \to \infty$? If so, how fast?

11.5.4. (*a*) Find an eigenfunction series representation of the fundamental solution for the heat equation $u_t = \Delta u$ on the unit square $\{0 \le x, y \le 1\}$ when subject to homogeneous Dirichlet boundary conditions. (*b*) Write the solution to the initial value problem $u(0,x,y) = f(x,y)$ in terms of the fundamental solution. (*c*) Discuss how your formula is related to the Fourier series solution (11.43).

11.5.5. Let $u(t,x,y)$ be a solution to the heat equation on all of $\mathbb{R}^2$ such that u and $\|\nabla u\| \to 0$ rapidly as $\|\mathbf{x}\| \to \infty$. (a) Prove that the total heat $H(t) = \iint u(t,x,y)\,dx\,dy$ is constant. (*b*) Explain how this can be reconciled with the statement that $u(t,x,y) \to 0$ as $t \to \infty$ at all points $(x,y) \in \mathbb{R}^2$.

♢ 11.5.6. Consider the initial value problem $u_t = \gamma\,\Delta u + H(t,x,y), \ u(0,x,y) = 0$, for the inhomogeneous heat equation on the entire (x,y)–plane, where $H(t,x,y)$ represents a time-varying external heat source. Derive an integral formula for its solution. *Hint*: Mimic the solution method in Section 8.1.

11.5.7. A flat plate of infinite extent with unit thermal diffusivity starts off at 0°. From then on, a unit heat source is continually applied at the origin. Find the resulting temperature distribution. Does the temperature eventually reach a steady state?
Hint: Use Exercise 11.5.6.

♡ 11.5.8. Building on Example 11.14, we model the "diffusion" of a set $D \subset \mathbb{R}^2$ as the solution $u(t,x,y)$ to the heat equation $u_t = \Delta u$ subject to the initial condition $u(0,x,y) = \chi_D(x,y)$, where $\chi_D(x,y) = \begin{cases} 1, & (x,y) \in D, \\ 0, & (x,y) \notin D, \end{cases}$ is the *characteristic function* of the set D.
(*a*) Write down a formula for the diffusion of the set D.
(*b*) *True or false*: At each t, the diffusion $u(t,x,y)$ is the characteristic function of a set D_t.
(*c*) Prove that $0 < u(t,x,y) < 1$ for all (x,y) and $t > 0$. (*d*) What is $\lim\limits_{t\to\infty} u(t,x,y)$?
(*e*) Write down a formula for the diffusion of a unit square $D = \{0 \le x, y \le 1\}$, and then plot the result at several times. Discuss what you observe.

11.5.9. (*a*) Explain why the delta function on $\mathbb{R}^2$ satisfies the scaling law $\delta(x,y) = \beta^2\,\delta(\beta x, \beta\,y)$, for $\beta \neq 0$. (*b*) Verify that the fundamental solution to the heat equation on $\mathbb{R}^2$ obeys the same scaling law: $F(t,x,y) = \beta^2\,F(\beta^2\,t, \beta\,x, \beta\,y)$. (*c*) Is the fundamental solution a similarity solution?

11.5.10. (a) Find the fundamental solution on $\mathbb{R}^2$ to the cable equation $u_t = \gamma\,\Delta u - \alpha\, u$, where $\alpha > 0$ is constant. (b) Use your solution to write down a formula for the solution to the general initial value problem $u(0,x,y) = f(x,y)$ for $(x,y) \in \mathbb{R}^2$.

11.5.11. (a) Prove that if $v(t,x)$ and $w(t,x)$ solve the dispersive wave equation (8.90), then their product $u(t,x,y) = v(t,x)\,w(t,y)$ solves the two-dimensional dispersive equation

$$u_t + u_{xxx} + u_{yyy} = 0.$$

(b) What is the fundamental solution on $\mathbb{R}^2$ of the latter equation? (c) Write down an integral formula for the solution to the initial value problem $u(0,x,y) = f(x,y)$ for $(x,y) \in \mathbb{R}^2$.

11.5.12. Define the two-dimensional convolution $f * g$ of functions $f(x,y)$ and $g(x,y)$ so that equation (11.135) is valid.

11.6 The Planar Wave Equation

Let us next consider the two-dimensional wave equation

$$\frac{\partial^2 u}{\partial t^2} = c^2 \Delta u = c^2 \left(\frac{\partial^2 u}{\partial x^2} + \frac{\partial^2 u}{\partial y^2} \right), \tag{11.137}$$

which models the unforced transverse vibrations of a homogeneous membrane, e.g., a drum. Here, $u(t,x,y)$ represents the vertical displacement of the membrane at time t and position $(x,y) \in \Omega$, where the domain $\Omega \subset \mathbb{R}^2$, assumed bounded, represents the undeformed shape. The constant $c^2 > 0$ encapsulates the membrane's physical properties — density, tension, stiffness, etc.; its square root, c, is called, as in the one-dimensional case, the *wave speed*, since it represents the speed of propagation of localized signals.

Remark: In this simplified model, we are only allowing small, transverse (vertical) displacements of the membrane. Large elastic vibrations lead to the nonlinear partial differential equations of elastodynamics, [**7**]. In particular, the bending vibrations of a flexible elastic plate are governed by a more complicated fourth-order partial differential equation.

The solution $u(t,x,y)$ to the wave equation will be uniquely specified once we impose suitable boundary and initial conditions. The Dirichlet conditions

$$u(t,x,y) = h(x,y), \qquad (x,y) \in \partial\Omega, \tag{11.138}$$

correspond to gluing our membrane to a fixed boundary — a rim; more generally, we can also allow h to depend on t, modeling a membrane attached to a moving boundary. On the other hand, the homogeneous Neumann conditions

$$\frac{\partial u}{\partial \mathbf{n}}(t,x,y) = 0, \qquad (x,y) \in \partial\Omega, \tag{11.139}$$

represent a free boundary where the membrane is not attached to any support — although in this model, its edge is allowed to move only in a vertical direction. Mixed boundary conditions attach part of the boundary and leave the remaining portion free to vibrate:

$$u = h \quad \text{on} \quad D \subsetneq \partial\Omega, \qquad \frac{\partial u}{\partial \mathbf{n}} = 0 \quad \text{on} \quad N = \partial\Omega \setminus D. \tag{11.140}$$

Since the wave equation is of second order in time, to uniquely specify the solution we need to impose two initial conditions,

$$u(0,x,y) = f(x,y), \qquad \frac{\partial u}{\partial t}(0,x,y) = g(x,y), \qquad (x,y) \in \Omega. \tag{11.141}$$

The first specifies the membrane's initial displacement, while the second prescribes its initial velocity.

Separation of Variables

Unfortunately, the d'Alembert solution method does not apply to the two-dimensional wave equation in any obvious manner. The reason is that, unlike the one-dimensional version (2.69), one cannot factorize the planar wave operator $\Box = \partial_t^2 - c^2\,\partial_x^2 - c^2\,\partial_y^2$, thus precluding any sort of reduction to a first-order partial differential equation. However, this is not the end of the story, and we will return to this issue at the end of Section 12.6.

We thus fall back on our universal solution tool for linear partial differential equations — separation of variables. According to the general framework established in Section 9.5, the separable solutions to the wave equation have the trigonometric form

$$u_k(t,x,y) = \cos(\omega_k\, t)\, v_k(x,y) \qquad \text{and} \qquad \widetilde{u}_k(t,x,y) = \sin(\omega_k\, t)\, v_k(x,y). \tag{11.142}$$

Substituting back into the wave equation, we find that $v_k(x,y)$ must be an eigenfunction of the associated Helmholtz boundary value problem

$$c^2\left(\frac{\partial^2 u}{\partial x^2} + \frac{\partial^2 u}{\partial y^2}\right) + \lambda_k\, v = 0, \tag{11.143}$$

whose eigenvalue $\lambda_k = \omega_k^2$ equals the square of the vibrational frequency. According to Theorem 9.47, on a bounded domain, there is an infinite number of such *normal modes* with progressively faster vibrational frequencies: $\omega_k \to \infty$ as $k \to \infty$. In addition, in the positive semi-definite case — which occurs under homogeneous Neumann boundary conditions — there is a single constant null eigenfunction, leading to the additional separable solutions

$$u_0(t,x,y) = 1 \qquad \text{and} \qquad \widetilde{u}_0(t,x,y) = t. \tag{11.144}$$

The first represents a stationary membrane that has been displaced to a fixed height, while the second represents a membrane that is moving off in the vertical direction with constant unit speed. (Think of the membrane moving in outer space unaffected by any external gravitational force.) As in Section 9.5, the general solution can be written as an infinite series in the eigensolutions (11.142). Unfortunately, as we know, the Helmholtz boundary value problem can be explicitly solved only on a rather restricted class of domains. Here we will content ourselves with investigating the two most important cases: rectangular and circular membranes.

Remark: The vibrational frequencies represent the tones and overtones one hears when the drum membrane vibrates. An interesting question is whether two drums of different shapes can have identical sounds — the exact same vibrational frequencies. Or, more descriptively, can one "hear" the shape of a drum? It was not until 1992 that the answer was shown to be no, but for quite subtle reasons. See [**47**] for a discussion and some examples of differently shaped drums that have the same vibrational frequencies.

Vibration of a Rectangular Drum

Let us first consider the vibrations of a membrane in the shape of a rectangle

$$R = \{0 < x < a,\ 0 < y < b\},$$

with side lengths a and b, whose edges are fixed to the (x, y)–plane. Thus, we seek to solve the wave equation

$$u_{tt} = c^2 \Delta u = c^2 (u_{xx} + u_{yy}), \qquad 0 < x < a, \qquad 0 < y < b, \tag{11.145}$$

subject to the initial and boundary conditions

$$\begin{aligned} u(t, 0, y) = u(t, a, y) = 0 = u(t, x, 0) = u(t, x, b), \qquad & 0 < x < a, \\ u(0, x, y) = f(x, y), \qquad u_t(0, x, y) = g(x, y), \qquad & 0 < y < b. \end{aligned} \tag{11.146}$$

As we saw in Section 11.2, the eigenfunctions and eigenvalues for the associated Helmholtz equation on a rectangle,

$$c^2 (v_{xx} + v_{yy}) + \lambda v = 0, \qquad (x, y) \in R, \tag{11.147}$$

when subject to the homogeneous Dirichlet boundary conditions

$$v(0, y) = v(a, y) = 0 = v(x, 0) = v(x, b), \qquad 0 < x < a, \qquad 0 < y < b, \tag{11.148}$$

are

$$v_{m,n}(x, y) = \sin \frac{m \pi x}{a} \, \sin \frac{n \pi y}{b}, \qquad \text{where} \qquad \lambda_{m,n} = \pi^2 c^2 \left(\frac{m^2}{a^2} + \frac{n^2}{b^2} \right), \tag{11.149}$$

with $m, n = 1, 2, \ldots$. The fundamental frequencies of vibration are the square roots of the eigenvalues, so

$$\omega_{m,n} = \sqrt{\lambda_{m,n}} = \pi c \sqrt{\frac{m^2}{a^2} + \frac{n^2}{b^2}}, \qquad m, n = 1, 2, \ldots \ . \tag{11.150}$$

The frequencies will depend upon the underlying geometry — meaning the side lengths — of the rectangle, as well as the wave speed c, which, in turn, is a function of the membrane's density and stiffness. The higher the wave speed, or the smaller the rectangle, the faster the vibrations. In layman's terms, (11.150) quantifies the observation that smaller, stiffer drums made of less-dense material vibrate faster.

According to (11.142), the normal modes of vibration of our rectangle are

$$\begin{aligned} u_{m,n}(t, x, y) &= \cos \left(\pi c \sqrt{\frac{m^2}{a^2} + \frac{n^2}{b^2}} \; t \right) \sin \frac{m \pi x}{a} \, \sin \frac{n \pi y}{b}, \\ \widetilde{u}_{m,n}(t, x, y) &= \sin \left(\pi c \sqrt{\frac{m^2}{a^2} + \frac{n^2}{b^2}} \; t \right) \sin \frac{m \pi x}{a} \, \sin \frac{n \pi y}{b}. \end{aligned} \tag{11.151}$$

The general solution can then be written as a double Fourier series

$$u(t, x, y) = \sum_{m,n=1}^{\infty} \left[a_{m,n} \, u_{m,n}(t, x, y) + b_{m,n} \, \widetilde{u}_{m,n}(t, x, y) \right]$$

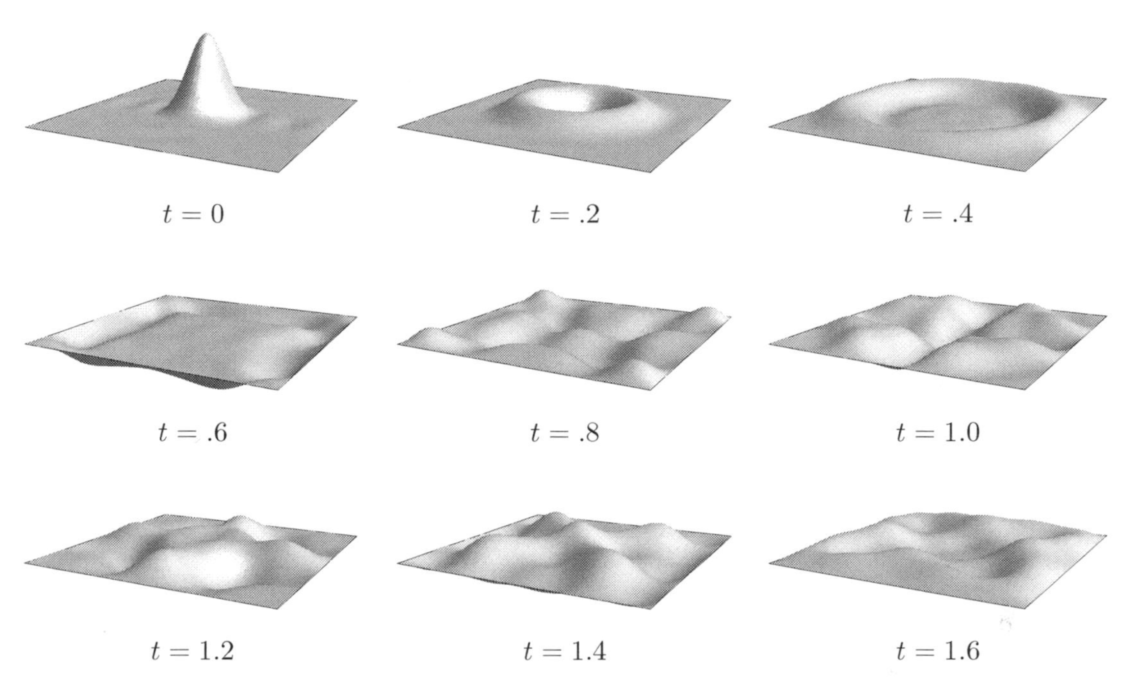

Figure 11.10. Vibrations of a square membrane.

in the normal modes. The coefficients $a_{m,n}, b_{m,n}$ are fixed by the initial displacement $u(0,x,y)=f(x,y)$ and the initial velocity $u_t(0,x,y)=g(x,y)$. Indeed, the usual orthogonality relations among the eigenfunctions imply

$$a_{m,n}=\frac{\langle\, v_{m,n}\,,f\,\rangle}{\|\,v_{m,n}\,\|^2}=\frac{4}{a\,b}\int_0^b\!\!\int_0^a f(x,y)\,\sin\frac{m\pi x}{a}\,\sin\frac{n\pi y}{b}\,dx\,dy, \tag{11.152}$$

$$b_{m,n}=\frac{\langle\, v_{m,n}\,,g\,\rangle}{\omega_{m,n}\,\|\,v_{m,n}\,\|^2}=\frac{4}{\pi\,c\,\sqrt{m^2b^2+n^2a^2}}\int_0^b\!\!\int_0^a g(x,y)\,\sin\frac{m\pi x}{a}\,\sin\frac{n\pi y}{b}\,dx\,dy.$$

Since the fundamental frequencies are not rational multiples of each other, the general solution is a genuinely quasiperiodic superposition of the various normal modes.

In Figure 11.10, we plot the solution resulting from the initially concentrated displacement†

$$u(0,x,y)=f(x,y)=e^{-100\,[\,(x-.5)^2+(y-.5)^2\,]}$$

at the center of a unit square, so $a=b=1$, with unit wave speed, $c=1$. Note that, unlike a concentrated displacement of a one-dimensional string, which remains concentrated at all subsequent times and periodically repeats, the initial displacement here spreads out in a radially symmetric manner and propagates to the edges of the rectangle, where it reflects

† The alert reader may object that the initial displacement $f(x,y)$ does not exactly satisfy the Dirichlet boundary conditions on the edges of the rectangle. But this does not prevent the existence of a well-defined (weak) solution to the initial value problem, whose initial boundary discontinuities will subsequently propagate into the square. However, here these are so tiny as to be unnoticeable in the solution graphs.

and then interacts with itself. Moreover, due to the quasiperiodicity of the solution, the drum's motion never exactly repeats, and the initially concentrated displacement never quite reforms.

Vibration of a Circular Drum

Let us next analyze the vibrations of a circular membrane of unit radius. In polar coordinates, the planar wave equation (11.137) takes the form

$$\frac{\partial^2 u}{\partial t^2} = c^2 \left(\frac{\partial^2 u}{\partial r^2} + \frac{1}{r}\frac{\partial u}{\partial r} + \frac{1}{r^2}\frac{\partial^2 u}{\partial \theta^2} \right). \tag{11.153}$$

We will again consider the homogeneous Dirichlet boundary value problem

$$u(t,1,\theta) = 0, \qquad t \geq 0, \qquad -\pi \leq \theta \leq \pi, \tag{11.154}$$

along with initial conditions

$$u(0,r,\theta) = f(r,\theta), \qquad \frac{\partial u}{\partial t}(0,r,\theta) = g(r,\theta), \tag{11.155}$$

representing the initial displacement and velocity of the membrane. As always, we build up the general solution as a quasiperiodic linear combination of the normal modes as specified by the eigenfunctions for the associated Helmholtz boundary value problem.

As we saw in Section 11.2, the eigenfunctions of the Helmholtz equation on a disk of radius 1, say, subject to homogeneous Dirichlet boundary conditions, are products of trigonometric and Bessel functions:

$$\begin{aligned} v_{0,n}(r,\theta) &= J_0(\zeta_{0,n} r), \\ v_{m,n}(r,\theta) &= J_m(\zeta_{m,n} r)\cos m\theta, \qquad\qquad m,n = 1,2,3,\ldots. \\ \widetilde{v}_{m,n}(r,\theta) &= J_m(\zeta_{m,n} r)\sin m\theta, \end{aligned} \tag{11.156}$$

Here r,θ are the usual polar coordinates, while $\zeta_{m,n} > 0$ denotes the nth (positive) root of the mth order Bessel function $J_m(z)$, cf. (11.118). The corresponding eigenvalue is its square, $\lambda_{m,n} = \zeta_{m,n}^2$, and hence the natural frequencies of vibration are equal to the Bessel roots scaled by the wave speed:

$$\omega_{m,n} = c\sqrt{\lambda_{m,n}} = c\,\zeta_{m,n}. \tag{11.157}$$

A table of their values (for the case $c = 1$) can be found in the preceding section. The Bessel roots do not follow any easily discernible pattern, and are not rational multiples of each other. This result, known as *Bourget's hypothesis*, [**119**; p. 484], was rigorously proved by the German mathematician Carl Ludwig Siegel in 1929, [**106**]. Thus, the vibrations of a circular drum are also truly quasiperiodic, thereby providing a mathematical explanation of why drums sound dissonant.

The frequencies $\omega_{0,n} = c\,\zeta_{0,n}$ correspond to simple eigenvalues, with a single radially symmetric eigenfunction $J_0(\zeta_{0,n} r)$, while the "angular modes" $\omega_{m,n}$, for $m > 0$, are double, each possessing two linearly independent eigenfunctions (11.156). According to the general

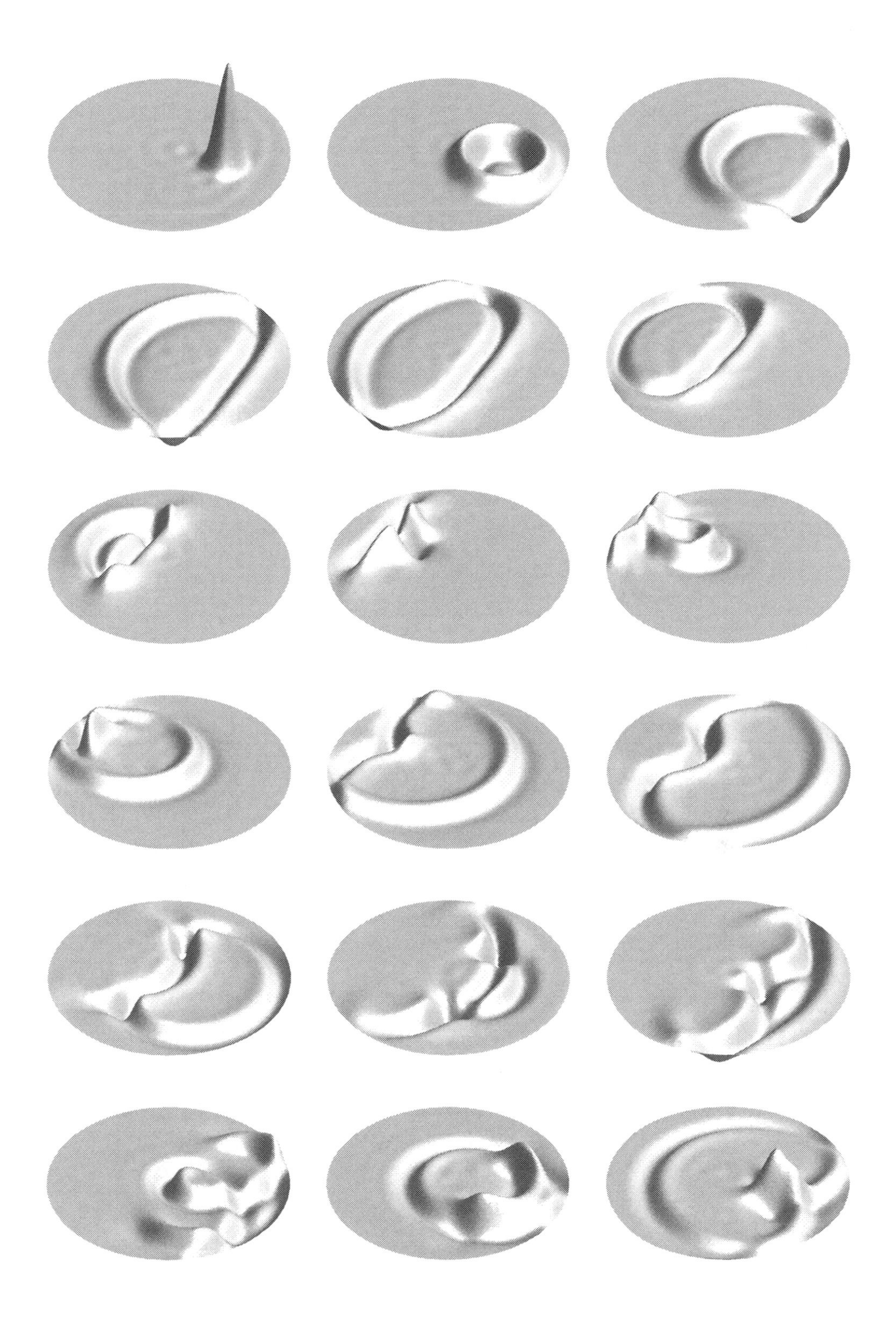

Figure 11.11. Vibration of a disk. 

formula (11.142), each eigenfunction engenders two independent normal modes of vibration, having the explicit forms

$$\begin{aligned} &\cos(c\,\zeta_{0,n}\,t)\;J_0(\zeta_{0,n}\,r), && \sin(c\,\zeta_{0,n}\,t)\;J_0(\zeta_{0,n}\,r),\\ &\cos(c\,\zeta_{m,n}\,t)\;J_m(\zeta_{m,n}\,r)\,\cos m\,\theta, && \sin(c\,\zeta_{m,n}\,t)\;J_m(\zeta_{m,n}\,r)\,\cos m\,\theta,\\ &\cos(c\,\zeta_{m,n}\,t)\;J_m(\zeta_{m,n}\,r)\,\sin m\,\theta, && \sin(c\,\zeta_{m,n}\,t)\;J_m(\zeta_{m,n}\,r)\,\sin m\,\theta. \end{aligned} \tag{11.158}$$

The general solution to (11.153–154) is then expressed as a Fourier–Bessel series:

$$\begin{aligned} u(t,r,\theta) = \frac{1}{2}\sum_{n=1}^{\infty} \big[\, a_{0,n}\cos(c\,\zeta_{0,n}\,t) + c_{0,n}\sin(c\,\zeta_{0,n}\,t)\,\big]\, J_0(\zeta_{0,n}\,r) \\ + \sum_{m,n=1}^{\infty} \big[\,\big(a_{m,n}\cos(c\,\zeta_{m,n}\,t) + c_{m,n}\sin(c\,\zeta_{m,n}\,t)\big)\cos m\,\theta \\ + \big(b_{m,n}\cos(c\,\zeta_{m,n}\,t) + d_{m,n}\sin(c\,\zeta_{m,n}\,t)\big)\sin m\,\theta\,\big]\, J_m(\zeta_{m,n}\,r), \end{aligned} \tag{11.159}$$

whose coefficients $a_{m,n}, b_{m,n}, c_{m,n}, d_{m,n}$ are determined, as usual, by the initial displacement and velocity of the membrane (11.155). In Figure 11.11, the vibrations due to an initially off-center concentrated displacement are displayed; the wave speed is $c = 1$, and the time interval between successive plots is $\Delta t = .3$. Again, the motion is only quasiperiodic and, no matter how long you wait, never quite returns to its original configuration.

Exercises

11.6.1. Use your physical intuition to decide whether the following statements are *true or false*. Then justify your answer.
(*a*) Increasing the stiffness of a membrane increases the wave speed.
(*b*) Increasing the density of a membrane increases the wave speed.
(*c*) Increasing the size of a membrane increases the wave speed

11.6.2. Two uniform membranes have the same shape, but are made out of different materials. Assuming that they are both subject to the same homogeneous boundary conditions, how are their vibrational frequencies related?

11.6.3. List the numerical values of the six lowest vibrational frequencies of a unit square with wave speed $c = 1$ when subject to homogeneous Dirichlet boundary conditions. How many linearly independent normal modes are associated with each of these frequencies?

♡ 11.6.4. The rectangular membrane $R = \{-1 < x < 1,\ 0 < y < 1\}$ has its two short sides attached to the (x,y)–plane, while its long sides are left free. The membrane is initially displaced so that its right half is one unit above, while its left half is one unit below the plane, and then released with zero initial velocity. (This discontinuous initial data serves to model a very sharp transition region.) Assume that the physical units are chosen so the wave speed $c = 1$. (*a*) Write down an initial-boundary value problem that governs the vibrations of the membrane. (*b*) What are the fundamental frequencies of vibration of the membrane? (*c*) Find the eigenfunction series solution that describes the subsequent motion of the membrane. (*d*) Is the motion (*i*) periodic? (*ii*) quasiperiodic? (*iii*) unstable? (*iv*) chaotic? Explain your answer.

11.6.5. Determine the solution to the following initial-boundary value problems for the wave equation on the rectangle $R = \{0 < x < 2,\ 0 < y < 1\}$:

(a) $\begin{cases} u_{tt} = u_{xx} + u_{yy}, & u(t,x,0) = u(t,x,1) = u(t,0,y) = u(t,2,y) = 0, \\ u(0,x,y) = \sin \pi y, & u_t(0,x,y) = \sin \pi y; \end{cases}$

(b) $\begin{cases} u_{tt} = u_{xx} + u_{yy}, & u(t,x,0) = u(t,x,1) = \dfrac{\partial u}{\partial x}(t,0,y) = \dfrac{\partial u}{\partial x}(t,2,y) = 0, \\ u(0,x,y) = \sin \pi y, & u_t(0,x,y) = \sin \pi y; \end{cases}$

(c) $\begin{cases} u_{tt} = u_{xx} + u_{yy}, & u(t,x,0) = u(t,x,1) = u(t,0,y) = u(t,2,y) = 0, \\ u(0,x,y) = \begin{cases} 1, & 0 < x < 1, \\ 0, & 1 < x < 2, \end{cases} & u_t(0,x,y) = 0; \end{cases}$

(d) $\begin{cases} u_{tt} = 2u_{xx} + 2u_{yy}, & u(t,x,0) = u(t,x,1) = u(t,0,y) = u(t,2,y) = 0, \\ u(0,x,y) = 0, & u_t(0,x,y) = \begin{cases} 1, & 0 < x < 1, \\ 0, & 1 < x < 2. \end{cases} \end{cases}$

11.6.6. *True or false*: The more sides of a rectangle that are tied down, the faster it vibrates.

11.6.7. Answer Exercise 11.6.3 when (*a*) two adjacent sides of the square are tied down and the other two are left free; (*b*) two opposite sides of the square are tied down and the other two are left free; (*c*) the membrane is freely floating in outer space.

11.6.8. A square drum has two sides fixed to a support and two sides left free. Does the drum vibrate faster if the fixed and free sides are adjacent to each other or on opposite sides?

11.6.9. Write down a periodic solution to the wave equation on a unit square, subject to homogeneous Dirichlet boundary conditions, that is *not* a normal mode. Does it vibrate at a fundamental frequency?

11.6.10. A rectangular drum with side lengths 1 cm by 2 cm and unit wave speed $c = 1$ has its boundary fixed to the (x, y)–plane while subject to a periodic external forcing of the form $F(t,x,y) = \cos(\omega t)\, h(x,y)$. (*a*) At which frequencies ω will the forcing incite resonance in the drum? (*b*) If ω is a resonant frequency, write down the condition(s) on $h(x,y)$ that ensure excitation of a resonant mode.

11.6.11. The right half of a rectangle of side lengths 1 by 2 is initially displaced, while the left half is quiescent. *True or false*: The ensuing vibrations are restricted to the right half of the membrane.

♡ 11.6.12. A torus (inner tube) can be obtained by gluing together each of the two pairs of opposite sides of a rubber rectangle. The (small) vibrations of the torus are described by the following periodic initial-boundary value problem for the wave equation, in which x, y represent angular variables:

$$\begin{aligned} &u_{tt} = c^2 \Delta u = c^2 (u_{xx} + u_{yy}), && u(0,x,y) = f(x,y), && u_t(0,x,y) = g(x,y), \\ &u(t,-\pi,y) = u(t,\pi,y), && u_x(t,-\pi,y) = u_x(t,\pi,y), && -\pi < x < \pi, \\ &u(t,x,-\pi) = u(t,x,\pi), && u_x(t,x,-\pi) = u_x(t,x,\pi), && -\pi < y < \pi. \end{aligned}$$

(*a*) Find the fundamental frequencies and normal modes of vibration. (*b*) Write down a series for the solution. (*c*) Discuss the stability of a vibrating torus. Is the motion (*i*) periodic; (*ii*) quasiperiodic; (*iii*) chaotic; (*iv*) none of these?

11.6.13. The *forced wave equation* $u_{tt} = c^2 \Delta u + F(x,y)$ on a bounded domain $\Omega \subset \mathbb{R}^2$ models a membrane subject to a constant external forcing function $F(x,y)$. Write down an eigenfunction series solution to the forced wave equation when the membrane is subject to homogeneous Dirichlet boundary conditions and initial conditions $u(0,x,y) = f(x,y)$, $u_t(0,x,y) = g(x,y)$. *Hint*: Expand the forcing function in an eigenfunction series.

11.6.14. A circular drum of radius $\zeta_{0,1} \approx 2.4048$ has initial displacement and velocity

$$u(0,x,y) = 0, \qquad \frac{\partial u}{\partial t}(0,x,y) = 2\, J_0\Big(\sqrt{x^2 + y^2}\,\Big).$$

Assuming that the circular edge of the drum is fixed to the (x, y)–plane, describe, both qualitatively and quantitatively, its subsequent motion.

11.6.15. Write out the integral formulae for the coefficients in the Fourier–Bessel series solution (11.159) to the wave equation in a circular disk in terms of the initial data

$$u(0,r,\theta) = f(r,\theta), \quad u_t(0,r,\theta) = g(r,\theta).$$

11.6.16. A circular drum at rest is struck with a concentrated blow at its center. Write down an eigenfunction series describing the resulting vibration.

♡ 11.6.17. (a) Set up and solve the initial-boundary value problem for the vibrations of a uniform circular drum of unit radius that is freely floating in space. (b) Discuss the stability of the drum's motion. (c) Are the vibrations slower or faster than when its edges are fixed to a plane?

11.6.18. A flat quarter-disk of radius 1 has its circular edge and one of its straight edges attached to the (x, y)–plane, while the other straight edge is left free. At time $t = 0$ the disk is struck with a hammer (unit delta function) at its midpoint, i.e., at radius $\frac{1}{2}$ and halfway between the straight edges. (a) Write down an initial-boundary value problem for the subsequent vibrations of the quarter-disk. *Hint*: Be careful with the form of the delta function in polar coordinates; see Exercise 6.3.6. (b) Assuming that the physical units are chosen so that the wave speed $c = 1$, determine the quarter-disk's vibrational frequencies. (c) Write down an eigenfunction series solution for the subsequent motion. (d) Is the motion unstable? periodic? If so, what is the period?

11.6.19. *True or false*: Assuming homogeneous Dirichlet boundary conditions, the fundamental frequencies of a vibrating half-disk are exactly twice those of the full disk of the same radius.

♡ 11.6.20. The edge of a circular drum is moved periodically up and down, so $u(t, 1, \theta) = \cos \omega t$. Assuming that the drum is initially at rest, discuss its response.

♣ 11.6.21. A drum is in the shape of a circular annulus with outer radius 1 meter and inner radius .5 meter. Find numerical values for its first three fundamental vibrational frequencies.

♡ 11.6.22. A homogeneous rope of length 1 and weight 1 is suspended from the ceiling. Taking x as the vertical coordinate, with $x = 1$ representing the fixed end and $x = 0$ the free end, the planar displacement $u(t, x)$ of the rope satisfies the initial-boundary value problem

$$\frac{\partial^2 u}{\partial t^2} = \frac{\partial}{\partial x}\left(x\,\frac{\partial u}{\partial x}\right), \qquad \begin{array}{ll} |\,u(t,0)\,| < \infty, & u(t,1) = 0, \\ u(0,x) = f(x), & \dfrac{\partial u}{\partial t}(0,x) = g(x), \end{array} \qquad t > 0, \quad 0 < x < 1.$$

(a) Find the solution. *Hint*: Let $y = \sqrt{x}$. (b) Are the vibrations periodic or quasiperiodic? (c) Describe the behavior of the rope when subject to uniform periodic external forcing $F(t, x) = a \cos \omega t$.

Scaling and Symmetry

Symmetry methods can also be effectively employed in the analysis of the wave equation. Let us consider the simultaneous rescaling

$$t \longmapsto \alpha t, \qquad x \longmapsto \beta x, \qquad y \longmapsto \beta y, \tag{11.160}$$

of time and space, whose effect is to change the function $u(t, x, y)$ into a rescaled version

$$U(t, x, y) = u(\alpha t, \beta x, \beta y). \tag{11.161}$$

The chain rule is employed to relate their derivatives:

$$\frac{\partial^2 U}{\partial t^2} = \alpha^2\,\frac{\partial^2 u}{\partial t^2}, \qquad \frac{\partial^2 U}{\partial x^2} = \beta^2\,\frac{\partial^2 u}{\partial x^2}, \qquad \frac{\partial^2 U}{\partial y^2} = \beta^2\,\frac{\partial^2 u}{\partial y^2}.$$

Therefore, if u satisfies the wave equation

$$u_{tt} = c^2\,\Delta u,$$

then U satisfies the rescaled wave equation

$$U_{tt} = \frac{\alpha^2 c^2}{\beta^2}\,\Delta U = C^2\,\Delta U, \quad \text{where the rescaled wave speed is} \quad C = \frac{\alpha\, c}{\beta}. \tag{11.162}$$

In particular, rescaling only time by setting $\alpha = 1/c$, $\beta = 1$, results in a unit wave speed $C = 1$. In other words, we are free to choose our unit of time measurement so as to fix the wave speed equal to 1.

If we set $\alpha = \beta$, scaling time and space in the same proportion, then the wave speed does not change, $C = c$, and so

$$t \longmapsto \beta\, t, \qquad x \longmapsto \beta\, x, \qquad y \longmapsto \beta\, y, \tag{11.163}$$

defines a *symmetry transformation* for the wave equation: If $u(t,x,y)$ is any solution to the wave equation, then so is its rescaled version

$$U(t,x,y) = u(\beta t, \beta x, \beta y) \tag{11.164}$$

for any choice of scale parameter $\beta \neq 0$. Observe that if $u(t,x,y)$ is defined on a domain Ω, then the rescaled solution $U(t,x,y)$ will be defined on the rescaled domain

$$\widetilde{\Omega} = \frac{1}{\beta}\,\Omega = \left\{ \left(\frac{x}{\beta}, \frac{y}{\beta} \right) \;\middle|\; (x,y) \in \Omega \right\} = \{\, (x,y) \mid (\beta x,\, \beta y) \in \Omega \,\}. \tag{11.165}$$

For instance, setting the scaling parameter $\beta = 2$ halves the size of the domain. The normal modes for the rescaled domain have the form

$$\begin{aligned} U_n(t,x,y) &= u_n(\beta t, \beta x, \beta y) = \cos(\beta\,\omega_n t)\; v_n(\beta x, \beta y), \\ \widetilde{U}_n(t,x,y) &= \widetilde{u}_n(\beta t, \beta x, \beta y) = \sin(\beta\,\omega_n t)\; v_n(\beta x, \beta y), \end{aligned}$$

and hence the rescaled vibrational frequencies are $\Omega_n = \beta\,\omega_n$. Thus, when $\beta < 1$, the rescaled membrane is larger by a factor $1/\beta$, and its vibrations are slowed down by the reciprocal factor β. For instance, a drum that is twice as large will vibrate twice as slowly, and hence have an octave lower overall tone. Musically, this means that all drums of a similar shape have the same pattern of overtones, differing only in their overall pitch, which is a function of their size, tautness, and density.

In particular, choosing $\beta = 1/R$ will rescale the unit disk into a disk of radius R. The fundamental frequencies of the rescaled disk are

$$\Omega_{m,n} = \beta\,\omega_{m,n} = \frac{c}{R}\,\zeta_{m,n}, \tag{11.166}$$

where c is the wave speed and $\zeta_{m,n}$ are the Bessel roots, defined in (11.118). Observe that the ratios $\omega_{m,n}/\omega_{m',n'}$ between vibrational frequencies remain the same, independent of the size of the disk R and the wave speed c. We define the *relative vibrational frequencies*

$$\rho_{m,n} = \frac{\omega_{m,n}}{\omega_{0,1}} = \frac{\zeta_{m,n}}{\zeta_{0,1}}, \qquad \text{in proportion to} \qquad \omega_{0,1} = \frac{c\,\zeta_{0,1}}{R} \approx 2.4\,\frac{c}{R}, \tag{11.167}$$

which is the drum's dominant, or lowest, vibrational frequency. The relative frequencies $\rho_{m,n}$ are independent of the size, stiffness or composition of the drum membrane. In the following table, we display a list of all relative vibrational frequencies (11.167) that are < 6. Once the lowest frequency $\omega_{0,1}$ has been determined — either theoretically, numerically, or experimentally — all the higher overtones $\omega_{m,n} = \rho_{m,n}\,\omega_{0,1}$ are simply obtained by rescaling.

Relative Vibrational Frequencies of a Circular Disk

$n \backslash m$	0	1	2	3	4	5	6	7	8	9	...
1	1.000	1.593	2.136	2.653	3.155	3.647	4.132	4.610	5.084	5.553	...
2	2.295	2.917	3.500	4.059	4.601	5.131	5.651	⋮	⋮	⋮	
3	3.598	4.230	4.832	5.412	5.977	⋮	⋮				
4	4.903	5.540	⋮	⋮	⋮						
⋮	⋮	⋮									

Exercises

11.6.23. *True or false*: Two rectangular membranes, made out of the same material and both subject to Dirichlet boundary conditions, have the same relative vibrational frequencies if and only if they are have similar shapes.

11.6.24. *True or false*: (*a*) The vibrational frequencies of a square with side lengths $a = b = 2$ are four times as slow as those of a square with side lengths $a = b = 1$.
(*b*) The vibrational frequencies of a rectangle with side lengths $a = 2$, $b = 1$, are twice as slow as those of a square with side lengths $a = b = 1$.

11.6.25. A vibrating rectangle of unknown size has wave speed $c = 1$ and is subject to homogeneous Dirichlet boundary conditions. How many of its lowest vibrational frequencies do you need to know in order to determine the size of the rectangle?

11.6.26. Answer Exercise 11.6.25 when the rectangle is subject to homogeneous Neumann boundary conditions.

♣ 11.6.27. A circular drum has the A above middle C, which has a frequency of 440 Hertz, as its lowest tone. What notes are the first five overtones nearest? Try playing these on a piano or guitar. Or, if you have a synthesizer, try assembling notes of these frequencies to see how closely it reproduces the dissonant sound of a drum.

11.6.28. In an orchestra, a circular snare drum of radius 1 foot sits near a second circular drum made out of the same material. Vibrations of the first drum are observed to excite an undesired resonant vibration in its partner. What are the possible radii of the second drum?

11.6.29. *True or false*: The relative vibrational frequencies of a half-disk, subject to Dirichlet boundary conditions, are a subset of the relative vibrational frequencies of a full disk.

11.6.30. *True or false*: If $u(t,x,y) = \cos(\omega t)\, v(x,y)$ is a normal mode of vibration for a unit square subject to homogeneous Dirichlet boundary conditions, then the function $\widehat{u}(t,x,y) = \cos(\omega t)\, v\left(\frac{1}{2}x, \frac{1}{3}y\right)$ is a normal mode of vibration for a 2×3 rectangle that is subject to the same boundary conditions, but with a possibly different wave speed. If true, how are the wave speeds of the two rectangles related?

11.6.31. Prove that if $u(t,x,y)$ is a solution to the two-dimensional wave equation, so is the translated function $U(t,x,y) = u(t - t_0, x - x_0, y - y_0)$, for any constants t_0, x_0, y_0.

◇ 11.6.32. (a) Prove that if $u(t,x,y)$ solves the wave equation, so does $U(t,x,y) = u(-t,x,y)$. Thus, unlike the heat equation, the wave equation is time-reversible, and its solutions can be unambiguously followed backwards in time. (b) Suppose $u(t,x,y)$ solves the initial value problem (11.141). Write down the initial value problem satisfied by $U(t,x,y)$.

11.6.33. (a) Prove that, on $\mathbb{R}^2$, the solution to the pure displacement initial value problem $u_{tt} = c^2 \Delta u$, $u(0,x,y) = f(x,y)$, $u_t(0,x,y) = 0$, is an even function of t.
(b) Prove that the solution to the pure velocity initial value problem $u_{tt} = c^2 \Delta u$, $u(0,x,y) = 0$, $u_t(0,x,y) = g(x,y)$, is an odd function of t.
Hint: Use Exercise 11.6.32 and uniqueness of solutions to the initial value problem.

11.6.34. Suppose $v(t,x)$ is any solution to the one-dimensional wave equation $v_{tt} = v_{xx}$. Prove that $u(t,x,y) = v(t, a\,x + b\,y)$, for any constants $(a,b) \neq (0,0)$, solves the two-dimensional wave equation $u_{tt} = c^2(u_{xx} + u_{yy})$ for some choice of wave speed. Describe the behavior of such solutions.

11.6.35. A *traveling-wave solution* to the two-dimensional wave equation has the form $u(t,x,y) = v(x - a\,t, y - a\,t)$, where a is a constant. Find the partial differential equation satisfied by the function $v(\xi,\eta)$. Is the equation hyperbolic?

11.6.36. Is the counterpart of Lemma 11.11 valid for the wave equation? In other words, if $v(t,x)$ and $w(t,x)$ are any two solutions to the one-dimensional wave equation, is their product $u(t,x,y) = v(t,x)\,w(t,y)$ a solution to the two-dimensional wave equation?

11.6.37. (a) How would you solve an initial-boundary value problem for the wave equation on a rectangle that is not aligned with the coordinate axes? (b) Apply your method to set up and solve an initial-boundary value problem on the square $R = \{\,|\,x+y\,| < 1,\ |\,x-y\,| < 1\,\}$.

Chladni Figures and Nodal Curves

When a membrane vibrates, its individual atoms typically move up and down in a quasiperiodic manner. As such, there is little correlation between their motions at different locations. However, if the membrane is set to vibrate in a pure eigenmode, say

$$u_n(t,x,y) = \cos(\omega_n t)\, v_n(x,y), \tag{11.168}$$

then all points move up and down at a common frequency $\omega_n = \sqrt{\lambda_n}$, which is the square root of the eigenvalue corresponding to the eigenfunction $v_n(x,y)$. The exceptions are the points where the eigenfunction vanishes:

$$v_n(x,y) = 0, \tag{11.169}$$

which remain stationary. The set of all points $(x,y) \in \Omega$ that satisfy (11.169) is known as the nth *Chladni figure* of the domain Ω, named in honor of the eighteenth-century German physicist and musician Ernst Chladni who first observed them experimentally by exciting a metal plate with his violin bow, [**43**]. The mathematical models governing such vibrating plates were formulated by the French mathematician Sophie Germain in the early 1800s. It can be shown that, in general, each Chladni figure consists of a finite system of *nodal curves*, [**34, 43**], that partition the membrane into disjoint *nodal regions*. As the membrane vibrates, the nodal curves remain stationary, while each nodal region is entirely either above or below the equilibrium plane, except, momentarily, when the *entire* membrane has zero displacement. As Chladni discovered in his original experiments, scattering small

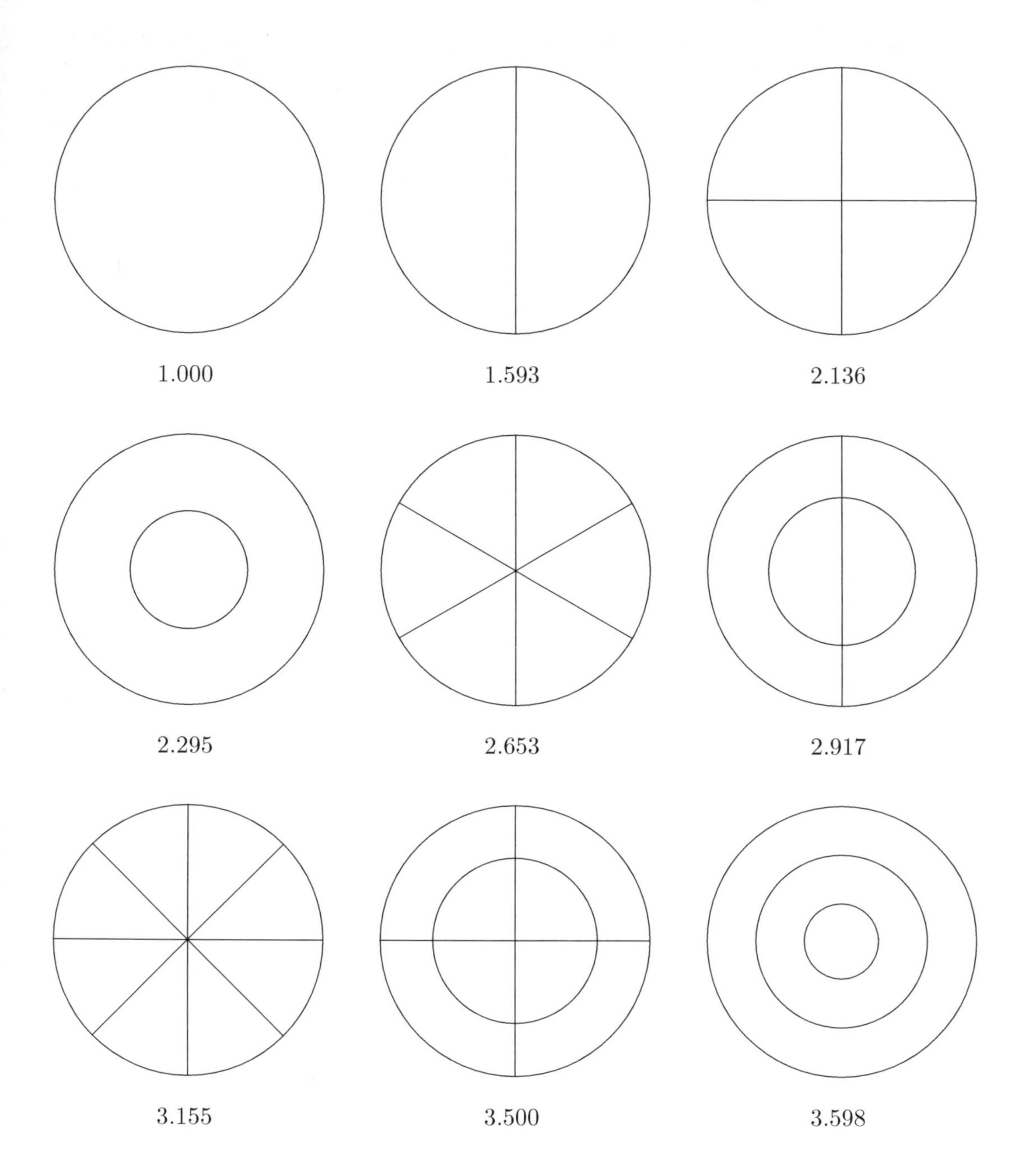

Figure 11.12. Nodal curves and relative vibrational frequencies of a circular membrane.

particles (e.g., fine sand) over a membrane or plate vibrating in an eigenmode will enable us to visualize the Chladni figure, because the particles will tend to accumulate along the stationary nodal curves. Adjacent nodal regions, lying on the opposite sides of a nodal curve, move in opposing directions — when one is up, its neighbors are down, and then they switch roles as the membrane becomes momentarily flat. Let us look at a couple of examples where the Chladni figures can be readily determined.

Example 11.15. *Circular Drums.* Since the eigenfunctions (11.156) for a disk are products of trigonometric functions in the angular variable and Bessel functions of the radius, the nodal curves for the normal modes of vibrations of a circular membrane are rays emanating from and circles centered at the origin. Consequently, the nodal regions are annular sectors. Chladni figures associated with the first nine normal modes, indexed by their relative frequencies, are plotted in Figure 11.12. Representative displacements of the membrane in each of the first twelve modes can be found earlier, in Figure 11.6. The dominant (lowest frequency) mode is the only one that has no nodal curves; it has the form of a radially symmetric bump where the entire membrane flexes up and down. The next lowest modes vibrate proportionally faster at a relative frequency $\rho_{1,1} \approx 1.593$. The most general solution with this vibrational frequency is a linear combination of the two eigensolutions: $\alpha\, u_{1,1} + \beta\, \widetilde{u}_{1,1}$. Each such combination has a single diameter as a nodal curve, whose angle with the horizontal depends on the ratio β/α. The two semicircular halves of the drum vibrate in opposing directions — when the top half is up, the bottom half is down and vice versa. The next set of modes have two perpendicular diameters as nodal curves; the four quadrants of the drum vibrate in tandem, with opposite quadrants moving in the same direction. Next in increasing order of vibrational frequency is a single mode, which has a circular nodal curve whose (relative) radius equals the ratio of the first two roots of the order zero Bessel function, $\zeta_{0,2}/\zeta_{0,1} \approx .43565$; see Exercise 11.6.39 for a justification. In this case, the inner disk and the outer annulus vibrate in opposing directions. And so on . . . .

Example 11.16. *Rectangular Drums.* For most rectangular drums, the Chladni figures are relatively uninteresting. Since the normal modes (11.151) are separable products of trigonometric functions in the coordinate variables x, y, the nodal curves are equally spaced straight lines parallel to the sides of the rectangle. The internodal regions are smaller rectangles, of identical size and shape, with adjacent rectangles vibrating in opposite directions.

More interesting figures appear when the rectangle admits multiple eigenvalues — so-called *accidental degeneracies*. Note that two of the eigenvalues (11.149) coincide, $\lambda_{m,n} = \lambda_{k,l}$, if and only if

$$\frac{m^2}{a^2} + \frac{n^2}{b^2} = \frac{k^2}{a^2} + \frac{l^2}{b^2}, \tag{11.170}$$

where $(m,n) \neq (k,l)$ are distinct pairs of positive integers. In such situations, the two eigenmodes happen to vibrate with a common frequency $\omega = \omega_{m,n} = \omega_{k,l}$. Consequently, any linear combination of the eigenmodes, e.g.,

$$\cos(\omega\, t) \left(\alpha \sin \frac{m \pi x}{a} \sin \frac{n \pi y}{b} + \beta \sin \frac{k \pi x}{a} \sin \frac{l \pi y}{b} \right), \qquad \alpha, \beta \in \mathbb{R},$$

is also a pure vibration, and hence qualifies as a normal mode. The associated nodal curves,

$$\alpha \sin \frac{m \pi x}{a} \sin \frac{n \pi y}{b} + \beta \sin \frac{k \pi x}{a} \sin \frac{l \pi y}{b} = 0, \qquad \begin{array}{l} 0 \leq x \leq a, \\ 0 \leq y \leq b, \end{array} \tag{11.171}$$

have a more intriguing geometry, which can change dramatically as the coefficients α, β vary.

For example, on the unit square $R = \{\, 0 < x, y < 1 \,\}$, an accidental degeneracy occurs whenever

$$m^2 + n^2 = k^2 + l^2 \tag{11.172}$$

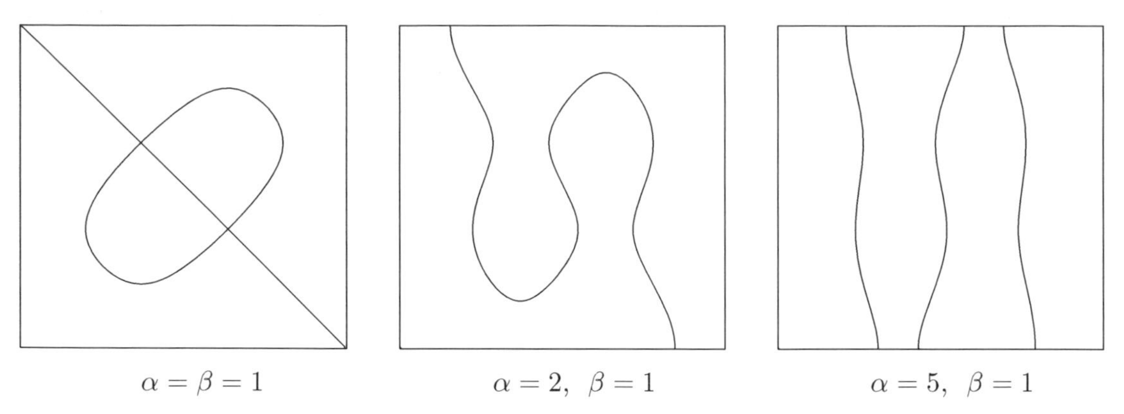

Figure 11.13. Some Chladni figures for a square membrane.

for distinct pairs of positive integers $(m, n) \neq (k, l)$. The simplest possibility arises whenever $m \neq n$, in which case we can merely reverse the order, setting $k = n$, $l = m$. In Figure 11.13 we plot three sample nodal curves

$$\alpha \sin 4\pi x \ \sin \pi y + \sin \pi x \ \sin 4\pi y = 0,$$

corresponding to three different linear combinations of the eigenfunctions with $m = l = 4$, $n = k = 1$. The associated vibrational frequency is, in all cases, $\omega_{4,1} = c\sqrt{17}\,\pi$, where c is the wave speed.

Classifying accidental degeneracies of rectangles takes us into the realm of number theory, [**9, 29**]. In the case of a square, equation (11.172) is asking us to locate all integer points $(m, n) \in \mathbb{Z}^2$ that lie on a common circle.

Remark: Bourget's hypothesis, mentioned after (11.157), implies that $\zeta_{m,n} \neq \zeta_{k,l}$ whenever $(m, n) \neq (k, l)$. This implies that a disk has no accidental degeneracies, and hence all its nodal curves are concentric circles and diameters.

Exercises

♢ 11.6.38. Suppose that a membrane is vibrating in a normal mode. Prove that the membrane lies instantaneously completely flat at regular time intervals.

♢ 11.6.39. For a vibrating disk of unit radius, determine the radius of the circular nodal curve for the next-to-lowest circular mode.

11.6.40. Order the five nodal circles displayed in Figure 11.12 according to their size.

11.6.41. Sketch the Chladni figures in a unit disk corresponding to the following vibrational frequencies. Determine numerical values for the radii of any circular nodal curves.
(a) $\omega_{4,0}$, (b) $\omega_{4,2}$, (c) $\omega_{2,4}$, (d) $\omega_{3,3}$, (e) $\omega_{1,5}$.

11.6.42. *True or false*: Any diameter of a circular disk is a nodal curve for some normal mode.

11.6.43. *True or false*: The nodal curves on a semicircular disk are all semicircles and rays emanating from the center.

11.6.44. (a) Find the smallest distinct pair of positive integers $(k,l) \neq (m,n)$ satisfying (11.172) that are not obtained by simply reversing the order, i.e., $(k,l) \neq (n,m)$. (b) Find the next-smallest example. (c) Plot two or three Chladni figures arising from such degenerate eigenfunctions.

♡ 11.6.45. Let R be a rectangle all of whose sides are fixed to the (x,y)–plane. Suppose that all its nodal curves are straight lines. What can you say about its side lengths a, b?

11.6.46. *True or false*: The nodal regions of a vibrating rectangle are similarly shaped rectangles.

◇ 11.6.47. Prove that any point of intersection (x_0, y_0) of two nodal curves associated with the same normal mode is a critical point of the associated eigenfunction: $\nabla v(x_0, y_0) = \mathbf{0}$.

11.6.48. *True or false*: The nodal curves on a domain do not depend on the choice of boundary conditions.

Chapter 12
Partial Differential Equations in Space

At last we have ascended to the ultimate rung of the dimensional ladder (at least for those of us living in a three-dimensional universe): partial differential equations in physical space. As in the one- and two-dimensional settings developed in the preceding chapters, the main protagonists are the Laplace and Poisson equations, modeling equilibrium configurations of solid bodies; the three-dimensional wave equation, governing vibrations of solids, liquids, and electromagnetic waves; and the three-dimensional heat equation, modeling spatial diffusion processes. To conclude this chapter — and the book — we will also analyze the particular three-dimensional Schrödinger equation that governs the hydrogen atom, and thereby characterizes atomic orbitals.

Fortunately, almost everything of importance has already appeared in the previous chapters, and appending a third dimension is, for the most part, simply a matter of appropriately adapting the constructions. We have already developed the principal solution techniques: separation of variables, Green's functions, and fundamental solutions. In three-dimensional problems, separation of variables is applicable in a variety of coordinate systems, including the usual rectangular, cylindrical, and spherical coordinates. The first two do not lead to anything fundamentally new, and are therefore relegated to the exercises. Separation in spherical coordinates requires spherical Bessel functions and spherical harmonics, which play essential roles in a wide variety of physical systems, both classical and quantum.

The Green's function for the three-dimensional Poisson equation in space can be identified as the classic Newton (Coulomb) $1/r$ gravitational (electrostatic) potential. The fundamental solution for the three-dimensional heat equation can be easily guessed from its one- and two-dimensional forms. The three-dimensional wave equation, surprisingly, has an explicit solution formula, named after Kirchhoff, of electrical fame, but originally due to Poisson. Counterintuitively, the best way to handle the two-dimensional wave equation is by "descending" from the simpler(!) three-dimensional Kirchhoff formula. Descent reveals a remarkable difference between waves in planar and spatial media. Huygens' Principle states that three-dimensional waves emanating from a localized initial disturbance remain localized as they propagate through space. In contrast, initially concentrated two-dimensional disturbances leave a slowly decaying remnant that never entirely disappears.

The final section is concerned with the Schrödinger equation for a hydrogen atom, that is, the quantum-dynamical system governing the spatial motion of a single electron around a positively charged nucleus. As we will see, the spherical harmonic eigensolutions account for the observed quantum energy levels of atoms that underly the periodic table and hence the foundations of molecular chemistry.

12.1 The Three–Dimensional Laplace and Poisson Equations

We begin our investigations, as usual, with systems in equilibrium, deferring dynamics until later. The prototypical equilibrium system is the three-dimensional Laplace equation

$$\Delta u = \frac{\partial^2 u}{\partial x^2} + \frac{\partial^2 u}{\partial y^2} + \frac{\partial^2 u}{\partial z^2} = 0, \tag{12.1}$$

in which $\mathbf{x} = (x, y, z)^T$ represents rectangular coordinates on $\mathbb{R}^3$. The solutions $u(x, y, z)$ continue to be known as *harmonic functions*. The Laplace equation models unforced equilibria; *Poisson's equation* is the inhomogeneous version

$$-\Delta u = f(x, y, z), \tag{12.2}$$

whose right-hand side represents some form of external forcing.

The basic boundary value problem for the Laplace and Poisson equations seeks a solution inside a bounded domain $\Omega \subset \mathbb{R}^3$ subject to either *Dirichlet boundary conditions*, prescribing the function values on the domain's boundary:

$$u = h \qquad \text{on} \qquad \partial\Omega, \tag{12.3}$$

or *Neumann boundary conditions*, prescribing its normal derivative or flux through the boundary:

$$\frac{\partial u}{\partial \mathbf{n}} = k \qquad \text{on} \qquad \partial\Omega, \tag{12.4}$$

or *mixed boundary conditions*, in which one imposes Dirichlet conditions on part of the boundary and Neumann conditions on the remainder. Keep in mind that the boundary of the solid domain Ω consists of one or more piecewise smooth closed surfaces, which will be oriented by use of the outward — meaning exterior to the domain — unit normal $\mathbf{n}$.

The boundary value problems for the three-dimensional Laplace and Poisson equations govern a wide variety of physical systems, including:

- *Heat conduction*: The solution u represents the equilibrium temperature in a solid body. The inhomogeneity f represents some form of internal heat source or sink. Dirichlet conditions correspond to fixing the temperature on the bounding surface(s), whereas homogeneous Neumann conditions correspond to an insulated boundary, i.e., one that does not allow any heat flux.
- *Ideal fluid flow*: Here the solution u to the Laplace equation represents the velocity potential for an incompressible, irrotational steady-state fluid flow inside a container governed by the velocity vector field $\mathbf{v} = \nabla u$. Homogeneous Neumann boundary conditions correspond to a solid boundary that the fluid cannot penetrate.
- *Elasticity*: In certain restricted contexts, u represents an equilibrium deformation of a solid body, e.g., the radial deformation of an elastic ball.
- *Electrostatics*: In applications to electromagnetism, u is the electric potential in a conducting medium; its gradient ∇u prescribes the electromotive force on a charged particle. The inhomogeneity f represents an external electrostatic force field.
- *Gravitation*: The Newtonian gravitational potential in flat empty space is also prescribed by the Laplace equation. (In contrast, Einstein's theory of general relativity requires a vastly more complicated nonlinear system of partial differential equations, [**75**].)

Self–Adjoint Formulation and Minimum Principle

The Laplace and Poisson equations naturally fit into the general self-adjoint equilibrium framework summarized in Chapter 9. We introduce the L^2 inner products

$$\begin{aligned}\langle\, u\,,\widetilde{u}\,\rangle &= \iiint_\Omega u(x,y,z)\,\widetilde{u}(x,y,z)\,dx\,dy\,dz,\\ \langle\, \mathbf{v}\,,\widetilde{\mathbf{v}}\,\rangle &= \iiint_\Omega \mathbf{v}(x,y,z)\cdot\widetilde{\mathbf{v}}(x,y,z)\,dx\,dy\,dz,\end{aligned} \tag{12.5}$$

between, respectively, scalar fields $u, \widetilde{u}$, and vector fields $\mathbf{v}, \widetilde{\mathbf{v}}$, which are defined on the domain $\Omega \subset \mathbb{R}^3$. We assume that the functions in question are sufficiently nice in order that these inner products be well defined; if Ω is unbounded, this, in essence, requires that they decay reasonably rapidly to zero at large distances.

When subject to suitable homogeneous boundary conditions, the three-dimensional Laplace equation can be placed in our standard self-adjoint form

$$-\,\Delta u = -\,\nabla\cdot\nabla u = \nabla^*\circ\nabla u. \tag{12.6}$$

This relies on the fact that the adjoint of the gradient operator with respect to the L^2 inner products (12.5) is minus the divergence operator:

$$\nabla^*\mathbf{v} = -\,\nabla\cdot\mathbf{v}. \tag{12.7}$$

As usual, the determination of the adjoint rests on an integration by parts formula, which, in three-dimensional space, is a consequence of the Divergence Theorem from multivariable calculus, [**8**, **108**]:

Theorem 12.1. *Let $\Omega \subset \mathbb{R}^3$ be a bounded domain whose boundary $\partial\Omega$ consists of one or more piecewise smooth simple closed surfaces. Let $\mathbf{n}$ denote the unit outward normal to the boundary of Ω. Let $\mathbf{v}$ be a C^1 vector field defined on Ω and continuous up to its boundary. Then the surface integral, with respect to surface area, of the normal component of $\mathbf{v}$ over the boundary of the domain equals the triple integral of its divergence over the domain*:

$$\iint_{\partial\Omega} \mathbf{v}\cdot\mathbf{n}\;dS = \iiint_\Omega \nabla\cdot\mathbf{v}\;\,dx\,dy\,dz. \tag{12.8}$$

Replacing $\mathbf{v}$ by the product $u\mathbf{v}$ of a scalar field u and a vector field $\mathbf{v}$ yields

$$\iiint_\Omega (u\,\nabla\cdot\mathbf{v} + \nabla u\cdot\mathbf{v})\,dx\,dy\,dz = \iiint_\Omega \nabla\cdot(u\,\mathbf{v})\,dx\,dy\,dz = \iint_{\partial\Omega} u\,(\mathbf{v}\cdot\mathbf{n})\,dS. \tag{12.9}$$

Rearranging the terms produces the desired *integration by parts formula* for triple integrals:

$$\iiint_\Omega (\nabla u\cdot\mathbf{v})\,dx\,dy\,dz = \iint_{\partial\Omega} u\,(\mathbf{v}\cdot\mathbf{n})\,dS - \iiint_\Omega u\,(\nabla\cdot\mathbf{v})\,dx\,dy\,dz. \tag{12.10}$$

The boundary surface integral will vanish, provided either $u = 0$ or $\mathbf{v}\cdot\mathbf{n} = \mathbf{0}$ at each point on $\partial\Omega$. When $u = 0$ on all of $\partial\Omega$, we have homogeneous Dirichlet conditions. Setting $\mathbf{v}\cdot\mathbf{n} = \mathbf{0}$ everywhere on $\partial\Omega$ results in the homogeneous Neumann boundary value problem owing to the identification of $\mathbf{v} = \nabla u$. Finally, the mixed boundary value problem takes $u = 0$ on part of $\partial\Omega$ and $\mathbf{v}\cdot\mathbf{n} = \mathbf{0}$ on the rest. Thus, subject to one of these choices, the integration by parts formula (12.10) reduces to

$$\langle\, \nabla u\,,\mathbf{v}\,\rangle = \langle\, u\,,-\,\nabla\cdot\mathbf{v}\,\rangle, \tag{12.11}$$

which suffices to establish the adjoint formula (12.7).

Remark: Adopting more general weighted inner products results in a more general elliptic boundary value problem. See Exercise 12.1.9 for details.

According to Theorem 9.20, the self-adjoint formulation (12.6) automatically implies positive semi-definiteness of the boundary value problem, with positive definiteness if $\ker \nabla = \{0\}$. Since, on a connected domain, only constant functions are annihilated by the gradient operator — see Lemma 6.16, which also applies to three-dimensional domains — both the Dirichlet and mixed boundary value problems are positive definite, while the Neumann boundary value problem is only positive semi-definite.

Finally, in the positive definite cases, Theorem 9.26 implies that the solution can be characterized by the three-dimensional version of the Dirichlet minimization principle (9.82).

Theorem 12.2. *The solution $u(x,y,z)$ to the Poisson equation* (12.2) *subject to homogeneous Dirichlet or mixed boundary conditions* (12.3) *is the unique function that minimizes the* Dirichlet integral

$$\tfrac{1}{2} \, ||| \, \nabla u \, |||^2 - \langle \, u \, , f \, \rangle = \iiint_\Omega \left[\tfrac{1}{2} \, (u_x^2 + u_y^2 + u_z^2) - f \, u \right] dx \, dy \, dz \tag{12.12}$$

among all C^2 *functions that satisfy the prescribed boundary conditions.*

As in the two-dimensional version discussed in Chapter 9, the Dirichlet minimization principle continues to hold in the case of the inhomogeneous Dirichlet boundary value problem. Modifications for the inhomogeneous mixed boundary value problem appear in Exercise 12.1.13.

Exercises

12.1.1. Find bases for the following: (*a*) the space of harmonic polynomials $u(x,y,z)$ of degree ≤ 2; (*b*) the space of homogeneous cubic harmonic polynomials $u(x,y,z)$.

12.1.2. *True or false*: (*a*) Every harmonic polynomial is homogeneous.
(*b*) Every homogeneous polynomial is harmonic.

12.1.3. Solve the Poisson boundary value problem $-\Delta u = 1$ on the unit ball $x^2 + y^2 + z^2 < 1$ with homogeneous Dirichlet boundary conditions. *Hint*: Look for a polynomial solution.

♢ 12.1.4. Prove that if $u(x,y,z)$ solves the Laplace equation, then so does the translated function $U(x,y,z) = u(x-a, y-b, z-c)$ for constants a, b, c.

♢ 12.1.5. (*a*) Prove that if $u(x,y,z)$ solves Laplace's equation, so does the rescaled function $U(x,y,z) = u(\lambda x, \lambda y, \lambda z)$ for any constant λ. (*b*) More generally, show that $U(x,y,z) = \mu u(\lambda x, \lambda y, \lambda z) + c$ solves Laplace's equation for any constants λ, μ, c.

♢ 12.1.6. Let A be a constant nonsingular 3×3 matrix, $u(\mathbf{x})$ a C^1 scalar field, and $\mathbf{v}(\mathbf{x})$ a C^1 vector field. Set $U(\mathbf{x}) = u(A\,\mathbf{x})$ and $\mathbf{V}(\mathbf{x}) = \mathbf{v}(A\,\mathbf{x})$. Prove that
(*a*) $\nabla U(\mathbf{x}) = A^T \nabla u(A\,\mathbf{x})$, (*b*) $\nabla \cdot \mathbf{V}(\mathbf{x}) = w(A\,\mathbf{x})$, where $w(\mathbf{x}) = \nabla \cdot (A\,\mathbf{v})(\mathbf{x})$.

♢ 12.1.7. Prove that every rotation and reflection is a symmetry of the Laplace equation. In other words, if Q is any 3×3 orthogonal matrix, so $Q^T Q = \mathrm{I}$, and $u(\mathbf{x})$ is a harmonic function, then so is $U(\mathbf{x}) = u(Q\,\mathbf{x})$. *Hint*: Use Exercise 12.1.6.

♢ 12.1.8. *The Weak Maximum Principle*: Let $\Omega \subset \mathbb{R}^2$ be a bounded domain. Let $u(x,y,z)$ solve the Poisson equation $-\Delta u = f(x,y,z)$, where $f(x,y,z) < 0$ for all $(x,y,z) \in \Omega$.
(*a*) Prove that the maximum value of u occurs on the boundary $\partial\Omega$.
Hint: Explain why u cannot have a local maximum at any interior point in Ω.
(*b*) Generalize your result to the case $f(x,y,z) \le 0$.
Hint: Look at $v_\varepsilon(x,y,z) = u(x,y,z) + \varepsilon\,(x^2+y^2+z^2)$ and let $\varepsilon \to 0^+$.

♢ 12.1.9. Find the equilibrium equations corresponding to minimizing $|||\, \nabla u \,|||^2$ subject to homogeneous Dirichlet boundary conditions, where the indicated norm is based on the weighted inner product

$$\langle\!\langle\, \mathbf{v}\,, \mathbf{w} \,\rangle\!\rangle = \iiint_\Omega \mathbf{v}(x,y,z) \cdot \mathbf{w}(x,y,z)\, \sigma(x,y,z)\, dx\, dy\, dz,$$

with $\sigma(x,y,z) > 0$ a positive scalar function.

♢ 12.1.10. Prove the following vector calculus identities:
(*a*) $\nabla \cdot (u\,\mathbf{v}) = \nabla u \cdot \mathbf{v} + u\,\nabla \cdot \mathbf{v}$, (*b*) $\nabla \times (u\,\mathbf{v}) = \nabla u \times \mathbf{v} + u\,\nabla \times \mathbf{v}$,
(*c*) $\nabla \cdot (\mathbf{v} \times \mathbf{w}) = (\nabla \times \mathbf{v}) \cdot \mathbf{w} - \mathbf{v} \cdot (\nabla \times \mathbf{w})$, (*d*) $\nabla \times (\nabla \times \mathbf{v}) = \nabla(\nabla \cdot \mathbf{v}) - \Delta \mathbf{v}$.
(In the final term, the Laplacian Δ acts component-wise on the vector field $\mathbf{v}$.)

♢ 12.1.11. Let Ω be a bounded domain with piecewise smooth boundary $\partial\Omega$. Prove the following identities: (*a*) $\displaystyle\iiint_\Omega \Delta u\, dx\, dy\, dz = \iint_{\partial\Omega} \frac{\partial u}{\partial \mathbf{n}}\, dS$,

(*b*) $\displaystyle\iiint_\Omega u\,\Delta u\, dx\, dy\, dz = \iint_{\partial\Omega} u\,\frac{\partial u}{\partial \mathbf{n}}\, dS - \iiint_\Omega \|\, \nabla u \,\|^2\, dx\, dy\, dz.$

12.1.12. Suppose the inhomogeneous Neumann boundary value problem (12.1, 4) has a solution. (*a*) Prove that $\displaystyle\iint_{\partial\Omega} k\, dS = 0$. (*b*) Is the solution unique? If not, what is the most general solution? (*c*) State and prove an analogous result for the inhomogeneous Poisson equation $-\Delta u = f(x,y,z)$. (*d*) Provide a physical explanation for your answers.

♢ 12.1.13. Find a minimization principle that characterizes the solution to the inhomogeneous mixed boundary value problem $-\Delta u = f$ on Ω, with $u = g$ on $D \subsetneq \partial\Omega$, and $\partial u/\partial \mathbf{n} = h$ on $N = \partial\Omega \setminus D$.

♡ 12.1.14. (*a*) Prove that, subject to suitable boundary conditions, the curl $\nabla\times$ defines a self-adjoint operator with respect to the L^2 inner product between vector fields. What kinds of boundary conditions do you need to impose for your integration by parts argument to be valid? *Hint*: Use the identity in Exercise 12.1.10(*c*). (*b*) What operator on vector fields is given by the self-adjoint composition $S = (\nabla\times)^* \circ (\nabla\times)$? (*c*) Choose a set of homogeneous boundary conditions that make S self-adjoint. Is the resulting boundary value problem $S[\mathbf{v}] = \mathbf{f}$ positive definite? If not, what does the Fredholm Alternative say about its solvability?

12.2 Separation of Variables for the Laplace Equation

In this section, we revisit the method of separation of variables in the context of the three-dimensional Laplace equation. As always, its applicability is unfortunately restricted to rather special, but important, geometric configurations, the simplest being rectangular, cylindrical, and spherical domains. Since the first two are straightforward extensions of their two-dimensional counterparts, we will discuss only spherically separable solutions in any detail.

The simplest domain to which the separation of variables method applies is a rectangular box:

$$B = \{0 < x < a,\ 0 < y < b,\ 0 < z < c\}.$$

For functions of three variables, one begins the separation process by splitting off one of them, by setting $u(x,y,z) = v(x)\,w(y,z)$, say. The function $v(x)$ satisfies a simple second-order ordinary differential equation, while $w(y,z)$ solves the two-dimensional Helmholtz equation (11.21), which is further separated by writing $w(y,z) = p(y)\,q(z)$. The resulting fully separated solutions $u(x,y,z) = v(x)\,p(y)\,q(z)$ are (mostly) products of trigonometric and hyperbolic functions. Implementation of the technique and analysis of the resulting series solutions are relegated to Exercise 12.2.34.

In the case that the domain is a cylinder, one passes to *cylindrical coordinates* r, θ, z, where

$$x = r\cos\theta, \qquad y = r\sin\theta, \qquad z = z, \tag{12.13}$$

to effect the separation. Writing $u(r,\theta,z) = v(r,\theta)\,w(z)$, one finds that $w(z)$ satisfies a simple second-order ordinary differential equation, while $v(r,\theta)$ solves the two-dimensional polar Helmholtz equation (11.51) on a disk. Applying a further separation to $v(r,\theta)$, as in Chapter 11, produces fully separable solutions $u(r,\theta,z) = p(r)\,q(\theta)\,w(z)$ as products of Bessel functions of the cylindrical radius r, trigonometric functions of the polar angle θ, and hyperbolic functions of z; see Exercise 12.2.40.

The most interesting case is that of spherical coordinates, which we proceed to analyze in detail in the following subsection.

Remark: These are just three of the many coordinate systems in which the three-dimensional Laplace equation separates. See [**78, 79**] for 37 additional exotic types, including ellipsoidal, toroidal, and parabolic spheroidal coordinates. The resulting separable solutions are written in terms of new classes of special functions that solve interesting second-order ordinary differential equations, all of Sturm–Liouville form (9.71).

Laplace's Equation in a Ball

Suppose a solid ball (e.g., the Earth) is subject to a specified steady temperature distribution on its spherical boundary. Our task is to determine the equilibrium temperature within the ball. We assume that the body is composed of an isotropic, uniform medium and, to slightly simplify the analysis, choose units in which its radius equals 1.

To find the equilibrium temperature within the ball, we must solve the Dirichlet boundary value problem

$$\begin{aligned} &\frac{\partial^2 u}{\partial x^2} + \frac{\partial^2 u}{\partial y^2} + \frac{\partial^2 u}{\partial z^2} = 0, \qquad && x^2 + y^2 + z^2 < 1, \\ &u(x,y,z) = h(x,y,z), && x^2 + y^2 + z^2 = 1, \end{aligned} \tag{12.14}$$

where h is prescribed on the bounding unit sphere. Problems in spherical geometries are most naturally analyzed in *spherical coordinates* r, φ, θ. Our convention is to set

$$x = r\sin\varphi\,\cos\theta, \qquad y = r\sin\varphi\,\sin\theta, \qquad z = r\cos\varphi, \tag{12.15}$$

where $-\pi < \theta \le \pi$ is the *azimuthal angle* or *longitude*, while $0 \le \varphi \le \pi$ is the *zenith angle* or *latitude* on the sphere of radius $r = \sqrt{x^2 + y^2 + z^2}$. In other words, φ measures

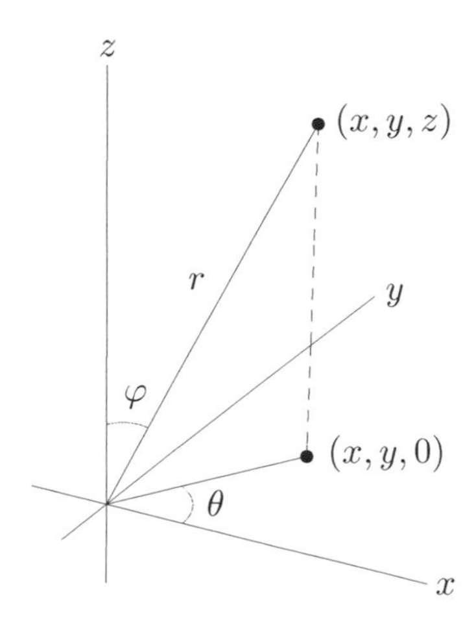

Figure 12.1. Spherical coordinates.

the angle between the vector $(x, y, z)^T$ and the positive z–axis, while θ measures the angle between its projection $(x, y, 0)^T$ on the (x, y)–plane and the positive x–axis; see Figure 12.1. On Earth, longitude θ is measured from the Greenwich prime meridian, while latitude is measured from the equator, and so equals $\frac{1}{2}\pi - \varphi$ (although the everyday units are degrees, not radians), whereby φ is sometimes referred to as the "co-latitude".

Warning: In many books, particularly those in physics, the roles of θ and φ are *reversed*, leading to much confusion when one is perusing the literature. We prefer the mathematical convention, since the azimuthal angle θ coincides with the cylindrical angle coordinate — as well as the polar coordinate on the (x, y)–plane — thus avoiding unnecessary confusion when going from one coordinate system to the other. You must be attentive to the convention being used when consulting any reference!

In spherical coordinates, the Laplace equation for $u(r, \varphi, \theta)$ takes the form

$$\Delta u = \frac{\partial^2 u}{\partial r^2} + \frac{2}{r}\frac{\partial u}{\partial r} + \frac{1}{r^2}\frac{\partial^2 u}{\partial \varphi^2} + \frac{\cos\varphi}{r^2 \sin\varphi}\frac{\partial u}{\partial \varphi} + \frac{1}{r^2 \sin^2\varphi}\frac{\partial^2 u}{\partial \theta^2} = 0. \tag{12.16}$$

This important formula is the final result of a fairly nasty chain rule computation, whose details are left to the motivated reader. (Set aside lots of paper and keep an eraser handy!)

To construct separable solutions to the spherical coordinate form (12.16) of the Laplace equation, we begin by separating off the radial part of the solution, setting

$$u(r, \varphi, \theta) = v(r)\, w(\varphi, \theta). \tag{12.17}$$

Substituting this ansatz into (12.16), multiplying the resulting equation through by $\dfrac{r^2}{v\,w}$, and then placing all the terms involving r on one side yields

$$\frac{1}{v}\left(r^2 \frac{d^2 v}{dr^2} + 2r\frac{dv}{dr} \right) = -\frac{1}{w}\Delta_S[w], \tag{12.18}$$

where

$$\Delta_S[w] = \frac{\partial^2 w}{\partial \varphi^2} + \frac{\cos\varphi}{\sin\varphi}\frac{\partial w}{\partial \varphi} + \frac{1}{\sin^2\varphi}\frac{\partial^2 w}{\partial \theta^2}. \tag{12.19}$$

The second-order differential operator Δ_S, which involves only the angular components of the full Laplacian operator Δ, is of particular significance. It is known as the *spherical Laplacian*, and governs the equilibrium and dynamics of thin spherical shells — see Example 12.15 below.

Returning to equation (12.18), our usual separation argument applies. The left-hand side depends only on r, while the right-hand side depends only on the angles φ, θ. This can occur only when both sides are equal to a common separation constant, which we denote by μ. As a consequence, the radial component $v(r)$ satisfies the ordinary differential equation

$$r^2 v'' + 2 r v' - \mu v = 0, \tag{12.20}$$

which is of Euler type (11.89), and hence can be readily solved. However, let us put this equation aside for the time being, and concentrate our efforts on the more complicated angular components.

The second equation in (12.18) assumes the form

$$\Delta_S[w] + \mu w = \frac{\partial^2 w}{\partial \varphi^2} + \frac{\cos\varphi}{\sin\varphi} \frac{\partial w}{\partial \varphi} + \frac{1}{\sin^2 \varphi} \frac{\partial^2 w}{\partial \theta^2} + \mu w = 0. \tag{12.21}$$

This second-order partial differential equation can be regarded as the eigenvalue equation for the spherical Laplacian operator Δ_S and is known as the *spherical Helmholtz equation*. To find explicit solutions, we adopt a further separation of angular variables,

$$w(\varphi, \theta) = p(\varphi)\, q(\theta), \tag{12.22}$$

which we substitute into (12.21). Dividing the result by the product $w = p\,q$, multiplying by $\sin^2 \varphi$, and then rearranging terms, we are led to the separated system

$$\frac{1}{p}\left(\sin^2 \varphi \frac{d^2 p}{d\varphi^2} + \cos\varphi \sin\varphi \frac{dp}{d\varphi}\right) + \mu \sin^2 \varphi = -\frac{1}{q}\frac{d^2 q}{d\theta^2} = \nu,$$

where, by our usual argument, ν is another separation constant. The spherical Helmholtz equation thereby splits into a pair of ordinary differential equations

$$\sin^2 \varphi \frac{d^2 p}{d\varphi^2} + \cos\varphi \,\sin\varphi \frac{dp}{d\varphi} + (\mu \sin^2 \varphi - \nu)\, p = 0, \qquad \frac{d^2 q}{d\theta^2} + \nu\, q = 0.$$

The equation for $q(\theta)$ is easy to solve. As one circumnavigates the sphere, the azimuthal angle θ increases from $-\pi$ to π, so $q(\theta)$ must be a 2π–periodic function. Thus, $q(\theta)$ solves the well-studied periodic boundary value problem treated, for instance, in (4.109). Up to a constant multiple, nonzero periodic solutions occur only when the separation constant assumes one of the values $\nu = m^2$, where $m = 0, 1, 2, \ldots$ is an integer, with

$$q(\theta) = \cos m\theta \qquad \text{or} \qquad \sin m\theta, \qquad\qquad m = 0, 1, 2, \ldots. \tag{12.23}$$

Each positive $\nu = m^2 > 0$ admits two linearly independent 2π–periodic solutions, while when $\nu = 0$, only the constant solutions are periodic.

The Legendre Equation and Ferrers Functions

With this information, we endeavor to solve the zenith differential equation

$$\sin^2 \varphi \frac{d^2 p}{d\varphi^2} + \cos\varphi \,\sin\varphi \frac{dp}{d\varphi} + (\mu \sin^2 \varphi - m^2)\, p = 0. \tag{12.24}$$

This is not so easy, and constructing analytic formulas for its solutions requires some ingenuity. The motivation behind the following steps may not be so apparent; indeed, they are the culmination of a long, detailed study of this important differential equation by mathematicians over the last 200 years.

As an initial simplification, let us get rid of the trigonometric functions, by invoking the change of variables

$$t = \cos\varphi, \qquad \text{with} \qquad p(\varphi) = P(\cos\varphi) = P(t). \tag{12.25}$$

Since

$$0 \le \varphi \le \pi, \qquad \text{we have} \qquad 0 \le \sqrt{1-t^2} = \sin\varphi \le 1.$$

According to the chain rule,

$$\begin{aligned} \frac{dp}{d\varphi} &= \frac{dP}{dt}\,\frac{dt}{d\varphi} = -\sin\varphi\,\frac{dP}{dt} = -\sqrt{1-t^2}\,\frac{dP}{dt}, \\ \frac{d^2p}{d\varphi^2} &= -\sin\varphi\,\frac{d}{dt}\left(-\sqrt{1-t^2}\,\frac{dP}{dt}\right) = (1-t^2)\,\frac{d^2P}{dt^2} - t\,\frac{dP}{dt}. \end{aligned}$$

Substituting these expressions into (12.24), we conclude that $P(t)$ must satisfy

$$(1-t^2)^2\,\frac{d^2P}{dt^2} - 2t\,(1-t^2)\,\frac{dP}{dt} + \left[\,\mu\,(1-t^2) - m^2\,\right] P = 0. \tag{12.26}$$

Unfortunately, the resulting differential equation is still not elementary, but at least its coefficients are polynomials. It is known as the *Legendre differential equation* of *order* m, having first been employed by Adrien–Marie Legendre to study the gravitational attraction of *ellipsoid*al bodies. In the cases of interest to us, the order parameter m is an integer, while the separation constant μ plays the role of an eigenvalue.

Power series solutions to the Legendre equation can be constructed by the standard techniques presented in Section 11.3. The most general solution is a new type of special function, called a *Legendre function*, [**86**]. However, it turns out that the solutions we are actually interested in can all be written in terms of elementary algebraic functions. First of all, since $t = \cos\varphi$, the solution only needs to be defined on the interval $-1 \le t \le 1$, the so-called *cut locus*. The endpoints of the cut locus, $t = 1$ and $t = -1$, correspond to the sphere's north pole, $\varphi = 0$, and south pole, $\varphi = \pi$, respectively. Both endpoints are singular points for the Legendre equation, since the coefficient $(1-t^2)^2$ of the leading-order derivative vanishes when $t = \pm 1$. In fact, both are regular singular points, as you are asked to show in Exercise 12.2.11. Since ultimately we need the separable solution (12.17) to be a well-defined function of x, y, z (even at points where the spherical coordinates degenerate, i.e., on the z–axis), we need $p(\varphi)$ to be well defined at $\varphi = 0$ and π, and this requires $P(t)$ to be bounded at the singular points:

$$|\,P(-1)\,| < \infty, \qquad\qquad |\,P(+1)\,| < \infty. \tag{12.27}$$

Let us begin our analysis with the Legendre equation of order $m = 0$

$$(1-t^2)\,\frac{d^2P}{dt^2} - 2t\,\frac{dP}{dt} + \mu\,P = 0. \tag{12.28}$$

In this case, the eigenfunctions, i.e., solutions to the Legendre boundary value problem (12.27–28), are the *Legendre polynomials*

$$P_n(t) = \frac{(-1)^n}{2^n\,n!}\,\frac{d^n}{dt^n}\,(1-t^2)^n, \tag{12.29}$$

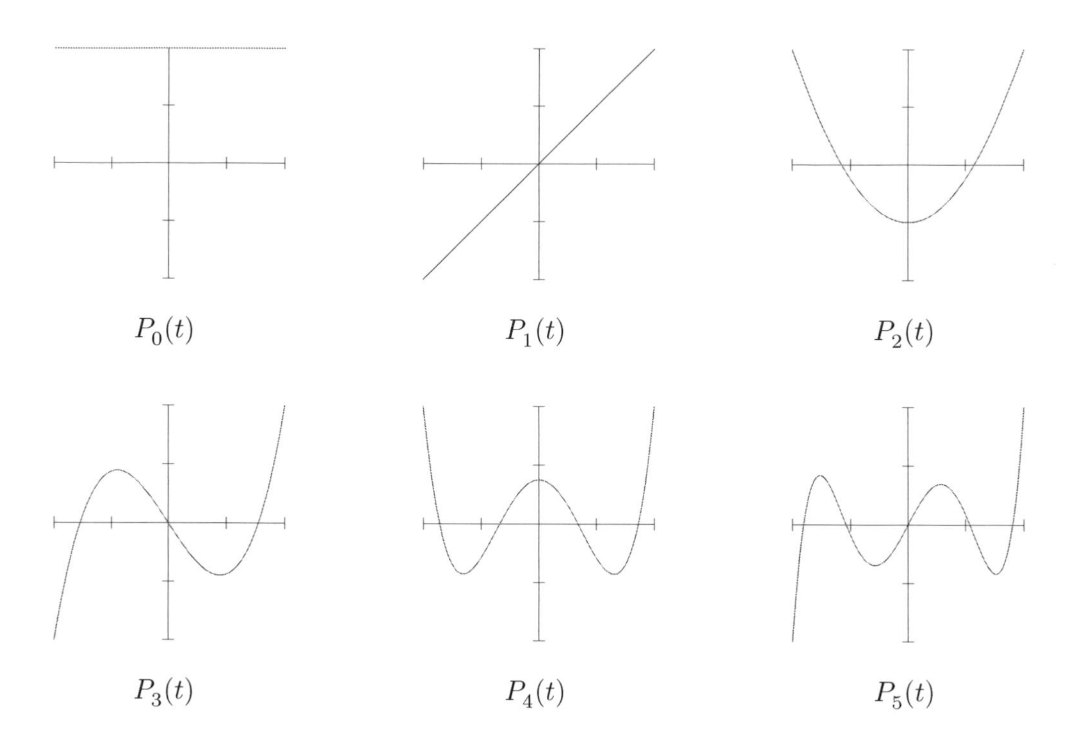

Figure 12.2. Legendre polynomials.

with corresponding eigenvalue parameter $\mu = n(n+1)$. (The initial factor is by common convention, [**86**]; see (12.64) for the explicit formula.) The first few are

$$P_0(t) = 1, \qquad P_1(t) = t, \qquad P_2(t) = \tfrac{3}{2}t^2 - \tfrac{1}{2}, \qquad P_3(t) = \tfrac{5}{2}t^3 - \tfrac{3}{2}t,$$
$$P_4(t) = \tfrac{35}{8}t^4 - \tfrac{15}{4}t^2 + \tfrac{3}{8}, \qquad P_5(t) = \tfrac{63}{8}t^5 - \tfrac{35}{4}t^3 + \tfrac{15}{8}t,$$

and are graphed in Figure 12.2.

Each Legendre polynomial clearly satisfies the boundary conditions (12.27). To verify that they are indeed solutions to the differential equation (12.28), we set

$$Q_n(t) = (1-t^2)^n.$$

By the chain rule, the derivative of $Q_n(t)$ is

$$Q_n' = -2nt(1-t^2)^{n-1}, \qquad \text{and hence} \qquad (1-t^2)Q_n' = -2nt(1-t^2)^n = -2ntQ_n.$$

Differentiating the latter formula yields

$$(1-t^2)Q_n'' - 2tQ_n' = -2ntQ_n' - 2nQ_n, \qquad \text{or} \qquad (1-t^2)Q_n'' = -2(n-1)tQ_n' - 2nQ_n.$$

A simple induction proves that the kth order derivative $Q_n^{(k)}(t) = \dfrac{d^kQ_n}{dt^k}$ satisfies

$$\begin{aligned}(1-t^2)\,Q_n^{(k+2)} &= -2(n-k-1)\,t\,Q_n^{(k+1)} - 2[n+(n-1)+\cdots+(n-k)]\,Q_n^{(k)} \\ &= -2(n-k-1)\,t\,Q_n^{(k+1)} - (k+1)(2n-k)\,Q_n^{(k)}.\end{aligned} \tag{12.30}$$

In particular, when $k = n$, this reduces to

$$(1-t^2)Q_n^{(n+2)} = 2tQ_n^{(n+1)} - n(n+1)Q_n^{(n)} = 0,$$

and so $\widehat{P}_n(t) = Q_n^{(n)}(t)$ satisfies

$$(1-t^2)\widehat{P}_n'' - 2t\widehat{P}_n' + n(n+1)\widehat{P}_n = 0,$$

which is precisely the order 0 Legendre equation (12.28) with $\mu = n(n+1)$. The Legendre polynomial P_n is a constant multiple of $\widehat{P}_n$, and hence it too satisfies the order 0 Legendre equation. According to Theorem 12.3 below, the Legendre polynomials form a complete system of eigenfunctions for the order 0 Legendre boundary value problem.

When the order $m > 0$, the eigenfunctions of the Legendre boundary value problem (12.26–27) are not always polynomials. They are known as the *Ferrers functions*, named after the nineteenth-century British mathematician Norman Ferrers, or, more generally, as *associated Legendre functions*. They have the explicit formula†

$$\begin{aligned} P_n^m(t) &= (1-t^2)^{m/2}\frac{d^m}{dt^m}P_n(t) \\ &= (-1)^n\frac{(1-t^2)^{m/2}}{2^n\, n!}\frac{d^{n+m}}{dt^{n+m}}(1-t^2)^n, \end{aligned} \qquad n = m, m+1, \ldots, \tag{12.31}$$

which generalizes the formula (12.29) for the Legendre polynomials. The eigenvalue parameter for $P_n^m(t)$ is also $\mu = n(n+1)$. In particular $P_n^0(t) = P_n(t)$. Here is a list of the first few Ferrers functions, which, for completeness, includes Legendre polynomials:

$$\begin{aligned}
&P_0^0(t) = 1, && P_1^0(t) = t, && P_1^1(t) = \sqrt{1-t^2}\,, \\
&P_2^0(t) = -\tfrac{1}{2} + \tfrac{3}{2}t^2, && P_2^1(t) = 3t\sqrt{1-t^2}\,, && P_2^2(t) = 3(1-t^2), \\
&P_3^0(t) = -\tfrac{3}{2}t + \tfrac{5}{2}t^3, && P_3^1(t) = \left(-\tfrac{3}{2} + \tfrac{15}{2}t^2\right)\sqrt{1-t^2}, && \\
&P_3^2(t) = 15t(1-t^2), && P_3^3(t) = 15(1-t^2)^{3/2}\,, && \\
&P_4^0(t) = \tfrac{3}{8} - \tfrac{15}{4}t^2 + \tfrac{35}{8}t^4, && P_4^1(t) = \left(-\tfrac{15}{2}t + \tfrac{35}{2}t^3\right)\sqrt{1-t^2}, && \\
&P_4^2(t) = \left(-\tfrac{15}{2} + \tfrac{105}{2}t^2\right)(1-t^2), && P_4^3(t) = 105t(1-t^2)^{3/2}\,, && P_4^4(t) = 105(1-t^2)^2.
\end{aligned} \tag{12.32}$$

When $m = 2k \leq n$ is an even integer, $P_n^m(t)$ is a polynomial function, while when $m = 2k+1 \leq n$ is odd, there is an extra factor of $\sqrt{1-t^2}$. Keep in mind that the square root is real and positive, since we are restricting our attention to the interval $-1 \leq t \leq 1$. If $m > n$, formula (12.31) reduces to the zero function and so is not included in the final tally.

Warning: Even though half of the Ferrers functions are polynomials, only those with $m = 0$, i.e., $P_n(t) = P_n^0(t)$, are called *Legendre polynomials*.

† *Warning*: Some authors include a $(-1)^m$ factor in the formula, resulting in the opposite sign when m is odd. Another source of confusion is that many tables define the associated Legendre functions using the alternative initial factor $(t^2-1)^{m/2}$. But this is unsuitable, since we are solely interested in values of t lying in the interval $-1 \leq t \leq 1$, and this convention would result in a complex-valued function when m is odd. Following [**86**], we use the term "Ferrers function" to refer to the restriction of the associated Legendre function to the cut locus $-1 \leq t \leq 1$.

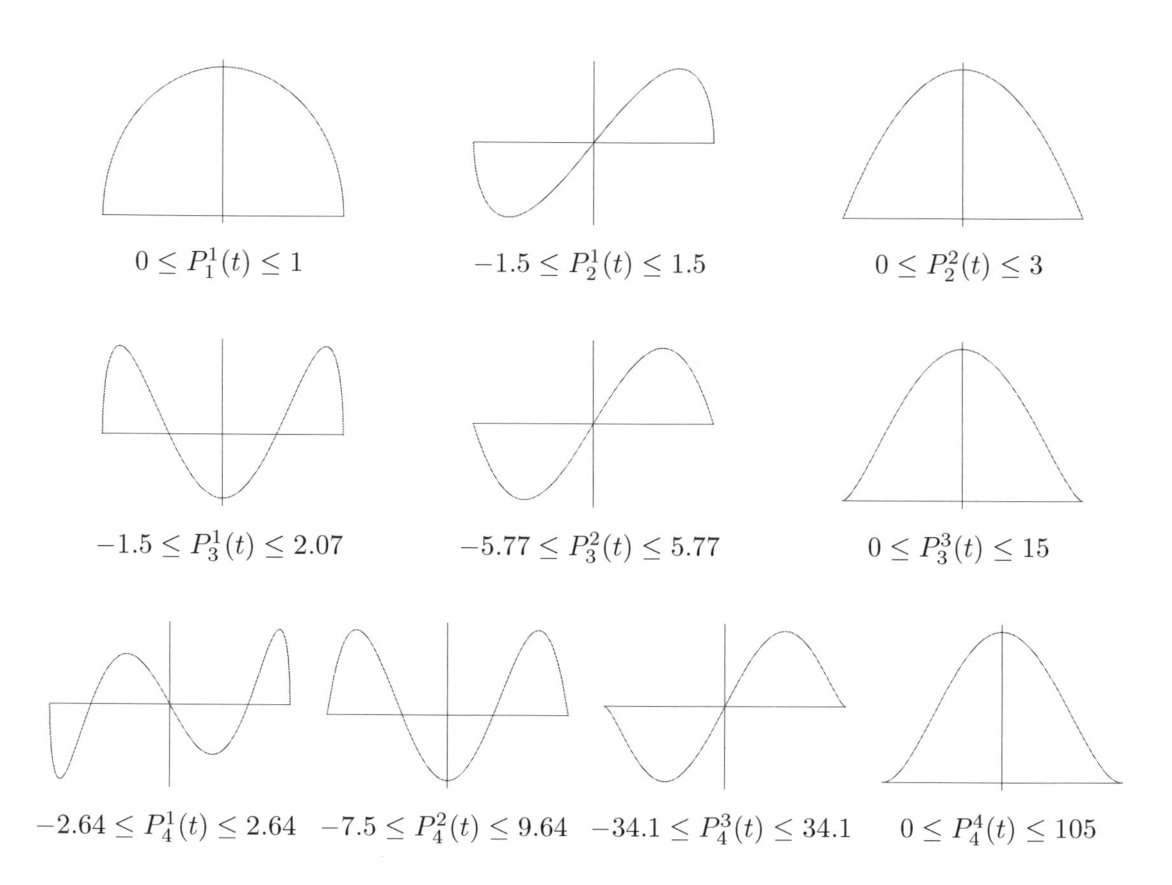

Figure 12.3. Ferrers functions.

Figure 12.3 displays graphs of the Ferrers functions $P_n^m(t)$ for $1 \le m \le n \le 4$. Pay particular attention to the fact that, owing to the choice of normalization factor, the graphs have very different vertical scales, as indicated by their minimum and maximum values (rounded to two decimal places) written below each — although one always has the freedom to rescale the eigenfunctions as desired, e.g., so as to be orthonormal.

To show that the Ferrers functions $P_n^m(t)$ satisfy the Legendre differential equation (12.26) of order m, we substitute $k = m + n$ in (12.30):

$$(1-t^2)\,\frac{d^2 R_n^m}{dt^2} - 2\,(m+1)\,t\,\frac{dR_n^m}{dt} + (m+n+1)\,(n-m)\,R_n^m = 0, \tag{12.33}$$

where

$$R_n^m(t) = Q_n^{(m+n)}(t).$$

This is *not* the order m Legendre equation, but it can be converted into it by setting

$$R_n^m(t) = (1-t^2)^{-m/2}\,S_n^m(t).$$

Differentiating, we obtain

$$\begin{aligned}\frac{dR_n^m}{dt} &= (1-t^2)^{-m/2}\,\frac{dS_n^m}{dt} - mt\,(1-t^2)^{-m/2-1}\,S_n^m,\\ \frac{d^2R_n^m}{dt^2} &= (1-t^2)^{-m/2}\,\frac{d^2S_n^m}{dt^2} - 2mt\,(1-t^2)^{-m/2-1}\,\frac{dS_n^m}{dt}\\ &\qquad + \left[\,m + m(m+1)t^2\,\right](1-t^2)^{-m/2-2}\,S_n^m.\end{aligned}$$

Therefore, after a little algebra, equation (12.33) takes the alternative form

$$\begin{aligned}(1-t^2)^{-m/2+1}\,\frac{d^2S_n^m}{dt^2} &- 2t\,(1-t^2)^{-m/2}\,\frac{dS_n^m}{dt}\\ &+ \left[\,n\,(n+1)\,(1-t^2) - m^2\,\right](1-t^2)^{-m/2-1}\,S_n^m = 0,\end{aligned}$$

which, when multiplied by $(1-t^2)^{m/2+1}$, is precisely the order m Legendre equation (12.26) with $\mu = n\,(n+1)$. Thus,

$$S_n^m(t) = (1-t^2)^{m/2}R_n^m(t) = (1-t^2)^{m/2}\,\frac{d^{n+m}}{dt^{n+m}}\,(1-t^2)^n,$$

which is a constant multiple of the Ferrers function $P_n^m(t)$, is a solution to the order m Legendre equation. Moreover, we note that

$$P_n^m(1) = P_n^m(-1) = 0, \qquad \text{when} \qquad m > 0, \tag{12.34}$$

and we conclude that $P_n^m(t)$ is an eigenfunction for the order m Legendre boundary value problem.

The following result states that the Ferrers functions provide a complete list of solutions to the Legendre boundary value problem (12.26–27).

Theorem 12.3. *Let $m \geq 0$ be a nonnegative integer. Then the order m Legendre boundary value problem prescribed by (12.26–27) has eigenvalues $\mu_n = n\,(n+1)$ for $n = 0, 1, 2, \ldots$, and associated eigenfunctions $P_n^m(t)$, where $m = 0, \ldots, n$. Moreover, the Ferrers eigenfunctions form a complete orthogonal system relative to the L^2 inner product on the cut locus $[-1, 1]$.*

Returning to the zenith variable φ via (12.25), Theorem 12.3 implies that our original boundary value problem

$$\sin^2\varphi\,\frac{d^2p}{d\varphi^2} + \cos\varphi\,\sin\varphi\,\frac{dp}{d\varphi} + (\mu\,\sin^2\varphi - m^2)\,p = 0, \qquad |\,p(0)\,|, \quad |\,p(\pi)\,| < \infty, \tag{12.35}$$

has its eigenvalues and eigenfunctions expressed in terms of the Ferrers functions:

$$\mu_n = n\,(n+1), \qquad p_n^m(\varphi) = P_n^m(\cos\varphi), \qquad \text{for} \qquad 0 \leq m \leq n. \tag{12.36}$$

Since $P_n^m(t)$ is either a polynomial or a polynomial multiplied by a power of $\sqrt{1-t^2}$, the eigenfunction $p_n^m(\varphi)$ is a trigonometric polynomial of degree n, which we call a *trigono-*

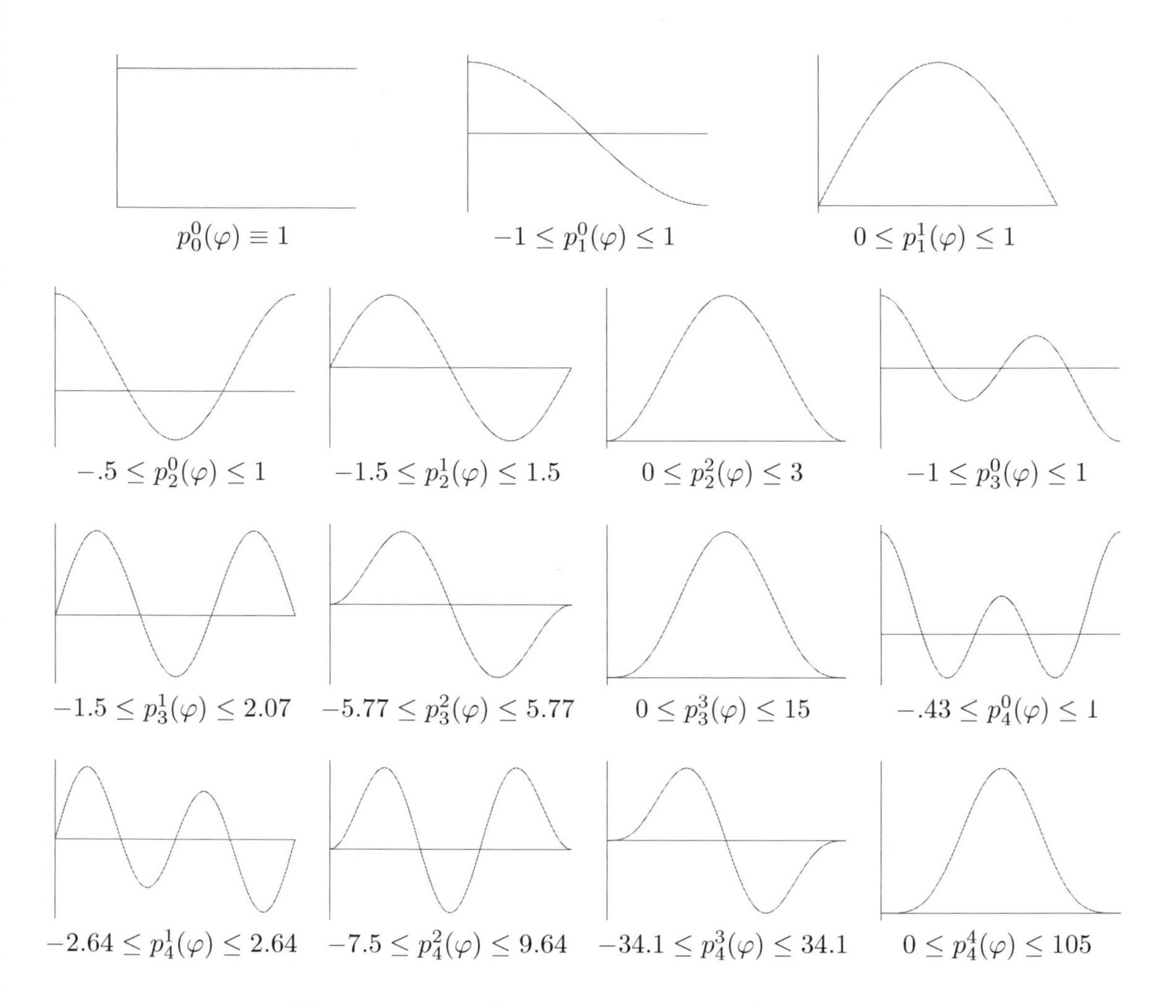

Figure 12.4. Trigonometric Ferrers functions.

metric Ferrers function. Here are the first few, written in Fourier form, as in (3.38):

$$\begin{aligned}
&p_0^0(\varphi) = 1, && p_1^0(\varphi) = \cos\varphi, && p_1^1(\varphi) = \sin\varphi, \\
&p_2^0(\varphi) = \tfrac{1}{4} + \tfrac{3}{4}\cos 2\varphi, && p_2^1(\varphi) = \tfrac{3}{2}\sin 2\varphi, && p_2^2(\varphi) = \tfrac{3}{2} - \tfrac{3}{2}\cos 2\varphi, \\
&p_3^0(\varphi) = \tfrac{3}{8}\cos\varphi + \tfrac{5}{8}\cos 3\varphi, && && p_3^1(\varphi) = \tfrac{3}{8}\sin\varphi + \tfrac{15}{8}\sin 3\varphi, \\
&p_3^2(\varphi) = \tfrac{15}{4}\cos\varphi - \tfrac{15}{4}\cos 3\varphi, && && p_3^3(\varphi) = \tfrac{45}{4}\sin\varphi - \tfrac{15}{4}\sin 3\varphi, \\
&p_4^0(\varphi) = \tfrac{9}{64} + \tfrac{5}{16}\cos 2\varphi + \tfrac{35}{64}\cos 4\varphi, && && p_4^1(\varphi) = \tfrac{5}{8}\sin 2\varphi + \tfrac{35}{16}\sin 4\varphi, \\
&p_4^2(\varphi) = \tfrac{45}{16} + \tfrac{15}{4}\cos 2\varphi - \tfrac{105}{16}\cos 4\varphi, && && p_4^3(\varphi) = \tfrac{105}{4}\sin 2\varphi - \tfrac{105}{8}\sin 4\varphi, \\
&p_4^4(\varphi) = \tfrac{315}{8} - \tfrac{105}{2}\cos 2\varphi + \tfrac{105}{8}\cos 4\varphi.
\end{aligned} \tag{12.37}$$

It is also instructive to plot the eigenfunctions in terms of the zenith angle φ; see Figure 12.4. As in Figure 12.3, the vertical scales are not the same, as indicated by the listed minimum and maximum values.

Spherical Harmonics

At this stage, we have determined both angular components of our separable solutions (12.22). Multiplying the two parts together results in the spherical angle functions

$$\begin{aligned} Y_n^m(\varphi,\theta) &= p_n^m(\varphi)\cos m\theta, & n &= 0,1,2,\ldots, \\ \widetilde{Y}_n^m(\varphi,\theta) &= p_n^m(\varphi)\sin m\theta, & m &= 0,1,\ldots,n, \end{aligned} \tag{12.38}$$

known as *spherical harmonics*. They satisfy the *spherical Helmholtz equation*

$$\Delta_S Y_n^m + n(n+1)\,Y_n^m \;=\; 0 \;=\; \Delta_S \widetilde{Y}_n^m + n(n+1)\,\widetilde{Y}_n^m, \tag{12.39}$$

and so are eigenfunctions for the spherical Laplacian operator, (12.19), with associated eigenvalues $\mu_n = n(n+1)$ for $n = 0,1,2,\ldots$. The nth eigenvalue μ_n admits a $(2n+1)$–dimensional eigenspace, spanned by the spherical harmonics

$$Y_n^0(\varphi,\theta),\quad Y_n^1(\varphi,\theta),\quad \ldots\,,\quad Y_n^n(\varphi,\theta),\quad \widetilde{Y}_n^1(\varphi,\theta),\quad \ldots\,,\quad \widetilde{Y}_n^n(\varphi,\theta).$$

(The omitted function $\widetilde{Y}_n^0(\varphi,\theta) \equiv 0$ is trivial, and so does not contribute.) In Figure 12.5 we plot the first few spherical harmonic surfaces $r = Y_n^m(\varphi,\theta)$. In these graphs, in view of the spherical coordinate formulae (12.15), points with a negative r coordinate appear on the opposite side of the origin from their positive r counterparts. Incidentally, the graphs of the other spherical harmonic surfaces $r = \widetilde{Y}_n^m(\varphi,\theta)$, when $m > 0$, are obtained by rotation around the z–axis by 90°; see Exercise 12.2.20. On the other hand, the graphs of Y_n^0 are cylindrically symmetric (why?), and hence unaffected by such a rotation.

Self-adjointness of the spherical Laplacian, as per Exercise 12.2.21, implies that the spherical harmonics are orthogonal with respect to the L^2 inner product

$$\langle f\,,g\rangle = \iint_{S_1} f\,g\,dS = \int_{-\pi}^{\pi}\int_0^{\pi} f(\varphi,\theta)\,g(\varphi,\theta)\sin\varphi\;d\varphi\,d\theta \tag{12.40}$$

given by integrating the product of the functions with respect to the surface area element $dS = \sin\varphi\;d\varphi\,d\theta$ on the unit sphere $S_1 = \{\,\|\,\mathbf{x}\,\| = 1\,\}$. More correctly, self-adjointness only guarantees orthogonality of the harmonics corresponding to distinct eigenvalues: $\mu_n \neq \mu_l$. However, the orthogonality relations

$$\begin{aligned} \langle Y_n^m\,,Y_l^k\rangle &= \iint_{S_1} Y_n^m\,Y_l^k\,dS = 0, &&\text{for} && (m,n)\neq(k,l), \\ \langle Y_n^m\,,\widetilde{Y}_l^k\rangle &= \iint_{S_1} Y_n^m\,\widetilde{Y}_l^k\,dS = 0, &&\text{for all} && (m,n),\ (k,l), \\ \langle \widetilde{Y}_n^m\,,\widetilde{Y}_l^k\rangle &= \iint_{S_1} \widetilde{Y}_n^m\,\widetilde{Y}_l^k\,dS = 0, &&\text{for} && (m,n)\neq(k,l), \end{aligned} \tag{12.41}$$

do, in fact, hold in full generality; Exercise 12.2.22 asks you to supply the details. Moreover, their norms can be explicitly computed:

$$\|\,Y_n^0\,\|^2 = \frac{4\pi}{2n+1}, \qquad \|\,Y_n^m\,\|^2 = \|\,\widetilde{Y}_n^m\,\|^2 = \frac{2\pi(n+m)!}{(2n+1)(n-m)!}, \qquad m = 1,\ldots,n. \tag{12.42}$$

Proofs of the latter formulae are outlined in Exercise 12.2.24.

With some further work, it can be shown that the spherical harmonics form a complete orthogonal system of functions on the unit sphere. This means that any reasonable (e.g.,

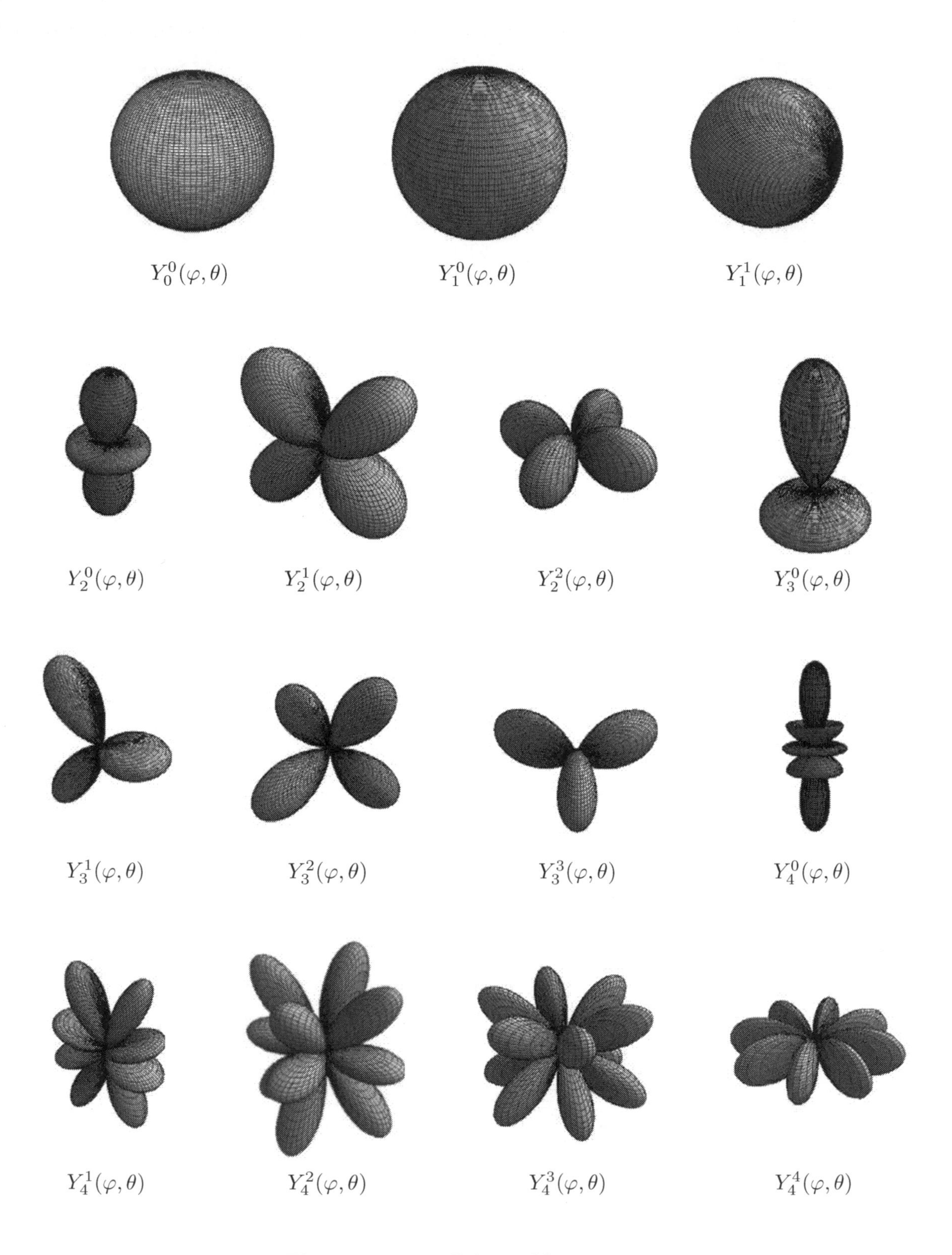

Figure 12.5. Spherical harmonics.

piecewise C^1 or even L^2) function $h\colon S_1 \to \mathbb{R}$, can be expanded into a convergent *spherical harmonic series*

$$h(\varphi,\theta) = \frac{c_{0,0}}{2} + \sum_{n=1}^{\infty} \left(\frac{c_{0,n}}{2}\, Y_n^0(\varphi) + \sum_{m=1}^{n} \left[c_{m,n} Y_n^m(\varphi,\theta) + \widetilde{c}_{m,n} \widetilde{Y}_n^m(\varphi,\theta) \right] \right). \tag{12.43}$$

Applying the orthogonality relations (12.41), we find that the spherical harmonic coefficients are given by the inner products

$$c_{0,n} = \frac{2\,\langle\, h\,, Y_n^0 \,\rangle}{\|\, Y_n^0 \,\|^2}, \qquad c_{m,n} = \frac{\langle\, h\,, Y_n^m \,\rangle}{\|\, Y_n^m \,\|^2}, \qquad \widetilde{c}_{m,n} = \frac{\langle\, h\,, \widetilde{Y}_n^m \,\rangle}{\|\, \widetilde{Y}_n^m \,\|^2}, \qquad \begin{array}{c} 0 \le n, \\ 1 \le m \le n, \end{array}$$

or, explicitly, using (12.40) and the formulae (12.42) for the norms,

$$\begin{aligned} c_{m,n} &= \frac{(2n+1)(n-m)!}{2\pi\,(n+m)!} \int_{-\pi}^{\pi} \int_0^{\pi} h(\varphi,\theta)\, p_n^m(\varphi)\, \cos m\theta\, \sin\varphi\; d\varphi\, d\theta, \\ \widetilde{c}_{m,n} &= \frac{(2n+1)(n-m)!}{2\pi\,(n+m)!} \int_{-\pi}^{\pi} \int_0^{\pi} h(\varphi,\theta)\, p_n^m(\varphi)\, \sin m\theta\, \sin\varphi\; d\varphi\, d\theta. \end{aligned} \tag{12.44}$$

As with an ordinary Fourier series, the extra $\frac{1}{2}$ was appended to the $c_{0,n}$ terms in (12.43) so that equations (12.44) remain valid for all values of m, n. In particular, the constant term in the spherical harmonic series is the mean of the function h over the unit sphere:

$$\frac{c_{0,0}}{2} = \frac{1}{4\pi} \iint_{S_1} h\, dS = \frac{1}{4\pi} \int_{-\pi}^{\pi} \int_0^{\pi} h(\varphi,\theta)\, \sin\varphi\, d\varphi\, d\theta. \tag{12.45}$$

Remark: Establishing uniform convergence of a spherical harmonic series (12.43) is more challenging than in the Fourier series case, because, unlike the trigonometric functions, the orthonormal spherical harmonics are not uniformly bounded. A recent survey of what is known in this regard can be found in [**10**].

Remark: An alternative approach is to replace the real trigonometric functions by complex exponentials, and work with the *complex spherical harmonics*†

$$\mathcal{Y}_n^m(\varphi,\theta) = Y_n^m(\varphi,\theta) + \mathrm{i}\, \widetilde{Y}_n^m(\varphi,\theta) = p_n^m(\varphi)\, e^{\,\mathrm{i}\, m\theta}, \qquad \begin{array}{c} n = 0,1,2,\ldots\,, \\ m = -n, -n+1, \ldots, n. \end{array} \tag{12.46}$$

The associated orthogonality and expansion formulas are relegated to the exercises.

Harmonic Polynomials

To complete our solution to the Laplace equation on the solid ball, we still need to solve the ordinary differential equation (12.20) for the radial component $v(r)$. In view of our analysis of the spherical Helmholtz equation, the original separation constant is $\mu = n(n+1)$ for some nonnegative integer $n \ge 0$, and so the radial equation takes the form

$$r^2 v'' + 2 r v' - n(n+1) v = 0. \tag{12.47}$$

† Here we use the convention that $Y_n^m = Y_n^{-m}$, $\widetilde{Y}_n^m = -\widetilde{Y}_n^{-m}$, and $\widetilde{Y}_n^0 \equiv 0$, which is compatible with their defining formulas (12.38).

To solve this Euler equation, we substitute the power ansatz $v(r) = r^\alpha$, and find that the exponent α must satisfy the quadratic indicial equation

$$\alpha^2 + \alpha - n(n+1) = 0, \qquad \text{and hence} \qquad \alpha = n \qquad \text{or} \qquad \alpha = -(n+1).$$

Therefore, the two linearly independent solutions are

$$v_1(r) = r^n \qquad \text{and} \qquad v_2(r) = r^{-n-1}. \tag{12.48}$$

Since we are currently interested only in solutions that remain bounded at $r = 0$ — the center of the ball — we will retain just the first solution $v(r) = r^n$ for our subsequent analysis.

At this stage, we have solved all three ordinary differential equations for the separable solutions. We combine (12.23, 38, 48) to produce the following spherically separable solutions to the Laplace equation:

$$\begin{aligned} H_n^m &= r^n\, Y_n^m(\varphi,\theta) = r^n\, p_n^m(\varphi)\cos m\,\theta, & n &= 0,1,2,\ldots, \\ \widetilde{H}_n^m &= r^n\, \widetilde{Y}_n^m(\varphi,\theta) = r^n\, p_n^m(\varphi)\sin m\,\theta, & m &= 0,1,\ldots,n. \end{aligned} \tag{12.49}$$

Although apparently complicated, these solutions are, perhaps surprisingly, elementary polynomial functions of the rectangular coordinates x, y, z, and hence are *harmonic polynomials*. The first few are

$$\begin{array}{llll} H_0^0 = 1, & H_1^0 = z, & H_2^0 = z^2 - \frac{1}{2}x^2 - \frac{1}{2}y^2, & H_3^0 = z^3 - \frac{3}{2}x^2 z - \frac{3}{2}y^2 z, \\ & H_1^1 = x, & H_2^1 = 3\,x\,z, & H_3^1 = 6\,x\,z^2 - \frac{3}{2}x^3 - \frac{3}{2}\,x\,y^2, \\ & \widetilde{H}_1^1 = y, & \widetilde{H}_2^1 = 3\,y\,z, & \widetilde{H}_3^1 = 6\,y\,z^2 - \frac{3}{2}x^2 y - \frac{3}{2}y^3, \\ & & H_2^2 = 3\,x^2 - 3\,y^2, & H_3^2 = 15\,x^2 z - 15\,y^2 z, \\ & & \widetilde{H}_2^2 = 6\,x\,y, & \widetilde{H}_3^2 = 30\,x\,y\,z, \\ & & & H_3^3 = 15\,x^3 - 45\,x\,y^2, \\ & & & \widetilde{H}_3^3 = 45\,x^2 y - 15\,y^3. \end{array} \tag{12.50}$$

The polynomials

$$H_n^0, \ H_n^1, \ \ldots\ , \ H_n^n, \ \widetilde{H}_n^1, \ \ldots\ , \ \widetilde{H}_n^n$$

are homogeneous of degree n. Orthogonality of the spherical harmonics implies that they form a basis for the vector space comprised of all homogeneous harmonic polynomials of degree n, which hence has dimension $2n+1$.

The harmonic polynomials (12.49) form a complete system, and therefore the general solution to the Laplace equation inside the unit ball can be written as a harmonic polynomial series:

$$u(x,y,z) = \frac{c_{0,0}}{2} + \sum_{n=1}^{\infty}\left(\frac{c_{0,n}}{2}\,H_n^0(x,y,z) + \sum_{m=1}^{n}\left[\,c_{m,n}H_n^m(x,y,z) + \widetilde{c}_{m,n}\widetilde{H}_n^m(x,y,z)\,\right]\right), \tag{12.51}$$

or equivalently, in spherical coordinates,

$$u(r,\varphi,\theta) = \frac{c_{0,0}}{2} + \sum_{n=1}^{\infty}\left(\frac{c_{0,n}}{2}\,r^n\,Y_n^0(\varphi) + \sum_{m=1}^{n}\left[\,c_{m,n}r^n\,Y_n^m(\varphi,\theta) + \widetilde{c}_{m,n}r^n\,\widetilde{Y}_n^m(\varphi,\theta)\,\right]\right). \tag{12.52}$$

The coefficients $c_{m,n}, \widetilde{c}_{m,n}$ are uniquely prescribed by the boundary conditions. Indeed, substituting (12.52) into the Dirichlet boundary conditions on the unit sphere $r = 1$ yields

$$u(1,\varphi,\theta) = \frac{c_{0,0}}{2} + \sum_{n=1}^{\infty}\left(\frac{c_{0,n}}{2}\,Y_n^0(\varphi) + \sum_{m=1}^{n}\left[\,c_{m,n}Y_n^m(\varphi,\theta) + \widetilde{c}_{m,n}\widetilde{Y}_n^m(\varphi,\theta)\,\right]\right) = h(\varphi,\theta). \tag{12.53}$$

Thus, the coefficients $c_{m,n}, \widetilde{c}_{m,n}$ are given by the inner product formulae (12.44). If the terms in the resulting series are uniformly bounded — which occurs for all piecewise continuous functions h, as well as all L^2 functions and many generalized functions such as the delta function — then the harmonic polynomial series (12.52) converges everywhere, and, in fact, uniformly on any smaller ball $\|\,\mathbf{x}\,\| = r \le r_0 < 1$.

Averaging, the Maximum Principle, and Analyticity

In rectangular coordinates, the nth summand of the series (12.51) is a homogeneous polynomial of degree n. Therefore, repeating the argument used in the two-dimensional situation (4.115), we conclude that the harmonic polynomial series is, in fact, a power series, and hence provides the *Taylor expansion for the harmonic function* $u(x,y,z)$ *at the origin*! In particular, its convergence for all $r < 1$ implies that the harmonic function $u(x,y,z)$ is analytic at $x = y = z = 0$.

The constant term in such a Taylor series can be identified with the value of the function at the origin: $u(0,0,0) = \frac{1}{2}\,c_{0,0}$. On the other hand, since $u = h$ on $S_1 = \partial\Omega$, the coefficient formula (12.45) tells us that

$$u(0,0,0) = \frac{c_{0,0}}{2} = \frac{1}{4\pi}\iint_{S_1} u\,dS. \tag{12.54}$$

Therefore, we have established the three-dimensional counterpart of Theorem 4.8: the value of a harmonic function u at the center of the sphere is equal to the average of its values on the sphere's surface. Moreover, each partial derivative $\dfrac{\partial^{i+j+k}u}{\partial x^i\partial y^j\partial z^k}(0,0,0)$ appears, up to a factor, as the coefficient of the terms $x^i y^j z^k$ in the Taylor series, and hence can be expressed as a certain linear combination of the coefficients $c_{m,n}, \widetilde{c}_{m,n}$, which are in turn given by the integral formulae (12.44).

More generally, the value of a harmonic function at the center of any ball contained within its domain equals the average of its values over the bounding sphere. As with the planar version in Theorem 4.8, it is preferable to give a direct proof that doesn't rely on the series expansion (12.51).

Theorem 12.4. *If $u(\mathbf{x})$ is a harmonic function defined on a domain $\Omega \subset \mathbb{R}^3$, then u is analytic inside Ω. Moreover, its value at any $\mathbf{x}_0 \in \Omega$ is obtained by averaging its values on any sphere $S_a = \{\,\|\,\mathbf{x} - \mathbf{x}_0\,\| = a\,\}$ centered at $\mathbf{x}_0$:*

$$u(\mathbf{x}_0) \;=\; \frac{1}{4\pi a^2}\iint_{S_a} u\;dS, \tag{12.55}$$

provided the enclosed ball lies within its domain of analyticity: $B_a = \{\,\|\,\mathbf{x} - \mathbf{x}_0\,\| \le a\,\} \subset \Omega$.

Proof: Let us denote the average of u over the sphere of radius a by

$$\begin{aligned} g(a) &= \frac{1}{4\pi a^2} \iint_{S_a} u \, dS \\ &= \frac{1}{4\pi} \int_{-\pi}^{\pi} \int_0^{\pi} u(x_0 + a \sin\varphi \, \cos\theta, y_0 + a \sin\varphi \, \sin\theta, z_0 + a\cos\varphi) \, \sin\varphi \, d\varphi \, d\theta. \end{aligned}$$

By continuity, as the radius $a \to 0$, the average of u on the sphere S_a tends to its value at the center: $g(a) \to u(\mathbf{x}_0)$.

On the other hand, since $u \in \mathrm{C}^2$ and harmonic in $B_a \subset \Omega$, the derivative

$$\begin{aligned} g'(a) &= \frac{1}{4\pi} \int_{-\pi}^{\pi} \int_0^{\pi} \left(\sin\varphi \, \cos\theta \, \frac{\partial u}{\partial x} + \sin\varphi \, \sin\theta \, \frac{\partial u}{\partial y} + \cos\varphi \, \frac{\partial u}{\partial z} \right) \sin\varphi \, d\varphi \, d\theta \\ &= \frac{1}{4\pi a^2} \iint_{S_a} \frac{\partial u}{\partial \mathbf{n}} \, dS = \frac{1}{4\pi a^2} \iiint_{B_a} \Delta u \, dx \, dy \, dz = 0, \end{aligned}$$

where $\mathbf{n}$ denotes the unit outwards normal to $S_a = \partial B_a$, and we used the divergence identity in Exercise 12.1.11(a). We conclude that $g(a)$ is constant, and hence $g(a) = u(\mathbf{x}_0)$ for any $a > 0$ provided $B_a \subset \Omega$. *Q.E.D.*

Arguing as in the planar case of Theorem 4.9, we readily establish the corresponding *Strong Maximum Principle* for harmonic functions of three variables.

Theorem 12.5. *A nonconstant harmonic function cannot have a local maximum or minimum at any interior point of its domain of definition. Moreover, its global maximum or minimum (if any) is located on the boundary of the domain.*

For instance, the Maximum Principle implies that the maximum and minimum temperatures in a solid body in thermal equilibrium are to be found only on its boundary. In physical terms, since heat energy must flow away from an internal maximum and towards an internal minimum, any local temperature extremum inside the body would preclude it from being in thermal equilibrium. The Maximum Principle immediately implies a Uniqueness Theorem for both the Laplace and Poisson equations, cf. Theorem 4.10, which in turn establishes the solution formula (12.51) and hence analyticity of every harmonic function.

Example 12.6. In this example, we shall determine the electrostatic potential inside a hollow sphere when the upper and lower hemispheres are held at different constant potentials. This device is called a *spherical capacitor* and is realized experimentally by separating the two charged conducting hemispherical shells by a thin insulating ring at the equator. A straightforward scaling argument allows us to choose our units so that the sphere has unit radius, while the potential is set equal to 1 on the upper hemisphere and equal to 0, i.e., grounded, on the lower hemisphere. The resulting electrostatic potential satisfies the Laplace equation

$$\Delta u = 0 \qquad \text{inside a solid ball} \qquad \| \mathbf{x} \| < 1, \tag{12.56}$$

and is subject to Dirichlet boundary conditions

$$u(x,y,z) = h(x,y,z) \equiv \begin{cases} 1, & z > 0, \\ 0, & z < 0, \end{cases} \qquad \text{on the unit sphere} \qquad \| \mathbf{x} \| = 1. \tag{12.57}$$

The solution will be prescribed by a harmonic polynomial series (12.51) whose coefficients are fixed by the boundary values (12.57). Before tackling the required computation, let us first note that since the boundary data does not depend upon the azimuthal angle θ, the solution $u = u(r,\varphi)$ will also be independent of θ. Therefore, we need only consider the θ-independent spherical harmonic polynomials (12.38), which are those with $m = 0$. Thus,

$$u(x,y,z) = \frac{1}{2}\sum_{n=0}^{\infty} c_n H_n^0(x,y,z) = \frac{1}{2}\sum_{n=0}^{\infty} c_n r^n P_n(\cos\varphi), \tag{12.58}$$

where we abbreviate $c_n = c_{0,n}$. The boundary conditions (12.57) require

$$u|_{r=1} = \frac{1}{2}\sum_{n=0}^{\infty} c_n P_n(\cos\varphi) = h(\varphi) = \begin{cases} 1, & 0 \le \varphi < \frac{1}{2}\pi, \\ 0, & \frac{1}{2}\pi < \varphi \le \pi. \end{cases}$$

The coefficients are given by (12.44), which, in the case $m = 0$, reduce to

$$c_n = \frac{2n+1}{2\pi}\iint_{S_1} h\, Y_n^0\, dS = (2n+1)\int_0^{\pi/2} P_n(\cos\varphi)\sin\varphi\, d\varphi = (2n+1)\int_0^1 P_n(t)\, dt, \tag{12.59}$$

since $h = 0$ when $\frac{1}{2}\pi < \varphi \le \pi$. The first few are $c_0 = 1$, $c_1 = \frac{3}{2}$, $c_2 = 0$, $c_3 = -\frac{7}{8}$, $c_4 = 0$. Therefore, the solution has the explicit Taylor expansion

$$\begin{aligned} u(x,y,z) &= \tfrac{1}{2} + \tfrac{3}{4}\, r\cos\varphi - \tfrac{21}{128}\, r^3\cos\varphi - \tfrac{35}{128}\, r^3\cos 3\varphi + \cdots \\ &= \tfrac{1}{2} + \tfrac{3}{4}\, z + \tfrac{21}{32}\,(x^2+y^2)\, z - \tfrac{7}{16}\, z^3 + \cdots . \end{aligned} \tag{12.60}$$

Note in particular that the value $u(0,0,0) = \frac{1}{2}$ at the center of the sphere is the average of its boundary values, in accordance with Theorem 12.4. The solution depends only on the cylindrical coordinates r, z, which is a consequence of the invariance of the Laplace equation under general rotations, coupled with the invariance of the boundary data under rotations around the z–axis.

Remark: The same solution $u(x,y,z)$ describes the thermal equilibrium in a solid sphere whose upper hemisphere is held at temperature 1° and lower hemisphere at 0°.

Example 12.7. A closely related problem is to determine the electrostatic potential *outside* a spherical capacitor. As in the preceding example, we take our capacitor of radius 1, with electrostatic charge of 1 on the upper hemisphere and 0 on the lower hemisphere. Here, we need to solve the Laplace equation $\Delta u = 0$ in the unbounded domain $\Omega = \{\,\|\,\mathbf{x}\,\| > 1\,\}$ — the exterior of the unit sphere — subject to the same Dirichlet boundary conditions (12.57). We anticipate that the potential will be vanishingly small at large distances away from the capacitor: $r = \|\,\mathbf{x}\,\| \gg 1$. Therefore, the harmonic polynomial solutions (12.49) will not help us solve this problem, since (except for the constant case) they become unboundedly large far away from the origin.

However, revisiting our original separation of variables argument will produce a different class of solutions having the desired decay properties. When we solved the radial equation (12.47), we discarded the solution $v_2(r) = r^{-n-1}$ because it had a singularity at the origin. In the present situation, the behavior of the function at $r = 0$ is irrelevant; our requirement is that the solution decay as $r \to \infty$, and $v_2(r)$ has this property. Therefore, we will utilize the *complementary harmonic functions*

$$\begin{aligned} K_n^m(x,y,z) &= r^{-2n-1} H_n^m(x,y,z) = r^{-n-1} Y_n^m(\varphi,\theta) = r^{-n-1} p_n^m(\varphi)\cos m\theta, \\ \widetilde{K}_n^m(x,y,z) &= r^{-2n-1} \widetilde{H}_n^m(x,y,z) = r^{-n-1} \widetilde{Y}_n^m(\varphi,\theta) = r^{-n-1} p_n^m(\varphi)\sin m\theta, \end{aligned} \tag{12.61}$$

for solving such exterior problems. For the capacitor problem, we need only those that are independent of θ, whereby $m = 0$. We write the resulting solution as a series

$$u(x,y,z) = \frac{1}{2} \sum_{n=0}^{\infty} c_n K_n^0(x,y,z) = \frac{1}{2} \sum_{n=0}^{\infty} c_n \, r^{-n-1} \, P_n(\cos\varphi). \tag{12.62}$$

The boundary conditions

$$u\big|_{r=1} = \frac{1}{2} \sum_{n=0}^{\infty} c_n P_n(\cos\varphi) = h(\varphi) \equiv \begin{cases} 1, & 0 \le \varphi < \frac{1}{2}\pi, \\ 0, & \frac{1}{2}\pi < \varphi \le \pi, \end{cases}$$

are identical to those in the previous example. Therefore, the coefficients are given by (12.59), leading to the series expansion

$$\begin{aligned} u(x,y,z) &= \frac{1}{2r} + \frac{3\cos\varphi}{4r^2} - \frac{21\cos\varphi + 35\cos 3\varphi}{128\,r^4} + \cdots \\ &= \frac{1}{2\sqrt{x^2+y^2+z^2}} + \frac{3z}{4\,(x^2+y^2+z^2)^{3/2}} + \frac{21\,(x^2+y^2)\,z - 14\,z^3}{32\,(x^2+y^2+z^2)^{7/2}} + \cdots . \end{aligned} \tag{12.63}$$

Observe that the higher-order terms become negligible at large distances, and hence the potential is asymptotic to that associated with a point charge concentrated at the origin of magnitude $\frac{1}{2}$, which is the average of the boundary potential over the sphere. This is indicative of a general fact, to be explored in Exercise 12.2.32.

Exercises

12.2.1. A solid ball of radius R has its upper hemispherical surface held at temperature T_1 and its lower hemispherical surface held at temperature T_0. Find the resulting equilibrium temperature.

12.2.2. A solid ball has its top hemispherical surface insulated and its bottom hemispherical surface held at a fixed temperature of 10°. Find its equilibrium temperature.

12.2.3. Find the potential inside a spherical capacitor of radius R when the upper hemisphere is at potential α and the lower is at β.

12.2.4. Find the potential $u(x,y,z)$ inside a unit spherical capacitor that has the indicated boundary values on the unit sphere $x^2+y^2+z^2=1$: (a) x, (b) x^2+y^2, (c) x^3. *Hint*: The potential is a polynomial.

12.2.5. Each point on the spherical boundary of a solid ball of radius 1 has temperature equal to its zenith angle φ. (*i*) Find the value of the equilibrium temperature at the center of the ball. (*ii*) Find the Taylor polynomial of degree 3, based at the origin, for the equilibrium temperature distribution.

12.2.6. Solve Exercise 12.2.5 when the boundary temperature equals (a) $\cos\varphi$, (b) $\cos\theta$, (c) θ.

12.2.7. A solid spherical container of radius 3 cm contains a hollow spherical cavity of radius 1 cm in its center. The inner cavity is filled with boiling water at 100°, while the entire container is immersed in an ice water bath at 0°. Assume that the container is in thermal equilibrium. *True or false*: The temperature at a point half-way between the container's inner and outer boundaries is 50°. If true, explain. If false, what is the temperature at such a point?

12.2.8. Find the electrostatic potential between two concentric spherical metal shells of respective radii 1 and 1.2, given that the inner shell is grounded, while the outer shell has potential equal to 1.

♢ 12.2.9. Use the chain rule to establish the formula (12.16) for the Laplacian in spherical coordinates.

♢ 12.2.10. (*a*) Prove that $t = \pm 1$ are both regular singular points for the order 0 Legendre differential equation (12.28). (*b*) Prove that the Legendre eigenvalue problem (12.27–28) is defined by a self-adjoint operator with respect to the L^2 inner product on the cut locus $[-1,1]$. (*c*) Discuss the orthogonality of the Legendre polynomials.

♢ 12.2.11. Solve Exercise 12.2.10 for the Legendre eigenvalue problem (12.26–27) of order m along with the relevant Ferrers eigenfunctions.

♢ 12.2.12. Suppose $m > 0$. (*a*) Find the Green's function for the boundary value problem

$$(1-t^2)\,\frac{d^2P}{dt^2} - 2t\,\frac{dP}{dt} - \frac{m^2}{1-t^2}\,P = f(t), \qquad |\,P(-1)\,|,\ |\,P(1)\,| < \infty.$$

Hint: The homogeneous differential equation has solutions $\left(\dfrac{1+t}{1-t}\right)^{\frac{m}{2}}$ and $\left(\dfrac{1-t}{1+t}\right)^{\frac{m}{2}}$.
(*b*) Use part (*a*) to prove completeness of the Ferrers functions of order $m > 0$ on $[-1,1]$.
(*c*) Explain why there is no Green's function in the order $m = 0$ case.
Remark: When $m = 0$, one can use the trick of Example 9.49 to prove completeness. Although the Green's function for the modified operator does not have an explicit elementary formula, one can prove that it has logarithmic singularities at the endpoints, and hence finite double L^2 norm. See [**120**; §43] for details.

12.2.13. What happens when $n < m$ in formula (12.31)?

♢ 12.2.14. Prove that the Legendre polynomial (12.29) has the explicit formula

$$P_n(t) = \sum_{0 \le 2m \le n} (-1)^m \frac{(2n-2m)!}{2^n\,(n-m)!\,m!\,(n-2m)!}\,t^{n-2m}. \tag{12.64}$$

♢ 12.2.15. Prove the following recurrence relation for the Ferrers functions:

$$P_n^{m+1}(t) = \sqrt{1-t^2}\,\frac{dP_n^m}{dt} + \frac{m\,t}{\sqrt{1-t^2}}\,P_n^m(t). \tag{12.65}$$

♡ 12.2.16. In this exercise, we determine the L^2 norms of the Ferrers functions. (*a*) First, prove that $\displaystyle\int_{-1}^{1}(1-t^2)^n\,dt = \frac{2^{2n+1}\,(n!)^2}{(2n+1)!}$. *Hint*: Set $t = \cos\theta$ and then integrate by parts repeatedly. (*b*) Prove that $\|\,P_n\,\|^2 = \dfrac{2}{2n+1}$. *Hint*: Integrate by parts repeatedly and then use part (*a*). (*c*) Prove that $\|\,P_n^{m+1}\,\|^2 = (n-m)(n+m+1)\,\|\,P_n^m\,\|^2$. *Hint*: Use (12.65) and an integration by parts. (*d*) Finally, prove that $\|\,P_n^m\,\|^2 = \dfrac{2}{2n+1}\,\dfrac{(n+m)!}{(n-m)!}$.

12.2.17. (*a*) Prove that $P_n^m(t)$ is an even or odd function according to whether $m+n$ is an even or odd integer. (*b*) Prove that its Fourier form, $p_n^m(\varphi)$, depends only on $\cos n\,\varphi$, $\cos(n-2)\varphi$, $\cos(n-4)\varphi$, ... if m is even, and only on $\sin n\,\varphi$, $\sin(n-2)\varphi$, $\sin(n-4)\varphi$, ... if m is odd.

12.2.18. Let m be fixed. Are the functions $p_n^m(\varphi)$ for $n = 0,1,2,\ldots$ mutually orthogonal with respect to the standard L^2 inner product on $[0,\pi]$? If not, is there an inner product that makes them orthogonal functions?

12.2.19. Prove that the surfaces defined by the first three spherical harmonics Y_0^0, Y_1^0, and Y_1^1, as in Figure 12.5, are all spheres. Find their centers and radii.

♢ 12.2.20. Explain why the surface defined by $r = \widetilde{Y}_n^m(\varphi,\theta)$ is obtained by rotating that defined by $r = Y_n^m(\varphi,\theta)$ around the z–axis by 90°.

◇ 12.2.21. Prove directly that the spherical Laplacian Δ_S is a self-adjoint linear operator with respect to the inner product (12.40).

◇ 12.2.22. (a) In view of Exercise 12.2.21, which orthogonality relations in (12.41) follow from their status as eigenfunctions of the spherical Laplacian?
(b) Prove the general orthogonality formulae by direct computation.

◇ 12.2.23. State and prove the orthogonality of the complex spherical harmonics (12.46). Then establish the following formula for their norms:

$$\| \mathcal{Y}_n^m \|^2 = \iint_{S_1} | \mathcal{Y}_n^m |^2 \, dS = \frac{4\pi (n+m)!}{(2n+1)(n-m)!} \qquad \begin{matrix} n = 0, 1, 2, \ldots, \\ m = -n, -n+1, \ldots, n. \end{matrix} \tag{12.66}$$

◇ 12.2.24. Prove the formulae (12.42) for the norms of the spherical harmonics.
Hint: Use Exercise 12.2.16.

◇ 12.2.25. Justify the formulas in (12.50) for (a) H_1^0, (b) H_2^0, (c) $\widetilde{H}_2^1$.

12.2.26. Find formulas for the following harmonic polynomials (*i*) in spherical coordinates; (*ii*) in rectangular coordinates: (a) H_4^0, (b) H_4^4, (c) $\widetilde{H}_4^4$.

12.2.27. Explain why every polynomial solution of the Laplace equation is a linear combination of the harmonic polynomials (12.49). *Hint*: Look at its Taylor series.

12.2.28. (a) Prove that if $u(x,y,z)$ is any harmonic polynomial, then so are $u(y,x,z)$, $u(z,x,y)$, and all other functions obtained by permuting the variables x, y, z. (b) Discuss the effect of such permutations on the basis harmonic polynomials $H_n^m(x,y,z)$ appearing in (12.50).

12.2.29. Find the formulas in rectangular coordinates for the following complementary harmonic functions: (a) K_0^0, (b) K_1^1, (c) K_2^0, (d) $\widetilde{K}_2^1$.

◇ 12.2.30. Let $u(x,y,z)$ be a harmonic function defined on the unit ball $r \leq 1$. Prove that its gradient at the center, $\nabla u(\mathbf{0})$, equals the average of the vector field $\mathbf{v}(\mathbf{x}) = \mathbf{x}\, u(\mathbf{x})$ over the unit sphere $r = 1$.

◇ 12.2.31. (a) Suppose $u(x,y,z)$ is a solution to the Laplace equation. Prove that the function $U(x,y,z) = r^{-1}\, u(x/r^2, y/r^2, z/r^2)$ obtained by *inversion* is also a solution. (b) Explain how inversion can be used to solve boundary value problems on the exterior of a sphere.
(c) Use inversion to relate the solutions to Examples 12.6 and 12.7.

◇ 12.2.32. Suppose $u(r,\varphi,\theta)$ is the potential exterior to a spherical capacitor of unit radius.
(a) Prove that $\lim_{r \to \infty} r\, u(r,\varphi,\theta)$ equals the average value of u on the sphere.
(b) Use Exercise 12.2.31 to deduce this result as a consequence of Theorem 12.4.

12.2.33. (a) Write out, using spherical coordinates, formulas for the L^2 inner product and norm for scalar fields $f(r,\varphi,\theta)$ and $g(r,\varphi,\theta)$ on a solid ball of unit radius centered at the origin.
(b) Let $f(x,y,z) = z$ and $g(x,y,z) = x^2 + y^2$. Find $\| f \|$, $\| g \|$ and $\langle f , g \rangle$.
(c) Verify the Cauchy–Schwarz and triangle inequalities for these two functions.

◇ 12.2.34. Use separation of variables to construct a Fourier series solution to the Laplace equation on a rectangular box, $B = \{ 0 < x < a,\ 0 < y < b,\ 0 < z < c \}$, subject to the Dirichlet boundary conditions $u(x,y,z) = \begin{cases} h(x,y), & z = 0, \quad 0 < x < a, \quad 0 < y < b, \\ 0, & \text{at all other points in } \partial B. \end{cases}$

12.2.35. Find the equilibrium temperature distribution inside a unit cube that has 100° temperature on its top face, 0° on its bottom face, while all four side faces are insulated.

12.2.36. Solve Exercise 12.2.35 when the top face of the cube has temperature

$$u(x,y,1) = \cos \pi x \, \cos \pi y.$$

♣ 12.2.37. A solid unit cube is in thermal equilibrium when subject to 100° temperature on its top face and 0° on all other faces. *True or false*: The temperature at the center equals the average temperature over the surface of the cube.

12.2.38. Solve the boundary value problem

$$- \frac{\partial^2 u}{\partial x^2} - \frac{\partial^2 u}{\partial y^2} - \frac{\partial^2 u}{\partial z^2} + u = \cos x \cos y, \qquad 0 < x, y, z < \pi,$$

$$u(x,y,0) = 1, \quad \frac{\partial u}{\partial z}(x,y,\pi) = \frac{\partial u}{\partial y}(x,0,z) = \frac{\partial u}{\partial y}(x,\pi,z) = \frac{\partial u}{\partial z}(0,y,z) = \frac{\partial u}{\partial x}(\pi,y,z) = 0.$$

12.2.39. Let C be the cylinder of height 1 and diameter 1 that sits on the (x,y)–plane centered on the z–axis. (a) Write out, in cylindrical coordinates, the explicit formula for the L^2 inner product and norm on C.
(b) Let $f(x,y,z) = z$ and $g(x,y,z) = x^2 + y^2$. Find $\| f \|$, $\| g \|$ and $\langle f , g \rangle$.
(c) Verify the Cauchy–Schwarz and triangle inequalities for these two functions.

♢ 12.2.40. (a) Write out the Laplace equation in cylindrical coordinates.
(b) Use separation of variables to construct a series solution to the Laplace equation on the cylinder $C = \{x^2 + y^2 < 1,\ 0 < z < 1\}$, subject to the Dirichlet boundary conditions

$$u(x,y,z) = \begin{cases} h(x,y), & z = 0, \quad x^2 + y^2 < 1, \\ 0, & \text{at all other points in } \partial C. \end{cases}$$

12.2.41. A cylinder of radius 1 and height 2 has 100° temperature on its top face, 0° on its bottom face, while its curved side is fully insulated. Find its equilibrium temperature distribution.

12.2.42. Solve Exercise 12.2.41 if the curved sides are kept at 0° instead.

12.3 Green's Functions for the Poisson Equation

We now turn to the inhomogeneous form of the three-dimensional Laplace equation: the *Poisson equation*

$$- \Delta u = f, \tag{12.67}$$

on a solid domain $\Omega \subset \mathbb{R}^3$. In order to uniquely specify the solution, we must impose appropriate boundary conditions: Dirichlet or mixed. (As in the planar version, Neumann boundary value problems have either infinitely many solutions or no solutions, depending upon whether the Fredholm conditions are satisfied or not.) We only need to discuss the case of homogeneous boundary conditions, since, by linear superposition, an inhomogeneous boundary value problem can be split into a homogeneous boundary value problem for the inhomogeneous Poisson equation along with an inhomogeneous boundary value problem for the homogeneous Laplace equation.

As in Chapter 6, we begin by analyzing the case of a delta function inhomogeneity that is concentrated at a single point in the domain. Thus, for each $\boldsymbol{\xi} = (\xi, \eta, \zeta) \in \Omega$, the *Green's function* $G(\mathbf{x}; \boldsymbol{\xi}) = G(x,y,z;\xi,\eta,\zeta)$ is the unique solution to the Poisson equation

$$- \Delta u = \delta(\mathbf{x} - \boldsymbol{\xi}) = \delta(x - \xi)\, \delta(y - \eta)\, \delta(z - \zeta) \qquad \text{for all} \qquad \mathbf{x} \in \Omega, \tag{12.68}$$

subject to the chosen homogeneous boundary conditions. The solution to the general Poisson equation (12.67) is then obtained by superposition: We write the forcing function

$$f(x,y,z) = \iiint_\Omega f(\xi,\eta,\zeta)\, \delta(x - \xi)\, \delta(y - \eta)\, \delta(z - \zeta)\, d\xi\, d\eta\, d\zeta \tag{12.69}$$

as a linear superposition of delta functions. By linearity, the solution

$$u(x,y,z) = \iiint_\Omega f(\xi,\eta,\zeta)\, G(x,y,z;\xi,\eta,\zeta)\, d\xi\, d\eta\, d\zeta \tag{12.70}$$

to the homogeneous boundary value problem for the Poisson equation (12.67) is then given as the corresponding superposition of the Green's function solutions.

The Green's function can also be used to solve the inhomogeneous Dirichlet boundary value problem

$$-\Delta u = 0, \qquad \mathbf{x} \in \Omega, \qquad\qquad u = h, \qquad \mathbf{x} \in \partial\Omega. \tag{12.71}$$

The same argument that was used in the two-dimensional situation produces the solution

$$u(\mathbf{x}) = -\iint_{\partial\Omega} \frac{\partial G}{\partial \mathbf{n}}(\mathbf{x};\boldsymbol{\xi})\, h(\boldsymbol{\xi})\, dS, \tag{12.72}$$

where the normal derivative is taken with respect to the variable $\boldsymbol{\xi} \in \partial\Omega$. In the case that Ω is a solid ball, this integral formula effectively sums the spherical harmonic series (12.51); see Theorem 12.12 below.

The Free–Space Green's Function

Only in a few specific instances is an explicit formula for the Green's function known. Nevertheless, certain general guiding features can be readily established. The starting point is to investigate the Poisson equation (12.68) when the domain $\Omega = \mathbb{R}^3$ is all of three-dimensional space. We impose boundary constraints by seeking a solution that goes to zero, $u(\mathbf{x}) \to 0$, at large distances, $\|\mathbf{x}\| \to \infty$. Since the Laplacian operator is invariant under translations, we can, without loss of generality, place our delta impulse at the origin, and concentrate on solving the particular case

$$-\Delta u = \delta(\mathbf{x}), \qquad\qquad \mathbf{x} \in \mathbb{R}^3.$$

Since $\delta(\mathbf{x}) = 0$ for all $\mathbf{x} \neq \mathbf{0}$, the desired solution will, in fact, be a solution to the homogeneous Laplace equation

$$\Delta u = 0, \qquad\qquad \mathbf{x} \neq \mathbf{0},$$

save, possibly, for a singularity at the origin.

The Laplace equation models the equilibria of a uniform isotropic medium, and so, as noted in Exercise 12.1.7, is also invariant under three-dimensional rotations. This suggests that, in any radially symmetric configuration, the solution should depend only on the distance $r = \|\mathbf{x}\|$ from the origin. Referring to the spherical coordinate form (12.16) of the Laplacian operator, if u is a function of r only, then its derivatives with respect to the angular coordinates φ, θ are zero, and so $u(r)$ solves the ordinary differential equation

$$\frac{d^2u}{dr^2} + \frac{2}{r}\frac{du}{dr} = 0. \tag{12.73}$$

This equation is, in effect, a first-order linear ordinary differential equation for $v = du/dr$ and hence is particularly easy to solve:

$$\frac{du}{dr} = v(r) = -\frac{b}{r^2}, \qquad \text{and hence} \qquad u(r) = a + \frac{b}{r},$$

where a, b are arbitrary constants. The constant solution $u(r) = a$ does not die away at large distances, nor does it have a singularity at the origin. Therefore, if our intuition is valid, the desired solution should be of the form

$$u = \frac{b}{r} = \frac{b}{\|\mathbf{x}\|} = \frac{b}{\sqrt{x^2+y^2+z^2}}. \tag{12.74}$$

Indeed, this function is harmonic — solves Laplace's equation — everywhere away from the origin and has a singularity at $\mathbf{x} = \mathbf{0}$.

The solution (12.74) is, up to a constant multiple, the three-dimensional Newtonian gravitational potential due to a point mass at the origin. Its gradient,

$$\mathbf{f}(\mathbf{x}) = \nabla \left(\frac{b}{\|\mathbf{x}\|} \right) = -\frac{b\,\mathbf{x}}{\|\mathbf{x}\|^3}, \tag{12.75}$$

defines the gravitational force vector at the point $\mathbf{x}$. When $b > 0$, the force $\mathbf{f}(\mathbf{x})$ points toward the mass at the origin. Its magnitude

$$\|\mathbf{f}\| = \frac{b}{\|\mathbf{x}\|^2} = \frac{b}{r^2}$$

is proportional to the reciprocal of the squared distance, which is the well-known inverse square law of three-dimensional Newtonian gravity. Formula (12.75) can also be interpreted as the electrostatic force due to a concentrated electric charge at the origin, with (12.74) giving the corresponding Coulomb potential. The constant b is positive when the charges are of opposite signs, leading to an attractive force, and negative in the repulsive case of like charges.

Returning to our problem, the remaining task is to fix the multiple b such that the Laplacian of our candidate solution (12.74) has a delta function singularity at the origin; equivalently, we must determine $a = 1/b$ such that

$$-\Delta(r^{-1}) = a\,\delta(\mathbf{x}). \tag{12.76}$$

This equation is certainly valid away from the origin, since $\delta(\mathbf{x}) = 0$ when $\mathbf{x} \neq \mathbf{0}$. To investigate near the singularity, we integrate both sides of (12.76) over a small solid ball $B_\varepsilon = \{ \|\mathbf{x}\| \le \varepsilon \}$ of radius ε:

$$-\iiint_{B_\varepsilon} \Delta(r^{-1})\; dx\,dy\,dz = \iiint_{B_\varepsilon} a\,\delta(\mathbf{x})\,dx\,dy\,dz = a, \tag{12.77}$$

where we used the definition of the delta function to evaluate the right-hand side. On the other hand, since $\Delta\, r^{-1} = \nabla \cdot \nabla\, r^{-1}$, we can use the divergence theorem (12.8) to evaluate the left-hand integral, whence

$$\iiint_{B_\varepsilon} \Delta(r^{-1})\,dx\,dy\,dz = \iiint_{B_\varepsilon} \nabla \cdot \nabla(r^{-1})\,dx\,dy\,dz = \iint_{S_\varepsilon} \frac{\partial}{\partial \mathbf{n}}\left(\frac{1}{r}\right) dS,$$

where the surface integral is over the bounding sphere $S_\varepsilon = \partial B_\varepsilon = \{ \|\mathbf{x}\| = \varepsilon \}$. The sphere's unit normal $\mathbf{n}$ points in the radial direction, and hence the normal derivative coincides with differentiation with respect to r; in particular,

$$\frac{\partial}{\partial \mathbf{n}}\left(\frac{1}{r}\right) = \frac{\partial}{\partial r}\left(\frac{1}{r}\right) = -\frac{1}{r^2}.$$

The surface integral can now be explicitly evaluated:

$$\iint_{S_\varepsilon} \frac{\partial}{\partial \mathbf{n}}\left(\frac{1}{r}\right) dS = -\iint_{S_\varepsilon} \frac{1}{r^2}\, dS = -\iint_{S_\varepsilon} \frac{1}{\varepsilon^2}\, dS = -4\pi,$$

since S_ε has surface area $4\pi\varepsilon^2$. Substituting this result back into (12.77), we conclude that

$$a = 4\pi, \qquad \text{and hence} \qquad -\Delta\, r^{-1} = 4\pi\,\delta(\mathbf{x}). \tag{12.78}$$

This is our desired formula! We conclude that a solution to the Poisson equation with a delta function impulse at the origin is

$$G(x,y,z) = \frac{1}{4\pi r} = \frac{1}{4\pi \, \| \mathbf{x} \|} = \frac{1}{4\pi \sqrt{x^2+y^2+z^2}}, \tag{12.79}$$

which is the three-dimensional Newtonian potential due to a unit point mass situated at the origin.

If the singularity is concentrated at some other point $\boldsymbol{\xi} = (\xi, \eta, \zeta)$, then we merely translate the preceding solution. This leads immediately to the *free-space Green's function*

$$G(\mathbf{x};\boldsymbol{\xi}) = G(\mathbf{x}-\boldsymbol{\xi}) = \frac{1}{4\pi \, \| \mathbf{x}-\boldsymbol{\xi} \|} = \frac{1}{4\pi \sqrt{(x-\xi)^2+(y-\eta)^2+(z-\zeta)^2}}. \tag{12.80}$$

The superposition principle (12.70) implies the following integral formula for the solutions to the Poisson equation on all of three-dimensional space.

Theorem 12.8. *Assuming that $f(\mathbf{x}) \to 0$ sufficiently rapidly as $\| \mathbf{x} \| \to \infty$, a particular solution to the Poisson equation*

$$-\Delta u = f, \qquad \text{for} \qquad \mathbf{x} \in \mathbb{R}^3, \tag{12.81}$$

is given by

$$u_\star(\mathbf{x}) = \frac{1}{4\pi} \iiint_{\mathbb{R}^3} \frac{f(\boldsymbol{\xi})}{\| \mathbf{x}-\boldsymbol{\xi} \|} \, d\boldsymbol{\xi} = \frac{1}{4\pi} \iiint_{\mathbb{R}^3} \frac{f(\xi,\eta,\zeta) \; d\xi \, d\eta \, d\zeta}{\sqrt{(x-\xi)^2+(y-\eta)^2+(z-\zeta)^2}}. \tag{12.82}$$

The general solution is $u(x,y,z) = u_\star(x,y,z) + w(x,y,z)$, where $w(x,y,z)$ is an arbitrary harmonic function.

Example 12.9. In this example, we compute the gravitational (or electrostatic) potential in three-dimensional space due to a uniform solid ball, e.g., a spherical planet such as the Earth. By rescaling, it suffices to consider the case in which the forcing function is equal to 1 inside a ball of radius 1 and zero outside:

$$f(\mathbf{x}) = \begin{cases} 1, & \| \mathbf{x} \| < 1, \\ 0, & \| \mathbf{x} \| > 1. \end{cases}$$

The particular solution to the resulting Poisson equation (12.81) is given by the integral

$$u(\mathbf{x}) = \frac{1}{4\pi} \iiint_{\| \boldsymbol{\xi} \| < 1} \frac{1}{\| \mathbf{x}-\boldsymbol{\xi} \|} \, d\xi \, d\eta \, d\zeta. \tag{12.83}$$

Clearly, since the forcing function is radially symmetric, the solution $u = u(r)$ is also radially symmetric. To evaluate the integral, then, we can take $\mathbf{x} = (0,0,z)$ to lie on the z–axis, so that $r = \| \mathbf{x} \| = |\, z \,|$. We use cylindrical coordinates $\boldsymbol{\xi} = (\rho\cos\theta, \rho\sin\theta, \zeta)$, so that

$$\| \mathbf{x}-\boldsymbol{\xi} \| = \sqrt{\rho^2 + (z-\zeta)^2}\,.$$

The integral in (12.83) can then be explicitly computed:

$$\begin{aligned} &\frac{1}{4\pi} \int_{-1}^{1} \int_{0}^{\sqrt{1-\zeta^2}} \int_{0}^{2\pi} \frac{\rho \, d\theta \, d\rho \, d\zeta}{\sqrt{\rho^2+(z-\zeta)^2}} \\ &\qquad = \frac{1}{2} \int_{-1}^{1} \left(\sqrt{1+z^2-2\,z\,\zeta} \; - \; |\, z-\zeta \,| \right) d\zeta = \begin{cases} \dfrac{1}{3\,|\, z \,|}, & |\, z \,| \geq 1, \\[2ex] \dfrac{1}{2} - \dfrac{z^2}{6}, & |\, z \,| \leq 1. \end{cases} \end{aligned}$$

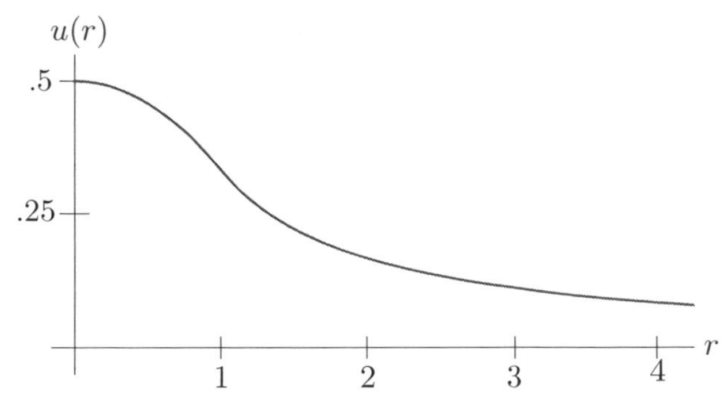

Figure 12.6. Solution to Poisson's equation in a solid ball.

Therefore, by radial symmetry, the solution is

$$u(\mathbf{x}) = \begin{cases} \dfrac{1}{3r}, & r = \|\,\mathbf{x}\,\| \geq 1, \\ \dfrac{1}{2} - \dfrac{r^2}{6}, & r = \|\,\mathbf{x}\,\| \leq 1, \end{cases} \tag{12.84}$$

plotted, as a function of $r = \|\,\mathbf{x}\,\|$, in Figure 12.6. Note that, outside the solid ball, the solution is a Newtonian potential corresponding to a concentrated point mass of magnitude $\frac{4}{3}\pi$ — the total mass of the planet. We have thus demonstrated a well-known result in gravitation and electrostatics: the exterior potential due to a spherically symmetric mass (or electrically charged body) is the same as if all the mass (charge) were concentrated at its center. In the darkness of outer space, if you cannot see a spherical planet, you can determine only its mass, not its size, by measuring its external gravitational force.

Bounded Domains and the Method of Images

Suppose we now wish to solve the inhomogeneous Poisson equation (12.67) on a bounded domain $\Omega \subset \mathbb{R}^3$. To construct the desired Green's function, we proceed as follows. The Newtonian potential (12.80) is a particular solution to the underlying inhomogeneous equation

$$-\,\Delta u = \delta(\mathbf{x} - \boldsymbol{\xi}), \qquad \mathbf{x} \in \Omega, \tag{12.85}$$

but it almost surely does not have the proper boundary values on $\partial\Omega$. By linearity, the general solution to such an inhomogeneous linear equation must take the form

$$u(\mathbf{x}) = \frac{1}{4\pi\,\|\,\mathbf{x} - \boldsymbol{\xi}\,\|} - v(\mathbf{x}), \tag{12.86}$$

where the first term is a particular solution, while $v(\mathbf{x})$ is an arbitrary solution to the homogeneous equation $\Delta v = 0$, i.e., an arbitrary harmonic function. The solution (12.86) satisfies the homogeneous boundary conditions, provided the boundary values of $v(\mathbf{x})$ match those of the Green's function. Let us explicitly state the result in the Dirichlet case.

Theorem 12.10. *The Green's function for the homogeneous Dirichlet boundary value problem*

$$-\,\Delta u = f \quad \textit{for} \quad \mathbf{x} \in \Omega, \qquad u = 0 \quad \textit{for} \quad \|\,\mathbf{x}\,\| \in \partial\Omega,$$

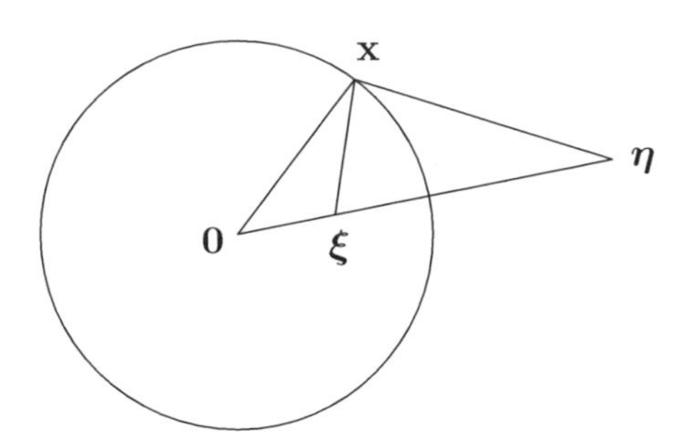

Figure 12.7. Method of Images for the unit ball.

for the Poisson equation in a domain $\Omega \subset \mathbb{R}^3$ *has the form*

$$G(\mathbf{x};\boldsymbol{\xi}) = \frac{1}{4\pi\,\|\,\mathbf{x}-\boldsymbol{\xi}\,\|} - v(\mathbf{x};\boldsymbol{\xi}), \qquad \mathbf{x},\boldsymbol{\xi} \in \Omega, \tag{12.87}$$

where $v(\mathbf{x};\boldsymbol{\xi})$ *is the harmonic function of* $\mathbf{x} \in \Omega$ *that satisfies*

$$v(\mathbf{x};\boldsymbol{\xi}) = \frac{1}{4\pi\,\|\,\mathbf{x}-\boldsymbol{\xi}\,\|} \qquad \textit{for all} \qquad \mathbf{x} \in \partial\Omega. \tag{12.88}$$

In this manner, we have reduced the determination of the Green's function to the solution to a particular family of Laplace boundary value problems, which are parametrized by the point $\boldsymbol{\xi} \in \Omega$. In certain domains with simple geometry, the Method of Images can be used to produce an explicit formula for the Green's function. As in Section 6.3, the idea is to match the boundary values of the free-space Green's function due to a delta impulse at a point inside the domain with one or more additional Green's functions corresponding to impulses at points outside the domain — the "image points".

The case of a solid ball of radius 1 with Dirichlet boundary conditions is the easiest to handle. Indeed, the *same* geometric construction that we used for a planar disk, redrawn in Figure 12.7, applies here. Although identical to Figure 6.13, we are re-interpreting it as a three-dimensional diagram, with the circle representing the unit sphere, while the lines remain lines. The required image point is given by *inversion*:

$$\boldsymbol{\eta} = \frac{\boldsymbol{\xi}}{\|\,\boldsymbol{\xi}\,\|^2}, \qquad \text{whereby} \qquad \|\,\boldsymbol{\xi}\,\| = \frac{1}{\|\,\boldsymbol{\eta}\,\|}.$$

By the similar triangles argument used before, we have

$$\frac{\|\,\boldsymbol{\xi}\,\|}{\|\,\mathbf{x}\,\|} = \frac{\|\,\mathbf{x}\,\|}{\|\,\boldsymbol{\eta}\,\|} = \frac{\|\,\mathbf{x}-\boldsymbol{\xi}\,\|}{\|\,\mathbf{x}-\boldsymbol{\eta}\,\|}, \qquad \text{and therefore} \qquad \|\,\mathbf{x}\,\| = 1.$$

As a result, the function

$$v(\mathbf{x},\boldsymbol{\xi}) = \frac{1}{4\pi}\,\frac{\|\,\boldsymbol{\eta}\,\|}{\|\,\mathbf{x}-\boldsymbol{\eta}\,\|} = \frac{1}{4\pi}\,\frac{\|\,\boldsymbol{\xi}\,\|}{\|\,\boldsymbol{\xi}-\|\,\boldsymbol{\xi}\,\|^2\,\mathbf{x}\,\|}$$

has the same boundary values on the unit sphere as the Newtonian potential:

$$\frac{1}{4\pi}\,\frac{\|\,\boldsymbol{\eta}\,\|}{\|\,\mathbf{x}-\boldsymbol{\eta}\,\|} = \frac{1}{4\pi\|\,\mathbf{x}-\boldsymbol{\xi}\,\|} \qquad \text{whenever} \qquad \|\,\mathbf{x}\,\| = 1.$$

We conclude that their difference

$$G(\mathbf{x};\boldsymbol{\xi}) = \frac{1}{4\pi}\left(\frac{1}{\|\,\mathbf{x}-\boldsymbol{\xi}\,\|} - \frac{\|\,\boldsymbol{\xi}\,\|}{\|\,\boldsymbol{\xi} - \|\,\boldsymbol{\xi}\,\|^2\,\mathbf{x}\,\|}\right) \tag{12.89}$$

has the required properties of the Green's function: it satisfies the Laplace equation inside the unit ball except at the delta function singularity $\mathbf{x} = \boldsymbol{\xi}$, and, moreover, $G(\mathbf{x};\boldsymbol{\xi}) = 0$ has homogeneous Dirichlet conditions on the spherical boundary $\|\,\mathbf{x}\,\| = 1$.

With the Green's function in hand, we can apply the general superposition formula (12.70) to arrive at a solution to the Dirichlet boundary value problem for the Poisson equation in the unit ball.

Theorem 12.11. *The solution to the Dirichlet boundary value problem*

$$-\Delta u = f \qquad \textit{for} \qquad \|\,\mathbf{x}\,\| < 1, \qquad\qquad u = 0 \qquad \textit{for} \qquad \|\,\mathbf{x}\,\| = 1,$$

on the unit ball is given by the integral

$$u(\mathbf{x}) = \frac{1}{4\pi}\iiint_{\|\,\boldsymbol{\xi}\,\|\le 1}\left(\frac{1}{\|\,\mathbf{x}-\boldsymbol{\xi}\,\|} - \frac{\|\,\boldsymbol{\xi}\,\|}{\|\,\boldsymbol{\xi} - \|\,\boldsymbol{\xi}\,\|^2\,\mathbf{x}\,\|}\right) f(\boldsymbol{\xi})\,d\xi\,d\eta\,d\zeta. \tag{12.90}$$

By the same token, formula (12.72) provides a solution to the inhomogeneous Dirichlet boundary value problem for the Laplace equation on a ball.

Theorem 12.12. *The solution to the Dirichlet boundary value problem*

$$-\Delta u = 0 \qquad \textit{for} \qquad \|\,\mathbf{x}\,\| < 1, \qquad\qquad u = h \qquad \textit{for} \qquad \|\,\mathbf{x}\,\| = 1,$$

on the unit ball is given by the following surface integral:

$$u(\mathbf{x}) = \frac{1}{4\pi}\iint_{\|\,\boldsymbol{\xi}\,\|=1}\frac{1-\|\,\mathbf{x}\,\|^2}{\|\,\boldsymbol{\xi}-\mathbf{x}\,\|^3}\,h(\boldsymbol{\xi})\,dS. \tag{12.91}$$

Proof: We start with the explicit formula (12.89) for the Green's function on the unit ball. Since the normal derivative on the unit sphere $\|\,\boldsymbol{\xi}\,\| = 1$ can be written as $\partial/\partial\mathbf{n} = \boldsymbol{\xi}\cdot\nabla_{\boldsymbol{\xi}}$, a short computation demonstrates that

$$\frac{\partial G}{\partial \mathbf{n}}(\mathbf{x};\boldsymbol{\xi}) = \frac{1}{4\pi}\left(\frac{\mathbf{x}\cdot\boldsymbol{\xi} - \|\,\boldsymbol{\xi}\,\|^2}{\|\,\mathbf{x}-\boldsymbol{\xi}\,\|^3} - \frac{\|\,\boldsymbol{\xi}\,\|^3\big(\mathbf{x}\cdot\boldsymbol{\xi} - \|\,\boldsymbol{\xi}\,\|^2\,\|\,\mathbf{x}\,\|^2\big)}{\|\,\boldsymbol{\xi} - \|\,\boldsymbol{\xi}\,\|^2\,\mathbf{x}\,\|^3}\right) = \frac{1}{4\pi}\frac{\|\,\mathbf{x}\,\|^2-1}{\|\,\boldsymbol{\xi}-\mathbf{x}\,\|^3}.$$

The solution formula (12.91) thus immediately follows from (12.72). *Q.E.D.*

For example, the series solution (12.60) to the spherical capacitor problem of Example 12.6 can thus be re-expressed as a surface integral:

$$\begin{aligned} u(x,y,z) &= \frac{1}{4\pi}\iint_{\{\xi^2+\eta^2+\zeta^2=1,\ \zeta>0\}}\frac{(1-x^2-y^2-z^2)\,dS}{\big[\,(\xi-x)^2+(\eta-y)^2+(\zeta-z)^2\,\big]^{3/2}} \\ &= \int_{-\pi}^{\pi}\int_0^{\pi/2}\frac{(1-x^2-y^2-z^2)\sin\varphi\,d\varphi\,d\theta}{\big[\,(\cos\theta\sin\varphi-x)^2+(\sin\theta\sin\varphi-y)^2+(\cos\varphi-z)^2\,\big]^{3/2}}. \end{aligned}$$

Exercises

12.3.1. Find the equilibrium temperature of a sphere of radius 1 whose boundary is held at 0° while a concentrated unit heat source is applied at (*a*) the center; (*b*) a point half-way between the center and the boundary.

12.3.2. A hot soldering iron is continually applied to the north pole of a solid spherical ball of radius 1. Find the equilibrium temperature.

12.3.3. Write down the gravitional potential — both external and internal — due to a spherical planet of radius R composed out of a uniform material with density ρ.

12.3.4. (*a*) Find the gravitational potential due to a spherical shell of unit density obtained by carving out a spherical cavity of radius a from a solid ball of radius $b > a$. *Hint*: Use the solution to Exercise 12.3.3. (*b*) What is the gravitational force inside the cavity? (*c*) Show that outside the shell, the gravitational potential is as if the entire mass were concentrated at the origin.

♣ 12.3.5. (*a*) Write down an integral formula for the gravitational potential and gravitational force field due to a mass of unit density in the shape of a solid unit cube that is centered at the origin. (*b*) Use numerical integration to determine the gravitational force vector at the points $(3, 0, 0)$ and $\left(\sqrt{3}, \sqrt{3}, \sqrt{3}\right)$. Before doing the calculation, see whether you can predict which experiences a stronger force, and then check your prediction numerically. (*c*) Suppose the mass is re-formed into a sphere. How does this affect the gravitational force at the two points? First predict whether it will increase, decrease, or stay the same. Then test your prediction by computing the values and comparing with those you computed in part (*b*).

12.3.6. A thin hollow metal sphere of unit radius is grounded. Find the electrostatic potential inside the sphere due to a small solid metal ball of radius $\rho < 1$ placed at its center, assuming unit charge density throughout the ball.

12.3.7. A thin straight rod of unit density and length 2ℓ is fixed on the z–axis centered at the origin. Find the induced (a) gravitational potential and (*b*) gravitational force experienced by a point (x, y, z) not on the rod.

♡ 12.3.8. (a) Find the gravitational force due to a thin, uniform straight rod of unit density and infinite length by letting $\ell \to \infty$ in your solution to Exercise 12.3.7(*b*). (*b*) Show that the force field of part (a) has a potential function that can be identified with the two-dimensional logarithmic gravitational potential due to a point mass at the origin. Thus, two-dimensional gravitation can be regarded as a cross-section of three-dimensional gravitation due to infinitely long vertical line masses. (*c*) Is your potential function the limit, as $\ell \to \infty$, of the potential function you found in Exercise 12.3.7(a)? Discuss.

12.3.9. Which well-known solutions to the Laplace equation comes from setting $m = n = 0$ in (12.61)?

12.3.10. Use the Fredholm Alternative to analyze the existence and uniqueness of solutions to the homogeneous Neumann boundary value problem for the Poisson equation on a bounded domain $\Omega \subset \mathbb{R}^3$.

◇ 12.3.11. Mimic the proof of Theorem 6.19 to establish the solution formula (12.72).

12.3.12. Use the Method of Images to find the Green's function for a solid hemisphere of unit radius subject to homogeneous Dirichlet boundary conditions.

12.4 The Heat Equation for Three–Dimensional Media

Thermal diffusion in a uniform isotropic solid body $\Omega \subset \mathbb{R}^3$ is modeled by the three-dimensional *heat equation*

$$\frac{\partial u}{\partial t} = \gamma\,\Delta u = \gamma\left(\frac{\partial^2 u}{\partial x^2} + \frac{\partial^2 u}{\partial y^2} + \frac{\partial^2 u}{\partial z^2}\right), \qquad (x,y,z) \in \Omega. \tag{12.92}$$

The positivity of the body's thermal diffusivity, $\gamma > 0$, is required on both physical and mathematical grounds. The physical derivation is exactly the same as that for the two-dimensional version (11.1), and does not need to be repeated in detail. Briefly, Fourier's law expresses the heat flux vector as a multiple of the temperature gradient, $\mathbf{w} = -\,\kappa\,\nabla u$, while energy conservation implies that its divergence is proportional to the rate of change of temperature: $\nabla \cdot \mathbf{w} = -\,\sigma\,u_t$. Combining these two physical laws and assuming uniformity, whereby κ and σ are constant, produces (12.92) with $\gamma = \kappa/\sigma$.

As always, we must impose suitable boundary conditions: either Dirichlet conditions $u = h$ that specify the boundary temperature; (homogeneous) Neumann conditions $\partial u/\partial \mathbf{n} = 0$ corresponding to an insulated boundary; or a mixture of the two. Given the body's temperature

$$u(t_0,x,y,z) = f(x,y,z) \tag{12.93}$$

at an initial time t_0, it can be proved, [**38**, **61**, **99**], that the resulting initial-boundary value problem is well-posed, which means that there is a unique classical solution $u(t,x,y,z)$, defined at all subsequent times $t > t_0$, that depends continuously on the initial data.

As in the one- and two-dimensional versions, we begin by restricting our attention to homogeneous boundary conditions. Separation of variables works as usual, and we quickly review the basic ideas. One begins by imposing an exponential solution ansatz

$$u(t,\mathbf{x}) = e^{-\lambda t}\,v(\mathbf{x}).$$

Substituting into the differential equation and canceling the exponentials, it follows that v satisfies the Helmholtz eigenvalue problem

$$\gamma\,\Delta v + \lambda\,v = 0,$$

subject to the relevant boundary conditions. For Dirichlet and mixed boundary conditions, the Laplacian is a positive definite operator, and hence the eigenvalues are all strictly positive,

$$0 < \lambda_1 \le \lambda_2 \le \cdots, \qquad \text{with} \qquad \lambda_n \longrightarrow \infty, \qquad \text{as} \qquad n \to \infty.$$

Moreover, on a bounded domain, the Helmholtz eigenfunctions are complete, and so linear superposition implies that the solution can be written as an eigenfunction series

$$u(t,\mathbf{x}) = \sum_{n=1}^{\infty} c_n\,e^{-\lambda_n t}\,v_n(\mathbf{x}). \tag{12.94}$$

The coefficients c_n are uniquely prescribed by the initial condition (12.93):

$$u(t_0,\mathbf{x}) = \sum_{n=1}^{\infty} c_n\,e^{-\lambda_n t_0}\,v_n(\mathbf{x}) = f(\mathbf{x}). \tag{12.95}$$

Self-adjointness of the boundary value problem implies orthogonality of the eigenfunctions, and hence the coefficients are obtained via the usual inner product formulae:

$$c_n = e^{\lambda_n t_0} \frac{\langle f, v_n \rangle}{\| v_n \|^2} = e^{\lambda_n t_0} \frac{\displaystyle\iiint_\Omega f(\mathbf{x})\, v_n(\mathbf{x})\, dx\, dy\, dz}{\displaystyle\iiint_\Omega v_n(\mathbf{x})^2\, dx\, dy\, dz}. \tag{12.96}$$

The resulting solution decays exponentially fast to thermal equilibrium, $u(t, \mathbf{x}) \to 0$ as $t \to \infty$, typically at a rate equal to the smallest positive eigenvalue $\lambda_1 > 0$, although special solutions, whose initial series coefficients vanish, will decay at a faster rate governed by a higher eigenvalue. Since the higher modes — the terms with $n \gg 0$ — go to zero extremely rapidly with increasing t, the solution can be well approximated by the first few terms in its eigenfunction expansion. As a consequence, the heat equation rapidly smooths out discontinuities and eliminates high-frequency noise in the initial data.

Unfortunately, explicit formulas for the eigenfunctions and eigenvalues are rare. Most explicit eigensolutions of the Helmholtz boundary value problem require a further separation of variables. In a rectangular box, one separates the solution into a product of functions depending upon the individual Cartesian coordinates, and the eigenfunctions are written as products of trigonometric functions; see Exercise 12.4.1 for details. In a cylindrical domain, the separation is effected in cylindrical coordinates, which leads to eigensolutions involving trigonometric and Bessel functions, as outlined in Exercise 12.4.5. The most interesting and enlightening case is a spherical domain, and we treat this particular problem in complete detail in the ensuing subsection.

Exercises

♢ 12.4.1. Let $B = \{0 < x < a,\ 0 < y < b,\ 0 < z < c\}$ be a solid box of size $a \times b \times c$. (a) Write down an initial-boundary value problem for the thermodynamics of the box when all its sides are all held at 0° and its initial temperature is $f(x, y, z)$. (b) Use separation of variables to construct the normal mode solutions. (c) Write down a series representing the general solution to the initial-boundary value problem. What are the formulas for the coefficients in your series? (d) What is the equilibrium temperature? How fast does the temperature in the box decay to equilibrium?

12.4.2. *True or false*: In the context of Exercise 12.4.1, among all boxes of a given volume V, a cube decays slowest to thermal equilibrium. What is the cube's decay rate?

12.4.3. Answer Exercises 12.4.1 and 12.4.2 when the top of the box, where $z = c$, is insulated.

12.4.4. A rectangular brick of size 1 cm × 2 cm × 3 cm made out of material with diffusion coefficient $\gamma = 6$ is insulated on five sides, while one of its small ends is held at temperature $u(x, y, 0) = \cos \pi x\, \cos 2\pi y$. (a) Find the eventual equilibrium temperature distribution. (b) If the brick is initially heated in an oven, how fast does it return to equilibrium?

♢ 12.4.5. Let $C = \left\{ 0 \le \sqrt{x^2 + y^2} < a,\ 0 < z < h \right\}$ be a solid cylinder of radius a and height h.
(a) Write down an initial-boundary value problem in cylindrical coordinates for the thermodynamics of the cylinder when its sides, top, and bottom are all held at 0°.
(b) Use separation of variables to write down a series representing the general solution to the initial-boundary value problem. What are the formulas for the coefficients in your series?

(c) What is the eventual equilibrium temperature?
(d) How fast does the temperature in the cylinder go to equilibrium?

12.4.6. Find the solution to the initial-boundary value problem in Exercise 12.4.5 when the initial temperature of the cylinder is uniformly 30°. *Hint*: Use (11.112) to evaluate the coefficients.

♡ 12.4.7. A cylindrical can that contains 355 ml of soda is removed from the refrigerator. Find the optimal cylindrical shape for such a can in order to keep the soda cold the longest. Is this the manufactured shape of a standard soda can in your country?

♡ 12.4.8. *True or false*: Among all solid cylinders of a given volume, the one that reaches thermal equilibrium the slowest, when subject to homogeneous Dirichlet boundary conditions, is the one that has the least surface area. Justify your answer.

♡ 12.4.9. Among all fully insulated solid cylinders of unit volume, which cools down (*i*) the slowest? (*ii*) the fastest?

♢ 12.4.10. Write down a series for the solution to the homogeneous Neumann boundary value problem for the heat equation on a bounded domain $\Omega \subset \mathbb{R}^3$, corresponding to the thermodynamics of a completely insulated solid body. What is the equilibrium temperature of the body? Does the solution decay to equilibrium? If so, how fast?

♢ 12.4.11. Suppose $u(t,x,y,z)$ is a solution to the heat equation on a fully insulated bounded domain $\Omega \subset \mathbb{R}^3$. Use the identities in Exercise 12.1.11 to prove the following:
(a) The total heat $H(t) = \iiint_\Omega u(t,x,y,z)\,dx\,dy\,dz$ is conserved, i.e., is constant. Explain how this can be used to determine the equilibrium temperature of the body.
(b) If u is a non-equilibrium solution, its squared L^2 norm $E(t) = \iiint_\Omega u(t,x,y,z)^2\,dx\,dy\,dz$ is a strictly decreasing function of t.
(c) Use part (b) to prove uniqueness of solutions to the initial value problem.

♢ 12.4.12. State and prove a Maximum Principle for the three-dimensional heat equation.

Heating of a Ball

Our goal is to study heat propagation in a solid spherical body, e.g., the Earth.† For simplicity, we take the diffusivity $\gamma = 1$, and consider the heat equation on a solid spherical ball of unit radius, $B_1 = \{\, \| \mathbf{x} \| < 1 \,\}$, that is subject to homogeneous Dirichlet boundary conditions. Once we know how to solve this particular case, an easy scaling argument, as outlined in Exercise 12.4.16, will allow us to find the solution for a ball of arbitrary radius and general diffusivity.

As usual, when dealing with a spherical geometry, we adopt spherical coordinates r, φ, θ, as in (12.15), in terms of which the heat equation takes the form

$$\frac{\partial u}{\partial t} = \Delta u = \frac{\partial^2 u}{\partial r^2} + \frac{2}{r}\frac{\partial u}{\partial r} + \frac{1}{r^2}\frac{\partial^2 u}{\partial \varphi^2} + \frac{\cos\varphi}{r^2 \sin\varphi}\frac{\partial u}{\partial \varphi} + \frac{1}{r^2 \sin^2\varphi}\frac{\partial^2 u}{\partial \theta^2}, \qquad (12.97)$$

where we have used our handy spherical coordinate formula (12.16) for the Laplacian. The

† In this admittedly simplistic model, we are assuming that the Earth is composed of a completely uniform and isotropic solid material.

standard diffusive separation of variables ansatz

$$u(t,r,\varphi,\theta) = e^{-\lambda t}\, v(r,\varphi,\theta)$$

requires us to analyze the spherical coordinate form of the Helmholtz equation

$$\Delta v + \lambda\, v = \frac{\partial^2 v}{\partial r^2} + \frac{2}{r}\frac{\partial v}{\partial r} + \frac{1}{r^2}\frac{\partial^2 v}{\partial \varphi^2} + \frac{\cos\varphi}{r^2 \sin\varphi}\frac{\partial v}{\partial \varphi} + \frac{1}{r^2 \sin^2\varphi}\frac{\partial^2 v}{\partial \theta^2} + \lambda\, v = 0 \qquad (12.98)$$

on the unit ball $\Omega = \{\, r < 1 \,\}$ under homogeneous Dirichlet boundary conditions. To make further progress, we invoke a second variable separation, splitting off the radial coordinate by setting

$$v(r,\varphi,\theta) = p(r)\, w(\varphi,\theta).$$

The function w must be 2π–periodic in θ and well defined on the z–axis, i.e., when $\varphi = 0, \pi$. Substituting this ansatz into (12.98), and separating all the r-dependent terms from those terms depending on the angular variables φ, θ leads to a pair of differential equations involving a separation constant, denoted by μ. The first is an ordinary differential equation

$$r^2 \frac{d^2 p}{dr^2} + 2\, r \frac{dp}{dr} + (\lambda\, r^2 - \mu) p = 0, \qquad (12.99)$$

for the radial component $p(r)$, while the second is a familiar partial differential equation

$$\Delta_S\, w + \mu\, w = \frac{\partial^2 w}{\partial \varphi^2} + \frac{\cos\varphi}{\sin\varphi}\frac{\partial w}{\partial \varphi} + \frac{1}{\sin^2\varphi}\frac{\partial^2 w}{\partial \theta^2} + \mu\, w = 0, \qquad (12.100)$$

for its angular counterpart $w(\varphi,\theta)$. The operator Δ_S is the *spherical Laplacian* from (12.19). In Section 12.2, we showed that its eigenvalues are

$$\mu_m = m(m+1) \qquad \text{for} \qquad m = 0, 1, 2, 3, \ldots .$$

The mth eigenvalue admits $2\,m + 1$ linearly independent eigenfunctions: the spherical harmonics $Y_m^0, \ldots, Y_m^m, \widetilde{Y}_m^1, \ldots, \widetilde{Y}_m^m$ defined in (12.38).

Spherical Bessel Functions

The radial ordinary differential equation (12.99) can be solved by setting

$$q(r) = \sqrt{r}\, p(r). \qquad (12.101)$$

We use the product rule to relate the derivatives of q and p, whereby

$$p = \frac{q}{r^{1/2}}, \qquad \frac{dp}{dr} = \frac{1}{r^{1/2}}\frac{dq}{dr} - \frac{q}{2\, r^{3/2}}, \qquad \frac{d^2 p}{dr^2} = \frac{1}{r^{1/2}}\frac{d^2 q}{dr^2} - \frac{1}{r^{3/2}}\frac{dq}{dr} + \frac{3\, q}{4\, r^{5/2}}.$$

Substituting these expressions back into (12.99) with $\mu = \mu_m = m(m+1)$ and multiplying the resulting equation by $\sqrt{r}$, we discover that $q(r)$ must solve the differential equation

$$r^2 \frac{d^2 q}{dr^2} + r \frac{dq}{dr} + \left[\, \lambda\, r^2 - \left(m + \tfrac{1}{2} \right)^2 \right] q = 0, \qquad (12.102)$$

which we recognize as the rescaled Bessel equation (11.56) of half-integer order $m + \frac{1}{2}$. Consequently, the solution to (12.102) that remains bounded at $r = 0$ is (up to a scalar multiple) the rescaled Bessel function

$$q(r) = J_{m+1/2}\left(\sqrt{\lambda}\, r\right).$$

The corresponding solution

$$p(r) = r^{-1/2} J_{m+1/2}(\sqrt{\lambda}\, r) \tag{12.103}$$

to (12.99) is important enough to warrant a special name.

Definition 12.13. The *spherical Bessel function* of order $m \geq 0$ is defined by the formula

$$S_m(x) = \sqrt{\frac{\pi}{2\,x}}\, J_{m+1/2}(x). \tag{12.104}$$

Remark: The multiplicative factor $\sqrt{\pi/2}$ is included in the definition so as to avoid annoying factors of $\sqrt{\pi}$ and $\sqrt{2}$ in the subsequent formulas.

Surprisingly, unlike the Bessel functions of integer order, the spherical Bessel functions are all elementary functions! Comparing (12.104) with (11.105), we see that the spherical Bessel function of order 0 is

$$S_0(x) = \frac{\sin x}{x}. \tag{12.105}$$

The corresponding explicit formulas for the higher-order spherical Bessel functions can be obtained through the general recurrence relation

$$S_{m+1}(x) = -\frac{dS_m}{dx} + \frac{m}{x}\, S_m(x), \tag{12.106}$$

which is a consequence of the Bessel function recurrence formula (11.111). Indeed,

$$\begin{aligned}
\frac{dS_m}{dx} &= \sqrt{\frac{\pi}{2x}}\,\frac{dJ_{m+1/2}}{dx} - \frac{1}{2}\sqrt{\frac{\pi}{2}}\,\frac{1}{x^{3/2}}\, J_{m+1/2}(x) \\
&= -\sqrt{\frac{\pi}{2x}}\left[J_{m+3/2}(x) + \frac{m+\frac{1}{2}}{x}\, J_{m+1/2}(x) \right] - \frac{1}{2}\sqrt{\frac{\pi}{2}}\,\frac{1}{x^{3/2}}\, J_{m+1/2}(x) \\
&= -\sqrt{\frac{\pi}{2x}}\, J_{m+3/2}(x) + \frac{m}{x}\sqrt{\frac{\pi}{2x}}\, J_{m+1/2}(x) = -S_{m+1}(x) + \frac{m}{x}\, S_m(x).
\end{aligned}$$

The next few spherical Bessel functions are, therefore,

$$\begin{aligned}
S_1(x) &= -\frac{dS_0}{dx} &&= -\frac{\cos x}{x} + \frac{\sin x}{x^2}, \\
S_2(x) &= -\frac{dS_1}{dx} + \frac{S_1}{x} &&= -\frac{\sin x}{x} - \frac{3\cos x}{x^2} + \frac{3\sin x}{x^3}, \\
S_3(x) &= -\frac{dS_2}{dx} + \frac{2\,S_2}{x} &&= \frac{\cos x}{x} - \frac{6\sin x}{x^2} - \frac{15\cos x}{x^3} + \frac{15\sin x}{x^4},
\end{aligned} \tag{12.107}$$

and so on. Figure 11.4 provides graphs of the first four spherical Bessel functions on the interval $0 \leq x \leq 20$; the vertical axes range from $-.5$ to 1.0. We note that

$$S_0(0) = 1, \qquad \text{whereas} \qquad S_m(0) = 0 \qquad \text{for} \qquad m > 0, \tag{12.108}$$

whose proof is the task of Exercise 12.4.26. Thus, our radial solution (12.103) is, apart from an inessential constant multiple, a rescaled spherical Bessel function of order m:

$$p(r) = S_m(\sqrt{\lambda}\, r).$$

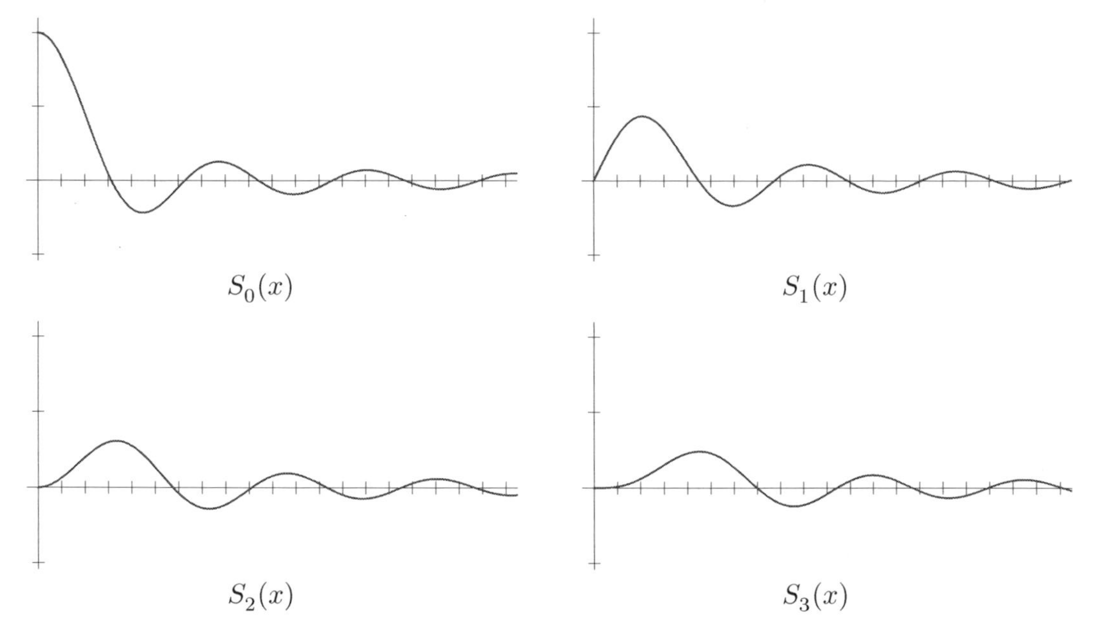

Figure 12.8. Spherical Bessel functions.

So far, we have not taken into account the (homogeneous) Dirichlet boundary condition at $r = 1$. This requires

$$p(1) = 0, \qquad \text{and hence} \qquad S_m(\sqrt{\lambda}) = 0.$$

Therefore, $\sqrt{\lambda}$ must be a root of the mth order spherical Bessel function. We introduce the notation

$$0 < \sigma_{m,1} < \sigma_{m,2} < \sigma_{m,3} < \cdots$$

to denote the successive (positive) *spherical Bessel roots*, satisfying

$$S_m(\sigma_{m,n}) = 0 \qquad \text{for} \qquad n = 1, 2, \ \ldots \ . \tag{12.109}$$

In particular the roots of the zeroth order spherical Bessel function $S_0(x) = x^{-1} \sin x$ are just the integer multiples of π:

$$\sigma_{0,n} = n\pi \qquad \text{for} \qquad n = 1, 2, \ \ldots \ .$$

The higher-order roots are not expressible in terms of known constants. A table of all spherical Bessel roots that are < 13 appears below. The columns of the table are indexed by m, the order, while the rows are indexed by n, the root number.

Re-assembling the individual constituents, we have now demonstrated that the separable eigenfunctions of the Helmholtz equation on a solid ball of radius 1, when subject to homogeneous Dirichlet boundary conditions, are products of spherical Bessel functions and spherical harmonics,

$$\begin{aligned} v_{k,m,n}(r,\varphi,\theta) &= S_m(\sigma_{m,n}\, r)\, Y_m^k(\varphi,\theta), \\ \widetilde{v}_{k,m,n}(r,\varphi,\theta) &= S_m(\sigma_{m,n}\, r)\, \widetilde{Y}_m^k(\varphi,\theta), \end{aligned} \qquad \begin{aligned} m &= 0,1,2,\ldots\,, \\ k &= 0,\ldots,m, \\ n &= 1,2,3,\ldots\,. \end{aligned} \tag{12.110}$$

Spherical Bessel Roots $\sigma_{m,n}$

$n \backslash m$	0	1	2	3	4	5	6	7	8	
1	3.1416	4.4934	5.7635	6.9879	8.1826	9.3558	10.5128	11.6570	12.7908	...
2	6.2832	7.7253	9.0950	10.4171	11.7049	12.9665	⋮	⋮	⋮	
3	9.4248	10.9041	12.3229	⋮	⋮	⋮				
4	12.5664	⋮	⋮							
⋮	⋮									

The corresponding eigenvalues

$$\lambda_{m,n} = \sigma_{m,n}^2, \qquad m = 0, 1, 2, \ldots, \qquad n = 1, 2, 3, \ldots, \tag{12.111}$$

are the squared spherical Bessel roots. Since there are $2m + 1$ independent spherical harmonics of order m, the eigenvalue $\lambda_{m,n}$ admits $2m + 1$ linearly independent eigenfunctions, namely $v_{0,m,n}, \ldots, v_{m,m,n}, \widetilde{v}_{1,m,n}, \ldots, \widetilde{v}_{m,m,n}$. In particular, the radially symmetric solutions are the eigenfunctions with $k = m = 0$:

$$v_n(r) = v_{0,0,n}(r) = S_0(\sigma_{0,n}\, r) = \frac{\sin n\pi\, r}{n\pi\, r}, \qquad n = 1, 2, \ldots . \tag{12.112}$$

Further analysis, cf. [**34**], demonstrates that the separable solutions (12.110) form a complete system of eigenfunctions for the Helmholtz equation on the unit ball with homogeneous Dirichlet boundary conditions.

We have thus completely determined the basic separable solutions to the heat equation on a solid unit ball subject to homogeneous Dirichlet boundary conditions. They are products of exponential functions of time, spherical Bessel functions of the radius, and spherical harmonics:

$$\begin{aligned} u_{k,m,n}(t, r, \varphi, \theta) &= e^{-\sigma_{m,n}^2 t}\, S_m(\sigma_{m,n}\, r)\, Y_m^k(\varphi, \theta), \\ \widetilde{u}_{k,m,n}(t, r, \varphi, \theta) &= e^{-\sigma_{m,n}^2 t}\, S_m(\sigma_{m,n}\, r)\, \widetilde{Y}_m^k(\varphi, \theta). \end{aligned} \tag{12.113}$$

The general solution can be written as an infinite "Fourier–Bessel–spherical harmonic" series in these fundamental modes:

$$\begin{aligned} u(t, r, \varphi, \theta) = \sum_{m=0}^{\infty} \sum_{n=1}^{\infty} e^{-\sigma_{m,n}^2 t}\, S_m(\sigma_{m,n}\, r) \Bigg(& \frac{c_{0,m,n}}{2}\, Y_m^0(\varphi, \theta) \\ & + \sum_{k=1}^{m} \Big[c_{k,m,n} Y_m^k(\varphi, \theta) + \widetilde{c}_{k,m,n} \widetilde{Y}_m^k(\varphi, \theta) \Big] \Bigg). \end{aligned} \tag{12.114}$$

The series' coefficients are uniquely prescribed by the initial data $u(0, r, \varphi, \theta) = f(r, \varphi, \theta)$, and their explicit formulae[†]

$$c_{k,m,n} = \frac{(2m+1)(m-k)!}{\pi\,(m+k)!\,S_{m+1}(\sigma_{m,n})^2} \int_{-\pi}^{\pi}\int_0^{\pi}\int_0^1 f(r,\varphi,\theta)\, v_{k,m,n}(r,\varphi,\theta)\, r^2 \sin\varphi\, dr\, d\varphi\, d\theta,$$

$$\widetilde{c}_{k,m,n} = \frac{(2m+1)(m-k)!}{\pi\,(m+k)!\,S_{m+1}(\sigma_{m,n})^2} \int_{-\pi}^{\pi}\int_0^{\pi}\int_0^1 f(r,\varphi,\theta)\, \widetilde{v}_{k,m,n}(r,\varphi,\theta)\, r^2 \sin\varphi\, dr\, d\varphi\, d\theta, \tag{12.115}$$

follow from the usual orthogonality relations among the eigenfunctions, combined with the formulas

$$\| v_{0,m,n} \| = \sqrt{\frac{2\pi}{2m+1}}\; S_{m+1}(\sigma_{m,n}),$$

$$\| v_{k,m,n} \| = \| \widetilde{v}_{k,m,n} \| = \sqrt{\frac{\pi(m+k)!}{(2m+1)(m-k)!}}\; S_{m+1}(\sigma_{m,n}), \qquad k > 0, \tag{12.116}$$

for their norms, to be established in Exercise 12.4.29. In particular, the slowest-decaying mode is the spherically symmetric function

$$u_{0,0,1}(t,r) = \frac{e^{-\pi^2 t}\sin \pi r}{\pi r}, \tag{12.117}$$

corresponding to the smallest eigenvalue $\lambda_{0,1} = \sigma_{0,1}^2 = \pi^2$. Therefore, typically, the decay to thermal equilibrium of a unit sphere is at an exponential rate of $\pi^2 \approx 9.8696$, or, to a very rough approximation, 10.

Exercises

12.4.13. It takes a solid ball of radius 1 cm ten minutes to return to (approximate) thermal equilibrium. How long does it take a similar ball of radius 2?

12.4.14. If a 200-gram potato served hot from the oven takes 15 minutes until its maximum temperature is less than 40° C, how long does it take a 300-gram potato of the same shape to cool off?

♡ 12.4.15. A uniform solid metal ball of radius 1 meter, with diffusion coefficient $\gamma = 2$, is taken from a 300° oven and immersed in a bucket of ice water. (a) Write down an initial-boundary value problem that describes the temperature of the ball. (*b*) Find a series solution for the temperature. (*c*) At what time is the temperature $\leq 50°$ throughout the ball?

◇ 12.4.16. Find the decay rate to thermal equilibrium of a solid spherical ball of radius R and diffusion coefficient γ when subject to homogeneous Dirichlet boundary conditions.

12.4.17. *True or false*: A heated solid hemisphere placed in a 0° environment cools down twice as fast as a solid sphere of the same radius made out of the same material.

12.4.18. A fully insulated solid spherical ball of radius 1 has initial temperature distribution $f(r,\varphi,\theta)$. (a) Write down a formula for the equilibrium temperature of the ball. (*b*) What is the rate of decay of the ball to thermal equilibrium?

[†] We use the spherical coordinate form of the L^2 inner product on the ball.

12.4.19. Which cools down to equilibrium faster: a fully insulated solid ball or one whose boundary is held fixed at $0°$? How much faster?

12.4.20. A solid sphere and solid cube are made out of the same material and have the same volume. Both are heated in an oven and then submerged in a large vat of water. Which will cool down faster? Explain and justify your answer.

12.4.21. Answer Exercise 12.4.20 when the two solids have the same surface area.

12.4.22. Suppose the solid spherical shell in Exercise 12.2.7 starts off at room temperature. Assuming that the water in the center remains at $100°$, find the rate at which the shell tends to thermal equilibrium.

♡ 12.4.23. The thermodynamics of a thin, uniform, spherical shell of unit radius is governed by the *spherical heat equation* $u_t = \gamma \Delta_S u$, $u(0, \varphi, \theta) = f(\varphi, \theta)$, in which Δ_S is the spherical Laplacian (12.19). The solution $u(t, \varphi, \theta)$ represents the temperature of the point on the unit sphere with angular coordinates φ, θ, while $f(\varphi, \theta)$ is the initial temperature distribution. (a) Find the eigensolutions. (b) Write down the solution to the initial value problem as a series in eigensolutions. (c) What is the final equilibrium temperature of the spherical shell? (d) What is its rate of decay to equilibrium? (e) Find the solution and the final equilibrium temperature when $f(\varphi, \theta) =$ (i) $\sin \varphi \cos \theta$; (ii) $\cos 2\varphi$.

12.4.24. A spherical potato, of radius $R = 7.5$ cm and thermal diffusivity $\gamma = .3\,\text{cm}^2/\text{sec}$, is initially at room temperature, $25°$ C, and is placed in a pot of boiling water at $100°$ C. The potato is cooked when it has reached the temperature of at least $90°$ C throughout. How long do you have to wait until the potato is done?

12.4.25. (a) Explain why the spherical Bessel function $S_1(x)$ is bounded at $x = 0$. What is $S_1(0)$? (b) Answer the same question for $S_2(x)$.

♢ 12.4.26. Prove the formulae (12.108).

♢ 12.4.27. (a) Find a recurrence relation expressing the spherical Bessel function $S_{m-1}(x)$ in terms of $S_m(x)$. (b) Prove that

$$\frac{d}{dx}\left[x^3 \big(S_m(x)^2 - S_{m-1}(x)\, S_{m+1}(x) \big) \right] = 2x^2\, S_m(x)^2.$$

♢ 12.4.28. Let $m \geq 0$ be a fixed integer. (a) Prove that the rescaled spherical Bessel functions $v_n(r) = S_m(\sigma_{m,n}\, r)$, $n = 1, 2, \ldots$, are mutually orthogonal under the inner product $\langle f, g \rangle = \int_0^1 f(r)\, g(r)\, r^2\, dr$. (b) Prove that $\| v_n \| = \frac{1}{\sqrt{2}} \, | S_{m+1}(\sigma_{m,n}) |$. *Hint*: Mimic the method outlined in Exercise 11.4.22, using the identity in Exercise 12.4.27(b).

♢ 12.4.29. (a) Use the result of Exercise 12.4.28 to prove the formulae (12.116) for the L^2 norms of the eigenfunctions (12.110). (b) Justify the formulae (12.115).

The Fundamental Solution to the Heat Equation in Space

For the heat equation (as well as more general diffusion equations), the fundamental solution measures the response of the body to an instantaneously applied concentrated unit heat source. Thus, given a point $\boldsymbol{\xi} = (\xi, \eta, \zeta) \in \Omega$ within the body, the *fundamental solution*

$$u(t, \mathbf{x}) = F(t, \mathbf{x}; \boldsymbol{\xi}) = F(t, x, y, z; \xi, \eta, \zeta)$$

solves the initial-boundary value problem

$$u_t = \Delta u, \qquad u(0, \mathbf{x}) = \delta(\mathbf{x} - \boldsymbol{\xi}), \qquad \text{for} \qquad \mathbf{x} \in \Omega, \quad t > 0, \qquad (12.118)$$

subject to the selected homogeneous boundary conditions — Dirichlet, Neumann, or mixed.

Explicit formulas for the fundamental solution are rare, although in bounded domains it is possible to construct it as an eigenfunction series, as described in Section 9.5. The one case amenable to a complete analysis is that in which the heat is distributed over all of three-dimensional space, so $\Omega = \mathbb{R}^3$. We recall that Lemma 11.11 showed how to construct solutions of the two-dimensional heat equation as products of one-dimensional solutions. In a similar manner, if $p(t,x)$, $q(t,x)$, and $r(t,x)$ are any three solutions to the one-dimensional heat equation $u_t = \gamma\, u_{xx}$, then their product

$$u(t,x,y,z) = p(t,x)\, q(t,y)\, r(t,z) \tag{12.119}$$

is a solution to the three-dimensional heat equation

$$u_t = \gamma\,(u_{xx} + u_{yy} + u_{zz}).$$

In particular, choosing

$$p(t,x) = \frac{e^{-(x-\xi)^2/4\gamma t}}{2\sqrt{\pi\gamma t}}, \qquad q(t,y) = \frac{e^{-(y-\eta)^2/4\gamma t}}{2\sqrt{\pi\gamma t}}, \qquad r(t,z) = \frac{e^{-(z-\zeta)^2/4\gamma t}}{2\sqrt{\pi\gamma t}},$$

to all be one-dimensional fundamental solutions, we are immediately led to the fundamental solution in the form of a three-dimensional *Gaussian filter*.

Theorem 12.14. *The fundamental solution*

$$F(t,\mathbf{x};\boldsymbol{\xi}) = F(t,\mathbf{x}-\boldsymbol{\xi}) = \frac{e^{-\,\|\,\mathbf{x}-\boldsymbol{\xi}\,\|^2/(4\gamma t)}}{8\,(\pi\gamma t)^{3/2}} \tag{12.120}$$

solves the three-dimensional heat equation $u_t = \gamma\,\Delta u$ *on* $\mathbb{R}^3$ *for* $t > 0$, *with an initial temperature equal to a delta function concentrated at the point* $\mathbf{x} = \boldsymbol{\xi}$.

Thus, the initially concentrated heat energy immediately begins to spread out in a spherically symmetric manner, with a minuscule, but nonzero effect that is felt immediately arbitrarily far away from the initial concentration. At each individual point $\mathbf{x} \in \mathbb{R}^3$, after an initial warm-up, the temperature decays back to zero at a rate proportional to $t^{-3/2}$ — more rapidly than in two dimensions, because, intuitively, there are more directions in which the heat energy can disperse.

To solve a more general initial value problem with the initial temperature distributed over all of space, we first write

$$u(0,\mathbf{x}) = f(\mathbf{x}) = \iiint f(\boldsymbol{\xi})\, \delta(\mathbf{x}-\boldsymbol{\xi})\, d\xi\, d\eta\, d\zeta$$

as a linear superposition of delta functions. By linearity, the solution to the initial value problem is given by the corresponding superposition

$$u(t,\mathbf{x}) = \frac{1}{8\,(\pi\gamma t)^{3/2}} \iiint f(\boldsymbol{\xi})\, e^{-\,\|\,\mathbf{x}-\boldsymbol{\xi}\,\|^2/(4\gamma t)}\, d\xi\, d\eta\, d\zeta \tag{12.121}$$

of the fundamental solutions. Since the fundamental solution has exponential decay as $\|\,\mathbf{x}\,\| \to \infty$, the superposition formula is valid even for initial temperature distributions that are moderately increasing at large distances. We remark that the integral (12.121) has the form of a three-dimensional convolution

$$u(t,\mathbf{x}) = F(t,\mathbf{x}) * f(\mathbf{x}) = \iiint f(\boldsymbol{\xi})\, F(t,\mathbf{x}-\boldsymbol{\xi})\, d\xi\, d\eta\, d\zeta \tag{12.122}$$

of the initial data with a one-parameter family of increasingly spread-out Gaussian filters. Thus, as before, convolution with a Gaussian filter has a smoothing effect on the initial temperature distribution.

Exercises

12.4.30. *True or false*: In a three-dimensional medium, heat energy propagates at infinite speed.

12.4.31. A solid spherical ball of radius 1 is heated to 100° and inserted into a three-dimensional medium filling the rest of $\mathbb{R}^3$ with uniform temperature 0°.
(*a*) Write down an integral formula for the subsequent temperature distribution over $\mathbb{R}^3$ at time $t > 0$, assuming a common diffusion coefficient $\gamma = 1$.
(*b*) Evaluate the resulting integral using spherical coordinates.

12.4.32. (*a*) Prove that $u(t, r)$ is a spherically symmetric solution to the three-dimensional heat equation if and only if $w(t, r) = r\, u(t, r)$ solves the one-dimensional heat equation: $w_t = w_{rr}$. (*b*) *True or false*: If $w(t, r)$ is the fundamental solution for the one-dimensional heat equation based at $r = 0$, then $u(t, r) = w(t, r)/r$ is the fundamental solution for the three-dimensional heat equation based at the origin.

12.4.33. Construct the solution to the initial value problem in Exercise 12.4.31 using radial symmetry and Exercise 12.4.32.

♡ 12.4.34. Suppose that, as Earth orbits the sun, its surface is subject to yearly periodic temperature variations $a \cos \omega\, t$, where the frequency ω is given by (4.56). (*a*) Assuming, for simplicity, that the Earth is a homogeneous solid ball, of radius R, formulate an initial-boundary value problem that governs the temperature fluctuations within the Earth due to its orbiting the sun. (*b*) At what depth does the temperature vary out of phase with the surface, i.e., is the warmest in winter and coldest in summer? Compare your answer with the root cellar computation at the end of Section 4.1. *Hint*: Use Exercise 12.4.32.

12.4.35. (*a*) Prove that if $u(t, x)$ is any (sufficiently smooth) solution to the heat equation, so is its time derivative $v = \partial u/\partial t$. (*b*) Write out the time derivative of the fundamental solution, and the initial value problem it satisfies.

12.4.36. Write down an explicit eigenfunction series for the fundamental solution $F(t, \mathbf{x}; \boldsymbol{\xi})$ to the heat equation in a unit cube with thermal diffusivity $\gamma = 1$ that is subject to homogeneous Dirichlet boundary conditions.

12.4.37. Write down an explicit eigenfunction series for the fundamental solution $F(t, \mathbf{x}; \boldsymbol{\xi})$ to the heat equation in a ball of radius 1 that has thermal diffusivity $\gamma = 1$ and is subject to homogeneous Dirichlet boundary conditions.

◇ 12.4.38. Justify the statement that formula (12.119) provides a solution to the three-dimensional heat equation.

12.4.39. Fill in the details of the proof of Theorem 12.14.

12.5 The Wave Equation for Three–Dimensional Media

The *three-dimensional wave equation*

$$u_{tt} = c^2 \Delta u = c^2 (u_{xx} + u_{yy} + u_{zz}), \tag{12.123}$$

in which $c > 0$ denotes the speed of light, governs the propagation of waves in a homogeneous isotropic three-dimensional medium, e.g., electromagnetic waves (light, X-rays, radio waves, etc.) in empty space. In this context, while the electric and magnetic vector fields $\mathbf{E}, \mathbf{B}$ are intrinsically coupled by the more complicated system of Maxwell's equations, each individual component satisfies the wave equation; see Exercise 12.5.14 for details.

The wave equation also models certain restricted classes of vibrations of a uniform solid body. The solution $u(t, \mathbf{x}) = u(t, x, y, z)$ represents a scalar-valued displacement of the body at time t and position $\mathbf{x} = (x, y, z) \in \Omega \subset \mathbb{R}^3$. For example, $u(t, \mathbf{x})$ might represent the radial displacement of the body. One imposes suitable boundary conditions, e.g., Dirichlet, Neumann, or mixed, on $\partial\Omega$, along with a pair of initial conditions

$$u(0, \mathbf{x}) = f(\mathbf{x}), \qquad \frac{\partial u}{\partial t}(0, \mathbf{x}) = g(\mathbf{x}), \qquad \mathbf{x} \in \Omega, \tag{12.124}$$

that specify the body's initial displacement and initial velocity. As long as the initial and boundary data are reasonably nice, there exists a unique classical solution to the initial-boundary value problem for all $-\infty < t < \infty$, cf. [**38, 61, 99**]. Thus, in contrast to the heat equation, one can follow solutions to the wave equation both forwards and backwards in time.

Let us focus our attention on the homogeneous boundary value problem. The fundamental vibrational modes are found by imposing our usual trigonometric ansatz

$$u(t, x, y, z) = \cos(\omega t)\, v(x, y, z) \qquad \text{or} \qquad \sin(\omega t)\, v(x, y, z).$$

Substituting into the wave equation (12.123), we discover (yet again) that $v(x, y, z)$ must be an eigenfunction for the associated Helmholtz eigenvalue problem

$$\Delta v + \lambda v = 0, \qquad \text{where} \qquad \lambda = \frac{\omega^2}{c^2}, \tag{12.125}$$

coupled to the relevant boundary conditions. In the positive definite cases, i.e., Dirichlet and mixed boundary conditions, the eigenvalues $\lambda_k = \omega_k^2/c^2 > 0$ are all positive. Each eigenfunction $v_k(x, y, z)$ yields two normal vibrational modes

$$u_k(t, x, y, z) = \cos(\omega_k t)\, v_k(x, y, z), \qquad \widetilde{u}_k(t, x, y, z) = \sin(\omega_k t)\, v_k(x, y, z),$$

of frequency $\omega_k = c\sqrt{\lambda_k}$ equal to the square root of the corresponding eigenvalue multiplied by the wave speed. The general solution is a quasiperiodic linear combination,

$$u(t, x, y, z) = \sum_{k=1}^{\infty} \big[\, a_k \cos(\omega_k t) + b_k \sin(\omega_k t) \,\big]\, v_k(x, y, z), \tag{12.126}$$

of the eigenmodes. The coefficients a_k, b_k are uniquely prescribed by the initial conditions (12.124). Thus,

$$u(0, x, y, z) = \sum_{k=1}^{\infty} a_k v_k(x, y, z) = f(x, y, z),$$

$$\frac{\partial u}{\partial t}(0, x, y, z) = \sum_{k=1}^{\infty} \omega_k b_k v_k(x, y, z) = g(x, y, z).$$

The explicit formulas follow immediately from the orthogonality of the eigenfunctions:

$$a_k = \frac{\langle f, v_k \rangle}{\| v_k \|^2} = \frac{\displaystyle\iiint_\Omega f\, v_k \, dx\, dy\, dz}{\displaystyle\iiint_\Omega v_k^2 \, dx\, dy\, dz}, \qquad b_k = \frac{1}{\omega_k} \frac{\langle g, v_k \rangle}{\| v_k \|^2} = \frac{\displaystyle\iiint_\Omega g\, v_k \, dx\, dy\, dz}{\displaystyle\omega_k \iiint_\Omega v_k^2 \, dx\, dy\, dz}. \tag{12.127}$$

In the positive semi-definite Neumann case, there is an additional zero eigenvalue $\lambda_0 = 0$ corresponding to the constant null eigenfunction $v_0(x, y, z) \equiv 1$. This results in two additional terms in the eigenfunction expansion — a constant term

$$a_0 = \frac{1}{\operatorname{vol} \Omega} \iiint_\Omega f(x, y, z)\, dx\, dy\, dz$$

that equals the average initial displacement, and an unstable mode $b_0 \, t$ that grows linearly in time, whose speed

$$b_0 = \frac{1}{\operatorname{vol} \Omega} \iiint_\Omega g(x, y, z)\, dx\, dy\, dz$$

is the average initial velocity over the entire body. Thus, the unstable mode will be excited if and only if there is a nonzero net initial velocity: $b_0 \neq 0$.

Most of the basic solution techniques we learned in the two-dimensional case apply here, and we will not dwell on the details. The case of a rectangular box is a particularly straightforward application of the method of separation of variables, and is outlined in the exercises. A similar analysis, now in cylindrical coordinates, can be applied to the case of a vibrating cylinder. The most interesting case is that of a solid spherical ball, which is the subject of the next subsection.

Vibration of Balls and Spheres

Let us focus on the radial vibrations of a solid ball, as modeled by the three-dimensional wave equation (12.123). The solution $u(t, x, y, z)$ represents the radial displacement of the "atom" that is situated at position (x, y, z) when the ball is at rest.

For simplicity, we look at the Dirichlet boundary value problem on the unit ball $B_1 = \{ \| \mathbf{x} \| < 1 \}$. The normal modes of vibration are governed by the Helmholtz equation (12.125) subject to homogeneous Dirichlet boundary conditions. According to (12.110), the eigenfunctions are

$$\begin{aligned} v_{0,m,n}(r, \varphi, \theta) &= S_m(\sigma_{m,n} r)\, Y_m^0(\varphi, \theta), & & n = 1, 2, 3, \ldots\,, \\ v_{k,m,n}(r, \varphi, \theta) &= S_n(\sigma_{n,m} r)\, Y_m^k(\varphi, \theta), & \text{for} \quad & m = 0, 1, 2, \ldots\,, \\ \widetilde{v}_{k,m,n}(r, \varphi, \theta) &= S_m(\sigma_{m,n} r)\, \widetilde{Y}_m^k(\varphi, \theta), & & k = 1, 2, \ldots, m. \end{aligned} \tag{12.128}$$

Here S_m denotes the mth order spherical Bessel function (12.104), $\sigma_{m,n}$ is its nth root, as in (12.109), while $Y_n^m, \widetilde{Y}_n^m$ are the spherical harmonics (12.38). Each eigenvalue

$$\lambda_{m,n} = \sigma_{m,n}^2, \qquad m = 0, 1, 2, \ldots, \qquad n = 1, 2, 3, \ldots,$$

corresponds to $2m + 1$ independent eigenfunctions, namely

$$v_{k,m,0}(r, \varphi, \theta), \ \ v_{k,m,1}(r, \varphi, \theta), \ \ldots\,, \ v_{k,m,m}(r, \varphi, \theta), \ \ \widetilde{v}_{k,m,1}(r, \varphi, \theta), \ \ldots\,, \ \widetilde{v}_{k,m,m}(r, \varphi, \theta).$$

Consequently, the fundamental vibrational frequencies of a solid ball

$$\omega_{m,n} = c\sqrt{\lambda_{m,n}} = c\,\sigma_{m,n}, \qquad m = 0,1,2,\ldots, \qquad n = 1,2,3,\ldots, \tag{12.129}$$

are equal to the spherical Bessel roots $\sigma_{m,n}$ multiplied by the wave speed. There is a total of $2(2m+1)$ independent vibrational modes associated with each distinct frequency (12.129), namely

$$\begin{aligned} u_{0,m,n}(t,r,\varphi,\theta) &= \cos(c\,\sigma_{m,n} t)\; S_m(\sigma_{m,n} r)\, Y_m^0(\varphi,\theta), \\ \widehat{u}_{0,m,n}(t,r,\varphi,\theta) &= \sin(c\,\sigma_{m,n} t)\; S_m(\sigma_{m,n} r)\, Y_m^0(\varphi,\theta), \\ u_{k,m,n}(t,r,\varphi,\theta) &= \cos(c\,\sigma_{m,n} t)\; S_m(\sigma_{m,n} r)\, Y_m^k(\varphi,\theta), \\ \widehat{u}_{k,m,n}(t,r,\varphi,\theta) &= \sin(c\,\sigma_{m,n} t)\; S_m(\sigma_{m,n} r)\, Y_m^k(\varphi,\theta), \\ \widetilde{u}_{k,m,n}(t,r,\varphi,\theta) &= \cos(c\,\sigma_{m,n} t)\; S_m(\sigma_{m,n} r)\, \widetilde{Y}_m^k(\varphi,\theta), \\ \widehat{\widetilde{u}}_{k,m,n}(t,r,\varphi,\theta) &= \sin(c\,\sigma_{m,n} t)\; S_m(\sigma_{m,n} r)\, \widetilde{Y}_m^k(\varphi,\theta), \end{aligned} \qquad \begin{aligned} n &= 1,2,3,\ldots\,, \\ m &= 0,1,2,\ldots\,, \\ k &= 1,2,\ldots,m. \end{aligned} \tag{12.130}$$

In particular, the radially symmetric modes of vibration have, according to (12.105), the elementary form

$$\begin{aligned} u_{0,0,n}(r,\varphi,\theta) &= \cos(c\,n\,\pi\,t)\; S_0(n\,\pi\,r) = \frac{\cos c\,n\,\pi\,t\;\sin n\,\pi\,r}{r}, \\ \widehat{u}_{0,0,n}(r,\varphi,\theta) &= \sin(c\,n\,\pi\,t)\; S_0(n\,\pi\,r) = \frac{\sin c\,n\,\pi\,t\;\sin n\,\pi\,r}{r}, \end{aligned} \qquad n = 1,2,3,\ldots\,. \tag{12.131}$$

Their vibrational frequencies, $\omega_{0,n} = c\,n\,\pi$, are integral multiples of the lowest frequency $\omega_{0,1} = c\,\pi$. Thus, intriguingly, if you excite only the radially symmetric modes, the resulting motion of the ball is periodic. However, more general vibrations are only quasiperiodic.

Adopting the same scaling argument as in (11.166), we conclude that the fundamental frequencies for a solid ball of radius R and wave speed c are given by $\omega_{m,n} = c\,\sigma_{m,n}/R$. The relative vibrational frequencies

$$\frac{\omega_{m,n}}{\omega_{0,1}} = \frac{\sigma_{m,n}}{\sigma_{0,1}} = \frac{\sigma_{m,n}}{\pi} \tag{12.132}$$

are independent of the size of the ball R or the wave speed c. In the accompanying table, we display all relative vibrational frequencies that are ≤ 4 in magnitude.

Relative Spherical Bessel Roots $\sigma_{m,n}/\sigma_{0,1}$

$n \backslash m$	0	1	2	3	4	6	7	8	…
1	1.0000	1.4303	1.8346	2.2243	2.6046	2.9780	3.3463	3.7105	…
2	2.0000	2.4590	2.8950	3.3159	3.7258	⋮	⋮	⋮	
3	3.0000	3.4709	3.9225	⋮	⋮				
4	4.0000	⋮	⋮						
⋮	⋮								

The purely radial modes of vibration (12.131) have individual frequencies

$$\omega_{0,n} = \frac{n \pi c}{R}, \qquad \text{so} \qquad \frac{\omega_{0,n}}{\omega_{0,1}} = n,$$

which appear in the first column of the table. The lowest frequency is $\omega_{0,1} = \pi c/R$, corresponding to a vibration with period $2\pi/\omega_{0,1} = 2R/c$. In particular, for the Earth, the radius $R \approx 6000$ km, and the wave speed in rock is, on average, $c \approx 5$ km/sec, so that the fundamental mode of vibration has period $2R/c \approx 2400$ seconds, or 40 minutes. Of course, we have suppressed almost all interesting terrestrial geology in this very crude approximation, which has been based on the assumption that the Earth is a uniform spherical body, globally vibrating only in its radial direction. A more realistic modeling of the vibrations of the Earth requires an understanding of the basic partial differential equations of linear and nonlinear elastodynamics, [**7**, **49**]. Nonuniformities in the Earth lead to scattering of the vibrational waves, which are then used to locate subterranean geological structures, e.g., oil and gas deposits. Localized vibrations of the Earth are also known as *seismic waves*, and, of course, earthquakes are their most severe manifestation. We refer the interested reader to [**5**] for an introduction to mathematical seismology. Understanding terrestrial vibrations is an issue of critical importance in geophysics and civil engineering, including the design of structures, buildings, and bridges, requiring the avoidance of potentially catastrophic resonant frequencies.

Example 12.15. The radial vibrations of a hollow thin spherical shell (e.g., an elastic balloon) are governed by the differential equation

$$\frac{\partial^2 u}{\partial t^2} = c^2 \, \Delta_S[u] = c^2 \left(\frac{\partial^2 u}{\partial \varphi^2} + \frac{\cos\varphi}{\sin\varphi} \frac{\partial u}{\partial \varphi} + \frac{1}{\sin^2\varphi} \frac{\partial^2 u}{\partial \theta^2} \right), \tag{12.133}$$

where Δ_S denotes the spherical Laplacian (12.19). The radial displacement $u(t,\varphi,\theta)$ of a point on the sphere depends only on time t and the angular coordinates φ, θ. The solution $u(t,\varphi,\theta)$ is required to be 2π–periodic in the azimuthal angle θ and bounded at the poles, where $\varphi = 0$ and π.

According to (12.38), the nth eigenvalue of the spherical Laplacian, $\lambda_n = n(n+1)$, possesses $2n+1$ linearly independent eigenfunctions, namely, the spherical harmonics

$$Y_n^0(\varphi,\theta), \ Y_n^1(\varphi,\theta), \ \ldots, \ Y_n^n(\varphi,\theta), \ \widetilde{Y}_n^1(\varphi,\theta), \ \ldots, \ \widetilde{Y}_n^n(\varphi,\theta).$$

As a consequence, the fundamental frequencies of vibration for a spherical shell are

$$\omega_n = c\sqrt{\lambda_n} = c\sqrt{n(n+1)}, \qquad n = 1, 2, \ldots . \tag{12.134}$$

The vibrational solutions are quasiperiodic combinations of the fundamental spherical harmonic modes

$$\begin{aligned} &\cos\bigl(\sqrt{n(n+1)}\,t\bigr)\, Y_n^m(\varphi,\theta), \qquad &&\sin\bigl(\sqrt{n(n+1)}\,t\bigr)\, Y_n^m(\varphi,\theta), \\ &\cos\bigl(\sqrt{n(n+1)}\,t\bigr)\, \widetilde{Y}_n^m(\varphi,\theta), &&\sin\bigl(\sqrt{n(n+1)}\,t\bigr)\, \widetilde{Y}_n^m(\varphi,\theta). \end{aligned} \tag{12.135}$$

Representative graphs can be seen in Figure 12.5. The smallest positive eigenvalue is $\lambda_1 = 2$, yielding a lowest tone of frequency $\omega_1 = c\sqrt{2}$. The higher-order frequencies are irrational multiples of the fundamental frequency, implying that a vibrating spherical bell sounds dissonant to our ears.

One further remark is in order. The spherical Laplacian operator is only positive semi-definite, since the lowest mode has eigenvalue $\lambda_0 = 0$, which corresponds to the constant null eigenfunction $v_0(\varphi,\theta) = Y_0^0(\varphi,\theta) \equiv 1$. Therefore, the wave equation (12.133) admits an unstable mode $b_{0,0}\,t$, corresponding to a uniform radial inflation; its coefficient

$$b_{0,0} = \frac{3}{4\pi} \iint_{S_1} \frac{\partial u}{\partial t}(0,\varphi,\theta)\,dS$$

represents the shell's average initial velocity. The existence of such an unstable mode is an artifact of the simplified linear model we are using, which fails to account for nonlinearly elastic effects that serve to constrain the inflation of a spherical balloon.

Exercises

12.5.1. Find the eigenfunction series solution to the initial-boundary value problem for the wave equation $u_{tt} = \Delta u$ on a unit cube $C = \{0 < x, y, z < 1\}$, subject to homogeneous Dirichlet boundary conditions and one of the following sets of initial conditions: (a) $u(0,x,y,z) = 1$, $u_t(0,x,y,z) = 0$; (b) $u(0,x,y,z) = 0$, $u_t(0,x,y,z) = 1$; (c) $u(0,x,y,z) = \sin \pi x \, \sin \pi y \, \sin \pi z$, $u_t(0,x,y,z) = 0$; (d) $u(0,x,y,z) = \sin 3\pi x$, $u_t(0,x,y,z) = \sin 2\pi y$; (e) $u(0,x,y,z) = 0$, $u_t(0,x,y,z) = x\,y\,z\,(1-x)(1-y)(1-z)$.

12.5.2. Suppose the cube in Exercise 12.5.1 is subject to homogeneous Neumann boundary conditions. Which of the preceding initial value problems leads to an unstable motion of the cube?

12.5.3. (a) Find the separable periodic vibrations of a unit cube subject to homogeneous Dirichlet boundary conditions. (b) Can you find a periodic mode that is not separable?

12.5.4. Answer Exercise 12.5.3 when one face of the cube is left free, while the other five faces are fixed.

12.5.5. Given a material with wave speed $c = 1.5$ cm/sec, find the natural vibrational frequencies of a solid rectangular box of size 1 cm × 2 cm × 3 cm whose sides are held fixed. List the lowest five such frequencies in order. Does the box vibrate periodically?

12.5.6. Find the natural vibrational frequencies of a solid cylinder of height 2, radius 1, and wave speed $c = 1$, when (a) all sides are fixed; (b) the top and bottom of the cylinder are free, while the curved side is fixed; (c) the curved side of the cylinder is free, while the top and bottom are fixed.

12.5.7. Among all solid cylinders of unit volume with fixed boundary, find the one that vibrates the slowest.

12.5.8. Does a solid spherical ball that is subject to homogeneous Neumann boundary conditions vibrate (i) faster, (ii) slower, or (iii) at the same rate as the same ball subject to homogeneous Dirichlet conditions. If your answer is (i) or (ii), estimate how much faster or slower.

12.5.9. A solid cube and solid sphere are made of the same material and have the same volume. Which vibrates faster when subject to homogeneous Dirichlet boundary conditions?

12.5.10. Assuming that they both have the same wave speed and fixed boundaries, which vibrates faster: a solid sphere or a circular membrane of the same radius?

12.5.11. A uniform, solid spherical planet is floating freely in outer space. Find its three slowest resonant frequencies.

12.5.12. *True or false*: Suppose we have two uniform solid bodies composed of the same material. If the first body cools down to thermal equilibrium the fastest, then it also vibrates the fastest. Explain your answer.

12.5.13. (a) Define what is meant by a nodal curve and a nodal region on a vibrating thin spherical shell. (b) *True or false*: All the nodal curves are arcs of circles.

♡ 12.5.14. The propagation of electromagnetic waves (including light) is governed by the electric field $\mathbf{E}(t,\mathbf{x})$ and magnetic field $\mathbf{B}(t,\mathbf{x})$, which are both time-dependent vector fields defined for $\mathbf{x} = (x,y,z)$ in a domain $\Omega \subset \mathbb{R}^3$. In empty space, *Maxwell's equations* (as formulated by Heaviside) are

$$\nabla \cdot \mathbf{E} = 0, \qquad \nabla \cdot \mathbf{B} = 0, \qquad \frac{\partial \mathbf{B}}{\partial t} = -\nabla \times \mathbf{E}, \qquad \frac{\partial \mathbf{E}}{\partial t} = \frac{1}{\mu_0\,\epsilon_0}\,\nabla \times \mathbf{B}, \tag{12.136}$$

where μ_0, ϵ_0 are, respectively, the *permeability* and *permittivity* constants. Prove that all individual components of $\mathbf{E}$ and $\mathbf{B}$ satisfy the scalar wave equation. What is the wave speed, i.e., the speed of light in empty space?

12.6 Spherical Waves and Huygens' Principle

For any dynamical partial differential equation, the fundamental solution measures the effect of applying an instantaneous concentrated unit impulse at a single point. Two representative physical effects to keep in mind are the light waves emanating from a sudden concentrated blast, e.g., a lightning bolt or a stellar supernova, and the sound waves due to an explosion or thunderclap, propagating in air at a much slower speed. Linear superposition utilizes the fundamental solution to build up more general solutions to initial value problems. For the wave and other second-order vibrational equations, the impulse can be applied either to the initial displacement or to the initial velocity, resulting in two distinct types of fundamental solution. The general solution to the initial value problem will be obtained by a double superposition. In this section, we derive explicit formulas for the two fundamental solutions for the three-dimensional wave equation on all of space, leading to Kirchhoff's formula for the solution to the general initial value problem. An important consequence is Huygens' Principle, which states that, in three-dimensional space, localized initial disturbances remain localized as they propagate. In the final subsection, we apply the method of descent to our three-dimensional solution formulas in order to solve the two-dimensional wave equation, for which, surprisingly, Huygens' Principle is no longer valid.

Spherical Waves

In a uniform isotropic medium, an initial concentrated blast results in a spherically expanding wave, moving away at the speed of light (or sound) in all directions. Invoking translation invariance, we will assume, without loss of generality, that the source of the disturbance is at the origin, and so the solution $u(t,\mathbf{x})$ should depend only on the distance $r = \|\,\mathbf{x}\,\|$ from the source. We adopt spherical coordinates and look for a solution $u = u(t,r)$ to the three-dimensional wave equation (12.123) with no angular dependence. Substituting the formula (12.16) for the spherical Laplacian and setting both angular derivatives to 0, we are led to the partial differential equation

$$\frac{\partial^2 u}{\partial t^2} = c^2 \left(\frac{\partial^2 u}{\partial r^2} + \frac{2}{r}\,\frac{\partial u}{\partial r} \right), \tag{12.137}$$

which governs the propagation of spherically symmetric waves in three-dimensional space. Surprisingly, we can explicitly solve (12.137). The secret is to multiply both sides of the equation by r:

$$\frac{\partial^2 (r\,u)}{\partial t^2} = r\,\frac{\partial^2 u}{\partial t^2} = c^2 \left(r\,\frac{\partial^2 u}{\partial r^2} + 2\,\frac{\partial u}{\partial r} \right) = c^2\,\frac{\partial^2}{\partial r^2}\,(r\,u).$$

Thus, the function

$$w(t,r) = r\,u(t,r)$$

satisfies the one-dimensional wave equation

$$\frac{\partial^2 w}{\partial t^2} = c^2\,\frac{\partial^2 w}{\partial r^2}\,. \tag{12.138}$$

According to Theorem 2.14, the general solution to the one-dimensional wave equation (12.138) can be written in d'Alembert form

$$w(t,r) = p(r - c\,t) + q(r + c\,t),$$

where $p(\xi)$ and $q(\eta)$ are arbitrary functions of a single characteristic variable. Therefore, spherically symmetric solutions to the three-dimensional wave equation assume the form

$$u(t,r) = \frac{p(r - c\,t)}{r} + \frac{q(r + c\,t)}{r}\,. \tag{12.139}$$

The first summand,

$$u(t,r) = \frac{p(r - c\,t)}{r}\,, \tag{12.140}$$

represents a wave moving at speed c in the direction of increasing r, and so describes the illumination from a variable light source that is concentrated at the origin, e.g., a pulsating quasar in interstellar space. To highlight this interpretation, let us concentrate on the case that $p(\xi) = \delta(\xi - a)$ is a delta function, keeping in mind that more general solutions can then be assembled by linear superposition. The induced solution

$$u(t,r) = \frac{\delta(r - c\,t - a)}{r} = \frac{\delta\big(r - c\,(t - t_0)\big)}{r}\,, \qquad \text{where} \qquad t_0 = -\,\frac{a}{c}\,, \tag{12.141}$$

represents a spherical wave propagating through space. At the instant $t = t_0$, the light is entirely concentrated at the origin, $r = 0$. The signal then moves away from the origin in all directions at speed c. At each later time $t > t_0$, the wave remains concentrated on the surface of a sphere of radius $r = c\,(t - t_0)$. Its intensity at each point on the sphere, however, has decreased by a factor $1/r$, and so, the farther the light travels away from the source, the dimmer it becomes. A stationary observer sitting at a fixed point in space will see only an instantaneous flash of light of intensity $1/r$ as the spherical wave passes by at time $t = t_0 + r/c$, where r is the observer's distance from the light source. A similar statement holds for sound waves — to an observer, the sound of a distant explosion will last momentarily. Thunder and lightning are the most familiar examples of this everyday phenomenon.

On the other hand, for $t < t_0$, the impulse is concentrated at a negative radius $r = c\,(t - t_0) < 0$. To interpret this, note that, for spherical coordinates (12.15), replacing r by $-r$ has the same effect as changing $\mathbf{x}$ to the antipodal point $-\mathbf{x}$. Thus, the solution (12.141) represents a concentrated spherically symmetric light wave arriving from the edges of the universe at speed c that strengthens in intensity as it collapses into the origin at $t = t_0$. After collapse, it immediately reappears and expands back out into the universe.

The second solution in the d'Alembert formula (12.139) has, in fact, exactly the same physical form under the antipodal identification. Indeed, if we set

$$\widetilde{r} = -r, \qquad \widetilde{p}(\xi) = -q(-\xi), \qquad \text{then} \qquad \frac{q(r+ct)}{r} = \frac{\widetilde{p}(\widetilde{r}-ct)}{\widetilde{r}}.$$

Thus, the second d'Alembert solution is redundant, and we only need to consider solutions of the form (12.140) from now on.

To effectively utilize such spherical wave solutions, we need to understand the nature of their originating singularity. For simplicity, we set $t_0 = 0$ in (12.141) and concentrate on the particular solution

$$u(t,r) = \frac{\delta(r-ct)}{r}, \tag{12.142}$$

which apparently has a bad singularity at the origin, $r = 0$, at the initial time $t = 0$. We need to pin down precisely which sort of distribution (generalized function) it represents. Invoking the limiting definition is tricky, and it will be easier to work with the dual characterization of a distribution as a linear functional. Thus, at a fixed time $t \geq 0$, we must evaluate the inner product†

$$\langle\, u(t,\cdot)\,,f\,\rangle = \iiint u(t,x,y,z)\, f(x,y,z)\, dx\, dy\, dz$$

of the solution with a smooth test function $f(\mathbf{x}) = f(x,y,z)$. We rewrite the triple integral in spherical coordinates, whereby

$$\langle\, u(t,\cdot)\,,f\,\rangle = \int_{-\pi}^{\pi}\int_0^{\pi}\int_0^{\infty} \frac{\delta(r-ct)}{r}\, f(r,\varphi,\theta)\, r^2 \sin\varphi\;dr\,d\varphi\,d\theta.$$

When $t \neq 0$, the r integration can be immediately evaluated, and so

$$\langle\, u(t,\cdot)\,,f\,\rangle = ct \int_{-\pi}^{\pi}\int_0^{\pi} f(ct,\varphi,\theta)\,\sin\varphi\;d\varphi\,d\theta = 4\pi c t\, \mathrm{M}_{ct}[\,f\,], \tag{12.143}$$

where

$$\mathrm{M}_{ct}[\,f\,] = \frac{1}{4\pi}\int_{-\pi}^{\pi}\int_0^{\pi} f(ct,\varphi,\theta)\,\sin\varphi\;d\varphi\,d\theta = \frac{1}{4\pi c^2t^2}\iint_{S_{ct}} f\;dS \tag{12.144}$$

is the *mean* or *average value* of the function f on the sphere $S_{ct} = \{\,\|\,\mathbf{x}\,\| = ct\,\}$ of radius $r = ct$ and, hence, surface area $4\pi c^2t^2$. In particular, in the limit as the sphere's radius $ct \to 0$, by continuity, the mean reduces to just the value of the function at the origin:

$$\lim_{t\to\infty} \mathrm{M}_{ct}[\,f\,] = \mathrm{M}_0[\,f\,] = f(\mathbf{0}). \tag{12.145}$$

Thus, (12.143) implies that

$$\lim_{t\to\infty}\,\langle\, u(t,\cdot)\,,f\,\rangle = \langle\, u(0,\cdot)\,,f\,\rangle = 0 \qquad \text{for } \textit{all} \text{ functions} \quad f,$$

† For fixed t, we use $u(t,\cdot)$ to indicate the real-valued function $(x,y,z) \mapsto u(t,x,y,z)$ on $\mathbb{R}^3$.

and hence $u(0,x,y,z) \equiv 0$ represents a zero initial displacement. In other words, there is, in fact, *no singularity* in the solution at $t = 0$!

In the absence of any initial displacement, how, then, can the solution (12.142) be nonzero? Clearly, this must be the result of a nonzero initial velocity. To evaluate $\partial u/\partial t$, we differentiate (12.143), whereby

$$\begin{aligned} \left\langle \frac{\partial u}{\partial t}, f \right\rangle &= \frac{\partial}{\partial t} \langle u(t,\cdot), f \rangle = \frac{\partial}{\partial t} \left(c\,t \int_{-\pi}^{\pi} \int_{0}^{\pi} f(c\,t,\varphi,\theta) \sin\varphi \; d\varphi \, d\theta \right) \\ &= c \int_{-\pi}^{\pi} \int_{0}^{\pi} f(c\,t,\varphi,\theta) \sin\varphi \; d\varphi \, d\theta + c^2 t \int_{-\pi}^{\pi} \int_{0}^{\pi} \frac{\partial f}{\partial r}(c\,t,\varphi,\theta) \sin\varphi \; d\varphi \, d\theta \\ &= 4\pi c \, \mathrm{M}_{ct}[\, f \,] + 4\pi c^2 t \, \mathrm{M}_{ct}\left[\frac{\partial f}{\partial r} \right]. \end{aligned} \tag{12.146}$$

The result is a linear combination of the means of f and its radial derivative f_r over the sphere of radius $c\,t$. In the limit, the second term goes to 0, and so, by (12.145),

$$\lim_{t \to 0} \langle u_t \, , f \rangle = 4\pi c \, \mathrm{M}_0[\, f \,] = 4\pi c \, f(\mathbf{0}).$$

Since this holds for all test functions f, we conclude that the initial velocity of our solution is a multiple of a delta function at the origin:

$$u_t(0,r) = 4\pi c \, \delta(\mathbf{x}).$$

Dividing through by $4\pi c$, we find that the spherical expanding wave

$$u(t,r) = \frac{\delta(r - c\,t)}{4\pi c\,r} \tag{12.147}$$

solves the initial value problem

$$u(0,\mathbf{x}) \equiv 0, \qquad \frac{\partial u}{\partial t}(0,\mathbf{x}) = \delta(\mathbf{x}),$$

corresponding to an initial unit-velocity impulse concentrated at the origin. This solution can be viewed as the three-dimensional version of the hammer-blow solution to the one-dimensional wave equation discussed in Exercise 6.3.28.

More generally, we use the translational symmetry of the wave equation to conclude that the function

$$G(t,\mathbf{x};\boldsymbol{\xi}) = \frac{\delta(\,\|\,\mathbf{x} - \boldsymbol{\xi}\,\| - c\,t\,)}{4\pi c\,\|\,\mathbf{x} - \boldsymbol{\xi}\,\|}, \qquad t \geq 0, \tag{12.148}$$

is the *fundamental solution* to the wave equation resulting from a unit-velocity impulse concentrated at the point $\boldsymbol{\xi}$ at the initial time $t = 0$:

$$G(0,\mathbf{x};\boldsymbol{\xi}) = 0, \qquad \frac{\partial G}{\partial t}(0,\mathbf{x};\boldsymbol{\xi}) = \delta(\mathbf{x} - \boldsymbol{\xi}). \tag{12.149}$$

With this in hand, we can apply linear superposition to solve the zero initial displacement initial value problem

$$u(0,x,y,z) = 0, \qquad \frac{\partial u}{\partial t}(0,x,y,z) = g(x,y,z). \tag{12.150}$$

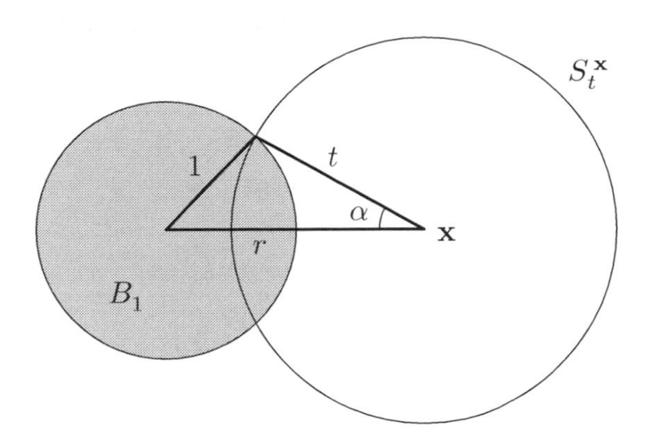

Figure 12.9. Cross-section of a sphere intersecting a ball.

Namely, we write the initial velocity

$$g(\mathbf{x}) = \iiint g(\boldsymbol{\xi})\, \delta(\mathbf{x} - \boldsymbol{\xi})\, d\xi\, d\eta\, d\zeta$$

as a superposition of impulses, and immediately conclude that the relevant solution is the selfsame superposition of spherical waves:

$$\begin{aligned} u(t, \mathbf{x}) &= \frac{1}{4\pi c} \iiint g(\boldsymbol{\xi})\, \frac{\delta(\,\|\,\mathbf{x} - \boldsymbol{\xi}\,\| - c\,t\,)}{\|\,\mathbf{x} - \boldsymbol{\xi}\,\|}\, d\xi\, d\eta\, d\zeta \\ &= \frac{1}{4\pi c^2 t} \iint_{\|\,\boldsymbol{\xi} - \mathbf{x}\,\| = c\,t} g(\boldsymbol{\xi})\, dS = t\, \mathrm{M}_{ct}^{\mathbf{x}}\,[\,g\,]\,. \end{aligned} \tag{12.151}$$

Thus, the value of our solution at position $\mathbf{x}$ and time $t > 0$ is equal to t times the *mean* of the initial velocity function g over the sphere of radius $r = c\,t$ centered at the point $\mathbf{x}$.

Example 12.16. Let us set the wave speed $c = 1$. Suppose that the initial velocity

$$g(\mathbf{x}) = \begin{cases} 1, & \|\,\mathbf{x}\,\| < 1, \\ 0, & \|\,\mathbf{x}\,\| > 1, \end{cases}$$

is 1 inside the unit ball B_1 centered at the origin and 0 outside. To solve the corresponding initial velocity problem, we must compute the average value of g over a sphere

$$S_t^{\mathbf{x}} = \{\,\boldsymbol{\xi} \mid \|\,\boldsymbol{\xi} - \mathbf{x}\,\| = t\,\}$$

of radius $t > 0$ centered at a point $\mathbf{x} \in \mathbb{R}^3$. Since $g = 0$ outside the unit ball, its average will be equal to the surface area of that part of the sphere that is contained inside the unit ball, namely $S_t^{\mathbf{x}} \cap B_1$, divided by the total surface area of $S_t^{\mathbf{x}}$, namely $4\pi\,t^2$.

To compute this quantity, let $r = \|\,\mathbf{x}\,\|$. If $t > r + 1$ or $0 < t < r - 1$, then the sphere of radius t lies entirely outside the unit ball, and so the average is 0; if $0 < t < 1 - r$, which requires $r < 1$ and so $\mathbf{x} \in B_1$, then the sphere lies entirely within the unit ball, and so the average is 1. Otherwise, referring to Figure 12.9 and Exercise 12.6.7, we see that the area of the spherical cap $S_t^{\mathbf{x}} \cap B_1$ is given by

$$2\pi\,t^2(1 - \cos\alpha) = 2\pi\,t^2 \left(1 - \frac{r^2 + t^2 - 1}{2\,r\,t}\right) = \frac{\pi\,t}{r}\,[\,1 - (t - r)^2\,]\,, \tag{12.152}$$

where α denotes the angle between the line joining the centers of the two spheres and the circle formed by their intersection, whose value is prescribed by the Law of Cosines. Assembling the different subcases, we conclude that

$$\mathrm{M}_{ct}^{\mathbf{x}}\,[\,g\,] = \begin{cases} 1, & 0 \le t \le 1 - r, \\ \dfrac{1 - (t-r)^2}{4\,r\,t}, & |\,r - 1\,| \le t \le r + 1, \\ 0, & 0 \le t \le r - 1 \quad \text{or} \quad t \ge r + 1. \end{cases} \tag{12.153}$$

The solution (12.151) is obtained by multiplying by t, and hence for $t \ge 0$,

$$u(t,\mathbf{x}) = \begin{cases} t, & 0 \le t \le 1 - \|\,\mathbf{x}\,\|, \\ \dfrac{1 - \big(t - \|\,\mathbf{x}\,\|\big)^2}{4\,\|\,\mathbf{x}\,\|}, & \big|\,\|\,\mathbf{x}\,\| - 1\,\big| \le t \le \|\,\mathbf{x}\,\| + 1, \\ 0, & 0 \le t \le \|\,\mathbf{x}\,\| - 1 \quad \text{or} \quad t \ge \|\,\mathbf{x}\,\| + 1. \end{cases} \tag{12.154}$$

The resulting function is not smooth at the interfaces $t = \big|\,\|\,\mathbf{x}\,\| - 1\,\big|$ and $\|\,\mathbf{x}\,\| + 1$, and hence does not qualify as a classical solution. Nevertheless, it can be shown that (12.154) is a bona fide weak solution to the initial value problem.

The first two rows of Figure 12.10 plot the solution as a function of time for several fixed values of $r = \|\,\mathbf{x}\,\|$. An observer sitting at the origin will see a linearly increasing light intensity followed by a sudden blackout. At other points inside the sphere, there is a similar linear increase, while the subsequent decrease follows a parabolic arc; if the observer is closer to the edge of the ball than its center, the parabolic portion will continue to increase for a while before eventually tapering off. On the other hand, an observer sitting outside the sphere will experience, after an initially dark period, a symmetric parabolic increase to a maximal intensity and then a decrease back to dark after a total time lapse of 2. The second two rows plot the solution as a function of r for various fixed times. Note that, up until time $t = 1$, the light spreads out while increasing in intensity near the origin, after which the solution is of gradually decreasing magnitude, supported within the domain lying between two concentric spheres of respective radii $t - 1$ and $t + 1$.

Returning to the general situation, we note that the solution formula (12.151) handles only nonzero initial velocities. What about solutions resulting from a nonzero initial displacement? Surprisingly, the answer comes from differentiation! The key observation is that if $u(t,\mathbf{x})$ is any (sufficiently smooth) solution to the wave equation, then so is its time derivative

$$v(t,\mathbf{x}) = \frac{\partial u}{\partial t}\,(t,\mathbf{x}). \tag{12.155}$$

This follows at once from differentiating both sides of the wave equation with respect to t and using the equality of mixed partial derivatives. Physically, this implies that the velocity of a wave obeys the same evolutionary principle as the wave itself, which is a manifestation of the linearity and time-independence (autonomy) of the equation.

Now, suppose u has initial conditions

$$u(0,\mathbf{x}) = f(\mathbf{x}), \qquad\qquad u_t(0,\mathbf{x}) = g(\mathbf{x}). \tag{12.156}$$

What are the initial conditions for its derivative $v = u_t$? Clearly, its initial displacement

$$v(0,\mathbf{x}) = u_t(0,\mathbf{x}) = g(\mathbf{x}) \tag{12.157}$$

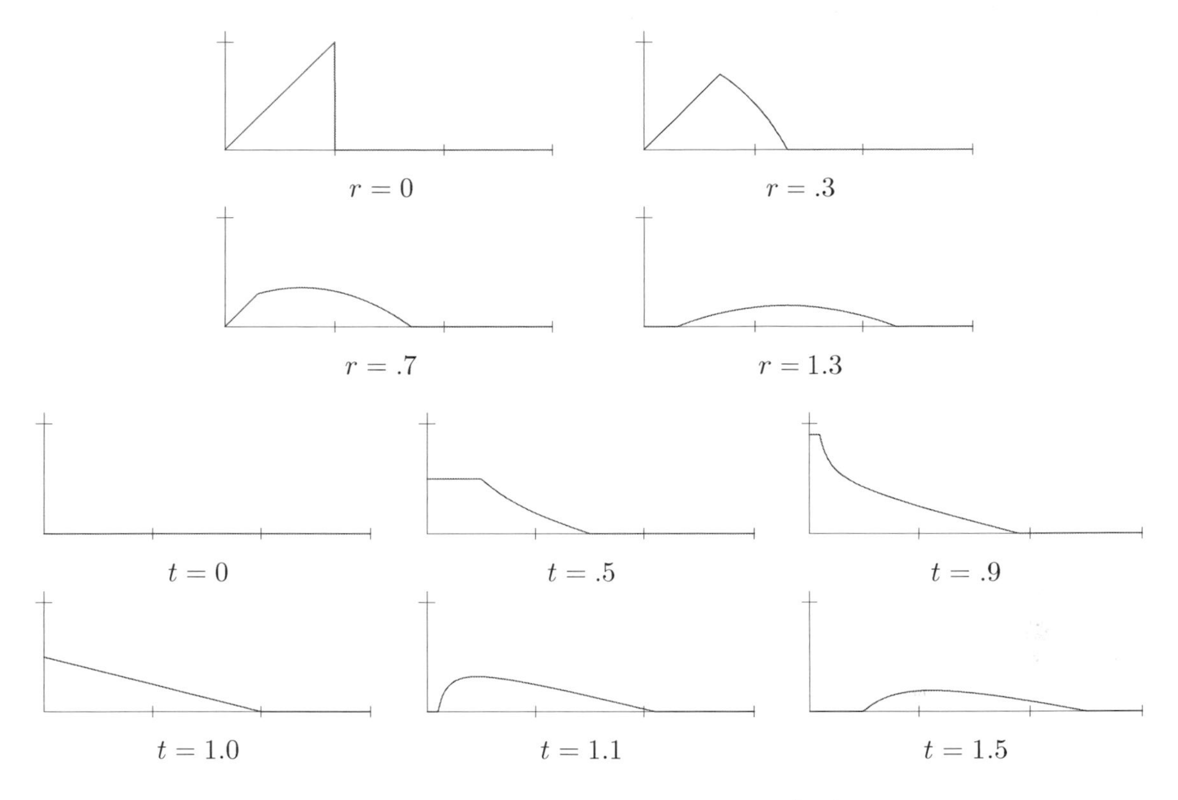

Figure 12.10. Wave equation solution $u(t, r)$ due to an initial velocity of the unit ball.

equals the initial velocity of u. As for its initial velocity, we have

$$\frac{\partial v}{\partial t} = \frac{\partial^2 u}{\partial t^2} = c^2 \, \Delta u,$$

because we are assuming that u solves the wave equation. Thus, at the initial time, the velocity,

$$\frac{\partial v}{\partial t}(0, \mathbf{x}) = c^2 \, \Delta u(0, \mathbf{x}) = c^2 \, \Delta f(\mathbf{x}), \tag{12.158}$$

equals c^2 times the Laplacian of the initial displacement f. In particular, if u satisfies the initial conditions

$$u(0, \mathbf{x}) = 0, \qquad u_t(0, \mathbf{x}) = g(\mathbf{x}), \tag{12.159}$$

then $v = u_t$ satisfies the initial conditions

$$v(0, \mathbf{x}) = g(\mathbf{x}), \qquad v_t(0, \mathbf{x}) = 0. \tag{12.160}$$

Thus, paradoxically, to solve the initial displacement problem we differentiate the initial velocity solution (12.151) with respect to t, and hence

$$v(t, \mathbf{x}) = \frac{\partial u}{\partial t}(t, \mathbf{x}) = \frac{\partial}{\partial t}\left(t \, \mathrm{M}^{\mathbf{x}}_{ct}[\, g \,] \right) = \mathrm{M}^{\mathbf{x}}_{ct}[\, g \,] + c \, t \, \mathrm{M}^{\mathbf{x}}_{ct}\left[\frac{\partial g}{\partial \mathbf{n}} \right], \tag{12.161}$$

where we have made use of our computation in (12.146). Therefore, $v(t,\mathbf{x})$ is a linear combination of the mean of the function g and the mean of its normal or radial derivative $\partial g/\partial \mathbf{n} = \partial g/\partial r$, taken over a sphere of radius $c\,t$ centered at the point $\mathbf{x}$. In particular, to obtain the solution corresponding to a concentrated initial displacement,

$$F(0,\mathbf{x};\boldsymbol{\xi}) = \delta(\mathbf{x}-\boldsymbol{\xi}), \qquad \frac{\partial F}{\partial t}(0,\mathbf{x};\boldsymbol{\xi}) = 0, \tag{12.162}$$

we differentiate the solution (12.148), resulting in

$$F(t,\mathbf{x};\boldsymbol{\xi}) = \frac{\partial G}{\partial t}(t,\mathbf{x};\boldsymbol{\xi}) = -\,\frac{\delta'(\,\|\,\mathbf{x}-\boldsymbol{\xi}\,\| - ct\,)}{4\pi\,\|\,\boldsymbol{\xi}-\mathbf{x}\,\|}\,, \tag{12.163}$$

which is the fundamental solution for the initial displacement problem. Thus, interestingly, a concentrated initial displacement spawns a spherically expanding doublet, cf. Figure 6.6, whereas a concentrated initial velocity spawns an expanding spherical singlet or delta wave.

Example 12.17. Let $c = 1$. Consider the initial conditions

$$u(0,\mathbf{x}) = f(\mathbf{x}) = \begin{cases} 1, & \|\,\mathbf{x}\,\| < 1, \\ 0, & \|\,\mathbf{x}\,\| > 1, \end{cases} \qquad \frac{\partial u}{\partial t}(0,\mathbf{x}) = 0, \tag{12.164}$$

modeling the effect of an instantaneously illuminated solid ball. To obtain the resulting solution, we differentiate (12.154) with respect to t, leading to

$$u(t,\mathbf{x}) = \begin{cases} 1, & 0 \le t < 1 - \|\,\mathbf{x}\,\|, \\ \dfrac{\|\,\mathbf{x}\,\| - t}{2\,\|\,\mathbf{x}\,\|}, & \big|\,\|\,\mathbf{x}\,\| - 1\,\big| \le t \le \|\,\mathbf{x}\,\| + 1, \\ 0, & 0 \le t < \|\,\mathbf{x}\,\| - 1 \quad \text{or} \quad t > 1 + \|\,\mathbf{x}\,\|. \end{cases} \tag{12.165}$$

As illustrated in the first two rows of Figure 12.11, an observer sitting at the center of the ball will see a constant light intensity until $t = 1$, at which time the solution suddenly goes dark. At other points inside the ball, $0 < r < 1$, the downwards jump in intensity arrives sooner, and even goes below 0, followed by a further linear decrease, and finally a jump back to quiescence. An observer placed outside the ball, at radius $r = \|\,\mathbf{x}\,\| > 1$, will experience, after an initially dark period, a sudden increase in the light intensity at time $t = r - 1$, followed by a linear decrease to negative, followed by a jump back up to darkness at time $t = r + 1$. The farther away from the source, the fainter the light. In the second two rows, we plot the same solution as a function of r for different values of t. Note the sudden appearance of a $1/r$ singularity at the origin at time $t = 1$, which is the result of a focusing of the initial discontinuities of $u(0,\mathbf{x}) = f(\mathbf{x})$ on the surface of the unit sphere. Afterwards, the residual radially symmetric disturbance moves off to ∞ while gradually decreasing in intensity. Again, the discontinuities imply that (12.165) is not a classical solution, but it does qualify as a weak solution to the initial value problem.

Kirchhoff's Formula and Huygens' Principle

Linearly combining the two solution formulas (12.151) and (12.161) establishes *Kirchhoff's formula* (first discovered by Poisson), which is the three-dimensional counterpart to d'Alembert's solution formula for the wave equation.

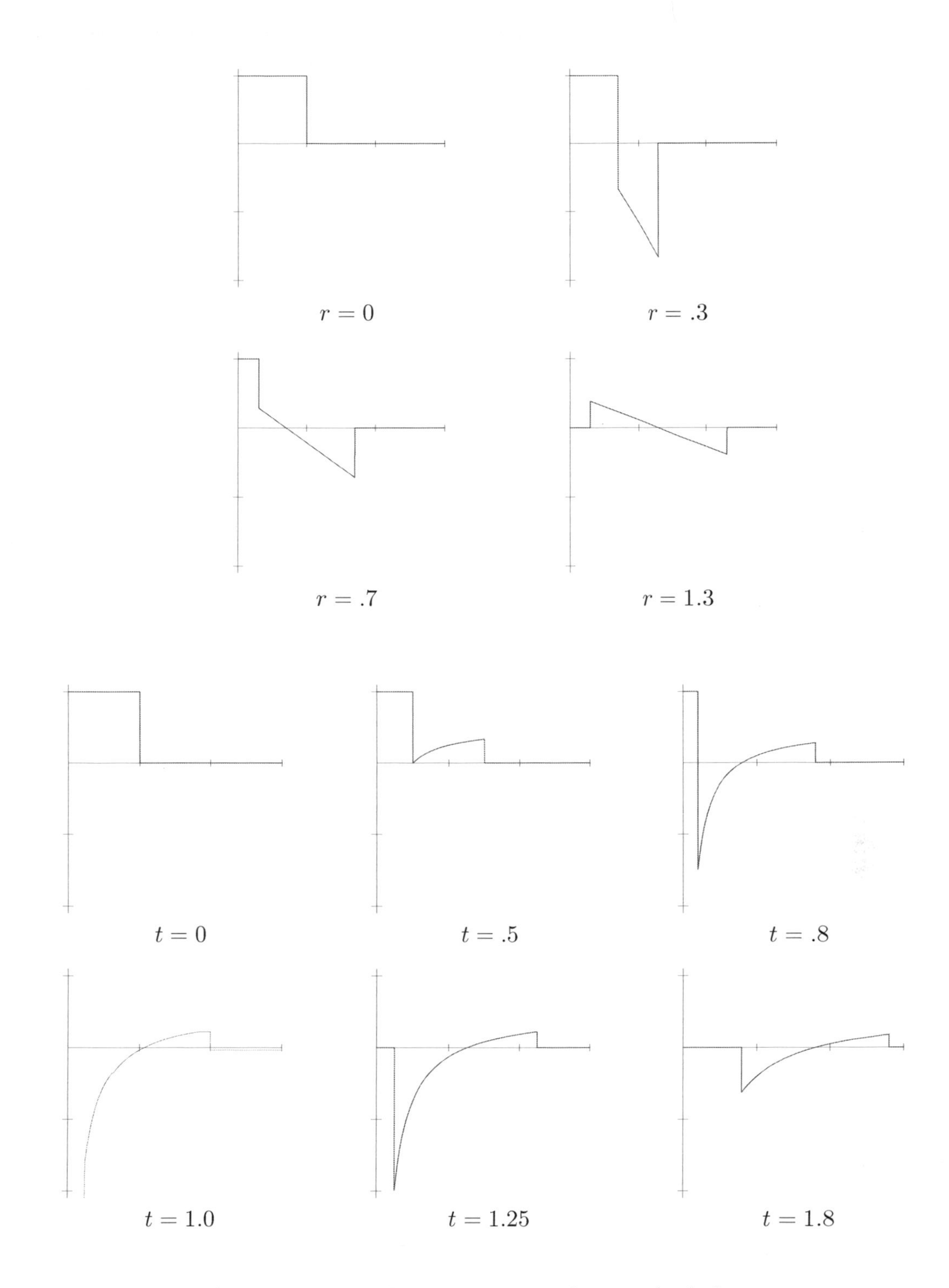

Figure 12.11. Wave equation solution $u(t, r)$ due to an initial displacement of the unit ball . ⊎

Theorem 12.18. *The solution to the initial value problem*

$$u_{tt} = c^2 \Delta u, \qquad u(0,\mathbf{x}) = f(\mathbf{x}), \qquad \frac{\partial u}{\partial t}(0,\mathbf{x}) = g(\mathbf{x}), \qquad \mathbf{x} \in \mathbb{R}^3, \tag{12.166}$$

for the wave equation in three-dimensional space is given by

$$u(t,\mathbf{x}) = \frac{\partial}{\partial t}\left(t\, \mathrm{M}^{\mathbf{x}}_{ct}[f]\right) + t\, \mathrm{M}^{\mathbf{x}}_{ct}[g] = \mathrm{M}^{\mathbf{x}}_{ct}[f] + ct\, \mathrm{M}^{\mathbf{x}}_{ct}\left[\frac{\partial f}{\partial \mathbf{n}}\right] + t\, \mathrm{M}^{\mathbf{x}}_{ct}[g], \tag{12.167}$$

where $\mathrm{M}^{\mathbf{x}}_{ct}[f]$ *denotes the average of the function* f *over the sphere* $S^{\mathbf{x}}_{ct} = \{\, \| \boldsymbol{\xi} - \mathbf{x} \| = ct \}$ *of radius* ct *centered at the point* $\mathbf{x}$.

A crucially important consequence of the Kirchhoff solution formula is a celebrated physical principle first set out by the pioneering seventeenth century Dutch scientist Christiaan Huygens.† Roughly, *Huygens' Principle* states that, in three-dimensional space, localized solutions to the wave equation remain localized. More concretely, (12.167) implies that the value of the solution at a point $\mathbf{x}$ and time t depends only on the values of the initial displacements and velocities at a distance ct away. Thus, all signals propagate along the relativistic light cone

$$c^2 t^2 = x^2 + y^2 + z^2$$

in four-dimensional Minkowski space-time. Physically, Huygens' Principle assures us that any light that we witness at time t arrived from points that lie a distance exactly $d = c(t - t_0)$ away at an earlier time $t_0 < t$. In particular, a localized initial signal, whether initial displacement or initial velocity, that is concentrated near a point produces a response that remains concentrated on an ever expanding sphere surrounding the point. In our three-dimensional universe, we witness the light from a sudden explosion or lightning bolt for only a brief moment, after which the view returns to darkness. Similarly, a sharp sound, e.g., a thunderclap, remains sharply concentrated with diminishing magnitude as it propagates through space. Huygens' Principle is responsible for the important astronomical fact that the light we now observe from a distant star was generated at a single past time that is directly proportional to the star's distance from the Earth. Remarkably, as we will show in the following subsection, Huygens' Principle *does not hold* in a two-dimensional universe! There, initially concentrated light and sound impulses will spread out as time progresses, and their effect will be experienced over an extended time range; see below for details.

Exercises

12.6.1. Solve the wave equation in three-dimensional space for the following initial conditions: (a) $u(0,x,y,z) = x + z$, $u_t(0,x,y,z) = 0$; (b) $u(0,x,y,z) = 0$, $u_t(0,x,y,z) = y$; (c) $u(0,x,y,z) = 1/(1 + x^2 + y^2 + z^2)$, $u_t(0,x,y,z) = 0$, (d) $u(0,x,y,z) = 0$, $u_t(0,x,y,z) = 1/(1 + x^2 + y^2 + z^2)$.

12.6.2. At what points in space-time does a three-dimensional wave vanish if it vanishes outside a sphere of radius R at the initial time $t = 0$?

† Don't even bother trying to pronounce his name correctly unless you are Dutch!

12.6.3. Consider the initial value problem

$$\frac{\partial^2 u}{\partial t^2} = \frac{\partial^2 u}{\partial x^2} + \frac{\partial^2 u}{\partial y^2} + \frac{\partial^2 u}{\partial z^2}, \qquad u(0,x,y,z) = 0, \qquad \frac{\partial u}{\partial t}(0,x,y,z) = \begin{cases} 1, & 0 < x,y,z < 1, \\ 0, & \text{otherwise}, \end{cases}$$

i.e., the initial velocity is 1 inside a unit cube and 0 outside the cube. We interpret the solution $u(t,x,y,z)$ as the intensity of light at a given point in space-time, measured in units that make the speed of light $c = 1$. (a) Write down an integral formula for $u(t,x,y,z)$. (b) Suppose a light sensor is placed at the point $(2,2,1)$. For which values of $t > 0$ will the sensor register a nonzero signal? Sketch a rough graph of what the sensor measures. (You do not need to find the precise formula, but explain how you obtained your graph.) (c) *True or false*: The solution $u(t,x,y,z) \geq 0$ at all points in space-time.

12.6.4. Is (12.151) a solution to the wave equation for $t < 0$? If not, write down a solution formula that is valid for negative t.

12.6.5. *True or false*: The function $u(t,x,y,z)$ defined by (12.154) is everywhere continuous.

12.6.6. A thermonuclear explosion occurs at the center of the Earth. Would you feel the effect first through a motion at the surface or a change in temperature at the surface? Discuss.

♢ 12.6.7. Prove that the area of the spherical cap $S_t^{\mathbf{x}} \cap B_1$ is given by formula (12.152).

Descent to Two Dimensions

So far, we have found explicit formulas for the solution to the wave equation on the one-dimensional line, and in three-dimensional space. The two-dimensional case

$$u_{tt} = c^2 \Delta u = c^2 (u_{xx} + u_{yy}) \tag{12.168}$$

is, counterintuitively, more complicated! For instance, seeking a radially symmetric solution $u(t,r)$ requires solving the partial differential equation

$$\frac{\partial^2 u}{\partial t^2} = c^2 \left(\frac{\partial^2 u}{\partial r^2} + \frac{1}{r}\frac{\partial u}{\partial r} \right), \tag{12.169}$$

which, unlike its three-dimensional counterpart (12.137), is not so easily integrated.

However, our solution to the three-dimensional problem can be adapted to construct a solution to the two-dimensional problem using the so-called *Method of Descent*. Observe that any solution $u(t,x,y)$ to the two-dimensional wave equation (12.168) can be viewed as a solution to the three-dimensional wave equation (12.123) that does not depend upon the vertical z coordinate, whence $\partial u/\partial z = 0$. Clearly, if the three-dimensional initial data does not depend on z, then the resulting solution $u(t,x,y)$ will also be independent of z.

Consider first the zero initial displacement condition

$$u(0,x,y) = 0, \qquad \frac{\partial u}{\partial t}(0,x,y) = g(x,y). \tag{12.170}$$

In the three-dimensional solution formula (12.151), if $g(x,y)$ does not depend on the z–coordinate, then the integrals over the upper and lower hemispheres

$$S_{ct}^{+} = \{ \, \| \boldsymbol{\xi} - \mathbf{x} \| = ct, \ \ \zeta \geq z \, \}, \qquad S_{ct}^{-} = \{ \, \| \boldsymbol{\xi} - \mathbf{x} \| = ct, \ \ \zeta \leq z \, \},$$

are identical. To evaluate these integrals, we parametrize the upper hemisphere as the graph of

$$\zeta = z + \sqrt{c^2t^2 - (\xi - x)^2 - (\eta - y)^2} \quad \text{over the disk} \quad D_{ct}^{\mathbf{x}} = \{ (\xi - x)^2 + (\eta - y)^2 \leq c^2 t^2 \},$$

concluding that

$$\begin{aligned} u(t,x,y) &= \frac{1}{4\pi c^2 t} \iint_{S_{ct}} g(\xi,\eta)\, dS = \frac{1}{2\pi c^2 t} \iint_{S_{ct}^+} g(\xi,\eta)\, dS \\ &= \frac{1}{2\pi c} \iint_{D_{ct}^{\mathbf{x}}} \frac{g(\xi,\eta)}{\sqrt{c^2t^2 - (\xi - x)^2 - (\eta - y)^2}}\, d\xi\, d\eta \end{aligned} \tag{12.171}$$

solves the initial value problem (12.170). In particular, if we take the initial velocity

$$\frac{\partial u}{\partial t}(0,x,y) = g(x,y) = \delta(x)\,\delta(y)$$

to be a unit impulse concentrated at the origin, then the resulting solution is

$$u(t,x,y) = \begin{cases} \dfrac{1}{2\pi c\sqrt{c^2t^2 - x^2 - y^2}}, & x^2 + y^2 < c^2t^2, \\ 0, & x^2 + y^2 > c^2t^2. \end{cases} \tag{12.172}$$

An observer sitting at distance $r = \| \mathbf{x} \| = \sqrt{x^2 + y^2}$ from the origin will first witness a concentrated displacement singularity at time $t = r/c$. However, in contrast to the three-dimensional solution, even after the impulse passes by, there will continue to be a decreasing, but nonzero, signal of magnitude roughly proportional to $1/t$. In Figure 12.12, we plot the solution (12.172) for unit wave speed $c = 1$. The first row plots intensity as a function of t at three different radii; note that the initial singularity, indicated by a spike in the graph, is followed by a progressively smaller residual displacement, which never entirely disappears. The second row shows the displacement at three different times as a function of $r = \| \mathbf{x} \|$.

As in the three-dimensional case, the solution to the initial displacement conditions

$$u(0,x,y) = f(x,y), \qquad \frac{\partial u}{\partial t}(0,x,y) = 0, \tag{12.173}$$

can then be obtained by differentiation of (12.171) with respect to t, and so

$$u(t,x,y) = \frac{1}{2\pi c}\,\frac{\partial}{\partial t} \iint_{D_{ct}^{\mathbf{x}}} \frac{f(\xi,\eta)}{\sqrt{c^2t^2 - (\xi - x)^2 - (\eta - y)^2}}\, d\xi\, d\eta. \tag{12.174}$$

As before, starting with a concentrated impulse, an observer will witness, after a time lapse $t = r/c$, an abrupt impulse passing by, followed by a progressively decaying residual wave. The general solution to the two-dimensional wave equation on all of $\mathbb{R}^2$ is a linear combination of these two types of solutions (12.171, 174).

As a consequence of these considerations, we discover that Huygens' Principle is *not* valid in a two-dimensional universe. The solution to the two-dimensional wave equation at a point $\mathbf{x}$ at time t depends on the initial displacement and velocity on the entire disk of radius $c\,t$ centered at the point, and not just on the points lying a distance $c\,t$ away. So a two-dimensional creature would experience not just an initial effect of a concentrated sound or light wave, but also an "afterglow" of slowly diminishing magnitude. It would be

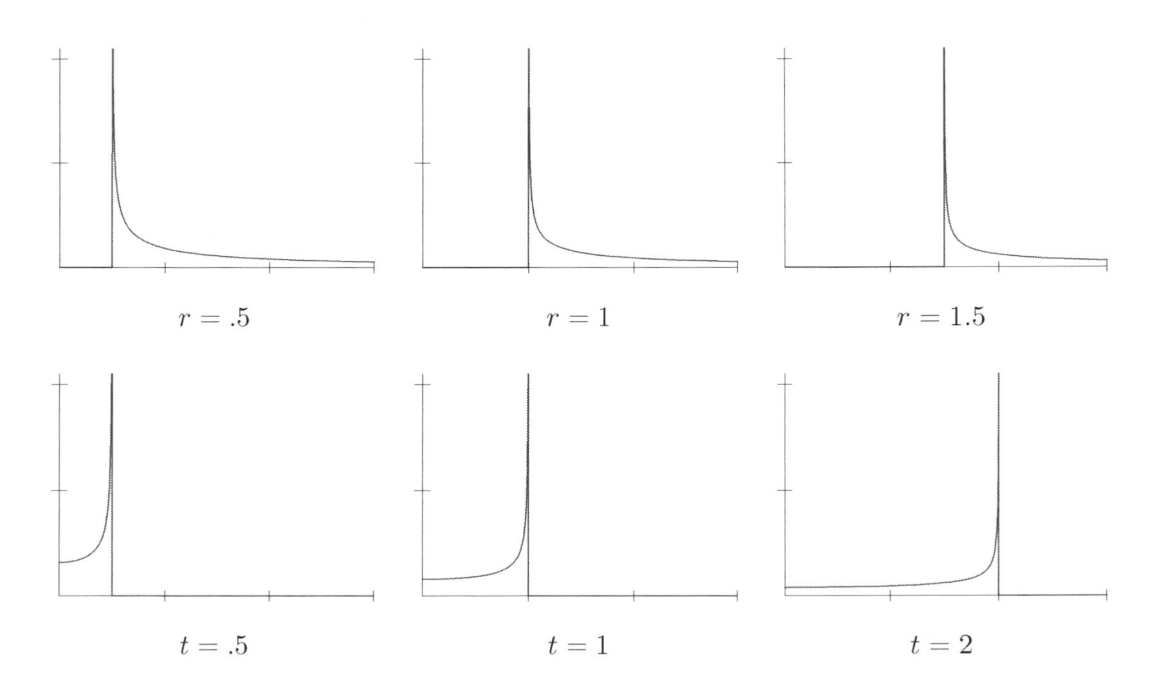

Figure 12.12. Solution to the two-dimensional wave equation for a concentrated impulse.

like living in a permanent echo chamber, and so understanding and acting upon sensory phenomena would be considerably more challenging. In general, it can be proved that Huygens' Principle for the wave equation is valid only in spaces of odd dimension $n = 2k + 1 \geq 3$; see also [**15**] for recent advances in the classification of partial differential equations that admit a Huygens' Principle.

Remark: Since the solutions to the two-dimensional wave equation can be interpreted as three-dimensional solutions with no z dependence, a concentrated delta impulse for the two-dimensional wave equation would correspond to a three-dimensional initial impulse that is concentrated along an entire vertical line, e.g., an instantaneous lightning bolt in the form of an infinite straight line. An observer fixed in space will first encounter the light flash arriving from the closest point on the line, but will subsequently experience the gradually decreasing effect of the light emitted by points that lie progressively farther away along the line. This accounts for the two-dimensional afterglow in formula (12.172).

Exercises

12.6.8. Solve initial value problem for the two-dimensional wave equation with the following initial data (*a*) $u(0, x, y) = x - y$, $u_t(0, x, y) = 0$; (*b*) $u(0, x, y) = 0$, $u_t(0, x, y) = y$.

12.6.9. (a) Prove that $u(t,x,y) = 1/\sqrt{x^2+y^2-c^2t^2}$ is a solution to the two-dimensional wave equation on the domain $\Omega = \{x^2+y^2 > c^2t^2\}$ exterior to the light cone passing through the origin. What is the corresponding initial data at $t = 0$? (b) Use part (a) to solve the initial value problem $u(0,x,y) = 0$, $u_t(0,x,y) = 1/\sqrt{x^2+y^2}$, on Ω.

12.6.10. Consider the two-dimensional wave equation on $\mathbb{R}^2$ with wave speed $c = 1$. Write down an integral formula for the solution to the following initial value problems. You need not evaluate the integrals. (a) $u(0,x,y) = x^3 - y^3$, $u_t(0,x,y) = 0$; (b) $u(0,x,y) = 0$, $u_t(0,x,y) = y^2$; (c) $u(0,x,y) = x^2 + y^2$, $u_t(0,x,y) = -x^2 - y^2$.

12.6.11. (a) Find the solution to the two-dimensional wave equation whose initial displacement is a concentrated delta impulse at the origin and whose initial velocity is zero.
(b) Is your expression a classical solution when $t > 0$?
(c) *True or false*: The solution tends to 0 uniformly as $t \to \infty$.

12.6.12. Use separation of variables to write down an eigenfunction series solution to the partial differential equation (12.169) when subject to homogeneous Dirichlet boundary conditions at $r = 1$ and bounded at $r = 0$.

◇ 12.6.13. Write down the fundamental solution for the one-dimensional wave equation with (a) a concentrated initial displacement at the origin; (b) a concentrated initial velocity at the origin. (c) Discuss the validity of Huygens' Principle in a one-dimensional universe.

12.6.14. Discuss how you can construct solutions to the one-dimensional wave equation by descent from the three-dimensional wave equation.

12.7 The Hydrogen Atom

A *hydrogen atom* consists of a single electron circling an atomic nucleus that contains a single proton, which, owing to its relatively tiny size, is assumed to be entirely concentrated at the origin. As a result of quantization of the corresponding classical Coulomb problem, the Schrödinger equation† governing the dynamical behavior of the electron moving around the nucleus takes the explicit form

$$\mathrm{i}\,\hbar\,\frac{\partial\psi}{\partial t} = -\frac{\hbar^2}{2\,M}\,\Delta\psi - \frac{\alpha^2}{r}\,\psi = -\frac{\hbar^2}{2\,M}\left(\frac{\partial^2\psi}{\partial x^2} + \frac{\partial^2\psi}{\partial y^2} + \frac{\partial^2\psi}{\partial y^2}\right) - \frac{\alpha^2\,\psi}{\sqrt{x^2+y^2+z^2}}. \tag{12.175}$$

Here $\psi(t,x,y,z)$ denotes the electron's time-dependent wave function, which, at each time t, prescribes its quantum probability density as it circles the nucleus. In the quantized Hamiltonian operator $K = -\frac{1}{2}(\hbar^2/M)\,\Delta - \alpha^2/r$, the coefficient of the Laplacian depends on Planck's constant $\hbar$ and the electron's mass M. The final term represents the three-dimensional electromagnetic (Coulomb) potential function $V(\mathbf{x}) = \alpha^2/r$ attracting the electron to the nucleus, with α representing the electron's (and proton's) charge, while $r = \|\,\mathbf{x}\,\|$ is its distance from the nucleus. Incidentally, the quantum-mechanical Schrödinger equation for multi-electron atoms or even molecules is not so difficult to write down, but its solution, even for, say, the helium atom, is *much* more difficult, and, in general, is still

† The reader is referred to (9.151) and the subsequent discussion for generalities regarding the Schrödinger equation and quantum mechanics.

a major challenge for numerical analysts, even on today's supercomputers, [**116**]. Thus, to keep matters as simple as possible, we will consider only the case of a single electron hydrogen atom here.

Bound States

According to the analysis in Section 9.5, the normal mode solutions to the Schrödinger equation are of the form

$$\psi(t,x,y,z) = e^{\,\mathrm{i}\,\lambda\,t/\hbar}\, v(x,y,z),$$

where v is an eigenfunction of the Hamiltonian operator with eigenvalue λ, and hence satisfies

$$\frac{\hbar^2}{2M}\,\Delta v + \left(\lambda + \frac{\alpha^2}{r}\right) v = 0. \tag{12.176}$$

The *bound states* of the atom, in which the electron remains trapped by the nucleus, are represented by the nonzero solutions to the eigenvalue problem (12.176) with unit L^2 norm:

$$\|\, v\,\|^2 = \iiint |\, v(x,y,z)\,|^2\, dx\, dy\, dz = 1.$$

The eigenvalue λ specifies the bound state's energy, and is necessarily negative: $\lambda < 0$. Since we are working on an unbounded domain, the bound states do *not* form a complete system of eigenfunctions, and so not every wave function $\varphi \in \mathrm{L}^2(\mathbb{R}^3)$ can be approximated by an eigenfunction series. The missing data are the so-called *scattering states* arising from the *continuous spectrum* of the Schrödinger operator; these represent electrons that scatter off the nucleus, and so do not remain trapped in an orbit. (For the classical Kepler problem of a planet circling a sun, the bound states would correspond to planets following bounded elliptic orbits, while the scattering states correspond to interstellar comets and the like moving along unbounded hyperbolic or parabolic trajectories.) We will leave the discussion of the quantum-mechanical scattering states and the associated continuous spectrum to a more advanced treatment of the subject, [**72, 95**].

To understand the bound states, we will apply the method of separation of variables. We begin by rewriting the eigenvalue problem (12.176) in spherical coordinates:

$$\frac{\hbar^2}{2M}\left(\frac{\partial^2 v}{\partial r^2} + \frac{2}{r}\,\frac{\partial v}{\partial r} + \frac{1}{r^2}\,\frac{\partial^2 v}{\partial \varphi^2} + \frac{\cos\varphi}{r^2\sin\varphi}\,\frac{\partial v}{\partial \varphi} + \frac{1}{r^2\sin^2\varphi}\,\frac{\partial^2 v}{\partial \theta^2}\right) + \left(\lambda + \frac{\alpha^2}{r}\right) v = 0. \tag{12.177}$$

We then separate off the radial coordinate, setting

$$v(r,\varphi,\theta) = p(r)\, w(\varphi,\theta).$$

The angular component satisfies the spherical Helmholtz equation

$$\Delta_S w + \mu\, w = \frac{\partial^2 w}{\partial \varphi^2} + \frac{\cos\varphi}{\sin\varphi}\,\frac{\partial w}{\partial \varphi} + \frac{1}{\sin^2\varphi}\,\frac{\partial^2 w}{\partial \theta^2} + \mu\, w = 0,$$

which we have already solved; see (12.21) and the ensuing discussion. The eigensolutions are spherical harmonics, which, because the quantum-mechanical solutions are intrinsically complex-valued, we take in their complex form (12.46). The associated eigenvalue

$$\mu = l\,(l+1), \qquad \text{where the integer} \qquad l = 0, 1, 2, \ldots \tag{12.178}$$

is known as the *angular quantum number*, admits a total of $2l+1$ linearly independent eigenfunctions

$$\mathcal{Y}_l^m(\varphi,\theta) = P_l^m(\cos\varphi)\, e^{\,i\,m\,\theta}, \qquad m = -l, -l+1, \ldots, l-1, l. \tag{12.179}$$

The radial equation with the separation constant (12.178) is

$$\frac{\hbar^2}{2M}\left(\frac{d^2p}{dr^2} + \frac{2}{r}\,\frac{dp}{dr}\right) + \left(\lambda + \frac{\alpha^2}{r} - \frac{l\,(l+1)}{r^2}\right) p = 0. \tag{12.180}$$

To eliminate the physical parameters, let us rescale the radial coordinate by setting

$$s = \sigma\, r, \qquad \text{where} \qquad \sigma = \frac{2\sqrt{-2\,M\,\lambda}}{\hbar}, \tag{12.181}$$

given that $\lambda < 0$. The resulting ordinary differential equation for the rescaled function

$$P(s) = p\left(\frac{s}{\sigma}\right)$$

is

$$\frac{d^2P}{ds^2} + \frac{2}{s}\,\frac{dP}{ds} - \left(\frac{1}{4} - \frac{n}{s} + \frac{l\,(l+1)}{s^2}\right) P = 0, \tag{12.182}$$

where

$$n = \frac{2\,M\,\alpha^2}{\sigma\,\hbar^2} = \frac{\alpha^2}{\hbar}\sqrt{-\frac{M}{2\,\lambda}}\,. \tag{12.183}$$

Equation (12.182) is a version of the *generalized Laguerre differential equation* — see Exercise 12.7.4 below — named after the nineteenth-century French mathematician Edmond Laguerre, who studied its solutions well before the appearance of quantum mechanics. Since we are searching for bound states, the relevant solutions should be defined on $0 \le s < \infty$, remain bounded at $s = 0$, and go to zero as $s \to \infty$:

$$\lim_{s\to 0+} P(s) < \infty, \qquad \lim_{s\to\infty} P(s) = 0. \tag{12.184}$$

The proof of the following key result is outlined in Exercises 12.7.4–5.

Theorem 12.19. *For each pair of nonnegative integers* $0 \le l < n$, *the boundary value problem* (12.182, 184) *has the eigensolution*

$$P_l^n(s) = s^l\, e^{-\,s/2}\, L_{n-l-1}^{2l+1}(s), \tag{12.185}$$

where

$$L_k^j(s) = \frac{s^{-j} e^s}{k!}\,\frac{d^k}{ds^k}\left[\, s^{j+k} e^{-\,s}\,\right] = \sum_{i=0}^{k} \frac{(-1)^i}{i!}\binom{j+k}{j+i} s^i, \qquad j,k = 0,1,2,\ldots\,, \tag{12.186}$$

are known as generalized[†] Laguerre polynomials.

[†] The ordinary *Laguerre polynomials* are $L_k(s) = L_k^0(s)$.

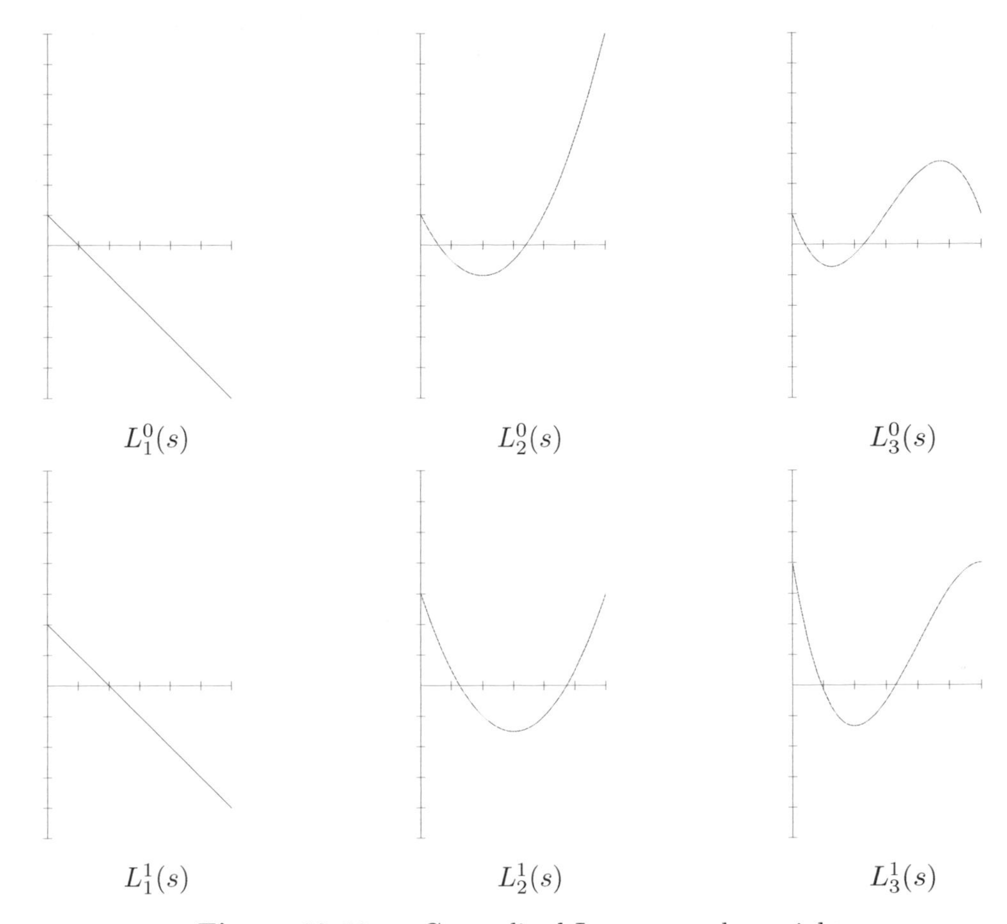

Figure 12.13. Generalized Laguerre polynomials.

The first few generalized Laguerre polynomials are

$$\begin{aligned}
&L_0^0(s) = 1, \quad L_1^0(s) = 1 - s, \quad L_2^0(s) = 1 - 2\,s + \tfrac{1}{2}\,s^2, \quad L_3^0(s) = 1 - 3\,s + \tfrac{3}{2}\,s^2 - \tfrac{1}{6}\,s^3,\\
&L_0^1(s) = 1, \quad L_1^1(s) = 2 - s, \quad L_2^1(s) = 3 - 3\,s + \tfrac{1}{2}\,s^2, \quad L_3^1(s) = 4 - 6\,s + 2\,s^2 - \tfrac{1}{6}\,s^3,\\
&L_0^2(s) = 1, \quad L_1^2(s) = 3 - s, \quad L_2^2(s) = 6 - 4\,s + \tfrac{1}{2}\,s^2, \quad L_3^2(s) = 10 - 10\,s + \tfrac{5}{2}\,s^2 - \tfrac{1}{6}\,s^3.
\end{aligned}$$

Note that $L_k^j(s)$ has degree k. A few graphs, on the interval $0 \leq t \leq 6$, appear in Figure 12.13. See [**86**] for details on their properties.

Atomic Eigenstates and Quantum Numbers

The integer n, whose physical value was noted in (12.183), is known as the *principal quantum number*. We further note that the scaling factor in (12.181) can be written as

$$\sigma = \frac{2\,M\,\alpha^2}{n\,\hbar^2} = \frac{2}{n\,a}, \qquad \text{where} \qquad a = \frac{\hbar^2}{M\,\alpha^2} \approx .529 \times 10^{-10} \text{ meter},$$

which approximates the radius of the electron's lowest energy level, is known as the *Bohr radius*, in honor of the pioneering Danish quantum physicist Niels Bohr. Reverting to physical coordinates, the bound state solutions (12.185) become, up to an inessential constant multiple, the *radial wave functions*

$$\beta_l^n(r) = \left(\frac{2r}{na}\right)^l e^{-r/(na)} L_{n-l-1}^{2l+1}\left(\frac{2r}{na}\right). \tag{12.187}$$

Combining them with the spherical harmonics (12.179) yields the *atomic eigenfunctions* or *eigenstates*

$$v_{lmn}(r,\varphi,\theta) = \sqrt{\frac{(2l+1)\,(l-m)!\,(n-l-1)!}{\pi\,a^3 n^4 (l+m)!\,(l+n)!}}\;\beta_l^n(r)\,\mathcal{Y}_l^m(\varphi,\theta), \tag{12.188}$$

where the initial factor is selected so as to make $\| v_{lmn} \| = 1$, and hence a bona fide wave function. (A proof of this fact is outlined in Exercise 12.7.8.) The eigenstates depend on three integers, which have the following physical designations:

- $n = 1, 2, 3, \ldots :$ the *principal quantum number*;
- $l = 0, 1, \ \ldots \ , n-1:$ the *angular quantum number*;
- $m = -l, -l+1, \ \ldots \ , l-1, l:$ the *magnetic quantum number*.

The energy is the associated eigenvalue:

$$\lambda_n = -\frac{\alpha^4 M}{2\,\hbar^2}\,\frac{1}{n^2} = -\frac{\alpha^2}{2a}\,\frac{1}{n^2}, \qquad n = 1, 2, 3, \ldots\,. \tag{12.189}$$

The fact that the ratios $\lambda_n/\lambda_1 = 1/n^2$ between the energy levels of an atom are inverse squares of integers was one of the key experimental discoveries that precipitated the discovery of quantum mechanics. Observe that the nth energy level has a total of

$$\sum_{l=0}^{n-1} (2l+1) = n^2 \tag{12.190}$$

linearly independent bound states (12.188). The dimension of the eigenspace corresponds to the number of orbital subshells in the atom for the corresponding energy level. The shells indexed by the angular quantum number, i.e., the order $l = 0, 1, 2, \ldots$ of the spherical harmonic, are traditionally labeled by a letter in the sequence $s, p, d, f, g, \ldots$, where each successive shell contains $2l+1$ individual subshells, indexed by the magnetic quantum number m.

The one missing ingredient in this simple model is the electron's *spin*. Since electrons can have one of two possible spins, the *Pauli Exclusion Principle*, first formulated by the Austrian physicist Wolfgang Pauli, tells us that each atomic energy shell can be occupied by at most two electrons. Consequently, the atomic shell with angular quantum number l may contain up to $2(2l+1)$ electrons. Keep in mind that, since $0 \le l < n$, the lth shell appears only when n is sufficiently large, so that, according to (12.190), the nth energy level contains up to $2n^2$ electrons.

The resulting atomic configuration of electronic energy shells is the explanation for Mendeleev's periodic table. Its rows are indexed by the principal quantum number n, while the columns are labeled by the angular and magnetic quantum numbers l, m, and the spin. As one moves up the periodic table, the electrons in each successive element's atom progressively fill up the lower energy levels, each new shell containing first a single electron, then two electrons with opposite spins. Thus, hydrogen (in its ground state) has a single electron in the $1s$ shell. Helium has two electrons in the $1s$ shell. Lithium has three electrons, with two of them filling the $1s$ shell and the third in the $2s$ shell. Neon has ten electrons filling the first two energy levels, with two electrons in the $1s$ shell, two in the $2s$ shell, and six in the $2p$ shell. And so on. The one complication is that, owing to the orbital's geometry, as prescribed by the associated spherical harmonic, the angular and, to a lesser extent, magnetic quantum numbers also affect the physically observed energy, and this can cause shells to fill later than might initially be expected. For example, in potassium and calcium, the $4s$ shell is successively filled, followed by scandium, which begins the process of filling the $3d$ subshells. The chemical properties of the elements are, to a very large extent, determined by the placement of their atom's electrons within the outermost energy level. The interested reader can consult, for example, [**67**, **79**] for further details.

Exercises

12.7.1. If the nucleus contains Z protons circled by a single electron, then its atomic potential $V(\mathbf{x})$ is rescaled accordingly, replacing α^2/r by $Z\alpha^2/r$. Discuss the induced effect on the energy levels of such an atomic ion.

♡ 12.7.2. (a) Write down the time-dependent wave function for a single electron atom when the electron is in its ground state, i.e., the lowest energy level. (*b*) What is the probability density of the electron? (*c*) What is the probability of finding the electron within 1 Bohr radius of the atom? (*d*) Find the distance d (measured in Bohr radii) so that there is a 95% probability of finding the electron within a distance d of the nucleus.

♢ 12.7.3. Prove that the two expressions for the Laguerre polynomials in (12.186) agree.

♢ 12.7.4. (a) Let $k = 0, 1, 2, \ldots$ be a nonnegative integer. The *Laguerre differential equation of order* k is

$$x\,u'' + (1-x)\,u' + k\,u = 0. \qquad (12.191)$$

Show that $x = 0$ is a regular singular point. Then prove that the Frobenius solution based at $x = 0$ is a polynomial of degree j that coincides with the Laguerre polynomial $L_k^0(x)$.
(*b*) Given nonnegative integers $j, k \geq 0$, use the Frobenius method to prove that the *generalized Laguerre differential equation*

$$x\,u'' + (j+1-x)\,u' + k\,u = 0 \qquad (12.192)$$

has a polynomial solution that can be identified with the generalized Laguerre polynomial $L_k^j(x)$ in (12.186).

♢ 12.7.5. Suppose that $P(s)$ solves the ordinary differential equation (12.182). Prove that $Q(s) = s^{-l}\,e^{s/2}\,P(s)$ solves the differential equation

$$s\,\frac{d^2Q}{ds^2} + [\,2(l+1) - s\,]\,\frac{dQ}{ds} + (n-l-1)Q = 0. \qquad (12.193)$$

Then apply the result of Exercise 12.7.4 to complete the proof of Theorem 12.19.

♡ 12.7.6. Suppose $f(x)$ is a polynomial, and let $L_k^j(s)$ denote the generalized Laguerre polynomials (12.186). (a) Prove that, for $j, k \geq 0$,

$$\int_0^\infty f(s)\,L_k^j(s)\,s^j\,e^{-s}\,ds = \frac{(-1)^k}{k!}\int_0^\infty f^{(k)}(s)\,s^{j+k}\,e^{-s}\,ds.$$

(*b*) For fixed j, prove that the generalized Laguerre polynomials $L_k^j(s)$, $k = 0, 1, 2, \ldots$, are orthogonal with respect to the weighted inner product $\langle\, f\,, g\,\rangle = \displaystyle\int_0^\infty f(s)\,g(s)\,s^j\,e^{-s}\,ds$.

(*c*) Prove the formula for their corresponding norms: $\|\, L_k^j \,\| = \sqrt{\dfrac{(j+k)!}{k!}}$.

♢ 12.7.7. (a) Prove that the generalized Laguerre polynomials satisfy the following recurrence relation:

$$(k+1)\, L^j_{k+1}(s) - (j+2k+1-s)\, L^j_k(s) + (j+k)\, L^j_{k-1}(s) = 0. \tag{12.194}$$

(b) Prove that

$$\int_0^\infty s^{j+1} e^{-s} \left[L^j_k(s) \right]^2 ds = \frac{(j+2k+1)\,(j+k)!}{k!}. \tag{12.195}$$

Hint: Use part (a) and Exercise 12.7.6.

♡ 12.7.8. Prove that the atomic eigenfunctions (12.188) form an orthonormal system of wave functions with respect to the L^2 inner product on $\mathbb{R}^3$.
Hint: Use Theorem 9.33 and equation (12.195).

Appendix A
Complex Numbers

The purpose of this short appendix is to review the basics of complex numbers and complex arithmetic, which are used throughout much of the text.

A *complex number* is an expression of the form $z = x + \mathrm{i}\, y$, where $x, y \in \mathbb{R}$ are real and $\mathrm{i} = \sqrt{-1}$ is the imaginary unit. The set of all complex numbers is denoted by $\mathbb{C}$. We call $x = \operatorname{Re} z$ the *real part* of z and $y = \operatorname{Im} z$ the *imaginary part* of $z = x + \mathrm{i}\, y$. (*Note*: The imaginary part is the real number y, *not* $\mathrm{i}\, y$.) A real number x is merely a complex number with zero imaginary part, $\operatorname{Im} z = 0$, and so we may regard $\mathbb{R} \subset \mathbb{C}$. Complex addition and multiplication are based on simple adaptations of the rules of real arithmetic to include the identity $\mathrm{i}^2 = -1$, and so

$$\begin{aligned} (x + \mathrm{i}\, y) + (u + \mathrm{i}\, v) &= (x + u) + \mathrm{i}\, (y + v), \\ (x + \mathrm{i}\, y)\, (u + \mathrm{i}\, v) &= (x\, u - y\, v) + \mathrm{i}\, (x\, v + y\, u). \end{aligned} \tag{A.1}$$

Complex numbers enjoy all the usual laws of real addition and multiplication, *including commutativity*: $z\, w = w\, z$.

We can identify a complex number $x + \mathrm{i}\, y$ with a vector $(x, y) \in \mathbb{R}^2$ in the real, two-dimensional plane. For this reason, $\mathbb{C}$ is sometimes referred to as the *complex plane*. (Although keep in mind that, as a complex vector space, $\mathbb{C}$ is only one-dimensional.) Based on this identification, we shall employ the standard terminology of planar vector calculus — domain, curve, etc. — without alteration. Complex addition (A.1) corresponds to vector addition, but the vector interpretation of complex multiplication is more obscure.

The *complex conjugate* of $z = x + \mathrm{i}\, y$ is $\overline{z} = x - \mathrm{i}\, y$. Note that $\operatorname{Re} \overline{z} = \operatorname{Re} z$, while $\operatorname{Im} \overline{z} = -\operatorname{Im} z$. Geometrically, the complex conjugate of z is obtained by reflecting the corresponding vector through the real axis, as illustrated in Figure A.1. In particular, $\overline{z} = z$ if and only if z is real. In general,

$$\operatorname{Re} z = \frac{z + \overline{z}}{2}, \qquad \operatorname{Im} z = \frac{z - \overline{z}}{2\,\mathrm{i}}. \tag{A.2}$$

Complex conjugation is compatible with complex arithmetic:

$$\overline{z + w} = \overline{z} + \overline{w}, \qquad \overline{z\, w} = \overline{z}\; \overline{w}.$$

In particular, the product of a complex number and its conjugate,

$$z\, \overline{z} = (x + \mathrm{i}\, y)\, (x - \mathrm{i}\, y) = x^2 + y^2, \tag{A.3}$$

is real and nonnegative. Its square root is known as the *modulus* or *norm* of the complex number $z = x + \mathrm{i}\, y$, and written

$$|\, z\, | = \sqrt{x^2 + y^2}\,. \tag{A.4}$$

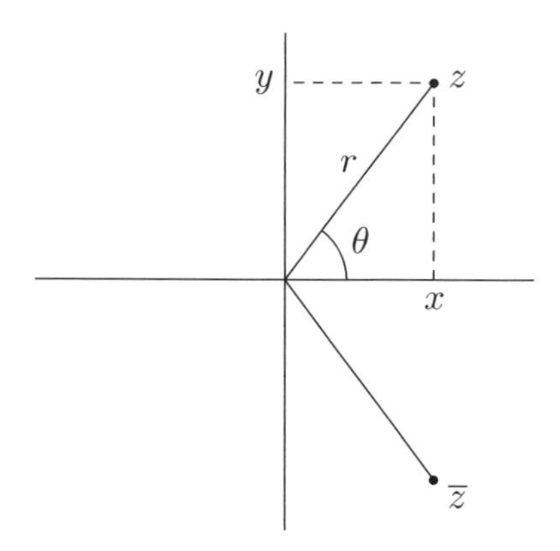

Figure A.1. Complex numbers.

Note that $|z| \geq 0$, with $|z| = 0$ if and only if $z = 0$. The modulus $|z|$ generalizes the absolute value of a real number and coincides with the standard Euclidean norm in the (x, y)–plane. This implies the validity of the triangle inequality

$$|z + w| \leq |z| + |w|. \tag{A.5}$$

Equation (A.3) can be rewritten in terms of the modulus as

$$z\,\overline{z} = |z|^2. \tag{A.6}$$

Rearranging the factors, we deduce the formula for the reciprocal of a nonzero complex number:

$$\frac{1}{z} = \frac{\overline{z}}{|z|^2}, \qquad z \neq 0, \qquad \text{or, equivalently,} \qquad \frac{1}{x + \mathrm{i}\,y} = \frac{x - \mathrm{i}\,y}{x^2 + y^2}. \tag{A.7}$$

The general formula for complex division,

$$\frac{w}{z} = \frac{w\,\overline{z}}{|z|^2} \qquad \text{or} \qquad \frac{u + \mathrm{i}\,v}{x + \mathrm{i}\,y} = \frac{(x\,u + y\,v) + \mathrm{i}\,(x\,v - y\,u)}{x^2 + y^2}, \tag{A.8}$$

is an immediate consequence.

The modulus of a complex number,

$$r = |z| = \sqrt{x^2 + y^2}\,,$$

is one component of its polar coordinate representation

$$x = r\cos\theta, \qquad y = r\sin\theta \qquad \text{or} \qquad z = r(\cos\theta + \mathrm{i}\,\sin\theta). \tag{A.9}$$

The polar angle θ, which measures the angle that the line connecting z to the origin makes with the horizontal axis, is known as the *phase*, and written

$$\theta = \operatorname{ph} z. \tag{A.10}$$

As such, the phase is defined only up to an integer multiple of 2π. The unique *principal value* of the phase is restricted to $-\pi < \operatorname{ph} z \leq \pi$. A more common term for the polar angle is the *argument* of z, written $\arg z = \operatorname{ph} z$. However, in conformity with [**85**, **86**], we prefer to use "phase" here, in part to avoid confusion with the argument z of a function $f(z)$.

Euler's celebrated formula for the complex exponential,

$$e^{\,\mathrm{i}\,\theta} = \cos\theta + \mathrm{i}\,\sin\theta, \tag{A.11}$$

can be used to compactly rewrite the polar form (A.9) of a complex number as

$$z = r\,e^{\,\mathrm{i}\,\theta}, \qquad \text{where} \qquad r = |\,z\,|, \qquad \theta = \operatorname{ph} z. \tag{A.12}$$

Consequently, the complex logarithm has the form

$$\log z = \log(r\,e^{\,\mathrm{i}\,\theta}) = \log r + \log e^{\,\mathrm{i}\,\theta} = \log r + \mathrm{i}\,\theta = \log|\,z\,| + \mathrm{i}\,\operatorname{ph} z. \tag{A.13}$$

More generally, the complex exponential is given by

$$e^{z} = e^{x}\cos y + \mathrm{i}\,e^{x}\sin y, \qquad \text{for} \qquad z = x + \mathrm{i}\,y. \tag{A.14}$$

We note that the modulus and phase of a product of complex numbers can be readily computed:

$$|\,z\,w\,| = |\,z\,|\,|\,w\,|, \qquad \operatorname{ph}(z\,w) = \operatorname{ph} z + \operatorname{ph} w, \tag{A.15}$$

the latter formula requiring that we allow multiply valued phases; the formula does *not* hold as stated for all z, w when the principal value of the phase is used. Similarly, the modulus and phase of the reciprocal of a nonzero complex number are

$$\left|\,\frac{1}{z}\,\right| = \frac{1}{|\,z\,|}, \qquad \operatorname{ph}\left(\frac{1}{z}\right) = -\operatorname{ph} z. \tag{A.16}$$

On the other hand, complex conjugation preserves the modulus, but negates the phase:

$$|\,\overline{z}\,| = |\,z\,|, \qquad \operatorname{ph}\overline{z} = -\operatorname{ph} z. \tag{A.17}$$

The latter formula is not valid for the principal value of the phase when z lies on the negative real axis.

Appendix B
Linear Algebra

In this appendix, we collect basic results and definitions from linear algebra that are used in our study of partial differential equations. The reader is referred to [**89**] for the proofs and further details.

B.1 Vector Spaces and Subspaces

Vector spaces and their ancillary structures provide the common language of linear algebra. The basic definition is modeled on the prototypical finite-dimensional example: the *Euclidean space* $\mathbb{R}^n$, which is the set of all real (column) vectors with n entries, equipped with the operations of vector addition and scalar multiplication. More generally:

Definition B.1. A (real) *vector space* is a set V equipped with two operations:

(i) *Addition*: adding any pair of elements $\mathbf{v}, \mathbf{w} \in V$ produces another vector $\mathbf{v} + \mathbf{w} \in V$.

(ii) *Scalar Multiplication*: multiplying an element $\mathbf{v} \in V$ by a scalar $c \in \mathbb{R}$ produces a vector $c\,\mathbf{v} \in V$.

These are subject to the following axioms: for all $\mathbf{u}, \mathbf{v}, \mathbf{w} \in V$ and all scalars $c, d \in \mathbb{R}$,

(a) *Commutativity of Addition*: $\mathbf{v} + \mathbf{w} = \mathbf{w} + \mathbf{v}$.

(b) *Associativity of Addition*: $\mathbf{u} + (\mathbf{v} + \mathbf{w}) = (\mathbf{u} + \mathbf{v}) + \mathbf{w}$.

(c) *Additive Identity*: There is a zero element $\mathbf{0} \in V$ satisfying $\mathbf{v} + \mathbf{0} = \mathbf{v} = \mathbf{0} + \mathbf{v}$.

(d) *Additive Inverse*: For each $\mathbf{v} \in V$ there is an element $-\mathbf{v} \in V$ such that $\mathbf{v} + (-\mathbf{v}) = \mathbf{0} = (-\mathbf{v}) + \mathbf{v}$.

(e) *Distributivity*: $(c + d)\,\mathbf{v} = (c\,\mathbf{v}) + (d\,\mathbf{v})$, and $c\,(\mathbf{v} + \mathbf{w}) = (c\,\mathbf{v}) + (c\,\mathbf{w})$.

(f) *Associativity of Scalar Multiplication*: $c\,(d\,\mathbf{v}) = (c\,d)\,\mathbf{v}$.

(g) *Unit for Scalar Multiplication*: the scalar $1 \in \mathbb{R}$ satisfies $1\,\mathbf{v} = \mathbf{v}$.

Complex vector spaces are defined in an identical manner, the only difference being that the scalars are allowed to be complex numbers. In this case, the prototype is the space $\mathbb{C}^n$ consisting of column vectors with n complex entries.

While finite-dimensional vector spaces play a significant role in the study of partial differential equations, particularly in the design of numerical solution schemes, for us the more important examples are infinite-dimensional vector spaces whose elements ("vectors") are functions. The main example is the following:

Example B.2. Let $I \subset \mathbb{R}$ be an interval. The *function space* $\mathcal{F} = \mathcal{F}(I)$, whose elements are all real-valued functions $f(x)$ defined for $x \in I$, has the structure of a vector space. Addition of functions in $\mathcal{F}$ is defined in the usual manner: $(f+g)(x) = f(x)+g(x)$ for all $x \in I$. Multiplication by scalars $c \in \mathbb{R}$ is the same as multiplication by constants, $(c\,f)(x) = c\,f(x)$. The zero element is the constant function that is identically 0 for all $x \in I$. With these operations, all the vector space axioms listed in Definition B.1 are valid, and hence $\mathcal{F}(I)$ is a real vector space.

More generally, if $\Omega \subset \mathbb{R}^n$ is any subset of n-dimensional Euclidean space, the function space $\mathcal{F}(\Omega)$ is defined as the set of all real-valued functions $f(x_1, \ldots, x_n)$ defined for all $x = (x_1, \ldots, x_n) \in \Omega$. Addition and scalar (constant) multiplication of functions are defined in the same manner.

A *subspace* of a vector space V is a subset $W \subset V$ that is a vector space in its own right. In particular, a subspace W *must* contain the zero element of V.

Proposition B.3. *A nonempty subset $W \subset V$ of a vector space is a subspace if and only if*

(a) *for every $\mathbf{v}, \mathbf{w} \in W$, the sum $\mathbf{v} + \mathbf{w} \in W$, and*

(b) *for every $\mathbf{v} \in W$ and every $c \in \mathbb{R}$, the scalar product $c\mathbf{v} \in W$.*

For example, a complete list of subspaces of $V = \mathbb{R}^3$ is (i) the origin $\{\mathbf{0}\}$; (ii) every line through the origin; (iii) every plane through the origin; (iv) all of $\mathbb{R}^3$.

Example B.4. Here are some examples of subspaces of the function space $\mathcal{F}(I)$.

(a) The space $\mathcal{P}^{(n)}$ of polynomials of degree $\leq n$.

(b) The space $\mathrm{C}^0(I)$ of all continuous functions on the interval I.

(c) The space $\mathrm{C}^n(I)$ consisting of all functions $f(x)$ that have n continuous derivatives $f'(x), f''(x), \ldots, f^{(n)}(x)$ on† I.

(d) The space $\mathrm{C}^\infty(I) = \bigcap_{n \geq 0} \mathrm{C}^n(I)$ of infinitely differentiable, or *smooth*, functions is also a subspace.

(e) The space $\mathcal{A}(I)$ of analytic functions. Recall that a function $f(x)$ is called *analytic* at a point a if it is smooth, and, moreover, its Taylor series

$$f(a) + f'(a)\,(x-a) + \tfrac{1}{2}\,f''(a)\,(x-a)^2 + \cdots = \sum_{n=0}^{\infty} \frac{f^{(n)}(a)}{n!}\,(x-a)^n \tag{B.1}$$

converges to $f(x)$ for all x sufficiently close to a. (The series is not required to converge on the entire interval I.) Not every smooth function is analytic, and so $\mathcal{A}(I) \subsetneq \mathrm{C}^\infty(I)$; see Exercise 11.3.21 for an explicit example.

B.2 Bases and Dimension

Definition B.5. Let $\mathbf{v}_1, \ldots, \mathbf{v}_k$ belong to a vector space V. A sum of the form

$$c_1\mathbf{v}_1 + c_2\mathbf{v}_2 + \cdots + c_k\mathbf{v}_k = \sum_{i=1}^{k} c_i\mathbf{v}_i, \tag{B.2}$$

† We use one-sided derivatives at any endpoint that belongs to the interval.

where the coefficients $c_1, c_2, \ldots, c_k$ are any scalars, is known as a *linear combination* of the elements $\mathbf{v}_1, \ldots, \mathbf{v}_k$. Their *span* is the subspace $W = \operatorname{span}\{\mathbf{v}_1, \ldots, \mathbf{v}_k\} \subset V$ consisting of all possible linear combinations.

Definition B.6. The elements $\mathbf{v}_1, \ldots, \mathbf{v}_k \in V$ are called *linearly dependent* if there exist scalars $c_1, \ldots, c_k$, *not all zero*, such that

$$c_1 \mathbf{v}_1 + \cdots + c_k \mathbf{v}_k = \mathbf{0}. \tag{B.3}$$

Elements that are not linearly dependent are called *linearly independent*.

In particular, a collection of functions $f_1(x), \ldots, f_n(x)$ is linearly dependent if and only if there exist constants $c_1, \ldots, c_n$, *not all zero*, such that the linear combination

$$c_1 f_1(x) + \cdots + c_n f_n(x) \equiv 0 \tag{B.4}$$

is identically zero. Conversely, if the only choice of constants for which (B.4) holds is $c_1 = \cdots = c_n = 0$, then the functions are linearly independent.

Definition B.7. A *basis* of a vector space V is a finite collection of elements $\mathbf{v}_1, \ldots, \mathbf{v}_n \in V$ that (a) spans V, and (b) is linearly independent.

The simplest example is the *standard basis* of $\mathbb{R}^n$, consisting of the n vectors

$$\mathbf{e}_1 = \begin{pmatrix} 1 \\ 0 \\ 0 \\ \vdots \\ 0 \\ 0 \end{pmatrix}, \qquad \mathbf{e}_2 = \begin{pmatrix} 0 \\ 1 \\ 0 \\ \vdots \\ 0 \\ 0 \end{pmatrix}, \qquad \ldots, \qquad \mathbf{e}_n = \begin{pmatrix} 0 \\ 0 \\ 0 \\ \vdots \\ 0 \\ 1 \end{pmatrix}, \tag{B.5}$$

so that $\mathbf{e}_i$ is the vector with 1 in the ith slot and 0's elsewhere. However, there are many other bases of $\mathbb{R}^n$; indeed, any n linearly independent vectors $\mathbf{v}_1, \ldots, \mathbf{v}_n \in \mathbb{R}^n$ form a basis.

Lemma B.8. *The elements $\mathbf{v}_1, \ldots, \mathbf{v}_n$ form a basis of V if and only if every $\mathbf{v} \in V$ can be written* uniquely *as a linear combination of the basis elements:*

$$\mathbf{v} = c_1 \mathbf{v}_1 + \cdots + c_n \mathbf{v}_n = \sum_{i=1}^{n} c_i \mathbf{v}_i. \tag{B.6}$$

The coefficients $(c_1, \ldots, c_n)$ are called the coordinates *of the vector $\mathbf{v}$ with respect to the given basis.*

Theorem B.9. *Suppose the vector space V has a basis $\mathbf{v}_1, \ldots, \mathbf{v}_n$. Then every other basis of V has the same number of elements in it. This number is called the* dimension *of V, and written $\dim V = n$.*

On the other hand, if the vector space contains infinitely many linearly independent elements, then it does not have a basis in the sense of Definition B.7, and is thus *infinite-dimensional*. All of the function spaces and subspaces listed above are infinite-dimensional vector spaces. An example of a finite-dimensional function space is the space $\mathcal{P}^{(n)} \subset \mathcal{F}(\mathbb{R})$ consisting of all polynomials $p(x) = a_0 + a_1 x + \cdots + a_n x^n$ of degree $\leq n$. The monomials $1, x, x^2, \ldots, x^n$ form a basis, and hence $\mathcal{P}^{(n)}$ has dimension $n+1$. (On the other hand, the vector space containing *all* polynomials is infinite-dimensional.)

B.3 Inner Products and Norms

The dot product on Euclidean space $\mathbb{R}^n$ plays an essential role in geometry, analysis, and mechanics. Its basic properties inspire the general definition of an inner product on a vector space.

Definition B.10. An *inner product* on the real vector space V is a pairing that takes two elements $\mathbf{v}, \mathbf{w} \in V$ and produces a real number $\langle \mathbf{v}, \mathbf{w} \rangle \in \mathbb{R}$, subject to the following three axioms for all $\mathbf{u}, \mathbf{v}, \mathbf{w} \in V$, and scalars $c, d \in \mathbb{R}$.

(*i*) *Bilinearity*:

$$\begin{aligned} \langle c\,\mathbf{u} + d\,\mathbf{v}, \mathbf{w} \rangle &= c \langle \mathbf{u}, \mathbf{w} \rangle + d \langle \mathbf{v}, \mathbf{w} \rangle, \\ \langle \mathbf{u}, c\,\mathbf{v} + d\,\mathbf{w} \rangle &= c \langle \mathbf{u}, \mathbf{v} \rangle + d \langle \mathbf{u}, \mathbf{w} \rangle. \end{aligned} \tag{B.7}$$

(*ii*) *Symmetry*:

$$\langle \mathbf{v}, \mathbf{w} \rangle = \langle \mathbf{w}, \mathbf{v} \rangle. \tag{B.8}$$

(*iii*) *Positivity*:

$$\langle \mathbf{v}, \mathbf{v} \rangle > 0 \qquad \text{whenever} \qquad \mathbf{v} \neq \mathbf{0}, \qquad \text{while} \qquad \langle \mathbf{0}, \mathbf{0} \rangle = 0. \tag{B.9}$$

Given an inner product, the associated *norm* of an element $\mathbf{v} \in V$ is defined as the positive square root of its inner product with itself:

$$\| \mathbf{v} \| = \sqrt{\langle \mathbf{v}, \mathbf{v} \rangle}\,. \tag{B.10}$$

Bilinearity of the inner product implies that

$$\| c\,\mathbf{v} \| = | c | \, \| v \|$$

for any scalar c. The positivity axiom implies that $\| \mathbf{v} \| \geq 0$ is real and nonnegative, and equals 0 if and only if $\mathbf{v} = \mathbf{0}$ is the zero element. A vector space norm induces a notion of *distance* between elements $\mathbf{v}, \mathbf{w} \in V$, with $\operatorname{dist}(\mathbf{v}, \mathbf{w}) = \| \mathbf{v} - \mathbf{w} \|$. In particular, $\operatorname{dist}(\mathbf{v}, \mathbf{w}) = 0$ if and only if $\mathbf{v} = \mathbf{w}$.

Example B.11. The most familiar example of an inner product is the *dot product*†

$$\langle \mathbf{v}, \mathbf{w} \rangle = \mathbf{v} \cdot \mathbf{w} = \mathbf{v}^T \mathbf{w} = v_1 w_1 + v_2 w_2 + \cdots + v_n w_n \tag{B.11}$$

on the Euclidean space $\mathbb{R}^n$. The associated *Euclidean norm*

$$\| \mathbf{v} \| \;=\; \sqrt{\mathbf{v} \cdot \mathbf{v}} \;=\; \sqrt{v_1^2 + v_2^2 + \cdots + v_n^2} \tag{B.12}$$

conforms to our usual notion of distance between points in Euclidean space.

To find the most general inner product on $\mathbb{R}^n$, we need to introduce the important class of positive definite matrices.

Definition B.12. An $n \times n$ matrix C is called *positive definite* if it satisfies the positivity condition

$$\mathbf{v}^T C\, \mathbf{v} > 0 \qquad \text{for all} \qquad \mathbf{0} \neq \mathbf{v} \in \mathbb{R}^n. \tag{B.13}$$

We will sometimes write $C > 0$ to mean that C is a positive definite matrix.

† The elements $\mathbf{v} \in \mathbb{R}^n$ are to be regarded as column vectors, while the transpose, written $\mathbf{v}^T$, is the corresponding row vector.

Warning: The condition $C > 0$ does *not* mean that all the entries of C are positive. For example, $\begin{pmatrix} 3 & -1 \\ -1 & 1 \end{pmatrix}$ is positive definite, whereas $\begin{pmatrix} 1 & 2 \\ 2 & 1 \end{pmatrix}$ is not.

Many authors, including [**89**], require that a positive definite matrix also be symmetric. We will *not* impose this condition here a priori. However, most of the positive definite matrices we will encounter in applications will be symmetric (or, more generally, self-adjoint — as in Example 9.15). For a symmetric matrix, the most useful test for positive definiteness is to perform Gaussian Elimination on the matrix C, which is positive definite if and only if no row interchanges are needed, and all the pivots are positive, [**89**].

Proposition B.13. *Every inner product on* $\mathbb{R}^n$ *is given by*

$$\langle \mathbf{v} , \mathbf{w} \rangle = \mathbf{v}^T C\, \mathbf{w} \qquad \text{for} \qquad \mathbf{v}, \mathbf{w} \in \mathbb{R}^n, \tag{B.14}$$

where $C > 0$ *is a symmetric positive definite matrix.*

The next example is of particular significance in Fourier analysis and partial differential equations.

Example B.14. Let $[a, b] \subset \mathbb{R}$ be a bounded closed interval. The integral

$$\langle f , g \rangle = \int_a^b f(x)\, g(x)\, dx \tag{B.15}$$

defines an inner product on the space $\mathrm{C}^0[a, b]$ of continuous functions. The associated norm

$$\| f \| = \sqrt{\int_a^b f(x)^2\, dx} \tag{B.16}$$

is known as the L^2 *norm* of the function f over the interval $[a, b]$. The positivity of the norm: $\| f \| > 0$ for $f \neq 0$, follows from the fact that the only continuous nonnegative function $g(x) \geq 0$ that satisfies $\displaystyle\int_a^b g(x)\, dx = 0$ is the zero function $g(x) \equiv 0$. Extending this construction to spaces containing discontinuous functions is trickier, since there are discontinuous functions that are not identically zero, but nevertheless have zero norm integral. An example is a function that is zero except at a single point. Further discussion can be found in Section 3.5.

The two most important inequalities in mathematical analysis apply to any inner product space.

Theorem B.15. *Every inner product satisfies the* Cauchy–Schwarz *and* triangle inequalities

$$|\langle \mathbf{v} , \mathbf{w} \rangle| \leq \| \mathbf{v} \|\, \| \mathbf{w} \|, \qquad \| \mathbf{v} + \mathbf{w} \| \leq \| \mathbf{v} \| + \| \mathbf{w} \|, \qquad \text{for all} \qquad \mathbf{v}, \mathbf{w} \in V. \tag{B.17}$$

Equality holds if and only if $\mathbf{v}$ *and* $\mathbf{w}$ *are parallel, i.e., scalar multiples of each other.*

Proof: We begin with the Cauchy–Schwarz inequality: $|\langle \mathbf{v} , \mathbf{w} \rangle| \leq \| \mathbf{v} \|\, \| \mathbf{w} \|$. The case $\mathbf{w} = \mathbf{0}$ is trivial, and so we assume $\mathbf{w} \neq \mathbf{0}$. Let $t \in \mathbb{R}$ be an arbitrary scalar. Using the three inner product axioms, we have

$$0 \leq \| \mathbf{v} + t\,\mathbf{w} \|^2 = \langle \mathbf{v} + t\,\mathbf{w} , \mathbf{v} + t\,\mathbf{w} \rangle = \| \mathbf{v} \|^2 + 2\,t\, \langle \mathbf{v} , \mathbf{w} \rangle + t^2\, \| \mathbf{w} \|^2, \tag{B.18}$$

with equality holding if and only if $\mathbf{v} = -t\,\mathbf{w}$, which requires $\mathbf{v}$ and $\mathbf{w}$ to be parallel vectors. We fix $\mathbf{v}$ and $\mathbf{w}$, and consider the right-hand side of (B.18) as a quadratic function of t. Its minimum value occurs when $t = \|\mathbf{w}\|^{-2}\langle \mathbf{v}, \mathbf{w}\rangle$. Substituting this value into (B.18), we obtain

$$0 \le \|\mathbf{v}\|^2 - 2\,\frac{\langle \mathbf{v}, \mathbf{w}\rangle^2}{\|\mathbf{w}\|^2} + \frac{\langle \mathbf{v}, \mathbf{w}\rangle^2}{\|\mathbf{w}\|^2} = \|\mathbf{v}\|^2 - \frac{\langle \mathbf{v}, \mathbf{w}\rangle^2}{\|\mathbf{w}\|^2},$$

and hence $\langle \mathbf{v}, \mathbf{w}\rangle^2 \le \|\mathbf{v}\|^2\,\|\mathbf{w}\|^2$, which, upon taking the square root, establishes the Cauchy–Schwarz inequality. Again, as noted above, equality holds if and only if $\mathbf{v}$ and $\mathbf{w}$ are parallel.

To establish the triangle inequality, we compute

$$\begin{aligned}\|\mathbf{v}+\mathbf{w}\|^2 = \langle \mathbf{v}+\mathbf{w}, \mathbf{v}+\mathbf{w}\rangle &= \|\mathbf{v}\|^2 + 2\,\langle \mathbf{v}, \mathbf{w}\rangle + \|\mathbf{w}\|^2 \\ &\le \|\mathbf{v}\|^2 + 2\,\|\mathbf{v}\|\,\|\mathbf{w}\| + \|\mathbf{w}\|^2 = \big(\|\mathbf{v}\| + \|\mathbf{w}\|\big)^2,\end{aligned}$$

where the middle inequality follows from the Cauchy–Schwarz inequality (which clearly also holds if the absolute value is removed.) Taking square roots of both sides completes the proof. *Q.E.D.*

We will also have occasion to use inner products on complex vector spaces. To ensure that the associated norm remains positive, the real definition must be modified. The complex conjugate of a complex scalar $c = a + \mathrm{i}\,b$, with $a, b \in \mathbb{R}$, will be indicated by an overbar: $\overline{c} = a - \mathrm{i}\,b$. When dealing with a complex inner product space, one must pay careful attention to complex conjugation.

Definition B.16. An *inner product* on the complex vector space V is a pairing that takes two vectors $\mathbf{v}, \mathbf{w} \in V$ and produces a complex number $\langle \mathbf{v}, \mathbf{w}\rangle \in \mathbb{C}$, subject to the following requirements, for $\mathbf{u}, \mathbf{v}, \mathbf{w} \in V$, and $c, d \in \mathbb{C}$:

(*i*) *Sesquilinearity*:

$$\begin{aligned}\langle c\,\mathbf{u} + d\,\mathbf{v}, \mathbf{w}\rangle &= c\,\langle \mathbf{u}, \mathbf{w}\rangle + d\,\langle \mathbf{v}, \mathbf{w}\rangle, \\ \langle \mathbf{u}, c\,\mathbf{v} + d\,\mathbf{w}\rangle &= \overline{c}\,\langle \mathbf{u}, \mathbf{v}\rangle + \overline{d}\,\langle \mathbf{u}, \mathbf{w}\rangle.\end{aligned} \tag{B.19}$$

(*ii*) *Conjugate Symmetry*:

$$\langle \mathbf{v}, \mathbf{w}\rangle = \overline{\langle \mathbf{w}, \mathbf{v}\rangle}. \tag{B.20}$$

(*iii*) *Positivity*:

$$\|\mathbf{v}\|^2 = \langle \mathbf{v}, \mathbf{v}\rangle \ge 0, \qquad \text{and} \qquad \langle \mathbf{v}, \mathbf{v}\rangle = 0 \quad \text{if and only if} \quad \mathbf{v} = \mathbf{0}. \tag{B.21}$$

Example B.17. The simplest example is the *Hermitian dot product*

$$\mathbf{z}\cdot\mathbf{w} = \mathbf{z}^T\,\overline{\mathbf{w}} = z_1\overline{w}_1 + z_2\overline{w}_2 + \cdots + z_n\overline{w}_n, \quad \text{for} \quad \mathbf{z} = \begin{pmatrix} z_1 \\ z_2 \\ \vdots \\ z_n \end{pmatrix}, \quad \mathbf{w} = \begin{pmatrix} w_1 \\ w_2 \\ \vdots \\ w_n \end{pmatrix}, \tag{B.22}$$

between complex vectors $\mathbf{v}, \mathbf{w} \in \mathbb{C}^n$.

Example B.18. Let $\mathrm{C}^0[-\pi, \pi]$ denote the complex vector space consisting of all complex-valued continuous functions $f(x) = u(x) + \mathrm{i}\,v(x)$ depending on the *real* variable $-\pi \le x \le \pi$. The L^2 *Hermitian inner product* on $\mathrm{C}^0[-\pi, \pi]$ is defined as

$$\langle f, g\rangle = \int_{-\pi}^{\pi} f(x)\,\overline{g(x)}\,dx\,, \tag{B.23}$$

i.e., the integral of f times the complex conjugate of g, with corresponding norm

$$\| f \| = \sqrt{\int_{-\pi}^{\pi} | f(x) |^2 \, dx} = \sqrt{\int_{-\pi}^{\pi} \left[u(x)^2 + v(x)^2 \right] dx} . \tag{B.24}$$

Inner products on complex vector spaces also satisfy the Cauchy–Schwarz and triangle inequalities (B.17). The proof is left as an exercise for the reader; see [**89**; Exercise 3.6.46].

B.4 Orthogonality

Definition B.19. Two elements $\mathbf{v}, \mathbf{w} \in V$ of an inner product space V are called *orthogonal* if their inner product vanishes: $\langle \mathbf{v} , \mathbf{w} \rangle = 0$.

For ordinary Euclidean space equipped with the dot product, two vectors are orthogonal if and only if they are perpendicular, i.e., meet at a right angle.

Definition B.20. A basis $\mathbf{u}_1, \ldots, \mathbf{u}_n$ of an inner product space V is called *orthogonal* if $\langle \mathbf{u}_i , \mathbf{u}_j \rangle = 0$ for all $i \neq j$. The basis is called *orthonormal* if, in addition, each vector has unit length: $\| \mathbf{u}_i \| = 1$, for all $i = 1, \ldots, n$.

For example, the standard basis vectors (B.5) form an orthonormal basis of $\mathbb{R}^n$ with respect to the dot product, but they are not orthonormal for any other inner product thereon.

Theorem B.21. *If $\mathbf{v}_1, \ldots, \mathbf{v}_n$ form an orthogonal basis, then the corresponding coordinates of a vector*

$$\mathbf{v} = a_1 \mathbf{v}_1 + \cdots + a_n \mathbf{v}_n \qquad \text{are given by} \qquad a_i = \frac{\langle \mathbf{v} , \mathbf{v}_i \rangle}{\| \mathbf{v}_i \|^2} . \tag{B.25}$$

Moreover, the vector's norm can be computed using the formula

$$\| \mathbf{v} \|^2 = \sum_{i=1}^{n} a_i^2 \, \| \mathbf{v}_i \|^2 = \sum_{i=1}^{n} \left(\frac{\langle \mathbf{v} , \mathbf{v}_i \rangle}{\| \mathbf{v}_i \|} \right)^2 . \tag{B.26}$$

Proof: We compute the inner product of (B.25) with one of the basis vectors. By orthogonality,

$$\langle \mathbf{v} , \mathbf{v}_i \rangle = \left\langle \sum_{j=1}^{n} a_j \mathbf{v}_j , \mathbf{v}_i \right\rangle = \sum_{j=1}^{n} a_j \langle \mathbf{u}_j , \mathbf{u}_i \rangle = a_i \, \| \mathbf{v}_i \|^2 .$$

To prove formula (B.26), we similarly expand

$$\| \mathbf{v} \|^2 = \langle \mathbf{v} , \mathbf{v} \rangle = \sum_{i,j=1}^{n} a_i \, a_j \langle \mathbf{v}_i , \mathbf{v}_j \rangle = \sum_{i=1}^{n} a_i^2 \, \| \mathbf{v}_i \|^2 . \qquad Q.E.D.$$

In the case of an orthonormal basis, the formulas (B.25–26) simplify to

$$\mathbf{v} = c_1 \mathbf{u}_1 + \cdots + c_n \mathbf{u}_n, \quad \text{where} \quad c_i = \langle \mathbf{v} , \mathbf{u}_i \rangle, \qquad \| \mathbf{v} \| = c_1^2 + \cdots + c_n^2 . \tag{B.27}$$

Example B.22. A particularly important orthogonal basis is provided by the following vectors lying in $\mathbb{C}^n$:

$$\begin{aligned}\boldsymbol{\omega}_k &= \left(1, \zeta^k, \zeta^{2k}, \zeta^{3k}, \ldots, \zeta^{(n-1)k}\right)^T \\ &= \left(1, e^{2k\pi \mathrm{i}/n}, e^{4k\pi \mathrm{i}/n}, \ldots, e^{2(n-1)k\pi \mathrm{i}/n}\right)^T,\end{aligned} \qquad k = 0, \ldots, n-1, \tag{B.28}$$

where

$$\zeta = e^{2\pi \mathrm{i}/n}. \tag{B.29}$$

Orthogonality relies on the fact that its powers, $\zeta^k = e^{2k\pi \mathrm{i}/n}$, $k = 0, \ldots, n-1$, are the complex roots of the elementary polynomial

$$z^n - 1 = (z-1)(1 + z + z^2 + \cdots + z^{n-1}), \tag{B.30}$$

while

$$\overline{\zeta} = e^{-2\pi \mathrm{i}/n} = \zeta^{-1}.$$

Since when $0 < k \le n-1$, the complex number $\zeta^k \neq 1$ is a root of the polynomial (B.30), it must also be a root of the second factor. This implies that

$$1 + \zeta^k + \zeta^{2k} + \zeta^{3k} + \cdots + \zeta^{(n-1)k} = \begin{cases} n, & k \equiv 0 \mod n, \\ 0, & k \not\equiv 0 \mod n, \end{cases}$$

where the former case $k \equiv 0 \mod n$ follows by direct substitution of $\zeta^k = 1$. Thus, the Hermitian inner products of the vectors (B.28) equal

$$\langle \boldsymbol{\omega}_k, \boldsymbol{\omega}_l \rangle = \sum_{j=0}^{n-1} \zeta^{jk}\, \overline{\zeta}^{jl} = \sum_{j=0}^{n-1} \zeta^{j(k-l)} = \begin{cases} n, & k = l, \\ 0, & k \neq l, \end{cases} \tag{B.31}$$

provided $0 \le k, l \le n-1$, thereby establishing orthogonality. These vectors are the discrete analogues of the orthogonal complex exponential functions that are used to construct complex Fourier series. They are the basis of the discrete Fourier transform, [**89**; §5.7], and their orthogonality is the key to modern signal processing.

B.5 Eigenvalues and Eigenvectors

The eigenvalues and eigenvectors of a matrix first appear when solving linear systems of ordinary differential equations. But their essential importance extends across all of mathematics and its manifold applications. Extensions of the eigenvalue method to linear operators on function spaces are critical to the analysis of partial differential equations.

Definition B.23. Let A be an $n \times n$ matrix. A scalar λ is called an *eigenvalue* of A if there is a *nonzero* vector $\mathbf{v} \neq \mathbf{0}$, called an associated *eigenvector*, such that

$$A\mathbf{v} = \lambda \mathbf{v}. \tag{B.32}$$

In particular, a matrix has $\lambda = 0$ as an eigenvalue if and only if it has a *null eigenvector* $\mathbf{v} \neq \mathbf{0}$, satisfying $A\mathbf{v} = \mathbf{0}$, and hence is a singular (non-invertible) matrix, with vanishing determinant: $\det A = 0$. An eigenvalue is called *simple* if it admits only one linearly independent eigenvalue; more generally, the *multiplicity* of an eigenvalue is defined as the dimension of the *eigenspace* consisting of all solutions to the eigenequation (B.32), including $\mathbf{0}$. Thus, a simple eigenvalue has multiplicity 1.

Even if A is a real matrix, we must allow the possibility of complex eigenvectors. Matrices with a "complete" set of eigenvectors are the most common, and also the easiest to deal with.

Definition B.24. An $n \times n$ real or complex matrix A is called *complete* if there exists a basis of $\mathbb{C}^n$ consisting of its (complex) eigenvectors.

It is not hard to show that eigenvectors corresponding to different eigenvalues are necessarily linearly independent. This means that matrices with all distinct (and hence simple) eigenvalues are necessarily complete:

Proposition B.25. *Any $n \times n$ matrix with n distinct eigenvalues is complete.*

Unfortunately, not all matrices with repeated eigenvalues are complete. For instance, $\begin{pmatrix} 1 & 0 \\ 0 & 1 \end{pmatrix}$ is complete, since, for instance, $\begin{pmatrix} 1 \\ 0 \end{pmatrix}$ and $\begin{pmatrix} 0 \\ 1 \end{pmatrix}$ form an eigenvector basis of $\mathbb{C}^2$, whereas $\begin{pmatrix} 1 & 1 \\ 0 & 1 \end{pmatrix}$ is not, since it has only one independent eigenvector, namely $\begin{pmatrix} 1 \\ 0 \end{pmatrix}$. Incomplete matrices are much more challenging to deal with, both theoretically and numerically. Fortunately, we can safely ignore the incomplete cases in this text.

The most common way for orthogonal bases to arise is as eigenvector bases of symmetric matrices. (Orthogonality is with respect to the standard dot product on $\mathbb{R}^n$.) The extension of this result to "self-adjoint" operators on function space forms the foundation of Fourier analysis and its generalizations.

Theorem B.26. *Let $A = A^T$ be a real symmetric $n \times n$ matrix. Then*

(a) *All the eigenvalues of A are real.*

(b) *Eigenvectors corresponding to distinct eigenvalues are orthogonal.*

(c) *There is an orthonormal basis of $\mathbb{R}^n$ consisting of n eigenvectors of A.*

Let us demonstrate orthogonality, leaving the remaining steps in the proof to [**89**; Theorem 8.20]. If

$$A\mathbf{v} = \lambda\mathbf{v}, \qquad A\mathbf{w} = \mu\,\mathbf{w},$$

where $\lambda \neq \mu$ are distinct real eigenvalues, then, by symmetry of A,

$$\lambda\,\mathbf{v}\cdot\mathbf{w} = (A\mathbf{v})\cdot\mathbf{w} = (A\mathbf{v})^T\mathbf{w} = \mathbf{v}^T A\mathbf{w} = \mathbf{v}\cdot(A\mathbf{w}) = \mathbf{v}\cdot(\mu\,\mathbf{w}) = \mu\,\mathbf{v}\cdot\mathbf{w},$$

and hence

$$(\lambda - \mu)\,\mathbf{v}\cdot\mathbf{w} = 0.$$

Since $\lambda \neq \mu$, this implies that the eigenvectors $\mathbf{v}, \mathbf{w}$ are necessarily orthogonal.

B.6 Linear Iteration

For numerical applications, we will require some basic results on iteration of linear systems. Consider first a *homogeneous linear iterative system* of the form

$$\mathbf{u}^{(k+1)} = A\,\mathbf{u}^{(k)}, \qquad \mathbf{u}^{(0)} = \mathbf{u}_0, \tag{B.33}$$

in which A is an $n \times n$ matrix and $\mathbf{u}_0 \in \mathbb{R}^n$ or $\mathbb{C}^n$. The solution to such a system is evidently obtained by repeatedly multiplying the initial vector $\mathbf{u}_0$ by the matrix A, and so

$$\mathbf{u}^{(k)} = A^k \mathbf{u}_0. \tag{B.34}$$

Definition B.27. A matrix A is called *convergent* if every solution to the homogeneous linear iterative system (B.33) tends to zero in the limit: $\mathbf{u}^{(k)} \to \mathbf{0}$ as $k \to \infty$. Equivalently, A is convergent if and only if its powers converge to the zero matrix: $A^k \to \mathrm{O}$ as $k \to \infty$.

The solution formula (B.34), while elementary, is not particularly enlightening. An alternative approach is to recognize that if λ_j is an eigenvalue of A and $\mathbf{v}_j$ a corresponding eigenvector, then

$$\mathbf{u}_j^{(k)} = \lambda_j^k \mathbf{v}_j \tag{B.35}$$

is a solution, since

$$A\mathbf{u}_j^{(k)} = \lambda_j^k A\mathbf{v}_j = \lambda_j^{k+1} \mathbf{v}_j = \mathbf{u}_j^{(k+1)}.$$

Moreover, linear combinations of such *eigensolutions* are also solutions. In particular, if A is complete, then we can write down the general solution to (B.33) as a linear combination of the independent eigensolutions:

$$\mathbf{u}^{(k)} = c_1 \lambda_1^k \mathbf{v}_1 + c_2 \lambda_2^k \mathbf{v}_2 + \cdots + c_n \lambda_n^k \mathbf{v}_n, \tag{B.36}$$

where $\{\mathbf{v}_1, \ldots, \mathbf{v}_n\}$ is the eigenvector basis. The coefficients $c_1, \ldots, c_n$ are uniquely determined by the initial conditions,

$$\mathbf{u}^{(0)} = c_1 \mathbf{v}_1 + c_2 \mathbf{v}_2 + \cdots + c_n \mathbf{v}_n = \mathbf{u}_0,$$

which relies on the fact that the eigenvectors $\mathbf{v}_1, \ldots, \mathbf{v}_n$ form a basis. Now, A is convergent if and only if all solutions $\mathbf{u}^{(k)} \to \mathbf{0}$. The individual eigensolution (B.35) goes to zero if and only if its associated eigenvalue is strictly less than 1 in modulus: $|\lambda_j| < 1$. This proves the following result for complete matrices. The proof in the incomplete case relies on the Jordan canonical form, [**89**; Chapter 10].

Theorem B.28. *The matrix A is convergent if and only if all its eigenvalues satisfy $|\lambda| < 1$.*

Definition B.29. The *spectral radius* of a matrix A is defined as the maximal modulus of all of its real and complex eigenvalues: $\rho(A) = \max\{ |\lambda_1|, \ldots, |\lambda_k| \}$.

Corollary B.30. *The matrix A is convergent if and only if $\rho(A) < 1$.*

Indeed, the spectral radius essentially governs the rate of convergence of the iterative system — the closer it is to 0, the faster the convergence rate.

Next, consider the *inhomogeneous linear iterative system*

$$\mathbf{v}^{(k+1)} = A\mathbf{v}^{(k)} + \mathbf{b}, \qquad \mathbf{v}^{(0)} = \mathbf{v}_0, \tag{B.37}$$

where $\mathbf{b}$ a fixed vector. A *fixed point* is a vector $\mathbf{v}^\star$ that satisfies

$$\mathbf{v}^\star = A\mathbf{v}^\star + \mathbf{b}, \qquad \text{or, equivalently,} \qquad (\mathrm{I} - A)\mathbf{v}^\star = \mathbf{b}, \tag{B.38}$$

where I is the *identity matrix* of the same size as A. Thus, if 1 is not an eigenvalue of A (which cannot happen when A is convergent), then $\mathrm{I} - A$ is nonsingular, and so the iterative system has a unique fixed point.

Theorem B.31. *Assume that* 1 *is not an eigenvalue of* A. *Then all solutions to* (B.37) *converge to the fixed point,* $\mathbf{v}^{(k)} \to \mathbf{v}^\star$ *as* $k \to \infty$ *if and only if* A *is a convergent matrix.*

Proof: Let $\mathbf{u}^{(k)} = \mathbf{v}^{(k)} - \mathbf{v}^\star$, so that $\mathbf{v}^{(k)} \to \mathbf{v}^\star$ if and only if $\mathbf{u}^{(k)} \to \mathbf{0}$. Now,

$$\mathbf{u}^{(k+1)} = \mathbf{v}^{(k+1)} - \mathbf{v}^\star = (A\mathbf{v}^{(k)} + \mathbf{b}) - (A\mathbf{v}^\star + \mathbf{b}) = A(\mathbf{v}^{(k)} - \mathbf{v}^\star) = A\mathbf{u}^{(k)},$$

and hence $\mathbf{u}^{(k)}$ solves the homogeneous version (B.33). Thus, the result is an immediate consequence of Definition B.27. *Q.E.D.*

B.7 Linear Functions and Systems

The most basic structural features of linear differential equations, both ordinary and partial, linear boundary value problems, etc., are founded on the concept of a linear function between vector spaces.

Definition B.32. Let U and V be real vector spaces. A function $L\colon U \to V$ is called *linear* if it obeys two basic rules:

$$L[\mathbf{u} + \mathbf{v}] = L[\mathbf{u}] + L[\mathbf{v}], \qquad\qquad L[c\,\mathbf{u}] = c\,L[\mathbf{u}], \tag{B.39}$$

for all $\mathbf{u}, \mathbf{v} \in U$ and all scalars c.

We will refer to U as the *domain space* of the function L, and V as the *target space*. The latter is to emphasize the fact that the *range* of L, namely

$$\operatorname{rng} L = \{\, \mathbf{v} \in V \mid \mathbf{v} = L[\mathbf{u}] \text{ for some } \mathbf{u} \in U \,\}, \tag{B.40}$$

may very well be a proper subspace of the target space V.

Theorem B.33. *Every linear function* $L\colon \mathbb{R}^n \to \mathbb{R}^m$ *is given by matrix multiplication,* $L[\mathbf{v}] = A\mathbf{v}$, *where* A *is an* $m \times n$ *matrix.*

Proving that matrix multiplication satisfies the linearity conditions (B.39) is easy. The converse is established by seeing what the linear function does to the basis vectors of $\mathbb{R}^n$; see [**89**; Theorem 7.5].

Corollary B.34. *Every linear function* $L\colon \mathbb{R}^n \to \mathbb{R}$ *is given by taking the dot product with a fixed vector* $\mathbf{a} \in \mathbb{R}^n$:

$$L[\mathbf{v}] = \mathbf{a} \cdot \mathbf{v}. \tag{B.41}$$

When U is a function space, a linear function is also referred to as a *linear operator* in order to avoid confusion with the elements of U. If the target space $V = \mathbb{R}$, then the term *linear functional* is also often used for $L\colon U \to \mathbb{R}$.

Here are some representative examples that arise in applications.

Example B.35. (*a*) Evaluation of a function at a point, namely $L[f] = f(x_0)$, defines a linear operator $L\colon \mathrm{C}^0[a,b] \to \mathbb{R}$.

(*b*) Integration,

$$I[f] = \int_a^b f(x)\,dx, \tag{B.42}$$

also defines a linear functional $I\colon \mathrm{C}^0[a,b] \to \mathbb{R}$.

(c) The operation $M_a[f(x)] = a(x)\,f(x)$ of multiplication by a continuous function a defines a linear operator $M_a\colon \mathrm{C}^0[a,b] \to \mathrm{C}^0[a,b]$.

(d) Differentiation of functions, $D[f] = f'$, serves to define a linear operator $D\colon \mathrm{C}^1[a,b] \to \mathrm{C}^0[a,b]$.

(e) A general *linear ordinary differential operator* of order n,

$$L = a_n(x)\,D^n + a_{n-1}(x)\,D^{n-1} + \cdots + a_1(x)\,D + a_0(x), \tag{B.43}$$

is obtained by summing such operators. If the coefficient functions $a_0(x), \ldots, a_n(x)$ are continuous, then

$$L[u] = a_n(x)\,\frac{d^n u}{dx^n} + a_{n-1}(x)\,\frac{d^{n-1} u}{dx^{n-1}} + \cdots + a_1(x)\,\frac{du}{dx} + a_0(x)u \tag{B.44}$$

defines a linear operator from $\mathrm{C}^n[a,b]$ to $\mathrm{C}^0[a,b]$.

Linear partial differential equations are based on linear partial differential operators, which are discussed in Chapter 1. They are particular examples of the general concept of a linear system.

Definition B.36. A *linear system* is an equation of the form

$$L[\mathbf{u}] = \mathbf{f}, \tag{B.45}$$

in which $L\colon U \to V$ is a linear function, $\mathbf{f} \in V$, while the desired solution $\mathbf{u} \in U$. The system is *homogeneous* if $\mathbf{f} = \mathbf{0}$; otherwise, it is called *inhomogeneous*.

Note that, by the definition (B.40) of the range of L, the linear system (B.45) will have a solution if and only if $\mathbf{f} \in \operatorname{rng} L$. In particular, a homogeneous linear system always has a solution, namely $\mathbf{u} = \mathbf{0}$. However, it may possibly admit other, nonzero, solutions.

Theorem B.37. *If $\mathbf{z}_1, \ldots, \mathbf{z}_k$ are all solutions to the same homogeneous linear system*

$$L[\mathbf{z}] = \mathbf{0}, \tag{B.46}$$

then any linear combination $c_1\,\mathbf{z}_1 + \cdots + c_k\,\mathbf{z}_k$, for any scalars $c_1, \ldots, c_k$, is also a solution.

In other words, the set of solutions to a homogeneous linear system (B.46) forms a subspace of the domain space U, known as the *kernel* of the linear function L:

$$\ker L = \{\, \mathbf{z} \in U \mid L[\mathbf{z}] = \mathbf{0} \,\}. \tag{B.47}$$

Theorem B.38. *If the inhomogeneous linear system $L[\mathbf{u}] = \mathbf{f}$ has a particular solution $\mathbf{u}^\star$, which requires $\mathbf{f} \in \operatorname{rng} L$, then the general solution is $\mathbf{u} = \mathbf{u}^\star + \mathbf{z}$, where $\mathbf{z} \in \ker L$ is any solution to the corresponding homogeneous system $L[\mathbf{z}] = \mathbf{0}$.*

The *Superposition Principle* for inhomogeneous linear systems allows us to combine solutions corresponding to different right-hand sides.

Theorem B.39. *Suppose that for each $i = 1, \ldots, k$, we know a particular solution $\mathbf{u}_i^\star$ to the inhomogeneous linear system $L[\mathbf{u}] = \mathbf{f}_i$ for some $\mathbf{f}_i \in \operatorname{rng} L$. Then, given scalars $c_1, \ldots, c_k$, a particular solution to the combined inhomogeneous system*

$$L[\mathbf{u}] = c_1\,\mathbf{f}_1 + \cdots + c_k\,\mathbf{f}_k \tag{B.48}$$

is the corresponding linear combination

$$\mathbf{u}^\star = c_1\,\mathbf{u}_1^\star + \cdots + c_k\,\mathbf{u}_k^\star \tag{B.49}$$

of particular solutions. The general solution to the inhomogeneous system (B.48) *is*

$$\mathbf{u} = \mathbf{u}^\star + \mathbf{z} = c_1\,\mathbf{u}_1^\star + \cdots + c_k\,\mathbf{u}_k^\star + \mathbf{z}, \tag{B.50}$$

where $\mathbf{z} \in \ker L$ *is an arbitrary solution to the associated homogeneous system* $L[\,\mathbf{z}\,] = \mathbf{0}$.

References

[**1**] Abdulloev, K. O., Bogolubsky, I. L., and Makhankov, V. G., One more example of inelastic soliton interaction, *Phys. Lett. A* **56** (1976), 427–428.

[**2**] Ablowitz, M. J., and Clarkson, P. A., *Solitons, Nonlinear Evolution Equations and the Inverse Scattering Transform*, L.M.S. Lecture Notes in Math., vol. 149, Cambridge University Press, Cambridge, 1991.

[**3**] Abraham, R., Marsden, J. E., and Ratiu, T., *Manifolds, Tensor Analysis, and Applications*, Springer–Verlag, New York, 1988.

[**4**] Airy, G. B., On the intensity of light in the neighborhood of a caustic, *Trans. Cambridge Phil. Soc.* **6** (1838), 379–402.

[**5**] Aki, K., and Richards, P. G., *Quantitative Seismology*, W. H. Freeman, San Francisco, 1980.

[**6**] Ames, W. F., *Numerical Methods for Partial Differential Equations*, 3rd ed., Academic Press, New York, 1992.

[**7**] Antman, S. S., *Nonlinear Problems of Elasticity*, Appl. Math. Sci., vol. 107, Springer–Verlag, New York, 1995.

[**8**] Apostol, T. M., *Calculus*, Blaisdell Publishing Co., Waltham, Mass., 1967–1969.

[**9**] Apostol, T. M., *Introduction to Analytic Number Theory*, Springer–Verlag, New York, 1976.

[**10**] Atkinson, K., and Han, W., *Spherical Harmonics and Approximations on the Unit Sphere: An Introduction*, Lecture Notes in Math., vol. 2044, Springer, Berlin, 2012.

[**11**] Bank, S. B., and Kaufman, R. P., A note on Hölder's theorem concerning the gamma function, *Math. Ann.* **232** (1978), 115–120.

[**12**] Batchelor, G. K., *An Introduction to Fluid Dynamics*, Cambridge University Press, Cambridge, 1967.

[**13**] Bateman, H., Some recent researches on the motion of fluids, *Monthly Weather Rev.* **43** (1915), 63–170.

[**14**] Benjamin, T. B., Bona, J. L., and Mahony, J. J., Model equations for long waves in nonlinear dispersive systems, *Phil. Trans. Roy. Soc. London A* **272** (1972), 47–78.

[**15**] Berest, Y., and Winternitz, P., Huygens' principle and separation of variables, *Rev. Math. Phys.* **12** (2000), 159–180.

[**16**] Berry, M. V., Marzoli, I., and Schleich, W., Quantum carpets, carpets of light, *Physics World* **14**(6) (2001), 39–44.

[**17**] Birkhoff, G., *Hydrodynamics — A Study in Logic, Fact and Similitude*, 2nd ed., Princeton University Press, Princeton, 1960.

[**18**] Birkhoff, G., and Rota, G.–C., *Ordinary Differential Equations*, Blaisdell Publ. Co., Waltham, Mass., 1962.

[**19**] Black, F., and Scholes, M., The pricing of options and corporate liabilities, *J. Political Economy* **81** (1973), 637–654.

[**20**] Blanchard, P., Devaney, R. L., and Hall, G. R., *Differential Equations*, Brooks–Cole Publ. Co., Pacific Grove, Calif., 1998.

[**21**] Boussinesq, J., Théorie des ondes et des remous qui se propagent le long d'un canal rectangulaire horizontal, en communiquant au liquide contenu dans ce canal des vitesses sensiblement pareilles de la surface au fond, *J. Math. Pures Appl.* **17** (2) (1872), 55–108.

[**22**] Boussinesq, J., Essai sur la théorie des eaux courants, *Mém. Acad. Sci. Inst. Nat. France* **23** (1) (1877), 1–680.

[**23**] Boyce, W. E., and DiPrima, R. C., *Elementary Differential Equations and Boundary Value Problems*, 7th ed., John Wiley & Sons, Inc., New York, 2001.

[**24**] Bradie, B., *A Friendly Introduction to Numerical Analysis*, Prentice–Hall, Inc., Upper Saddle River, N.J., 2006.

[**25**] Bronstein, M., *Symbolic integration I: Transcendental Functions*, Springer–Verlag, New York, 1997.

[**26**] Burgers, J. M., A mathematical model illustrating the theory of turbulence, *Adv. Appl. Mech.* **1** (1948), 171–199.

[**27**] Cantwell, B. J., *Introduction to Symmetry Analysis*, Cambridge University Press, Cambridge, 2003.

[**28**] Carleson, L., On the convergence and growth of partial sums of Fourier series, *Acta Math.* **116** (1966), 135–157.

[**29**] Carmichael, R., *The Theory of Numbers*, Dover Publ., New York, 1959.

[**30**] Chen, G., and Olver, P. J., Dispersion of discontinuous periodic waves, *Proc. Roy. Soc. London* **469** (2012), 20120407.

[**31**] Coddington, E. A., and Levinson, N., *Theory of Ordinary Differential Equations*, McGraw–Hill, New York, 1955.

[**32**] Cole, J. D., On a quasilinear parabolic equation occurring in aerodynamics, *Q. Appl. Math.* **9** (1951), 225–236.

[**33**] Courant, R., Friedrichs, K. O., and Lewy, H., Über die partiellen Differenzengleichungen der mathematischen Physik, *Math. Ann.* **100** (1928), 32–74.

[**34**] Courant, R., and Hilbert, D., *Methods of Mathematical Physics*, vol. I, Interscience Publ., New York, 1953.

[**35**] Courant, R., and Hilbert, D., *Methods of Mathematical Physics*, vol. II, Interscience Publ., New York, 1953.

[**36**] Drazin, P. G., and Johnson, R. S., *Solitons: An Introduction*, Cambridge University Press, Cambridge, 1989.

[**37**] Dym, H., and McKean, H. P., *Fourier Series and Integrals*, Academic Press, New York, 1972.

[**38**] Evans, L. C., *Partial Differential Equations*, Grad. Studies Math. vol. 19, Amer. Math. Soc., Providence, R.I., 1998.

[**39**] Feller, W., *An Introduction to Probability Theory and Its Applications*, 3rd ed., J. Wiley & Sons, New York, 1968.

[**40**] Fermi, E., Pasta, J., and Ulam, S., Studies of nonlinear problems. I., preprint, Los Alamos Report LA 1940, 1955; in: *Nonlinear Wave Motion*, A. C. Newell, ed., Lectures in Applied Math., vol. 15, American Math. Soc., Providence, R.I., 1974, pp. 143–156.

[**41**] Forsyth, A. R., *The Theory of Differential Equations*, Cambridge University Press, Cambridge, 1890, 1900, 1902, 1906.

[**42**] Fourier, J., *The Analytical Theory of Heat*, Dover Publ., New York, 1955.

[**43**] Gander, M. J., and Kwok, F., Chladni figures and the Tacoma bridge: motivating PDE eigenvalue problems via vibrating plates, *SIAM Review* **54** (2012), 573–596.

[**44**] Garabedian, P., *Partial Differential Equations*, 2nd ed., Chelsea Publ. Co., New York, 1986.

[**45**] Gardner, C. S., Greene, J. M., Kruskal, M. D., and Miura, R. M., Method for solving the Korteweg–deVries equation, *Phys. Rev. Lett.* **19** (1967), 1095–1097.

[**46**] Gonzalez, R. C., and Woods, R. E., *Digital Image Processing*, 2nd ed., Prentice–Hall, Inc., Upper Saddle River, N.J., 2002.

[**47**] Gordon, C., Webb, D. L., and Wolpert, S., One cannot hear the shape of a drum, *Bull. Amer. Math. Soc.* **27** (1992), 134–138.

[**48**] Gradshteyn, I. S., and Ryzhik, I. W., *Table of Integrals, Series and Products*, Academic Press, New York, 1965.

[**49**] Gurtin, M. E., *An Introduction to Continuum Mechanics*, Academic Press, New York, 1981.

[**50**] Haberman, R., *Elementary Applied Partial Differential Equations*, 3rd ed., Prentice–Hall, Inc., Upper Saddle River, NJ, 1998.

[**51**] Hairer, E., Lubich, C., and Wanner, G., *Geometric Numerical Integration*, Springer–Verlag, New York, 2002.

[**52**] Hale, J. K., *Ordinary Differential Equations*, 2nd ed., R.E. Krieger Pub. Co., Huntington, N.Y., 1980.

[**53**] Henrici, P., *Applied and Computational Complex Analysis*, vol. 1, J. Wiley & Sons, New York, 1974.

[**54**] Hille, E., *Ordinary Differential Equations in the Complex Domain*, John Wiley & Sons, New York, 1976.

[**55**] Hobson, E. W., *The Theory of Functions of a Real Variable and the Theory of Fourier's Series*, Dover Publ., New York, 1957.

[**56**] Hopf, E., The partial differential equation $u_t + u u_x = \mu u$, *Commun. Pure Appl. Math.* **3** (1950), 201–230.

[**57**] Howison, S., *Practical Applied Mathematics: Modelling, Analysis, Approximation*, Cambridge University Press, Cambridge, 2005.

[**58**] Hydon, P. E., *Symmetry Methods for Differential Equations*, Cambridge Texts in Appl. Math., Cambridge University Press, Cambridge, 2000.

[**59**] Ince, E. L., *Ordinary Differential Equations*, Dover Publ., New York, 1956.

[**60**] Iserles, A., *A First Course in the Numerical Analysis of Differential Equations*, Cambridge University Press, Cambridge, 1996.

[**61**] Jost, J., *Partial Differential Equations*, Graduate Texts in Mathematics, vol. 214, Springer–Verlag, New York, 2007.

[**62**] Kamke, E., *Differentialgleichungen Lösungsmethoden und Lösungen*, vol. 1, Chelsea, New York, 1971.

[**63**] Keller, H. B., *Numerical Methods for Two-Point Boundary-Value Problems*, Blaisdell, Waltham, MA, 1968.

[**64**] Knobel, R., *An Introduction to the Mathematical Theory of Waves*, American Mathematical Society, Providence, RI, 2000.

[**65**] Korteweg, D. J., and de Vries, G., On the change of form of long waves advancing in a rectangular channel, and on a new type of long stationary waves, *Phil. Mag.* (5) **39** (1895), 422–443.

[**66**] Landau, L. D., and Lifshitz, E. M., *Quantum Mechanics* (*Non-relativistic Theory*), Course of Theoretical Physics, vol. 3, Pergamon Press, New York, 1977.

[**67**] Levine, I. N., *Quantum Chemistry*, 5th ed., Prentice–Hall, Inc., Upper Saddle River, N.J., 2000.

[68] Lighthill, M. J., *Introduction to Fourier Analysis and Generalised Functions*, Cambridge University Press, Cambridge, 1970.

[69] Lin, C. C., and Segel, L. A., *Mathematics Applied to Deterministic Problems in the Natural Sciences*, SIAM, Philadelphia, 1988.

[70] McOwen, R. C., *Partial Differential Equations: Methods and Applications*, Prentice–Hall, Inc., Upper Saddle River, N.J., 2002.

[71] Merton, R. C., Theory of rational option pricing, *Bell J. Econ. Management Sci.* **4** (1973), 141–183.

[72] Messiah, A., *Quantum Mechanics*, John Wiley & Sons, New York, 1976.

[73] Miller, W., Jr., *Symmetry and Separation of Variables*, Encyclopedia of Mathematics and Its Applications, vol. 4, Addison–Wesley Publ. Co., Reading, Mass., 1977.

[74] Milne–Thompson, L. M., *The Calculus of Finite Differences*, Macmillan and Co., Ltd., London, 1951.

[75] Misner, C. W., Thorne, K. S., and Wheeler, J. A., *Gravitation*, W. H. Freeman, San Francisco, 1973.

[76] Miura, R. M., Gardner, C. S., and Kruskal, M. D., Korteweg–deVries equation and generalizations. II. Existence of conservation laws and constants of the motion, *J. Math. Phys.* **9** (1968), 1204–1209.

[77] Moon, F. C., *Chaotic Vibrations*, John Wiley & Sons, New York, 1987.

[78] Moon, P., and Spencer, D. E., *Field Theory Handbook*, Springer-Verlag, New York, 1971.

[79] Morse, P. M., and Feshbach, H., *Methods of Theoretical Physics*, McGraw–Hill, New York, 1953.

[80] Morton, K. W., and Mayers, D. F., *Numerical Solution of Partial Differential Equations*, 2nd ed., Cambridge University Press, Cambridge, 2005.

[81] Murray, J. D., *Mathematical Biology*, 3rd ed., Springer-Verlag, New York, 2002–2003.

[82] Oberhettinger, F., *Tables of Fourier Transforms and Fourier Transforms of Distributions*, Springer-Verlag, New York, 1990.

[83] Øksendal, B., *Stochastic Differential Equations: An Introduction with Applications*, Springer–Verlag, New York, 1985.

[84] Okubo, A., *Diffusion and Ecological Problems: Mathematical Models*, Springer-Verlag, New York, 1980.

[85] Olver, F. W. J., *Asymptotics and Special Functions*, Academic Press, New York, 1974.

[86] Olver, F. W. J., Lozier, D. W., Boisvert, R. F., and Clark, C. W., eds., *NIST Handbook of Mathematical Functions*, Cambridge University Press, Cambridge, 2010.

[87] Olver, P. J., *Applications of Lie Groups to Differential Equations*, 2nd ed., Graduate Texts in Mathematics, vol. 107, Springer–Verlag, New York, 1993.

[88] Olver, P. J., Dispersive quantization, *Amer. Math. Monthly* **117** (2010), 599–610.

[89] Olver, P. J., and Shakiban, C., *Applied Linear Algebra*, Prentice–Hall, Inc., Upper Saddle River, N.J., 2005.

[90] Oskolkov, K. I., A class of I. M. Vinogradov's series and its applications in harmonic analysis, *in*: *Progress in Approximation Theory* , Springer Ser. Comput. Math., 19, Springer, New York, 1992, pp. 353–402.

[91] Pinchover, Y., and Rubinstein, J., *An Introduction to Partial Differential Equations*, Cambridge University Press, Cambridge, 2005.

[92] Pinsky, M. A., *Partial Differential Equations and Boundary–Value Problems with Applications*, 3rd ed., McGraw–Hill, New York, 1998.

[**93**] Polyanin, A. D., and Zaitsev, V. F., *Handbook of Exact Solutions for Ordinary Differential Equations*, 2nd ed., Chapman & Hall/CRC, Boca Raton, Fl., 2003.

[**94**] Press, W. H., Teukolsky, S. A., Vetterling, W. T., and Flannery, B. P., *Numerical Recipes: The Art of Scientific Computing*, 3rd ed., Cambridge University Press, Cambridge, 2007.

[**95**] Reed, M., and Simon, B., *Methods of Modern Mathematical Physics*, Academic Press, New York, 1972.

[**96**] Royden, H. L., and Fitzpatrick, P. M., *Real Analysis*, 4th ed., Pearson Education Inc., Boston, MA, 2010.

[**97**] Rudin, W., *Principles of Mathematical Analysis*, 3rd ed., McGraw–Hill, New York, 1976.

[**98**] Rudin, W., *Real and Complex Analysis*, 3rd ed., McGraw–Hill, New York, 1987.

[**99**] Salsa, S., *Partial Differential Equations in Action: From Modelling to Theory*, Springer–Verlag, New York, 2008.

[**100**] Sapiro, G., *Geometric Partial Differential Equations and Image Analysis*, Cambridge University Press, Cambridge, 2001.

[**101**] Schrödinger, E., *Collected Papers on Wave Mechanics*, Chelsea Publ. Co., New York, 1982.

[**102**] Schumaker, L. L., *Spline Functions: Basic Theory*, John Wiley & Sons, New York, 1981.

[**103**] Schwartz, L., *Théorie des distributions*, Hermann, Paris, 1957.

[**104**] Scott Russell, J., On waves, *in*: *Report of the 14th Meeting*, British Assoc. Adv. Sci., 1845, pp. 311–390.

[**105**] Sethares, W. A., *Tuning, Timbre, Spectrum, Scale*, Springer–Verlag, New York, 1999.

[**106**] Siegel, C. L., Über einige Anwendungen diophantischer Approximationen, *in*: *Gesammelte Abhandlungen*, vol. 1, Springer–Verlag, New York, 1966, pp. 209–266.

[**107**] Smoller, J., *Shock Waves and Reaction–Diffusion Equations*, 2nd ed., Springer-Verlag, New York, 1994.

[**108**] Stewart, J., *Calculus: Early Transcendentals*, vols. 1 & 2, 7th ed., Cengage Learning, Mason, OH, 2012.

[**109**] Stokes, G. G., On a difficulty in the theory of sound, *Phil. Mag.* **33**(3) (1848), 349–356.

[**110**] Stokes, G. G., *Mathematical and Physical Papers*, Cambridge University Press, Cambridge, 1880–1905.

[**111**] Stokes, G. G., *Mathematical and Physical Papers*, 2nd ed., Johnson Reprint Corp., New York, 1966.

[**112**] Strang, G., *Introduction to Applied Mathematics*, Wellesley Cambridge Press, Wellesley, Mass., 1986.

[**113**] Strang, G., and Fix, G. J., *An Analysis of the Finite Element Method*, Prentice–Hall, Inc., Englewood Cliffs, N.J., 1973.

[**114**] Strauss, W. A., *Partial Differential Equations: An Introduction*, John Wiley & Sons, New York, 1992.

[**115**] Thaller, B., *Visual Quantum Mechanics*, Springer–Verlag, New York, 2000.

[**116**] Thijssen, J., *Computational Physics*, Cambridge University Press, Cambridge, 1999.

[**117**] Titchmarsh, E. C., *Theory of Functions*, Oxford University Press, London, 1968.

[**118**] Varga, R. S., *Matrix Iterative Analysis*, 2nd ed., Springer–Verlag, New York, 2000.

[**119**] Watson, G. N., *A Treatise on the Theory of Bessel Functions*, Cambridge University Press, Cambridge, 1952.

[**120**] Weinberger, H. F., *A First Course in Partial Differential Equations*, Dover Publ., New York, 1995.

[**121**] Wiener, N., *I Am a Mathematician*, Doubleday, Garden City, N.Y., 1956.

[**122**] Whitham, G. B., *Linear and Nonlinear Waves*, John Wiley & Sons, New York, 1974.

[**123**] Wilmott, P., Howison, S., and Dewynne, J., *The Mathematics of Financial Derivatives*, Cambridge University Press, Cambridge, 1995.

[**124**] Yong, D., Strings, chains, and ropes, *SIAM Review* **48** (2006), 771–781.

[**125**] Zabusky, N. J., and Kruskal, M. D., Interaction of "solitons" in a collisionless plasma and the recurrence of initial states, *Phys. Rev. Lett.* **15** (1965), 240–243.

[**126**] Zienkiewicz, O. C., and Taylor, R. L., *The Finite Element Method*, 4th ed., McGraw–Hill, New York, 1989.

[**127**] Zwillinger, D., *Handbook of Differential Equations*, Academic Press, Boston, 1992.

[**128**] Zygmund, A., *Trigonometric Series*, 3rd ed., Cambridge University Press, Cambridge, 2002.

Symbol Index

Symbol	Meaning	Page(s)
$c+d$	addition of scalars	575
$z+w$	complex addition	571
$A+B$	addition of matrices	575
$\mathbf{v}+\mathbf{w}$	addition of vectors	575
$f+g$	addition of functions	575
zw	complex multiplication	571
z/w	complex division	572
$c\mathbf{v},\ cA,\ cf$	scalar multiplication	575
$\overline{z}$	complex conjugate	571
$\overline{\Omega}$	closure of subset or domain	243
$\mathbf{0}$	zero vector	xvi, 575
>0	positive definite	355, 578
≥ 0	positive semi-definite	355
f^{-1}	inverse function	xvi
A^{-1}	inverse matrix	xvi
$f(x^+),\ f(x^-)$	one-sided limits	xvi
$n!$	factorial	163, 453
$\binom{n}{k}$	binomial coefficient	163
$\lvert\cdot\rvert$	absolute value, modulus	94, 225, 571
$\lVert\cdot\rVert$	norm	73, 89, 106, 284, 356, 578, 579, 581
$\lVert\cdot\rVert$	double norm	380
$\lvert\lvert\lvert\cdot\rvert\rvert\rvert$	norm	356
$\mathbf{v}\cdot\mathbf{w}$	dot product	578
$\mathbf{z}\cdot\mathbf{w}$	Hermitian dot product	580
$\langle\cdot\rangle$	expected value	287
$\langle\cdot,\cdot\rangle$	inner product	73, 89, 107, 285, 341, 578, 579, 581
$\langle\langle\cdot,\cdot\rangle\rangle$	inner product	341
$[0,1]$	closed interval	xvi
$\{f \mid C\}$	set	xvi
$\in$	element of	xvi

$\notin$	not element of	xvi
$\subset$, $\subsetneq$	subset	xvi
$\cup$	union	xvi
$\cap$	intersection	xvi
$\setminus$	set theoretic difference	xvi
$:=$	definition of symbol	xvi
$\equiv$	identical equality of functions	xvi
$\equiv$	equivalence in modular arithmetic	xvi
$\circ$	composition	xvi
$*$	convolution	95, 281
L^*	adjoint operator	341
$\sim$	Fourier series representation	74
$\sim$	asymptotic equality	300
$f\colon X \to Y$	function	xvi
$x_n \to x$	convergent sequence	xvi
$f_n \rightharpoonup f$	weak convergence	230
$f(x^+),\ f(x^-)$	one-sided limits	41, 79
$u', u'', \ldots$	space derivativcs	xvii
$\dot{u}, \ddot{u}, \ldots$	time derivatives	xvii
$u_x, u_{xx}, u_{tx}, \ldots$	partial derivatives	xvii, 1
$\frac{du}{dx}, \frac{d^2u}{dx^2}, \ldots$	ordinary derivatives	xvii, 1
∂	partial derivative	xvii, 1
∂	boundary of domain	5, 152, 504
$\frac{\partial u}{\partial x}, \frac{\partial^2 u}{\partial x^2}, \frac{\partial^2 u}{\partial t\, \partial x}, \ldots$	partial derivatives	xvii, 1
$\partial_x, \frac{\partial}{\partial x}$	partial derivative operator	2
$\frac{\partial}{\partial \mathbf{n}}$	normal derivative	153, 244, 504
∇	gradient	150, 242, 345, 505
$\nabla\cdot$	divergence	242, 347, 505
$\nabla\times$	curl	242
∇^2	Laplacian	243
$\Box$	wave operator	50
$\sum_{i=1}^{n}$	summation	xvi
$\int f(x)\, dx$	indefinite integral	xvii
$\int_a^b f(x)\, dx$	definite integral	xvii

$\fint_{-\infty}^{\infty} f(x)\,dx$	principal value integral	283
$\iint_\Omega f(x,y)\,dx\,dy$	double integral	243
$\iiint_\Omega f(x,y,z)\,dx\,dy\,dz$	triple integral	505
$\int_C f(s)\,ds$	line integral with respect to arc length	244
$\int_C \mathbf{v}\,d\mathbf{x}$	line integral	243
$\oint_C \mathbf{v}\,d\mathbf{x}$	line integral around closed curve	243
$\iint_{\partial\Omega} f\,dS$	surface integral	505
a	Bohr radius	567
$\mathcal{A}$	space of analytic functions	576
a_k	Fourier coefficient	74, 89
Ai	Airy function	327, 460
arg	argument (see phase)	xvi, 573
$\mathbf{b}$	finite element vector	401
$\mathbf{B}$	magnetic field	551
b_k	Fourier coefficient	74, 89
Bi	Airy function of the second kind	462
c	wave speed	19, 24, 50, 486, 546
$\mathbf{c}$	finite element coefficient vector	401
$\mathbb{C}$	complex numbers	xv, 571
c_g	group velocity	331
c_k	complex Fourier coefficient	89
c_k	eigenfunction series coefficient	378
c_p	phase velocity	330
C^0	space of continuous functions	108, 576
C^n	space of differentiable functions	5, 576
C^∞	space of smooth functions	576
$\mathbb{C}^n$	n-dimensional complex space	xv, 575
coker	cokernel	350
cos	cosine	6, 89
cosh	hyperbolic cosine	88
coth	hyperbolic cotangent	91, 317
csc	cosecant	230
curl	curl (see also $\nabla\times$)	242
d	ordinary derivative	xvii, 1
D	derivative operator	342, 585
D	domain	5

det	determinant	582
dim	dimension	577
div	divergence (see also $\nabla\cdot$)	242
ds	arc length element	244
dS	surface area element	505
e	base of natural logarithm	xvi
E	energy	61, 132, 151
$\mathbf{E}$	electric field	551
e^x	exponential	5
e^z	complex exponential	573
$\mathbf{e}_i$	standard basis vector	216, 577
erf	error function	55
erfc	complementary error function	302
$\widetilde{f}$	periodic extension	77
$\mathcal{F}$	function space	575
$\mathcal{F}$	Fourier transform	264
$\mathcal{F}^{-1}$	inverse Fourier transform	265
$F(t,x;\xi)$	fundamental solution	292, 387, 481, 543
$G(x;\xi),\ G_\xi(x)$	Green's function	234, 240, 248, 527
$G(t,x;\tau,\xi)$	general fundamental solution	297
h	step size	182
$\hbar$	Planck's constant	6, 287, 394
H_n	Hermite polynomial	311
$H_n^m,\ \widetilde{H}_n^m$	harmonic polynomial	520
$\mathrm{i}=\sqrt{-1}$	imaginary unit	571
I	identity matrix	575
Im	imaginary part	571
J_m	Bessel function	468
k	frequency variable	264
k	wave number	330
K	finite element matrix	401
$K[u]$	right hand side of evolution equation	291
k_{ij}^ν	elemental stiffness	417
$K_n^m,\ \widetilde{K}_n^m$	complementary harmonic function	523
ker	kernel	350, 577
l	angular quantum number	568
L^2	Hilbert space	106, 284
L_k	Laguerre polynomial	566
L_k^j	generalized Laguerre polynomial	566
$L[u]$	linear function/operator	10, 64, 585
$\lim\limits_{x\to a},\ \lim\limits_{n\to\infty}$	limits	xvi

$\lim_{x \to a^-}$, $\lim_{x \to a^+}$	one-sided limits	xvi
log	natural or complex logarithm	xvi, 573
m	mass	6
m	magnetic quantum number	568
M	electron mass	564
M_r, $\mathrm{M}_r^{\mathbf{x}}$	spherical mean	553
max	maximum	xvi
min	minimum	xvi
mod	modular arithmetic	xvi
n	principal quantum number	568
$\mathbf{n}$	unit normal	153, 244, 505
$\mathbb{N}$	natural numbers	xv
O	zero matrix	575
$\mathrm{O}(h)$	Big Oh notation	182
p	pressure	3
p	option exercise price	299
P	Péclet number	311
P_n	Legendre polynomial	511, 525
p_n^m	trigonometric Ferrers function	515
P_n^m	Ferrers (associated Legendre) function	513
$\mathcal{P}^{(n)}$	space of polynomials of degree $\leq n$	577
ph	phase (argument)	xvi, 572
$Q[u]$	quadratic function(al)	362
r	radial coordinate	xv, 3, 160, 572
r	cylindrical radius	xv, 3, 508
r	spherical radius	xv, 3, 508
r	interest rate	299
$\mathbb{R}$	real numbers	xv
$\mathbb{R}^n$	n-dimensional Euclidean space	xv, 575
$R[u]$	Rayleigh quotient	375
Re	real part	571
rng	range	576
s	arc length	244
S	surface area	505
S_m	spherical Bessel function	539
s_n	partial sum	75, 113
S_r, $S_r^{\mathbf{x}}$	sphere of radius r	553, 555
sech	hyperbolic secant	334
sign	sign function	94, 225
sin	sine	6, 89
sinh	hyperbolic sine	13, 88

span	span	576
supp	support	407
t	time	xv, 3
T	conserved density	38, 256
A^T	transpose of matrix	341, 578
T_ν	finite element triangle	411
tan	tangent	1
tanh	hyperbolic tangent	135
u	dependent variable	xv, 3
$u_x, u_{xx}, \ldots$	partial derivative	1
v	dependent variable	xv, 3
v	eigenvector/eigenfunction	371
$\mathbf{v}$	vector	xv, 575
$\mathbf{v}$	eigenvector	66, 582
$\mathbf{v}$	vector field	3, 242
V	vector space	575
V	potential function	6
$\mathbf{v}^\perp$	perpendicular vector	244
v_{lmn}	atomic eigenfunction	568
V_λ	eigenspace	371
w	dependent variable	xv, 3
w	heat flux	122
$\mathbf{w}$	heat flux vector	437
x	Cartesian space coordinate	xv, 3, 152, 504
x	real part of complex number	571
X	flux	38, 256
y	Cartesian space coordinate	xv, 3, 152, 504
y	imaginary part of complex number	571
Y	flux	256
Y_m	Bessel function of the second kind	470
$Y_n^m,\ \widetilde{Y}_n^m$	spherical harmonic	517
$\mathcal{Y}_n^m$	complex spherical harmonic	519
z	Cartesian space coordinate	xv, 3, 504
z	cylindrical coordinate	xv, 3, 508
z	complex number	571
$\mathbb{Z}$	integers	xv
α	electron charge	564
β_l^n	radial wave function	568
γ	thermal diffusivity	124, 438, 535
γ	Euler–Mascheroni constant	471
Γ	gamma function	454

$\delta,\ \delta_\xi$	delta function	217, 219, 246, 247, 527
$\widetilde{\delta}$	periodically extended delta function	229
$\delta',\ \delta'_\xi$	derivative of delta function	225, 226
Δ	Laplacian	4, 152, 161, 243, 504, 509
Δ	discriminant	172, 173
Δx	step size	186
Δx	variance	287
Δ_S	spherical Laplacian	509
ε	thermal energy density	122, 437
ϵ_0	permittivity constant	551
$\zeta_{m,n}$	Bessel root	474
η	characteristic variable	51
θ	polar angle	xv, 3, 160, 572
θ	cylindrical angle	xv, 3, 508
θ	azimuthal angle	xv, 3, 508
ζ	root of unity	582
κ	thermal conductivity	65, 123, 437
κ	stiffness or tension	49
λ	eigenvalue	66, 371, 573
λ	magnification factor	189
μ_0	permeability constant	551
ν	viscosity	3
ξ	characteristic variable	19, 25, 32, 51
π	area of unit circle	5
ρ	density	49, 122, 438
ρ	spectral radius	584
$\rho,\ \rho_\xi$	ramp function	91, 223
$\rho_n,\ \rho_{n,\xi}$	n^{th} order ramp function	95, 223
$\rho_{m,n}$	relative vibrational frequency	495
σ	shock position	41
σ	heat capacity	65, 122, 438
σ	volatility	299
$\sigma,\ \sigma_\xi$	unit step function	61, 80, 222
$\sigma_{m,n}$	spherical Bessel root	540
φ	zenith angle	xv, 3, 508
φ	wave function	286
φ_k	orthogonal or orthonormal system	109
φ_k	basis for finite element subspace	401
χ	specific heat capacity	122, 431
χ_D	characteristic function	485

ψ	time-dependent wave function	394, 564
ω	frequency	59, 330
Ω	domain	152, 242, 504

Author Index

Abdulloev, K. O. 337, [**1**]
Ablowitz, M. J. 283, 292, 324, 333, 337, 338, [**2**]
Abraham, R. 161, [**3**]
Airy, G. B. 281, 327, 334, [**4**]
Aki, K. 549, [**5**]
Ames, W. F. 181, 213, 400, 410, [**6**]
Antman, S. S. 324, 486, 549, [**7**]
Apostol, T. M. 5, 20, 76, 87, 100, 105, 169, 182, 236, 242, 245, 267, 312, 437, 500, 505, [**8**], [**9**]
Atkinson, K. 519, [**10**]

Bank, S. B. 455, [**11**]
Batchelor, G. K. 3, [**12**]
Bateman, H. 315, [**13**]
Benjamin, T. B. 337, [**14**]
Berest, Y. 563, [**15**]
Bernoulli, J. xvii, 452
Berry, M. V. 329, [**16**]
Bessel, F. W. 111, 452
Birkhoff, G. ix, 2, 11, 29, 67, 298, 305, 309, 457, [**17**], [**18**]
Black, F. 299, [**19**]
Blanchard, P. ix, 2, 11, 22, 25, 29, 65, 68, [**20**]
Bogolubsky, I. L. 337, [**1**]
Bohr, N. H. D. 567
Boisvert, R. F. xvi, 55, 310, 327, 364, 435, 452, 468, 511, 512, 513, 567, 573, [**86**]
Bona, J. L. 337, [**14**]
Bourget, J. 490
Boussinesq, J. 292, 333, 335, [**21**], [**22**]
Boyce, W. E. ix, 2, 11, 22, 25, 29, 65, 67, 68, 162, 169, 263, 298, 300, 309, 466, [**23**]
Bradie, B. ix, 135, 185, 407, 453, [**24**]
Bronstein, M. 267, [**25**]
Burgers, J. M. 315, [**26**]

Cantor, G. F. L. P. xvii, 64
Cantwell, B. J. 305, [**27**]
Carleson, L. 117, [**28**]
Carmichael, R. 500, [**29**]
Cauchy, A. L. xvii, 175, 215
Chen, G. 329, [**30**]
Chladni, E. F. F. 497
Clark, C. W. xvi, 55, 310, 327, 364, 435, 452, 468, 511, 512, 513, 567, 573, [**86**]
Clarkson, P. A. 283, 292, 324, 333, 337, 338, [**2**]
Coddington, E. A. 2, [**31**]
Cole, J. D. 318, [**32**]
Courant, R. xvii, 4, 176, 177, 179, 197, 246, 255, 340, 377, 436, 440, 449, 477, 497, 541, [**33**], [**34**], [**35**]
Crank, J. 192

d'Alembert, J. L. R. xvii, 15, 50, 140, 149, 558
de Broglie, L. V. P. R. 287
de Coulomb, C.–A. 252, 503
Devaney, R. L. ix, 2, 11, 22, 25, 29, 65, 68, [**20**]
de Vries, G. 333, [**65**]
Dewynne, J. 299, [**123**]
DiPrima, R. C. ix, 2, 11, 22, 25, 29, 65, 67, 68, 162, 169, 263, 298, 300, 309, 466, [**23**]
Dirac, P. A. M. 217
Dirichlet, J. P. G. L. 7, 368
Drazin, P. G. 38, 292, 324, 333, 337, 338, [**36**]
du Bois–Reymond, P. D. G. 430
Duhamel, J. M. C. 298
Dym, H. 76, 99, 107, 115, 117, 263, 265, 275, 286, 344, [**37**]

Einstein, A. 19, 31, 63, 149, 504
Euler, L. xvii, 3, 454, 461, 573
Evans, L. C. xvii, 4, 314, 340, 350, 427, 436, 535, 546, [**38**]

Feller, W. 55, 295, [**39**]
Fermi, E. 333, [**40**]
Ferrers, N. M. 513
Feshbach, H. 170, 508, 569, [**79**]
Fitzpatrick, P. M. 76, 100, 102, 107, 108, 119, 217, 219, 344, [**96**]
Fix, G. J. 399, 400, 410, 431, [**113**]
Flannery, B. P. ix, 135, 181, [**94**]
Forsyth, A. R. 317, [**41**]
Fourier, J. xvii, 63, 64, 71, 114, 123, 149, 437, 452, 535, [**42**]
Fredholm, E. I. 350

Friedrichs, K. O. 197, [**33**]
Frobenius, F. G. 464

Galilei, G. 20
Gander, M. J. 497, [**43**]
Garabedian, P. xvii, 4, 173, 176, 246, 340, 350, 376, 377, 427, [**44**]
Gardner, C. S. 336, 338, [**45**], [**76**]
Gauss, J. C. F. 63
Germain, M.–S. 497
Gibbs, J. W. 84
Gonzalez, R. C. 442, [**46**]
Gordon, C. 487, [**47**]
Gradshteyn, I. S. 334, [**48**]
Green, G. 3, 215, 243
Greene, J. M. 336, [**45**]
Gregory, J. 78
Gurtin, M. E. 256, 549, [**49**]

Haberman, R. xvii, [**50**]
Hairer, E. 181, [**51**]
Hale, J. K. ix, 2, 11, 29, 67, [**52**]
Hall, G. R. ix, 2, 11, 22, 25, 29, 65, 68, [**20**]
Han, W. 519, [**10**]
Heaviside, O. 215, 217, 551
Heisenberg, W. K. 286, 288
Henrici, P. 256, [**53**]
Hilbert, D. xvii, 4, 106, 176, 177, 179, 246, 255, 340, 368, 377, 436, 440, 449, 477, 497, 541, [**34**], [**35**]
Hille, E. ix, 2, 453, 457, 459, 463, [**54**]
Hobson, E. W. 329, [**55**]
Hopf, E. 318, [**56**]
Howison, S. vii, viii, 43, 46, 299, [**57**], [**123**]
Hugoniot, P. H. 40
Huygens, C. 560
Hydon, P. E. 305, [**58**]

Ince, E. L. ix, 2, 453, 457, 459, 463, 472, [**59**]
Iserles, A. ix, 181, 185, 453, [**60**]

Johnson, R. S. 38, 292, 324, 333, 337, 338, [**36**]
Jost, J. xvii, 4, 246, 255, 314, 340, 350, 427, 535, 546, [**61**]

Kamke, E. 453, [**62**]
Kaufman, R. P. 455, [**11**]
Keller, H. B. 185, 355, 364, [**63**]
Kelvin, L. (Thomson, W.) 3, 41, 139, 331
Kirchhoff, G. R. 503, 558
Knobel, R. 317, [**64**]
Korteweg, D. J. 333, [**65**]
Kovalevskaya, S. V. 175
Kruskal, M. D. 333, 336, 338, [**45**], [**76**], [**125**]
Kwok, F. 497, [**43**]

Lagrange, J.–L. xvii, 182
Laguerre, E. N. 566
Landau, L. D. 108, 278, 288, 383, 394, [**66**]
Laplace, P.–S. xvii, 152
Lebesgue, H. L. 107, 112
Legendre, A.–M. 454, 511
Leibniz, G. W. 1, 78
Levine, I. N. 569, [**67**]
Levinson, N. 2, [**31**]
Lewy, H. 197, [**33**]
Lie, M. S. 305
Lifshitz, E. M. 108, 278, 288, 383, 394, [**66**]
Lighthill, M. J. 76, 233, 263, 286, [**68**]
Lin, C. C. vii, viii, [**69**]
Liouville, J. 363
Lozier, D. W. xvi, 55, 310, 327, 364, 435, 452, 468, 511, 512, 513, 567, 573, [**86**]
Lubich, C. 181, [**51**]

Mahony, J. J. 337, [**14**]
Makhankov, V. G. 337, [**1**]
Marsden, J. E. 161, [**3**]
Marzoli, I. 329, [**16**]
Maxwell, J. C. 551
Mayers, D. F. 181, 200, 213, 453, [**80**]
McKean, H. P. 76, 99, 107, 115, 117, 263, 265, 275, 286, 344, [**37**]
McOwen, R. C. xvii, 56, 122, 179, 246, 255, [**70**]
Mendeleev, D. I. 568
Merton, R. C. 299, [**71**]
Messiah, A. 108, 278, 288, 383, 394, 565, [**72**]
Miller, W., Jr. 305, [**73**]
Milne–Thompson, L. M. 185, [**74**]
Minkowski, H. 56, 560
Misner, C. W. 56, 504, [**75**]
Miura, R. M. 336, 338, [**45**], [**76**]
Moon, F. C. 60, [**77**]
Moon, P. 170, 508, [**78**]
Morse, P. M. 170, 508, 569, [**79**]
Morton, K. W. 181, 200, 213, 453, [**80**]
Murray, J. D. 438, [**81**]

Navier, C. L. M. H. 3
Neumann, C. G. 7
Newton, I. vii, 49, 63, 135, 149, 182, 252, 388, 503
Nicolson, P. 192

Oberhettinger, F. 273, [**82**]
Okubo, A. 438, [**84**]
Olver, F. W. J. xvi, 55, 310, 327, 331, 364, 435, 452, 453, 461, 462, 468, 470, 472, 474, 511, 512, 513, 567, 573, [**85**], [**86**]
Olver, P. J. ix, xi, xv, xviii, 11, 27, 38, 39, 65, 66, 67, 73, 75, 98, 110, 162, 191, 210, 212, 300, 305, 308, 310, 311, 329, 338, 350, 357, 363, 372, 378, 390, 400, 402, 407, 408, 411, 575, 579, 581, 582, 583, 584, 585, [**30**], [**87**], [**88**], [**89**]
Oskolkov, K. I. 332, [**90**]
Øksendal, B. viii, 299, [**83**]

Parseval des Chênes, M.–A. 114
Pasta, J. 333, [**40**]
Pauli, W. E. 568
Pinchover, Y. xvii, [**91**]
Pinsky, M. A. xvii, [**92**]
Plancherel, M. 114
Planck, M. K. E. L. 287, 394
Poisson, S. D. 31, 152, 503, 558
Polyanin, A. D. 453, [**93**]
Press, W. H. ix, 135, 181, [**94**]

Rankine, W. J. M. 40
Ratiu, T. 161, [**3**]
Rayleigh, L. (Strutt, J.W.) 40, 375
Reed, M. 344, 374, 383, 565, [**95**]
Richards, P. G. 549, [**5**]
Richardson, L. F. 194
Riemann, G. F. B. xvii, 31, 63, 87, 112
Robin, V. G. 123
Rota, G.–C. ix, 2, 11, 29, 67, 298, 309, 457, [**18**]
Royden, H. L. 76, 100, 102, 107, 108, 119, 217, 219, 344, [**96**]
Rubinstein, J. xvii, [**91**]
Rudin, W. 5, 76, 100, 102, 105, 107, 108, 119, 169, 182, 217, 219, 256, 263, 267, 344, [**97**], [**98**]
Ryzhik, I. W. 334, [**48**]

Salsa, S. xvii, 4, 122, 340, 350, 427, 436, 535, 546, [**99**]
Sapiro, G. 442, [**100**]
Schleich, W. 329, [**16**]
Scholes, M. 299, [**19**]
Schrödinger, E. 394, [**101**]
Schumaker, L. L. 210, 400, 408, [**102**]
Schwartz, L. 217, [**103**]
Scott Russell, J. 334, [**104**]
Segel, L. A. vii, viii, [**69**]
Sethares, W. A. 144, [**105**]
Shakiban, C. ix, xv, xviii, 11, 27, 65, 66, 67, 73, 75, 98, 110, 162, 191, 210, 212, 300, 350, 357, 363, 372, 378, 390, 400, 402, 407, 408, 411, 575, 579, 581, 582, 583, 584, 585, [**89**]
Siegel, C. L. 490, [**106**]
Simon, B. 344, 374, 383, 565, [**95**]
Smoller, J. 317, 427, 434, 438, [**107**]
Spencer, D. E. 170, 508, [**78**]
Stewart, J. 5, 20, 105, 236, 242, 245, 312, 407, 437, 505, [**108**]
Stokes, G. G. 3, 41, 331, [**109**], [**110**], [**111**]
Strang, G. xvii, 357, 399, 400, 410, 431, [**112**], [**113**]
Strauss, W. A. xvii, [**114**]
Strutt, J.W. *see* Rayleigh, L.
Sturm, F. O. R. 363

Talbot, W. H. F. 329
Taylor, B. 75
Taylor, R. L. 400, 410, 431, [**126**]
Teukolsky, S. A. ix, 135, 181, [**94**]
Thaller, B. 108, 329, 394, [**115**]
Thijssen, J. 565, [**116**]
Thomson, W. *see* Kelvin, L.
Thorne, K. S. 56, 504, [**75**]
Titchmarsh, E. C. 263, 265, 275, 286, [**117**]

Ulam, S. 333, [**40**]

Varga, R. S. 213, 402, 411, [**118**]
Vetterling, W. T. ix, 135, 181, [**94**]
Victoria, Q. 139
von Helmholtz, H. L. F. 374
von Neumann, J. 190

Wanner, G. 181, [**51**]
Watson, G. N. 452, 470, 472, 474, 490, [**119**]
Webb, D. L. 487, [**47**]
Weierstrass, K. T. W. xvii, 100, 329
Weinberger, H. F. xvii, 50, 447, 525, [**120**]
Wheeler, J. A. 56, 504, [**75**]
Whitham, G. B. 31, 44, 46, 179, 315, 316, 317, 324, 331, 427, 434, [**122**]
Wiener, N. 254, [**121**]
Wilmott, P. 299, [**123**]
Winternitz, P. 563, [**15**]
Wolpert, S. 487, [**47**]
Woods, R. E. 442, [**46**]

Yong, D. 50, [**124**]

Zabusky, N. J. 333, [**125**]

Zaitsev, V. F. 453, [**93**]
Zienkiewicz, O. C. 400, 410, 431, [**126**]
Zwillinger, D. 453, [**127**]
Zygmund, A. 76, 99, 102, 105, 115, 117, [**128**]

Subject Index

absolute
  convergence 101
  value 86, 94, 105, 225
  zero 139
abstraction 339
acceleration 7, 49
accidental degeneracy 499
acoustics 2, 15, 31
acoustic wave 15
Acta Numerica 181
addition 571, 575
  identity 575
  inverse 575
adjoint 339, 341, 344, 428, 438, 505
  formal 344
  system 350
  weighted 342
advection 322
aerodynamics 173
affine 404
  element 414, 416
  piecewise 404, 411
afterglow 562, 563
air 15, 174, 551
airplane 15, 173, 174
Airy
  differential equation 281, 459
  function 327, 364, 435, 461
    second kind 462
algebra 263, 273, 275, 453
  linear viii, ix, 63, 215, 221, 234, 339, 353, 400, 575
  numerical linear ix, 399
algebraic
  differential equation 455
  equation 1, 8, 11, 215, 428
  function 511
  multiplicity 373
algorithm 399
altitude vector 418
amplitude 334
analysis 63, 76, 99, 400, 578
  complex 31, 175, 256, 263
  functional 340, 350, 362
  numerical ix, 181
  real ix, 219
  vector viii
analytic 76, 105, 158, 169, 175, 181, 521, 576
  function 98, 456, 463
  solution 431
angle 509
  azimuthal 508, 522, 549
  cylindrical 509
  polar 572
  right 73, 581
  zenith 508, 515
angular
  coordinate 130
  quantum number 566, 568
animal population 2, 435, 485
anisotropic 442
annulus 170, 171, 415, 474, 480, 494, 499
ansatz 66, 124, 161, 330, 390, 394, 466, 475, 535, 538
  exponential 67, 330
  power 162, 464, 520
  trigonometric 546
approximation vii, 9, 110, 181, 363, 406
arbitrage 299
arbitrary function 6, 21
arc 551, 556
arc-length 244
area 39, 244, 413, 414, 415, 477
  equal 40, 46, 47, 431
  surface 517, 529, 537, 543, 553, 555
argument (*see* phase) xvi, 573
arithmetic 184, 571
  floating-point ix, 184
  single-precision 184
asset 299
associated Legendre function 513
associativity 281, 575
astronomy 560
asymptotics 69, 284, 453, 468
atom 37, 279, 394, 497, 547, 568
  eigenfunction 568, 570
  orbital 503
audio 63
autonomous 24, 29, 556
average 40, 92, 131, 167, 252, 285, 521, 523, 553, 560
  weighted 213

axis
  horizontal 18
  vertical 18
azimuthal angle 508, 522, 549

B spline 283
back substitution 212
backward difference 182
backwards heat equation 129, 299, 442
bacteria 438, 485
ball 362, 467, 508, 528, 530, 533, 534, 537, 540, 542, 545, 547, 550, 555
balloon 549
bank 299
bar 2, 15, 49, 64, 65, 96, 122, 124, 132, 134, 138, 140, 144, 234, 239, 241, 292, 293, 298, 307, 344, 351, 357, 405, 440
barge 334
barrier 15, 173
basis ix, 112, 350, 401, 405, 430, 577, 583
  eigenvector 66, 583, 584
  orthogonal 73, 112, 372, 581
  solution 458
  standard 216, 577, 581
bath 123, 134, 154, 436
beam 2, 146, 393
  equation 146, 396
  light 286
bell curve 295
bend 486
Benjamin–Bona–Mahony equation 337
Bessel
  boundary value problem 475, 480
  differential operator 366, 374
  equation 364, 452, 467, 474, 538
    modified 473
  function 109, 364, 374, 435, 445, 452, 468, 474, 490, 499, 508, 536, 538
    second kind 470, 474
    spherical 435, 503, 539, 540, 543
  inequality 111, 113, 379, 380, 441
  operator 367
  root 474, 490, 499
    spherical 540, 541, 548
best approximation 110
bi-affine function 415
bidiagonal 211
bidirectional 5, 324
big Oh 182
biharmonic equation 361
bilinearity 281, 578
binomial
  coefficient 164
  formula 163
biology vii, 4, 15, 63, 121, 130
black 441
Black–Scholes
  equation vii, viii, 291, 299
  formula 302
block 144
  bidiagonal 211
  tridiagonal 211
blow-up 23, 32, 37
blurring 129, 442
body 121, 341
  solid 2, 15, 65, 144, 440, 503, 504, 522, 546
Bohr radius 567, 569
boiling water 138, 170, 524, 543
boost 20, 311
boundary ix, 5, 7, 152, 220, 243, 245, 254, 436, 437, 522
  coefficient matrix 425
  condition 4, 7, 68, 217, 293, 306, 339, 400, 521
    Dirichlet 7, 123, 141, 147, 153, 166, 186, 201, 343, 345, 347, 359, 364, 368, 404, 412, 436, 439, 441, 446, 486, 488, 504, 521, 522, 527, 535, 537, 540, 544, 546
    impedance 154
    mixed 124, 343, 345, 347, 359, 364, 368, 436, 439, 441, 486, 504, 527, 535, 544, 546
    Neumann 7, 123, 144, 153, 343, 345, 347, 359, 364, 368, 374, 436, 439, 441, 486, 504, 527, 535, 544, 546
    periodic 69, 124, 130, 292, 343, 345, 361
    Robin 123, 134, 154, 361, 436, 439
  curved 411, 412
  data 207
  element 424
  free 486
  integral 347
  moving 486
  node 188, 207, 411, 418, 422, 425
  point 153, 169
  polygonal 411, 417
  solid 154, 504
  term 343
  value problem ix, 2, 6, 7, 74, 121, 152, 172, 215, 217, 234, 237, 255, 263, 278, 336, 339, 350, 355, 371, 376, 386, 399, 401, 402, 410, 585
    Bessel 475, 480
    Chebyshev 385

boundary value problem (*continued*)
Dirichlet 125, 132, 160, 207, 213, 245, 254, 383, 410, 416, 442, 450, 474, 508, 528, 531, 533, 547
elliptic 207, 216, 256, 506
Helmholtz 160, 441, 443, 445, 475, 487, 490
Legendre 511, 515
mixed 7, 134, 154, 214, 241, 245, 505, 507
Neumann 148, 238, 245, 352, 387, 534
periodic 69, 373, 383, 387, 510
regular 379
singular 379
Sturm–Liouville 363, 382
bounded ix, 152, 345, 359, 504, 519, 531
bound state 278, 337, 383, 565, 568
Bourget's hypothesis 490, 500
bow 497
bowl 362
box 508, 526, 536, 547, 550
function 60, 88, 149, 266, 283, 296
brick 536
bridge 549
Brownian motion 438
building 549
bump 169
Burgers' equation 2, 9, 291, 315, 317
inviscid 315
potential 318

C^n 5, 456, 463, 576
piecewise 80–2, 94, 223, 233
cable equation 139, 304
calcium 569
calculus vii, viii, 63, 76, 98, 182, 221, 223, 233, 263, 275, 452
of finite differences 182, 185
multivariable 215, 437, 505
of variations 255
vector 242, 507, 571
call option 304
can 537
canal 334
cap 555, 561
capacity 65
capacitor 522, 523, 526, 533
Carleson's Theorem 117
Cartesian coordinate viii, 242, 446, 536
Cauchy data 175
–Kovalevskaya Theorem 175
problem 175
–Schwarz inequality ix, 107, 175, 285, 526, 579, 581
sequence 107, 119
causality 15, 43, 433
cavity 524, 534
CD 63
ceiling 494
cell phone 63
cellar 136, 137, 545
Celsius 306
center 252
centered difference 184, 198
CFL condition 197, 205
chain 333, 452
rule viii, 20, 24, 32, 57, 161, 167, 171, 300, 305, 308, 333, 494, 509, 511, 524
change of variables viii, 51, 58, 174, 179, 318, 511
chaos 60, 393, 492
characteristic 56, 175
coordinate 57
curve 15, 24, 30, 31, 47, 49, 176–9
lifted 49
equation 177
function 485
line 21, 45, 195
variable 20, 22, 25, 30, 32, 51, 52, 552
charge 249, 252, 256, 529, 531, 564
Chebyshev
boundary value problem 385
equation 385, 473
polynomial 473
chemical 2
diffusion 123
reaction vii, 31, 438
chemistry vii, 435, 569, 503
Chladni figure 497
chromatography 15, 31
circle 167, 258, 450, 500, 551
civil engineering 549
clarinet 144
classical
mechanics 394, 503
solution 5, 7, 17, 51, 144, 255, 399, 410, 427, 428, 432, 535, 546, 556, 558
clockwise 243
closed ix
curve 152, 243
surface 505
coefficient
constant ix, 2, 9
diffusion 129, 307
Fourier 71, 72, 75, 85, 102, 107, 228, 264, 441
matrix 188, 191, 350, 386
transfer 156

cokernel 350
cold 436
collision 54, 292, 324, 336, 337, 438
column vector 216, 578
comb 229, 233
comet 565
commutativity 575
compact ix
  support 230, 430, 432
complementary
  error function 277, 302, 321
  harmonic function 523
complete 66, 107, 113, 119, 340, 371, 372, 378, 379, 381, 382, 383, 447, 477, 480, 517, 535, 583
  system 520, 541, 565
complex
  addition 571
  analysis 31, 175, 256, 263
  arithmetic 571
  change of variables 175
  conjugate 571, 580
  division 572
  eigenvalue 372
  exponential ix, 88, 107, 109, 136, 181, 189, 265, 271, 285, 519, 573
  Fourier coefficient 89, 229, 284
  Fourier series 89, 582
  function 284, 324
  inner product 118
  integral 266, 461
  logarithm 164, 573
  multiplication 571
  number viii, ix, xv, 571
  plane xv, 571
  solution 6
  spherical harmonic 519, 526, 565
  variable 163, 571
  vector space 289, 571, 575, 580
  zero 87
compressible 37
compression wave 36, 44
computer 184, 190, 400
  algebra 273, 453
  arithmetic ix
  graphics 63
concentration 34
conditionally stable 190
conduction 504
conducting medium 504
conductivity 124
  thermal 65, 123, 437
conductor 123
cone 56, 560, 564
conformal mapping 256
conic 172
conjugate symmetry 580
connected 17, 141, 245, 345, 359
  simply 243
conservation law 15, 38, 46, 131, 201, 255, 256, 295, 304, 332, 337, 431, 437
  energy 61, 151, 535
  heat energy 122, 304
  mass 41, 47, 360
conserved density 38, 46, 201, 256
constant
  coefficient ix, 2, 9
  Euler–Mascheroni 471
  function 285, 506
  Planck 6, 287, 394, 564
  separation 141, 156, 446, 510
constitutive assumption 122
contaminant 123
continuous viii, 7, 63, 80, 82, 94, 102, 231
  dependence 130
  function 99, 108, 117, 219, 220, 344, 576, 579
  Lipschitz 29
  media 390
  piecewise 79, 81, 108, 127, 441
  spectrum 337, 340, 374, 383, 565
continuum mechanics/physics viii, 38
contract 299
control 9, 263
convection 438
  -diffusion equation 139, 311, 314, 438
convective flow 311
convergence viii, xvi, 63, 72, 75, 76, 98, 109, 213, 383, 400, 410
  in norm 98, 99, 109, 110 nonuniform 267
  pointwise 99, 109, 115, 117, 231, 285, 378
  radius of 453, 458
  test
    integral viii, 105
    ratio viii, 75, 462, 468
    root viii, 75
  theorem 82
  uniform 99, 100–102, 104, 378, 519
  weak 99, 230, 270, 327, 429
convergent 584, 585
convolution 242, 281, 282, 295, 484, 544
  integral 301
  periodic 95
  summation 284
  theorem 284

coordinate
  angular 130
  Cartesian viii, 242, 446, 536
  characteristic 57
  curvilinear 364, 435
  cylindrical viii, 3, 503, 508, 523, 527, 530, 536, 547
  ellipsoidal 508
  moving 15, 19
  parabolic 174
  parabolic spheroidal 508
  polar viii, 3, 62, 160, 161, 171, 250, 383, 451, 477, 479, 490, 509, 572
  rectangular viii, 161, 503
  separable 508
  space 3
  spherical viii, 3, 503, 508, 520, 524, 528, 537, 551, 553, 565
  toroidal 508
corner 81, 193, 441
  node 209
corpse 129
cosine transform 274
Coulomb
  potential 503, 529, 564
  problem 564
counterclockwise 243, 415
crack 256, 427
Cramer's Rule 413
Crank–Nicolson method 192
crest 331
critical point 501
cross product viii
cube 526, 534, 536, 543, 545, 550, 561
cubic 283, 408
curl viii, 13, 242, 243, 349, 507
curve 172, 176, 254, 571
  bell 295
  characteristic 15, 24, 30, 31, 47, 49, 176, 177, 178
    lifted 49
  closed 152, 243
    simple 152
  nodal 497, 551
curved boundary 411, 412
curvilinear coordinate 364, 435
cusp 254
cut locus 511, 515, 525
cylinder 450, 452, 467, 508, 527, 536, 537, 547, 550
cylindrical
  angle 509
  coordinates viii, 3, 503, 508, 523, 527, 530, 536, 547
  shell 450
cylindrical (*continued*)
  symmetry 517
cymbal 144

d'Alembert
  formula 16, 121, 140, 146, 260, 427, 552, 558
  solution 53, 201, 324, 487
dam 61
damped 62, 200
  heat equation 65
  transport equation 49
  wave equation 207
data 363, 410
de Broglie relation 287
death 129
decay 22, 105, 276, 285, 387, 441, 443, 479, 482, 536, 544
  entropic 291
  exponential 127
decimal expansion 108
deep water 331
deflection 249
deformable body 341
deformation 121
degree 306, 509
delta
  comb 229, 233
  distribution 215, 217
  function 63, 215, 217, 221, 223, 225, 228, 233, 246, 249, 270, 276, 277, 280, 281, 292, 293, 321, 326, 358, 379, 405, 441, 479, 481, 483, 485, 521, 544, 552, 554
    three-dimensional 527
    two-dimensional 246, 255
  impulse 221, 234, 248, 291, 292, 387
  wave 558
denoise 128, 296, 441
dense 107, 344
  subspace 344, 346, 371
density 49, 122, 124, 132, 142, 253, 344, 357, 438, 486, 488, 492, 495
  conserved 38, 46, 201, 256
  momentum 61
  probability 108, 286, 564
dependent variable 3
deposit 299
depression 31
derivative 1, 5, 105, 182, 223, 236, 275, 342, 576
  of delta function 233, 277
  of Fourier series 94
  left-hand 81
  logarithmic 336

derivative (*continued*)
  normal 7, 153, 245, 250, 436, 504, 528, 529
  one-sided 576
  partial viii, 1, 3, 521
    mixed 5, 8, 50, 242, 556
  radial 554
  right-hand 81
  of series 101
  operator 354
  ordinary 1, 3
  space 291
descent 503, 551, 561, 564
determinant ix
  Jacobian 58, 255
deviation 287, 295
diameter 499, 500
difference
  backward 182
  centered 184, 198
  finite vii, 9, 181, 182, 185, 186, 213, 214, 399, 407, 411, 422
  forward 182
  quotient 182
differentiable 80, 285
  infinitely 5, 75, 105, 128, 158
  nowhere 329
differential
  equation 1, 181, 342, 350, 427, 428, 585
    stochastic viii, 299
    system of 2, 66
    zenith 510
  see also ordinary differential equation, partial differential equation
  geometry 63
  operator 9, 64, 339, 350, 371
    Bessel 366, 374
    partial 2, 50
    Sturm–Liouville 364, 365, 480
differentiation viii, 233, 263, 276, 556
  implicit 49
  numerical 181
  operator 286
diffusion 172, 299, 311, 315, 385, 388, 435, 436, 503, 535
  chemical 123
  coefficient 129, 307
  equation 123, 315, 340, 395, 438, 439, 543
  nonlinear 2, 122, 291, 437, 442
  of set 485
  process 121, 129, 386
diffusive transport equation 194
diffusivity 307, 537
  thermal 124, 134, 186, 293, 298, 438, 535
Dilation Theorem 271, 274
dimension 2, 112, 577
  eigenspace 583
  finite ix, 11, 98, 109, 215, 220, 400, 410, 430
  infinite ix, 11, 99, 109, 215, 340, 342, 371, 400, 577
Dirac
  comb 229
  delta function 217
  equation vii
direct method 255
Dirichlet
  boundary condition 7, 123, 141, 147, 153, 166, 186, 201, 343, 345, 347, 359, 364, 368, 404, 412, 436, 439, 441, 446, 486, 488, 504, 521, 522, 527, 535, 537, 540, 544, 546
  boundary value problem 125, 132, 160, 207, 213, 245, 254, 383, 410, 416, 442, 450, 474, 508, 528, 531, 533, 547
  eigenvalue 373, 377
  functional 410, 416, 424
  integral 368, 506
  principle 368, 400, 443, 506
disconnected 17
discontinuity 37, 193, 223, 441
  jump 80, 81, 82, 96, 164, 223, 233, 236, 405, 432
  removable 80
discontinuous 15, 76, 102, 427
  initial data 292
discrete Fourier transform 582
discriminant 172, 173, 174
disease 123
disk 121, 160, 167, 249, 251, 253, 374, 415, 427, 444, 445, 450, 467, 479, 490, 499, 500, 508
  half 260, 480, 494, 496
  metal 166, 479
  quarter 494
  semi-circular 170, 418
  unit 155, 166
dislocation 256, 427
dispersion vii, 292, 324, 330
  relation 330
dispersive 329, 396
  equation 328, 486
  medium 2
  quantization 328, 329
  tail 337
  wave 2, 324, 459

displacement 49, 216, 341, 351, 486
  initial 53, 59, 145, 487, 546, 547, 551, 554, 556, 557, 560, 561, 562
  radial 546
dissipative 315
dissonant 144, 490, 496, 549
distance 307, 578
distribution ix, 215, 217, 553
  Gaussian 295
distributive 575
disturbance 15, 22, 179, 292
diverge 98
divergence viii, 13, 242, 244, 359, 437, 439, 505, 535
  operator 347
  theorem viii, 505, 529
division 263
domain ix, 5, 17, 152, 207, 243, 245, 248, 253, 339, 341, 345, 359, 486, 504, 571
  of dependence 59, 197, 199
  of influence 56
  irregular 213
  rescaled 495
  space 401, 585, 586
dominant
  frequency 495
  mode 499
dot product viii, 73, 341, 346, 354, 372, 578, 585
  Hermitian 580
double
  Fourier series 488
  integral viii, 58, 243, 245, 248, 251, 346, 422, 432, 437, 448
  weighted L^2 norm 381
  L^2 norm 383, 525
doublet 558
doubly infinite series 89, 91
driver 44
drum 63, 144, 152, 153, 214, 486, 487, 490, 495, 496
  circular 154, 160, 490, 494, 496, 499
  rectangular 499
  square 493
du Bois–Reymond lemma 431, 434
duality 221, 247, 286, 553
Duhamel's principle 298
DVD 63
dynamical 3
  partial differential equation 291, 551
  process 7, 15, 172
  system 340, 385
dynamics 46, 47, 49, 340, 435
  fluid 291, 315
  gas 15, 31, 315
  shock 15, 431
ear 144
Earth 136, 137, 508, 530, 537, 545, 549, 560, 561
earthquake 549
echo chamber 563
ecology 15
economics vii, 4
eigenequation 66, 67, 371, 439, 583
eigenfunction 67, 74, 110, 125, 190, 340, 371, 374, 375, 376, 378, 387, 395, 439, 445, 515, 517, 535, 540, 546, 547, 549, 565, 566
  atomic 568, 570
  expansion 340, 379
  Fourier–Bessel 477
  null 70, 132, 145, 386, 387, 389, 441, 487, 547, 550
  series 109, 371, 378, 386, 435, 441, 443, 535, 544
  Sturm–Liouville 382
eigenmode 389, 497, 499, 546
eigensolution 66, 67, 125, 140, 389, 395, 447, 475, 487, 566, 584
  series 440
eigenspace 371, 382, 517, 568, 583
eigenstate 568
eigenvalue ix, 2, 66, 67, 125, 190, 336, 340, 371, 376, 378, 387, 389, 395, 487, 517, 535, 541, 546, 549, 565, 568, 582–4
  complex 372
  equation 66, 67, 371, 439, 582
  Dirichlet 373, 377
  Helmholtz 377, 383, 446, 474, 535, 546
  multiplicity 582
  null 131, 439, 582
  problem 130, 371, 446
  simple 372, 391, 582
  zero 131, 439, 582
eigenvector ix, 66, 371, 372, 375, 582, 583, 584
  basis 66, 583, 584
  null 582
eikonal equation vii
Einstein equation vii
elastic 550
  ball 504
  bar 15, 49, 234, 241, 351, 357
  beam 146, 393
  media 427
  plate 324, 486

elastic (*continued*)
vibration 486
wave 121
elasticity vii, 2, 121, 175, 504
elastodynamics 486, 549
elastomechanics 341
electric
charge 256, 529, 531, 564
field 341, 504, 546, 551
potential 504
electromagnetic 254, 395
potential 2, 564
vibration 2
wave 15, 121, 388, 503, 546, 551
electromagnetism vii, 121, 154, 341, 504
electromotive force 504
electron 6, 108, 249, 279, 395, 503, 564, 567, 569
electronic music 63
electrostatic 242, 249, 531
attraction 279
force 504, 529
potential 152, 249, 252, 256, 503, 522, 534
element 413, 568
affine 414, 416
boundary 424
finite viii, 9, 181, 207, 340, 362, 399, 410, 411, 416, 427, 430, 431
zero 428
elemental stiffness 417, 418
elementary function xvi, 9, 55, 267, 310, 337, 435, 445, 452, 453, 459, 467, 539
elevation 31
ellipse 173
ellipsoid 511
ellipsoidal coordinates 508
elliptic 2, 122, 172, 173, 178, 212, 399, 404, 410
boundary value problem 207, 216, 256, 506
orbit 565
energy 39, 286, 292, 324, 565, 568, 569
conservation 61, 151, 535
density 61, 122
heat 122, 246, 304, 435–7, 482
kinetic 61
level 395, 503, 567
operator 394
potential 6, 61, 152, 242, 318, 340, 362
thermal 121, 122, 132, 134, 139, 169, 295, 304
total 61, 151
engineering vii, 2, 4, 63, 217, 263, 305, 549
enhancement 129, 442
entropic decay 291
entropy 317
condition 43, 46
envelope 230
Equal Area Rule 40, 46, 47, 431
equation
Airy 281, 459
algebraic 1, 8, 428
algebraic differential 455
backwards heat 129, 299, 442
beam 146, 396
Benjamin–Bona–Mahony 337
Bessel 364, 452, 467, 474, 538
modified 473
biharmonic 361
Black–Scholes vii, viii, 291, 299
Burgers 2, 9, 291, 315, 317
inviscid 315
potential 318
cable 139, 304
characteristic 177
Chebyshev 385, 473
convection-diffusion 139, 311, 314, 438
damped heat 65
damped transport 49
damped wave 207
differential 1, 181, 342, 350, 427, 428, 585
diffusion 123, 194, 315, 340, 395, 438, 439, 543
forced 388
diffusive transport 194
Dirac vii
dispersive 328, 486
eigenvalue 66, 67, 371, 439, 582
eikonal vii
Einstein vii
equilibrium 3, 5, 386
Euler vii, 3, 162, 170, 300, 315, 452, 463, 465, 510, 520
evolution 2, 64, 67, 291, 298, 324, 325, 385, 439, 442
Fitz-Hugh–Nagumo vii
functional 454
Ginzburg–Landau vii
Hamilton–Jacobi 305

equation (*continued*)
heat vii, 2, 5, 9, 11, 12, 64, 69, 121, 124, 140, 152, 172, 173, 178, 181, 186, 291, 292, 295, 301, 305, 309, 311, 315, 318, 324, 325, 339, 340, 371, 374, 385, 387, 435, 438, 443, 445, 446, 467, 474, 503, 535, 537, 543, 544
backwards 129, 299, 442
forced 134, 296, 312, 314, 442
generalized 65
lossy 139
spherical 543
Helmholtz 178, 370, 374, 378, 439, 488, 490, 508, 538, 540, 547
polar 451, 508
spherical 510, 517, 519, 565
Hermite 462
integral 350
implicit 26
indicial 464, 465, 520
integral 9, 284
Kadomstev–Petviashvili vii
Kolmogorov–Petrovsky–Piskounov vii
Korteweg–de Vries vii, 2, 292, 324, 333, 336
modified 337
Laguerre 364, 566, 569
Lamé vii
Laplace 2, 8, 13, 121, 152, 155, 166, 169, 172, 173, 175, 178, 181, 207, 213, 243, 249, 256, 291, 311, 312, 339, 399, 422, 435, 467, 503, 504, 507, 509, 520, 522, 527
Legendre 364, 511, 514, 525
linear 9, 11, 215, 216, 428
Mathieu 409
Maxwell vii, 546, 551
Maxwell–Bloch vii
minimal surface 153, 175
Navier–Stokes vii, 3, 9, 315
ordinary differential ix, 2, 6, 8, 11, 22, 65, 67, 121, 130, 140, 185, 215, 217, 263, 291, 308, 333, 363, 399, 435, 445, 452, 455, 463, 585
first-order ix, 297
linear 11, 585
second-order ix, 297, 363, 459, 463
separable ix
partial differential vii, ix, 2, 63, 67, 130, 175, 215, 263, 305, 317, 339, 350, 394, 399, 429, 487, 582
dynamical 291, 551
elliptic 2, 122, 172, 173, 178, 212, 399, 404, 410
first-order 176, 179, 195

equation, partial differential (*continued*)
fourth-order 486
fully nonlinear 174
hyperbolic 2, 122, 172, 173, 178, 179, 195, 385, 497
ill-posed 129, 175, 436, 442
nonlinear 2, 12, 31, 174, 222, 305, 315, 333, 486
parabolic 2, 122, 172, 173, 178, 193, 291, 315, 385, 438
quasilinear 174
second-order 121, 173
third-order 2, 324
type 172, 173
well-posed 136, 395, 535
Poisson 2, 121, 152, 169, 172, 173, 181, 207, 215, 242, 245, 248, 253, 255, 256, 339, 352, 359, 361, 368, 383, 399, 401, 410, 422, 435, 442, 503, 504, 507, 527, 530, 532, 533, 534
Poisson–Darboux 62
quadratic 172, 464, 465
reaction-diffusion 438
Schrödinger vii, 6, 146, 304, 340, 372, 374, 383, 385, 394, 396, 397, 503, 564
nonlinear vii
stochastic differential viii, 299
Sturm–Liouville 336, 364, 370, 404, 508
telegrapher 62, 146, 393
transcendental 1, 134
transport 2, 19, 31, 51, 65, 181, 195, 291, 315, 324, 330, 434
Tricomi 173, 178, 462
vibration 340, 388, 390
wave 2, 8, 13, 15, 50, 64, 121, 140, 149, 151, 152, 169, 172, 173, 178, 181, 195, 201, 291, 315, 324, 339, 340, 371, 374, 385, 389, 427, 434, 435, 467, 486, 488, 494, 503, 545, 551, 552, 554, 556, 558, 561
forced 56, 58, 393, 493
nonlinear 317, 333
vibration 340, 388
Vlasov vii
von Karman vii
zenith 510
equator 509, 522
equilateral triangle 260, 415, 417, 419, 424, 426
equilibrium 2, 131, 132, 172, 216, 239, 399, 435, 442, 503, 504
equation 3, 5, 386
mechanics 121

equilibrium (*continued*)
  solution 132, 443
  system 339, 340
  temperature 133, 154, 214, 246, 435, 537
  thermal 127, 152, 153, 169, 441, 448, 479, 522, 536
equivalence class 108
error 182, 184
  bound 331
  function 55, 277, 296, 310, 435
  oscillatory 193
  round-off ix, 184, 190
Euclidean
  norm 169, 572, 578
  space ix, 575, 578
Euler
  equation 162, 170, 300, 452, 463, 465, 510, 520
  equations vii, 3, 315
  –Mascheroni constant 471
  formula ix, 6, 88, 267, 573
European call option 300, 301
even 85, 149, 229, 268, 275, 497, 525
evolution equation 64, 67, 291, 298, 325, 385, 439, 442
    third-order 2, 324
exercise
  price 299
  time 302
existence ix, 2, 3, 29, 130, 246, 254, 350, 355, 368, 376, 390, 436
  theorem 67, 364, 457
expected value 286, 287
explicit scheme 188
explosion 551, 560, 561
exponential 445, 452
  ansatz 67, 330
  complex ix, 88, 107, 109, 136, 181, 189, 265, 271, 285, 519, 573
  decay 127
  function 66, 90, 140
  integral 267
  pulse 275
external
  force 12, 56, 58, 152, 215, 216, 234, 291, 292, 351, 386, 390, 504
  heat source 12, 152, 246, 296, 297, 312, 436, 437, 442, 485

factorial 453, 454, 468
factorization 15, 50, 62, 243, 324, 487
Fahrenheit 306, 311
feature 128
Fermi–Pasta–Ulam problem 333
Ferrers function 435, 513, 515, 525
  trigonometric 516
field
  electric 341, 504, 546, 551
  force 6, 252
  gravitational 341
  magnetic 341, 546, 551
  scalar 242, 345, 359, 439, 505
  vector 242, 243, 346, 359, 360, 439, 505, 507, 526
figure
  Chladni 497
filter 291, 295, 482, 544, 545
final
  condition 300
  value problem 300
finance vii, 63, 291, 299
fingerprint 63
finite difference vii, 9, 181, 182, 185, 186, 213, 214, 399, 407, 411, 422
finite-dimensional ix, 11, 98, 109, 215, 220
  subspace 400, 410, 430
finite element viii, 9, 181, 207, 340, 362, 399, 410, 411, 427, 430
  linear system 431
  matrix 418
  subspace 416
fire 254
first-order 19, 182
  ordinary differential equation ix, 297
  partial differential equation 176, 179, 195
Fitz-Hugh–Nagumo equation vii
fixed point 25, 29, 30, 584
floating-point ix, 184
flood wave 15, 31
flow 3, 19, 22, 139, 242, 243, 244, 304, 311, 360, 504
  compressible 37
  gradient 386
  ideal 504
  incompressible 3, 243, 244, 504
fluctuation 137
fluid 2, 15, 37, 152, 292, 388, 438, 504
  dynamics 291, 315
  flux 153, 244
  mechanics vii, 3, 121, 152, 153, 305, 315, 324, 341
  source 360
  velocity 341
flute 144
flux 38, 46, 201, 256, 504
  fluid 153, 244
  heat 7, 122, 123, 139, 153, 154, 242, 246, 436, 437, 504, 535

focusing 179
food 136
force 49, 216, 218
  electromotive 504
  electrostatic 504, 529
  external 12, 56, 58, 152, 215, 216, 234, 291, 292, 351, 386, 390, 504
  field 6, 252, 504
  gravitational 154, 242, 487, 529, 531
  restoring 169
forced
  diffusion equation 388
  heat equation 134, 296, 312, 314, 442
  vibrational equation 390
  wave equation 56, 58, 393, 493
forcing 4, 206
  frequency 390
  function 215, 255, 350, 357, 422
  periodic 304, 390, 397
forensics 129
formal adjoint 344
formula
  binomial 163
  Black–Scholes 302
  d'Alembert 16, 121, 140, 146, 260, 427, 552, 558
  Euler ix, 6, 88, 267, 573
  Green 244, 250, 251, 254, 432
  integral 468, 521, 530
  Kirchhoff 503, 551, 558
  Lagrange 182
  Parseval 114, 118, 285, 287, 289
  Plancherel 113, 118, 286, 287, 289, 379
  Poisson 165, 171, 259
  superposition 248, 291, 297
  Taylor 75
forward
  difference 182
  substitution 212
foundations 64
Fourier
  analysis vii, ix, 11, 181, 233, 579
  –Bessel eigenfunction 477
  –Bessel series 477, 492, 493
  –Bessel–spherical harmonic series 541
  coefficient 71, 72, 75, 85, 102, 107, 228, 264, 441
    complex 89, 229, 284
  integral 284, 331, 383
  law 123, 437, 535
  mode 142, 189, 193
Fourier (*continued*)
  series viii, 71, 74, 76, 82, 102, 109, 113, 121, 131, 163, 228, 231, 233, 263, 285, 328, 340, 371, 378, 402, 519
    complex 89, 582
    cosine 85, 87, 144, 148, 233
    derivative of 94
    double 488
    generalized 109, 378
    rescaled 495
    sine 85, 87, 126, 142, 147, 157, 233, 382
  transform 263, 264, 265, 269, 273, 274, 278, 282, 284, 285, 293, 297, 325, 330, 337, 340, 374, 462
    discrete 582
    inverse 263, 265, 269
    Table 272
    two-dimensional 274, 278
fractal 329, 332
Fredholm Alternative 239, 339, 350, 356, 390, 507, 527, 534
free
  boundary 486
  end 239
  -space Green's function 249, 530
freezer 304, 449
freezing 436
frequency 144, 263, 265, 292, 330, 546
  dominant 495
  forcing 390
  high 127, 128, 193, 231, 296, 331, 442, 536
  low 128
  resonant 60, 279, 304, 391, 397, 493, 549
  space 263, 284
  variable 286
  vibrational 389, 395, 487, 495, 548
friction 62, 315, 317, 388
Frobenius series 464, 569
fully
  nonlinear 174
  weak 429, 430
function xvi, 63, 76, 108, 215, 220, 263, 575
  Airy 327, 364, 435, 461
    second kind 462
  algebraic 327, 364, 435, 461
  analytic 98, 456, 463
  arbitrary 6, 21
  associated Legendre 513

function (*continued*)
Bessel 109, 364, 374, 435, 445, 452, 468, 474, 490, 499, 508, 536, 538
second kind 470, 474
spherical 435, 503, 539, 540, 543
bi-affine 415
box 60, 88, 149, 266, 283, 296
C^{∞} 456, 463
characteristic 485
complementary error 277, 302, 321
complementary harmonic 523
complex 284, 324
constant 285, 506
continuous 99, 108, 117, 219, 220, 344, 576, 579
delta 63, 215, 217, 221, 223, 225, 228, 233, 246, 249, 270, 276, 277, 280, 281, 292, 293, 321, 326, 358, 379, 405, 441, 479, 481, 483, 485, 521, 544, 552, 554
three-dimensional 527
two-dimensional 246, 255
elementary xvi, 9, 55, 267, 310, 337, 435, 445, 452, 453, 459, 467, 539
error 55, 277, 296, 310, 435
even 85, 149, 229, 268, 275, 497, 525
exponential 66, 90, 140
Ferrers 435, 513, 515, 525
trigonometric 516
forcing 215, 255, 350, 357, 422
gamma 454, 455, 461, 468
Gaussian 227, 247, 273, 277, 326, 334
generalized ix, 63, 215, 223, 228, 233, 263, 270, 275, 553
Green's vii, 3, 215, 234, 236, 239, 240, 242, 248, 253, 256, 258, 280, 291, 292, 340, 358, 360, 371, 379, 383, 388, 480, 503, 525, 527, 528, 531, 533
free-space 249, 530
modified 358, 381, 480
harmonic 121, 152, 156, 168, 169, 383, 504, 521, 530, 531
hat 88, 227, 274, 283, 405, 407, 412
hyperbolic 88, 91, 157, 241, 322, 334, 508
hypergeometric 364
inverse xvi
L^2 286, 344
Legendre 364, 435, 511, 513
linear 220, 339, 341, 344, 585
odd 85, 93, 147, 268, 275, 497, 525
parabolic cylinder 310
pyramid 412, 414, 418
quadratic 340, 362, 401, 416
ramp 82, 91, 95, 223, 235, 239

function (*continued*)
rational 218, 279, 452
rescaled 96, 506
rotated 155
sawtooth 78
series of 100
sign 94, 225, 270
space 63, 74, 99, 109, 215, 220, 340, 342, 344, 386, 400, 576
special vii, 2, 9, 327, 364, 435, 453, 455, 472, 508, 511
step 29, 40, 42, 61, 105, 276, 320
unit 80, 83, 90, 102, 222, 232, 266, 329, 396
symmetric 236
trigonometric 60, 63, 70, 72, 74, 89, 109, 113, 125, 130, 189, 231, 264, 273, 374, 445, 451, 452, 457, 499, 508, 511, 519, 536
wave 108, 286, 288, 394, 396, 564, 565, 568
weight 379
zero 114, 219, 231, 281, 456, 463
zeta 87
functional 220, 399
analysis 340, 350, 362
Dirichlet 368, 400, 410, 416, 424, 443, 506
equation 454
linear 220, 553, 585
quadratic 340, 362, 400, 402
fundamental
solution 291, 292, 294, 301, 304, 326, 328, 387, 388, 435, 481, 482, 503, 543, 551, 554, 564
general 297
theorem viii, 16, 39, 223, 236, 243, 245, 267
future 130

Galilean boost 20, 311
gamma function 454, 455, 461, 468
gas 37, 123, 129, 549
dynamics 15, 31, 315
Gauss–Seidel method 212, 402, 411
Gaussian
distribution 295
Elimination ix, 210, 213, 351, 402, 407, 413, 416, 579
filter 291, 295, 482, 544, 545
function 227, 247, 273, 277, 326, 334
kernel 283
general
fundamental solution 297
relativity 31, 63, 504
solution 12, 67, 253, 390

generalized
  Fourier series 109, 378
  function ix, 63, 215, 223, 228, 233, 263, 270, 275, 553
  heat equation 65
  Laguerre equation 566, 569
  Laguerre polynomial 567, 569
geology 549
geometric
  multiplicity 371, 373, 382
  sum 116
geometry 31, 63, 121, 156, 181, 423, 488, 578
geophysics 549
Gibbs phenomenon 84, 88, 99, 267
Ginzburg–Landau equation vii
glacier 15
glass 123
gradient viii, 13, 152, 172, 242, 245, 252, 278, 345, 359, 400, 416, 438, 504, 505, 526, 529
  flow 386
  temperature 341, 437, 535
Gram–Schmidt process 372, 378
gravitation 341, 362, 504, 511, 529
  field 341
  force 154, 242, 487, 529, 531
  potential 152, 249, 252, 503, 529, 534
gray-scale 441, 442
Green's
  formula 244, 250, 251, 254, 432
  function vii, 3, 215, 234, 236, 239, 240, 242, 248, 253, 256, 258, 280, 291, 292, 340, 358, 360, 371, 379, 383, 388, 480, 503, 525, 527, 528, 531, 533
    free-space 249, 530
    modified 358, 381, 480
  identity 244
  theorem viii, 3, 168, 215, 243, 437
Greenwich 509
Gregory's series 78, 85, 86
grid 414, 421, 422
grounded 524
group 305, 308
  velocity 292, 331
guitar 496

half
  disk 260, 480, 494, 496
  plane 260
Hamilton–Jacobi equation 305
Hamiltonian 394, 395, 564
hammer 55, 261, 494, 554
handbook 453
harmonic
  function 121, 152, 156, 168, 169, 383, 504, 521, 530, 531
  oscillator 397
  part 256
  polynomial 163, 171, 506, 520, 522, 526
  spherical 109, 435, 503, 517, 519, 528, 538, 540, 549, 565, 568
harmonious 144
hat function 88, 227, 274, 283, 405, 407, 412
heat 63
  bath 123, 134, 154, 436
  capacity 65
  conduction 504
  energy 122, 246, 304, 435–7, 482
  equation vii, 2, 5, 9, 11, 12, 64, 69, 121, 124, 140, 152, 172, 173, 178, 181, 186, 291, 292, 295, 301, 305, 309, 311, 315, 318, 324, 325, 339, 340, 371, 374, 385, 387, 435, 438, 443, 445, 446, 467, 474, 503, 535, 537, 543, 544
    backwards 129, 299, 442
    forced 312, 314, 442
    generalized 65
    lossy 139
    spherical 543
  flux 7, 122, 123, 139, 153, 154, 242, 246, 436, 437, 504, 535
  sensor 303
  sink 504
  source 292, 438, 442, 504, 543
    external 12, 152, 246, 296, 297, 312, 436, 437, 442, 485
  specific 122, 124, 132, 438
  total 444, 485, 537
Heisenberg Uncertainty Principle 286–7
helium 564, 568
helix 164
Helmholtz
  boundary value problem 160, 441, 443, 445, 475, 487, 490
  eigenvalue 377, 383, 446, 474, 535, 546
  equation 178, 370, 374, 378, 439, 488, 490, 508, 538, 540, 547
    polar 451, 508
    spherical 510, 517, 519, 565
hemisphere 522, 523, 534, 542, 561
Hermite
  equation 462
  polynomial 311, 462

Hermitian
  dot product 580
  inner product 89, 107, 118, 271, 285, 361, 372, 394, 580
  norm 89, 394
Hessian matrix 443
high frequency 127, 128, 193, 231, 296, 331, 442, 536
highway 38
Hilbert
  space ix, 106, 108, 112, 119, 263, 284, 286, 368, 394
  transform 283
hole 214, 243, 286, 426
homogeneous 2, 9, 10, 172, 356, 390, 583, 586
  polynomial 169, 520
Hopf–Cole transformation 318
Hôpital's Rule 117, 229
horizontal axis 18
horse 334
Huygens' Principle vii, 503, 551, 560, 562, 564
hydrodynamics 305
hydrogen 383, 503, 564, 568
hyperbola 173
hyperbolic 2, 122, 172, 173, 178, 179, 195, 385, 497
  cosine 241
  cotangent 91, 322
  function 88, 508
  secant 334
  sine 157, 241
  trajectory 565
hypergeometric function 364
hypersurface 175

ice 138, 524, 542
ideal fluid flow 504
identity
  additive 575
  matrix 584
  operator 383, 384
ill-posed 129, 175, 436, 442
image xvi, 63, 441
  enhancement 129, 442
  method 256, 532, 534
  mirror- 143, 148, 149
  point 532
  processing 2, 63, 129, 263
imaginary
  part ix, 571
  unit 571
impedance 154
implicit
  differentiation 49
  equation 26
  function theorem 49
  scheme 191
impulse 221, 234, 248, 291, 292, 387
  boundary value problem 236
  point 236
  unit 215–218, 234, 280, 551, 562
incomplete matrix 583, 584
incompressible 3, 243, 244, 504
independent variable 3, 6, 15, 291, 300
index 464
indicial equation 464, 465, 520
inelastic 337
inequality
  Bessel 111, 113, 379, 380, 441
  Cauchy–Schwarz ix, 107, 175, 285, 526, 579, 581
  triangle ix, 107, 526, 572, 579, 581
infinite-dimensional ix, 11, 99, 109, 215, 340, 342, 371, 400, 577
infinitely differentiable 5, 75, 105, 128, 158
inflation 550
inhomogeneous 2, 12, 133, 215, 217, 234, 350, 424, 427, 442, 507, 584, 586
initial
  -boundary value problem 7, 121, 172, 293, 297, 306, 314, 339, 386, 440, 535, 546
    periodic 328, 493
  condition 4, 7, 53, 291, 292, 297, 306, 389, 453, 458, 487, 535
  data 127, 175, 292, 295, 436, 481, 535
  displacement 53, 59, 145, 487, 546, 547, 551, 554, 556, 557, 560, 561, 562
  position 50
  temperature 131, 544
  value problem ix, 2, 6, 19, 50, 53, 172, 175, 263, 292, 325, 328, 371, 458, 544, 551, 560
  velocity 50, 55, 59, 145, 487, 546, 547, 551, 554, 557, 560, 562
inner product ix, 107, 112, 118, 114, 339, 341, 371, 379, 400, 428, 578, 580
  Hermitian 89, 107, 118, 271, 285, 361, 372, 394, 580
  L^2 73, 220, 230, 247, 342, 346, 358, 359, 371, 439, 477, 505, 515, 517, 525
  space 350, 354, 400, 428, 429, 578
    complex 118, 580

inner product (*continued*)
  weighted 341, 344, 354, 358, 359, 365, 378, 438, 506, 569
inner tube 361, 493
insulated 122, 153, 154, 436, 441, 443, 504, 522, 535
insurance 299
integer xv, 453
integrable 127, 292, 324, 333, 337
  square- 106, 108, 231, 284, 285, 293, 394
integral viii, 219, 236, 264, 406
  boundary 347
  complex 266, 461
  convergence test viii, 105
  convolution 301
  double viii, 58, 243, 245, 248, 251, 346, 422, 432, 437, 448
  equation 9, 284, 350
  exponential 267
  formula 468, 521, 530
  Fourier 284, 331, 383
  line viii, 39, 243, 245, 251, 256, 433, 437
  Poisson 165, 171, 259
  of series 101
  surface viii, 505, 522, 529, 533
  triple viii, 505, 553
integrating factor 22
integration viii, 63, 92, 101, 263, 276, 585
  by parts viii, 226, 245, 288, 342, 346, 428, 430, 432, 454, 461, 505
  Lebesgue ix, 107, 217, 219, 263, 270
  numerical 185, 407, 422, 453
  of series 92, 101
  Riemann 217, 219, 270
interest 299, 304
interior
  maximum 443
  node 188, 207, 209, 404, 411, 412, 418, 422, 425
  point 169
interpolation 27, 210, 400
interstellar space 552
interval 17, 73, 95, 102, 217, 220, 264, 284, 576
invariant 42, 291
inverse
  additive 575
  Fourier transform 263, 265, 269
  function xvi
  matrix xvi, 217
  scattering 337
  square law 252, 529
  tangent 277
inversion 258, 526, 532
investment 291, 299
inviscid
  Burgers' equation 315
  limit 292, 315, 317, 320
irrational 60, 107, 144, 389
  time 292, 329, 332
irregular
  domain 213
  singular point 453, 463
irreversibility 130
irrotational 242, 360, 504
isosceles 413, 415, 417, 421
isotropic 435, 436, 437, 485, 508, 528, 535, 537
iterative 187, 453, 583
  numerical method ix, 203, 212, 411
    Gaussian–Seidel 212, 402, 411
    Jacobi 212
    successive over-relaxation (SOR) 212, 402, 411
  system 190, 402, 416

Jacobi method 212
Jacobian determinant 58, 255
Jordan canonical form 67, 584
joule 287
jump 441
  discontinuity 80, 81, 82, 96, 164, 223, 233, 236, 405, 432
  magnitude 80, 223
  rope 148, 393

Kadomstev–Petviashvili equation vii
Kelvin 311
Kepler problem 565
kernel 350, 356, 400, 586
  Gaussian 283
  Poisson 165, 259
kinetic energy 61
Kirchhoff's formula 503, 551, 558
Kolmogorov–Petrovsky–Piskounov equation vii
Korteweg–de Vries equation vii, 2, 292, 324, 333, 336
  modified 337

L^2
  function 286, 344
  Hermitian inner product 89, 107, 580
  Hermitian norm 89
  inner product 73, 220, 230, 247, 342, 346, 358, 359, 371, 439, 477, 505, 515, 517, 525
  norm 89, 106, 284, 383, 394, 445, 477, 525, 579
ladder 503

lag 137
Lagrange's formula 182
Laguerre
  equation 364, 566, 569
  polynomial 109, 566, 567, 569
Lamé equation vii
Laplace
  equation 2, 8, 13, 121, 152, 155, 166, 169, 172, 175, 178, 181, 207, 213, 243, 249, 256, 291, 311, 312, 339, 399, 422, 435, 467, 503, 504, 507, 509, 520, 522, 527
  transform 263
Laplacian 4, 13, 152, 171, 243, 250, 339, 359, 374, 447, 451, 507, 525, 528, 535, 537, 557, 564
  spherical 510, 517, 525, 538, 543, 549, 551
laser 292
latitude 508
law
  of cosines 258, 556
  Fourier 123, 437, 535
  inverse square 252, 529
  Newton's second 49
Lax–Wendroff scheme 200
least squares 110, 363
Lebesgue integration ix, 107, 217, 219, 263, 270
left-hand
  derivative 81
  limit 79, 81
Legendre
  boundary value problem 511, 515
  equation 364, 511, 514, 525
  function 364, 435, 511, 513
  polynomial 511, 513, 525
lemma
  du Bois–Reymond 431, 434
  Riemann–Lebesgue 112, 117, 231, 276
Lie group 305
lifted characteristic curve 49
light 15, 149, 546, 551, 552, 560, 562
  beam 286
  cone 56, 560, 564
  ray 179
  sensor 561
  speed 295, 551
  wave 551
lightning 551, 552, 560, 563
limit viii, 63, 98, 107, 218, 219, 221, 247
  inviscid 292, 315, 317, 320
  left-hand 79, 81
  one-sided viii
  right-hand 79, 81
line 32, 396, 404, 563, 576
  characteristic 21, 45, 195
  integral viii, 39, 243, 245, 251, 256, 433, 437
  shock 40, 173
  spectral 395
  real xv, 107, 284
linear 10, 64, 306, 394, 404, 585
  algebra viii, ix, 63, 215, 221, 234, 339, 353, 400, 575
    numerical ix, 399
  combination 11, 12, 156, 216, 221, 401, 413, 430, 577
  dependence 577
  differential operator 9, 10, 64, 339, 350, 371, 586
  equation 9, 11, 215, 216, 428
  function 220, 339, 341, 344, 585
  functional 220, 553, 585
  independence ix, 112, 401, 577
  operator 110, 284, 339, 340, 350, 354, 355, 386, 400, 428, 585
  superposition 11, 51, 59, 67, 71, 234, 242, 293, 328, 330, 476, 481, 483, 535, 544, 551, 552, 554
  system ix, 187, 191, 216, 339, 350, 386, 400, 428, 582, 586
  transformation 285
linearization 317, 318
Lipschitz continuity 29
liquid 121, 123, 324, 438, 503
  crystal 427
lithium 568
local maximum 507
localized 333, 503, 551
  disturbance 179, 292
  solution 560
logarithm 250, 452, 464
  complex 164, 573
logarithmic
  derivative 336
  potential 253, 256, 383
  singularity 383, 525
  term 468
logic 64
longitude 508
lossy
  convection-diffusion equation 139
  diffusion equation 194
  heat equation 139
low frequency 128
lower triangular matrix 211

M–test 100, 105
Mach number 174

magnetic
  field 341, 546, 551
  quantum number 568, 569
magneto-hydrodynamics vii, 315
magnification factor 189, 192, 199, 205
magnitude 80, 221, 223
manufactured solution 210
MAPLE 273, 453
mass 6, 37, 39, 49, 95, 216, 333, 396, 531, 564
  conservation 41, 47, 360
  flux 38, 47
  point 249, 252, 256, 529
mass-spring system 95
materials science 175
MATHEMATICA 273, 453
mathematics vii, ix, 2, 16, 64, 435
Mathieu equation 409
matrix 66, 216, 339, 582
  boundary coefficient 425
  coefficient 188, 191, 350, 386
  convergent 584, 585
  finite element 418
  Hessian 443
  identity 584
  incomplete 583, 584
  inverse xvi, 217
  lower triangular 211
  multiplication 341, 344, 356, 585
  orthogonal 506
  positive definite 342, 353, 362
  self-adjoint 354, 373
  symmetric 66, 74, 342, 353, 354, 362, 372, 401
  tridiagonal 188, 191, 211, 407
  upper triangular 211
  weighted adjoint 342
  zero 583
maximum 169, 507, 522
  principle vii, 169, 213, 291, 312, 318, 443, 537
    strong 169, 314, 522
    weak 314, 507
Maxwell's equations vii, 546, 551
Maxwell–Bloch equation vii
mean 92, 286, 287, 519, 553, 555
  temperature 136
  zero 78, 92
measure ix, 63, 102
  zero 102, 104, 108, 114, 117
mechanics viii, 435, 578
  classical 394, 503
  continuum viii, 38
  equilibrium 121
mechanics (*continued*)
  fluid vii, 3, 121, 152, 153, 305, 315, 324, 341
  Newtonian 7
  quantum vii, viii, x, 6, 108, 149, 263, 278, 284, 286, 288, 329, 337, 340, 383, 386, 394, 503, 564, 566, 568
  solid 121, 292
medium 2, 64, 390, 427, 435, 504
membrane 15, 121, 152, 153, 154, 158, 169, 179, 249, 361, 435, 486, 487, 490, 492, 550
meridian 509
mesh 186, 195, 207, 411, 414, 418, 421, 423
metal
  ball 534, 542
  bar 64, 138, 292
  disk 166, 479
  pipe 170
  plate 214, 418, 426, 435, 481, 497
  shell 524
  sphere 534
method
  of images 256, 532, 534
  Newton ix, 135
  Perron 255
  Richardson 194
  Runge–Kutta ix
  stationary phase 331
microwave 15
minimal surface 153, 175
minimization principle 255, 340, 371, 399, 427, 429, 507
minimizer 399, 400, 401, 443
minimum 168, 522
Minneapolis 137
minute 307
mirror-image 143, 148, 149
mixed
  boundary condition 124, 343, 345, 347, 359, 364, 368, 436, 439, 441, 486, 504, 527, 535, 544, 546
  boundary value problem 7, 134, 154, 214, 241, 245, 505, 507
  partial derivatives 5, 8, 50, 242, 556
mode
  Fourier 142, 189, 193
  normal 389, 475, 487, 488, 489, 492, 499, 546, 565
  unstable 145, 340, 547
modeling viii

modified
  Bessel equation 473
  Dirichlet integral 368
  Green's function 358, 381, 480
  Korteweg–de Vries equation 337
modulus ix, 571
molecule 438, 503, 564
momentum 39, 61, 286, 287
money 299
monomial 577
monotone 29, 30
Monte Carlo 181
movie xv, 63
moving
  boundary 486
  coordinate 15, 19
multigrid 181
multiplication 263
  complex 571
  matrix 341, 344, 356, 585
  operator 286, 586
  scalar 575
  unit 575
multiplicity 371, 373, 382
multiply valued solution 36, 37
multipole 181
multiscale 442
multi-soliton solution 337
multivariable calculus 215, 437, 505
music 63, 144

N–wave 48
natural number xv
Nature 340
Navier–Stokes equations vii, 3, 9, 315
negative semi-definite 443
neon 569
Neumann
  boundary condition 7, 123, 144, 153, 343, 345, 347, 359, 364, 368, 374, 436, 439, 441, 486, 504, 527, 535, 544, 546
  boundary value problem 148, 238, 245, 352, 387, 534
neuron 139
neuroscience vii
Newton
  mechanics 7
  potential 503, 504, 529, 530, 532
  method ix, 135
  Second Law 49
nodal
  curve 497, 551
  region 497, 551
node 186, 187, 404, 411
  boundary 188, 207, 411, 418, 422, 425
  corner 209
  interior 188, 207, 209, 404, 411, 412, 418, 422, 425
no-flux 153
noise 193, 442, 536
non-characteristic 176
nonlinear vii, 60, 222, 324, 428, 549
  diffusion 2, 122, 291, 437, 442
  fully 174
  partial differential equation 2, 12, 31, 222, 305, 315, 333, 486
  Schrödinger equation vii
  transport 38, 46, 130, 292, 308, 315, 317, 427, 431
  wave equation 317, 333
nonlinearizing 318
nonsingular 211, 216
nonsmooth 129, 399, 427
nonuniform 10
  convergence 267
  transport 24
norm ix, 98, 109, 344, 356, 400, 571, 578
  convergence in 98, 99, 109, 110, 231
  double weighted L^2 381
  Euclidean 169, 572, 578
  L^2 89, 106, 284, 383, 394, 445, 477, 525, 579
  -preserving 286
  unit 108, 286, 394
normal ix, 295
  derivative 7, 153, 245, 250, 436, 504, 528, 529
  mode 389, 475, 487, 488, 489, 492, 499, 546, 565
  unit 153, 244, 433, 436, 437, 504, 505
north pole 511, 534
nowhere differentiable 329
nucleus 279, 394, 395, 503, 564, 569
null
  eigenfunction 70, 132, 145, 386, 387, 389, 441, 487, 547, 550
  eigenvalue 131, 439, 582
  eigenvector 582
number
  complex viii, ix, xv, 571
  Mach 174
  natural xv
  Péclet 311
  quantum 566–9
  rational 59, 107
  real xv, 98, 571

number (*continued*)
  theory 31, 121, 500
  wave 330
numerical
  algorithm 399
  analysis ix, 181
  approximation vii, 9, 181, 406
  Crank–Nicolson method 192
  differentiation 181
  domain of dependence 197, 199
  explicit scheme 188
  geometric method 181
  implicit scheme 191
  integration 185, 407, 422, 453
  iterative method ix, 203, 212, 411
    Gaussian–Seidel 212, 402, 411
    Jacobi 212
    successive over-relaxation (SOR) 212, 402, 411
  Lax–Wendroff scheme 200
  linear algebra ix, 399
  multigrid method 181
  multipole method 181
  Monte Carlo method 181
  pseudo-spectral method 181
  Richardson's method 194
  Runge–Kutta method ix
  simulation 324
  spectral method 181
  stability 181, 190, 205
  symplectic method 181
  upwind scheme 200

observer 552, 556, 558, 562
obtuse 414
odd 85, 93, 147, 268, 275, 497, 525
oil 129, 549
one-sided
  derivative 576
  limit viii
open ix, 152
operator
  Bessel 367
  derivative 354
  differential 9, 64, 339, 350, 371
    partial 2, 50
  divergence 347
  identity 383, 384
  linear 110, 284, 339, 340, 350, 354, 355, 386, 400, 428, 585
  multiplication 286, 586
  Schrödinger 565
  Sturm–Liouville 364, 365, 480
  wave 50, 487
optics vii, 179, 292, 329, 459
option 291, 299, 300–2, 304
orbit 452, 503, 565
order 1, 3, 366, 467, 511, 569
ordinary
  derivative 1, 3
  differential equation ix, 2, 6, 8, 11, 22, 65, 67, 121, 130, 140, 185, 215, 217, 263, 291, 308, 333, 363, 399, 435, 445, 452, 455, 463, 585
    first-order ix, 297
    linear 11, 585
    second-order ix, 297, 363, 459, 463
    separable ix
orientation 243, 504
orthogonal ix, 73, 271, 286, 340, 350, 371, 378, 387, 428, 440, 441, 447, 462, 477, 489, 517, 525, 536, 547, 569, 581, 583
  basis 73, 112, 372, 581
  matrix 506
  system 73, 113, 441, 515
orthonormal 109, 378, 514
  basis 112, 113, 581, 583
  system 73, 113, 570
oscillation 230, 267, 397, 474
oscillatory
  error 193
  wave 292, 327
outer space 487, 493, 531, 550
out of phase 137
oval 418, 423, 425
oven 138, 449, 480, 542
overdamped 146
overtone 63, 487, 495, 496

parabola 173, 309
parabolic 2, 122, 172, 173, 178, 193, 291, 315, 385, 438
  arc 556
  coordinates 174
  cylinder function 310
  spheroidal coordinates 508
  trajectory 565
parallel 579
parameter viii, 66
  relaxation 213
Parseval's formula 114, 118, 285, 287, 289
partial
  derivative viii, 1, 3, 521
    mixed 5, 8, 50, 242, 556
  differential equation vii, ix, 2, 63, 67, 130, 176, 215, 263, 305, 317, 339, 350, 394, 399, 429, 487, 582
    dynamical 291, 551
    elliptic 2, 122, 172, 173, 178, 212, 399, 404, 410

partial differential equation (*continued*)
first-order 176, 179, 195
fourth-order 486
fully nonlinear 174
hyperbolic 2, 122, 172, 173, 178, 179, 195, 385, 497
ill-posed 129, 174, 436, 442
nonlinear 2, 12, 31, 174, 222, 305, 315, 333, 486
parabolic 2, 122, 172, 173, 178, 193, 291, 315, 385, 438
quasilinear 174
second-order 121, 172
third-order 2, 324
type 172, 173
well-posed 136, 395, 535
differential operator 2, 50
fraction 279
sum 75, 100, 110, 113, 229
particle 149, 286
quantum 6, 396
subatomic 108, 149, 279
-wave duality 55, 149
particular solution 12, 253, 390
passenger 19, 20
path-independent 256
Pauli Exclusion Principle 568
Péclet number 311
percussion 144
periodic 59, 82, 144, 147, 389, 393, 395, 489, 548
boundary condition 69, 124, 130, 292, 343, 345, 361
boundary value problem 69, 373, 383, 387, 510
convolution 95
eigenvalue problem 130
extension 77, 96, 102, 229
force 304, 390, 397
function 60, 130, 278
initial-boundary value problem 328, 493
Table 503, 568
permeability 551
permittivity 551
perpendicular 244, 581
Perron method 255
petri dish 438, 485
phase ix, xvi, 137, 572
lag 137
shift 292, 324, 334
stationary 331
velocity 330
photoelectric effect 149
photon 108, 149, 286, 395
physical
space 263, 284, 503
variable 286
physicist 217
physics vii, viii, x, 37, 38, 55, 63, 305, 324, 339, 341, 394, 435, 509
piano 63, 144, 145, 261, 393, 496
piecewise
affine 404, 411
continuous 79, 81, 108, 127, 441
cubic 408
smooth 254, 427, 505
C^1 80, 82, 94, 223, 233
C^2 94
C^n 81
pie wedge 170
pipe 31, 170
piston 37
pitch 495
pivot 579
Plancherel formula 113, 118, 286, 287, 289, 379
Planck's constant 6, 287, 394, 564
plane 411, 571, 576
complex xv, 571
curve 172
half 260
quarter 24
square 255
planet 452, 530, 534, 550, 565
plasma vii, 2, 292, 388, 315
plate 2, 65, 152, 153, 246, 253, 324, 435, 441, 442, 486, 497
rectangular 154, 445
semi-circular 155
point
boundary 153, 169
charge 249, 252
critical 501
fixed 25, 29, 30, 584
floating ix, 184
image 532
interior 169
mass 249, 252, 256, 529
regular 453, 457, 463
set topology viii
singular 2, 283, 363, 452, 457, 463
irregular 453, 463
regular 453, 463, 511, 569
pointwise convergence 99, 109, 115, 117, 285, 378

Poisson
 –Darboux equation 62
 equation 2, 121, 152, 169, 172, 173, 181, 207, 215, 242, 245, 248, 253, 255, 256, 339, 352, 359, 361, 368, 383, 399, 401, 410, 422, 435, 442, 503, 504, 507, 527, 530, 532, 533, 534
 integral 165, 171, 259
 kernel 165, 259
polar
 angle 572
 coordinates viii, 3, 62, 160, 161, 171, 250, 383, 451, 477, 479, 490, 509, 572
 Helmholtz equation 451, 508
pole 511, 534, 549
pollutant 2, 15, 19, 23, 139, 304, 435
polygon
 boundary 411, 417
 vertex 414, 423
polyhedral surface 411
polynomial 1, 50, 75, 119, 400, 402, 576, 577
 Chebyshev 473
 Laguerre 109, 566, 567, 569
 harmonic 163, 171, 506, 520, 522, 526
 Hermite 311, 462
 homogeneous 169, 520
 Laguerre 109, 566, 569
 Legendre 511, 513, 525
 quadratic 316
 trigonometric 71, 75, 91, 108, 111, 119, 400, 402
pond 292, 324, 331
population 2, 121, 123, 435, 438, 485
portfolio 291, 300
posed
 ill 129, 175, 436, 442
 well 136, 395, 535
position 50, 286, 287, 396
positive 578, 580
 definite 339, 340, 355, 356, 362, 363, 371, 372, 373, 375, 376, 378, 383, 386, 387, 389, 400, 401, 429, 431, 439, 441, 442, 474, 506, 535, 546, 578
  matrix 342, 353, 362
 semi-definite 339, 355, 356, 371, 372, 378, 386, 387, 389, 441, 487, 506, 547, 549
potassium 569
potato 542, 543
potential 341
 Burgers' equation 318
 Coulomb 503, 529, 564
 electric 504
 electromagnetic 2, 564
 electrostatic 152, 249, 252, 256, 503, 522, 534
 energy 6, 61, 152, 242, 318, 340, 362
 gravitational 152, 249, 252, 503, 529, 534
 logarithmic 253, 256, 383
 Newtonian 503, 504, 529, 530, 532
 theory 121
 velocity 360, 504
power
 ansatz 162, 464, 520
 series viii, 75, 92, 98, 169, 453, 456, 458, 464, 511, 521
  solution 445, 452, 463
pressure 3, 35
price 299
prime meridian 509
principal
 quantum number 567, 568
 value 283, 572
principle
 Dirichlet 368, 400, 410, 416, 424, 443, 506
 Duhamel 298
 Heisenberg Uncertainty 286, 287
 Huygens vii, 503, 551, 560, 562, 564
 maximum vii, 169, 213, 291, 312, 318, 443, 537
  strong 169, 314, 522
  weak 314, 507
 minimization 255, 340, 371, 399, 427, 429, 507
 Pauli Exclusion 568
 superposition 9, 11, 12, 215, 217, 235, 335, 586
 symmetry 269, 271, 275
probability 109, 121, 181, 263, 394
 density 108, 286, 564
process
 diffusion 121, 129, 386
 dynamical 7, 15, 172
 Gram–Schmidt 372, 378
product 282
 convolution 281
 cross viii
 dot viii, 73, 341, 346, 354, 372, 578, 580, 585

product (*continued*)
inner ix, 107, 112, 118, 114, 339, 341, 371, 379, 400, 428, 578
Hermitian 89, 107, 118, 271, 285, 361, 372, 394, 580
L^2 73, 220, 230, 247, 342, 346, 358, 359, 371, 439, 477, 505, 515, 517, 525
weighted 341, 344, 354, 358, 359, 365, 378, 438, 506, 569
rule viii
profit 299, 302
program 453
proof ix
proton 564, 569
pseudo-spectral 181
pulse 54, 266, 267, 275
put option 300, 304
pyramid function 412, 414, 418

quadrant 499
quadratic
equation 172, 464, 465
function 340, 362, 401, 416
functional 340, 362, 400, 402
polynomial 316
quadrilateral 416
quantization 287, 394, 564
dispersive 328, 329
quantum
-dynamical system 503
mechanics vii, viii, x, 6, 55, 108, 149, 263, 278, 284, 286, 288, 329, 337, 340, 383, 386, 394, 503, 564, 566, 568
number 566–9
particle 6, 396
quasar 552
quasilinear 174
quasiperiodic 60, 144, 389, 393, 395, 489, 492, 497, 546, 548, 549
quotient
Rayleigh 340, 371, 375, 376, 377, 381, 384, 387
rule viii
radial
derivative 554
displacement 546
Gaussian function 247
symmetry 249, 479, 528, 531, 541, 548
variable 249
vibration 547, 549
wave function 568
radian 509
radiation 65
radio 15, 63, 546
radioactive 22
radius
Bohr 567, 569
of convergence 453, 458
spectral 584
rainbow 459
ramp function 82, 91, 95, 223, 235, 239
higher order 95, 223
random 60, 129
range 585, 586
Rankine–Hugoniot condition 40, 46, 47, 317, 321, 431, 433
rarefaction wave 34, 43, 44, 309, 320, 323, 433
ratio test viii, 75, 462, 468
rational 59, 389
function 218, 279, 452
number 59, 107
time 292, 329
ray 179, 257, 499
Rayleigh quotient 340, 371, 375, 376, 377, 381, 384, 387
reaction vii, 31, 438
-diffusion equation 438
real 6, 371, 571
analysis ix, 219
line xv, 107, 284
number xv, 98, 571
part ix
reciprocal 572
rectangle 121, 156, 214, 248, 255, 312, 361, 374, 415, 418, 445, 482, 488, 493, 496, 497, 499
rectangular
box 508, 526, 536, 547, 550
coordinates viii, 161, 503
drum 499
grid 422
mesh 186, 195, 207, 411, 418
plate 154, 445
pulse 266
recurrence relation 454, 456, 459, 467, 471, 525, 539, 543, 570
reduced finite element matrix 418
reflection 506, 571
refraction 179
refrigeration 136, 537
regular 457
boundary value problem 379
mesh 414
point 453, 457, 463
singular point 453, 463, 511, 569

regularization 317, 442
relative vibrational frequency 495, 548
relativity vii, 2, 31, 56, 63, 295, 504, 560
relaxation parameter 213
remainder term 182
removable discontinuity 80
reparametrization 179
repulsive 529
rescaled
  domain 495
  Fourier series 96, 264
  function 96, 506
reservoir 123
resonant 60, 279, 304, 391, 397, 493, 549
restoring force 169
reverse shock 433
Richardson's method 194
Riemann
  hypothesis 87
  geometry 31
  integration 217, 219, 270
  –Lebesgue Lemma 112, 117, 231, 276
  sum 265
  zeta function 87
right
  angle 73, 581
  triangle 413, 417, 419, 423
right-hand
  derivative 81
  limit 79, 81
rim 486
ring 69, 124, 130, 146
ripple 292
river 31
Robin boundary condition 123, 134, 154, 361, 436, 439
rock 292, 324, 549
rod 534
root ix, 316, 464, 474, 582
  Bessel 474, 490, 499
    spherical 540, 541, 548
  cellar 136, 137, 545
  test viii, 75
rope 148, 393, 494
rotated function 155
rotation 305, 506, 517, 528
rough 284, 296
round-off error ix, 184, 190
row vector 216, 578
rubber 493
rule
  chain viii, 20, 24, 32, 57, 161, 167, 171, 300, 305, 308, 333, 494, 509, 511, 524
  product viii
  quotient viii
  trapezoid 407
Runge–Kutta method ix

sample 27, 187, 189
sand 498
sawtooth function 78
saxophone 144
scalar
  field 242, 345, 359, 439, 505
  multiplication 575
scaling 155, 305, 306, 307, 308, 323, 494, 521
  symmetry 291, 307, 308, 327, 337, 479
scandium 569
scattering 337, 383, 549, 565
Schrödinger
  equation vii, 6, 146, 304, 340, 372, 374, 383, 385, 394, 396, 397, 503, 564
  operator 565
science vii, 2, 4
screen 286
sea 332
secant 182, 185
second-order
  iterative scheme 203
  ordinary differential equation ix, 297, 363, 459, 463
  partial differential equation 121, 173
  ramp function 95
segment 404
seismology 63, 129, 549
self-adjoint 74, 110, 237, 243, 249, 339, 340, 354, 356, 362, 371, 378, 386, 400, 429, 438, 447, 505, 517, 525, 536, 579
  matrix 354, 373
semi-
  circle 480, 499
  circular disk 170, 418
  circular plate 155
  infinite bar 136
  weak formulation 429
semi-definite
  negative 443
  positive 339, 355, 356, 371, 372, 378, 386, 387, 389, 441, 487, 506, 547, 549
sensor 23, 303, 561

separable
  coordinates 508
  ordinary differential equation ix
  solution 68, 121, 124, 141, 156, 163, 306, 320, 386, 388, 439, 463, 482, 509, 517, 520
separation
  constant 141, 156, 446, 510
  of variables vii, ix, 25, 68, 121, 140, 155, 161, 215, 256, 305, 340, 364, 386, 435, 447, 451, 467, 487, 503, 507, 527, 535, 538, 547, 565
sequence viii, xvi, 98, 104
  Cauchy 107, 119
series viii, 2, 9, 11, 67, 68, 75, 100, 121, 156, 256, 435, 455
  complex Fourier 89, 582
  differentiated 101
  doubly infinite 89, 91
  eigenfunction 109, 340, 371, 378, 379, 386, 435, 441, 443, 535, 544
  eigensolution 440
  Fourier viii, 71, 74, 76, 82, 102, 109, 113, 121, 131, 163, 228, 231, 233, 263, 285, 328, 340, 371, 378, 402, 519
    –Bessel 477, 492, 493
    –Bessel–spherical harmonic 541
    complex 89, 582
    cosine 85, 87, 144, 148, 233
    derivative of 94
    double 488
    generalized 109, 378
    rescaled 495
    sine 85, 87, 126, 142, 147, 157, 233, 382
  of functions 100
  Frobenius 464, 569
  Gregory 78, 85, 86
  integrated 101
  power viii, 75, 92, 98, 169, 453, 456, 458, 464, 511, 521
    solution 445, 452, 463
  Taylor 75, 169, 171, 182, 183, 193, 453, 458, 463, 521, 576
sesquilinear 372, 394, 580
set
  theory 64
  diffusion of 485
shallow water 292, 333, 435
sharpen 442
shell 324, 450, 524, 568
  spherical 510, 534, 543, 549, 551
shift
  phase 292, 324, 334
  theorem 271, 274, 284
shock 39, 45, 47
  dynamics 15, 431
  line 40, 173
  reverse 433
  wave vii, 2, 5, 15, 31, 37, 130, 195, 292, 315, 316, 317, 320, 324, 399, 427, 431, 433
shore 332
signal 15, 22, 23, 128, 179, 295, 486, 552, 560
  processing 63, 263, 582
sign function 94, 225, 270
silver 123
similar triangle 257, 532
similarity
  solution vii, 42, 291, 305, 308, 323, 327, 337, 485
  transformation 307
  variable 308
simple
  closed curve 152
  eigenvalue 372, 391, 582
simply
  connected 243
  supported 146, 393
simulation 324
sine transform 274
single-precision 184
singlet 558
singular 173, 582
  boundary value problem 379
  point 2, 283, 363, 452, 457, 463
    irregular 453, 463
    regular 453, 463, 511, 569
singularity 251, 256, 427, 454, 553
  logarithmic 383, 525
sink 360, 438, 442, 504
slope 21, 37, 182, 185
Smith Prize 3
smooth 5, 284, 441, 536, 576
  piecewise 254, 427, 505
smoothing 128, 296, 441, 545
smoothness 7, 105, 276, 402, 410, 428, 430
snare drum 496
soap 153
soda 537
soil 137
soldering iron 292, 297, 304, 481, 534
solid 121, 388
  body 2, 15, 65, 144, 440, 503, 504, 522, 546
  boundary 154, 504
  geometry 423
  mechanics 121, 292

solitary wave 324, 334
soliton 292, 324, 336, 337
solute 22, 121, 438
solution 5, 6, 11, 181, 291, 350, 452
  analytic 431
  basis 458
  classical 5, 7, 17, 51, 144, 255, 399, 410, 427, 428, 432, 535, 546, 556, 558
  complex 6
  d'Alembert 53, 201, 324, 487
  equilibrium 132, 443
  fundamental 291, 292, 294, 297, 301, 304, 326, 328, 387, 388, 435, 481, 482, 503, 543, 551, 554, 564
  general 12, 67, 253, 390
  localized 560
  manufactured 210
  multiply valued 36, 37
  multi-soliton 337
  particular 12, 253, 390
  power series 445, 452, 463
  separable 68, 121, 124, 141, 156, 163, 306, 320, 386, 388, 439, 463, 482, 509, 517, 520
  similarity vii, 42, 291, 305, 308, 323, 327, 337, 485
  weak 5, 43, 53, 144, 399, 427, 429, 431, 432, 433, 489, 556, 558
sonic boom 173
SOR 212, 402, 411
sound 560, 562
  barrier 15, 173
  speed 174
  wave 121, 551, 552
source 194, 292, 360, 438, 442, 504, 543, 552
space 6, 15, 18, 121, 291, 305, 494, 504
  coordinates 3
  derivative 291
  Euclidean ix, 575, 578
  frequency 263, 284
  function 63, 74, 99, 109, 215, 220, 340, 342, 344, 386, 400, 576
  Hilbert ix, 106, 108, 112, 119, 263, 284, 286, 368, 394
  inner product 118, 350, 354, 400, 428, 429, 578, 580
  outer 487, 493, 531, 550, 552
  physical 263, 284, 503
  three-dimensional 503, 528, 544, 552
  -time 19, 56, 560
  two-dimensional 252
  vector ix, 11, 98, 112, 220, 341, 371, 575, 578
span ix, 112, 401, 413, 577
sparse 402, 407, 410
special
  function vii, 2, 9, 327, 364, 435, 453, 455, 472, 508, 511
  relativity 56
specific heat 122, 124, 132, 438
spectral 181
  line 395
  radius 584
spectrum 395
  continuous 337, 340, 374, 383, 565
speech 63
speed 331, 334
  of light 295, 551
  of sound 174
  wave 19, 22, 24, 50, 51, 65, 140, 195, 197, 292, 330, 334, 396, 486, 488, 492, 495, 549
sphere 108, 508, 511, 517, 521, 522, 523, 526, 532, 534, 552, 555, 560
spherical
  Bessel function 435, 503, 539, 540, 543
  Bessel root 540, 541, 548
  cap 555, 561
  capacitor 522, 523, 526, 533
  cavity 524, 534
  coordinates viii, 3, 503, 508, 520, 524, 528, 537, 551, 553, 565
  harmonic 109, 435, 503, 517, 519, 528, 538, 540, 549, 565, 568
    complex 519, 526, 565
    Fourier–Bessel 541
  heat equation 543
  Helmholtz equation 510, 517, 519, 565
  Laplacian 510, 517, 525, 538, 543, 549, 551
  shell 510, 534, 543, 549, 551
  symmetry 531, 552
  wave 552, 554
spike 218, 224, 229
spin 568
spiral 164
spline 27, 210, 283, 400, 408
spring 95, 154, 216, 333, 473
square 369, 385, 415, 426, 444, 449, 479, 489, 492, 496, 499
  drum 493
  grid 414, 421
  hole 426
  -integrable 106, 108, 231, 284, 285, 293, 394
  least 110, 363

square (*continued*)
mesh 421, 423
plate 255
unit 155, 158, 260, 485
stable 181, 190, 205, 340
conditionally 190
unconditionally 192, 193
von Neumann criterion 198, 199, 205
standard
basis 216, 577, 581
deviation 295
standing wave 55
star 560
state space 286
stationary 20
phase 331
point 331
wave 17
statistics vii
steepest decrease 437
step
function 29, 40, 42, 61, 105, 276, 320
unit 80, 83, 90, 102, 222, 232, 266, 329, 396
size 182, 187, 406
time 197
stiffness 49, 142, 234, 344, 357, 473, 486, 488, 492, 495
elemental 417, 418
stochastic differential equation viii, 299
stone 331
stopping time 128
St. Elmo's fire 254
strain 341
string 2, 13, 15, 49, 62, 96, 140, 142, 261, 391, 489
Strong Maximum Principle 169, 314, 522
structure 549
Sturm–Liouville
boundary value problem 363, 382
eigenfunction 382
equation 336, 364, 370, 404, 508
operator 364, 365, 480
subatomic particle 108, 149, 279
subsolution 255
subsonic 173, 174
subspace 576
dense 344, 346, 371
finite-dimensional 400, 410, 430
finite element 416
subterranean 129
successive over-relaxation (SOR) 212, 402, 411
sum
convolution 284
geometric 116
partial 75, 100, 110, 113, 229
Riemann 265
trigonometric 116
summer 136, 137, 545
sun 545, 565
supercomputer 565
superconductivity vii
supernova 551
superposition 11, 51, 59, 67, 71, 234, 242, 293, 328, 330, 476, 481, 483, 535, 544, 551, 552, 554
formula 248, 291, 297
principle 9, 11, 12, 215, 217, 235, 335, 586
supersonic 15, 173, 174
support 284, 407, 411, 414
compact 230, 430, 432
surface viii, 324, 504
area 517, 529, 537, 543, 553, 555
curve 505
integral viii, 505, 522, 529, 533
minimal 153, 175
polyhedral 411
wave 292, 333
surfer 332
symbolic program 453
symmetric 188, 211, 249, 358, 579, 583
function 236
matrix 66, 74, 342, 353, 354, 362, 372, 401
symmetry vii, ix, 170, 237, 263, 281, 291, 305, 358, 360, 494, 578
conjugate 580
cylindrical 517
principle 269, 271, 275
radial 249, 479, 528, 531, 541, 548
rotational 305
scaling 291, 307, 308, 327, 337, 479
spherical 531, 552
transformation 495
weighted 358, 380
symplectic 181
synthesizer 63, 496
system
adjoint 350
complete 520, 541, 565
of differential equations 2, 66
dynamical 340, 385
equilibrium 339, 340

system (*continued*)
linear ix, 187, 191, 216, 339, 350, 386, 400, 428, 582, 586
finite element 431
orthogonal 73, 113, 570
quantum-dynamical 503

table
of Fourier transforms 272
periodic 503, 568
tail 337
Talbot effect 292, 329, 396
tangent 153, 182, 277
target xvi, 339, 341, 401, 428, 585
tautness 495
Taylor
series 75, 169, 171, 182, 183, 193, 453, 458, 463, 521, 576
formula 75
theorem viii, 203
telegrapher's equation 62, 146, 393
television 63
temperature 7, 68, 122, 129, 136, 152, 153, 291, 312, 341, 436, 449, 504, 508, 522, 535
equilibrium 133, 154, 214, 246, 435, 537
fluctuation 137
gradient 341, 437, 535
initial 131, 544
maximum 169
minimum 169
mean 136
tension 49, 142, 486
test
function 220, 225, 230, 429, 553
integral viii, 105
ratio viii, 75, 462, 468
root viii, 75
Weierstrass M 100, 105
tetrahedron 423
theorem ix
Carleson 117
Cauchy–Kovalevskaya 175
convergence 82
convolution 284
dilation 271, 274
divergence viii, 505, 529
existence 67, 364, 457
fundamental viii, 16, 39, 223, 236, 243, 245, 267
Green viii, 3, 168, 215, 243, 437
implicit function 49
shift 271, 274, 284
Taylor viii, 203
uniqueness 62, 67, 151, 169, 522

thermal
conductivity 65, 123, 437
diffusivity 124, 134, 186, 293, 298, 438, 535
energy 121, 122, 132, 134, 139, 169, 295, 304
equilibrium 127, 152, 153, 169, 441, 448, 479, 522, 536
reservoir 123
thermodynamics vii, viii, 2, 7, 12, 123, 130, 139, 242, 295, 312, 341, 436, 452
thermomechanics 121, 153
thermometer 306
thermonuclear explosion 561
third-order evolution equation 324
three-dimensional
delta function 527
space 503, 528, 544, 552
thunder 551, 552, 560
timbre 63
time 2, 3, 6, 7, 15, 18, 39, 121, 130, 286, 291, 299, 305, 307, 395, 442, 494
exercise 302
irrational 292, 329, 332
rational 292, 329
-reversible 497
space- 19, 56, 560
step 197
stopping 128
tone 63, 487
topology viii
toroidal
coordinates 508
membrane 361
torus 493
total
energy 61, 151
heat 444, 485, 537
traffic 15, 31, 38, 44
train 19, 20
trajectory 565
transatlantic cable 139
transcendental equation 1, 134
transfer coefficient 156
transform
cosine 274
Fourier 263, 264, 265, 269, 273, 274, 278, 282, 284, 285, 293, 297, 325, 330, 337, 340, 374, 462
discrete 582
inverse 263, 265, 269
table 272
two-dimensional 274, 278

transform (*continued*)
  Hilbert 283
  Laplace 263
  sine 274
transformation
  Hopf–Cole 318
  linear 285
  similarity 307
  symmetry 495
translation 21, 155, 291, 295, 305, 328, 506, 528, 551
transport
  equation 2, 19, 31, 51, 65, 181, 195, 291, 315, 324, 330, 434
  nonlinear 38, 46, 130, 292, 308, 315, 317, 427, 431
  nonuniform 24
transpose 216, 339, 341, 353, 578
transverse vibration 393
trapezoid rule 407
traveling wave 21, 51, 195, 291, 292, 305, 316, 330, 333, 334, 497
triangle 40, 58, 144, 197, 199, 205, 210, 257, 411, 414, 415, 416, 418
  equilateral 260, 415, 417, 419, 424, 426
  inequality ix, 107, 526, 572, 579, 581
  isosceles 413, 415, 417, 421
  obtuse 414
  right 413, 417, 419, 423
  similar 257, 532
  wave 44, 48, 322
triangulation 411, 414, 417
Tricomi equation 173, 178, 462
tridiagonal 188, 191, 211, 407
trigonometric 340, 371, 388, 487, 490
  ansatz 546
  Ferrers function 516
  function 60, 63, 70, 72, 74, 89, 109, 113, 125, 130, 189, 231, 264, 273, 374, 445, 451, 452, 457, 499, 508, 511, 519, 536
  polynomial 71, 75, 91, 108, 111, 119, 400, 402
  sum 116
triple integral viii, 505, 553
trumpet 63, 144
two-dimensional
  delta function 246, 255
  Fourier transform 274, 278
  space 252
type 172, 173

Uncertainty Principle 286, 287
unconditionally stable 192, 193
underdamped 146
unidirectional 19, 324
uniform 50, 99, 104, 435, 436, 438, 485, 508, 528, 535, 537
  convergence 99, 100–102, 104, 378, 519
uniqueness ix, 2, 29, 130, 245, 314, 340, 355, 400, 429, 458
  theorem 62, 67, 151, 169, 522
unit 217, 281
  circle 167, 258
  disk 155, 166
  imaginary 571
  impulse 215–218, 234, 280, 551, 562
  multiplicative 575
  norm 108, 286, 394
  normal 153, 244, 433, 436, 437, 504, 505
  sphere 108
  square 155, 158, 260, 485
  step function 80, 83, 90, 102, 222, 232, 266, 329, 396
unitary 286
universe 130, 503, 553, 560, 564
unstable 136, 190, 389
  mode 145, 340, 547
upper triangular matrix 211
upwind scheme 200

value of option 302
variable 1
  change of viii, 51, 58, 175, 179, 318, 511
  characteristic 20, 22, 25, 30, 32, 51, 52, 552
  complex 163, 571
  dependent 3
  frequency 286
  independent 3, 6, 15, 291, 300
  physical 286
  radial 249
  separation of vii, ix, 25, 68, 121, 140, 155, 161, 215, 256, 305, 340, 364, 386, 435, 447, 451, 467, 487, 503, 507, 527, 535, 538, 547, 565
  similarity 308
variance 287
vector 66, 98, 571, 575
  addition 571
  analysis viii
  calculus 242, 507, 571
  column 216, 578
  field 242, 243, 346, 359, 360, 439, 505, 507, 526
    electric 341, 546, 551
    magnetic 341, 546, 551
    velocity 3, 152, 244, 504

vector (*continued*)
  row 216, 578
  space ix, 11, 98, 112, 220, 341, 371, 575, 578
    complex 289, 571, 575, 580
    finite-dimensional ix, 11, 98, 109, 215, 220
    infinite-dimensional ix, 11, 99, 109, 215, 340, 342, 371, 400, 577
vehicle 38, 44
velocity 7, 19, 35
  fluid 341
  group 292, 331
  initial 50, 55, 59, 145, 487, 546, 547, 551, 554, 557, 560, 562
  phase 330
  potential 360, 504
  vector field 3, 152, 244, 504
vertex 411, 415, 416
  polygon 414, 423
vertical axis 18
vibration 10, 15, 49, 60, 121, 140, 142, 149, 172, 340, 385, 389, 503, 546
  elastic 486
  electromagnetic 2
  equation 340, 388, 390
  frequency 389, 395, 487, 495, 548
  radial 547, 549
  transverse 393
video 63
violin 13, 15, 49, 63, 144, 497
viscosity 3, 291, 315, 317
Vlasov equation vii
volatility 299, 302, 304
volume 243, 543
von Karman equation vii
von Neumann stability 198, 199, 205

water 2, 3, 15, 121, 149, 243, 331
  deep 331
  shallow 292, 333, 435
wave 2, 15, 54, 121, 149, 292, 324, 388
  acoustic 15
  compression 36, 44
  delta 558
  dispersive 2, 324, 459
  elastic 121
  electromagnetic 15, 121, 388, 503, 546, 551
  equation 2, 8, 13, 15, 50, 64, 121, 140, 149, 151, 152, 169, 172, 173, 178, 181, 195, 201, 291, 315, 324, 339, 340, 371, 374, 385, 389, 427, 434, 435, 467, 486, 488, 494, 503, 545, 551, 552, 554, 556, 558, 561
    damped 207

wave (*continued*)
  flood 15, 31
  function 108, 286, 288, 394, 396, 564, 565, 568
  light 551
  N– 48
  number 330
  operator 50, 487
  oscillatory 292, 327
  packet 331
  -particle duality 55, 149
  rarefaction 34, 43, 44, 309, 320, 323, 433
  shock vii, 2, 5, 15, 31, 37, 130, 195, 292, 315, 316, 317, 320, 324, 399, 427, 431, 433
  solitary 324, 334
  sound 121, 551, 552
  speed 19, 22, 24, 50, 51, 65, 140, 195, 197, 292, 330, 334, 396, 486, 488, 492, 495, 549
  spherical 552, 554
  standing 55
  surface 292, 333
  traveling 21, 51, 195, 291, 292, 305, 316, 330, 333, 334, 497
  triangle 44, 48, 322
weak
  convergence 99, 230, 270, 327, 429
  formulation 428, 429, 430, 432
  maximum principle 314, 507
  solution 5, 43, 53, 144, 399, 427, 429, 431, 432, 433, 489, 556, 558
weakening spring 473
wedge 35, 43
Weierstrass M–test 100, 105
weight function 379
weighted
  adjoint matrix 342
  average 213
  inner product 341, 344, 354, 358, 359, 365, 378, 438, 506, 569
  Sturm–Liouville differential operator 365
  symmetry condition 358, 380
well-posed 136, 395, 535
white 441
width 334
wind instrument 15, 144
winter 136, 137, 545
wire 153, 158, 160, 169, 393
wolf 438

X-ray 546
xylophone 144

yard 307

zenith
  angle 508, 515
  differential equation 510
zero
  complex 87
  eigenvalue 131, 439, 582
  element 428
  function 114, 219, 231, 281, 456, 463
  matrix 583
  mean 78, 92
  measure 102, 104, 108, 114, 117
zeta function 87

Manufactured by Amazon.ca
Bolton, ON